HEAT PUMP SYSTEMS

HEAT PUMP SYSTEMS

HARRY J. SAUER, JR.
RONALD H. HOWELL
University of Missouri—Rolla

A Wiley-Interscience Publication

JOHN WILEY & SONS

New York   Chichester   Brisbane   Toronto   Singapore

Library of Congress Cataloging in Publication Data:

Sauer, Harry J., 1935–
  Heat pump systems.

  "A Wiley-Interscience publication."
  Includes bibliographical references and index.
  1. Heat pumps. I. Howell, Ronald H. (Ronald
Hunter), 1935–   . II. Title.
TJ262.S28   1983   621.402′5   83-1113
ISBN 0-471-08178-7

Printed in the United States of America

10 9 8 7 6 5 4 3 2 1

PREFACE

This book is designed to provide a complete, comprehensive treatment of heat pumps . . . their fundamentals, performance, design, cost, and selection. It is an outgrowth of a manual developed for an intensive seminar on electric heat pumps sponsored by the American Gas Association (AGA) in 1979. Although the manual has been completely rewritten and much material added, the authors are greatly indebted to AGA for providing both initial incentive as well as considerable material for this book. A large debt of appreciation is also owed the American Society of Heating, Refrigerating and Air-Conditioning Engineers for allowing us to draw heavily on the methodology in the ASHRAE Handbooks and from *Environmental Control Principles,* the educational supplement to the *1981 Fundamentals Handbook,* developed for ASHRAE by the authors.

Although it is impossible to acknowledge all the people and companies who have contributed to this book, we hope that the references at the end of each chapter provide adequate recognition to the major sources of material. However, with a combined total of over 45 years of teaching, the authors have benefitted from many semesters of using the texts of Jennings, Stoecker, and Threlkeld (refrigeration/air-conditioning), Van Wylen and Sonntag (thermodynamics), Holman (heat transfer), and many others. Undoubtedly, this presentation has been influenced by these experiences but the sources of specific ideas, perspectives, examples, approaches, and the like have been forgotten and only a general "thank you" can be offered. And, of course, very special thanks go to our families for all their encouragement and patience.

Over the years many excellent technical texts and reference books have appeared dealing with refrigeration. However, the heating capabilities of such equipment have failed to receive much attention. With the value of energy today, the heat pump has become a more timely concept. The early days of the heat pump showed that three essential characteristics had to exist in order to generate consumer interest in heat pumps. First, a need for cooling as well as heating was required in order the justify the high initial equipment cost. Second, relatively low electrical rates were necessary for the heat pump to compete favorably in terms of operating costs with fossil fuel fired furnaces. Third, the heat pump had to have satisfactory reliability and low maintenance requirements. Achievement of this combination of events has led to a period of explosive growth and the industry moved from a level of 82,000 heat pumps shipped in 1971 to over 540,000 units in 1981.

Accompanying this increase in interest and use, the need for knowledge of heat pump principles and technology has greatly increased. This book attempts to be a complete reference/text book on heat pumps linking the fundamental principles of economics, thermodynamics, and heat transfer with current design, selection, and operating practices. Emphasis is placed on specific, practical methods and solutions—not broad generalizations. All the essential procedures are presented in step-by-step fashion to enable HVAC engineers, contractors, architects, and builders to design, size, and select heat pump systems and components. Example problems are included throughout which demonstrate fundamental principles, techniques of calculational procedures, and overall system analysis, design, and/or selection.

The early chapters provide a combination of historical background and technical introduction to heat pumps, emphasizing the more common vapor-compression types, but including gas-fired versions, solar and geothermal units, and hybrid systems. The reader is then provided the necessary tools for sizing heat pumps, for estimating the energy use by such systems, and for determining life cycle costs. The results of many studies and experiences in actual residential, commercial, and industrial installations are presented in detail as relate to energy consumption, owning and operating costs, reliability and maintenance, realistic expectations, and environmental impacts. The book concludes with a look at developments underway today as well as possible future technologies.

We hope it makes interesting and informative reading.

HARRY J. SAUER, JR.
RONALD H. HOWELL

Rolla, Missouri
April 1983

CONTENTS

9. ECONOMIC ANALYSIS FUNDAMENTALS   512

1

INTRODUCTION

1.1. HEATING AND COOLING SYSTEMS

Our "new" energy sources, solar and geothermal, were undoubtedly the earliest forms of energy to warm the human environment. The sun kept people warm when it was out, and when it wasn't, cave homes and earth shelters provided warmth from geothermal heat. The most primitive method of heating was the open fire of the cave man, who thousands of years ago modified his personal climate by nurturing the first fire. Then the Romans ingeniously engineered ventilation and panel heating into their baths. One of the first notable steps in the evolution of the heating system consisted of the addition of a chimney to carry away the products of combustion, from which it was only a short step to the fireplace and then to the stove. An important later innovation in the development of the furnace system was the addition of a fan to provide a mechanical means of forcing air through the duct system. The remarkable Leonardo da Vinci had built a ventilating fan at the end of the 15th century.

On the cooling side, in both Egypt and India, the evaporative cooling process supported by radiation to clear skies at night furnished ice for the royal tables as early as 500 BC. The Scottish physician Dr. William Cullen in 1775 evacuated a vessel of water to make ice. A few years later our own Benjamin Franklin wrote his treatise on Pennsylvania fireplaces, detailing their construction, installation, and operation with elaborate illustrations.

In the United States, by 1800, the techniques of warming and ventilating dwellings were developing well. Fans, boilers, and radiators had been invented and were in common use. Refrigeration technology was not very far behind. Dr. John Gorrie, a physician in Charleston, South Carolina, invented a dense air compression machine in 1849 and in 1851 was granted

1

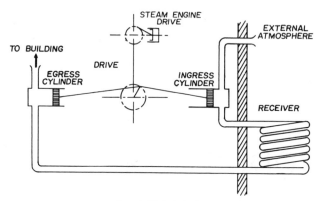

Fig. 1.1. Lord Kelvin's heat engine.

U.S. Patent 8080. This was the first commercial machine in the world built and used for refrigeration and air-conditioning, and used compressed air in an open cycle. In France, Ferdinand Carré in 1851 designed the first ammonia absorption unit. Alexander Catlin Twining of New Haven, Connecticut, received U.S. Patent 10,221 in 1853 for the first compression machine in the world to make ice commercially by the vapor compression system.

The growth in understanding of physical processes in the 19th century led to interest in the possibility of pumping heat energy to a higher temperature. Joule demonstrated the principle of changing the temperature of a gas by altering its pressure, and the theoretical concept of the heat pump was described in an 1824 book written by a young French army officer, Sadi Carnot. Professor William Thomson (later to be Lord Kelvin) was the first to propose the heat pump, or "heat multiplier" as he called it. Thomson published a paper in 1852 [1]* describing a system in which, using a linked compressor and expander, air was moved to and from a reservoir that also acted as a heat exchanger (Fig. 1.1). This open-cycle unit could be used for either heating or cooling buildings. In his paper Thomson foresaw the closed-cycle vapor compression machine, but neither the refrigerants nor the drive motors were available to enable him to design anything really resembling the modern heat pump.

Development of refrigeration equipment using these ideas progressed rapidly in the 1870s. A number of cold-air refrigeration machines were produced in response to the needs of the international frozen meat trade. The cold-air refrigeration machines were ousted by carbon dioxide machines, and by the 1920s ammonia compression had become established. Smaller refrigeration equipment used methyl chloride in the 1930s, and by the early 1940s the first of the modern halocarbon refrigerants, R-12, was available. By 1940 and the beginning of World War II, the art of air cooling was introduced in an

*Numbers in brackets refer to references at the end of the chapter.

ever-increasing number of applications. Air cooling of department stores, restaurants, hotels and hospitals became an accepted practice, and extended research was underway in the preferred methods of cooling automobiles, airplanes, and homes.

Over this period of refrigeration equipment development, the development of heat pumps lagged behind. While refrigeration met an established need, heat pump development depended on energy costs and availability and on the alternative heat generators available. Heat pumps are refrigeration systems in which the rejected heat is at least as important as the refrigeration effect. Heat pumps have been referred to as *reverse-cycle refrigeration systems*.

A heat pump, in the common thermodynamic sense, is a system in which refrigeration components (compressors, condensers, evaporators, and expansion devices) are used in such a manner as to take heat from a source (air, water, ground, etc.) and give it up to a heat sink (air, water, ground, etc.) that is at a higher temperature than the source. For many applications, the heat pump is designed in such a manner as to reverse the cooling and heating functions. This allows the use of the same equipment for both heating and cooling an individual structure. Other applications are designed to simultaneously utilize both the heating and cooling effects obtained from the cycle in the same structure.

First heat pump applications were considered in the 1920s, with restatements of and improvements on Thomson's paper by Krauss [2] and Morley [3]. Although there were no heat pumps as such in existence, it was possible to examine their feasibility by analyzing the performance of the rapidly increasing amount of refrigeration equipment that was being installed. This was done by Haldane [4], who analyzed data from a number of refrigerating plants operating between 1891 and 1926. From these results, Haldane was able to recommend that reversible heat pumps should be considered for cooling and heating buildings.

Not content with theoretical calculations, Haldane constructed an experimental heat pump in the mid-1920s to provide space heating and water heating in his home in Scotland. He used both outdoor air and mains water as heat sources, a low-temperature, hot-water radiator heat-distribution system, and an electrically driven refrigeration compressor. The performance of the domestic unit showed a worthwhile coefficient of performance (COP)* and, apart from a "little" noise difficulty, appears to have been effective.

Probably the first large-scale heat pump application was in the Los Angeles offices of the Southern California Edison Company [5], where in 1930–1931 refrigerating equipment was used for heating purposes. A COP of between 1.5 and 2 was obtained.

In the early 1930s many manufacturers became interested in the development of economical heat pump systems. Several custom-designed systems

---

*COP = units of heating energy per unit of work input.

were developed for individual buildings and put into operation. These projects were limited in number and were privately financed. They confirmed the heat pump principle proposed by Carnot and further developed by Kelvin. Enough actual performance data were collected from these systems to promote renewed interest and development in the 1940s [6].

The economic difficulties of the 1930s provided the necessary spur for heat pump development in Europe, and an appreciable number of large-scale applications were in existence by 1943. The first major heat pump installation in Europe was commissioned during the period 1938–1939 in Zurich. A unit that used river water as the heat source utilized a rotary compressor with R-12 as the working fluid [7]. Used to heat the Town Hall, the output of the Zurich heat pump was 175 kW, delivering water at a temperature of 60°C for space heating. A thermal storage system was incorporated in the circuit. Water temperature could be boosted by electric heating at periods of peak demand. The system was also arranged so that it could provide cooling in summer months. Kemler and Oglesby [8] list 15 commercial applications installed by 1940 in the United States, the majority using well water as their heat source.

This activity was interrupted by World War II, which diverted the technical skills of industry to more urgent matters. Interest resumed after the war, and there were many more demonstration projects installed in the late 1940s. It became evident at that time that if there was to be wide acceptance of heat pumps for comfort heating, products based on the unitary concept would have to be developed. A unitary refrigeration system is one that is factory-engineered and factory-built and then shipped to the field in one or two assemblies.

In the United States, development of smaller heat pump units, basically reversible air conditioners for domestic use providing either heating or cooling, had progressed sufficiently by 1948 for a field test procedure [9] to be formulated and for field tests to be performed by electric utility companies [10]. Possible problems of electricity demand characteristics were examined closely, and the dangers of excessive heat losses from air-distribution ductwork were noted.

The use of heat pumps for the heating and cooling of the Equitable Building in Portland, Oregon, initiated in 1948, was a pioneering achievement in the Western Hemisphere. In 1980 the system was designated a National Historic Mechanical Engineering Landmark by the American Society of Mechanical Engineers.

Around 1950 work was done in both the United States and Britain on domestic heat pumps using ground coils as a heat source. Baker [11] designed a reversible unit incorporating an antifreeze storage bath, and reported a coefficient of performance greater than 3 on average over the winter of 1950–1951.

In the late 1940s and into the early 1950s, development work continued on unitary heat pumps for residential and small commercial installations. These are factory-engineered and factory-assembled units that, like conventional domestic boilers, could be installed easily and cheaply in the home or in small commercial premises. In 1952, heat pumps developed along these lines were offered in quantity to the market [6].

The early days of the heat pump showed that three essential characteristics had to exist in order to generate consumer interest in heat pumps. First, there had to be a need for cooling as well as heating to justify the high initial equipment cost. Second, relatively low electricity rates were necessary for the heat pump to compete favorably in terms of operating costs with furnaces fired with fossil-fuels. Third, the heat pump had to have satisfactory reliability in order to minimize service calls. Figures 1.2 and 1.3 illustrate some of the claims and studies made by proponents and opponents (respectively) of the electric heat pump and offer conflicting conclusions as to the energy effectiveness, reliability, and market acceptance of the electric heat pump.

Since the high first cost of a heat pump precludes using it for heating only, the growth of the heat pump market closely paralleled the post–World War II growth in demand for comfort cooling. Comfort cooling, for obvious reasons, first spread in regions with a hot and humid climate. Heat pumps were first marketed on a commercial scale along the Gulf Coast and in the Mississippi River valley. Approximately 1000 units were sold in 1952. By 1954 sales were still only at 2000 units. In 1957, 10,000 units were shipped (Fig. 1.4).

In the early 1960s, reversible domestic air-to-air heat pumps established an appreciable sales success in the United States. In 1963 the number of units shipped was 76,000. Unfortunately they also established a poor reputation for reliability, as it had not been fully appreciated that a reversible heat pump needs to be more than just an air conditioner with a refrigerant flow-reversing valve added. While reputable manufacturers realized that a heat pump would have to operate many times longer than a typical air conditioner, and under conditions that would subject the heat pump components, especially the compressor, to high stress and wear, many other manufacturers marketed products that were little more than standard air conditioners with reversing valves. The heat pump, although generally only slightly more difficult to install and maintain than a central air-conditioning system, is much less tolerant of improper installation or poor servicing.

The early heat pump models lacked sufficient durability as outdoor winter temperatures dropped. During periods of low winter temperatures, stresses were very severe on the heat pump components (compressors, reversing valves, and control hardware). Many of the early heat pump models had a high mortality rate due to the severe conditions under which they had to operate. These experiences during the 1950s and early 1960s almost destroyed the heat pump industry.

Given some equipment that was of poor quality to begin with, or good equipment that was misapplied or serviced by poorly trained refrigeration

The heat pumps are here!

Save energy without sacrificing comfort

ELECTRIC HEAT PUMP

The heat pump —best hope for home comfort

"It's clean; it doesn't produce soot; I don't have to put up a chimney. And it's quiet"

Now may be the time for ground water heat pumps

...with a heat pump.

Some enthusiasts predict a one-million-unit year by 1980

Here Come the Heat Stealers

In a time of energy shortages, the heat pump, using Du Pont refrigerants, is emerging as an alternative to conventional heating systems

Augments heat pumps
New nocturnal cooling system dissipates heat to night air

25% saturation
1978: year of heat pumps in new homes

30-60 Pct. Savings Possible With Electric Heat Pump

Heat Pump Is Super Efficient

Solar collector and heat pump: Is it a marriage made in heaven?

Is This New-Fangled Thing Worth The Extra Cost?
The Magical (Almost) Heat Pump

Stage Reset for
HEAT PUMP BOOM

Heat pumps save energy in engineering center

Fig. 1.2.   Some favorable statements on Heat Pump. . . .

mechanics, by the early 1960s the product had earned a sufficiently poor reputation that the entire heat pump industry faced impending disaster in the form of plummeting sales and a rapidly growing volume of complaints. By 1964, reliability problems were sufficiently severe for U.S. military authorities to ban heat pumps from military housing, a ban that lasted until 1975.

During the 1960s, electric rates continued to decline, and electric furnaces offered the consumer competitive operating rates compared to the heat pump, along with a higher reliability. During this decade, many manufactur-

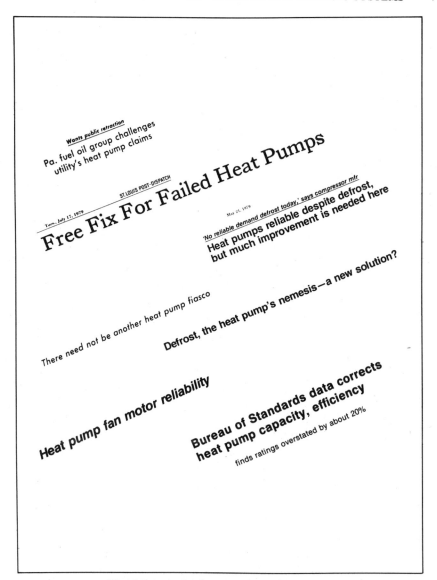

Fig. 1.3.   . . . and some not so favorable.

ers dropped heat pumps from their product lines while others curtailed sales in northern regions of the United States. These occurrences caused a slow-down in heat pump sales, which persisted at a flat rate between 1963 and 1971.

But the industry persevered. During this period, improved heat pump designs were developed that included more reliable compressors and lubri-cating systems, improved reversing valves, and refined control systems. In

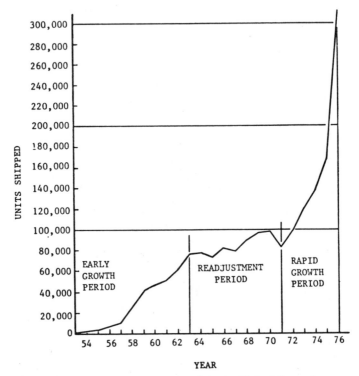

Fig. 1.4.   Manufacture of unitary heat pumps in the United States between 1954 and 1978 [6].

addition, electric utilities set up programs to improve heat pump installations. These programs included training programs for installers, certification of installers and maintenance personnel, and collection and dissemination of service and reliability information.

In 1963, the Edison Electric Institute (EEI) initiated a heat pump improvement research program to improve the reliability and performance of these appliances. The contracts for development of a prototype 3-ton single-phase residential and 5-ton three-phase commercial unit were awarded to Westinghouse and Carrier, respectively, who built 200 units of each for field testing. Improved compressors able to withstand the lubrication and cooling problems associated with reversing the cycle were produced, and new operating cycles were devised. Several million units have now been sold. Also, to add integrity to manufacturer's performance claims, the industry trade association, the Air-Conditioning and Refrigeration Institute, initiated in 1964 a program that certified the basic performance characteristics and the cooling and heating capacity of the product.

Following these American developments, heat pump applications increased worldwide in the late 1960s and early 1970s. Small domestic units

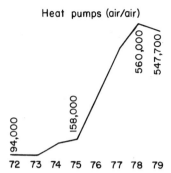

Fig. 1.5.   Heat pump shipment 1972–1979.

using outdoor air as a heat source were produced (notably in Japan, Sweden, and France), and larger units were increasingly incorporated into integrated heat-recovery designs for larger commercial and public buildings, particularly in the United Kingdom and in Germany. Many of the smaller units were reversible for applications with appreciable cooling loads.

Following the oil price increases of the early 1970s, heat pump applications where cooling was not required began to be economically more attractive, and national energy use surveys in Britain and Sweden showed how heat pumping could reduce domestic space-heating energy requirements economically. In many areas of the United States, curtailments were placed on the use of some fossil fuels for heating, and the only alternatives were oil furnaces, electric furnaces, and heat pumps.

The reliability of improved designs had been established, and while questions about the reputation earned in the 1950s still remained, the evidence indicated that the heat pump industry was ready. The combination of these events had led to a period of explosive growth, and the industry moved from a level of 82,000 units shipped in 1971 to over 560,000 units in 1978, which was truly the year of the heat pump, with its installation in 42% of the 797,000 new homes built with central air-conditioning that year. Although in 1979 heat pumps recorded their first plateau since the previous low in 1971, the 547,694 unitary heat pumps shipped that year showed graphically that the renewed availability of natural gas did not sound the heat pump's death knell (Fig. 1.5). In fact, if the increases in water-source heat pumps, packaged terminal air conditioners (ptac's), field-erected systems, and room-unit heat pumps were added, the overall heat pump total in 1979 would have shown a healthy increase instead of a 2% decline from the 560,000 units shipped in 1978. Unitary heat pump sales declined in 1980 as a reflection of the disruption of the new-home market by expensive money rather than the heat pump's inability to compete in a market once more dominated by natural gas.

If we consider an average life of 15 years, there are somewhat over 3 million unitary heat pumps in use today. In the interests of energy conservation, development started in the mid 1970s on nonelectric heat pumps operating directly from the combustion of fuel [12], and it is possible that such equipment, using a gas-fired Stirling engine, might be marketed in the United States shortly.

Throughout the 1970s and the early 1980s especially, a considerable literature has grown on the subject of heat pumps, although, of course, a number of articles on the subject date back to much earlier. A very brief discussion follows of selected technical literature describing prior work. This review makes no attempt to be comprehensive; rather, its purpose is to give the reader some additional source material on the topic of heat pumps.

A survey of the technical literature on heat pumps properly begins with the well-known 1966 monograph by Ambrose [13]. It discusses the thermodynamics of the Carnot and Rankine refrigeration cycle, heat sources and sinks for the heat pump, typical configurations and applications, energy-estimation methods, and residential and commercial diversified demand characteristics. Additional discussions of heat pump fundamentals may be found in several of the *ASHRAE Handbook and Product Directory* volumes [14–16].

For nonresidential buildings, projections by the U.S. Department of Energy (DOE) indicate electric heat pump sales were as high as 581,000 units in 1981 (Table 1.1). Potential market growth through the year 2000 is set at 2% per year. The potential market in terms of area, tonnage, and units for selected years is shown in Table 1.2.

The gas heat pump growth pattern is more limited than the electric heat pump market, because not all areas of the United States are served by natural gas. But nonetheless the DOE predicts that nonresidential building market potential for gas heat pumps is estimated to be 479,000 units in 1981 (Table 1.3). Area, tonnage, and units estimated for selected years are given in Table 1.4.

Early efforts at improving electric heat pump efficiency and reliability are described in Edison Electric Institute's final report on the Heat Pump Improvement Research Project [17]. A study by Hise [18], of Oak Ridge National Laboratory, discusses seasonal performance in detail and presents a critical review of some of the literature data on comparative energy consumption. Reliability is discussed in an appendix to the Hise report.

Studies sponsored by the Federal Energy Agency (FEA) and the DOE have been conducted and reported [19, 20] by Gordian Associates on the reliability, life-cycle cost, market prospects, institutional factors, and primary or resource energy efficiency of heat pumps for residential and commercial applications.

Heat pumps have been the subject of a number of technical conferences. Oklahoma State University has organized annual conferences on heat pump technology since 1975 at Stillwater, Oklahoma. In 1975, The Pennsylvania State University organized a conference on solar energy heat pump systems for heating and cooling of buildings. Proceedings of these conferences have been assembled or published. In addition, recent books on heat pumps have been written or edited by Heap [21], Reay and Macmichael [22], and Collie [23].

TABLE 1.1. Potential Nonresidential Building Market for Electric Heat Pumps (1981)

Building Type	New Construction, 1981 (ft² × 10⁻⁶)	Light Service Factor[e]	New Construction (ft² × 10⁻⁶)	Potential Market Conversions per Replacements[f] (ft² × 10⁻⁶)	ft² × 10⁻⁶	Total Potential Market Tonnage[g] (× 10⁻³)	Units[g] (× 10⁻³)
Office[a]	502	50	251	25	276	920	193
Commercial[b]	420	75	316	32	348	1160	242
Institutional[c]	618	31	193	19	212	707	167
Total[d]	1540	—	760	76	836	2787	581

[a] Includes miscellaneous and public buildings.

[b] Includes stores, warehouses, and shopping centers.

[c] Includes hospitals, educational, and religious buildings.

[d] Gordian estimate excluding industrial construction.

[e] Estimated percent of building type that is air conditioned and a potential user of an electric heat pump up to 25 tons capacity. (*The institutional factor is weighted.)

[f] Estimated 10% of new construction market.

[g] Number of units based on 1 ton cooling requirement for every 300 ft², with each unit having an average 4.8-ton capacity. SOURCE: Gordian Associates, Inc.

11

TABLE 1.2.   Potential Nonresidential Electric Heat Pump Market

Year	Building Square Footage ($\times 10^{-6}$)	Heat Pump Tonnage ($\times 10^{-3}$)	Heat Pump Units ($\times 10^{-3}$)
1977	770	2567	535
1981	836	2787	581
1985	903	3010	627
1990	997	3323	692
1995	1100	3667	764
2000	1215	4050	844

SOURCE: Gordian Associates, Inc.

1.2.   FACTS ABOUT ENERGY AND POWER

1.2.1.   Forms of Energy

Generally speaking, all energy on earth comes from the sun. Energy exists in different forms: chemical, nuclear, sound, light, mechanical, heat, and so on. Energy can be converted from one form to another, and all energy eventually converts into heat.

Energy can be defined as the ability to do work, and work is defined as a force acting through a distance, for example, the lifting of an object to a higher elevation. If a box weighing 10 lb is lifted from the floor onto a table that is 3 ft above the floor, the work required is 10 lb times 3 ft, or 30 ft·lb. To do this work will theoretically require 30 ft·lb of energy.

There are three important aspects of the concept of energy.

1. All matter and all things have energy.
2. The energy of the whole is the sum of the energies of the parts.
3. Energy is conserved.

The first recognizes that energy is a property of matter; the molecules in a certain chunk of matter, the electromagnetic waves in a certain field of radiation, or the cells in a certain living organism have energy. The second aspect tells us that the amount of energy in a complex system is the sum of the energies in its various parts. This may seem trivial, but matter has other properties (for example, temperature) that do not behave in this manner. The third states that the energy of a system that does not interact in any way with anything else is constant. In other words, the universe always contains

TABLE 1.3. Potential Nonresidential Building Market for Gas Heat Pump (1981)

Building Type	New Construction, 1981 (ft² × 10⁻⁶)	Light Service Factor[d]	New Construction (ft² × 10⁻⁶)	Potential Market Conversions per Replacement[e] (ft² × 10⁻⁶)	ft² × 10⁻⁶	Total Potential Market[f]	
						Tonnage (× 10⁻³)	Units (× 10⁻³)
Office[a]	502	41.3	207	21	228	760	158
Commercial[b]	420	62.0	261	26	287	957	199
Institutional[c]	618	25.6*	159	16	175	583	122
Total	1540	—	627	63	690	2300	479

[a] Includes miscellaneous and public buildings.

[b] Includes stores, warehouses, and shopping centers.

[c] Includes hospitals, educational and religious buildings.

[d] Estimated percent of building type that is air conditioned and a potential user of an electric heat pump up to 25 ton capacity. (*The institutional factor is weighted.)

[e] Estimated 10 percent of new construction market.

[f] Number of units based on 1 ton cooling requirement for every 300 ft², with each unit having an average 4.8 ton capacity.

SOURCE: Gordian Associates, Inc.

13

TABLE 1.4. Potential Nonresidential Gas Heat Pump Market

Year	Building Area $(ft^2 \times 10^{-6})$	Heat Pump Tonnage $(\times 10^{-3})$	Heat Pump Units $(\times 10^{-3})$
1981	690	2,300	479
1985	743	2,427	516
1990	814	2,713	565
1995	893	2,977	620
2000	922	3,240	675

SOURCE: Gordian Associates, Inc.

the same amount of energy (although perhaps the *form* of the energy might change). We cannot really tell anybody what energy is without invoking these three ideas; they are part of the basic concept of energy. They are as fundamental as energy itself.

1.2.2. Conversion of Energy

A law of physics known as the *law of conservation of energy* states that energy can be neither created nor destroyed. This law is very important in understanding how energy is used and how it can be conserved. The 30 ft·lb of energy that was required to lift the box onto the table is not used up and lost; it is stored in the box as "potential energy." If at a later time the box were lowered from the table back to the floor, the 30 ft·lb of stored potential energy would be available to do some other work, such as lifting an object to another height, assuming box and object are connected to a system of frictionless, massless cords and pulleys. An example of such a device is the double-hung window with counterweights in the window jamb. When the upper window is pulled down, counterweights rise within the jamb; force expended by the person to pull down the window is stored as potential energy in the counterweights. When the window is again raised, energy stored in the counterweights provides most of the required work.

A simple equation, called the "energy balance," describes this conservation of energy for any system:

$$E_{in} = E_{out} + \Delta E_{stored}$$

Here E stands for energy; E_{in} represents the amount of energy that enters the system in the time period under consideration, E_{out} the amount that leaves the system in this period, and ΔE_{stored} the change in the amount of energy stored within the system. In the language of mathematics, the symbol Δ is used to denote "final minus initial" (which of course is the "change"), and so

$$\Delta E_{\substack{\text{stored}}} = E_{\substack{\text{final} \\ \text{stored}}} - E_{\substack{\text{initial} \\ \text{stored}}}$$

Aside from the direct work energy, there are other forms of energy. Most people have heard of the Boy Scout's technique of starting a fire by rubbing two sticks together. This technique uses different forms of energy. It takes work (and thus energy) to rub the sticks together because of a resistance to the rubbing, which is called *friction*. That is, the sticks are not completely free to move against one another when they are pressed together; it takes energy or work to make them slide. But this energy is not stored as was the energy used to lift the box. What happens is that the sticks get hot, and work energy is thus converted to heat energy. Usually heat energy is measured not in foot-pounds, but in British thermal units, Btu. It requires 778 ft·lb of work to obtain 1 Btu of heat energy. In this manner, the work energy of rubbing sticks is converted to heat energy. When enough heat energy is accumulated, it could ignite some combustible materials, such as wood chips, and a fire would be started by a process of combustion. In the combustion process, chemical energy stored in the wood chips would combine with oxygen in the air and release thermal or heat energy. The chemical energy stored in the wood is useless until it is released through this combustion process in the form of heat. Once the heat is released, energy can be utilized for space and water heating, cooking, and other processes.

Gasoline used in an automobile engine ignited by spark plugs in a mixture with air (containing 21% oxygen) converts chemical energy to heat and work energy, providing the mechanical force to move the car.

1.2.3. Fuels

The materials used to release stored chemical energy by combustion are called fuels; most common fuels contain as basic ingredients carbon and hydrogen, or hydrocarbons. There are two common classifications of fuels: biofuels and fossil fuels.

1. Bio-fuels are products of growing matter such as trees and plants. These matters receive energy from the sun which is combined with other chemicals of the earth as they grow over a relatively short time span. At some stage of growth, these bio-fuels can be harvested and used for fuel with replacements planted for a later harvest, for example, a forest cut and replanted every 50 years.

2. Fossil fuels are generally considered nonreplenishable and therefore depleting. Fossil fuels also derived their energy from the sun, but it took millions of years for them to form their present state. Common fossil fuels are natural gas, petroleum, and coal.

The amount of energy in the form of heat that can be obtained by burning a fuel is called the *heating value* of the fuel. Some of the most common materials and their approximate heating values are given in Table 1.5.

1.2.4. Electrical Energy

Another familiar form of energy is electrical energy. Electrical energy provides a very useful method or means of moving energy from one place to another, or of transferring from one form to another. Electrical energy can be obtained from chemical energy (fuel cell and batteries), nuclear energy, heat energy (thermoelectric and thermionic), magneto-hydrodynamic energy, mechanical energy (hydro), and thermal-mechanical energy. Over 90% of the electrical energy used in this country is generated by thermal-mechanical methods, driving a rotary electrical generator with a steam turbine. The steam turbine derives its energy from expanding steam produced by burning fossil fuel. In this energy conversion process, chemical energy is converted first to heat energy, then to mechanical energy, and finally to electrical energy. Electrical energy is then sent through transmission lines to distant locations where it is converted to other useful forms, such as mechanical energy (running a fan), heating energy (cooking), and lighting energy (lighting a task).

Electrical energy is usually measured in kilowatt-hours (kW·hr or kWh) where 1 kWh is equivalent to 3413 Btu. Another energy unit commonly used is "horsepower-hour" (hp·hr) where 1 hp·hr is the equivalent of 2545 Btu.

1.2.5. Energy Units

There are several important measures of energy, or energy units, in common use today. These include

1. The *calorie* (cal),* which is roughly the amount of energy that must be added to 1 g of water to increase its temperature by 1°C.
2. The *British thermal unit* (Btu), which is roughly the amount of energy that must be added to 1 lb of water to increase its temperature 1°F. 1 Btu = 252 cal.
3. The *joule* (J), which is precisely the amount of energy that is used by a 1-W light bulb in 1 s. 1055 J = 1 Btu.

The British thermal unit (Btu) is the most common unit for energy measurement or for a common base of energy conversion. However, it is a very

---

*1000 cal = 1 kcal (kilocalorie). Food values are usually stated in kilocalories, often denoted simply as Calories (note the capital C, which is often left out inadvertently by dietitians).

TABLE 1.5. Typical Heating Values of Energy Sources

Material	Heating Value "As Fired"
Solids	Btu/lb
Anthracite coal	13,000
Bituminous coal	12,000
Subbituminous coal	9,000
Lignite coal	6,900
Coke	11,000
Newspaper	8,000
Brown paper	7,300
Corrugated board	7,000
Magazines	5,300
Waxed milk cartons	11,400
Asphalt or tar	17,000
Typical urban refuse	5,000
Corn cobs	8,000
Rags	7,500
Wood	9,000
Liquids	Btu/gal
Fuel oil	
grade 1	135,000
grade 2	140,000
grade 6	154,000
Kerosene	133,000
Gasoline	111,000
Methyl alcohol	68,000
Ethyl alcohol	88,000
LPG	91,000
Gases	Btu/ft^3
Natural gas	1,000
Commercial propane	2,500
Commercial butane	3,200
Acetylene	1,500
Methane	950
Bio-gas	500
Mass per se	Btu/lb
Uranium (1000 ppm U in ore)	
in LWRs	175,000
in breeder reactors	17,800,000
Fusion (deuterium)	145,000,000,000
Complete "mass-energy" conversion	39,000,000,000,000

TABLE 1.6. Energy Equivalent

Form	Unit of Measurement	Btu Equivalent
Heat	British thermal unit (Btu)	—
Electrical	Kilowatt-hour (kWh)	3413
Mechanical	Horsepower-hour (hp·hr)	2545
Chemical (fuel)	Pounds, gallons, cubic feet	See Table 1.5

small unit compared to the amount usually required in building systems. In building energy evaluation, one must constantly deal with astronomically large numbers. Frequently, therms (1 therm = 10^5 Btu), mega-Btus (10^6 Btu), giga-Btu (10^9 Btu), quads (10^{12} Btu), or Q's (10^{15} Btu) are used for expressing annual energy consumption.

Forms of energy common in building systems and their units of measurement are given in Table 1.6.

1.2.6. Power versus Energy

Power has often been confused with energy. In discussing energy conservation and energy management, it is very important to understand the difference between these two terms.

An important aspect of systems that use or produce energy is the *rate* of energy use or production. A battery that could deliver 20×10^8 J would be ideal for an urban electric automobile, but the battery would be useless unless this energy could be delivered in a few hours (nobody wants to take three days to go to the supermarket). The *power* of the device is the characteristic that describes the *rate* at which it supplies or uses energy. Power is sometimes referred to as the *demand*.

Power is the rate of consumption or conversion of energy; that is, power is an expression of how long or how quickly a given amount of energy is consumed or converted. An example: if 100 hp·hr of energy is used in 1 hr, the power required would be 100 hp·hr divided by 1 hr, or 100 hp. A motor or engine providing energy at this rate would be a 100-hp motor. However, if the same amount of energy were used over a 10-hr period, it would require only a 10-hp motor. Denoting power by P, energy by E, and the time interval by t, the relation between power and energy is

$$\text{Power} = \frac{\text{energy}}{\text{time}} \quad \text{or} \quad P = \frac{E}{t}$$

The power used by a toaster is about 1000 J/s. The unit combination *joule per second* is called the *watt* (W) after a famous contributor to energy technology. 1000 W is termed a kilowatt ("kilo" = 1000), abbreviated kW. So,

TABLE 1.7. Power Units

Form	Unit of Measurement	Btu/hr Equivalent
Heat	Btu per hour	1
Electrical	Kilowatt	3,413
Mechanical	Horsepower (hp)	2,545
Cooling	Tons of refrigeration	12,000
Chemical (fuel)	Btu per hour	1

your toaster power is about 1 kW. Power units are often given in kilowatts, which is why the "toaster calibration" is so useful. Another unit of power that is in common use is the *horsepower,* abbreviated hp. One horsepower is equivalent to about ¾ kW.

Since electrical power is usually measured in kilowatts, electrical energy is usually given in kilowatt-hours (kWh); power times time equals energy. Most appliances, lights, etc., have power requirements in kilowatts or watts stamped on them.

1.2.7. Power Units

If each energy term in Table 1.6 is divided by time, then the power unit for heat, mechanical, and electrical energy will be Btu/hr, hp, and kW, respectively.

To cool a building, it is necessary to remove heat (energy) from the space. This cooling procedure is more commonly called "air-conditioning." The rate at which heat is removed can be expressed as Btu/hr (also Btuh) cooling, or tons of refrigeration. Common power units and their equivalent values in Btu/hr are given in Table 1.7.

The units of power in any system can be found by dividing work units by time units. In the fps gravitational system,* power may be expressed as foot-pounds per second or foot-pounds per minute. Since the days of James Watt and his steam engine, the horsepower has been a common unit of power and is numerically equal to 550 ft·lbf/s or 33,000 ft·lbf/min. The SI unit of power is the watt, where 746 W = 1 hp.

$$
\begin{aligned}
1 \text{ hp} &= 550 \text{ ft·lbf/s} \\
&= 33,000 \text{ ft·lbf/min} \\
&= 746 \text{ W} \\
&= 0.746 \text{ kW}
\end{aligned}
$$

---

*System of units in which the basic units are the foot, pound, and second. Now largely displaced by other systems, such as S1.

EXAMPLE 1.1

A box weighing 1100 lbf (see Fig. E1.1) is lifted 15 ft in 3 s. How much power is necessary?

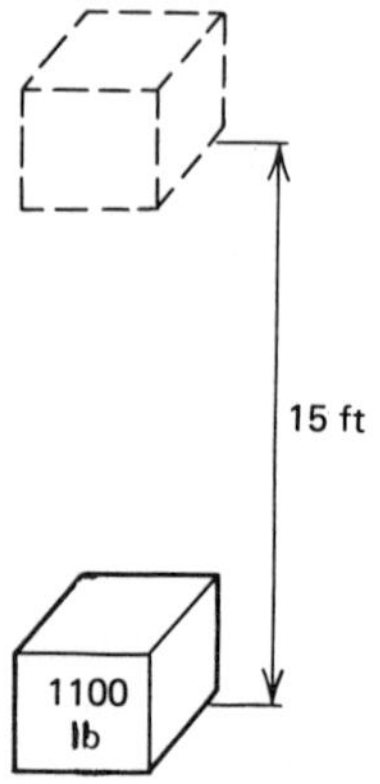

Solution.

$$\text{Power} = \frac{\text{work}}{\text{time}}$$

$$\text{Power} = \frac{(1100)\text{lbf}(15)\text{ft}}{3 \text{ s}} = 5500 \text{ ft·lbf/s}$$

$$\text{Horsepower} = \frac{\text{work in ft·lbf}}{(\text{time in s})(550)} = \frac{(1100)(15)}{(3)(550)} = 10 \text{ hp}$$

$$\text{Power} = 746 \frac{\text{W}}{\text{hp}} (10 \text{ hp}) = 7460 \text{ W}$$

15 ft

1100 lb

Fig. E1.1.

EXAMPLE 1.2

How much will it cost to operate a 150-W electric light for 2.5 hr when the utility company charges are 6.5 cents per kilowatt-hour?

Solution.

Work (or energy) in kWh = (power in kW)(time in hr)

$$\text{Energy} = \left(\frac{150}{1000}\right)(2.5)$$
$$= 0.375 \text{ kWh}$$
$$\text{Cost of electric work (or energy)} = (\text{kWh})(\text{cost per kWh})$$
$$= (0.375)(6.5)$$
$$= 2.43 \text{ cents}$$

The horsepower unit is used for motors, pumps, and other devices that produce or use energy. Your kitchen refrigerator probably has about a 1-hp motor; your heart uses about 0.01 hp; your automobile engine may be rated at 300 hp but probably seldom actually produces more than 100 hp, even under drag-strip conditions. Large jet engines produce about 25,000 hp. The largest modern steam turbines in central power stations produce about a million horsepower. Common conversion factors are given in Appendix D.

1.2.8. Effect of Power on Costs

Consider the case of the 100 hp · hr of energy consumed in the earlier example. If it is feasible to use a 10-hp motor for 10 hr per day, it would certainly be better than using a 100-hp motor for only 1 hr per day because the investment cost of a 100-hp motor is much higher than the 10-hp motor even though they both consume the same amount of energy (100 hp · hr) per day. It can therefore be said that power is related to investment cost, and energy is related to operating cost.

Electrical consumption for a building is generally charged for by the amount of energy (kWh) consumed per month. However, for large users, the utility company may also base its charge on power demand, which is normally determined by the maximum demand during any 15-min interval in a month or year. In this case power demand is also related to operating cost for the user. The utility company, however, must charge for the demand (power) to offset their investment cost in the generating plant and distribution network.

1.2.9.   Energy Required in Buildings

Energy is generally used in buildings to perform functions of heating, lighting, mechanical drives, cooling, and special applications. Figure 1.6 indi-

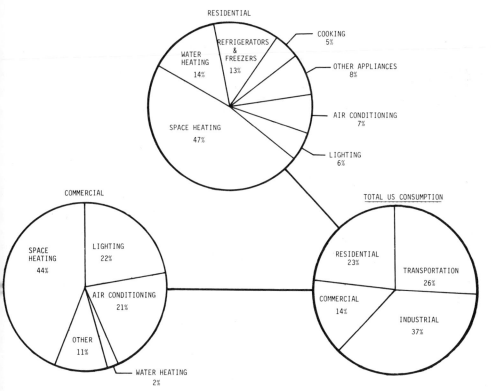

Fig. 1.6.   End use of energy in the United States.

TABLE 1.8.   Typical Building Design Heat Losses/Gains

Building Type	Air Conditioning (ft^2/ton)	Heating [(Btu/hr)/ft^3]
Apartment	350–450	4.9
Bank	200–250	3.2
Department store	200–250	0.75–1.2
Dormitory	350–450	4.9
House	600–700	3.2
Medical center	250–300	4.9
Night club	100–250	3.2
Office		
Interior	300–350	3.2
Exterior	225–275	3.2
Post office	200–250	3.2
Restaurant	100–250	3.2
School	225–275	3.2
Shopping center	200–250	3.2

cates the relation of some of these uses of energy to total U.S. energy consumption. The energy is available to the building in limited forms, such as electricity, fossil fuels, and solar energy, and these energy forms must be converted within the building to serve the end use of the various functions. A loss of energy is associated with any conversion process. In energy-conservation efforts, there are two avenues of approach: reducing the amount of use and/or reducing conversion losses. For example, the furnace that heats the building produces unusable and toxic flue gas, which must be vented to the outside of the building, and thus part of the energy is lost. (The "lost" energy is not destroyed, it simply ends up as low-grade heat energy.) Table 1.8 presents typical values for building heat losses and gains at design conditions.

The ultimate source of energy, as stated earlier, is the sun. It is plentiful and nondepletable for millions of years. However, from the perspective of current technology and the social systems, the major energy resource is still fossil fuel, which is a depletable commodity. Even though electricity can be converted within buildings to other forms of end use with relatively low losses, it has already undergone a conversion process in which approximately two-thirds of the fossil-fuel energy was lost, unless it was generated by a hydroelectric plant, where energy is obtained from a drop in water level (potential energy).

Under the National Energy Act of 1978, a building's total annual energy use will be expressed in Btu's per gross square foot and energy cost per gross square foot. The following conversion factors will be used:

Electricity	11,600 Btu/kWh
Natural gas	1030 Btu/ft^3
Distillate fuel oil	138,690 Btu/gal
Residual fuel oil	149,690 Btu/gal
Coal	24.5 million Btu per standard short ton
LP gases, including propane and butane	95,475 Btu/gal
Steam	1390 Btu/lb

The Department of Energy said the conversion factors result "from comparisons of buildings which indicate that using an index of Btu/ft^2 can more accurately indicate conservation potential if electricity used is converted to the Btu at the point of generation."

REFERENCES

1. Thomson, W., "On the Economy of the Heating or Cooling of Buildings by Means of Currents of Air," *Glasgow Phil. Soc. Proc.* **1852,** pp. 269–272.
2. Krauss, F., "The Heat Pump in Theory and Practice," *Power* **54,** 298–300 (1921).
3. Morley, T. B., "The Reversed Heat Engine as a Means of Heating Buildings," *Engineer,* 10 February, 145–146, 1922.
4. Haldane, T. G. N., "The Heat Pump—An Economical Method of Producing Low-Grade Heat from Electricity," *J.I.E.E.E.,* 666–675 (1930).
5. Doolittle, H. L., "Heating with Refrigerating Equipment Cuts Current Costs," *Power,* **76,** 29–31 (1932).
6. Pietsch, J. A., "The Unitary Heat Pump Industry—25 Years of Progress," *ASHRAE J.,* **19**(7), 15–18 (1977).
7. Egle, M., "The Heating of the Zurich Town Hall by the Heat Pump," *SEV Bull.* **29,** 261–273 (1978).
8. Kemler, E. N., and Oglesby, S., *Heat Pump Applications,* McGraw-Hill, New York, 1950.
9. Joint AEIC-EEI Heat Pump Committee, "Suggested Field Test Procedure for Determination of Coefficient of Performance and Performance Factor of an Electric Heat Pump While Operating on the Heating Cycle," *Edison Electric Inst. Bull.* **16,** 341–348 (1948).
10. Sporn, P. and Ambrose, E. R., "Tests Show How Heat Pumps Perform in Different Climates," *Electrical World* **135**(4), 97–100 (1951).
11. Baker, M., "Design and Performance of a Residential Earth Heat Pump," *ASHVE Trans.* **59,** 371–394 (1953).
12. Sarkes, L. A., Nicholls, J. A., and Menzer, M. S., "Gas Fired Heat Pumps: An Emerging Technology," *ASHRAE J.,* **19**(3), 36–41 (1977).
13. Ambrose, E. R., *Heat Pumps and Electric Heating,* Wiley, New York, 1966.
14. American Society of Heating, Refrigerating and Air Conditioning Engineers, *Handbook and Product Directory, 1977, Fundamentals,* ASHRAE, New York, 1977.

15. American Society of Heating, Refrigerating and Air Conditioning Engineers, *Handbook and Product Directory, 1979, Equipment,* ASHRAE, New York, 1979.

16. American Society of Heating, Refrigerating and Air Conditioning Engineers, *Handbook and Product Directory, 1980, Systems,* ASHRAE, New York, 1980.

17. Edison Electric Institute, Final Report, *Heat Pump Improvement Research Project* (RP69), Publ. No. 71-901, May 1971.

18. Hise, E. C., *Seasonal Fuel Utilization Efficiency of Residential Heating Systems* (ORNL Report No. ORNL-NSF-EP-82, April 1975).

19. Gordian Associates, Inc., *Evaluation of the Air-to-Air Heat Pump for Residential Space Conditioning,* Report prepared for FEA, NTIS No. PB-255-652, April 1976.

20. Gordian Associates, Inc., *Heat Pump Technology, A Survey of Technical Development Market Prospects,* Report prepared for DOE, HCP/M2121-01, June 1978.

21. Heap, R. D., *Heat Pumps,* Halsted Press, John Wiley & Sons, London, 1979.

22. Reay, D. A., and Macmichael, D. B. A., *Heat Pumps: Design and Application,* Pergamon Press, Oxford, 1979.

23. Collie, M. J. (Ed.), "Heat Pump Technology for Saving Energy," Energy Technology Review No. 39, Noyes Data Corporation, Park Ridge, NJ, 1979.

PROBLEMS

1. Calculate the energy equivalent of 0.1 Q (~annual U.S. energy use) in:
   (a) Tons of coal.
   (b) Gallons of #2 fuel oil.
   (c) Gallons of LPG.
   (d) Cubic feet of natural gas.
   (e) Barrels of oil.

2. The kilowatt-hour meter for a residence had the following readings:

March 1   19,389

April 1   20,458

Determine the amount of energy, in kWh and in Btu, used for the month, the average electrical power, and the cost if the following rate schedule applies.

Electric Rate Breakdown

Minimum bill (up to 25 kWh)	$5.00
Next 200 kWh	@5.45¢
Next 275 kWh	@3.15¢
Next 1000 kWh	@2.60¢
Next 3000 kWh	@2.15¢
All over 4500 kWh	@2.55¢

3. A factory has the following electrical power–time history:

8 A.M.–11 A.M.	760 kW
11 A.M.–2 P.M.	430 kW
2 P.M.–4 P.M.	1870 kW
4 P.M.–8 P.M.	319 kW
8 P.M.–8 A.M.	199 kW

Determine the daily energy usage, in kWh.

4. Prepare a table of power and energy requirements as shown in Table P.1.4.

TABLE P.1.4.   Actual and Predicted Power and Energy Requirements for the United States

Year	1980[a]	1985	2000	2050
Total energy, Q				
Electrical Energy, kWh				
Peak Electrical Power, kW				
Per capita energy, Btu/yr				
Per capita electrical energy, kWh/yr				
Per capita electrical power, kW				

[a] Or most recent year for which actual values are available.

5. A 110-V household circuit is fused at 30 A. What is the maximum power that can be drawn from this circuit?

6. A household hot-water system heats water only as it is needed. The water enters the electric heater at 60°F and is heated to 150°F. For a flow rate of 2 gal/min determine (a) required power rating of the heater (watts), and (b) for operation on 230 V, the size breaker (amps) needed.

7. A 10-kW heater is used to heat each of the following from 50 to 200°F. How long does the process take?
   (a)   1 ft^3 of water (density = 62.4 lb/ft^3).
   (b)   1 ft^3 of air at 1 atm (volume held constant).
   (c)   1 ft^3 of concrete (density = 100 lb/ft^3; specific heat = 0.2 Btu/lb·°F)

2

THERMODYNAMICS

2.1. SOME DEFINITIONS AND BASIC CONCEPTS*

What is thermodynamics?† Let's look at the word itself. Breaking it up into two pieces we find that it is made up of a couple of Greek or Latin words—*thermo,* which means "hot" or "hot stuff," and *dynamics,* which has to do with action or motion. Thermodynamics is the study of heat in motion.

What is the greatest boon to mankind today? What is the one single thing that separates civilized man from uncivilized man? The indoor bathroom. And what makes the indoor bathroom so wonderful? The plumbing put in by a bunch of civil engineers? No. Its warmth. Thermo.

And in case you have gotten the idea that thermo just goes around making it hot for people, consider for a moment the refrigerator. Now this is also a very important application of thermo. This comes from the law of thermodynamics which comes after the first law which says that heat runs downhill. This law says that you cannot make a refrigerator work, but engineers have figured out a way. It works because you use that motor in the refrigerator to turn the hill upside down so that when the heat runs down it is really running up and then it drips out into that pan underneath on the old-style refrigerators, or in the new types it is taken out by the thing they call an automatic defroster.

Seriously, what is thermodynamics?

Very briefly, thermodynamics is the *study of energy and its transformations.*

---

*The *serious* student should proceed directly to Section 2.2.
† Description of thermodynamics based on a speech by Professor Dwight A. Nesmith presented at the annual Spring Banquet, Pi Tau Sigma Honorary Mechanical Engineering Fraternity, Kansas State University, April 1957.

We can also say immediately that all of thermodynamics is contained within four apparently simple statements known as the *zeroth, first, second, and third laws of thermodynamics*.

The Zeroth Law of Thermodynamics.   The zeroth law states that when two bodies have equality of temperature with a third body, they in turn have equality of temperature with each other.

This law is really the basis of temperature measurement, and although obvious to us it is not derivable from other laws and in the logical presentation of thermodynamics precedes the first law.

The First Law of Thermodynamics.   The first law says that energy is conserved. You don't get something for nothing. Energy can neither be created nor destroyed, but it can be transformed.

The Second Law of Thermodynamics.   The second law says: "You can't beat city hall," or "The house always takes its cut." Entropy cannot be destroyed, but it can be created.

Entropy is that "house cut" and it is computed by a complicated formula which needs calculus or can be looked up in a steam table. There is one thing that is as sure as death and taxes and that is that you are always going to have more entropy than you started with.

Two classical statements of this law are:

"No process is possible whose sole result is the extraction of heat from a single reservoir and the performance of an equivalent amount of work." (Kelvin-Planck)

"No process is possible whose sole result is the removal of heat from a reservoir at one temperature and the absorption of an equal quantity of heat by a reservoir at a higher temperature." (Clausius)

The second law involves the fact that processes proceed in a certain direction, and not in the opposite direction, e.g. gasoline is used as a car drives up a hill, but as it coasts down the hill the fuel in the tank cannot be restored. The second law of thermodynamics is nothing more or less than a generalized statement of such phenomena.

The Third Law of Thermodynamics.   A statement of the third law, attributed to Nernst and frequently known as the Nernst theorem, is as follows: "The absolute entropy of a pure crystalline substance in complete internal equilibrium is zero at zero degrees absolute." Although there is some question about the possibility of nuclear-spin energies at a temperature of zero degrees absolute, it appears that the Nernst theorem is valid. Even though the entropy of a pure crystalline substance may be zero at zero degrees absolute, this condition may not be true for other substances. Planck suggested that the entropy of a substance whose temperature is absolute zero is a function of the logarithm of the number of configurations in which the substance exists. For a pure crystalline substance, there is only one configuration, and hence the entropy is zero. It may be possible to supercool a certain type of solution to a temperature close to absolute zero. However, the entropy of such a solution cannot approach absolute zero, since a solution must possess the entropy caused by mixing.

In the final analysis, the entropy of any substance will not be zero unless the molecules of the system have been so arranged that they have their highest possible degree of order.

2.2. THERMODYNAMIC FUNDAMENTALS

Thermodynamics is the science devoted to the study of energy, its transformations, and its relation to states of matter. Since every engineering operation involves an interaction between energy and materials, the principles of thermodynamics can be found to apply to all engineering activities. Thermodynamics may be studied or undertaken from either a microscopic or macroscopic point of view. The microscopic point of view would consider matter to be composed of molecules and would concern itself with the actions of these individual molecules. The macroscopic point of view is concerned with effects of the action of many molecules and is not concerned with the action of the individual molecules. The macroscopic approach, then, actually considers the average properties of a large group of molecules. When the macroscopic view of matter is pursued, the field is called *classical thermodynamics* or quite often just *thermodynamics*. When the view is microscopic, the science is called *statistical thermodynamics* and includes kinetic theory, statistical mechanics, quantum mechanics, and wave mechanics.

Thermodynamics is a physical theory of great generality impinging on practically every phase of human experience. It may be called the description of the behavior of matter in equilibrium and its changes from one equilibrium state to another. Thermodynamics operates with two master concepts and two great principles. The concepts are *energy* and *entropy*, and the principles are called the first and second laws of thermodynamics, although they are not really laws in the strict physical sense, since they do not describe regularities in experience directly. They are hypotheses whose use is justified by the agreement of their consequences with experience. The idea of energy is the embodiment of the attempt to find in the physical universe an invariant, something that remains constant in the midst of the obvious flux of change. It is in the transformation process that nature appears to exact a penalty, and this is where the second principle makes its appearance. Every naturally occurring transformation of energy is accompanied, somewhere, by a loss in the availability of energy for the future performance of work.

The German physicist Rudolf Clausius (1822–1888) invented the concept of *entropy* to describe quantitatively the loss in available energy in all naturally occurring transformations. Thus, although the natural tendency is for heat to flow from a hot to a colder body with which it is placed in contact, corresponding to an increase in entropy, it is perfectly possible to make heat flow from the colder body to the hot body, as is done every day in a refrigerator. But it does not take place naturally or without some extra effort exerted somewhere.

As Clausius epitomized the fundamental principles of thermodynamics, the energy of the world stays constant and the entropy of the world increases without limit. If the essence of the first principle in everyday life is that we cannot get something for nothing, the second principle emphasizes that every time we do get something we reduce by a measurable amount the opportunity to get that something in the future, until ultimately there will be no more "getting." This is the "heat death" envisioned by Clausius and Boltzmann, when the whole universe will have reached a dead level of temperature. Although the total amount of energy will be the same as ever, there will be no means of making it available; the entropy will have reached its maximum value.

Like all sciences, thermodynamics is based on experimental observation. In thermodynamics these findings have been formalized into certain basic laws. In the sections that follow, we shall present these laws and the thermodynamic properties related to them, and apply them to a number of representative examples. The examples, and the problems at the end of each chapter, are presented to help the reader gain a thorough understanding of the fundamentals and an ability to apply these fundamentals to thermodynamic problems. It is not necessary to memorize numerous equations, for problems are best solved by the application of the definitions and laws of thermodynamics.

Thermodynamic reasoning is deductive in character rather than inductive. The reasoning is always from the general law to the specific case. To illustrate the elements of thermodynamic reasoning, the analytical processes may be arbitrarily divided into two steps:

1. Idealize or substitute an analytical model for a real system. This step is taken in all engineering sciences. Idealizations are fairly easy to make after a little experience, and skill in making them is an essential part of the engineering art.
2. Apply deductive reasoning from the first and second laws of thermodynamics.

Included within these steps are: an energy balance, a suitable properties relation, and an accounting of entropy changes.

2.2.1. System and Surroundings

Most applications of thermodynamics require the definition of a system and its surroundings. A system can be any object, any quantity of matter, any region of space, or similar entity selected for study and set apart mentally from everything else, which then becomes the surroundings. The systems of interest in thermodynamics are finite, and the point of view taken is macroscopic rather than microscopic. That is to say, no account is taken of the detailed structure of matter, and only the coarse characteristics of the system, such as its temperature and pressure, are regarded as thermodynamic coordinates. These are advantageously dealt with because they have a direct relation to our sense perceptions and are measurable.

A *thermodynamic system* is defined as a region in space or quantity of matter bounded by a closed surface upon which attention is focused for study. Everything external to the system is the surroundings, and the system is separated from the surroundings by the system boundaries. These boundaries may be either movable or fixed, and either real or imaginary.

An *isolated system* cannot exchange either matter or energy with its surroundings. If a system is not isolated, its boundaries may permit either matter or energy or both to be exchanged with its surroundings. If the exchange of matter is allowed, the system is said to be *open;* if only energy and not matter may be exchanged, the system is *closed* (but not isolated), and its mass is constant.

2.2.2. Properties and State

A property of a system is any observable characteristic of the system. A listing of a sufficient number of independent properties makes up a complete definition of the state of a system. The more common thermodynamic properties are: temperature, pressure, specific volume or density, internal energy, enthalpy, and entropy.

The *state* of a system is its condition or configuration described in sufficient detail that one state may be distinguished from all other states. The state may be identified or described by specifying properties, such as temperature, pressure, and density. Each property at a given state has only one value, and properties always have the same value for that state.

Thermodynamic properties are divided into two classes, intensive and extensive properties. An *intensive property* is independent of the mass; the value of an *extensive property* varies directly with the mass. If a quantity of matter is divided into two equal parts, the intensive properties of each part will have the same value as the original, and the extensive properties will have half the value. Pressure, temperature, and density are intensive properties. Mass and total volume are extensive properties.

The state of a system is the condition of the system characterized by the values of its properties. Attention will be directed toward what are known as equilibrium states. The word *equilibrium* is used here in its generally accepted sense, the equality of forces or the state of balance. In future discussion the term *state* will refer to an equilibrium state unless otherwise noted. The concept of equilibrium is an important one, since it is only in an equilibrium state that thermodynamic properties have any real meaning. By definition:

A system is in **thermodynamic equilibrium** if it is not capable of a finite, spontaneous change to another state without a finite change in the state of the surroundings.

There are many types of equilibrium, all of which must be met to fulfill the condition of thermodynamic equilibrium. Included among these are thermal, mechanical, and chemical equilibria. If a system is in thermal equilibrium, the system is at the same temperature as the environment, and the temperature will be the same throughout the whole system. If a system is in mechanical equilibrium, no part of the system is accelerating ($\Sigma F = 0$), and the pressure within the system is the same as in the environment. If a system is in chemical equilibrium, the system does not tend to undergo a chemical reaction.

When a system is isolated, it is not affected by its surroundings. Nevertheless, changes may occur within the system that can be detected with measuring devices such as thermometers and pressure gauges. However, such changes are observed to cease after a period of time, and the system is said to have reached a condition of internal equilibrium such that it has no further tendency to change.

For a closed system, which may exchange energy with its surroundings, a final static condition may also eventually be reached such that the system is not only internally at equilibrium but also in external equilibrium with its surroundings.

An equilibrium state represents a particularly simple condition of a system and is subject to precise mathematical description because in such a state the system exhibits a set of identifiable, reproducible properties. Indeed, the word *state* represents the totality of macroscopic properties associated with a system. Certain properties are readily detected with instruments. The existence of other properties, such as internal energy, is recognized only indirectly. The number of properties that may be arbitrarily set at given values in order to fix the state of a system (that is, to fix all properties of the system) depends on the nature of the system. This number is generally small and is the number of properties that may be selected as independent variables for a particular system. These properties then represent one set of thermodynamic coordinates for the system.

To the extent that a system exhibits a set of identifiable properties it has a thermodynamic state, whether or not the system is at equilibrium. Moreover, the laws of thermodynamics have general validity, and their application is not limited to equilibrium states.

The importance of equilibrium states in thermodynamics derives from the fact that a system at equilibrium exhibits a set of fixed properties which are independent of time and which may therefore be measured with precision. Moreover, such states are readily reproduced from time to time and from place to place. Since any property of a thermodynamic system has a fixed value in a given equilibrium state, regardless of how the system arrives at that state, the change that occurs in the value of a property when a system is altered from one equilibrium state to another is always the same. This is true regardless of the method used to bring about a change between the two end states. The converse of this statement is equally true. If a measured quantity always has the same value between two given states, that quantity is a measure in the change in a property. This latter assertion will be useful to us in connection with the conservation of energy principle introduced in the next section.

The uniqueness of a property value for a given state can be described mathematically in the following manner. The integral of an exact differential dY is given by

$$\int_1^2 dY = Y_2 - Y_1 = \Delta Y$$

Thus the value of the integral depends solely on the initial and final states. But the change in the value of a property likewise depends only on the end states. Hence the differential change dY in a property Y is an exact differential. Throughout this text the infinitesimal variation of a property will be signified by the differential symbol d preceding the property symbol. For example, an infinitesimal change in the pressure P of a system is given by dP. A finite change in a property is denoted by the symbol Δ (capital delta), for example, ΔP. The change in a property value ΔY always represents the final value minus the initial value. This convention must be kept in mind.

Use of the symbol δ, instead of the usual differential operator d, is intended as a reminder that some quantities depend on the process and are not a property of the system. δQ represents a small quantity of heat, not a differential. δm represents a small quantity of matter. The same qualifications for δ hold in the case of thermodynamic work. There being no exact differential dW, small quantities of W similar in magnitude to differentials are expressed as δW.

2.2.3. Processes and Cycles

A process is a change in state that can be described as "any change in the properties of a system." A process is described in part by the series of states passed through by the system. Often, but not always, some sort of interaction between the system and surroundings occurs during a process; the specification of this interaction completes the description of the process.

A description of a process typically involves specification of the initial and final equilibrium states, the path (if identifiable), and the interactions that take place across the boundaries of the system during the process. *Path* in thermodynamics refers to the series of states through which the system passes. Of special significance in thermodynamics is a quasistatic process or path. During this process the system internally must always be infinitesimally close to a state of equilibrium. That is, the path of a quasistatic process is a series of equilibrium steps. Although a quasistatic process is an idealization, many actual processes closely approximate quasistatic conditions. For nonequilibrium processes, certain intermediate information during the process is missing. One is always able to predict various overall effects, even though a detailed description is not possible as long as the initial and final states are equilibrium states.

For some processes, there are special names:

Process Characteristic	Descriptive Term
Pressure remains constant	Isobaric
Temperature remains constant	Isothermal
Volume remains constant	Isometric
No heat is transferred to or from the system	Adiabatic
Entropy remains constant	Isentropic

A cycle is a process or a series of processes wherein the initial and final states of the system are the same. Therefore, at the conclusion of a cycle all the properties have the same value as at the beginning.

2.2.4. Reversibility

All naturally occurring changes or processes occur in one direction only. In nature they are irreversible. Like a clock they tend to "run down" and cannot "rewind" themselves. Familiar examples are the transfer of heat with a finite temperature difference, the mixing of two gases, a waterfall, a chemical reaction. However, *all of these changes can be reversed.* We can transfer heat from a region of low temperature to one of higher temperature; we can separate a gas into its components; we can cause water to flow uphill. The important point is, we can do these things *only at the expense of some other system,* which itself becomes run down.

A process is said to be reversible if its direction can be reversed at any stage by an infinitesimal change in external conditions. If we consider a connected series of equilibrium states, each representing only an infinitesimal displacement from the adjacent one but with the overall result a finite change, then we have a reversible process.

All actual processes can be made to approach more or less closely to a reversible process by suitable choice of conditions. However, like the absolute zero of temperature, the strictly reversible process is purely a concept that aids in the analysis of certain problems. The approach of actual processes to this ideal limit can be made almost as close as we please. The closeness of approach is usually limited by economic factors rather than physical factors. The truly reversible process would require an infinite time for its completion, but the concept of the reversible process establishes a standard for the comparison of actual processes. The reversible process is one that yields the greatest amount of work or requires the least amount of work to bring about a given change. It tells us the maximum efficiency of the process. With the reversible process as a standard, we know at once whether an actual process is already highly efficient or whether it is very inefficient and therefore capable of considerable improvement.

Since the reversibile process represents a succession of equilibrium states each of which is only a differential step from its neighbor, the reversible process can be represented as a continuous line on a state diagram [pressure-volume (P-v), temperature-entropy (T-s), and so on]. On the other hand, the irreversible process cannot be so represented. One can note the terminal states and indicate the general direction of change, but it is inherent in the nature of the irreversible process that the complete path of the change is indeterminate and therefore cannot be presented as a line in a thermodynamic diagram.

Irreversibility always lowers the efficiency of a process. Its effect in this respect is identical with that of friction, which is one cause of irreversibility. Conversely, no process more efficient than a reversible process can even be imagined. The reversible process is an abstraction, an idealization, that is never achieved in practice. It is, however, of enormous utility because it allows calculation of work from knowledge of the system properties alone. In addition, it represents a standard of perfection that cannot be exceeded because (1) it places an upper limit on the work that may be obtained for a given work-producing process, and (2) it places a lower limit on the work input for a given work-requiring process.

2.2.5. Conservation of Mass

For most practical purposes, there is *no* change in mass when the energy changes. Einstein's equation of relativity

$$E = mc^2 \qquad (2.1)$$

where c = velocity of light (9.83×10^8 ft/s) and E = energy when applied to the burning of one pound of gasoline shows that the loss of mass (amount converted into energy) is only 3.2×10^{-11} lbm.

The mass rate of flow (m) of a fluid passing through a cross-sectional area A is

$$m = \frac{AV}{v} \qquad (2.2)$$

where V is the average velocity of the fluid in a direction normal to the plane of the area A, and v is the specific volume of the fluid. For steady flow with fluid entering a system at section 1 and leaving at section 2,

$$m_1 = m_2 = \frac{A_1 V_1}{v_1} = \frac{A_2 V_2}{v_2}$$

This is the continuity equation of steady flow.

2.3. PHASES AND CHANGES IN PHASE OF MATTER

A pure substance is one that has a homogeneous and constant chemical composition. It may exist in more than one phase, but the chemical composition is the same in all phases. Thus, liquid water, a mixture of liquid water and water vapor (steam), or a mixture of solid water (ice) and liquid water are all pure substances. However, a mixture of liquid air and gaseous air is not a pure substance, since the composition of the liquid is different from that of the vapor.

Sometimes a mixture of gases, such as air, is considered a pure substance as long as there is no change of phase. Strictly speaking, this is not true, but rather, as we shall see later, we should say that a mixture of gases such as air exhibits some of the characteristics of a pure substance as long as there is no change of phase.

Consider as a system the water contained in the piston-cylinder arrangement of Fig. 2.1. Suppose that the piston maintains a pressure of 14.7 lbf/in.^{2} in the cylinder containing water, and that the initial temperature is 60°F. As heat is transferred to the water, the temperature increases appreciably, the specific volume increases slightly, and the pressure remains constant. When the temperature reaches 212°F, additional heat transfer results in a change of phase. That is, some of the liquid becomes vapor, and during this process both the temperature and pressure remain constant, but the specific volume increases considerably. When the last drop of liquid has vaporized, further transfer of heat results in an increase in both temperature and specific volume of the vapor.

The term *saturation temperature* designates the temperature at which vaporization takes place at a given pressure. Conversely, the pressure is called the *saturation pressure* for the given temperature. Thus for water at 212°F the saturation pressure is 14.7 lbf/in.^{2}, and for water at 14.7 lbf/in.^{2} the saturation temperature is 212°F.

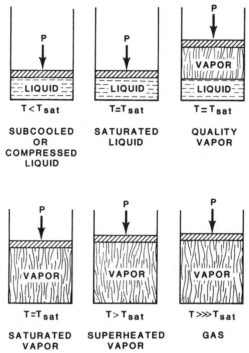

Fig. 2.1. Thermodynamic fluid states. Reprinted by permission from *ASHRAE Handbook of Fundamentals 1981.*

If a substance exists as liquid at the saturation temperature and pressure, it is called *saturated liquid.* If the temperature of the liquid is lower than the saturation temperature for the existing pressure, it is called either a subcooled liquid (indicating that the temperature is below the saturation temperature for the given pressure) or a *compressed liquid* (indicating that the pressure is greater than the saturation pressure). When a substance exists as part liquid and part vapor at the saturation temperature, its *quality* is defined as the ratio of the mass of vapor to the total mass and is given the symbol x. Quality has meaning only when the substance is at saturation pressure and temperature.

When a substance exists as vapor at the saturation temperature, it is called *saturated vapor.* When the vapor is at a temperature greater than the saturation temperature, it is said to be a *superheated vapor.* The pressure and temperature of superheated vapor are independent properties, since the temperature may increase while the pressure remains constant. The substances we call gases are highly superheated vapors.

The entire range of phases may be summarized by the diagram of Fig. 2.2, which shows how the solid, liquid, and vapor phases may exist together in equilibrium. Along the sublimation line the solid and vapor phases are in

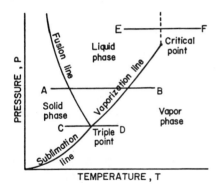

Fig. 2.2.   The pure substance.

equilibrium, along the fusion line the solid and liquid phases are in equilibrium, and along the vaporization line the liquid and vapor phases are in equilibrium. The only point at which all three phases may exist in equilibrium is the *triple point*. The vaporization line ends at the critical point because there is no distinct change from the liquid phase to the vapor phase above the critical point.

Consider the solid phase in Fig. 2.2. When the temperature is increased while the pressure, if less than the triple point pressure, is constant, the substance passes directly from the solid to the vapor phase. Along the constant-pressure line *AB*, the substance first passes from the solid to the liquid phase at one temperature, and then from the liquid to the vapor phase at a higher temperature. Constant-pressure line *CD* passes through the triple point, and it is only at the triple point that the three phases may exist together in equilibrium. At a pressure above the critical pressure, such as *EF*, there is no sharp distinction between the liquid and vapor phases.

One important reason for introducing the concept of pure substance is that the state of a pure substance is defined by two independent properties.

To illustrate the significance of the term *independent property,* consider the saturated-liquid and saturated-vapor states of a pure substance. These two states have the same pressure and the same temperature but are definitely different. In a saturation state, therefore, pressure and temperature are not independent properties. Two independent properties such as pressure and specific volume, or pressure and quality, are required to specify a saturation state.

The change of a substance from liquid to vapor, or from solid to liquid as in the case of melting ice, is called a change of phase. All matter can exist in three phases: solid, liquid, or vapor. Heat must be added to or taken from a substance to change its phase. The heat required to change a liquid to a vapor is called the *latent heat of vaporization*. The heat required to change a solid to a liquid is called the *latent heat of fusion*.

Suppose a warmed liquid is to be cooled back to its original temperature. Exactly the same quantity of heat must be removed as was originally used to heat it. Similarly, to condense steam, heat must be removed; the amount is exactly equal to the latent heat that went into the water to change it to steam. In freezing water, the heat that must be removed is exactly equal in amount to the latent heat that was absorbed in changing the ice to water. Heat that changes the temperature of a substance is called *sensible heat* because our senses give evidence of its presence. Heat supplied to a substance during a phase change is called *latent heat*. Figure 2.3 displays graphically the energy change with either temperature change or phase change for water.

By definition, a saturated vapor exists in the presence of its liquid. Another way of putting it is to say that a saturated vapor is in the two-phase condition. This is an absolutely accurate expression because, in a saturated condition, vapor is constantly condensing into a liquid, to be replaced by liquid that is evaporating into vapor. As a result, there is an equilibrium in the percent of vapor at a given pressure. In the liquid–vapor mixture, the vapor may be superheated and/or the liquid may be subcooled.

A mechanical refrigeration system depends totally on the process of evaporation (the change of state from liquid to gas) and condensation (the change of state from gas to liquid) in order to function efficiently. The importance of this change of state can be illustrated by recalling the number of Btu's required to change the state of water from a liquid to a gas. It required only 180 Btu to raise the temperature of 1 lbm of water from 32 to 212°F, but it required 970 Btu to change state from water to gas, or almost 5½ times the heat energy.

An efficient air-conditioning system must have the capability of readily absorbing and rejecting large amounts of heat at the normal operating temperatures of the equipment. Also it must be possible to continuously repeat the process of evaporation and condensation with the same substance. While water could be used for this purpose, it boils at temperatures too high for ordinary cooling and it freezes at temperatures too high for the low-temperature conditions. Therefore, some very special liquids were developed to accomplish efficient cooling. These refrigerants also had to have certain other characteristics. They had to be

1. High in density so small volumes could be used effectively.
2. Capable of operating at low differences in pressure.
3. Nonflammable.
4. Nonexplosive in either gas or liquid form.
5. Noncorrosive.
6. Nontoxic.
7. Able to carry oil in solution.
8. Highly resistant to electricity.

Most initial substances used for refrigeration—ammonia, sulfur dioxide, methyl chloride, propane, methane, and others—were missing at least one

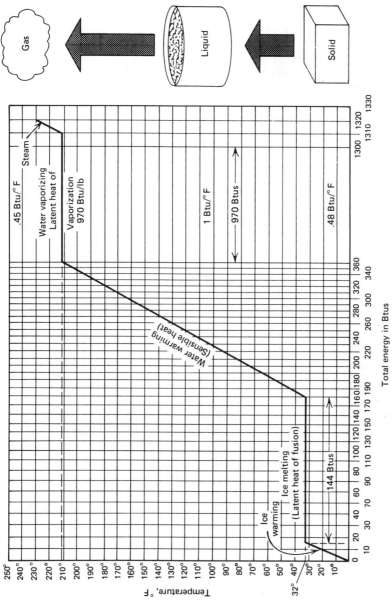

Fig. 2.3. Phase change from solid to liquid to gas.

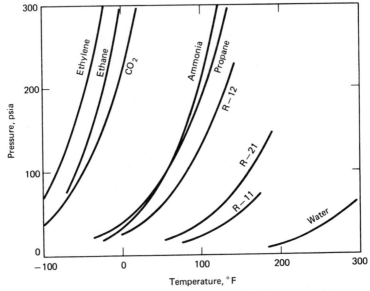

Fig. 2.4. Saturation pressure–temperature curves for some refrigerants.

of these qualities. These liquids have been almost completely replaced by the specialized refrigerants that have been developed over the years to avoid their undesirable qualities.

One of the first refrigerants developed specifically for the air-conditioning and refrigeration industry was Refrigerant 12, usually referred to as R-12. It is widely used in household refrigerators, commercial refrigeration, and window coolers but it is uncommon in residential air-conditioning.

A later development was Refrigerant 22, or R-22, which is similar in many characteristics to R-12. It does have a much higher latent heat of evaporation, 86 Btu/lb, which allows the absorption of greater heat content, and it has a higher density than R-12. Therefore, more pounds of refrigerant can be pumped through the same size compressor, or an equivalent mass of refrigerant can be pumped through a smaller compressor. This has resulted in smaller compressors with greater cooling capacities than were previously available. R-22 is a nontoxic, odorless, clear liquid. It will not burn or explode, and it can be used for many varied applications. Figure 2.4 presents the variation of boiling point with pressure for some refrigerants.

2.4. PROPERTIES OF A SUBSTANCE

2.4.1. Specific Volume and Density

The *specific volume* of a substance is defined as the volume per unit mass, and is given the symbol v. The *density* of a substance is defined as the mass per unit volume, and is therefore the reciprocal of the specific volume.

Fig. 2.5. Volume change with phase change.

Density is designated by the symbol ρ (Greek rho). The volume of a given mass of substance may be dramatically different in different phases, as illustrated in Fig. 2.5.

2.4.2. Pressure

The pressure at a given point in a stationary fluid is the same in all directions, and pressure is defined as the normal component of force per unit area.

In most thermodynamic investigations we are concerned with absolute (actual) pressure. Most pressure and vacuum gauges, however, read the difference between the absolute pressure and the atmospheric pressure existing at the gauge, as shown graphically in Fig. 2.6.

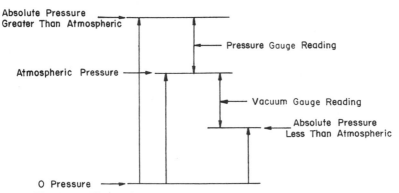

Fig. 2.6. Illustration of terms used in pressure measurement.

EXAMPLE 2.1

A pressure gauge reads 40.7 lbf/in.^{2}, and the barometer reads 29.2 in.Hg. Calculate the absolute pressure in lbf/in.^{2} and atm.

Solution.

$$P_{abs} = P_{gauge} + P_{atm} = 40.7 \frac{lbf}{in.^2} + .491 \frac{psi}{in.Hg} \times 29.2 \text{ in.Hg} = 55.0 \text{ psia}$$

$$55.0/14.7 = 3.74 \text{ std atm}$$

EXAMPLE 2.2

A mercury manometer which is used to measure a vacuum reads 29.3 in., and the barometer reads 29.7 in.Hg. Determine the pressure in lbf/in.^{2}

Solution.

$$P_{abs} = P_{atm} - \text{vacuum} = 29.7 \text{ in.Hg} - 29.3 \text{ in.Hg} = 0.4 \text{ in.Hg}$$

$$0.4 \text{ in.Hg} \times .491 \frac{psi}{in.Hg} = 0.1964 \text{ psia}$$

2.4.3. Temperature

Although temperature is a property with which everyone is familiar, a definition of it is difficult. We are aware of "temperature" first of all as a feeling of hotness or coldness when one touches an object. Because of the difficulties in defining temperature, we define equality of temperature. We may say that two bodies have equality of temperature when no change in any observable property occurs when they are in thermal communication, that is, thermally coupled.

The zeroth law of thermodynamics states that when two bodies have the same temperature as a third body, they in turn have the same temperature as each other. This law is the basis of temperature measurement, since numbers can be placed on the thermometers, and every time a body has equality of temperature with the thermometer, we can say that the body has the temperature indicated by the thermometer.

The two commonly used scales for measuring temperature are the Fahrenheit (after Gabriel Fahrenheit, 1686–1735) and the Celsius scale. The Celsius scale was formerly called the centigrade scale, but since there are other centigrade scales it is now named after Anders Celsius (1701–1744), the Swedish astronomer who devised it. Figure 2.7 gives a comparison of the two scales.

The absolute scale related to the Celsius scale is referred to as the Kelvin scale. One division of the Kelvin scale is called a kelvin (K). The relation between these scales is

$$K = {}^\circ C + 273.15$$

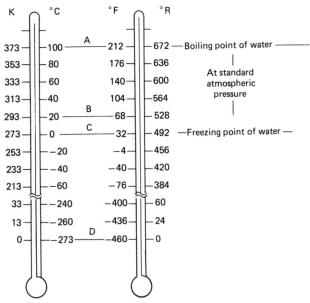

Fig. 2.7. Fahrenheit, Celsius, Rankine, and Kelvin thermometer scales. *A*, Boiling temperature of water; *B*, standard conditions temperature; *C*, freezing temperature of water; *D*, absolute zero.

The absolute scale related to the Fahrenheit scale is referred to as the Rankine scale; one division is 1 °R. The relation between these scales is

$$°R = °F + 459.67$$

2.4.4. Internal Energy

Internal energy is the energy *possessed* by a material due to the motion and/or position of the molecules. It may consist of (1) kinetic internal energy, which is due to the velocity of the molecules, and (2) potential internal energy, which is due to the attractive forces existing between molecules. Changes in the average velocity of the molecules are indicated by temperature changes of the system, while variations in position are denoted by changes in phase of the system. Internal energy is an extensive property, since it depends upon the mass of the system.

The symbol U is used to designate the internal energy of a given mass of a substance. Following the convention used with other extensive properties, the symbol u designates the internal energy per unit mass. We could speak of u as the specific internal energy, as we do in the case of specific volume. However, since the context will usually make it clear whether u or U is referred to, we will simply use the term internal energy to refer to both internal energy per unit mass and total internal energy.

2.4.5. Enthalpy

In analyzing specific types of processes, we frequently encounter certain combinations of thermodynamic properties, which are therefore also properties of the substance undergoing the change of state. One such combination is $U + PV$. Therefore, we find it convenient to define a new extensive property, called the *enthalpy*, which we denote H and define as

$$H \equiv U + PV$$

or, per unit mass, $\hspace{10cm}$ (2.3)

$$h \equiv u + Pv$$

As in the case of internal energy, we could speak of specific enthalpy, h, and total enthalpy, H. However, we will simply refer to both as enthalpy, since the context will make it clear which is referred to.

2.4.6. Entropy

Entropy, S, is a measure of molecular disorder or of the probability of a given state. The more completely shuffled any system is, the greater is its entropy. Conversely, an orderly or unmixed configuration is one of low entropy. Entropy is an absolute quantity, and if a substance reaches a state in which all randomness has disappeared, it should then have zero entropy.

By application of the theory of probability to molecular systems, to which it should apply rigorously because of the enormous number of individuals concerned, Boltzmann showed that there was a simple relationship between the entropy of a given system of molecules and the probability of its occurrence. This relationship is given as

$$S = k \ln W$$

where k is a constant called the Boltzmann constant and W is the thermodynamic probability. The derivation of this relationship and its further elaboration to permit the calculation of absolute entropies are beyond the scope of this book.

2.4.7. Specific Heats

The constant-volume specific heat and the constant-pressure specific heat are useful functions for thermodynamic calculations, particularly for gases. The constant-volume specific heat c_v is defined by the relation

$$c_v \equiv \left(\frac{\partial u}{\partial T} \right)_v \hspace{6cm} (2.4)$$

The constant-pressure specific heat c_p is defined by the relation

$$c_p \equiv \left(\frac{\partial h}{\partial T}\right)_p \tag{2.5}$$

Note that each of these quantities is defined in terms of properties, and therefore the constant-volume and constant-pressure specific heats are thermodynamic properties of a substance.

The *heat capacity* is sometimes used instead of the *specific heat*.

The amount of heat that must be added to a closed system to accomplish a given change of state depends on how the process is carried out. Only for a reversible process where the path is fully specified is it possible to relate the heat to a property of the system. On this basis we define heat capacity in general by

$$C_X = \left(\frac{\delta Q}{dT}\right)_X \tag{2.6}$$

where X indicates that the process is reversible and the path is fully specified. We could define a number of heat capacities according to this prescription; however, only C_v and C_p are in common use. In both cases the system is presumed closed and to be of constant composition.

By definition, C_v is the amount of heat required to increase the temperature by dT when the system is held at constant volume, and may therefore also be expressed as

$$C_v = \left(\frac{\delta Q}{dT}\right)_v$$

Similarly, C_p is the amount of heat required to increase the temperature by dT when the system is heated in a reversible process at constant pressure, and is given by the equation

$$C_p = \left(\frac{\delta Q}{dT}\right)_p$$

2.4.8. Tables and Graphs of Thermodynamic Properties

Tables of thermodynamic properties of many substances are available, and they generally all have the same form. In this section we shall refer to the tables for water and R-12, as well as their respective Mollier diagrams, the *h-s* chart for steam and the *p-h* diagram for R-12.

Table 2, Chapter 6 of the *ASHRAE Handbook and Product Directory, 1977 Fundamentals,* * gives the values of the thermodynamic properties of water at saturation and is reproduced in part as Fig. 2.8. Other versions of the "steam tables" are almost identical.

---

*Hereafter referred to simply as *ASHRAE 1977 Fundamentals*.

Fahr. Temp. t(F)	Absolute Pressure p_s		Specific Volume, cu ft per lb			Enthalpy, Btu per lb			Entropy, Btu per (Lb) (°F)			Fahr. Temp. t(F)
	Lb/Sq In.	In. Hg	Sat. Liquid v_f	Evap. v_{fg}	Sat. Vapor v_g	Sat. Liquid h_f	Evap. h_{fg}	Sat. Vapor h_g	Sat. Liquid s_f	Evap. s_{fg}	Sat. Vapor s_g	
34	0.095999	0.19546	0.01602	3061.7	3061.7	2.01	1074.03	1076.04	0.00409	2.1755	2.1796	34
35	0.099908	0.20342	0.01602	2947.8	2947.8	3.02	1073.46	1076.48	0.00612	2.1700	2.1761	35
36	0.10396	0.21166	0.01602	2838.7	2838.7	4.02	1072.90	1076.92	0.00815	2.1644	2.1726	36
37	0.10815	0.22020	0.01602	2734.1	2734.1	5.03	1072.33	1077.36	0.01018	2.1589	2.1691	37
38	0.11249	0.22904	0.01602	2633.8	2633.8	6.03	1071.77	1077.80	0.01220	2.1535	2.1657	38
39	0.11699	0.23819	0.01602	2537.6	2537.6	7.01	1071.20	1078.24	0.01422	2.1480	2.1622	39
40	0.12164	0.24767	0.01602	2445.4	2445.4	8.04	1070.64	1078.68	0.01623	2.1426	2.1588	40
41	0.12648	0.25748	0.01602	2356.9	2356.9	9.05	1070.06	1079.11	0.01824	2.1372	2.1554	41
42	0.13145	0.26763	0.01602	2272.0	2272.0	10.05	1069.50	1079.55	0.02024	2.1318	2.1520	42
43	0.13660	0.27813	0.01602	2190.5	2190.5	11.05	1068.91	1079.99	0.02224	2.1265	2.1487	43
44	0.14194	0.28899	0.01602	2112.3	2112.3	12.06	1068.37	1080.43	0.02423	2.1211	2.1453	44
45	0.14746	0.30023	0.01602	2037.3	2037.3	13.06	1067.81	1080.87	0.02622	2.1158	2.1420	45
46	0.15317	0.31185	0.01602	1965.2	1965.2	14.06	1067.24	1081.30	0.02820	2.1105	2.1387	46
47	0.15907	0.32387	0.01602	1896.0	1896.0	15.06	1066.68	1081.74	0.03018	2.1052	2.1354	47
48	0.16517	0.33629	0.01602	1829.5	1829.5	16.07	1066.11	1082.18	0.03216	2.0999	2.1321	48
49	0.17148	0.34913	0.01602	1765.7	1765.7	17.07	1065.55	1082.62	0.03413	2.0947	2.1288	49
50	0.17799	0.36240	0.01602	1704.3	1704.3	18.07	1064.99	1083.06	0.03610	2.0895	2.1256	50
51	0.18473	0.37611	0.01602	1645.4	1645.4	19.07	1064.42	1083.49	0.03806	2.0842	2.1223	51
52	0.19169	0.39028	0.01602	1588.7	1588.7	20.07	1063.86	1083.93	0.04002	2.0791	2.1191	52
53	0.19888	0.40492	0.01603	1534.3	1534.3	21.07	1063.30	1084.37	0.04197	2.0739	2.1159	53
54	0.20630	0.42003	0.01603	1481.9	1481.9	22.08	1062.72	1084.80	0.04392	2.0688	2.1127	54
55	0.21397	0.43564	0.01603	1431.5	1431.5	23.08	1062.16	1085.24	0.04587	2.0637	2.1096	55
56	0.22188	0.45176	0.01603	1383.1	1383.1	24.08	1061.60	1085.68	0.04781	2.0586	2.1064	56
57	0.23006	0.46840	0.01603	1336.5	1336.5	25.08	1061.04	1086.12	0.04975	2.0535	2.1033	57
58	0.23849	0.48558	0.01603	1291.7	1291.7	26.08	1060.47	1086.55	0.05168	2.0485	2.1002	58
59	0.24720	0.50330	0.01603	1248.6	1248.6	27.08	1059.91	1086.99	0.05361	2.0434	2.0970	59
60	0.25618	0.52160	0.01603	1207.1	1207.1	28.08	1059.34	1087.42	0.05553	2.0385	2.0940	60
61	0.26545	0.53047	0.01604	1167.2	1167.2	29.08	1058.78	1087.86	0.05746	2.0334	2.0909	61
62	0.27502	0.55994	0.01604	1128.7	1128.7	30.08	1058.22	1088.30	0.05937	2.0284	2.0878	62
63	0.28488	0.58002	0.01604	1091.7	1091.7	31.08	1057.65	1088.73	0.06129	2.0235	2.0848	63
64	0.29505	0.60073	0.01604	1056.1	1056.1	32.08	1057.09	1089.17	0.06320	2.0186	2.0818	64
65	0.30554	0.62209	0.01604	1021.7	1021.7	33.08	1056.52	1089.60	0.06510	2.0136	2.0787	65
66	0.31636	0.64411	0.01604	988.63	988.65	34.07	1055.97	1090.04	0.06700	2.0087	2.0757	66
67	0.32750	0.66681	0.01605	956.76	956.78	35.07	1055.40	1090.47	0.06890	2.0039	2.0728	67
68	0.33900	0.69021	0.01605	926.06	926.08	36.07	1054.84	1090.91	0.07080	1.9990	2.0698	68
69	0.35084	0.71432	0.01605	896.47	896.49	37.07	1054.27	1091.34	0.07269	1.9941	2.0668	69
70	0.36304	0.73916	0.01605	867.95	867.97	38.07	1053.71	1091.78	0.07458	1.9893	2.0639	70
71	0.37561	0.76476	0.01605	840.45	840.47	39.07	1053.14	1092.21	0.07646	1.9845	2.0610	71
72	0.38856	0.79113	0.01606	813.95	813.97	40.07	1052.58	1092.65	0.07834	1.9797	2.0580	72
73	0.40190	0.81829	0.01606	788.38	788.40	41.07	1052.01	1093.08	0.08022	1.9749	2.0551	73
74	0.41564	0.84626	0.01606	763.73	763.75	42.06	1051.48	1093.52	0.08209	1.9701	2.0522	74
75	0.42979	0.87506	0.01606	739.95	739.97	43.06	1050.89	1093.95	0.08396	1.9654	2.0494	75
76	0.44435	0.90472	0.01606	717.01	717.03	44.06	1050.32	1094.38	0.08582	1.9607	2.0465	76
77	0.45935	0.93524	0.01607	694.88	694.90	45.05	1049.76	1094.82	0.08769	1.9560	2.0437	77
78	0.47478	0.96666	0.01607	673.52	673.54	46.06	1049.19	1095.25	0.08954	1.9513	2.0408	78
79	0.49066	0.99900	0.01607	652.91	652.93	47.06	1048.62	1095.68	0.09140	1.9466	2.0380	79
80	0.50701	1.0323	0.01607	633.01	633.03	48.05	1048.07	1096.12	0.09325	1.9419	2.0352	80
81	0.52382	1.0665	0.01608	613.80	613.82	49.05	1047.50	1096.55	0.09510	1.9373	2.0324	81
82	0.54112	1.1017	0.01608	595.25	595.27	50.05	1046.93	1096.98	0.09694	1.9328	2.0297	82
83	0.55892	1.1380	0.01608	577.34	577.36	51.05	1046.37	1097.42	0.09878	1.9281	2.0269	83
84	0.57722	1.1752	0.01608	560.01	560.06	52.05	1045.80	1097.85	0.10062	1.9236	2.0242	84
85	0.59604	1.2136	0.01609	543.33	543.35	53.05	1045.23	1098.28	0.10246	1.9189	2.0214	85
86	0.61540	1.2530	0.01609	527.19	527.21	54.04	1044.67	1098.71	0.10429	1.9144	2.0187	86
87	0.63530	1.2935	0.01609	511.60	511.62	55.04	1044.10	1099.14	0.10611	1.9099	2.0160	87
88	0.65575	1.3351	0.01610	496.52	496.54	56.04	1043.53	1099.58	0.10794	1.9051	2.0133	88
89	0.67678	1.3779	0.01610	481.96	481.98	57.01	1042.97	1100.01	0.10976	1.9008	2.0106	89
90	0.69838	1.4219	0.01610	467.88	467.90	58.01	1042.40	1100.44	0.11158	1.8963	2.0079	90
91	0.72059	1.4671	0.01610	454.26	454.28	59.03	1041.84	1100.87	0.11339	1.8919	2.0053	91
92	0.74340	1.5136	0.01611	441.10	441.12	60.03	1041.27	1101.30	0.11520	1.8874	2.0026	92
93	0.76684	1.5613	0.01611	428.38	428.40	61.03	1040.70	1101.73	0.11701	1.8830	2.0000	93
94	0.79091	1.6103	0.01611	416.06	416.09	62.04	1040.13	1102.16	0.11881	1.8786	1.9974	94
95	0.81564	1.6607	0.01612	404.17	404.19	63.03	1039.56	1102.59	0.12061	1.8741	1.9947	95
96	0.84103	1.7124	0.01612	392.65	392.67	64.02	1039.00	1103.02	0.12241	1.8698	1.9922	96
97	0.86711	1.7655	0.01612	381.51	381.53	65.02	1038.43	1103.45	0.12420	1.8654	1.9896	97
98	0.89388	1.8200	0.01612	370.73	370.75	66.02	1037.86	1103.88	0.12600	1.8610	1.9870	98
99	0.92137	1.8759	0.01613	360.30	360.32	67.02	1037.29	1104.31	0.12778	1.8566	1.9844	99
100	0.94959	1.9334	0.01613	350.20	350.22	68.02	1036.72	1104.74	0.12957	1.8523	1.9819	100
101	0.97854	1.9923	0.01614	340.42	340.44	69.01	1036.16	1105.17	0.13135	1.8480	1.9793	101
102	1.0083	2.0529	0.01614	330.98	330.98	70.01	1035.58	1105.59	0.13313	1.8437	1.9768	102
103	1.0388	2.1140	0.01614	321.80	321.82	71.01	1035.01	1106.02	0.13490	1.8394	1.9743	103

Fig. 2.8. Thermodynamic properties of water at saturation. Reprinted by permission from *ASHRAE Handbook of Fundamentals 1977.*

In Fig. 2.8, the first two columns after the temperature give the corresponding saturation pressure in pounds force per square inch and in inches of mercury. The next three columns give specific volume in cubic feet per pound mass. The first of these gives the specific volume of the saturated liquid, v_f; the third column gives the specific volume of saturated vapor, v_g. The difference between these two, $v_g - v_f$, represents the increase in specific volume when the state changes from saturated liquid to saturated vapor and is designated v_{fg}.

The specific volume of a substance having a given quality can be found by utilizing the definition of quality, the ratio of the mass of vapor to total mass of liquid plus vapor when a substance is in a saturation state. Let us consider a mass of 1 lb having a quality x. The specific volume is the sum of the volume of the liquid and the volume of the vapor. The volume of the liquid is $(1 - x)v_f$, and the volume of the vapor is xv_g. Therefore the specific volume v is

$$v = xv_g + (1 - x)v_f \qquad (2.7)$$

Since $v_f + v_{fg} = v_g$, Eq. (2.7) can also be written

$$v = v_f + xv_{fg}$$

The same procedure is followed for determining the enthalpy and entropy for quality conditions:

$$h = xh_g + (1 - x)h_f \quad \text{and} \quad s = xs_g + (1 - x)s_f \qquad (2.8)$$

Internal energy can then be obtained from the definition of enthalpy as $u = h - Pv$.

If the substance is a compressed or subcooled liquid, the thermodynamic properties of specific volume, enthalpy, internal energy, and entropy are strongly dependent on temperature (rather than pressure) and thus may be approximated, if compressed liquid tables are not available, by the corresponding values for saturated liquid (v_f, h_f, u_f, and s_f) at the existing temperature.

In the superheat region, thermodynamic properties must be obtained from superheat tables or a plot of the thermodynamic properties, commonly called a Mollier diagram. An example is shown as Fig. 2.9.

The thermodynamic properties of the refrigerants used in vapor-compression systems are found in similar tables. Figure 2.10 reproduces a section of Table 2 (Refrigerant 12, properties of liquid and saturated vapor) from Chapter 16 of ASHRAE *1977 Fundamentals*. However, for these refrigerants the common Mollier plot is the *p-h* diagram as illustrated in Fig. 2.11. More complete tables for both R-12 and R-22 are given in Appendixes A and B.

For fluids used in absorption refrigeration systems, the thermodynamic properties are commonly found on a different type of plot, the enthalpy–concentration diagram, as illustrated by Fig. 2.12 for water-ammonia mixture.

2.4.9. Property Equations for Ideal Gases

At this point certain relations for the internal energy, enthalpy, entropy, and the constant-pressure and constant-volume specific heats of an ideal gas can be examined. An *ideal gas* is defined as a gas at sufficiently low density that intermolecular forces and the associated energy are negligibly small. As a

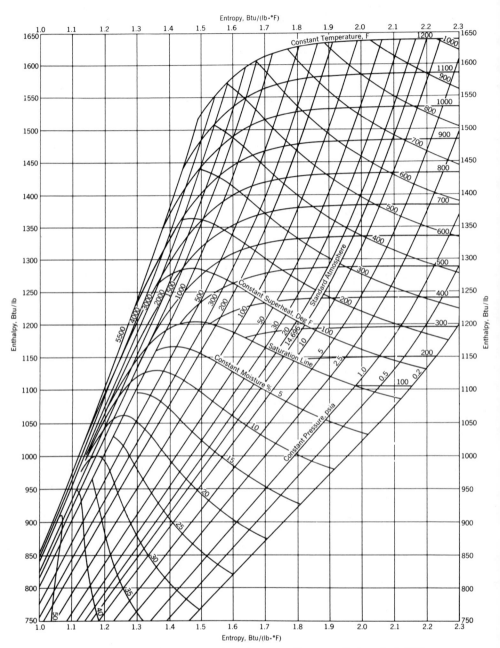

Fig. 2.9. A plot of the properties of steam. Courtesy Babcock & Wilcox.

Refrigerant 12 Properties of Saturated Liquid and Saturated Vapor

TEMP.	PRESSURE		VOLUME cu ft/lb		DENSITY lb/cu ft		ENTHALPY Btu/lb			ENTROPY Btu/(lb)(° R)		TEMP.
°F	PSIA	PSIG	LIQUID v_f	VAPOR v_g	LIQUID $1/v_f$	VAPOR $1/v_g$	LIQUID h_f	LATENT h_{fg}	VAPOR h_g	LIQUID s_f	VAPOR s_g	°F
15	32.415	17.719	0.011227	1.2050	89.070	0.82986	11.771	67.090	78.861	0.026243	0.16758	15
16	33.060	18.364	0.011241	1.1828	88.962	0.84544	11.989	66.977	78.966	0.026699	0.16750	16
17	33.714	19.018	0.011254	1.1611	88.854	0.86125	12.207	66.864	79.071	0.027154	0.16742	17
18	34.378	19.682	0.011268	1.1399	88.746	0.87729	12.426	66.750	79.176	0.027608	0.16734	18
19	35.052	20.356	0.011282	1.1191	88.637	0.89356	12.644	66.636	79.280	0.028062	0.16727	19
20	35.736	21.040	0.011296	1.0988	88.529	0.91006	12.863	66.522	79.385	0.028515	0.16719	20
21	36.430	21.734	0.011310	1.0790	88.419	0.92679	13.081	66.407	79.488	0.028968	0.16712	21
22	37.135	22.439	0.011324	1.0596	88.310	0.94377	13.300	66.293	79.593	0.029420	0.16704	22
23	37.849	23.153	0.011338	1.0406	88.201	0.96098	13.520	66.177	79.697	0.029871	0.16697	23
24	38.574	23.878	0.011352	1.0220	88.091	0.97843	13.739	66.061	79.800	0.030322	0.16690	24
25	39.310	24.614	0.011366	1.0039	87.981	0.99613	13.958	65.946	79.904	0.030772	0.16683	25
26	40.056	25.360	0.011380	0.98612	87.870	1.0141	14.178	65.829	80.007	0.031221	0.16676	26
27	40.813	26.117	0.011395	0.96874	87.760	1.0323	14.398	65.713	80.111	0.031670	0.16669	27
28	41.580	26.884	0.011409	0.95173	87.649	1.0507	14.618	65.596	80.214	0.032118	0.16662	28
29	42.359	27.663	0.011424	0.93509	87.537	1.0694	14.838	65.478	80.316	0.032566	0.16655	29
30	43.148	28.452	0.011438	0.91880	87.426	1.0884	15.058	65.361	80.419	0.033013	0.16648	30
31	43.948	29.252	0.011453	0.90286	87.314	1.1076	15.279	65.243	80.522	0.033460	0.16642	31
32	44.760	30.064	0.011468	0.88725	87.202	1.1271	15.500	65.124	80.624	0.033905	0.16635	32
33	45.583	30.887	0.011482	0.87197	87.090	1.1468	15.720	65.006	80.726	0.034351	0.16629	33
34	46.417	31.721	0.011497	0.85702	86.977	1.1668	15.942	64.886	80.828	0.034796	0.16622	34
35	47.263	32.567	0.011512	0.84237	86.865	1.1871	16.163	64.767	80.930	0.035240	0.16616	35
36	48.120	33.424	0.011527	0.82803	86.751	1.2077	16.384	64.647	81.031	0.035683	0.16610	36
37	48.989	34.293	0.011542	0.81399	86.638	1.2285	16.606	64.527	81.133	0.036126	0.16604	37
38	49.870	35.174	0.011557	0.80023	86.524	1.2496	16.828	64.406	81.234	0.036569	0.16598	38
39	50.763	36.067	0.011573	0.78676	86.410	1.2710	17.050	64.285	81.335	0.037011	0.16592	39
40	51.667	36.971	0.011588	0.77357	86.296	1.2927	17.273	64.163	81.436	0.037453	0.16586	40
41	52.584	37.888	0.011603	0.76064	86.181	1.3147	17.495	64.042	81.537	0.037893	0.16580	41
42	53.513	38.817	0.011619	0.74798	86.066	1 3369	17.718	63.919	81.637	0.038334	0.16574	42
43	54.454	39.758	0.011635	0.73557	85.951	1.3595	17.941	63.796	81.737	0.038774	0.16568	43
44	55.407	40.711	0.011650	0.72341	85.836	1.3823	18.164	63.673	81.837	0.039213	0.16562	44
45	56.373	41.677	0.011666	0.71149	85.720	1.4055	18.387	63.550	81.937	0.039652	0.16557	45
46	57.352	42.656	0.011682	0.69982	85.604	1.4289	18.611	63.426	82.037	0.040091	0.16551	46
47	58.343	43.647	0.011698	0.68837	85.487	1.4527	18.835	63.301	82.136	0.040529	0.16546	47
48	59.347	44.651	0.011714	0.67715	85.371	1.4768	19.059	63.177	82.236	0.040966	0.16540	48
49	60.364	45.668	0.011730	0.66616	85.254	1.5012	19.283	63.051	82.334	0.041403	0.16535	49
50	61.394	46.698	0.011746	0.65537	85.136	1.5258	19.507	62.926	82.433	0.041839	0.16530	50
51	62.437	47.741	0.011762	0.64480	85.018	1.5509	19.732	62.800	82.532	0.042276	0.16524	51
52	63.494	48.798	0.011779	0.63444	84.900	1.5762	19.957	62.673	82.630	0.042711	0.16519	52
53	64.563	49.867	0.011795	0.62428	84.782	1.6019	20.182	62.546	82.728	0.043146	0.16514	53
54	65.646	50.950	0.011811	0.61431	84.663	1.6278	20.408	62.418	82.826	0.043581	0.16509	54
55	66.743	52.047	0.011828	0.60453	84.544	1.6542	20.634	62.290	82.924	0.044015	0.16504	55
56	67.853	53.157	0.011845	0.59495	84.425	1.6808	20.859	62.162	83.021	0.044449	0.16499	56
57	68.977	54.281	0.011862	0.58554	84.305	1.7078	21.086	62.033	83.119	0.044883	0.16494	57
58	70.115	55.419	0.011879	0.57632	84.185	1.7352	21.312	61.903	83.215	0.045316	0.16489	58
59	71.267	56.571	0.011896	0.56727	84.065	1.7628	21.539	61.773	83.312	0.045748	0.16484	59
60	72.433	57.737	0.011913	0.55839	83.944	1.7909	21.766	61.643	83.409	0.046180	0.16479	60
61	73.613	58.917	0.011930	0.54967	83.823	1.8193	21.993	61.512	83.505	0.046612	0.16474	61
62	74.807	60.111	0.011947	0.54112	83.701	1.8480	22.221	61.380	83.601	0.047044	0.16470	62
63	76.016	61.320	0.011965	0.53273	83.580	1.8771	22.448	61.248	83.696	0.047475	0.16465	63
64	77.239	62.543	0.011982	0.52450	83.457	1.9066	22.676	61.116	83.792	0.047905	0.16460	64
65	78.477	63.781	0.012000	0.51642	83.335	1.9364	22.905	60.982	83.887	0.048336	0.16456	65
66	79.729	65.033	0.012017	0.50848	83.212	1.9666	23.133	60.849	83.982	0.048765	0.16451	66
67	80.996	66.300	0.012035	0.50070	83.089	1.9972	23.362	60.715	84.077	0.049195	0.16447	67
68	82.279	67.583	0.012053	0.49305	82.965	2.0282	23.591	60.580	84.171	0.049624	0.16442	68
69	83.576	68.880	0.012071	0.48555	82.841	2.0595	23.821	60.445	84.266	0.050053	0.16438	69
70	84.888	70.192	0.012089	0.47818	82.717	2.0913	24.050	60.309	84.359	0.050482	0.16434	70

Fig. 2.10. Sample refrigerant property table. Copyright by DuPont. Reproduced by permission.

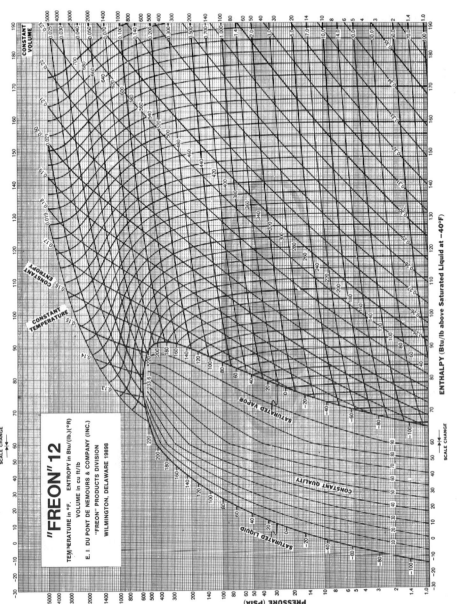

Fig. 2.11. Typical pressure–enthalpy plot for a refrigerant. Copyright by DuPont. Reproduced by permission.

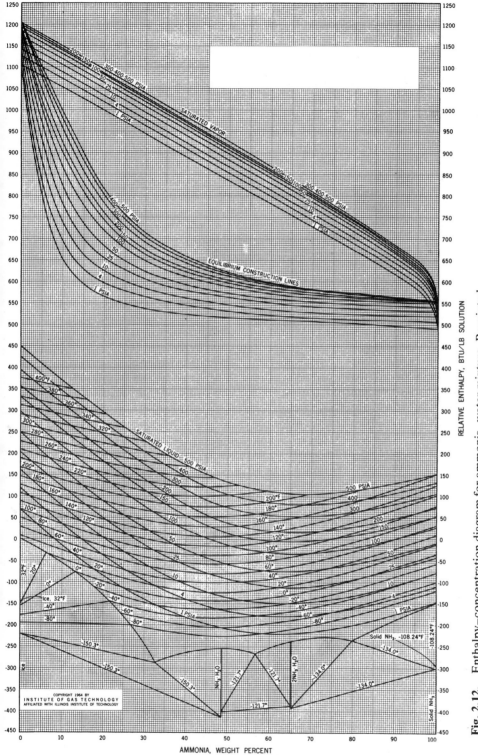

Fig. 2.12. Enthalpy–concentration diagram for ammonia–water mixture. Reprinted by permission from American Gas Association.

51

result, an ideal gas has the equation of state

$$Pv = RT \tag{2.9}$$

It can be shown that the internal energy of an ideal gas is a function of temperature only. This means that an ideal gas at a given temperature has a certain definite specific internal energy u, regardless of the pressure.

The relation between the internal energy u and the temperature T can be established by using the definition of constant-volume specific heat given by $c_v = (\partial u / \partial T)_v$. Since its internal energy is not a function of volume, for an ideal gas we can write

$$c_v = \frac{du}{dT}$$

so that

$$du = c_v dT \tag{2.10}$$

Equation (2.10) is always valid for an ideal gas regardless of what kind of process is considered.

From the definition of enthalpy and the equation of state of an ideal gas, it follows that

$$h = u + Pv = u + RT$$

Since R is a constant and u is a function of temperature only, it follows that the enthalpy h of an ideal gas is also a function of temperature only. The relation between enthalpy and temperature is found from the constant-pressure specific heat as defined by $c_p = (\partial h / \partial T)_p$. Since the enthalpy of an ideal gas is a function of temperature only and is independent of the pressure, it follows that

$$c_p = \frac{dh}{dT}$$

and therefore

$$dh = c_p dT \tag{2.11}$$

Equation (2.11) is always valid for an ideal gas regardless of what kind of process is considered.

Entropy, however, remains a function of both temperature and pressure:

$$ds = c_p \frac{dT}{T} - R \frac{dP}{P} \tag{2.12}$$

The ratio of heat capacities is often denoted by

$$k = \frac{c_p}{c_v} \tag{2.13}$$

and is a useful quantity in calculations for ideal gases. Table 2.1 gives the ideal gas values for some common gases.

TABLE 2.1. Properties of Gases

Gas	Chemical Formula	Molecular Weight	R ft·lbf lbm-°R	c_p Btu lbm-°R	c_p kJ kg-K	c_v Btu lbm-°R	c_v kJ kg-K	k
Air	—	28.97	53.34	0.240	1.0	0.171	0.716	1.400
Argon	Ar	39.94	38.66	0.125	0.523	0.075	0.316	1.667
Carbon dioxide	CO_2	44.01	35.10	0.203	0.85	0.158	0.661	1.285
Carbon monoxide	CO	28.01	55.16	0.249	1.04	0.178	0.715	1.399
Helium	He	4.003	386.0	1.25	5.23	0.753	3.153	1.667
Hydrogen	H_2	2.016	766.4	3.43	14.36	2.44	10.22	1.404
Methane	CH_4	16.04	96.35	0.532	2.23	0.403	1.69	1.32
Nitrogen	N_2	28.016	55.15	0.248	1.04	0.177	0.741	1.400
Oxygen	O_2	32.000	48.28	0.219	0.917	0.157	0.657	1.395
Steam	H_2O	18.016	85.76	0.445	1.863	0.335	1.402	1.329

The ideal gas is, of course, an idealization, and no real gas exactly satisfies these equations over any finite range of temperature and pressure. However, all real gases approach ideal behavior at low pressure and in the limit as $P \to 0$ do in fact meet the requirements. Thus the equations for an ideal gas provide good approximations to real gas behavior at low pressures, and because of their simplicity they are very useful.

From a practical point of view in the solution of problems, two things should be noted. First, at very low pressures, ideal gas behavior can be assumed with good accuracy, regardless of the temperature. Second, at temperatures that are double the critical temperature or above, ideal gas behavior can be assumed with good accuracy to pressures of at least 1000 lbf/in.^{2}. When the temperature is less than twice the critical temperature, and the pressure above a very low value, say atmospheric pressure, we are in the superheated vapor region, and the deviation from ideal gas behavior may be considerable. In this region it is preferable to use tables of thermodynamic properties or charts for a particular substance.

2.4.10. Mixtures

Up to this point we have limited our consideration primarily to pure substances. A large number of thermodynamic problems involve mixtures of pure substances.

A pure substance has been defined as a substance that is homogeneous and unchanging in chemical composition. Homogeneous mixtures of gases that do not react with each other are therefore pure substances, and the properties of such mixtures can be determined, correlated, and tabulated or fitted by equations just like the properties of any other pure substance. This has been done for common mixtures such as air and certain combustion products, but, since an unlimited number of mixtures is possible, properties of all of them cannot be determined experimentally and tabulated. It is therefore important to be able to calculate the properties of any mixture from the properties of its constituents. This section pertains to such calculations, first for gas mixtures and then for gas-vapor mixtures.

Mixtures of different gases are encountered in many engineering applications. Air is a good example of such a mixture. Since the individual gases can often be approximated as perfect gases, the study of mixtures of perfect gases and their properties is of considerable importance.

The first properties we shall examine are pressure, volume, and temperature.

Each constituent gas in a mixture has its own pressure, called the *partial pressure* of the particular gas. The Gibbs-Dalton law states that in a mixture of ideal gases, the pressure of the mixture is equal to the sum of the partial pressures of the individual constituent gases. In equation form,

$$p_m = p_A + p_B + p_C \tag{2.14}$$

where p_m is the total pressure of the mixture of gases A, B, and C, and p_A, p_B, and p_C are the partial pressures. In a mixture of ideal gases the partial pressure of each constituent equals the pressure which that constituent would exert if it existed alone at the temperature and volume of the mixture.

With regard to volume, on the other hand, we know from experience and experiment that generally, in mixtures of gases, each constituent gas behaves as though the other gases were not present. Each gas occupies the total volume of the mixture at the mixture temperature. If V_m is the volume of the mixture, then

$$V_m = V_A = V_B = V_C \qquad (2.15)$$

where V_A, V_B, and V_C are the volumes of the constituents.

However, the volume of a mixture of ideal gases equals the sum of the volumes of its constituents if each existed alone at the temperature and pressure of the mixture. This is known as Amagat's Law, Leduc's Law, or the law of additive volumes. Like Dalton's Law it is strictly true only for ideal gases but holds approximately for real-gas mixtures even in some ranges of pressure and temperature where $pv = RT$ is inaccurate. When the temperature of a real-gas mixture is well above the critical temperatures of all its constituents, the additive volume law is usually more accurate than the additive pressure law.

For ideal-gas mixtures, volumetric analyses are frequently used. The volume fraction is defined as

$$\text{Volume fraction of A} = \frac{V_A(p_m, T_m)}{V_m}$$

$$= \frac{\text{volume of A existing alone at } p_m, T_m}{\text{volume of the mixture at } p_m, T_m}$$

Notice that in a gas mixture each constituent occupies the total volume, and so volume fraction is not defined as the ratio of a constituent volume to the mixture volume because this ratio is always unity.

Avogadro's law states

Equal volumes of perfect gases held under exactly the same temperature and pressure have equal numbers of molecules.

If T_m is the temperature of the mixture, then the temperature relationship is

$$T_m = T_A = T_B = T_C$$

The analysis of a gas mixture based on mass is called a gravimetric analysis. It is based on the fact that the mass of a mixture is equal to the sum of the masses of its constituents:

$$m_m = m_A + m_B + m_C \qquad (2.16)$$

where the subscript m refers to the mixture and the subscripts A, B, and C refer to individual constituents of the mixture. The ratio m_A/m_m is called the mass fraction of constituent A.

The total number of moles, N_m, in a mixture is defined as the sum of the number of moles of its constituents:

$$N_m \equiv N_A + N_B + N_C + \cdots \tag{2.17}$$

The mole fraction x is defined as N/N_m, and

$$M_m = x_A M_A + x_B M_B + x_C M_C \tag{2.18}$$

where M_m is called the apparent (or average) molecular weight of the mixture.

The second part of the Gibbs-Dalton law can be taken as a basic definition:

Internal energy, enthalpy, and entropy of a gas mixture are equal to the sums of those properties for each component when it occupies the total volume by itself.

$$U_m = U_A + U_B + U_C \tag{2.19}$$

$$H_m = H_A + H_B + H_C \tag{2.20}$$

$$S_m = S_A + S_B + S_C \tag{2.21}$$

The constituent entropies must be evaluated at the temperature and volume of the mixture or at the mixture temperature and the constituent partial pressures. The entropy of any constituent at the volume and temperature of the mixture (and hence at its partial pressure) is greater than its entropy at the pressure and temperature of the mixture (and hence at its partial volume).

Consider the constituents as perfect gases:

$$R_m = \frac{m_A R_A + m_B R_B + m_C R_C}{m_m} \tag{2.22}$$

$$c_{v_m} = \frac{m_A c_{v_A} + m_B c_{v_B} + m_C c_{v_C}}{m_m} \tag{2.23}$$

$$c_{p_m} = \frac{m_A c_{p_A} + m_B c_{p_B} + m_C c_{p_C}}{m_m} \tag{2.24}$$

2.4.11. Psychrometrics: Moist Air Properties

Let us consider a simplification, which in many cases is a reasonable one, of the problem involving a mixture of ideal gases in contact with a solid or liquid phase of one of the components. The most familiar example is a mixture of air and water vapor in contact with liquid water or ice, such as the problems encountered in air-conditioning or drying. We are all familiar with the condensation of water from the atmosphere when it is cooled on a summer day.

Performance of air-conditioning phenomena and a number of similar processes can be analyzed quite simply and with considerable accuracy if the following assumptions are made:

1. The solid or liquid phase contains no dissolved gases.
2. The gaseous phase can be treated as a mixture of ideal gases.
3. When the mixture and the condensed phase are at a given pressure and temperature, the equilibrium between the condensed phase and its vapor is not influenced by the presence of the other component. This means that when equilibrium is achieved the partial pressure of the vapor will be equal to the saturation pressure corresponding to the temperature of the mixture.

Since this approach is used extensively and with considerable accuracy, let us give some attention to the terms that have been defined and the type of problems for which this approach is valid and relevant.

If the vapor is at the saturation pressure and temperature, the mixture is referred to as a saturated mixture, and for a air-water vapor mixture, the term "saturated air" is used.

Psychrometrics is the science involving thermodynamic properties of moist air and the effect of atmospheric moisture on materials and human comfort. As it applies in this text, the definition must be broadened to include the method of controlling the thermal properties of moist air.

When moist air is considered to be a mixture of two independent perfect gases, dry air and water vapor, each is assumed to obey the perfect gas equation of state:

$$\text{Dry air:} \qquad p_a V = n_a R T$$
$$\text{Water vapor:} \quad p_w V = n_w R T$$

where p_a is the partial pressure of dry air, p_w is the partial pressure of water vapor, V is the total mixture volume, n_a is the number of moles of dry air, R is the universal gas constant (1545.32 ft·lbf/lb·mole), n_w is the number of moles of water vapor, and T is the absolute temperature (°R).

The following definitions are useful when dealing with air/water vapor mixtures:

Dry-bulb temperature, t, is the temperature of air as registered by an ordinary thermometer.

Thermodynamic wet-bulb temperature, t^*, is the temperature at which water (liquid or solid) evaporating into moist air at a given dry-bulb temperature t and humidity ratio W can bring the air to saturation adiabatically while the pressure p is maintained constant.

The wet-bulb temperature of air is commonly measured with an ordinary thermometer whose glass bulb is covered by a wet cloth or gauze. The temperature is recorded after the thermometer has been moved rapidly in the air. Evaporation of water from the "wet sock" causes a cooling of the bulb. The temperature spread between the dry bulb and wet bulb depends upon the moisture content of the air.

Humidity ratio, W, of a given moist air sample is defined as the ratio of the mass of water vapor to the mass of dry air contained in the sample:

$$W = \frac{m_w}{m_a} \tag{2.25}$$

$$W = 0.62198 \frac{p_w}{p - p_w} \tag{2.26}$$

$$W = \frac{(1093 - 0.556t^*)W_s^* - 0.240(t - t^*)}{1093 + 0.444t - t^*} \tag{2.27}$$

Relative humidity, ϕ, is the ratio of the mole fraction of water vapor x_w in a given moist air sample to the mole fraction x_{ws} in an air sample that is saturated at the same temperature and pressure:

$$\phi = \frac{x_w}{x_{ws}} = \frac{p_w}{p_{ws}} \tag{2.28}$$

The term p_{ws} represents the saturation pressure of water vapor at the given temperature t.

Dew-point temperature, t_d, is the temperature of moist air that is saturated at the same pressure p and has the same humidity ratio W as that of the given sample of moist air.

The *volume v* of a moist air mixture is expressed in terms of a unit mass of dry air, with the relation $p = p_a + p_w$,

$$v = \frac{R_a T}{p - p_w} \tag{2.29}$$

where $R_a = R/28.9645 = 53.352$ (ft·lbf)/(lbm·°R).

The *enthalpy* of a mixture of perfect gases is equal to the sum of the individual partial enthalpies of the components. The enthalpy of moist air is then

$$h = h_a + Wh_g$$

where h_a is the specific enthalpy for dry air and h_g is the specific enthalpy for saturated water vapor at the temperature of the mixture. Approximately,

$$h_a = 0.240t \quad \text{Btu/lbm}$$

and

$$h_g = 1061 + 0.444t \quad \text{Btu/lbm}$$

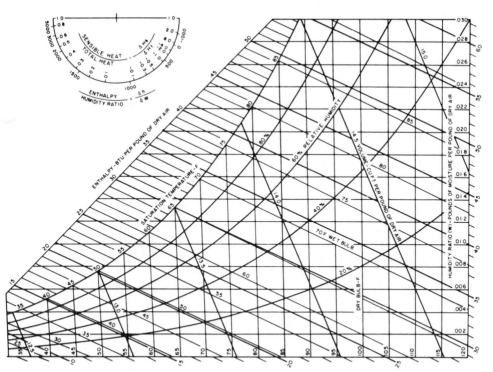

Fig. 2.14.   ASHRAE normal psychrometric chart. Reprinted by permission from *ASHRAE Handbook of Fundamentals 1977*.

where t is the dry-bulb temperature, °F. The moist air enthalpy then becomes

$$h = 0.240t + W(1061 + 0.444t) \tag{2.30}$$

where h has units of Btu/lbm dry air.

The primary data sources for moist air properties are the reports of an ASHRAE cooperative research project conducted at the University of Pennsylvania by John A. Goff and Serge Gratch. Table 1 in Chapter 6 of *1977 Fundamentals* (illustrated in Fig. 2.13) is a tabulation of the thermodynamic properties for the temperature range of -160 to $200°F$ and standard atmospheric pressure.

2.4.12.   The Psychrometric Chart

The ASHRAE psychrometric chart may be used conveniently in the solution of numerous process problems with moist air. Processes performed with air can be plotted on the chart for quick visualization as well as for determining changes in significant properties such as temperature, humidity ratio, and enthalpy for the process. Figure 2.14 is an abridgment of ASHRAE Psychro-

Thermodynamic Properties of MOIST AIR[a] (Standard Atmospheric Pressure, 29.921 in. Hg) (Continued)

Fahr. Temp. t(F)	Humidity Ratio $W_s \times 10^3$	Volume cu ft/lb dry air			Enthalpy Btu/lb dry air			Entropy Btu per (°F) (lb dry air)			Condensed Water			Fahr. Temp. t(F)
		v_a	v_{as}	v_s	h_a	h_{as}	h_s	s_a	s_{as}	s_s	Enthalpy Btu/Lb h_w	Entropy Btu per (°F)(Lb) s_w	Vap. Press In. Hg $p_s \times 10^2$	
7	1.130	11.756	0.021	11.777	1.681	1.202	2.883	0.00364	0.00271	0.00635	−155.61	−0.3172	5.4022	7
8	1.189	11.781	0.022	11.803	1.922	1.266	3.188	0.00415	0.00285	0.00700	−155.13	−0.3162	5.6832	8
9	1.251	11.806	0.024	11.830	2.162	1.332	3.494	0.00467	0.00299	0.00766	−154.65	−0.3152	5.9776	9
10	1.315	11.831	0.025	11.856	2.402	1.401	3.803	0.00518	0.00314	0.00832	−154.17	−0.3141	6.2858	10
11	1.383	11.857	0.026	11.883	2.642	1.474	4.116	0.00569	0.00330	0.00899	−153.69	−0.3131	6.6085	11
12	1.454	11.882	0.028	11.910	2.882	1.550	4.432	0.00620	0.00346	0.00966	−153.21	−0.3121	6.9462	12
13	1.528	11.907	0.029	11.936	3.123	1.620	4.753	0.00671	0.00363	0.01034	−152.73	−0.3111	7.2997	13
14	1.606	11.933	0.030	11.963	3.363	1.713	5.076	0.00721	0.00380	0.01101	−152.24	−0.3100	7.6696	14
15	1.687	11.958	0.032	11.990	3.603	1.800	5.403	0.00772	0.00399	0.01171	−151.76	−0.3090	8.0565	15
16	1.772	11.983	0.034	12.017	3.843	1.892	5.735	0.00822	0.00418	0.01240	−151.27	−0.3080	8.4612	16
17	1.861	12.009	0.035	12.044	4.083	1.988	6.071	0.00873	0.00438	0.01311	−150.78	−0.3070	8.8843	17
18	1.953	12.034	0.038	12.072	4.324	2.088	6.412	0.00923	0.00459	0.01382	−150.29	−0.3059	9.3267	18
19	2.051	12.059	0.040	12.099	4.564	2.192	6.756	0.00973	0.00481	0.01454	−149.80	−0.3049	9.7889	19
20	2.152	12.084	0.042	12.126	4.804	2.302	7.106	0.01023	0.00504	0.01527	−149.31	−0.3039	10.272	20
21	2.258	12.110	0.044	12.154	5.044	2.416	7.460	0.01073	0.00528	0.01601	−148.82	−0.3029	10.777	21
22	2.369	12.135	0.046	12.181	5.284	2.536	7.820	0.01123	0.00553	0.01676	−148.33	−0.3018	11.305	22
23	2.485	12.160	0.049	12.209	5.525	2.661	8.186	0.01173	0.00579	0.01752	−147.84	−0.3008	11.856	23
24	2.606	12.186	0.051	12.237	5.765	2.792	8.557	0.01223	0.00607	0.01830	−147.34	−0.2998	12.431	24
25	2.733	12.211	0.054	12.265	6.005	2.929	8.934	0.01273	0.00635	0.01908	−146.85	−0.2988	13.032	25
26	2.865	12.236	0.057	12.293	6.245	3.072	9.317	0.01322	0.00665	0.01987	−146.35	−0.2977	13.659	26
27	3.003	12.262	0.059	12.321	6.485	3.221	9.706	0.01372	0.00696	0.02068	−145.85	−0.2967	14.313	27
28	3.147	12.287	0.062	12.349	6.726	3.377	10.103	0.01421	0.00728	0.02149	−145.36	−0.2957	14.966	28
29	3.297	12.312	0.065	12.377	6.966	3.540	10.506	0.01470	0.00761	0.02231	−144.86	−0.2947	15.709	29
30	3.454	12.338	0.068	12.406	7.206	3.709	10.915	0.01519	0.00796	0.02315	−144.36	−0.2936	16.452	30
31	3.617	12.363	0.071	12.434	7.446	3.887	11.333	0.01568	0.00832	0.02400	−143.86	−0.2926	17.227	31
32	3.788	12.388	0.075	12.463	7.686	4.072	11.758	0.01617	0.00870	0.02487	−143.36	−0.2916	18.035	32
32*	3.788	12.388	0.075	12.463	7.686	4.072	11.758	0.01617	0.00870	0.02487	0.01	0.0000	18.037	32*
33	3.944	12.413	0.079	12.492	7.927	4.242	12.169	0.01666	0.00904	0.02570	1.05	0.0020	18.778	33
34	4.107	12.438	0.082	12.520	8.167	4.418	12.585	0.01715	0.00940	0.02655	2.06	0.0041	19.546	34
35	4.275	12.464	0.085	12.549	8.407	4.601	13.008	0.01764	0.00977	0.02741	3.06	0.0061	20.342	35
36	4.450	12.489	0.089	12.578	8.647	4.791	13.438	0.01812	0.01016	0.02828	4.07	0.0081	21.166	36
37	4.631	12.514	0.093	12.607	8.887	4.987	13.874	0.01861	0.01056	0.02917	5.07	0.0102	22.020	37

Fahr. Temp. t(F)	Humidity Ratio $W_s \times 10^3$	Volume cu ft/lb dry air			Enthalpy Btu/lb dry air			Entropy Btu per (°F) (lb dry air)			Condensed Water			Fahr. Temp. t(F)
		v_a	v_{as}	v_s	h_a	h_{as}	h_s	s_a	s_{as}	s_s	Enthalpy Btu/Lb h_w	Entropy Btu per (°F)(Lb) s_w	Vap. Press In. Hg p_s	
38	4.818	12.540	0.097	12.637	9.128	5.191	14.319	0.01909	0.01097	0.03006	6.08	0.0122	0.22904	38
39	5.012	12.565	0.101	12.666	9.368	5.403	14.771	0.01957	0.01139	0.03096	7.08	0.0142	0.23819	39
40	5.213	12.590	0.105	12.695	9.608	5.622	15.230	0.02005	0.01183	0.03188	8.09	0.0162	0.24767	40
41	5.421	12.616	0.109	12.725	9.848	5.849	15.697	0.02053	0.01228	0.03281	9.09	0.0182	0.25748	41
42	5.638	12.641	0.114	12.755	10.088	6.084	16.172	0.02101	0.01275	0.03376	10.09	0.0202	0.26763	42
43	5.860	12.666	0.119	12.785	10.329	6.328	16.657	0.02149	0.01323	0.03472	11.10	0.0222	0.27813	43
44	6.091	12.691	0.124	12.815	10.569	6.580	17.149	0.02197	0.01373	0.03570	12.10	0.0242	0.28899	44
45	6.331	12.717	0.129	12.846	10.809	6.841	17.650	0.02245	0.01425	0.03670	13.10	0.0262	0.30023	45
46	6.578	12.742	0.134	12.876	11.049	7.112	18.161	0.02293	0.01478	0.03771	14.10	0.0282	0.31185	46
47	6.835	12.767	0.140	12.907	11.289	7.391	18.680	0.02340	0.01534	0.03874	15.11	0.0302	0.32386	47
48	7.100	12.792	0.146	12.938	11.530	7.681	19.211	0.02387	0.01591	0.03978	16.11	0.0321	0.33629	48
49	7.374	12.818	0.151	12.969	11.770	7.981	19.751	0.02434	0.01650	0.04084	17.11	0.0341	0.34913	49
50	7.658	12.843	0.158	13.001	12.010	8.291	20.301	0.02481	0.01711	0.04192	18.11	0.0361	0.36240	50
51	7.952	12.868	0.164	13.032	12.250	8.612	20.862	0.02528	0.01774	0.04302	19.11	0.0381	0.37611	51
52	8.256	12.894	0.170	13.064	12.491	8.945	21.436	0.02575	0.01839	0.04414	20.11	0.0400	0.39028	52
53	8.569	12.919	0.178	13.097	12.731	9.289	22.020	0.02622	0.01906	0.04528	21.12	0.0420	0.40492	53
54	8.894	12.944	0.185	13.129	12.971	9.644	22.615	0.02669	0.01976	0.04645	22.12	0.0439	0.42004	54
55	9.229	12.970	0.192	13.162	13.211	10.01	23.22	0.02716	0.02047	0.04763	23.12	0.0459	0.43565	55
56	9.575	12.995	0.200	13.195	13.452	10.39	23.84	0.02762	0.02121	0.04883	24.12	0.0478	0.45176	56
57	9.934	13.020	0.208	13.228	13.692	10.79	24.48	0.02809	0.02197	0.05006	25.12	0.0497	0.46840	57
58	10.30	13.045	0.216	13.261	13.932	11.19	25.12	0.02855	0.02276	0.05131	26.12	0.0517	0.48558	58
59	10.69	13.071	0.224	13.295	14.172	11.61	25.78	0.02902	0.02357	0.05259	27.12	0.0536	0.50330	59
60	11.08	13.096	0.233	13.329	14.413	12.05	26.46	0.02948	0.02441	0.05389	28.12	0.0555	0.52159	60
61	11.49	13.121	0.242	13.363	14.653	12.50	27.15	0.02994	0.02527	0.05521	29.12	0.0574	0.54047	61
62	11.91	13.147	0.251	13.398	14.893	12.96	27.85	0.03010	0.02616	0.05656	30.12	0.0594	0.55994	62
63	12.35	13.172	0.261	13.433	15.134	13.44	28.57	0.03086	0.02708	0.05794	31.12	0.0613	0.58002	63
64	12.80	13.197	0.271	13.468	15.374	13.94	29.31	0.03132	0.02803	0.05935	32.12	0.0632	0.60073	64
65	13.26	13.222	0.282	13.501	15.614	14.45	30.06	0.03177	0.02901	0.06078	33.11	0.0651	0.62209	65
66	13.74	13.247	0.292	13.539	15.855	14.98	30.83	0.03223	0.03002	0.06225	34.11	0.0670	0.64411	66
67	14.24	13.273	0.303	13.576	16.095	15.53	31.62	0.03269	0.03106	0.06375	35.11	0.0689	0.66681	67
68	14.75	13.298	0.315	13.613	16.335	16.09	32.42	0.03314	0.03213	0.06527	36.11	0.0708	0.69019	68
69	15.28	13.323	0.327	13.650	16.576	16.67	33.25	0.03360	0.03323	0.06683	37.11	0.0727	0.71430	69

[a] Compiled by John A. Goff and S. Gratch.
* Extrapolated to represent metastable equilibrium with undercooled liquid.

Fig. 2.13. Air–water vapor table illustration. Reprinted by permission from *ASHRAE Handbook of Fundamentals 1977.*

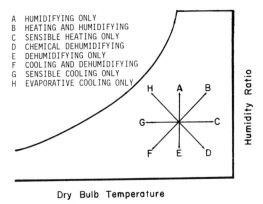

A HUMIDIFYING ONLY
B HEATING AND HUMIDIFYING
C SENSIBLE HEATING ONLY
D CHEMICAL DEHUMIDIFYING
E DEHUMIDIFYING ONLY
F COOLING AND DEHUMIDIFYING
G SENSIBLE COOLING ONLY
H EVAPORATIVE COOLING ONLY

Dry Bulb Temperature

Fig. 2.15. Psychrometric representations of basic air-conditioning processes.

metric Chart No. 1 for ready reference. Figure 2.15 shows some of the basic air-conditioning processes.

Sensible heating only *(C)* and sensible cooling only *(G)* show a change in dry-bulb temperature with no change in humidity ratio. For either sensible heat change process, the temperature changes but the moisture content of the air does not. Humidifying only *(A)* and dehumidifying only *(E)* show a change in humidity ratio with no change in dry-bulb temperature. For these latent heat processes, the moisture content of the air is changed but not the temperature.

Cooling and dehumidifying *(F)* result in a reduction of both the dry-bulb temperature and the humidity ratio. Cooling coils generally perform this type of process. Heating and humidifying *(B)* result in an increase of both the dry-bulb temperature and the humidity ratio.

Chemical dehumidifying *(D)* is a process in which moisture from the air is absorbed or adsorbed by a hygroscopic material. Generally, the process essentially occurs at constant enthalpy.

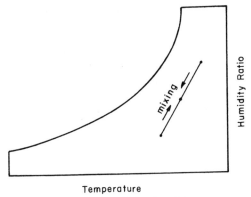

Temperature

Fig. 2.16. Adiabatic mixing.

ASHRAE PSYCHROMETRIC CHART NO. 1
NORMAL TEMPERATURE
BAROMETRIC PRESSURE 29.921 INCHES OF MERCURY
Copyright 1963
AMERICAN SOCIETY OF HEATING, REFRIGERATING AND AIR-CONDITIONING ENGINEERS, INC.

Evaporative cooling only *(H)* is an adiabatic heat transfer process in which the wet-bulb temperature of the air remains constant but the dry-bulb temperature drops as the humidity rises.

Adiabatic mixing of air at one condition with air at some other condition is represented on the psychrometric chart by a straight line drawn between the points representing the two air conditions as shown on Fig. 2.16.

ASHRAE Psychrometric Chart No. 1, "Normal Temperature," is reproduced as Fig. 2.17.

EXAMPLE 2.3

Complete the following table using the psychrometric chart.

Dry Bulb (°F)	Wet Bulb (°F)	Dew Point (°F)	Humidity Ratio m_w/m_a	Relative Humidity %	Enthalpy (Btu/lb)	Specific Volume (ft^3/lb)
90	75					
105					35	
		65		30		
			.022			14.5
45	45					

Solution.

Dry Bulb (°F)	Wet Bulb (°F)	Dew Point (°F)	Humidity Ratio m_w/m_a	Relative Humidity (%)	Enthalpy (Btu/lb)	Specific Volume (ft^3/lb)
90	75	68.8	0.0154	50	38.6	14.2
105	71.6	53.4	0.0088	18.5	35	14.4
102.7	76.2	65	0.0133	30	39.3	14.5
96	83.3	79.5	.022	59	47.2	14.5
45	45	45	0.0064	100	17.7	12.7

2.5. FORMS OF ENERGY

2.5.1. Energy

Energy is the capacity for producing an effect. Thermodynamics is founded on the law of conservation of energy, which says in effect that energy can neither be created nor destroyed. Heat and work are transitory forms of

Fig. 2.17. ASHRAE psychrometric chart. Reprinted by permission of ASHRAE.

energy; they lose their identity as soon as they are absorbed by the body or region to which they are delivered. *Work and heat are not possessed* by a system and therefore are not properties. Thus if there is a net transfer of energy across the boundary from a system (as heat and/or work), from where did this energy come? The only answer is that it must have come from a store of energy in the given system. These stored forms of energy may be assumed to reside within the bodies or regions with which they are associated. In thermodynamics, accent is placed on the *changes* of stored energy rather than on absolute quantities.

2.5.2. Stored Forms of Energy

Energy may be stored in many forms, such as thermal (internal), mechanical, electrical, chemical, and atomic (nuclear).

Thermal (Internal) Energy, U.   Internal energy relates to the energy possessed by a material due to the motion and/or position of the molecules. This form of energy may be divided into two parts: (1) kinetic internal energy, which is due to the velocity of the molecules, and (2) potential internal energy which is due to the attractive forces existing between molecules. Changes in the velocity of molecules are indicated by temperature changes of the system, while variations in position are denoted by changes in phase of the system.

Potential Energy, PE.   Potential energy is the energy possessed by the system due to the elevation or position of the system. This potential energy is equivalent to the work required to lift the medium from the arbitrary datum (zero elevation) to its elevation z in the absence of friction.

$$F = \frac{ma}{g_c} = \frac{mg}{g_c}$$

$$\text{PE} = W = \int_0^x F \, dx = \int_0^z m \, \frac{g}{g_c} \, dx = m \, \frac{g}{g_c} \, z \qquad (2.31)$$

Kinetic Energy, KE.   Kinetic energy is the energy possessed by the system as a result of motion. It is equal to the work that could be done in bringing to rest a medium that is in motion with a velocity V in the absence of gravity.

$$F = \frac{ma}{g_c} = -\frac{m}{g_c} \frac{dV}{d\tau}$$

$$\text{KE} = W = \int_0^x F \, dx = -\int_V^0 \frac{m}{g_c} \frac{dV}{d\tau} \, dx = -\int_V^0 \frac{m}{g_c} \, V \, dV = \frac{mV^2}{2g_c}$$

$$(2.32)$$

Chemical Energy, E_c.   Chemical energy is possessed by the system because of the arrangement of the atoms composing the molecules. Reactions that liberate energy are termed *exothermic,* and those that absorb energy are termed *endothermic.*

Nuclear (Atomic) Energy, E_a.   Nuclear energy is possessed by the system due to the cohesive forces holding the protons and neutrons together in the nucleus of the atom.

Stored energy is concerned with (1) the molecules of the system (internal energy); (2) the system as a unit (kinetic and potential energy); (3) the arrangement of the atoms (chemical energy), and (4) the nucleus of the atom (nuclear energy). Molecular stored energy is associated with the relative position and velocity of the molecules, and the total effect is called internal or thermal energy. It is called thermal energy because it cannot be readily converted into work. The stored energy associated with the velocity of the system is called kinetic energy, while the stored energy associated with the position of the system is called potential energy. These are both forms of mechanical energy since they can be converted readily and completely into work. Chemical, electrical, and atomic energy would be included in any accounting of stored energy; however, engineering thermodynamics frequently confines itself to systems not undergoing changes in these forms of energy. If the basic principles are understood, it is a simple matter to include them.

2.5.3.   Transient Forms of Energy

Heat, Q.   Heat is the mechanism in which energy is transferred across the boundary between systems by reason of the difference in temperature of the two systems. The transfer is always in the direction of the lower temperature. Being transitory, *heat is not a property.* It is redundant to speak of heat as being transferred, for the term "heat" signifies in itself "energy in transit." However, we will refer to heat as being transferred in keeping with common usage.

Although a body or system cannot "contain" heat, it will be useful in discussing many processes to speak of heat received or heat rejected so that the direction of heat transfer relative to the system is immediately obvious. This should not be construed as the treatment of heat as though it were a substance.

Further, heat transferred to a system is denoted $+Q$, and heat transferred from a system, $-Q$. Thus, positive heat represents energy transferred *to* a system, and negative heat represents energy transferred *from* a system. In an adiabatic process, $Q = 0$.

From a mathematical perspective, heat is a path function and is recognized as an inexact differential. That is, the amount of heat transferred when a system undergoes a change of state from state 1 to state 2 depends on the path that the system follows during the change of state. Since heat is an inexact differential, the differential is written δQ. On integrating, we write

$$\int_1^2 \delta Q = {}_1Q_2$$

In words, ${}_1Q_2$ is the heat transferred during the given process between state 1 and state 2.

The rate at which heat is transferred to a system is designated by the symbol $\dot{Q}$.

$$\dot{Q} \equiv \frac{\delta Q}{d\tau}$$

It is also convenient to speak of the heat transfer per unit mass of the system, q, which is defined as

$$q \equiv \frac{Q}{m}$$

Work, W.   Work is the mechanism by which energy is transferred across the boundary between systems by reason of the difference in pressure (or force of any kind) of the two systems, and is in the direction of the lower pressure. If the total effect produced in the system can be reduced to the raising of a weight, then nothing but work has crossed the boundary. Work, like heat, is not possessed by the system but occurs only as energy being transferred.

Work is, by definition, the energy resulting from a force applied through a distance. If the force varies with distance x, work may be expressed as $\delta W = F\, dx$ or

$$W = \int_0^x F\, dx \tag{2.33}$$

In thermodynamics one often finds work done by a force distributed over an area, by a pressure p acting through a volume V, as in the case of a fluid pressure exerted on a piston. In this event,

$$\delta W = p\, dV \tag{2.34}$$

where p is an external pressure exerted on the system.

Mechanical or Shaft Work, W.   This is the energy delivered or absorbed by a mechanism, such as a turbine, air compressor, or internal combustion engine. Shaft work can always be evaluated from the basic relation for work.

Work done *by* a system is considered positive and work done *on* a system is considered negative. The symbol W designates the work done by a system.

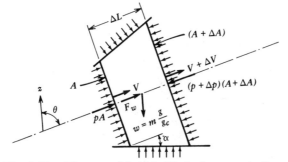

Fig. 2.18. Element of fluid in frictionless steady flow.

There are a variety of ways work may be done on or by a system. In addition to mechanical work, work may be done due to surface tension, the flow of electricity, magnetic fields, and in many other ways.

For *nonflow* processes, the form of mechanical work most frequently encountered is that done at the moving boundary of a system, such as the work done in moving the piston in a cylinder, and may be expressed in equation form for *reversible* processes as $W = \int p \, dV$. In general for the nonflow process, we can express work as follows:

$$W = \int p \, dV + \cdots \tag{2.35}$$

where the dots indicate other ways in which work can be done by or on the system and would include the work loss due to irreversibility.

A useful expression for the work of a frictionless steady-flow process will now be derived. The derivation procedure is

1. Make a free-body diagram of an element of fluid (Fig. 2.18).
2. Evaluate the external forces on the free body.
3. Relate the sum of the external forces to the mass and acceleration of the free body.
4. Solve the resulting relation for the force by which work is done on the fluid.
5. Apply the definition of work as $\int F \, dx$.

Applying Newton's second law of motion, the sum of the external forces on the fluid element must equal ma/g_c. The mass of the element is $\rho(A + \Delta A/2)\Delta L$, and the acceleration is approximately $\Delta V/\Delta\tau$. Thus,

$$\Sigma F = \frac{ma}{g_c} = \frac{\rho}{g_c}\left(A + \frac{\Delta A}{2}\right)\Delta L \frac{\Delta V}{\Delta\tau}$$

The sum or resultant of the forces is

$$\Sigma F = PA - (p + \Delta p)(A + \Delta A) - m\,\frac{g}{g_c}\cos\theta + \left(p + \frac{\Delta p}{2}\right)\Delta A + F_w$$

$$= -A\,\Delta p - \frac{\Delta p\,\Delta A}{2} - m\,\frac{g}{g_c}\cos\theta + F_w$$

$$\text{Work}_{\text{in}} = F_w\Delta L = mv\,\Delta p + \frac{mV\,\Delta V}{g_c} + \frac{mg}{g_c}\,\Delta z$$

and per unit mass

$$\text{Work}_{\text{in}} = v\,\Delta p + \frac{V\,\Delta V}{g_c} + \frac{g}{g_c}\,\Delta z$$

Now, if ΔL is made to approach dL, then the other differences also approach differentials, and the work (per unit mass) done on the fluid in the distance dL is

$$\delta\text{Work}_{\text{in}} = v\,dp + \frac{V\,dV}{g_c} + \frac{g}{g_c}\,dx \qquad (2.36)$$

or for flow between sections a finite distance apart,

$$\text{Work}_{\text{in}} = \int v\,dp + \Delta\left(\frac{V^2}{2g_c}\right) + \frac{g}{g_c}\,\Delta z \qquad (2.37)$$

This is an important relation for the mechanical work done on a unit mass of fluid in a frictionless steady-flow process.

For the *open system,* in addition to commonly encountering work done at a moving boundary, there is always flow work to be considered.

Flow Work.   This consists of the energy carried into or transmitted across the system boundary as a result of the fact that a pumping process occurs somewhere outside the system, causing fluid to enter the system. It might be more easily conceived as the work done by the fluid just outside the system on the adjacent fluid entering the system to force or "push" it into the system. Flow work also occurs as fluid leaves the system. This time the fluid in the system does work on the fluid just leaving the system. Although the energy was provided by a pump or other mechanical means outside of the system, we consider the direct effect, the work of fluid on fluid.

$$\text{Flow work (per unit mass)} = \int F\,dx = \int pA\,dx$$

$$= p\int_0^v dV = pv \qquad (2.38)$$

where v is the specific volume, or the volume displaced per unit mass.

Flow work is also called flow energy or displacement energy. Disagree-

TABLE 2.2.   Forms of Energy

Transient Forms of Energy	
Work	Potential : force
Heat	Potential : temperature
Stored Forms of Energy	
By substances as entities	
Potential	Manifested by elevation
Kinetic	Manifested by velocity
Internal or intrinsic	
Molecular	
Molecular kinetic	Manifested by temperature
Molecular potential	Manifested by phase
Chemical	Manifested by change in molecular composition
Nuclear	Manifested by changes in atomic composition

ment on the name used for this quantity results from the fact that the pv term is generally derived as a work quantity, yet it is unlike other work quantities in that it is expressed in terms of a point function. Because it is so expressed, some engineers prefer to group it with stored energy quantities and speak of it as "transported energy" or "convected energy" instead of work. However, it must be remembered that pv can be treated as energy only when a fluid is crossing a system boundary. For a closed system, pv does not represent any form of energy.

Table 2.2 summarizes the forms of energy and their sources.

2.6.   THE FIRST LAW OF THERMODYNAMICS (CONSERVATION OF ENERGY)

Having completed our consideration of basic definitions and concepts, we are ready to proceed to a discussion of the first law of thermodynamics. Often this law is called the law of conservation of energy.

From the first law or the law of conservation of energy, we can conclude that for any system, open or closed, there is an "energy balance" as

$$\begin{bmatrix} \text{Net amount of energy} \\ \text{added to system} \end{bmatrix} = \begin{bmatrix} \text{net increase in stored} \\ \text{energy of system} \end{bmatrix}$$

or

$$\text{Energy in} - \text{energy out} = \text{increase in energy in system} \qquad (2.39)$$

With both open and closed systems, energy can be added to the system or taken from it by means of heat and work. In the case of an open system, there is an additional mechanism for increasing or decreasing the stored energy of the system. When mass enters a system, the stored energy of the system is increased by the stored energy of the entering mass. Conversely, the stored energy of a system is decreased whenever mass leaves the system. If we distinguish this transfer of stored energy of mass crossing the system boundary from heat and work,

$$\begin{bmatrix} \text{Stored energy} \\ \text{of mass en-} \\ \text{tering system} \end{bmatrix} - \begin{bmatrix} \text{stored energy} \\ \text{of mass leav-} \\ \text{ing system} \end{bmatrix} + \begin{bmatrix} \text{net amount of} \\ \text{energy added} \\ \text{to system as} \\ \text{heat and all} \\ \text{forms of work} \end{bmatrix} = \begin{bmatrix} \text{net increase} \\ \text{in stored} \\ \text{energy of} \\ \text{system} \end{bmatrix}$$

The net exchange of energy between the system and its surroundings must be balanced by the change in energy of the system. Within exchange of energy we include our definition of energy in transition being either work or heat. However, we must describe further what is meant by the energy of the system and the energy associated with any matter entering or leaving the system.

The energy E of the system is a property of the system and consists of all of the various forms in which energy is characteristic of a system. These forms include: the potential energy (due to position), kinetic energy (due to any motion), electrical energy (due to charge), etc. Note that since work and heat are energy in transition and are not characteristic of the system, they are not included here.

All of the energy of a system—exclusive of kinetic and potential energy— is called internal energy.

The symbol for internal energy per unit mass is u. The symbol for internal energy contained in a mass of m pounds is U. Therefore,

$$U = mu$$

We must now describe precisely what is meant by the energy associated with mass entering or leaving the system. Each pound of mass that flows into or out of the system carries with it the energy characteristic of that pound of mass. This energy includes the internal energy, u, plus the kinetic and potential energy.

If we investigate the flow of mass across the boundary of a system, we find that work is always done on or by a system where fluid flows across the system boundary. Therefore, the work term in an energy balance for an open system is usually separated into two parts: (1) the work required to push a fluid into or out of the system, and (2) all other forms of work.

For each unit mass crossing the boundary of a system, the flow work is pv. If the pressure or specific volume or both vary as a fluid flows across a

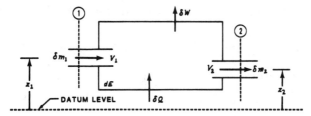

Fig. 2.19. Energy flows in a general thermodynamic system. Reprinted by permission from *ASHRAE Handbook of Fundamentals 1981*.

system boundary, the flow work is calculated as $\int pv\ \delta m$, where δm is an infinitesimal mass crossing the boundary. (The symbol δm is used instead of dm because the amount of mass crossing the boundary is not a property. The mass within the system is a property, so the infinitesimal change in mass within the system is properly represented as dm.)

The term work, W, without modifiers is conventionally understood to stand for all other forms of work except flow work, and the complete two-word name is always used when referring to flow work.

An equation representing the first law can now be written utilizing the symbols we have defined. As shown in Fig. 2.19, we will let δm_1 be the mass entering the system and δm_2 be the mass leaving. The first law in differential or incremental form becomes

$$[m(e + pv)]_{\text{in}} - [m(e + pv)]_{\text{out}} + Q - W = dE$$

$$\delta m_1 \left(u_1 + p_1 v_1 + \frac{V_1^2}{2g_c} + z_1 \frac{g}{g_c} \right) \tag{2.40}$$

$$- \delta m_2 \left(u_2 + p_2 v_2 + \frac{V_2^2}{2g_c} + z_2 \frac{g}{g_c} \right) + \delta Q - \delta W = dE$$

where δQ and δW are the increments of heat and work, respectively, and dE is the differential change in the energy of the system.

As properties of the system, E or U (or e or u) are treated like any other property. Tables of properties of various substances are available, which give values for internal energy u along with other properties.

Since the combination of properties, $u + pv$, is also a property, we use enthalpy for simplicity and speed in obtaining numerical values for this combination.

We let the symbol H stand for the total enthalpy associated with a mass m, and h for specific enthalpy, or enthalpy per unit mass. In equation form,

$$h = u + pv$$

h is tabulated in the literature along with other properties such as u, p, v, and T. The pv term represents flow energy for an open system, but it is merely the product of two properties of a closed system.

In terms of enthalpy, the first-law equation becomes

$$\delta m_1\left(h_1 + \frac{V_1^2}{2g_c} + \frac{g}{g_c}z_1\right) - \delta m_2\left(h_2 + \frac{V_2^2}{2g_c} + \frac{g}{g_c}z_2\right) + \delta Q - \delta W = dE$$

or in integrated form,

$$\int_0^{m_1}\delta m_1\left(h_1 + \frac{V_1^2}{2g_c} + \frac{g}{g_c}z_1\right) - \int_0^{m_2}\delta m_2\left(h_2 + \frac{V_2^2}{2g_c} + \frac{g}{g_c}z_2\right) + Q - W$$

$$= E_{\text{final}} - E_{\text{initial}} \qquad (2.41)$$

Dividing through by the same interval $\Delta\tau$, we have

$$\frac{\delta m_1}{\Delta\tau}\left(h_1 + \frac{V_1^2}{2g_c} + z_1\frac{g}{g_c}\right) - \frac{\delta m_2}{\Delta\tau}\left(h_2 + \frac{V_2^2}{2g_c} + z_2\frac{g}{g_c}\right)$$

$$+ \frac{\delta Q}{\Delta\tau} - \frac{\delta W}{\Delta\tau} = \frac{dE}{\Delta\tau}$$

as $\Delta\tau \to 0$,

$$\frac{\delta Q}{\Delta\tau} \to \dot Q, \quad \frac{\delta W}{\Delta\tau} \to \dot W, \quad \frac{\delta m_1}{\Delta\tau} \to \dot m_1, \quad \frac{\delta m_2}{\Delta\tau} \to \dot m_2, \quad \frac{dE}{\Delta\tau} \to \frac{dE}{d\tau}$$

and

$$\dot m_1\left(h_1 + \frac{V_1^2}{2g_c} + z_1\frac{g}{g_c}\right) - \dot m_2\left(h_2 + \frac{V_2^2}{2g_c} + z_2\frac{g}{g_c}\right) + \dot Q - \dot W = \frac{dE}{d\tau}$$

$$(2.42)$$

where $\dot Q$ and $\dot W$ are the rates of heat flow and work, respectively. $\dot W$ is recognized as power.

Integration of the first-law equation for the most general case in which a prior integration is possible, that is, where

1. The properties of the fluids crossing the boundary remain constant at each point on the boundary.
2. The flow rate at each section where mass crosses the boundary is constant. (The flow rate cannot change as long as all properties, including velocity, at each point remain constant.)
3. All interactions with the surroundings occur at a steady rate.

yields

$$\sum m_{\text{in}}\left(u + \frac{Pv}{J} + \frac{V^2}{2g_cJ} + \frac{g}{g_c}\frac{z}{J}\right)_{\text{in}}$$

$$- \sum m_{\text{out}}\left(u + \frac{Pv}{J} + \frac{V^2}{2g_cJ} + \frac{g}{g_c}\frac{z}{J}\right)_{\text{out}} + Q - W$$

$$= \left[m_f \left(u + \frac{V^2}{2g_c J} + \frac{g}{g_c} \frac{z}{J} \right)_f \right.$$

$$\left. - m_i \left(u + \frac{V^2}{2g_c J} + \frac{g}{g_c} \frac{z}{J} \right)_i \right]_{\text{system}} \tag{2.43}$$

where J is the conversion factor, 778 ft·lbf/Btu. Subscripts i and f are for initial and final, respectively.

A special case of considerable importance in engineering applications is that of the steady-flow process, a process in which all quantities associated with the system remain constant over time. Consequently,

$$\sum_{\substack{\text{all streams} \\ \text{leaving}}} \dot{m} \left(h + \frac{V^2}{2g_c} + \frac{g}{g_c} z \right) - \sum_{\substack{\text{all streams} \\ \text{entering}}} \dot{m} \left(h + \frac{V^2}{2g_c} + \frac{g}{g_c} z \right) + \dot{Q} - \dot{W} = 0 \tag{2.44}$$

A second common application is the closed stationary system, for which the first-law equation reduces to

$$Q - W = [m(u_f - u_i)]_{\text{system}} \tag{2.45}$$

2.7. THE SECOND LAW OF THERMODYNAMICS

2.7.1. The Second Law from Classical Thermodynamics

In its broader significance the second law involves the fact that processes proceed in a certain direction and not in the opposite direction.

A system that undergoes a series of processes and always returns to its initial state is said to have gone through a cycle. For the closed system undergoing a cycle, from the first law of thermodynamics,

$$\oint \delta Q = \oint \delta W$$

The symbol $\oint$ stands for the cyclical integral of the increment of heat or work. Any heat supplied to a cycling system must be balanced by an equivalent amount of work done by the system. Or, conversely, any work done on the cycling system would result in an equivalent amount of heat being given off.

Many examples can be given of work being completely converted into heat. However, a cycling system that completely converts heat into work has never been observed, although such complete conversion would not be a violation of the first law. The fact that heat cannot be completely converted into work is the basis for the second law of thermodynamics. The justification for the second law is empirical.

The second law has been stated in a number of equivalent ways. We shall discuss two of them.

The Kelvin-Planck statement of the second law is:

It is impossible for any cycling device to exchange heat with only a single reservoir and produce work.

In this context, a *reservoir* is a body whose temperature remains constant regardless of how much heat is added to or taken from it. In other words, the Kelvin-Planck statement says that heat cannot be continuously and completely converted into work; a fraction of the heat must be rejected to another reservoir at a lower temperature. The second law thus places a restriction on the first law in relation to the way energy is transferred. Work can be continuously and completely converted into heat, but not vice versa.

If the Kelvin-Planck statement were not true and heat could be completely converted into work, the heat might be obtained from a low-temperature source, converted into work, and the work converted back into heat in a region of higher temperature. The net result of this series of events would be the flow of heat from a low-temperature region to a high-temperature region with no other effect. This phenomenon has never been observed and is contrary to all our experience.

The Clausius statement of the second law is:

No process is possible whose sole result is the removal of heat from a reservoir at one temperature and the absorption of an equal quantity of heat by a reservoir at a higher temperature.

Some of the corollaries of the second law are given below.

Corollary A.   No engine operating between two given reservoirs can have a greater efficiency than a reversible engine operating between the same two reservoirs.

Corollary B.   All reversible engines operating between the same temperature limits have the same efficiency.

Corollary C.   It is theoretically impossible to reduce the temperature of any system to absolute zero by any series of finite processes.

Corollary D.   The thermodynamic temperature scale. Define the ratio of two temperatures as the ratio of the heat absorbed by a Carnot engine to the heat rejected, when the engine is operated between reservoirs at these temperatures. Thus the equality $Q_2/Q_1 = T_2/T_1$ becomes a matter of definition, and the fundamental problem of thermometry, that of establishing a temperature scale, reduces to a problem in calorimetry.

Corollary E.   The inequality of Clausius. When a system is carried around a cycle and the heat δQ added to it at every point is divided by its temperature at that point, the sum of all such quotients is less than zero for irreversible cycles and in the limit is equal to zero for reversible cycles.

$$\oint \frac{\delta Q}{T} < 0$$

Corollary F.   *Principle of the increase of entropy*. In any process what-ever between two equilibrium states of a system, the sum of the increase in entropy of the system and the increase in entropy of its surroundings is equal to or greater than zero.

2.7.2.   The Second Law from Statistical Thermodynamics

To help our understanding of the significance of the second law of thermody-namics, let us consider briefly the molecular nature of matter. Although a sample of a gas may be "at rest," its molecules are not at rest. They are in a state of continuous random motion with an average speed of the same order of magnitude as the speed of sound waves in the gas. For air, this is about 1100 ft/s at room temperature. Some of the molecules move more rapidly than this and some more slowly, and as a result of collisions with one another and with the walls of the containing vessel, the velocity of any one molecule is continually being changed in magnitude and direction. The num-ber of molecules traveling in a given direction with a given speed, however, remains constant. If the gas as a whole is at rest, the molecular velocities are distributed randomly in direction. We would not expect the molecules of a gas, flying about in all directions and with a wide range of speeds, to get together in a cooperative effort and all acquire simultaneously a common velocity component in the same direction, although such a process would be entirely possible from the standpoint of conservation of energy, that is, from the first law. Nevertheless, the possibility cannot be excluded that purely as a result of chance such events do occur and are easily observed. These phenomena have been observed and are described by the general term, *fluctuations*. Hence, while the dogmatic statement cannot be made that a process never occurs whose sole result is the flow of heat from a heat reservoir and the performance of an equivalent amount of work, it is a fact that such processes do not occur with sufficient frequency or with objects of sufficient size to make them of any practical utility. Thus for a completely accurate statement of the second law, the term "impossible" would have to be replaced with "improbable."

The second law is merely a statement of the improbability of the spon-taneous passage of the system from a highly probable state (random or disordered) to one of lower probability. To make this clearer, consider the simple analogy of a deck of cards arranged in four hands of 13 cards each. There is a total of 635,000,000,000 different possible arrangements, or hands. There are only 4 all-one-suit arrangements possible, but there are 1.37×10^{11} possible 4-4-3-2 arrangements. The chances, therefore, of getting 13 cards of one suit from a well-shuffled deck as compared to a 4-4-3-2 hand are only 1 in 3.4×10^{10}. The principle is the same with shuffled molecules, and because the numbers are so very much greater, the chances of an orderly arrangement arising from a random mixture are infinitesimal.

2.7.3. Physical Meaning of Entropy

What is the relation of all this to entropy? Entropy is a measure of "mixed-upness" or of the probability of a given state. The more completely shuffled any system is, the greater is its entropy; and conversely, an orderly or unmixed configuration is one of low entropy. If the entropy of the universe has been irretrievably increased in an irreversible process; so what? There has been no loss of energy in the process. From the engineer's point of view, something has been "lost" when an irreversible process takes place. What is lost, however, is *not energy,* but *opportunity*—the opportunity to convert internal energy to mechanical energy.

Entropy is not conserved, except in reversible processes. When a beaker of hot water is mixed with a beaker of cold water, the heat lost by the hot water equals the heat gained by the cold water and energy is conserved. On the other hand, while the entropy of the hot water decreased, the entropy of the cold water increased a greater amount, and the total entropy of the system is greater at the end of the process than it was at the beginning. Entropy was created in the process. "Energy can neither be created nor destroyed," states the first law. "Entropy cannot be destroyed, but it can be created," states the second law.

2.7.4. Entropy Equation of the Second Law of Thermodynamics

For the general case of an open system, the second law can be written

$$dS_{\text{system}} = \left(\frac{\delta Q}{T}\right)_{\text{rev}} + \delta m_i s_i - \delta m_e s_e + dS_{\text{irr}} \qquad (2.46)$$

where $\delta m_i s_i$ is the entropy increase due to the mass entering, $\delta m_e s_e$ is the entropy decrease due to the mass leaving, $\delta Q/T$ is the entropy change due to reversible heat transfer between system and surroundings, and dS_{irr} is the entropy created due to irreversibilities. The equation is merely an accounting for all the entropy changes in the system.

Rearranging Eq. (2.46), we obtain

$$\delta Q = T[\delta m_e s_e - \delta m_i s_i) + dS_{\text{sys}} - dS_{\text{irr}}]$$

In integrated form, subject to the restrictions that inlet and outlet properties, mass flow rates, and interactions with the surroundings do not vary with time, the general equation for the second law is

$$(S_f - S_i)_{\text{system}} = \int_{\text{rev}} \frac{\delta Q}{T} + \sum (ms)_{\text{in}} - \sum (ms)_{\text{out}} + \Delta S_{\text{produced}} \qquad (2.47)$$

2.8. APPLICATION OF THERMODYNAMICS TO REFRIGERATION

Continuous refrigeration can be accomplished by several different processes. In the great majority of applications, and almost exclusively in the smaller horsepower range, the vapor compression system, commonly termed the simple compression cycle, is used for the refrigeration process. However, absorption systems and steam-jet vacuum systems are being successfully used in many applications. In larger equipment, centrifugal systems are basically an adaptation of the compression cycle.

A larger number of working fluids (refrigerants) are used in vapor-compression refrigeration systems than in vapor power cycles. Ammonia and sulfur dioxide were first used as vapor-compression refrigerants. Today, the main refrigerants are the halogenated hydrocarbons, which are marketed under trade names such as Freon and Genetron. Two important considerations in selecting a refrigerant are the temperature at which refrigeration is desired and the type of equipment to be used.

Refrigerants used in most mechanical refrigeration systems are R-22, which boils at $-41.4°F$, and R-12, which boils at $-21.6°F$, at atmospheric pressure.

Figure 2.20 illustrates schematically the basic vapor-compression cycle.

Cool low-pressure liquid refrigerant enters the evaporator and evaporates. As it does, it absorbs heat from some other substance, air or water, thereby accomplishing refrigeration. The refrigerant then leaves the evaporator as a cool low-pressure gas and proceeds to the compressor. Here, its pressure and temperature are increased, and it is discharged as a hot high-pressure gas to the condenser, where it is condensed into a liquid. The condensing agent, air or water, is at a temperature lower than the refrigerant gas. The hot high-pressure liquid flows from the condenser through the

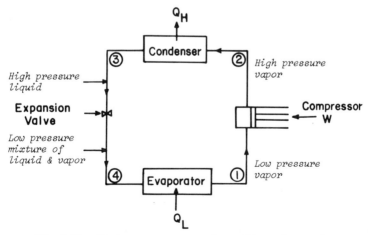

Fig. 2.20. Basic vapor-compression refrigeration cycle.

expansion valve to the evaporator. The expansion valve meters the liquid flow, reducing the hot high-pressure liquid to a cool low-pressure liquid as it enters the evaporator.

The refrigeration cycle can be plotted on the pressure–enthalpy diagram as shown in Figs. 2.21 and 2.22. Subcooled liquid, at point *A*, begins to lose pressure as it goes through the metering valve, located at the point where the vertical liquid line meets the saturation curve. As it leaves the metering point, some of the liquid flashes into vapor and cools the liquid entering the evaporator at point *B*. Notice that there is additional reduction in pressure from the metering point to point *B*, but no change in enthalpy.

As it passes from point *B* to *C*, the remaining liquid picks up heat and changes from a liquid to a gas but does not increase in pressure. Enthalpy, however, does increase. Superheat is added between point *C,* where the vapor passes the saturation curve, and point *D*. About 1 Btu is added to the enthalpy, due to superheat.

As it passes through the compressor, *D* to *E*, the temperature and the pressure are markedly increased, as is the enthalpy, due to the heat of compression. Line *E-F* indicates that the vapor must be de-superheated within the condenser before it attains a saturated condition and begins to condense. Line *F-G* represents the change from vapor to liquid within the condenser. Line *G-A* represents subcooling within the liquid line or capillary tube, prior to flow through the metering device.

Note that the pressure remains essentially constant as the refrigerant passes through the evaporator, but its temperature is increased beyond the saturation point, due to superheat, before it enters the compressor. The pressure likewise remains constant as the refrigerant enters the condenser as a vapor and leaves as a liquid. While the temperature is constant through the condenser, it is reduced as the liquid is subcooled before entering the metering valve.

The change in enthalpy, that is, the heat content, as the refrigerant passes through the evaporator is almost all latent heat since the temperature does

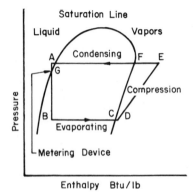

Fig. 2.21.   Simplified pressure–enthalpy curve.

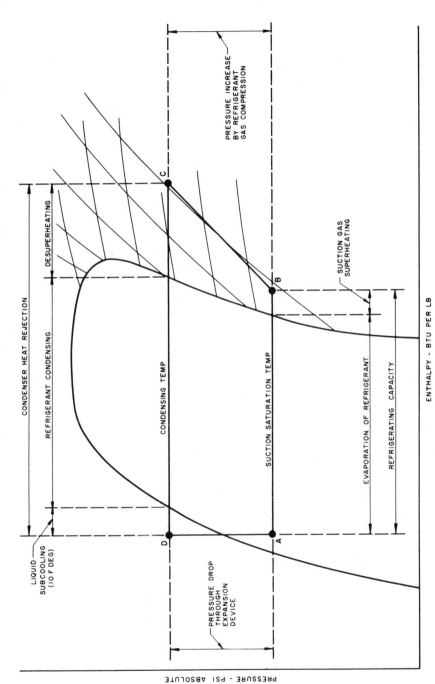

Fig. 2.22. Typical pressure–enthalpy (Mollier) diagram for complete refrigeration cycle. Reprinted by permission from *ASHRAE 1969 Equipment Handbook*.

not change appreciably. At 45°F (normal evaporator operating temperature) and in a liquid state, 1 lb of R-22 contains 23 Btu and in the vapor state it contains 109 Btu. The heat absorbed by vaporization of R-22 therefore is 86 Btu. This means it takes 86 Btu of latent heat to convert R-22 from a liquid to a gas.

Applying the first law of thermodynamics to this system as a whole requires that the sum of all energies in must equal the sum of all energies out when the unit is operating at a steady-state rate; hence,

$$Q_L + W = Q_H, \quad \text{Btu/hr}$$

The rate of heat being rejected at the condenser, Q_H, is seen to be greater numerically than the rate at which work is delivered to drive the compressor, W, and is also greater than the rate of heat absorption into the evaporator, Q_L.

It is apparent from this relation that every refrigeration cycle operates at all times as a heat pump. The household refrigerator absorbs a quantity of heat Q_L at a low temperature (e.g., 0°F) in the vicinity of the ice-making section (the evaporator) and rejects heat Q_H at a higher temperature (e.g., 100°F) to the air in the room where it is located. The rate of heat rejection Q_H is greater than the rate of absorption Q_L by W, the power input to drive the compressor.

The air-conditioning applications, for example, for a room air conditioner, the desired effect is cooling (the heat is absorbed at the evaporator), and the evaporator is located inside the air-conditioned space for the benefit of the occupants. Heat is rejected through the condenser, which is outside the conditioned space.

The name *heat pump* is reserved for this cycle when the desired effect is heating and the condenser is located inside the building. In this case, the evaporator is located outside the building and absorbs heat from outside air. Figure 2.23 depicts the refrigeration cycle in both modes.

2.8.1. Energy Relations for the Basic Refrigeration Cycle

The first law can be applied to each component of the system individually, since energy must be conserved by each component as well as by the entire system.

Compressor. Figure 2.24 shows the mass and major energy flows for a compressor. The rate of energy inflow must equal the rate of energy outflow during steady-state operation. Hence,

$$mh_1 + W = mh_2$$

or

$$W = m(h_2 - h_1)$$

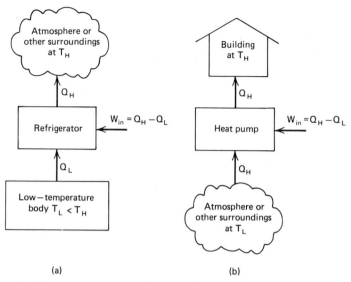

Fig. 2.23. Refrigeration cycle application (a) as a refrigerator and (b) as a heat pump.

where m = rate of refrigerant flow, lbm/hr

h_1, h_2 = enthalpies of the refrigerant at inlet and outlet of the compressor, respectively

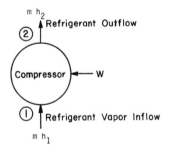

Fig. 2.24. Compressor flows.

Condenser. Figure 2.25 shows the mass and energy flows for a condenser, where

$$mh_2 = mh_3 + Q_R$$

or

$$Q_R = m(h_2 - h_3), \quad \text{Btu/hr}$$

Typically, heat Q_R is rejected from the condensing refrigerant to another fluid. The heat energy leaving the condensing refrigerant, Q_R, must equal the

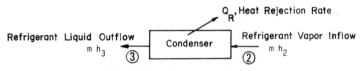

Fig. 2.25.   Condenser flows.

heat energy Q absorbed by the fluid that receives it. So, from the viewpoint of the other fluid,

$$Q_R = m_{\text{fluid}}\, c_{p_{\text{fluid}}} (t_{\text{out}} - t_{\text{in}})_{\text{fluid}}$$

Expansion Device.   This device is essentially a flow restrictor, a small valve seat opening or a long length of small-bore tubing, that prevents either work or any significant amount of heat transfer from occurring. Hence,

$$mh_3 = mh_4, \quad \text{Btu/hr}$$

Dividing both sides by m gives

$$h_3 = h_4, \quad \text{Btu/lbm}$$

Evaporator.   Figure 2.26 shows major mass and energy flows for an evaporator, where

$$mh_4 + Q_A = mh_1$$

or

$$Q_A = m(h_1 - h_4), \quad \text{Btu/hr}$$

Typically, the evaporator receives the heat flow quantity Q_A by heat transfer from another fluid, usually water or air. From the viewpoint of that other fluid,

$$Q_A = m_{\text{fluid}} c_{p_{\text{fluid}}} (t_{\text{in}} - t_{\text{out}})_{\text{fluid}}$$

Figure 2.27 summarizes the results of applying the laws of thermodynamics to the basic vapor-compression refrigeration system.

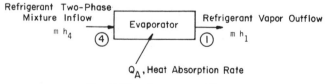

Fig. 2.26.   Evaporator flows.

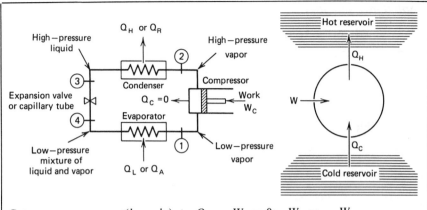

Compressor: $m(h_1 - h_2) + {_1Q_2} - {_1W_2} = 0;$ $W_C = -{_1W_2}$

Condenser: $m(h_2 - h_3) + {_2Q_3} = 0;$ $Q_R = -{_2Q_3}$

Expansion device: $h_3 - h_4 = 0$

Evaporator: $m(h_4 - h_1) + {_4Q_1} = 0;$ $Q_A = {_4Q_1}$

Overall: $Q_A + W_c = Q_c + Q_R$

or $({_1Q_2} + {_2Q_3} + {_4Q_1}) - ({_1W_2}) = 0$

Coefficient of performance for cooling:

$$COP_c = \frac{\text{Useful effect}}{\text{Input that costs}} = \frac{Q_L}{W} = \frac{1}{Q_H/Q_L - 1}$$

$$COP_{c,\max} = \frac{1}{T_H/T_L - 1}$$

Coefficient of performance for heating:

$$COP_h = \frac{Q_H}{W}$$

$$COP_{h,\max} = \frac{1}{1 - T_L/T_H}$$

Fig. 2.27.   The vapor-compression system.

EXAMPLE 2.4

A vapor-compression refrigeration system uses Freon-12 as the refrigerant and operates between a condensing temperature of 110°F and an evaporating temperature of 5°F. For a cooling effect of 24,000 Btu/hr, determine: (a) the minimum size motor (hp) required to drive the compressor, and (b) the maximum coefficient of performance for cooling for this vapor-compression system.

Solution. See Fig. E2.4.

(1) $T_1 = 5°F$, $x_1 = 1$; $h_1 = h_{g_{5°F}} = 77.805$, $S_1 = 0.16842$

$P_1 = 26.483$

(2) $P_2 = P_{\text{sat}_{110°F}} = 151.11$ psia

$S_L = S_1 = 0.16842$

$T_2 \approx 126°F$; $h_2 \approx 90.1$ Btu/lbm

(3) $P_3 = P_2$; $h_3 = h_{f_{110°F}} = 33.531$; $S_3 = 0.067451$

(4) $h_4 = h_3$ (1st law) $= 33.531 = x(77.805) + (1 - x)(9.61)$

$x_4 = .351$

$S_4 = 0.351(0.16833) + 0.649(0.02211) = 0.0734$

Evaporator: $m(h_1 - h_4) + Q = 0$

$$m = \frac{2400}{(77.805 - 33.531)} = 542.1 \text{ lb/hr}$$

Compressor: $m(h_1 - h_2) - w = 0$

(a) $w = \dfrac{542.1(77.805 - 90.1)}{2545} = -2.62$ hp

(b) COP $= \dfrac{Q_c}{w} = \dfrac{24,000}{2.62 \times 2545} = 3.6$

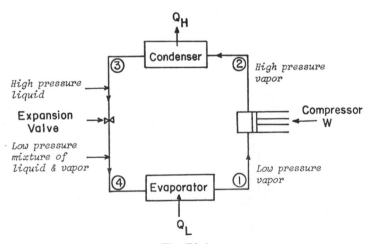

Fig. E2.4.

EXAMPLE 2.5

The following data are for the Freon-12 refrigeration cycle shown in Fig. E2.5.

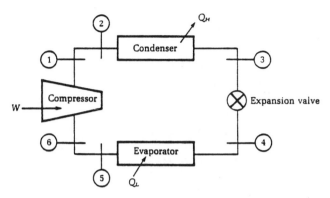

Fig. E2.5.

$P_1 = 180$ lbf/in.^{2}, $T_1 = 240°F$
$P_2 = 178$ lbf/in.^{2}, $T_2 = 220°F$
$P_3 = 175$ lbf/in.^{2}, $T_3 = 100°F$
$P_4 = 29$ lbf/in.^{2},
$P_5 = 27$ lbf/in.^{2}, $T_5 = 20°F$
$P_6 = 25$ lbf/in.^{2}, $T_6 = 40°F$
Rate of flow of Freon $= 200$ lbm/hr
Power input to compressor $= 2.5$ hp

Calculate

(a) The heat transfer per hour from the compressor.
(b) The heat transfer per hour from the Freon in the condenser.
(c) The heat transfer per hour to the Freon in the evaporator.
(d) COP (cooling).
(e) EER (cooling).
(f) COP (heating).
(g) EER (heating).

Solution.

(a) Compressor:

$$\dot{m}(h_6 - h_1) + \dot{Q} - \dot{W} = 0$$

$$200(83.01 - 110.48) + \dot{Q} - (-2.5 \times 2545) = 0$$

$$\dot{Q}_{cp} = -860 \text{ Btuh}$$

(b) Condenser:

$$\dot{m}(h_2 - h_3) + \dot{Q} = 0; \quad 200(106.98 - 31.1) + \dot{Q} = 0$$

$$\dot{Q}_c = -15,200 \text{ Btuh}$$

(c) Evaporator:

$$\dot{m}(h_4 - h_5) + \dot{Q} = 0; \quad 200(31.1 - 79.96) + \dot{Q} = 0$$

$$\dot{Q}_e = 9770 \text{ Btuh}$$

(d) $\text{COP}_c = \dfrac{\dot{Q}_e}{\dot{W}} = \dfrac{9770}{2.5(2545)} = 1.54$

(e) $\text{EER}_c = \dfrac{\dot{Q}_e}{\dot{W}} = \dfrac{9770}{2.5(2545)/3.413} = 5.26 \text{ Btuh/W}$

(f) $\text{COP}_h = \dfrac{\dot{Q}_c}{\dot{W}} = \dfrac{15,200}{2.5(2545)} = 2.39$

(g) $\text{EER}_h = \dfrac{\dot{Q}_c}{\dot{W}} = 2.39(3.413) = 8.16 \text{ Btuh/W}$

2.9. APPLICATION OF THERMODYNAMICS TO HEAT PUMPS

A heat pump is a thermodynamic device that operates in a cycle that requires work and accomplishes the objective of transferring heat from a low-temperature body to a high-temperature body. The heat-pump cycle is identical to a refrigeration cycle in principle except that the primary purpose of the heat pump is to supply heat rather than remove it from an enclosed space.

A heat pump accomplishes its task of transferring energy from a low-temperature source to a high-temperature receptacle through the use of a secondary fluid (usually Freon-22) which has a boiling point several degrees below 0°F. Space heating (in the winter) can be accomplished by transferring energy from the low-temperature outside air to the even lower-temperature secondary fluid in the liquid phase. The secondary liquid is evaporated in an outdoor heat exchanger and is converted to a cool secondary vapor. Work is then done on the secondary vapor through the use of a compressor. Following the compression process the secondary vapor is at high pressure and at a corresponding high temperature as defined by the equation of state for the fluid. The vapor temperature at this point is higher than the temperature of the indoor air. By condensing the secondary vapor with an indoor heat exchanger and fan, energy is transferred from the high-temperature secondary vapor to the indoor air, thus heating the indoor air. The secondary fluid then moves to the outdoor unit, at which point the high pressure is relieved

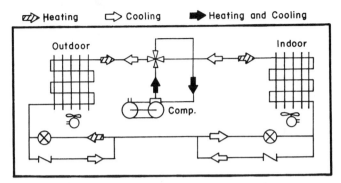

Fig. 2.28. Basic heat pump cycle. Reprinted by permission from *ASHRAE Systems Handbook 1973*.

by capillary tubes or an expansion valve, the secondary fluid returns to the evaporator in its liquid state at very low temperature, and the cycle continues. The net result is that heat is transferred from a low-temperature source to a high-temperature receptacle with the addition of work input to the cycle. The advantage of this type of heating system is that more energy is made available for space heating than the work required to operate the heat pump.

A heat pump cycle can be reversed to provide space cooling during the summer months. That is, most heat pumps provide a four-way valve, as shown in Fig. 2.28, that effectively switches the indoor and outdoor heat exchangers so that the indoor exchanger becomes the evaporator and the outdoor exchanger becomes the condenser. The heat pump then operates normally except that heat is removed instead of supplied to the space.

Figure 2.29 shows the four basic components of a heat pump: the compressor, the condenser, the expansion device, and the evaporator. The thermodynamic operating cycle for a heat pump is identical to the conventional vapor-compression refrigeration cycle, as shown in Fig. 2.29. The compressor takes superheated refrigerant vapor with low pressure and temperature at state 1 and compresses it to a much higher pressure and temperature at state 2. The high-pressure, high-temperature gas is then passed through the condenser (indoor coil of a heat pump), where it gives up heat to the high-temperature environment and changes from vapor to liquid at high pressure. The refrigerant exits from the condenser, usually as a subcooled liquid at state 3. Next, the refrigerant passes through an expansion device, where it drops in pressure. The drop in pressure is accompanied by a drop in temperature such that the refrigerant leaves the expansion device and enters the evaporator (outdoor coil of a heat pump) as a low-pressure, low-temperature mixture of liquid and vapor at state 4. Finally, the refrigerant passes through the evaporator, where it picks up heat from the low-temperature environment, changing to all vapor and exiting at state 1.

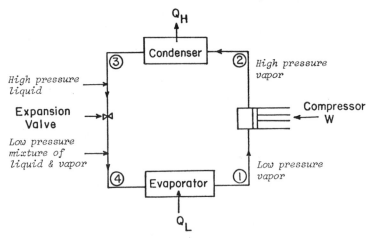

Fig. 2.29.   Basic heat pump components.

A simple energy balance on the system shown in Fig. 2.29 gives

$$Q_H = Q_L + W$$

where Q_H = heat energy rejected to the high-temperature environment
Q_L = heat energy taken from the low-temperature environment
W = input energy required to move the quantity of heat Q_L from the low-temperature environment to the high-temperature environment

The coefficient of performance (COP) is then equal to the heat output divided by the work input:

$$COP = \frac{Q_H}{W} = \frac{Q_L + W}{W} = 1 + \frac{Q_L}{W} \qquad (2.48)$$

It is thus seen that the COP of a heat pump is always greater than 1. That is, a heat pump always produces more heat energy than work energy consumed, because there is a net gain of energy Q_L transferred from the low-temperature to the high-temperature environment.

The heat pump is a reverse heat engine and is therefore limited by the Carnot cycle COP:

$$COP_{Carnot} = \frac{1}{1 - T_L/T_H} \qquad (2.49)$$

where T_L = low temperature in cycle and T_H = high temperature in cycle.

The maximum possible COP for a heat pump, maintaining a fixed temperature in the heated space, is hence a function of source temperature, as shown in Fig. 2.30. However, any real heat-transfer system must have finite temperature differences across the heat exchangers. Also shown in Fig. 2.30 are the Carnot COP for a typical air-to-air heat pump, accounting for ΔT

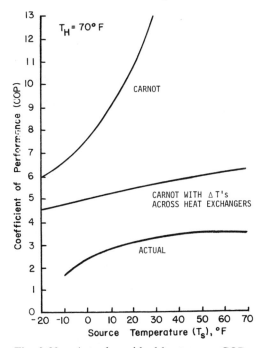

Fig. 2.30.   Actual vs. ideal heat pump COP.

values across the heat exchangers, and the actual COP for the same heat pump, accounting for compressor efficiencies and other effects. It is evident that the influence of temperature difference across the heat exchangers on COP is significant, causing a major portion of the discrepancy between actual and ideal COPs at higher source temperatures. The remaining difference between actual and Carnot COPs is a result of real working fluids, flow losses, and compressor efficiency.

EXAMPLE 2.6

A heat pump using Refrigerant-12 (shown in Fig. E2.6) is to be used for winter space heating of a residence. The building heat loss is 65,000 Btu/hr. The compressor process ideally will be reversible and adiabatic.

(a)   Determine the work required by the compressor, in hp.

(b)   Determine the COP for heating.

(c)   If the same work is supplied to the compressor in the summer and the operating conditions of the refrigeration system remain unchanged, determine the rating of the unit as an air-conditioner (for cooling), Btu/hr.

(d)   Determine the COP for cooling.

(e)   Determine the EER for cooling, Btu/(hr·W)

Solution.

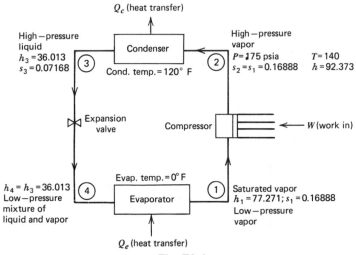

Fig. E2.6.

$$m(h_2 - h_3) + Q_c = 0; \quad m(92.373 - 36.013) + 65,000 = 0$$
$$m = 1153 \text{ lb/hr}$$

(a) $m(h_1 - h_2) - \dot{W} = 0;$

$$\dot{W} = \frac{1153(77.271 - 92.373)}{2545} = -6.84 \text{ hp}$$

(b) $\text{COP}_h = \dfrac{Q_{out}}{\dot{W}} = \dfrac{65,000}{17,413} = 3.73$

(c) $m(h_4 - h_1) + Q_e = 0; \quad Q_e = 1153(77.271 - 36.013) = 47,600 \text{ Btu/hr}$

(d) $\text{COP}_c = \dfrac{Q_{in}}{\dot{W}} = \dfrac{47,600}{17,413} = 2.73$

(e) $\text{EER}_c = \dfrac{47,600}{17,413/3.413} = 9.3 \text{ Btu/(hr·W)}$

2.10. AMMONIA-ABSORPTION REFRIGERATION CYCLE

The ammonia-absorption refrigeration cycle differs from the vapor compression cycle in the manner in which compression is achieved. Figure 2.31 is a schematic of such a system. The low-pressure ammonia vapor leaving the evaporator enters the absorber, where it is absorbed in the weak ammonia solution. Since the temperature is slightly above that of the surroundings,

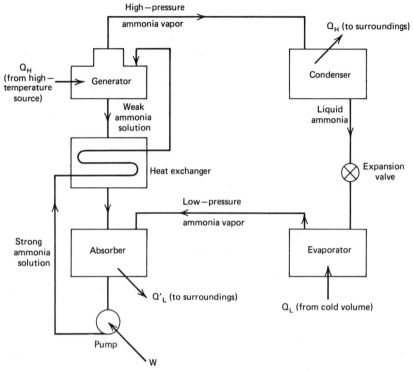

Fig. 2.31. The ammonia-absorption refrigeration cycle.

heat must be transferred to the surroundings during this process. The strong ammonia solution is then pumped through a heat exchanger to the generator at a higher pressure and temperature. The ammonia evaporates from the solution as a result of heat transfer. The ammonia vapor goes to the condenser (as in a vapor-compression system) and then to the expansion valve and evaporator. The weak ammonia solution is returned to the absorber through the heat exchanger.

The distinctive feature of the absorption system is that very little work input is required because the pumping process involves a liquid. In addition, more equipment is involved in an absorption system than in the vapor-compression cycle. Thus, it is economically feasible only in those cases where a source of heat is available that would otherwise be wasted.

2.11. APPLICATION OF THERMODYNAMICS TO HEATING, VENTILATING, AND AIR-CONDITIONING (HVAC) PROCESSES

A complete air-conditioning system is given schematically as Fig. 2.32 and shows various space heat and moisture transfers that may be present. The

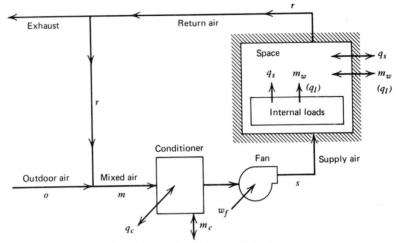

Fig. 2.32. Schematic of air-conditioning system.

symbol q_S represents a sensible heat transfer rate, while m_w represents a moisture transfer rate. The symbol q_L designates the transfer of energy that accompanies the moisture transfer and is given by $\Sigma m_w h_w$, where h_w is the specific enthalpy of the moisture added (or removed). Solar radiation and internal loads are always gains upon the space. Heat transmitted through solid construction components due to a temperature difference and energy transfers due to infiltration may represent gains or losses.

Referring to the conditioner of Fig. 2.32, it is important to note that the energy (q_c) and moisture (m_c) transfers at the conditioner cannot be determined from the space heat and moisture transfers alone. The effect of the outdoor ventilation air must also be included as well as other system load components. The designer must recognize that items such as fan energy, duct transmission, roof and ceiling transmissions, heat of lights, bypass and leakage, type of return air system, location of main fans, and actual vs.

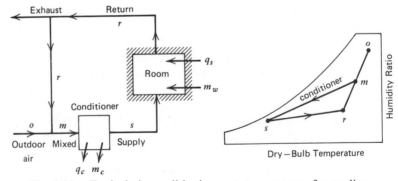

Fig. 2.33. Typical air-conditioning system processes for cooling.

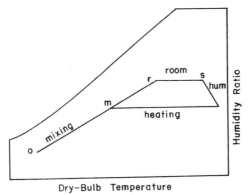

Fig. 2.34.   Psychrometric representation of heating/humidifying process.

design room conditions are all related to one another, to component sizing, and to system arrangement.

The most powerful analytical tools of the air-conditioning design engineer are the first law of thermodynamics, or energy balance, and the conservation of mass, or mass balance. These conservation laws are the basis for the analysis of moist air processes. The following sections demonstrate the application of these laws to specific HVAC processes.

In many air-conditioning systems, air is taken from the room and returned to the air-conditioning apparatus, where it is reconditioned and supplied again in the room. In most systems, the return air from the room is mixed with outdoor air required for ventilation.

Figure 2.33 shows a typical air-conditioning system and the corresponding psychrometric chart representation of the process for cooling conditions. Outdoor air (o) is mixed with return air (r) from the room and enters the apparatus (m). Air flows through the conditioner and is supplied to the space (s). The air supplied to the space picks up heat q_s and moisture m_w, and the cycle is repeated.

A typical psychrometric representation of the same type of system operating under conditions of heating followed by humidification is given in Fig. 2.34.

2.11.1.   Absorption of Space Heat and Moisture Gains

The problem of air-conditioning a space usually reduces to the determination of the quantity of moist air that must be supplied and the necessary condition which it must have in order to remove given amounts of energy and water from the space and be withdrawn at a specified condition.

Figure 2.35 schematically shows a space with incident rates of energy and moisture gains. The quantity q_s denotes the net sum of all rates of heat gain arising from transfers through boundaries and from sources within the space. This heat gain involves addition of energy alone and does not include

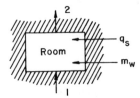

Fig. 2.35.   Space process.

energy contributions due to addition of water (or water vapor). It is usually called the *sensible heat gain*. The quantity m_w denotes the net sum of all rates of moisture gain arising from transfers through boundaries and from sources within the space. Each unit mass of moisture injected into the space adds an amount of energy equal to its specific enthalpy.

Assuming steady-state conditions, the governing equations are

$$m_a h_1 + m_w h_w - m_a h_2 + q_s = 0$$

and

$$m_a W_1 + m_w = m_a W_2$$

2.11.2.   Heating or Cooling of Air

When air is heated or cooled without the loss or gain of moisture, the process yields a straight horizontal line on the psychrometric chart, since the humidity ratio is constant. Such processes can occur when moist air flows through a heat exchanger. Figure 2.36 is a schematic of a device used to heat or cool air.

For steady flow conditions, the governing equations are

$$m_a h_1 - m_a h_2 + q = 0$$

and

$$W_2 = W_1$$

2.11.3.   Cooling and Dehumidifying of Air

When moist air is cooled to a temperature below its dew point, some of the water vapor will condense and leave the air stream. Figure 2.37 is a schematic of a cooling and dehumidifying device.

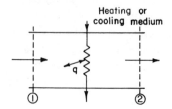

Fig. 2.36.   Schematic heating or cooling device.

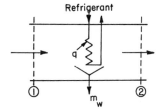

Fig. 2.37. Schematic cooling and dehumidifying device.

Although the actual process path will vary considerably, depending on the type of surface, surface temperature, and flow conditions, the heat and mass transfer can be expressed in terms of the initial and final states.

Although water may be separated at various temperatures ranging from the initial dew point to the final saturation temperature, it is assumed that condensed water is cooled to the final air temperature t_2 before it drains from the system.

For the system of Fig. 2.37, the steady flow energy and material balance equations are

$$m_a h_1 = m_a h_2 + {}_1 q_2 + m_w h_{w2}$$

$$m_a W_1 = m_a W_2 + m_w$$

Thus,

$$m_w = m_a(W_1 - W_2)$$

$${}_1 q_2 = m_a[(h_1 - h_2) - (W_1 - W_2)h_{w2}]$$

The cooling and dehumidifying process involves both sensible and latent heat transfer, where sensible heat transfer is associated with the decrease in dry-bulb temperature and the latent heat transfer is associated with the decrease in humidity ratio. These quantities may be expressed as

$$q_s = m_a c_p(t_1 - t_2)$$

and

$$q_1 = m_a(W_1 - W_2)h_{fg}$$

2.11.4. Heating and Humidifying Air

A device to heat and humidify moist air is shown in Fig. 2.38. This process is generally required during the cold months of the year. An energy balance on the device yields

$$m_a h_1 + q + m_w h_w = m_a h_2$$

and a mass balance on the water gives

$$m_a W_1 + m_w = m_a W_2$$

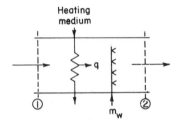

Fig. 2.38. Schematic heating and humidifying device.

2.11.5.   Adiabatic Mixing of Two Streams of Air

A common process involved in air-conditioning systems is the adiabatic mixing of two streams of moist air. Figure 2.39 schematically shows the problem. If the mixing is adiabatic, it must be governed by the three equations:

$$m_{a1}h_1 + m_{a2}h_2 = m_{a3}h_3$$

$$m_{a1} + m_{a2} = m_{a3}$$

$$m_{a1}W_1 + m_{a2}W_2 = m_{a3}W_3$$

2.11.6.   Adiabatic Mixing of Moist Air with Injected Water

Injection of steam or liquid water into a moist air stream to raise the humidity ratio of the moist air, a frequent air-conditioning process, is schematically shown in Fig. 2.40. If the mixing is adiabatic, the following equations apply:

$$m_a h_1 + m_w h_w = m_a h_2$$

$$m_a W_1 + m_w = m_a W_2$$

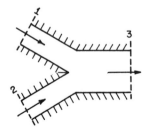

Fig. 2.39. Adiabatic mixing of two streams of moist air.

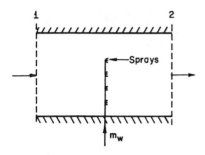

Fig. 2.40.   Schematic injection of water into moist air.

2.11.7.   Moving Air

In all HVAC systems, there must be a fan or blower to move the air. Under steady-flow conditions for the fan shown schematically in Fig. 2.41, the conservation equations are

$$m_a h_1 - m_a h_2 - W_k = 0$$

and

$$W_1 = W_2$$

2.11.8.   Approximate Equations Using Volume Flow Rates

Since the specific volume of air varies appreciably with temperature, all calculations should be made with the mass of air instead of volume. However, volume values, usually in cubic feet per minute (cfm), are required for selection of coils, fans, ducts, and so on.

A practical method of using volume values, while still actually working in mass so that accurate results are obtained, is the use of volume values based on measurement at standard conditions. The value taken as standard is 0.075 lb of dry air per cubic foot (13.33 ft³/lb dry air). This is a condition corresponding to about 60°F at saturation, and 69°F dry (at 14.7 psia). Thus, in the range at which the air usually passes through the coils, fans, ducts, and so on, it is at a density close to standard, and the accuracy desired is not likely to require correction. When the air flow is to be measured at any particular condition or particular point, such as a coil entering or leaving condition, the corresponding specific volume can be taken from the psychrometric chart

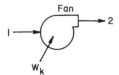

Fig. 2.41.   Air moving.

and the standard volume multiplied by the actual specific volume divided by 13.33.

Air-conditioning design often requires calculation of:

1. Sensible heat gain corresponding to the change of dry-bulb temperature (Δt) for a given air flow (standard conditions). Sensible heat change in Btu/hr:

$$q_s = (\text{cfm})(60)(0.075)(0.24 + 0.45W)\Delta t$$

where 0.24 = specific heat of dry air, Btu/lb
       W = humidity ratio, lb water/lb dry air
    0.45 = specific heat of water vapor, Btu/lb

(The specific heats are for a range from about -100 to $+200°F$.)

The value of $(60)(0.075)(0.24 + 0.45\,W)$ varies with W. When $W = 0$, the value is 1.08; when $W = 0.01$, the value is 1.10; when $W = 0.02$, the value is 1.12; and when $W = 0.03$, the value is 1.14.

Since a value of $W = 0.01$ approximates conditions found in many air-conditioning problems, the sensible heat change (Btu/hr) can be found from the relation

$$q_S = (\text{cfm})(1.10)(\Delta t) \qquad \text{in Btu/hr} \qquad (2.50)$$

2. Latent heat gain corresponding to the change of humidity ratio (W) for given air flow (standard conditions). Latent heat gain is

$$q_L = (\text{cfm})(60)(0.075)(1076)(\Delta W)$$

where 1076 is the approximate heat content of 50% relative humidity vapor at 75°F, less the heat content of water at 50°F. The 50% Rh at 75°F is a common design condition for the space, and 50°F is normal condensate temperature from cooling and dehumidifying coils. Combining the three values, the latent heat change (Btu/hr) can be found.

$$q_L = (\text{cfm})(4840)(\Delta W) \qquad (2.51)$$

3. Total heat, in Btu/hr, corresponds to the change of a given cfm (standard) through an enthalpy difference Δh.

$$\text{Total heat change} = (\text{cfm})(60)(0.075)(\Delta h)$$

where 60 = min/hr and 0.075 = lb dry air/ft^3.

If the product of the two constants is used as a single number (4.5), the total heat change is

$$q_T = 4.5(\text{cfm})(\Delta h) \qquad (2.52)$$

EXAMPLE 2.7

A fan is used to provide fresh air to the welding area in an industrial plant. The fan takes in outside air at 80°F and 14.7 psia at the rate of 1200 cfm with negligible inlet velocity. In the 10-ft^2 duct leaving the fan, air pressure is 1 psig. If the process is assumed to be reversible, adiabatic, determine the size motor needed to drive the fan.

Solution.

$$\dot{m} = \frac{P_1 V_1}{RT_1} = \frac{(14.7)(144)(1200)}{(53.34)(540)} = 88.189 \text{ lbm/min}$$

$$T_2 = T_1 \left(\frac{P_2}{P_1}\right)^{(k-1)/k} = (540)\left(\frac{15.7}{14.7}\right)^{(1.4-1)/1.4} = 550.25°R$$

$$v_2 = \frac{RT_2}{P_2} = \frac{(53.34)(550.25)}{(15.7)(144)} = 12.98 \text{ ft}^3/\text{lbm}$$

$$V_2 = \frac{mv_2}{A_2} = \frac{(88.189)(12.98)}{(10)} = 114.47 \text{ ft/min}$$

$$\dot{W} = m\left[h_1 - \left(h_2 + \frac{V_2^2}{2g_cJ}\right)\right] = m\left[c_p(T_1 - T_2) - \frac{V_2^2}{2g_cJ}\right]$$

$$= (88.189)(60)\left[.240(540 - 550.25) - \frac{(114.47/60)^2}{2(32.2)(778)}\right] = -13,020 \text{ Btu/hr}$$

$$= 5.1 \text{ hp}$$

EXAMPLE 2.8

Find the heat transfer rate required to warm 1500 cfm (ft^3/min) air at 60°F and 90% relative humidity to 120°F without the addition of moisture.

Solution.

$$\dot{m}_a = \frac{V_1 A_1}{v_1}$$

$$\dot{m}_a = \frac{1500(60)}{13.31} = 6762 \text{ lb/hr}$$

$$\dot{q} = \dot{m}_a(h_2 - h_1)$$

$$\dot{q} = 6762(40.0 - 25.3) = 99,400 \text{ Btu/hr}$$

or

$$\dot{q} = \dot{m}_a c_p(t_2 - t_1)$$

$$\dot{q} = (6762)(0.245)(120 - 60) = 99,400 \text{ Btu/hr}$$

EXAMPLE 2.9

Moist air at 80°F db and 67°F wb is cooled to 58°F db and 80% relative humidity. The volume flow rate is 2000 cfm and the condensate leaves at 60°F. Find the heat transfer rate.

Solution.   See Fig. E2.9.

$$\dot{m}_a = \frac{2000(60)}{13.85} = 8664 \text{ lb/hr}$$

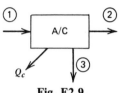

Then

$$\dot{q} = \dot{m}_a[(h_1 - h_2) - (W_1 - W_2)h_3]$$

$$\dot{q} = 8664[(31.6 - 22.9) - (0.0112 - 0.0082)28.08]$$

$$\dot{q} = 8664[(8.7) - (0.084)]$$

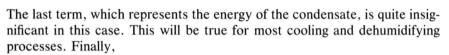

Fig. E2.9.

The last term, which represents the energy of the condensate, is quite insignificant in this case. This will be true for most cooling and dehumidifying processes. Finally,

$$\dot{q} = 74,649 \text{ Btu/hr}$$

A ton of refrigeration is 12,000 Btu/hr. Then

$$\dot{q} = 6.22 \text{ tons}$$

EXAMPLE 2.10

A fan in an air-conditioning system is drawing 1.25 hp at 1760 rpm. The capacity through the fan is 1800 cfm of 75°F air and the inlet and outlet duct are 12 in. in diameter. What is the temperature rise of air due to this fan?

Solution.   See Fig. E2.10.

Steady flow:

$$\dot{V} = 1800 \text{ cfm at 75°F and 14.7 psia}$$

$$P = \rho RT$$

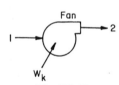

Fig. E2.10.

$$\rho = \frac{14.7(144)}{53.3(535)} = 0.075 \text{ lb/ft}^3$$

$$\dot{m} = 1800(.075)(60) \text{ lb/min}$$

$$\dot{m}(h_1 - h_2) + \cancel{Q}^0 - \dot{W} = 0$$

$$\dot{m}c_p(T_1 - T_2) = \dot{W}$$

$$1800(.075)(60)(0.24)(T_1 - T_2) = -1.25(2545)$$

$$T_1 - T_2 = -1.64°F \quad \text{or} \quad T_2 - T_1 = 1.64°F$$

EXAMPLE 2.11

In an air-conditioning unit, 71,000 cfm at 80°F dry bulb, 60% relative humidity, and standard atmospheric pressure, enter the unit. The leaving condition of the air is 57°F db and 90% humidity.

Calculate:

- (a) the cooling capacity of the air-conditioning unit, in Btu/hr
- (b) the rate of water removal from the unit
- (c) the sensible heat load on the conditioner, in Btu/hr
- (d) the latent heat load on the conditioner, in Btu/hr
- (e) the dew point of the air leaving the conditioner

Solution. See Fig. E2.11.

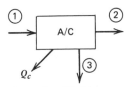

Fig. E2.11.

(1)	(2)	(3)
$t = 80°F$	$t = 57°F$	$h \approx 25$
$\phi = 60\%$	$\phi = 90\%$	(h_f at 57°F)
71,000 cfm	$v = 13.2$	
$v = 13.9$	$h = 23.5$	
$h = 33.7$	$W = 0.009$	
$W = 0.0132$		

$h_c = 29.0$

$$\dot{m}_a = \frac{\dot{V}}{v} = \frac{(71,000)(60)}{(13.9)} = 306,475 \text{ lbm air/hr}$$

$$\text{scfm} = 71,000 = \frac{13.3}{13.9} = 67,900$$

- (a) $\dot{m}_a h_1 - \dot{m}_a h_2 - \dot{m}_a(W_1 - W_2)h_3 = -Q$

  $-Q = (306,475)[33.7 - 23.5 - (0.0132 - 0.009)(25)]$

  $= -3,093,900 \text{ Btu/hr} = 257.8 \text{ tons removed}$

- (b) $\dot{m}_w = \dot{m}_a(W_1 - W_2) = (306,475)(0.0132 - 0.009) = 1287 \text{ lbm/hr}$

- (c) $Q_s \cong \dot{m}_a(h_c - h_2) = (306,475)(29 - 23.5) = 1,685,600 \text{ Btu/hr} = $
  140.5 tons

  or

  $Q_s \cong \dot{m}_a c_p(t_1 - t_2) = (306,475)(0.244)(80 - 57) = 1,719,900 \text{ Btu/hr} = $
  143.3 tons

or

$$Q_s \approx 1.10(\text{scfm})(t_1 - t_2) = 67{,}900(1.10)(23) = 1{,}718{,}000 \text{ Btu/hr} =$$
143.2 tons

(d) $Q_L \cong \dot{m}_a(h_1 - h_c) = (306{,}475)(33.7 - 29) = 1{,}440{,}400 \text{ Btu/hr} =$
120.0 tons

or

$$Q_L \cong \dot{m}_a(W_1 - W_2)h_{fg} = (306{,}475)(0.0132 - 0.009)(1076) = 1{,}385{,}000$$
= 115.4 tons

or

$$Q_L \approx 4840 \, (\text{cfm})\Delta W$$
$$= 4840(67{,}900)(0.0132 - 0.009) = 1{,}380{,}000 \text{ Btuh} = 115.0 \text{ tons}$$

(e) $t_d = 54.2°F$

EXAMPLE 2.12

An air-conditioned room with an occupancy of 20 people has a sensible heat load of 200,000 Btu/hr, and a latent load of 50,000 Btu/hr, and is maintained at 76°F db and 64°F wb. On a mass basis, 25% outside air is mixed with return air. Outside air is at 95°F db and 76°F wb. Conditioned air leaves the apparatus and enters the room at 60°F db. Neglect any temperature change due to the fan.

(a) Draw and label the schematic flow diagram for the complete system.

(b) Complete the following table:

Point	T, db (°F)	φ (%)	h (Btu/lb)	W (lb/lb)	m_a (lb/hr)	SCFM	CFM
OA							
r							
m							
s							

(c) Plot and draw all processes on a psychrometric chart.

(d) Specify the fan size, scfm

(e) Determine the size refrigeration unit needed, Btu/hr and tons

(f) What percent of the required refrigeration is for (i) sensible cooling and (ii) for dehumidification?

(g) What percent of the required refrigeration is due to outside air load?

Solution.

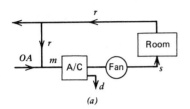

(a)   **Fig. E2.12A.**

(b)

Point	T, db (°F)	φ (%)	h (Btu/lb)	W (lb/lb)	m_a (lb/hr)	SCFM	CFM
OA	95	42	39.4	0.015	12,810	2,846	3,053
r	76	52	29.3	0.010	38,420	8,536	8,773
m	81	50	31.8	0.0113	51,230	11,380	11,783
s	60	82	24.4	0.0091	51,230	11,380	11,356

$$\left.\begin{array}{l} .75(29.3) + .25(39.4) = h_m = 31.8 \\ .75(0.010) + .25(0.015) = W_m = 0.0113 \end{array}\right\} \begin{array}{l} T = 81°F \\ \phi = 50\% \end{array}$$

$$m_{a(\text{supply})} = \frac{200,000}{.244(76 - 60)} = 51,230 \text{ lb/hr}$$

$$W_s = W_r - \frac{m_s}{m_a} = 0.010 - \frac{50,000/1100}{51,230} = 0.0091$$

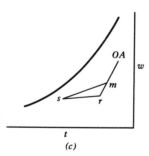

(c)   **Fig. E2.12C.**

(d) Fan size = 11,400 scfm

(e) $m_a(h_m) - m_a(h_s) - m_a(W_m - W_s)h_d + Q = 0$

   $51,230\ [31.8 - 24.4 - (0.0113 - 0.0091)28] + Q = 0$

   $Q = -375,900 \text{ Btu/hr} = 31.3 \text{ tons}$

(f)   % sensible $= \dfrac{51{,}230(0.244)(81 - 60)}{375{,}900} = .698$ or 69.8% sensible,

30.2% latent

(g)   % outside air $= \dfrac{2846[1.10(95 - 76) + 4840(.015 - .010)]}{375{,}900} = 0.341$ or

34.1%

EXAMPLE 2.13

1.   On summer days, only the cooling coil of the air-conditioning system (Fig. E2.13) is operating. At summer design conditions, the following conditions exist:

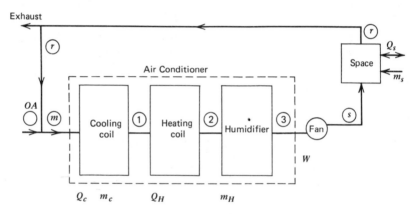

Fig. E2.13.

$r: t = 75°F$ db

$1,2,3: t = 55°F$ db, $\phi = 100\%$

$OA: t = 95°F$ db, $t^* = 78°F$ wb

$s: t = 56°F$ db

10% by weight outide air is required for ventilation.
Space sensible heat gain, $Q_s = 129{,}000$ Btu/hr
Space moisture gain, $m_s = 55.7$ lbf/hr

Determine:

(a)   Summer air flow rate to space, m_a, lb$_{da}$/hr; da = dry air.
(b)   Size of cooling unit required, Btu/hr and tons
(c)   Sensible load on cooling coil, Btu/hr
(d)   Latent load on cooling coil, Btu/hr

2. On winter days, the humidifier and heating coil components of the air-conditioning system (Fig. E2.13) are operating. At winter design conditions, the following conditions exist:

r: $t = 75°F$ db, $\phi = 25\%$

OA: $t = 0°F$, $\phi = 100\%$

2: $t = 135°F$ db

s: $t = 135.5°F$ db

10% by weight outside air is required for ventilation.
Space sensible heat loss, $Q_s = 214{,}000$ Btu/hr
Space moisture gain, $m_s = 8.3$ lb/hr

Determine:
(a) Winter air flow rate to space, m_a, lb_{da}/hr
(b) Supply humidity ratio to space, W_s, lb/lb_{da}
(c) Size of heating unit required, Btu/hr
(d) Size of humidifier required, lb/hr

Solution.

1. (a) $Q_s = m_a c_p (t_r - t_s) = 129{,}000 = m_a(0.244)(75 - 56)$

   $m_a = 27{,}800$ lb$_a$/hr

   (b) 1,2,3: $t = 55°F$, $\phi = 100\% = h = 23.4$, $W = .0092$

   s: $t = 56°F$, $W = 0.0092 = h = 23.6$

   r: $t = 75°F$, $W = 0.0092 + \dfrac{55.7}{27{,}800} = 0.01120$, $h = 30.2$

   OA: $t = 95°F$, $t^* = 78°F = h = 41.4$, $w = 0.0168$

   m: $m_{OA}h_{OA} + m_r h_r = m_m h_m$

   $m_{OA}W_{OA} + m_r W_r = m_m W_m$

   $.1(41.4) + .9(30.2) = h_m$

   $h_m = 31.3$

   $.1(.0168) + .9(.0120) = W_m$

   $W_m = 0.01176$

   $t_m = 76.7°F$

   $m_a[h_m - h_1 - (W_m - W_1)h_f] + Q = 0$

   $27{,}800[31.3 - 23.4 - (0.01176 - 0.0092)(23)] = -Q$

   $Q_c = 218{,}000$ Btu/hr $= 18.2$ tons

(c) $Q_s = m_a c_p (t_m - t_1) = 27,800(0.244)(76.7 - 55) = 147,200$ Btu/hr

(d) $Q_L = m_a (W_m - W_1)(1076) = 27,800(0.01176 - 0.0092)(1076) = 76,600$ Btu/hr

2. (a) $Q_s = m_a c_p (t_s - t_r) = 214,000 = m_a(0.244)(135.5 - 75)$
    $m_a = 14,500$ lb$_a$/hr

(b) $W_r = W_s + \dfrac{m_s}{m_a}$

   $0.0046 = W_s + \dfrac{8.3}{14,500};\quad W_s = 0.00403$

(c) r: $t = 75°F$; $\phi = 25\%$; $W = .0046$; $h = 23.0$

   s: $t = 135.5°F$, $W = .00403$

   OA: $t = 0°F$; $\phi = 100\%$; $W = 0.0007872$; $h = 0.835$

   $m,1$: $h_m = .1(.835) + .9(23.0) = 20.78$

      $W_m = .1(.00079) + .9(.0046) = 0.00422$

      $t_m = 68°F$ db

   2: $t = 135°F$; $W_2 = W_m = 0.00422$

   $m_a[h_1 - h_2] + Q = 0$

   $14,500(0.244)(68 - 135) = -Q$

   $Q_H = 237,000$ Btu/hr

(d) $m_H = m_a(W_3 - W_2)$

      $= 14,500(0.00403 - 0.00422) = -2.8$ (no humidification needed)

PROBLEMS

1. Determine the specific volume of a gas at 500 kPa and 20°C. Assume that $v = RT/p$ and $R = 287$ N·m/(kg·K).

2. A pressure gauge reads 31.2 lbf/in.^{2}, and the barometer reads 29.92 in. Hg. Calculate the absolute pressure in psia, psfa, and atm.

3. A mercury manometer is used to measure a vacuum and reads 26.8 in. when the barometer reads 29.5 in. Hg. Determine the absolute pressure in in. Hg, psia, atm, and microns of Hg.

4. A water manometer used to measure the pressure rise across a fan reads 1.1 in. H_2O when the density of the water is 62.1 lbm/ft³. Determine the pressure difference in lb/in.².

5. A thermometer reads 72°F. Specify the temperature in °C, K, and °R.

6. One gallon of fuel oil having a heating value of 139,000 Btu/gal burns in a home furnace. Determine the mass loss (converted to energy) per gallon of fuel burned.

7. Air with a density of 0.075 lb/ft³ enters a steady-flow system through a 12-in. diameter duct with a velocity of 10 ft/sec. It leaves with a specific volume of 5.0 ft³/lb through a 4-in. diameter duct. Determine (a) the mass flow rate in lb/hr and (b) the outlet velocity in ft/sec.

8. Air is compressed in a cylinder by a piston. At the initial pressure of 60 psia, the temperature is 100°F. The following table describes the path of the process:

Pressure (lbf/in.²)	Volume (in.³)
60	80.0
80	67.5
100	60.0
120	52.5
140	45.0
160	37.5
180	32.5

Determine the work required for compression of the air assuming a reversible process.

9. Complete the following table:

Substance	T (°F)	P (psia)	v (ft³/lbm)	Condition (x, °SH, or °SC)
H_2O	600	140		
H_2O		2000	0.018439	
Air	735	60		
Freon-12	120	35		
Freon-12	120			$x = 0.62$ lb$_v$/lbm

10. Complete the following table:

Substance	T (°F)	P (psia)	v (ft³/ lbm)	u (Btu/ lbm)	h (Btu/ lbm)	s (Btu/ lbm·°R)	Condition (°SC; x; °SH)
	50						x = 0.80
Ammonia	80		1.43				
		80	4.75				
	20				22.83		
Freon-12		50	0.6				
		50					75°SH
	100	1000					
Water	80		20				
		1000			1123		

11. A hot water heater has 2.0 gal/min entering at 50°F and 40 psig. The water leaves the heater at 160°F and 39 psig. Determine (a) the change in enthalpy per pound and (b) the gal/min of water leaving if the heater is operating under steady flow conditions.

12. As the pressure in a steam line reaches 100 psia, the safety valve opens, releasing steam to the atmosphere in a constant-enthalpy process across the valve. The temperature of the escaping steam (after the valve) was measured as 250°F. Determine the temperature of the steam in the line as well as its specific volume and condition.

13. A room of dimensions 4 m × 6 m × 2.4 m contains an air–water vapor mixture at a total pressure of 100 kPa and a temperature of 25°C. The partial pressure of the water vapor is 1.4 kPa. Calculate:
(a) The humidity ratio
(b) The dew point
(c) The total mass of water vapor in the room

14. An air-conditioning coil cools 2000 cfm of air at 14.7 psia and 80°F to 45°F. Determine the rating of the air-conditioner, in Btu/hr.

15. Air is heated to 80°F without the addition of water, from 60°F db and 50°F wb. By use of the psychrometric chart, find:
(a) Relative humidity of the original mixture.
(b) Original dew point temperature.

(c) Original specific humidity.

(d) Initial enthalpy.

(e) Final enthalpy.

(f) Heat added.

(g) Final relative humidity.

16. Air is used for cooling an electronics compartment. Atmospheric air enters at 60°F, and the maximum allowable air temperature is 100°F. If the equipment in the compartment dissipates 3600 W of energy to the air, determine the necessary air flow rate in (a) lbm/hr and (b) cfm at inlet conditions.

17. Air is heated as it flows through a constant-diameter tube in a steady flow. The air enters the tube at 50 psia, 80°F and has a velocity of 10 ft/s at entrance. The air leaves at 45 psia and 255°F.

(a) Determine the velocity of the air (ft/s) at the exit.

(b) If 23 lbm/min of air is to be heated, what diameter (in.) tube must be used?

18. Air undergoes a steady-flow reversible adiabatic process. The initial state is 1400 kPa, 815°C, and the final pressure is 140 kPa. Changes in kinetic and potential energy are negligible. Determine:

(a) Final temperature.

(b) Final specific volume.

(c) Change in specific internal energy.

(d) Change in specific enthalpy.

(e) Specific work.

19. Air is compressed in a reversible steady-state, steady-flow process from 15 lbf/in.^{2}, 80°F to 120 lbf/in.^{2}. Calculate the work of compression per pound, the change in entropy, and the heat transfer per pound of air compressed, assuming the process is:

(a) Isothermal.

(b) Polytropic, $n = 1.25$.

(c) Adiabatic.

20. Air enters an air-conditioning duct at a rate of 2000 ft^3/min at 40°F, 14.9 psia. The air discharges from the duct at 60°F, 14.7 psia. Determine:

(a) Mass flow rate of air, lb/hr.

(b) Volume flow rate at discharge, cfm.

(c) Change in enthalpy of air between inlet and outlet, Btu/hr.

21. The heat from students, lights, conduction through the walls, and so on, to the air moving through a classroom is 21,000 Btu/hr. Air is

supplied to the room from the air-conditioner at 55°F. The air leaves the room at 78°F. Specify:

(a) Air flow rate, lb/hr.

(b) Air flow rate, cfm, at inlet conditions.

(c) Duct diameter, in., for air velocity of 600 ft/min.

22. Water at 30 psig is heated at the rate of 5 gal/min from 62°F to 164°F. If electric heating elements are used, determine the (a) wattage required and (b) the current (A), if a single-phase 220-V circuit is used.

23. What *minimum* size motor (hp) would be necessary for a pump that handles 85 gal/min of city water while increasing the water pressure from 15 psia to 90 psia?

24. The water level in College Hills is 400 ft below the surface. It is desired to install a well pump that will deliver 15 gal/min of water (8.33 lb/gal and 0.016 ft³/lb) at a pressure of 30 psig. What hp motor should be used?

25. A booster pump is used to move water from the basement equipment room to the 13th floor of an apartment building at the rate of 800 lbm/min. Elevation change is 130 ft between basement and the 13th floor. Determine the minimum size pump (in hp) required.

26. In order to produce liquid nitrogen, nitrogen initially at 3000 psia and 80°F is expanded isentropically to atmospheric pressure. For every pound initially, determine

(a) Amount (lbm) of liquid nitrogen obtained.

(b) Internal energy change, Btu/lbm.

(c) Final temperature, °F.

27. Refrigerant-12 is compressed in a piston–cylinder system having an initial volume of 80 in.³. Initial pressure and temperature are 20 psia, 140°F. The process is *isentropic* to a final pressure of 175 psia. Determine:

(a) Final temperature, °F.

(b) Mass of R-12, lbm.

(c) Change in enthalpy, Btu/lbm.

(d) Change in internal energy, Btu.

28. In an ideal low-temperature refrigeration unit, Refrigerant-12 is compressed isentropically from saturated vapor at 15.3 psia to a pressure of 200 psia. Determine the change in internal energy across the compressor per pound of the Refrigerant-12.

29. Refrigerant-12 vapor enters a compressor at 25 psia and 40°F, and the mass rate of flow is 15 lbm/min. What is the smallest diameter tubing (in.) that can be used if the velocity of refrigerant must not exceed 20 ft/s?

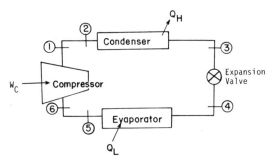

Fig. P2.33.

30. Refrigerant-12 is compressed in a residential air-conditioner from saturated vapor at 40°F to superheated vapor at 100 psia having an entropy of 0.170 Btu/lbm·°R. Determine the change in enthalpy for this compression process.

31. In a household refrigerator, Refrigerant-12 enters the compressor as saturated vapor at 30°F. If the process across the compressor is isentropic and the discharge pressure is 150 psia, determine the refrigerant temperature at the compressor outlet.

32. In the standard home freezer refrigeration unit, a capillary tube is often used to produce a throttling (constant-enthalpy) process. In one such system, R-12 is throttled from saturated liquid at 151 psia to a pressure of 12 psia. Determine:
   (a) Initial temperature, °F.
   (b) Final temperature, °F.
   (c) Final condition [°SC, x, or °SH].
   (d) The change in specific volume, ft³/lbm.

33. A refrigeration unit employing Refrigerant-12 is shown in Fig. P2.33. Condensing pressure is 216 psia. Evaporator temperature is −10°F. The unit is rated at 66,000 Btu/hr for cooling. If the velocity in the line between evaporator and compressor is not to exceed 5 ft/s, determine the inside diameter (in.) of the tubing to be used.

34. Ammonia at 15 psia, 20°F, is compressed polytropically, $pv^{1.35} = C$, in a steady-flow process to 50 psia. Determine the *minimum* work of compression *and* the corresponding heat transfer, both in Btu/lbm.

35. Determine the coefficient of performance for each of the following:
   (a) An ideal heat pump using Freon-12 and operating between pressures of 35.7 and 172.4 psia,
   (b) An actual refrigerator providing 4500 Btu/hr of cooling while drawing 585 W.

36. A heat pump is used in place of a furnace for heating a house. In winter when the outside air temperature is 10°F, the heat loss from the house is 60,000 Btu/hr if the inside is maintained at 70°F. Determine the *minimum* electric power required to operate the heat pump (in kW).

37. A heat pump is used in place of a furnace for heating a house. In winter when the outside air temperature is $-10°C$, the heat loss from the house is 200 kW if the inside is maintained at 21°C. Determine the minimum electric power required to operate the heat pump.

38. Solar energy is to be used to warm a large "collector plate." This energy would, in turn, be transferred as heat to a fluid within a heat engine, and the engine would reject energy as heat to the atmosphere. Experiments indicate that about 200 Btu/(hr·ft²) of energy can be "collected" when the plate is operating at 190°F. Estimate the minimum collector area that would be required for a plant producing 1 kW of useful shaft power, when the atmospheric temperature is 70°F.

39. The load on a residential air conditioner is 36,000 Btuh when the outdoor air temperature is 95°F and the indoor temperature is maintained at 75°F. Determine the *minimum* power requirement (kW) to operate the air-conditioner.

40. Tests performed on a residential air-conditioner system yielded the following data:

Refrigerant	R-12
Evaporating pressure	50 psia
Condensing pressure	200 psia
Actual air-cooling effect	32,450 Btu/hr
Power meter reading	5.76 kW

Determine both *actual* and *ideal* performance: (a) COP; (b) EER; (c) hp/ton.

41. A refrigerator utilizes R-12 as the refrigerant and handles 200 lbm/hr. Condensing temperature is 110°F, and evaporating temperature is 5°F. For a cooling effect of 11,000 Btu/hr, determine the minimum size motor (hp) required to drive the compressor.

42. Refrigerant-12 enters the condenser of a vapor-compression refrigeration system at 175 psia and 140°F and leaves as saturated liquid at 120°F. Mass flow rate of the refrigerant is 4.8 lbm/min. The heat is rejected to the surrounding air, which is at 90°F. Determine:
  (a) Heat rejection rate, Btu/hr.
  (b) Separate the overall entropy changes per hour (Freon and surroundings).

43. Refrigerant-12 enters the evaporator of a freezer at $-20°F$ with a quality of 85%, and leaves as saturated vapor. Determine the heat transfer per pound of refrigerant by (a) use of the first law of thermodynamics and (b) by use of the second law.

44. A refrigeration unit employing R-12 is shown in Fig. P2.44. Condensing pressure is 216 psia. Evaporator temperature is $-10°F$. The unit is rated at 66,000 Btu/hr for cooling. Determine: (a) the *minimum* size motor required to drive the compressor, (b) the corresponding COP_c of the unit, and (c) the output of the system as a heat pump (Btu/hr).

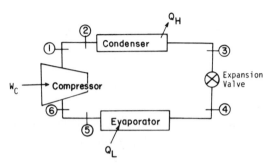

Fig. P2.44.

45. The cycle shown in Fig. P2.45 is used for air-conditioning aircraft and uses air as the working fluid. Considering the compression process as ideal, determine:

(a)   Net work required (in hp) per ton of refrigeration (12,000 Btu/hr).

(b)   Heat rejected at the heat exchanger, Btu/hr.

(c)   Turbine efficiency, %.

(d)   COP.

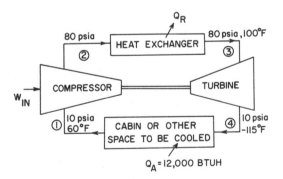

Fig. P2.45.

46. A solar-operated heat pump is designed to operate as shown in Fig. P2.46. Solar energy is used as the heat source for the boiler in a Rankine power system operating with R-22 as the working fluid. The turbine

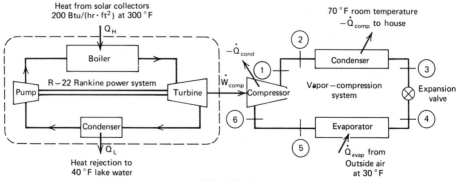

Fig. P2.46.

output drives the compressor in a regular R-12 heat-pump system. For conditions shown, determine the *minimum* square footage of solar collectors if the heat loss from the house is 45,000 Btu/hr.

47. In a mixing process of two streams of air, 10,000 cfm of air at 75°F and 50% relative humidity mix with 4000 cfm of air at 98°F dry bulb and 78°F wet bulb temperature. Calculate the following conditions after mixing at atmospheric pressure:

    (a) Dry bulb temperature.
    (b) Humidity content.
    (c) Relative humidity.
    (d) Enthalpy.
    (e) Dew point temperature.

48. In an auditorium maintained at a temperature not to exceed 77°F, and at a relative humidity not to exceed 55%, a sensible-heat load of 350,000 Btu and 1,000,000 grains of moisture per hour must be removed. Air is supplied to the auditorium at 67°F.

    (a) How many pounds of air per hour must be supplied?
    (b) What is the dew-point temperature of the entering air, and what is its relative humidity?
    (c) How much latent-heat load is picked up in the auditorium?
    (d) What is the sensible heat ratio?

49. A space in an industrial building has a winter sensible heat loss of 200,000 Btu/hr and a negligible latent heat load (latent losses to the outside are made up by latent gains within the space). The space is to be maintained precisely at 75°F and 50% relative humidity. Due to the nature of the process, 100% outside air is required for ventilation. The outdoor air conditions can be taken as saturated air at 20°F. The amount of ventilation air is 7000 scfm, and the air is to be preheated, humidified with an adiabatic saturator to the desired humidity, and then

reheated. The temperature out of the adiabatic saturator is to be maintained at 60°F db. Determine the following:

(a) Temperature of the air entering the space to be heated, °F.

(b) Heat supplied to preheat coil, Btu/hr.

(c) Heat supplied to reheat coil, Btu/hr.

(d) Amount of humidification, gal/min.

50. An office building in Jefferson City, MO, using the HVAC system shown in Fig. P2.50, has space design loads as follows:

Summer	Winter
314,000 Btu/hr sensible (gain)	407,000 Btu/hr sensible (loss)
81,000 Btu/hr latent (gain)	Negligible latent

Minimum outside air for meeting the ventilation requirements of the 410 occupants is used.

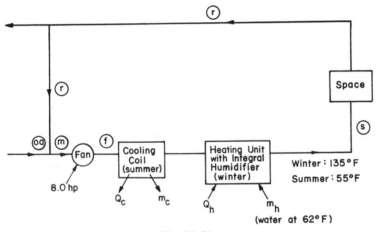

Fig. P2.50.

(a) Specify:
  (1) Summer air flow to space, scfm.
  (2) Cooling coil rating, Btu/hr.
  (3) Sensible cooling coil load, Btu/hr.
  (4) Latent cooling coil load, Btu/hr.
  (5) Actual relative humidity of air leaving space, %.
  (6) Winter air flow to space, scfm.
  (7) Heating unit rating, Btu/hr.
  (8) Humidifier rating, gal/hr.
  (9) Latent load on heating/humidifying unit, Btu/hr.

(b) Plot all summer points and the summer cycle, with complete labeling, on a psychrometric chart.

3

HEAT PUMP SYSTEMS AND APPLICATIONS

A heat pump, in the common thermodynamic sense, is a system in which refrigeration components (compressors, condensers, evaporators, and expansion devices) are used in such a manner as to take heat from a source (air, water, ground, etc.) and give it up to a heat sink (air, water, ground, etc.) that is at a higher temperature than the source. For many applications, the heat pump is designed in such a manner as to reverse the cooling and heating functions. This allows the use of the same equipment for both heating and cooling requirements in a single structure. Other applications are designed to simultaneously utilize both the heating and cooling effects obtained from the cycle in the same structure.

3.1. BASIC ARRANGEMENTS

There are two basic arrangements of heat pump components that provide for simple transition between the cooling and heating modes of operation. One controls the directions of flow of the ambient and room air by a complex arrangement of ducts so that the condenser and the evaporator do not reverse their roles during the cooling and heating modes of operation. Figure 3.1 is a simplified schematic presentation of such an air-to-air system. In the summer, all four air dampers are in the upper position, so that recirculated room air is passed over the evaporator coils for cooling. Outside air, passing over the condenser, removes the energy picked up by the working fluid in the evaporator plus the work of the compressor. For operation of the system

116

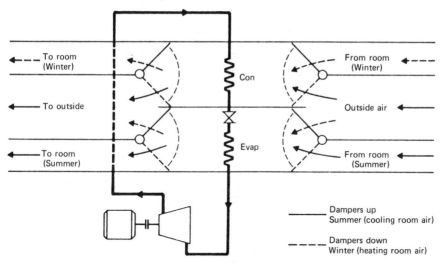

Fig. 3.1.   Air-to-air heat pump for summer cooling and winter heating.

as a heat pump in the winter, all four dampers are moved to the lower position. Heat transfer from the outside air, even at relatively low temperatures, can then be accomplished at the evaporator coils, while the heat rejected from the condenser coils is delivered to the recirculated room air. Controlled mixing of fresh outside air with recirculated room air can be provided for in the ducting arrangement.

A more practical embodiment of an air-to-air heat-pump system is presented in Fig. 3.2.

First, consider the conventional refrigeration cooling cycle shown in Fig. 3.3. Heat is absorbed by the indoor evaporator and discharged by the outdoor air-cooled condenser. If we could physically reverse these components and absorb heat from the outdoor air and, by means of the refrigerant, discharge it into the indoor space, then we would have created heating. This is exactly what a reverse-cycle heat pump does, except that it does not actually physically reverse the evaporator and condenser. By means of a reversing valve, it can direct the refrigerant flow alternately to make the process heat or cool. Thus the heat pump cycle looks like Fig. 3.2, where the coils are relabeled as indoor and outdoor. The reversing valve directs the discharge and suction gas as shown by the arrows. Check valves are installed in the outlets of both coils.

The one additional element of a heat pump refrigerant cycle that is most important is a suction-line accumulator that protects the compressor from refrigerant floodback during the changeover cycles (heating to cooling and vice versa) and during the defrost cycle. The physical volume of this element depends on the capacity of the unit. Its shape is a function of the brand.

COOLING

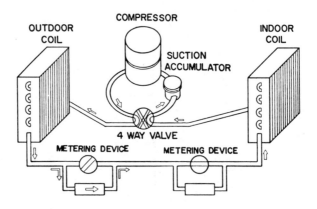

HEATING

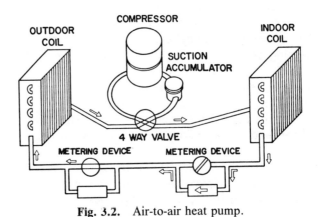

Fig. 3.2.   Air-to-air heat pump.

3.2.   TYPICAL CLASSIFICATIONS

Heat pumps for building heating and cooling systems are classified according to (1) type of heat source and sink, (2) heating and cooling distribution fluid, (3) type of thermodynamic cycle, (4) type of building structure, and (5) size and configuration.

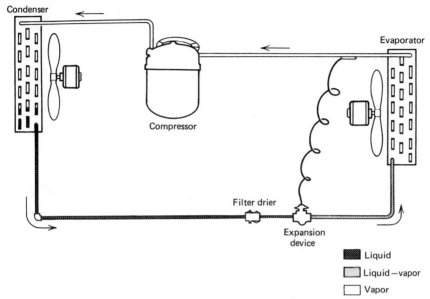

Fig. 3.3. Conventional cooling cycle.

The four basic heat-pump designs for space heating and cooling may be tabulated as follows:

Heat Source-Sink	Heating and Cooling Medium
Air	Air
Air	Water
Water	Air
Water	Water

Each of these basic designs can supply the required heating and cooling effect by changing the direction of the refrigerant flow or by maintaining a fixed refrigerant circuit and changing the direction of the heat source-sink medium. A third alternative is to incorporate an intermediate transfer fluid in the design. In this case the direction of the fluid is changed to obtain heating or cooling, and both the refrigerant and heat source-sink circuits are fixed. The fixed refrigerant circuit designs, generally referred to as the indirect type of application, are becoming increasingly popular, particularly in the larger capacities.

3.2.1. Sources and Sinks

The air-to-air heat pump is the most common type in use today. It is particularly suitable for factory-built or unitary heat pumps and is widely used in residential and light commercial applications. The first diagram in Table 3.1 is typical of the refrigeration cycle or circuit used in air-to-air heat pumps. A few installations have been made in which the forced-convection indoor coil has been replaced by a radiant heating panel.

In the air-to-air heat-pump system shown in the second diagram of Table 3.1, the air circuits may be interchanged by means of dampers (motor-driven or manually operated) to produce either heated or cooled air, which is then used for the conditioned space. With this system, one heat-exchanger coil is always the evaporator and the other is always the condenser. Therefore, refrigerant reversing valves are not needed. The conditioned air will pass over the evaporator during the cooling cycle, and the outdoor air will pass over the condenser. The change from cooling to heating is accomplished by changing the appropriate dampers.

A water-to-air heat pump uses water as a heat source and sink and uses air to move the heat to or from the conditioned space. Almost any water can be used as the source: river water, lake water, ground or well water, waste water, and so on.

Air-to-water heat pumps are commonly used in large buildings where zone control is necessary, and are also quite commonly used for the production of hot or cold water for industrial applications.

Earth-to-air heat pumps may employ direct expansion of the refrigerant in an embedded coil, as illustrated in Table 3.1, or they may be of the indirect type, where heat is exchanged by circulating water through underground coils.

A water-to-water heat pump uses water as the heat source and sink for both cooling and heating operation. Heating-cooling changeover may be accomplished in either the refrigerant circuit or in the water circuits as shown in Table 3.1.

An earth-to-water heat pump may be like the earth-to-air type shown in Table 3.1 except for the substitution of a refrigerant-to-water heat exchanger for the finned coil shown on the indoor side. It may also take a form similar to the water-to-water system shown, when a secondary-fluid earth coil is used.

Internal-source heat pumps utilize the high internal cooling load generated in modern buildings either directly or with a storage system. The storage means is usually water, which is kept in an insulated tank.

Some heat pumps that use the earth as the heat source-sink are essentially of the water-to-air type. An antifreeze solution is pumped through a circuit consisting of the chiller-condenser and a pipe coil buried in the earth. Earth source-sink systems are seldom used today because of the difficulties in maintaining their capacity and reliability.

TABLE 3.1. Common Heat Pump Types [1][a]

HEAT SOURCE AND SINK	DISTR FLUID	THERMAL CYCLE	DIAGRAM
AIR	AIR	REFRIGERANT* CHANGEOVER	
AIR	AIR	AIR CHANGEOVER*	
WATER	AIR	REFRIGERANT* CHANGEOVER	
AIR	WATER		
EARTH	AIR	REFRIGERANT* CHANGEOVER	
WATER	WATER	WATER CHANGEOVER*	

*ALL SINGLE STAGE COMPRESSION

[a]Reprinted by permission, from *ASHRAE Systems Handbook 1973* p. 11.2.

121

Other types of heat pumps in addition to those shown in Table 3.1 are possible. An example is one that utilizes solar energy as a heat source; its refrigerant circuit may resemble the water-to-air, air-to-air, or other types, depending on the form of solar collector and the means of heating and cooling distribution employed. Solar-assisted heat pumps have received a great deal of attention in the last five years.

Another variation is the use of more than one heat source. Some heat pumps have utilized air as the primary heat source, but are changed over to extract heat from water (e.g., from a well or storage tank) during periods of peak load. The use of solar energy requires another heat source during periods of insufficient solar radiation. Quite often, electric coils or gas-fired units are used for the auxiliary heat source.

Any thermodynamic cycle that is capable of producing a cooling effect may theoretically be used as a heat pump. Other than the ordinary vapor-compression cycle, possible cycles include: (1) the heat-operated absorption cycle, (2) the ejector cycle, (3) gas cycles, both open and closed, and (4) the thermoelectric cycle. It is not currently practicable nor economically viable to use any of these cycles as heat pumps, due to limitations in efficiency, cost, or size.

Table 3.2 shows the principal media currently being used with heat pumps as a heat source for heating and as a heat sink for cooling. The most practical choice for a particular application will be influenced primarily by geographic location, climatic conditions, initial cost, availability, and type of structure. Various characteristics that should be considered in the selection of a source or sink are given in Table 3.2.

3.2.2. Applied, Unitary, Package, and Split Systems

Applied heat-pump system is a term that generally refers to a system that is individually designed or constructed from separate components for a specific application. These systems generally have a cooling capacity in the range of 30–1000 tons. *Unitary heat pumps* are defined as consisting of one or more factory-assembled units, which normally include an indoor conditioning coil, compressor, and outdoor coil and controls or means to provide both cooling and heating. Capacities of unitary heat pumps range between $\frac{3}{4}$ ton and 50 tons. The classification of unitary air-source heat pumps is shown in Table 3.3, taken from ARI Standard 240-77 [2]. *Package heat pumps* have all components in a single package, whereas *split systems* will have both an indoor package or unit and an outdoor package or unit. Figure 3.4 illustrates a package type of heat pump; Fig. 3.5 illustrates a split system.

Four types of heat-pump systems are in common use today: (1) single-package heat pumps using an air source, (2) split-system heat pumps using an air source, (3) single-package heat pumps using a water source, and (4) split-system heat pumps using a water source.

An air-source unitary heat pump consists of one or more factory-made

TABLE 3.2. Heat Pump Heat Sources and Sinks [1]

Heat Source	Air	City Water	Well Water	Surface Water	Waste Water	Earth	Solar
Source classification	Primary	Primary or auxiliary	Primary	Primary	Primary or auxiliary	Primary or auxiliary	Auxiliary
Suitability as heat sink	Good	Good	Good	Good	Variable with source	Usually poor	May be used to dissipate heat to air
Availability (location)	Universal	Cities	Uncertain	Rare	Limited	Extensive	Universal
Availability (time)	Continuous	Continuous-except local shortages	Continuous-check water table	Continuous	Variable	Continuous, temperature drops as heat is removed, slowly rises when pump stops	Intermittent, unpredictable, except over extended time
Expense (original)	Low, less than earth & water sources except city	Usually lowest	Variable, depending on cost of drilling well	Low	Variable	High	High
Expense (operating)	Relatively low	High, usually prohibitive	Low to moderate	Relatively low	Low	Relatively moderate	Unexplored. Promising as auxiliary for reducing operating cost
Temperature (level)	Favorable 75-95% of time in most of U.S.	Usually satisfactory	Satisfactory	Satisfactory	Usually good	Initially good-drops with time and rate of heat withdrawal	Excellent
Temperature (variation)	Extreme	Variable with location (10 to 25 F deg)	Small	Moderate	Usually moderate	Large-less than for air, however	Extreme
Size of equipment	Moderate	Small	Small (except for well)	Small	Variable (usually moderate)	Small (except ground coils)	Available in some areas
Special problems	Least heat available when demand greatest. Coil frosting requires extra capacity, alternate source, or standby heat. May require ductwork.	Scale on coils. Local use restrictions during shortages. Disposal. Water temperature may become too low to permit further heat removal.	Corrosion, scale may form on heat transfer surface. Disposal may require second well. Water location, temperature, composition usually unknown until well drilled. Well may run dry.	Water may cause scale, corrosion, and algae fouling.	Usually scale forming or corrosive. Often insufficient supply. Very limited application, hence required individual design. Freeze-up hazards.	Limited by local geology and climate. Installation costs difficult to estimate. Requires considerable ground area, may damage lawns, gardens. Leaks difficult to repair.	Probably will require heat storage equipment at either evaporator or condenser side.

SOURCE: Reprinted by permission from *ASHRAE Systems Handbook 1973.*

TABLE 3.3.   Classification of Unitary Heat Pumps

Designation	ARI Type — Heating and Cooling	Heating Only	Arrangement
Single package	HSP-A	HOSP-A	Fan | Comp / Indoor coil | Outdoor coil
Remote outdoor coil	HRC-A-CB	HORC-A-CB	Fan / Indoor coil | Outdoor coil / Comp
Remote outdoor coil with no indoor fan	HRC-A-C	HORC-A-C	Indoor coil | Outdoor coil / Comp
Split system	HRCU-A-CB	HORCU-A-CB	Fan | Comp / Indoor coil | Outdoor coil
Split system with no indoor fan	HRCU-A-C	HORCU-A-C	Comp / Indoor coil | Outdoor coil

SOURCE: ARI Standard 240-77.

assemblies, which normally include an indoor conditioning coil, compressor(s), and outdoor coil, including means of providing a heating function, and may optionally include a cooling function. When such equipment is provided in more than one assembly, the separated assemblies are designed to be used together, and the requirements of rating outlined in the standard are based upon the use of matched assemblies.

Packaged terminal heat pumps are usually not equipped with a means for automatically defrosting the outdoor coil because of the difficulty of disposing of the melted frost, so the units employ an automatic means for limiting the operation of the refrigeration cycle in the heating mode to outdoor ambient conditions above the range that produces significant amounts of frost on the outdoor coil. Below this point, heating is done by built-in supplementary heaters.

In recent years, air-to-air heat pumps designed to be added to gas- or oil-fired warm-air furnaces have been introduced. Typically, they are operated as conventional heat pumps except during extreme winter weather. Either automatically or manually (at the owner's option), the heat pump may be turned off and the warm-air furnace turned on to become the source of heat. These add-on heat pumps share the same air-distribution system with the furnace. They may be arranged to operate in either a series or parallel air-flow arrangement with the furnace.

Fig. 3.4. Horizontal package heat pump. Courtesy Lennox Industries Inc.

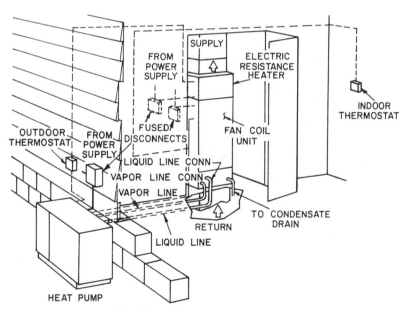

Fig. 3.5. Split-system heat pump.

3.2.3.   Installation Illustrations

Just as in an air-conditioning system, the outdoor unit of a heat pump may be mounted in the wall, on a slab, or on the roof. A typical installation for a residential split system is shown in Fig. 3.6. The indoor unit may be located in the basement, in an attic, or in an interior space such as a closet. The air flow through the indoor unit may be up-flow, down-flow, or horizontal. In all cases, the coil is upstream of the blower, and the auxiliary heat is down-stream.

Duct systems, including grilles, registers, and diffusers are sized and located in the same manner as for a straight air-conditioning system.

A packaged heat-pump unit is adaptable to a variety of installations and can be installed economically in an attic, basement, garage, breezeway, or utility room, or in commercial locations, as shown pictorially in Fig. 3.7 and schematically in Fig. 3.8. Only a provision to allow the escape of the moisture and heat-filled condenser air must be made.

Figure 3.9a depicts a valance-type heat pump system requiring no duct-work. In this system, finned tubing is run round the perimeter of the home. Figure 3.9b portrays a nonresidential roof mounted heat-pump system.

3.3.   TYPICAL CHARACTERISTICS

3.3.1.   Air-Source Heat Pumps

Outdoor air offers a universal heat-source, heat-sink medium for the heat pump. Extended-surface, forced-convection heat-transfer coils are normally employed to transfer the heat between the air and the refrigerant. Typically, these surfaces are 50–100% larger than the corresponding surface on the indoor side of heat pumps using air as the distributive medium. The volume of outdoor air handled is also usually greater in about the same proportions. The temperature difference during heating operation between the outdoor air and the evaporating refrigerant is generally in the range of 10–25°F.

As the outdoor temperature goes down, the heating capacity of an air-source heat pump decreases. Selecting the equipment for a given outdoor heating design temperature is therefore more critical than for a fuel-fired system. Care must be exercised to size the equipment to provide as much heating as possible without having excessive and unnecessary cooling capacity during the summer periods.

When the surface temperature of an outdoor air coil is 32°F or lower, frost may form, and if allowed to continue to form it will interfere with heat transfer. Research has shown that with a nominal amount of frost deposit (typically about 2–3 lb per square foot of coil face area), the heat-transfer capacity of the coil is not substantially affected. The number of defrosting operations will be influenced by the climate, the air-coil design, and the

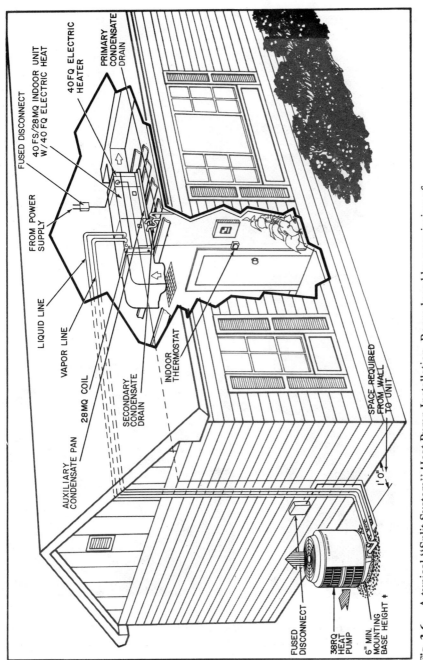

FUSED DISCONNECT

40 FS/28MQ INDOOR UNIT
W/40 FQ ELECTRIC HEAT

40FQ ELECTRIC
HEATER

PRIMARY
CONDENSATE
DRAIN

FROM POWER
SUPPLY

LIQUID LINE

VAPOR LINE

28MQ COIL

AUXILIARY
CONDENSATE PAN

SECONDARY
CONDENSATE
DRAIN

INDOOR
THERMOSTAT

SPACE REQUIRED
FROM WALL
TO UNIT

1'0"

FUSED
DISCONNECT

38RQ HEAT
PUMP

6" MIN.
MOUNTING
BASE HEIGHT ✳

Fig. 3.6. A typical "Split System" Heat Pump Installation. Reproduced by permission of Carrier Corporation.

Fig. 3.7. Heat pump installations. (a) Window unit, courtesy Carrier; (b) Through the wall unit, courtesy Westinghouse: (c) Packaged terminal unit, courtesy General Electric Co.

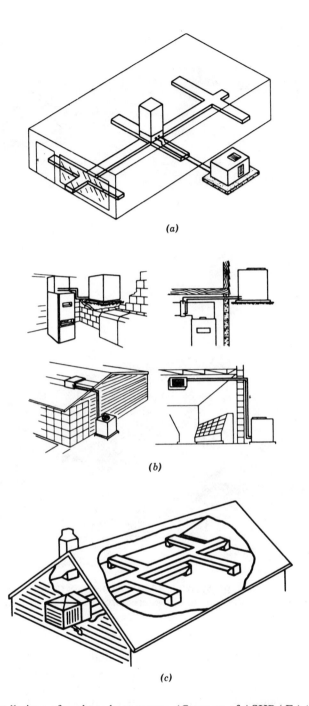

Fig. 3.8. Installations of package heat pumps. (*Courtesy* of ASHRAE.) (*a*) Outdoor split-system air-cooled condensing unit with indoor coil, type RCU-A-C, and down-flow furnace. (*b*) Outdoor split-system air-cooled condensing units with coil, type RCU-A-C, and upflow furnace, or with indoor flower coils, type RCU-A-CB. (*c*) Through-the-wall installation of air-cooled single-package unit, type SP-A.

1 ECONOMIZER SECTION
2 FILTER SECTION
3 HEATING/COOLING COIL
4 FANS AND BLOWERS
5 COMPRESSORS
6 WATER TO REFRIGERANT HEAT EXCHANGER
7 REVERSING VALVE

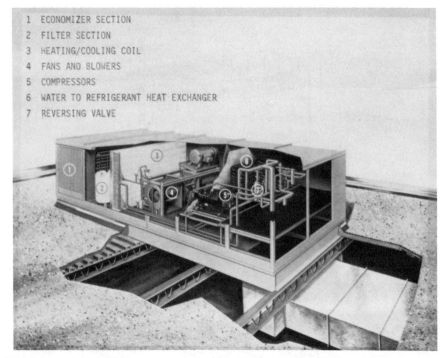

Fig. 3.9. (*a*) Valance-type system. Reproduced by permission of Conservation Technologies, Inc. (*b*) Roof-mounted water-source heat pump. From American Air Filter Co. by permission.

hours of operation. Experience has shown that generally little defrosting is required below 20°F if the relative humidity does not exceed 60%. This may be confirmed by psychrometric analysis using the principles given in the section on psychrometrics. However, it should be noted that under very humid conditions, when small suspended water droplets may be present in the air, the rate of frost deposit may be about three times as great as would be predicted from psychrometric theory. Under such conditions, a heat pump may require defrosting after as little as 20 min of operation. In apply-

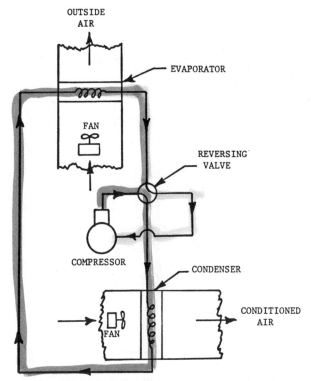

Fig. 3.10. Air-to-air heat pump—heating mode.

ing an air-source heat pump, the effects of this condition on available heating capacity should be taken into account.

Figure 3.10 shows an air-to-air heat pump operating in the heating mode. The hot gas leaves the compressor, passes through the reversing valve, and is condensed within the inside coil, providing heat for the conditioned air. The condensed liquid is then expanded through a valve and boils in the evaporator, gaining heat from the outside air. The cool gas then enters the compressor for continuation of the cycle.

In Fig. 3.11 the air-to-air heat pump is depicted operating in the cooling mode. The reversing valve now directs the hot gas from the compressor to the outside coil, where heat is removed from the refrigerant. Cooling of the conditioned air now takes place in the evaporator or indoor coil.

Table 3.4 lists selected manufacturers of air-to-air heat pumps and the unit sizes available. There are at least 26 manufacturers producing heat pumps in the size range 1½–45 tons. This list was compiled from the ARI *Directory of Certified Unitary Heat Pumps* [3] and is presented and discussed by Christian [4]. The space requirements for typical units are shown in Table 3.5.

Nine heat pump models were selected from the list of manufacturers in Table 3.4 and have been analyzed with respect to performance and cost [4].

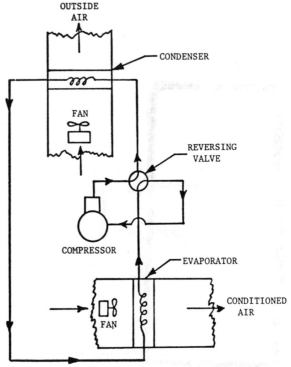

OUTSIDE
AIR

CONDENSER

FAN

REVERSING
VALVE

COMPRESSOR

EVAPORATOR

CONDITIONED
AIR

FAN

Fig. 3.11. Air-to-air heat pump—cooling mode.

These nine models are listed in Table 3.6 with their nominal heating COP and nominal cooling EER. The heating capacities of these nine models of air-to-air heat pumps were integrated into a nominal value at rated conditions [2]. The indoor air is at 70°F with a flow rate of 450 cfm/ton, and the outdoor air at rated performance is at 47°F and 85% relative humidity. Figure 3.12 depicts the performance of the nine models at various outdoor air temperatures. The shaded area represents the deviation in capacity between the nine models at each temperature. These differences are caused by variations in one or more of the following variables: heat-transfer area, efficiency of the compressor, the instrumentation and control philosophy used by the manufacturer, type of refrigerant used, and the refrigerant circulating in the two coils.

The heating COPs for the same nine models are compared in Fig. 3.13. Again significant variations are noted in the coefficient of performance. The variation at rated outdoor temperature (47°F) was 2.4–2.9 (see Table 3.6). These variations are again determined by unit design and operating characteristics. Changes in the heating capacity and the COP will occur as the indoor temperature changes and also as the indoor air-flow rate changes [4].

The cooling capacities of the nine air-to-air heat-pump models are de-

TABLE 3.4. Selected Heat Pump Manufacturers and Unit Sizes [3]

Company	Available Sizes (Tons)
Addison Products Co.	2.0–5.0
Airtemp Corporation	2.5–3.0
Amana Refrigeration, Inc.	2.0–5.0
Bard Manufacturing Co.	1.5–5.0
Bryant Air Conditioning	2.0–5.0
Carrier Air Conditioning	1.5–45
Fedders Corp.	2.0–4.5
Fraser & Johnston Co.	2.0–5.0
Friedrich Air Conditioning & Refrigerating Co.	2.5–5.0
General Electric Co.	2.0–20
Goettle Bros. Metal Products, Inc.	1.5–10
Heil-Quaker Corp.	2.0–5.0
Henry Furnace Co.	2.0–5.0
Lennox Industries, Inc.	2.0–15
Luxaire, Inc.	1.5–5.0
Mueller Climatrol Corp.	2.0–4.5
Payne Air Conditioning	1.5–5.0
Rheem Manufacturing Co.	1.5–5.0
The Ruud Co.	1.5–5.0
Singer Co.	2.0–5.0
Stewart-Warner Corp.	2.0–5.0
Tappan Co.	2.0–5.0
Trane Co.	2.0–5.0
Westinghouse Electric Corp.	1.5–5.0
Whirlpool Heating and Cooling Products	2.0–5.0
Williamson Co.	2.0–3.0

picted in Figure 3.14. The common or nominal point is at 95°F, where the cooling function of the heat pump is rated. The indoor conditions for rating purposes are at 80°F db and 67°F wb with 450 cfm/ton air flow. Again, significant differences in cooling capacities exist between different models.

The energy efficiency ratios (EER) for the nine heat-pump models are shown in Fig. 3.15 as a function of the outdoor air temperature. The EER is the ratio of the cooling effect (Btu/hr) and the total electrical power (W) required to produce that cooling effect. Changes in the cooling capacity and the EER will occur as the indoor dry-bulb and wet-bulb temperatures are changed.

The heat pump is normally sized to meet the cooling load so that proper humidity can be maintained during the cooling season. This procedure results in the heat pump being unable to supply all of the heat required during

TABLE 3.5.   Unitary Heat Pump Space Requirements [4]

Heat Pump	Length × Width × Height (ft)	Weight (lb)
3-Ton Split System		
Indoor unit		
Horizontal	3.5 × 5.0 × 3.5	100
Vertical	4.0 × 3.5 × 5.0	100
Outdoor unit	7.0 × 7.0 × 4.0	225
15-Ton Split System		
Indoor unit		
Horizontal	7.0 × 6.5 × 3.5	740
Vertical	7.0 × 4.5 × 5.5	795
Outdoor unit	8.5 × 7.5 × 3.5	1300
15-Ton Package Unit	22.0 × 8.5 × 4.0	2200
45-Ton Split System		
Indoor unit		
Horizontal	12.0 × 5.0 × 8.0	840
Outdoor unit	14.6 × 26.0 × 5.0	3030
(3–15 ton units)		

the heating season. This requires the use of some form of auxiliary heat in colder climates when the outdoor temperature is below 30–40°F. The colder the climate, the more auxiliary heat is required. Figure 3.16 shows the seasonal performance factor (SPF) of air-to-air heat pumps as a function of the number of degree days. The seasonal COP is the total heating output divided by the sum of the total input power to the heat pump and the auxiliary heat for the complete heating season. It is seen from Fig. 3.16 that the heat pump suffers from significant reductions in SPF in northern climates.

TABLE 3.6.   Heating and Cooling Performance of Nine Heat-Pump Models [4]

Nominal Capacity Size Range (Tons; 47°F)	Nominal Heating COP	Nominal Cooling EER
1½–2	2.8	7.1
2–3	2.4	7.1
3–4	2.4	6.5
4–5	2.8	8.1
5–10	2.6	7.4
10–15	2.7	8.6
15–20	2.9	8.7
20–25	2.8	8.6
25–45	2.6	8.6

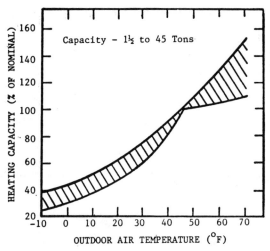

Fig. 3.12.   Heating capacity of nine air-to-air heat pump models [4].

3.3.2. Water-Source Heat Pumps

Water may represent a satisfactory, and in many cases an ideal, heat source for heat pumps. Well water is particularly attractive because of its relatively high and nearly constant temperature, generally about 50°F in northern areas and 60°F and higher in the south. Frequently, sufficient water may be available from wells, but the condition of the water often will either cause corrosion in heat exchangers or induce scale formation. Other considerations are the costs of drilling, piping, and pumping and means of disposing of used water. Information on well water availability, temperature, and chemical and physical analysis is generally available from U.S. Geological Survey offices located in many major cities. Industry specialists point out that groundwater is available to residents in more than 75% of the United States.

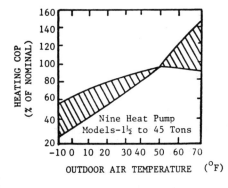

Fig. 3.13.   Heating coefficient of performance of nine heat pump models [14].

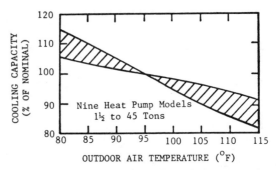

Fig. 3.14.   Cooling capacity of nine air-to-air heat pump models [4].

Rain is the major supplier of well-water systems. After reaching the earth's surface, rain water seeps into the ground until a nonporous layer of clay or rock is reached. This natural underground barrier forms a reservoir to contain groundwater. The upper surface of water is referred to as the *water table,* and unlike surface water, which is level, the underground water table is not uniform. Figure 3.17 shows a typical cross section of an unconfined reservoir.

Aside from a few extreme cases, most scaling with the groundwater-source heat pump is related to the air-conditioning mode when large amounts of heat must be rejected to the condenser water. This temperature-induced scaling results from the high temperature differential produced between compressor superheat (usually 240°F or slightly higher) and the cooling water temperature. In this situation, the potential for scaling will increase as the temperature of the condenser cooling water increases.

Scaling can seriously affect system efficiency. The thermal conductivity (k) of carbonate scale may range from 0.2 to 2.06 Btu/hr·ft·°F. With a thermal

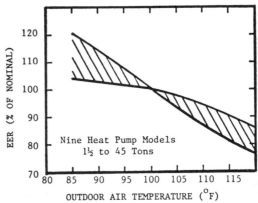

Fig. 3.15.   Cooling EER of nine air-to-air heat pump models [4].

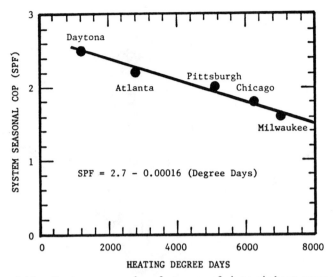

Fig. 3.16. System seasonal performance of air-to-air heat pumps.

conductivity this low, a $\frac{1}{16}$-in. layer of scale could effectively render a water-source system less efficient than an air-source heat pump.

The groundwater heat pump is rather new to the heating, ventilating, and air-conditioning industry; however, it has gained steadily in popularity during recent years. Figure 3.18 illustrates the groundwater heat-pump system, sometimes referred to as the "geothermal" heat pump.

While air-to-air heat pumps labor in extremely hot or extremely cold weather, the groundwater-to-air heat pump runs smoothly, at constant efficiency, on water as cold as 40°F. Instead of extracting heat from air during the winter, the groundwater heat pump extracts heat from water. During the summer, the unit expels heat to the groundwater. In Figs. 3.19 and 3.20 the unit heating and cooling modes are illustrated for 70°F groundwater.

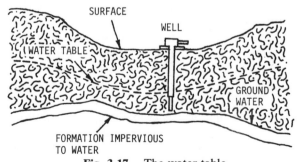

Fig. 3.17. The water table.

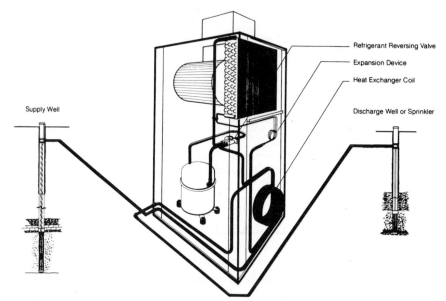

Fig. 3.18.   A groundwater ''geothermal'' heat pump. Reproduced by permission of Friedrich Air Conditioning and Refrigeration Co.

Efficiencies obtained from groundwater heat pumps appear impressive. COPs up to 3.5 are commonplace. One manufacturer has a line of heat pumps with EERs up to 10.5 and COPs up to 3.7. Another manufacturer claims custom-engineered units with COPs of 3.8–4.2.

The two biggest problems impeding the growth of water-to-air heat pumps are legal and technological. Currently, groundwater heat pumps are used mostly in rural areas where owners require a well for domestic water. However, they are also being installed in surburban areas where residents are willing to drill a well due to high electricity costs even though they have city water. Some of the legal problems encountered include laws against withdrawing water for any purpose other than for drinking water, fire fighting, or agriculture; restrictions against discharging spent water into the sewers; prohibitions on the discharge of any liquid into the ground. The biggest technological problem of groundwater heat pumps is designing the unit, at a reasonable cost, to operate with 40°F water. Some groundwater heat pumps have a variable flow rate determined by the entering groundwater temperature. At 60°F, 1.75 gpm/ton enters the unit, while at 34°F, 4 gpm/ton enters. The heat-exchanger coil is typically oversized by 30% for low entering water temperatures. Some units have a defrost cycle (reverse refrigerant flows) to remove any ice buildup that occurs at low inlet temperatures.

A third problem with the groundwater heat pump is the initial cost. The well required for the heat pump could cost several thousand dollars in addition to the cost of the heat pump itself. However, according to the National

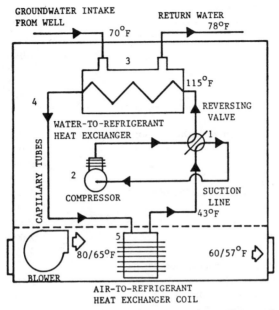

GROUNDWATER INTAKE
FROM WELL
70°F
RETURN WATER
78°F

3

115°F

WATER-TO-REFRIGERANT
HEAT EXCHANGER

4

CAPILLARY TUBES

REVERSING
VALVE

1

2

COMPRESSOR

SUCTION
LINE
43°F

5

80/65°F

60/57°F

BLOWER

AIR-TO-REFRIGERANT
HEAT EXCHANGER COIL

Fig. 3.19. Groundwater heat pump in cooling mode.

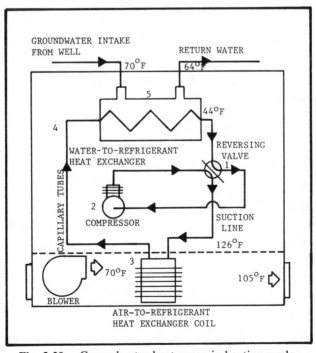

GROUNDWATER INTAKE
FROM WELL
70°F
RETURN WATER
64°F

5

44°F

WATER-TO-REFRIGERANT
HEAT EXCHANGER

4

CAPILLARY TUBES

REVERSING
VALVE

1

2

COMPRESSOR

SUCTION
LINE

126°F

3

70°F

105°F

BLOWER

AIR-TO-REFRIGERANT
HEAT EXCHANGER COIL

Fig. 3.20. Groundwater heat pump in heating mode.

139

TABLE 3.7. Typical Water Use Rates

Water Uses	Peak Demand Allowance for Pump (gpm)	Individual Fixture Flow Rate (gpm)
Household Uses		
Bathtub or tub-and-shower combination	2.00	8.0
Shower only	1.00	4.0
Lavatory	0.50	2.0
Toilet, flush tank	0.75	3.0
Sink, kitchen, including garbage disposal	1.00	4.0
Dishwasher	0.50	2.0
Laundry sink	1.50	6.0
Clothes washer	2.00	8.0
Irrigation, Cleaning, and Miscellaneous		
Lawn irrigation (per sprinkler)	2.50	5.0
Garden irrigation (per sprinkler)	2.50	5.0
Automobile washing	2.50	5.0
Tractor and equipment washing	2.50	5.0
Flushing driveways and walkways	5.00	10.0
Cleaning milking equipment and milk storage tank	4.00	8.0
Hose cleaning barn floors, ramps, etc.	5.00	10.0
Swimming pool (initial filling)	2.50	5.0

Water Well Association, a water-source heat pump system should pay for itself in 2 years if a well is already in place. If a new well must be drilled, the payback period should be somewhere from 4 to 6 years.

The increased demand for groundwater heat pumps requires knowledge concerning their water requirements and appropriate sizing.

Most groundwater heat pumps are being installed in homes where groundwater is the source for all domestic uses (bathing, drinking, laundry, etc.). Therefore, in calculating average daily water use, we will assume that the water system must supply all domestic water needs, including the operation of the heat pump. Table 3.7 lists flow rates for certain plumbing, household, and farm fixtures. The average per-capita consumption in a home is usually about 70 gal/day. In a home with many water-using appliances, the daily per-capita consumption may be as high as 100 gal.

The water demand of a groundwater heat pump varies with size and design. Typical water consumption falls between 1 and 3 gpm per 12,000 Btu output. For example, a 5-ton (or 60,000-Btu) heat pump would require between 5 and 15 gpm.

Table 3.7 lists the more common domestic uses of water along with the water demand each puts on a pump. USDA studies show that the demand on the pump is only about one-fourth of the individual fixture flow rates.

TABLE 3.8. Residential Water Use Example

Water Fixture	Peak Demand Allowance
Tub and shower	2.00
Two lavatories (0.5 gpm each)	1.00
Toilet	0.75
Kitchen sink	1.00
Dishwasher	0.50
Clothes washer	2.00
Heat pump	12.00
Total gpm	19.25

To determine the pump capacity needed for a particular installation, list all domestic water uses and their peak demand allowances, as shown in the example in Table 3.8. To determine the ideal capacity of the pump, add the demand allowances for each fixture. In the case of Table 3.8, the ideal capacity of the pump would be 19.25 gpm.

If a pump provides less water than required during peak demand, selecting the right type and size of water-storage facility is critical. The storage facility must be large enough to meet peak demand needs while allowing some reserve. Tank size may vary considerably, depending upon daily climatic conditions. If prolonged severe weather is expected, larger storage tanks may be desirable. Such tanks may be sized to meet 12 hr or more of continuous heat pump demand.

Water temperature is an important factor in a heat-pump installation. A typical water-source heat pump requires 1–3 gpm per ton of capacity. With some exceptions, the deeper the well, the warmer the water. Water obtained at a depth of 200 ft is normally about 5°F warmer than the local average annual air temperature. At depths between 30 and 60 ft, the water is usually 2–3°F warmer than the annual air temperature. As a rule of thumb, add 1°F to the average annual air temperature for every 64 ft of well depth to find the temperature of the well water. Except in cases where streams or recharging wells are close by, well water may be assumed to remain constant the year around for depths over 30 ft.

There are no federal controls on groundwater use that will have a significant impact on heat pump development. Various state regulations have the potential to affect groundwater heat pump utilization. These include well construction standards, water use restrictions, and effluent disposal regulations. There are thousands of county and municipal regulations throughout the United States that relate to the development of groundwater-source heat pumps for domestic use. These local controls often impede heat-pump development more than applicable federal or state legislation.

The use of a groundwater-source heat pump for residential heating requires a well that can provide up to 15 gal water per minute depending on the

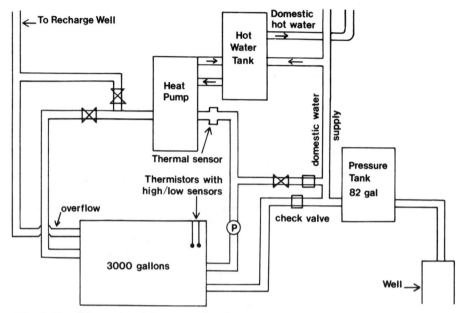

Fig. 3.21. Schematic of the water supply system at an NWWA facility. Reprinted by permission of the Water Well Journal Publishing Co. © Copyright 1980. All rights reserved.

size of the house, the type and efficiency of the system, and climatic conditions. However, the installation or use of the type of well needed to supply a groundwater heat pump is prohibited in some areas of the country. Development of groundwater heat pumps will not be possible in these areas.

Most groundwater heat pumps discharge water at relatively moderate temperatures. The temperature of the water used is altered by only a few degrees from natural groundwater temperatures, and the resulting effluent is not extremely hot or cold. It is unlikely, therefore, that most heat-pump water disposal will cause an impermissible alteration of surface water temperatures. The possibility does exist, however, that in instances of low surface-water flow, large-volume discharges, or extremely narrow temperature restrictions, the disposal of surface-water heat pump effluents will be restricted or prohibited.

One method of conserving water and cutting down on storage requirements is to recycle some of the water leaving the heat pump back to the storage tank. Figure 3.21 illustrates such a system now operating at a National Water Well Association test facility in Sunbury, Ohio. In this system, water is pumped from the wells into a pressure tank. Some of it is used to meet the needs of the house, and the rest goes to the storage tank. When the heat pump requires water, a circulating pump withdraws it from the storage

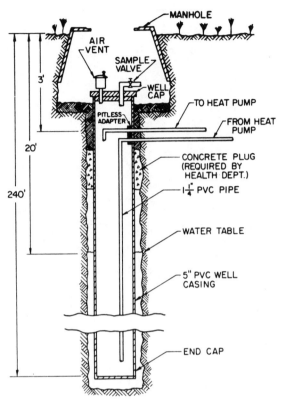

Fig. 3.22. Geothermal well design. Courtesy of J.E. Bose, Oklahoma State University.

tank and pumps it under pressure to the heat pump. The water from the heat pump is then returned to the storage tank.

During periods of the year when heating and cooling loads are light, the storage tank can meet the full demand of the heat pump. However, when heating and cooling needs increase, there must be some input of groundwater. Thermistors in the tank activate controls that permit the flow of water from the pressure tank into the heat-pump system or the storage tank. In the winter, this added water will keep the temperature of the water in storage from falling too low. Conversely, in the summer this system will prevent the water in the storage tank from becoming too warm. Water added to the heat-pump system displaces some water in storage. This displaced water flows out to a recharge well.

Under some circumstances, it may be desirable to pump the water directly from the well to the heat pump and directly out to the recharge well. Valving in the system will permit operation in this manner.

Bose [5] reports results on the geothermal coil under study at Oklahoma State University shown in Fig. 3.22. The coil is placed with a conventional

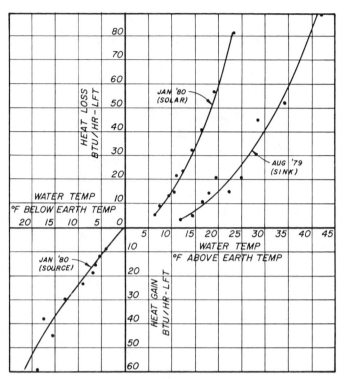

Fig. 3.23. Heat flow as a past function of well history. Courtesy of J. E. Bose, Oklahoma State University.

water well–drilling machine. The vertical heat exchanger consists of a 5-in. plastic pipe that is capped at both ends and pressurized at approximately 15 psig. This sealed and pressurized heat exchanger allows a low-power pump to circulate water through both the heat pump and vertical heat-exchanger system. Depending upon the type of test being studied, the water may be discharged to the top or bottom, and the discharge tube may be insulated.

The rate of heat gain or loss in the vertical heat exchanger is dependent upon past well performance. Figure 3.23 is a plot of three tests showing the performance of the well as a heat sink (August 1979), a solar storage device (January 1980), and a heat source for the heat pump (January 1980). The test data taken in 1979 indicate that the earth has been heat saturated from heat rejected during the cooling cycle. The January 1980 test data indicate that the earth around the heat exchanger has "cooled," and the ability to store or retain heat is less than that occurring a year earlier. The same generalized statements can be made about the well as a heat source.

Surface or stream water may be utilized, but under reduced winter temperatures the cooling spread between inlet and outlet must be limited to prevent freezeup in the water chiller that is absorbing the heat. Some large

TABLE 3.9. Selected Manufacturers of Water-to-Air Heat Pumps [3]

Company	Size Range (ton)
American Air Filter Co.	0.75–4.25
Dunham-Bush, Inc.	1–5
Command-Aire Corp.	0.75–26
FHP Manufacturing Corp.	0.75–10
Friedrick Air Conditioning and Refrigeration Co.	0.5–10
Heat Exchangers, Inc.	0.75–5.3
Singer Company	0.75–4.3
York Division	1.5–3.6

heat pump systems have been in operation in Europe in areas where winter stream temperatures of about 35°F have permitted inlet-outlet temperature spreads of only 1 or 2°F.

Under certain industrial circumstances, waste process water, such as spent warm water in laundries and warm condenser water, may be a source for specialized heat pump operations.

Utilization of water during the cooling operation generally follows conventional practice with water-cooled condensers.

Water-refrigerant heat exchangers generally take the form of direct-expansion water coolers, either of the shell-and-coil type or of the shell-and-tube type. They are circuited to permit use as a refrigerant condenser during the heating cycle and as a refrigerant evaporator during the cooling cycle.

Table 3.9 lists selected manufacturers of water-to-air heat pumps and the ranges of sizes that are available. These units are available from at least eight manufacturers in a size range 0.5–26 tons. In Table 3.10 is listed the space

TABLE 3.10. Selected Water-to-Air Heat Pump Space Requirements [2]

Water-to-Air Heat Pump	Length × Width × Height (ft)	Weight (lb)
1½-ton vertical unit	6.0 × 2.0 × 3.0	180
3-ton vertical unit	3.0 × 5.0 × 4.0	250
3-ton horizontal unit	6.0 × 2.0 × 3.5	250
5-ton vertical unit	6.0 × 2.5 × 4.5	330
11-ton vertical unit	6.0 × 3.5 × 3.0	720
26-ton vertical unit	6.0 × 5.0 × 3.5	1550

TABLE 3.11.   Performance of Nine Representative Heat Pump Models at ARI Standard 240-75 Test Conditions[a,b]

Manufacturer	Heating Capacity (Mbh)	COP	Cooling Capacity (Mbh)	EER [Btu/(W·hr)]	Water Flow (gpm/ton)
1. Singer CC-100	9.5	2.50	10.0	8.33	3.5
2. Dunham-Bush AQM-22	22.0	2.74	21.5	8.60	3.3
3. Singer CC and FV 400	41.0	2.73	40.0	9.09	3.5
4. American Air Filter VW 40	46.5	2.89	40.0	8.69	3.3
5. American Air Filter VW 51	58.3	2.91	51.0	8.64	3.3
6. Command-Aire SWP/H-800	126.0	3.1	100.0	8.50	3.2
7. Command-Aire SWP/H-1000	170.0	3.4	128.0	8.50	3.2
8. Command-Aire SWP/2000	274.0	3.1	210.0	8.5	3.2
9. Command-Aire SWP/2500	408.0	3.1	315.0	8.5	3.2

[a]	Test	Entering Indoor Air		Water Source	
		db	wb	In	Out
	Cooling	80°F	67°F	85°F	95°F
	Heating	70°F	60°F*	60°F	—†

*Maximum.
†Existing water temperature depends on the water flow rate as determined in the cooling test.

[b] Future water-to-air heat-pump ratings will be based on ARI Standard 320-76, which calls for a 70°F inlet water temperature instead of the 60°F called for in Standard 240-75.

required for typical or representative water-to-air heat pumps. These units appear to require less space and have less weight than air-to-air heat pumps.

Listed in Table 3.11 are nine representative models of water-to-air heat pumps in the full range of sizes available to the consumer. From these results one notes a trend of increasing heating COP with size that was not present in the air-to-air heat pump. The EER for cooling performance does not show this trend for water-to-air heat pumps, whereas it did increase with size for air-to-air heat pumps. The rating conditions for these heat pumps are given in the footnotes to Table 3.11. Significant changes in output and performance are obtained if the source water temperature changes.

3.3.3. Ground-Coupled Heat Pumps

Earth as a heat source and sink, by heat transfer through buried coils, has not been extensively used. This may be attributed to high installation expense, ground area requirements, and the difficulty and uncertainty of predicting performance.

Soil composition varies widely from wet clay to sandy soil and has a predominant effect on thermal properties and attendant overall performance. The heat-transfer process in the soil is primarily one of bulk temperature change with time (e.g., transient heat flow). The moisture content has an influence, since energy in the ground may also be transported via moisture travel in the soil.

In general, earth coils, usually arranged to be spaced horizontally 3–6 ft apart are submerged 3–6 ft below the surface. Although a lower depth might be preferred, excavation cost requires a compromise. Mean undisturbed ground temperatures generally follow the mean annual climatic temperature. Many individual studies involving theoretical as well as practical considerations have been conducted to determine earth heat transfer with respect to heat pump possibilities. While the earth provides a constant heat source and is of such volume that the temperature changes only in the immediate area of the heat exchanger, that area will stabilize to its original temperature within a short period of time.

Emphasized consistently is the fact that sufficient heat transfer is dependent upon wetted soil contacting the heat-exchange pipe. Efficiency tails off as the soil dries.

3.3.4. Solar-Assisted Heat Pumps

Renewed interest is being directed today at the solar-assisted heat pump— by architects, engineers, manufacturers, building owners, and in the marketplace in general. The principal advantage of employing solar radiation as a heat-pump heat source is that, when available, it provides heat at a higher temperature level than other sources, thus resulting in an increase in the COP. As compared to a solar heating system without a heat pump, the collector efficiency and capacity are materially increased due to the lower collector temperature required.

Research and development in the solar-source heat pump field has been concerned with two basic types of systems, direct and indirect. In the direct system, refrigerant evaporator tubes are embodied in a solar collector, usually of the flat-plate type. Research has shown that when the collector has no glass cover plates, the same collector surface can also function to extract heat from the outdoor air. The same surface may then be employed as a condenser, using outdoor air as a heat sink for cooling. The refrigeration circuit employed may resemble that shown in Table 3.1 for an earth-to-air heat pump.

An indirect system employs another fluid, either water or air, which is circulated through the solar collector. When air is used, the first system shown in Table 3.1 for an air-to-air heat pump may be employed, the collector being added in such a way that (1) the collector can serve as an outdoor-air preheater; (2) the outdoor-air loop can be closed so that all source heat is derived from the sun; or (3) the collector may be disconnected and the outdoor air used as the source or sink. When water is circulated through the collector, the heat pump circuit may be of either the water-to-air or water-to-water type illustrated in Table 3.2.

Solar-assisted heat pumps for residential applications have received a great deal of attention since 1975. They combine a solar heating system with a heat pump to provide the necessary heating and cooling requirements. Both air-to-air and water-to-air heat pumps can be used in this system. The solar system basically collects the energy for the heating requirements, and the heat pump supplies this energy at the proper time and temperature and also helps to improve the efficiency of the solar collector in the water-to-air operating mode.

Heat pumps complement solar collectors and heat storage in solar-boosted heat pump systems in several ways:

1. Heat pumps are more efficient and can provide more heat for a given heat pump size if their evaporators can operate from a warm source. Thermal energy storage, heated by solar collectors, can provide that warming.

2. Solar collectors operate more efficiently if they collect heat at lower temperatures. If the collected heat can be stored at a lower temperature because it is used to warm the evaporator of a heat pump, the collector is more efficient, and therefore less collector area is needed to collect a required amount of heat.

3. Heat pumps are the most efficient way to use electricity for backup heat for a solar heating system, even if there is no direct thermal connection of the heat pump with the solar collector and storage system.

4. Heat pumps can allow sensible-heat storage units (water, rock beds) to operate over wide temperature ranges, because stored heat down to 5°C can be used in conjunction with heat pumps to heat a building.

In Fig. 3.24 several possible combinations of solar-assisted heat pumps are depicted. A *parallel* system would use solar energy in the heat exchanger first and then use the air-to-air heat pump when the solar coil is unable to provide the necessary heat. A *series* system would use the solar energy in the heat exchanger first and then use the water-to-air heat pump when the solar coil is unable to provide the necessary heat. *Dual-source* heat pumps are currently in the development stage. A dual-source system would have the choice of either an air-heated coil when warm air is available or a water-

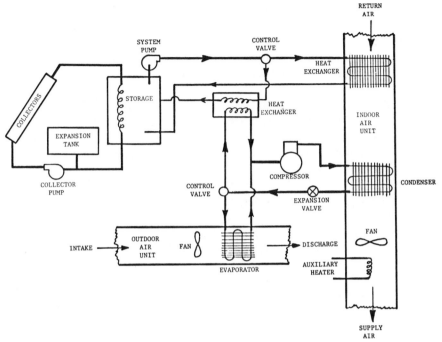

Fig. 3.24. Solar-assisted heat pump combinations.

heated coil when warm water is available. This type of capability would be advantageous in the solar-assisted heat pump, the ACES system (discussed in Chapter 12) and applications where well water is available as a source during the heating season. It would also alleviate the need for a cooling tower when the heat pump operates in the cooling mode.

It is not unusual for the solar-assisted heat pump to save up to 20% in operating costs during the heating season compared to an air-to-air heat pump. This represents a heating COP of almost 3. It has been reported that parallel systems can save up to 50% of the energy required when compared to electrical resistance heating. Series systems can save up to 60% of the energy normally required. Dual-source systems can save up to 70% of the energy normally required.

In the competition for commercial viability, all solar heating systems must compete for some version of life-cycle economic advantage with alternative nonsolar heating systems. In going to solar heating, an economic tradeoff is made between the high first cost of solar heating or cooling equipment and the cost savings realized by reduced use of fuel or electricity. One way to compare nonsolar and solar heating systems is to determine the cost of electricity at which they have the same overall life-cycle cost. A rough comparison of this break-even cost of electricity shows that many solar-assisted systems will be more cost effective than the conventional air-to-air heat pump available in the United States today only when electricity costs become greater than 10–12¢/kWh. The current national average cost of residential electricity is about 4¢/kWh, with a range up to about 10¢/kWh in some major urban utility systems. With the tax credits allowed by the U.S. government since 1980 the break-even cost, for the parallel system for example, becomes 6¢/kWh.

A wide variety of options exist to provide air-conditioning along with solar heating of buildings. The efficacy of coupling these options with a solar heating system depends strongly on which system is to be used and whether one is considering systems that can already be built reliably or ones that promise optimum performance in the future following extensive development efforts. Comly [6] has discussed the relative merits of several solar-boosted heat pump heating systems with regard to their ability to provide air-conditioning.

Several solar-boosted heat pump systems will be described to illustrate their relationship with air-conditioning: simple parallel system; series system; three-coil system; and cascade system.

Simple Parallel System.   Equipment is available today on the commercial market to build the parallel solar-boosted heat pump system illustrated in Fig. 3.25(*a*). On the right is shown a simple solar heating system. The collector heats water when sunlight is available and hot water is stored in a tank. When the thermostat calls for heat, water from the tank is circulated through a water-to-air heat exchanger to heat the building air. If the tank temperature is too low (below 40°C) when the house calls for heat, a controller shifts to an air-to-air heat pump, which heats the house in the conventional way. This system can provide air-conditioning using the normal function of a reversible heat pump. The equipment is already available on the commercial market and no development is required. The air-conditioning efficiency is the same as for a central air-conditioning unit.

Series System.   Figure 3.25(*b*) shows a series solar-boosted heat pump system. The solar collector heats water, which is then stored in a tank. The tank provides heat directly to the house if the tank temperature is above 40°C and the heat pump draws heat from the tank when temperature of the tank is between 40 and 5°C. This system has the advantage that the tank can be operated at lower temperatures when required, which allows the collector to operate with high efficiency, and the tank heat storage capacity can be increased by the amount of sensible heat stored between 5 and 40°C. The disadvantage of the system is that when the tank temperature finally drops to 5°C, the heat pump cannot be used further without danger of freeze-up in the water tank. The system shown will not provide air-conditioning, because there is no way to exhaust the waste heat to outside air. It would be possible to provide air-conditioning if there were another heat-exchanger loop between the storage tank and outside air, but this would add to the cost of the heat exchanger and its associated circulating pump. A development effort is required on the heat pump in this system because residential heat pumps are not currently designed to operate efficiently with evaporator temperatures over 20°C, and this one would have to operate up to 40°C.

Three-Coil System. This system, illustrated in Fig. 3.25(c), combines the advantages of the parallel and series systems and overcomes their disadvantages. The solar collector, heat storage, and building heat exchanger all continue to work as a simple solar system as long as the tank temperature remains above 40°C. At temperatures below that, the heat pump system is called upon by a microprocessor controller, and a decision is made as to whether it is better to draw the heat pump's heat from the tank or from the outside air. The control strategy options are numerous for this system; it can be operated to optimize savings of electricity or to reduce peak loads. It would be the most efficient of all the systems described here. It has the disadvantage that the required heat pump is complex and will be costly to manufacture. Considerable development work will be required to assure that the unit can be operated in all sequences of all of its modes with full reliability. With this system, air-conditioning is accomplished by operating the unit as a simple air-to-air heat pump in the cooling mode, using only the outside heat exchanger as a condenser. The system is not capable of using the storage tank to reduce air-conditioning peak loads in the summer, because the heat pump cannot be used to cool the tank to the outside air.

Cascade System. As Fig. 3.25(d) shows, the cascade system allows complete access to the storage tank in both the heating and air-conditioning mode. Two heat pumps are used, one between the outside air and the tank, and the other between the tank and the house. The inside heat pump allows the system to operate in the series heat pump mode, providing heat to the house when the tank temperature drops below 40°C. When supplemental heat is required, the outside heat pump charges the tank with heat removed from the outside air. In air-conditioning, the inside heat pump cools the house on demand, depositing the heat in storage; the outside heat pump then removes heat from the tank to outside air during off-peak periods. Both of the heat pumps used in the cascade system require some development effort. The outside heat pump must be optimized for operation between outside air temperatures and the cool water temperatures of the storage tank. The inside heat pump is the same as the series unit described above, except that it must include the ability to air-condition as well as heat. Required development effort would probably be less than for a three-coil unit. This system is probably the most expensive of those described here, because of the cost of heat pump equipment. When operating with the outside heat pump as a supplementary heat source, it will be less efficient than the three-coil or

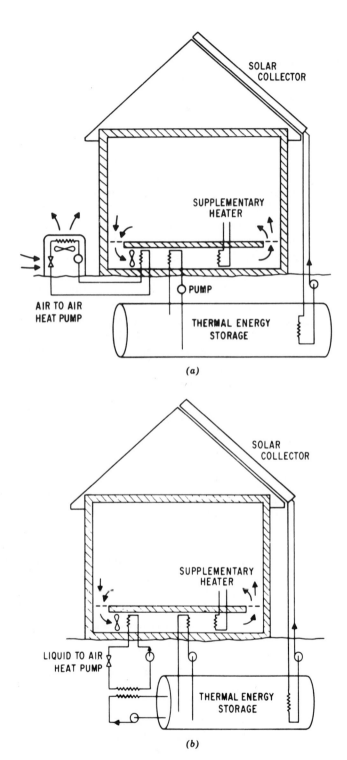

Fig. 3.25. Solar-assisted heat pump combinations. Courtesy of General Electric Co.

152

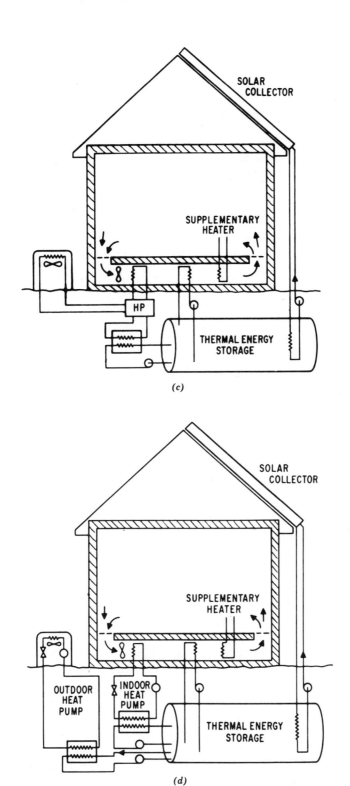

(c)

(d)

153

parallel systems, because the heat must go through a second heat pump in heating the house. On the other hand, it is the most flexible system, allowing full heating and air-conditioning and the use of the storage tank to store heat for off-peak use (solar, or electrical utility) during the heating season and to reduce air-conditioning peak loads during the cooling season.

It is possible to view much of the work of the last two years on solar-assisted heat pump (SAHP) systems as an attempt to address a single fundamental question: What does one do when the sun is not shining and storage is depleted? Studies based on the assumption that electrical resistance is used as the backup heat source have generally concluded that the SAHP is not a good idea. However, it is now possible to identify four general paths that have been taken to avoid this difficulty. These are:

1. Bimodal SAHP systems
2. Direct-expansion solar-collector heat pump systems
3. Volume-dominated ground-coupled systems
4. Area-dominated ground-coupled systems

Each type of system has its own special characteristics. These are described below.

Bimodal System.   The term *bimodal* is used to describe a system in which the heat pump processes solar-derived heat from storage as long as the storage temperature is above a preset minimum, and some type of backup heat source takes over when the storage temperature falls below this minimum. This is the basic SAHP system, with the proviso that the backup heat source need not be electrical resistance, with a coefficient of performance (COP) equal to 1, as has been assumed by most previous studies of SAHP systems.

A recent study [7] has shown that a backup heat source with a COP of 1 (electrical resistance) is a poor backup to a SAHP system. It has also been shown that a relationship exists between the COP of the backup and the optimum temperature at which solar energy should be collected and used. This temperature can be controlled by setting the minimum storage temperature below which the system goes to auxiliary. Backup with a COP greater than 1 can be achieved either by using the heat pump to process an alternative source of low-grade heat or, on a primary energy basis, by using a fossil-fueled burner. Figure 3.26 shows the relationship between optimum storage temperature and backup COP. For this example, which was done for Madison, Wisconsin, an auxiliary COP of 2 results in optimum storage temperatures of from −5 to 11°C in midwinter, the range depending on collector characteristics. If the auxiliary has a COP of 3, the optimum storage temperature is 21–39°C. In the first case, a heat pump is needed to process the heat to the load; in the second case the optimums are close enough to what is

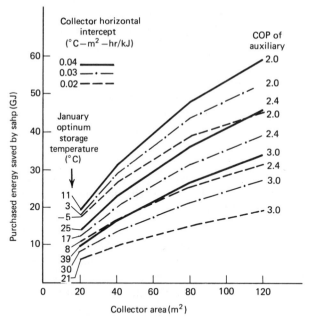

Fig. 3.26. Purchased energy saved by SAHP relative to use of auxiliary only, without SAHP (Courtesy of ASHRAE.)

required for direct heating that it probably pays to go to direct heating. Thus, the optimum system configuration can depend in detail on what is used as backup. A second result seen from Fig. 3.26 is that as the efficiency of the auxiliary increases, the energy that can be saved by the SAHP system decreases relative to use of the auxiliary only. There are two reasons for this. First, with a more efficient auxiliary there is simply less room for conservation, since the auxiliary is now by itself relatively energy efficient. Second, the SAHP, in order to compete with the auxiliary, must operate at higher source temperatures to provide COPs that are attractive relative to the auxiliary. Operating at higher source and collection temperatures, the SAHP system will now collect and use less solar energy than before. Bimodal systems can be of interest to electric utilities because they avoid the impact of everyone going to peak electrical resistance at the same time during very cold ambient conditions, as can happen with air-to-air heat pumps. Of course, bimodal air-to-air heat pump systems with fossil-fueled backup are also possible, and the bimodal SAHP must compete with them.

Direct-Expansion Solar Collector Heat Pump. In this type of system, exemplified by the Sigma Research, Inc. system, the collector itself is used as the heat pump evaporator. Refrigerant in the vapor-compression loop is passed through finned tubes exposed to the sunlight; absorbed solar energy evaporates the refrigerant directly, without intermediate storage or heat exchangers. These systems are seen as especially suited for areas such as the Pacific Northwest, with high humidity and rainfall, moderate winter temperatures, and relatively low insolation rates.

Volume-Dominated Ground-Coupled SAHP.   In this system concept a large volume of water thermally coupled to the ground is used as the storage element with the SAHP. For example, such a system may use a large uninsulated buried tank or a swimming pool as the storage element. Advantages of this type of system are:

1. The cost of storage in dollars per gallon is much less than that of conventional insulated storage. In the case of the swimming pool, whose cost is justified by its recreational value, the cost of storage is close to zero.

2. The potential exists for long-term storage (weeks to months) of low-grade heat. Energy collected at times when the solar collectors are underutilized may be saved for times when they are overworked. The extent of this potential is still under investigation.

3. The system can be designed so that no electrical resistance backup is required. This has great advantage to the load factor of the electric utility that is coupled to the system, since heat is always supplied by the heat pump at a high COP, meaning that more customers can be served by the same electric power-generating capacity.

Area-Dominated Ground-Coupled SAHP.   In this system a relatively small amount of heat-transfer fluid is spread out to present a large area, relative to its volume, to the ground. The typical configuration is a serpentine coil of buried plastic pipe, although others are possible. Here the ground is used as a source of heat (or heat sink during the cooling season). Such systems have been installed both in the United States and Europe and have performed well. Even without solar input, they have provided SPFs in the heating mode in the range 2.5–3.0. In colder climates the heat-transfer water that is passed through the pipes must contain antifreeze, and during the depths of winter the ground will be frozen. Minimum source temperatures to the heat pump of $\sim -5°C$ are encountered. If the heat pump and ground coil are sized to meet the design heating load at the minimum ground-source temperature, high SPFs are obtained by avoiding the energy-consuming defrost cycle and the need for supplemental resistance heating, both of which seriously degrade the performance of air-to-air heat pumps.

If it is desired to add solar energy to such a system to improve performance further, it must be done in such a way that the added cost of the solar components is justified by the additional energy saved. Since the nonsolar ground-coupled system is already quite efficient, this is a far more severe test than competing with electrical resistance. There are, however, at least three different avenues to coupling solar energy to such a system, and there is hope that one or more will prove competitive. These are:

1. Use of an active solar system to provide space heat and hot water, with the ground-coupled heat pump used as backup and for cooling. The solar-derived heat could be provided directly to the load, it could be processed through the heat pump at lower source temperatures, or it could be used to preheat the incoming room air, with the ground-coupled heat pump boosting the temperature of the air stream to the service value.
2. Use of the most inexpensive passive solar concepts in the building design, with the ground-coupled heat pump providing backup heat, hot water, and cooling.
3. Use of photovoltaic cells to provide electricity to drive the ground-coupled heat pump.

A number of solar-assisted heat pumps are commercially available either as complete systems or as component packages. The following descriptions of commercial units are provided only as examples of what is available.

Solar-assisted EnerCon water-source heat pumps manufactured by American Air Filter Co., Inc. (AAF), have helped the New Brunswick Electric Power Commission reduce energy consumption for heating and cooling a two-story district office and warehouse building. The heat pumps are integrated with solar collectors into a system that reportedly uses less than 12 kWh per square foot per year. This power consumption is 25% below the target of 16 $kWh/(ft^2 \cdot yr)$ set by the New Brunswick Power Commission for all its buildings, according to AAF. The HVAC system consists of two heat pumps (one for each floor of the office), electrical resistance in-duct heaters, a cooling tower, and supply and return air ducts. Data monitored since July 1977 indicated that the solar input to heating the 7000-ft^2 office area was approximately 45%. The 5000-ft^2 warehouse is heated separately with electrical resistance heaters.

Eight Daystar solar collectors in conjunction with a solar-assisted dual-source (water and air) heat pump system supply space heating and cooling and domestic hot water for a 2200-ft^2 home in Largo, Maryland (in the Washington, DC, area)—one of three homes in a program sponsored by Solar Thermal Systems, Division Exxon Enterprises, Inc. The system schematic is given here as Fig. 3.27. The program was developed to study solar energy used in conjunction with energy-efficient design and construction. Each home offers a distinctive solar energy system and numerous passive energy-saving features that use today's construction techniques and building materials. Some common features are additional ceiling, wall, and foundation insulation; exterior vented fireplace; and vestibule airlock entrance.

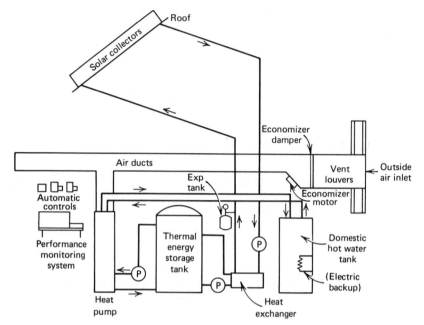

Fig. 3.27. Residential dual-source heat pump system. Courtesy Daystar Corporation—Exxon Enterprises.

The schematic for another residential solar-assisted heat pump is given as Fig. 3.28.

Northrup, Inc., is promoting a broad line of water-to-air heat pumps to complement its heating, air-conditioning and solar energy products.

The heat pumps for solar-assist systems cover a capacity range from 18,000 to 36,000 Btu/hr and are available in vertical configurations. The line is specially designed to operate with the higher water temperatures that result from the solar assist. Northrup, Inc. provides various schematics to show the interface of a heat-pump system with solar domestic hot water systems, one of which is presented here as Fig. 3.29. In winter, an additional source of heat is provided by the Northrup freeze-protected closed-loop solar domestic hot-water system. When the system is in its summer cooling mode, the heat of rejection is utilized for preheating the potable hot-water supply. The domestic hot water is then provided by the closed-loop solar domestic hot-water system. The equipment schedule for this system includes the Northrup FPIG solar collector; Northrup VWTA-19 water-to-air heat pump, 4 gpm flow; Northrup DWH storage, DWH 120; Grundios VP26-64 circulating pump, 8 gpm; an auxiliary water heater; TACO 4406s heat exchanger; and Northrup DWH 120 heat pump storage. Other system designs are available for use with groundwater and solar systems and for using the swimming pool as a heat sink.

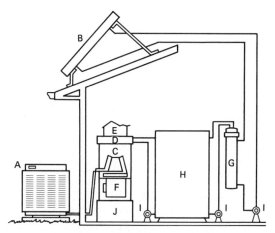

Fig. 3.28. A commercially available solar-assisted heat pump based on the General Electric Weathertron: A, Weathertron outdoor section; B, solar heat collection panel; C, indoor heat exchanger; D, hydronic coil; E, supply duct; F, flower; G, heat exchanger; H, heat storage tank; I, pumps; J, return duct. (Courtesy of General Electric Co.)

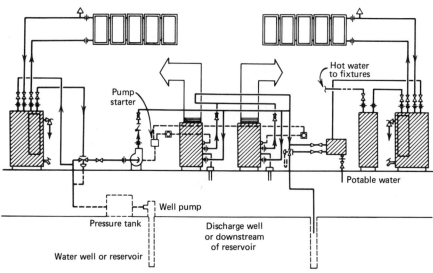

Fig. 3.29. Northrup solar-assisted heat pump system and components. (Courtesy of Northrup, Inc.)

159

3.3.5. Commercial Heat Pump Systems

Many types of heat pump systems have been applied to commercial structures during the last 30 years. Some are used in buildings that require simultaneous heating and cooling, while others have storage integrated into the
system or are designed for heat recovery from the building. The following
sections provide brief descriptions of several types.

Simultaneous Heating and Cooling.   Figures 3.30, 3.31, and 3.32 show three
of the several possible operating cycles for simultaneous heating and cooling. Figure 3.30 illustrates one method of using water as the heat source and
sink or as the heat source and heating and cooling medium. The compressor,
chiller, condenser, refrigerant piping, and accessories are essentially standard and factory preassembled. The cycle is very flexible, and the heating or
cooling medium is instantly available at all times.

Heating may be provided exclusively to the zone conditioners by closing
valves 2 and 3 and opening valves 1 and 4. With the valves in these positions,
the water will be divided into two separate circuits. The warm-water circuit
consists of the condenser (where the heat is supplied by the high-
temperature refrigerant), valve 1, zone conditioners, and a circulating pump.
The cold-water circuit consists of the chiller (where heat is taken from the
water by the low-temperature refrigerant), valve 4, exchanger (where heat is

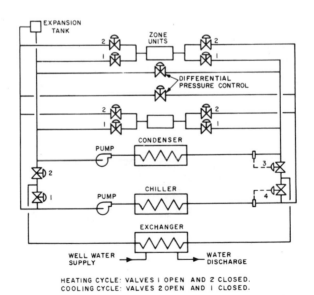

Fig. 3.30. Water-to-water heat pump cycle [1]. Reprinted by permission from
ASHRAE Systems Handbook 1973.

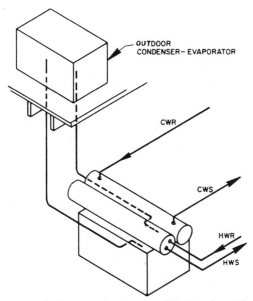

Fig. 3.31. Air-to-water single-stage heat pump [1]. Reprinted by permission from *ASHRAE Systems Handbook 1973*.

taken from the well water), valve 1, and a circulating pump. The refrigerating compressor is operated to maintain the desired temperature for water exiting from the condenser.

Similarly, cooling may be obtained exclusively in the cycle of Fig. 3.30 by opening valves 2 and 3 and closing valves 1 and 4. With this arrangement, the cold-water circuit will consist of the chiller (where heat is removed from the water by the low-temperature refrigerant), valve 2, zone conditioners, and a circulating pump. The warm-water circuit consists of the condenser (which receives the heat from the refrigerant), valve 3, exchanger (where heat is rejected to the well water), valve 2, and a circulating pump. The refrigerating compressor is operated to maintain the desired temperature of water leaving the chiller.

During the intermediate season, simultaneous heating and cooling may be provided by a cycle such as that in Fig. 3.30 by modulating valves 3 and 4 when valves 1 and 2 are open. Valve 3 is usually adjusted to maintain 85–120°F water in the condenser circuit and valve 4 to maintain 45–50°F in the chiller circuit. The excess heating or cooling effect is wasted to the exchanger, which in turn passes it on to the well water. The well water can be supplied directly to the condenser and chiller instead of by means of an exchanger, thus eliminating one heat-transfer surface. This direct supply of well water is generally used in small systems but may be used in larger systems, providing the well water is chemically acceptable and does not unduly contaminate the heat-transfer surfaces.

Figure 3.31 shows an air-source air-to-water single-stage heat pump work-

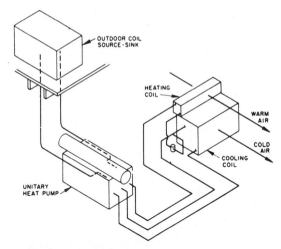

Fig. 3.32. Air-to-air heat pump [1]. (Courtesy of ASHRAE.) Reprinted by permission from *ASHRAE Systems Handbook 1973*.

ing with a four-pipe system for simultaneous heating and cooling. The refrigerant cycle of this package is also factory preassembled. However, the outside summer condenser or winter evaporator must be field connected to the compressor unit. In such heat pumps, which are available in packages up to 150 tons of cooling, the only operational reversal takes place in the outdoor condenser-evaporator. The shell-and-tube chiller and condenser re-

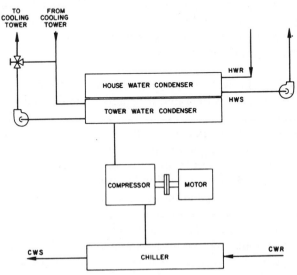

Fig. 3.33. Heat-transfer heat pump with double-bundle condenser [1]. Reprinted by permission from *ASHRAE Systems Handbook 1973*.

main as cooler and heater throughout, respectively. Most commercial units of this type use the flooded principle. The air-to-air heat pump (Fig. 3.32) is somewhat more efficient than the air-to-water system, as the conditioned air is heated or cooled directly by the refrigerant without the addition of another media. No shell-and-tube heat exchangers are required, reducing initial costs somewhat. All previously described heat pumps are also available without the simultaneous feature.

The double-bundle condenser working with a reciprocating or centrifugal compressor is quite often used as a simultaneous heating and cooling device. Its application is illustrated in Fig. 3.33. Heat is usually available in the range of 100–130°F.

Storage System Heat Pump. Figure 3.34 shows a system similar to the double-bundle condenser except that a storage tank has been added. This enables the system to store heat during the occupied hours by raising the temperature of the water in the tank. During the hours when the building is unoccupied, the water from the tank is used as input to the chiller to provide a load on the chiller, which raises the temperature of the condenser water.

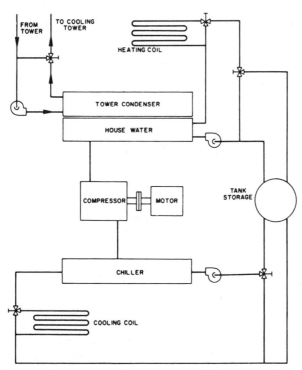

Fig. 3.34. Heat-transfer system with storage tank [1]. Reprinted by permission from *ASHRAE Systems Handbook 1973.*

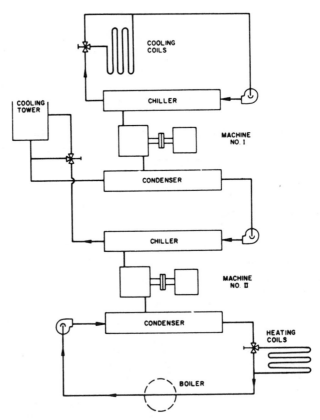

Fig. 3.35. Multistage (cascade) heat-transfer system [1]. Reprinted by permission from *ASHRAE Systems Handbook 1973*.

The high-temperature condenser water then heats the building during the unoccupied hours.

Cascade System. Figure 3.35 is another transfer system capable of generating 130–140°F warm water during occupied hours by cascading two compressors hydronically. In this configuration, one chiller can be considered as a chiller only and the second one as a heater.

Closed-Loop System. Another cycle that combines transfer characteristics with an actual water-to-air heat pump unit is shown in Fig. 3.36. Each module is furnished with one or more water-to-air heat pumps, the units being connected hydronically with a two-pipe system. Cooling is accomplished by each unit in a conventional manner, supplying cool air to the individual modules and rejecting the heat thus removed to the two-pipe system by means of a shell-and-tube condenser. The total heat gathered by the two-pipe system will be rejected to an evaporative cooler, which is usually mounted on the roof. If and when some of the modules, particularly

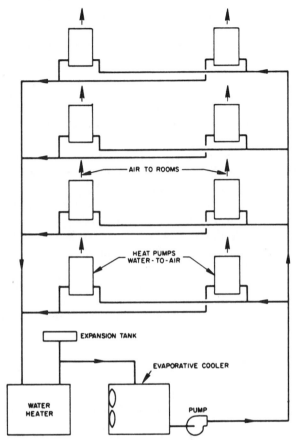

Fig. 3.36. Heat-transfer system using water-to-air unitary heat pump [1]. Reprinted by permission from *ASHRAE Systems Handbook 1973*.

on the northern side, require heat, the individual units will switch (by means of four-way refrigerant valves) into the heating cycle, deriving their heat source from the two-pipe water loop thereby basically obtaining heat from a relatively high source, the condenser water of the other units. Naturally, when only heating is required, all units will be in the heating cycle, and consequently a boiler of some nature must be added to provide 100% heating capability. The water loop is usually between 70 and 90°F and therefore needs no insulation.

Note: Some aspects of these cycles may be of proprietary nature and covered by patents, so they should not be used without appropriate investigation.

Heat Recovery System.   On many larger buildings, positive means of exhausting air are necessary to obtain good air distribution and sufficient ventilation. If the air is exhausted at one or more central locations, heat may be extracted from it economically, thereby reducing the outside heat source capacity required. One way to extract heat is to apply direct-expansion or water-cooling coils in the exhaust air stream, to remove heat from the air with the heat-pump compression system and dissipate it to offset building heat losses.

3.4. SPECIFICATIONS AND PERFORMANCE DATA

Heat pumps come in a wide variety of sizes and are normally rated in terms of the amount of cooling (in tons or Btu) they can produce. They also come in a range of voltages, both single-phase and three-phase to suit almost any application. Typical operating characteristics are shown in Fig. 3.37.

Typical manufacturer's ratings are illustrated by Figs. 3.38, 3.39, and 3.40. Note that the manufacturer's ratings shown in Fig. 3.38 are based on 80°F air entering the cooling coil, but the indoor design temperature normally selected is 75°F. Also, the desired capacity is usually based on a 3°F temperature swing, or indoor temperature of 78°F. The equipment capacity with 78°F entering air will be within a small percentage of the capacity with 80°F entering air. Therefore, it is highly unlikely that the capacity at outdoor

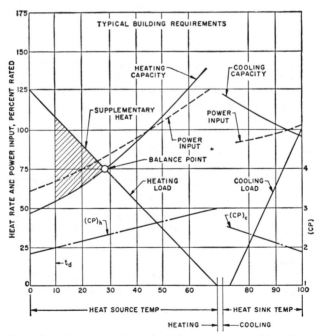

Fig. 3.37. Operating characteristics of single-stage unmodulated heat pump [8]. Reprinted by permission from *ASHRAE 1975 Equipment Handbook*.

XYZ Heat Pump Total Cooling Capacities (Btu/h)

Model No.	Indoor Coil Airflow (ft³/min)*	Temp. of Air Entering Outdoor Coil			
		85F	95F	105F	115F
18H	565	18,150	17,280	15,710	13,420
	625	18,520	17,640	16,040	13,720
	675	18,900	18,000	16,360	13,990
24H	700	24,500	23,330	21,200	18,110
	785	25,310	24,100	21,910	18,740
	920	25,820	24,590	22,350	19,110

Fig. 3.38. Example of manufacturer's cooling capacity ratings [9]. Airflow includes allowance for a 1-in. disposable filter in place. Entering air at 80°F db and 67°F wb. Reprinted by permission of Air Conditioning Contractors of America.

air design conditions will fall outside the 95–115% recommended selection limits.

Inspection of the airflow rates versus capacity ratings in Figs. 3.38 and 3.39 reveals that capacity increases with increasing airflow rate. This is usually the case, but not always. Note also in Fig. 3.40 that the indicated minimum airflow rate for heating with a model 24H heat pump is 700 ft³/min. Reference back to Fig. 3.38 shows that the ratio of 700 ft³/min airflow rate to the corresponding cooling capacity of 23,330 Btu/hr at 95°F outdoor air temperature is equivalent to an airflow rate per 12,000 Btu/hr of 360 ft³/min.

	Indoor Coil Airflow (ft³/min)**					
Outdoor Air Temp. (deg. F)	700		785		920	
	Cap'y.	Power	Cap'y.	Power	Cap'y.	Power
70	35,300	3230	36,300	3210	36,900	3140
60	30,900	2940	31,200	2950	32,100	2920
50	25,000	2680	26,500	2710	27,500	2700
40	20,900	2450	22,400	2480	23,500	2470
30	17,900	2270	19,200	2290	20,200	2280
20	14,600	2090	15,800	2110	16,600	2100
10	11,200	1920	12,300	1940	13,100	1930
0	8,500	1790	9,300	1810	9,900	1800

XYZ Model 24H Heat Pump Heating Capacity * (Btu/h) and Power (W) Input Ratings

Fig. 3.39. Example of manufacturer's heat pump heating capacity ratings [9]. Reprinted by permission of Air Conditioning Contractors of America.

XYZ Heat Pump Supplementary Electric Heater Ratings					
Heat Pump Model	Heater Package	Circuits		Total kW* Output	Min. Airflow (ft³/min)
		No.	Amps (each)		
18H	18-1	1	12.5	3	500
	18-2	1	25.0	6	
	18-3	2	12.5. 25.0	9	
24H	24-1	1	12.5	3	650
	24-2	1	25.0	6	
	24-2	2	12.5. 25.0	9	

Fig. 3.40. Example of manufacturer's supplementary resistance heater data [9]. Reprinted by permission of Air Conditioning Contractors of America.

Similarly, the airflow-to-capacity ratios for 785 and 920 ft³/min are 391 and 449 ft³/min per 12,000 Btu/hr, respectively. Airflow rates, capacities, and airflow-to-capacity ratios are design factors that are entirely under the control of the manufacturer. In order to publish a certified ARI Standard rating for a heat pump, however, a manufacturer must qualify its heat pump under certain specified airflow and air temperature conditions. Among the specified conditions is the requirement that the standard rating shall be determined at the indoor-side airflow delivered when operating against a minimum (specified) external resistance but not to exceed 37.5 ft³/min per 1000 Btu/hr of rated capacity (or 450 ft³/min per 12,000 Btu/hr). Thus, it is apparent that the manufacturer has a great deal of latitude in determining the relationship between airflow rate and unit capacity. There is no standard airflow-to-capacity ratio for heat pumps.

As shown in Fig. 3.40, XYZ Model 24H heat pumps have supplementary resistance heaters available in packages with capacity increments of 3 kW at 240 V. It is necessary to determine which heater package must be selected to ensure adequate heating down to the outside design temperature. However, in some locales there is the requirement that supplementary heat must satisfy 80–100% of design heat loss.

In a study conducted for EPRI, Blatt and Erickson [10] contacted 29 manufacturers of heat pumps requesting information on heat pump integrated performance, a heat-pump/fossil-furnace control scenario, heat pump specifications, furnace specifications, furnace efficiency, defrost operation, component descriptions, and unique heat pump or furnace sizing procedures. A complete set of data was obtained from eight manufacturers. The manufacturer's literature was excerpted to obtain the information contained in Fig. 3.41 on the heat pump steady-state integrated COP versus temperature for nominal 3-ton heat pumps.

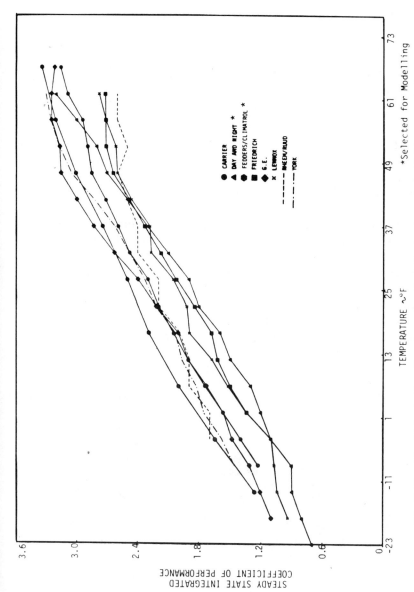

Fig. 3.41. Coefficient of performancy (heating) for heat pumps with 3-ton nominal cooling capacity [10]. Courtesy of M. H. Blatt, R. C. Erickson and the Electric Power Institute.

3.5.   SIZING AND SELECTING THE HEAT PUMP

Generally, the size of the heat pump is determined from the cooling load. However, in some areas, local utilities may *stipulate* a "balance point," which then will determine how the unit is sized in relation to the load. If the balance point has not been stipulated and the need for heating is much greater than cooling, then oversizing on the cooling load (25–35%) may be considered. However, with oversizing, there exists the higher first cost of equipment, higher installation cost (larger main service, etc.), and less-than-ideal summer humidity control. Humidity control can be improved by running the fan at low speed in cooling and at high speed in heating. The number of heating hours and the cost of local and future power will determine the payoff period required to offset the initial cost.

The map given as Fig. 3.42 is divided into several annual COP or SPF (seasonal performance factor) bands based on average industry heat pump performance. These SPFs assume the heat pump is sized on the basis of the cooling load and includes supplemental strip heating where required. This map may be used as a quick reference for comparing the operating cost advantage of a heat pump system versus a resistance electric heat system, which always has an SPF of 1.0. A heat pump in the 2.0-SPF band would have an operating cost approximately half that of the resistance heat system. The map may be used as a rule-of-thumb reference as to which load, heating or cooling, to use as a base for sizing the pump and, further, how much oversizing might reasonably be tolerated.

Very few sections of the country will allow an exact match of the building-

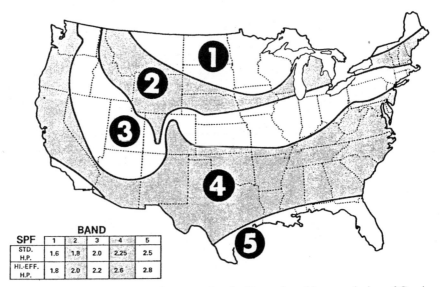

BAND					
SPF	1	2	3	4	5
STD. H.P.	1.6	1.8	2.0	2.25	2.5
HI.-EFF. H.P.	1.8	2.0	2.2	2.6	2.8

Fig. 3.42. Map of similar performance bands. Reproduced by permission of Carrier Copyright © Carrier Corporation 1978.

load requirements to both the heating and cooling capacities of any single heat pump. Heat-to-cool ratios vary widely throughout the country. In the south, cooling loads tend to be much greater than heating loads. In the north, the opposite occurs. It follows, then, that the further north, the more difficult it is to select a heat pump that fulfills the heating requirements when the equipment is selected on the basis of the cooling load.

On the other hand, the heating capacity of a heat pump drops as the outdoor temperature falls, and since northern winter design temperatures are lower, longer operating hours are likely to be needed for heating. The desire to react to these two circumstances may lead one to select a larger unit than is required to reduce heating costs or ensure enough heat.

As a general rule, in bands with an SPF of 2.25 or higher, heat pump systems should be sized to match the cooling load (or balance point designated by the local utilities). In the areas with an SPF of 2 and lower, slight oversizing may be considered. This procedure is simple enough. In map bands with an SPF of 2.0 or lower, calculate the cooling load and select a pump that has a cooling capacity in Btu/hr a maximum of 35% higher than the calculated cooling load.

When the pump has been selected in this manner, the heating capacity of the machine should be plotted against the heat-loss line for the house to locate the balance point. *Balance point* is defined as the outdoor temperature at which the heating capacity of the heat pump compressor unit and the heating requirements of the conditioned space are equal or balanced. When outdoor temperatures are at or above the balance point, the heat pump compressor unit alone provides the total heating required.

At temperatures below the balance point, the heat pump compressor unit must be supplemented. Figure 3.43 gives an example of the most accurate method of determining the balance point of a specific heat pump under specific heating load conditions. A curve is plotted of the heat pump unit's heating capacity, and a second curve is plotted of the structure's heat loss over a usual range of outdoor temperature. The point at which these curves cross is the balance point.

In the southeast, a balance point of 35°F or lower will produce good heating operation economy for the heat pump. Less than 10% of its heating kilowatt-hours are generally used below this point, and the effect of the supplemental heater use is nominal. Some heat pumps automatically shut off the compressor at 15°F and operate solely on supplemental electrical resistance heaters. As operating hours below 15°F are relatively few, this increases operating costs very little.

The following example on the balance point, illustrated in Fig. 3.43, also provides a simple graphic method of determining the sizing and staging of the increments of supplemental electric heat. By standard calculation methods the heating load at a 10°F design temperature has been determined to be 52,000 Btu/hr. From this data, the building structure's heat load is plotted as line 1 in Fig. 3.43.

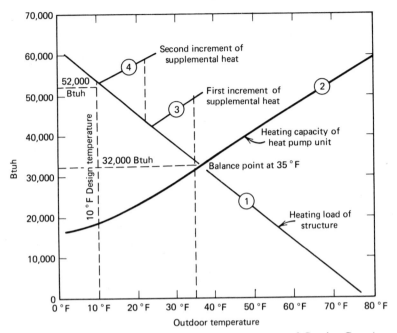

Fig. 3.43.   Example heat pump sizing. (Courtesy of Carrier Corp.)

The size of the heat pump to be used is generally based on a building's cooling needs. If the pump is too large, it will cycle on and off frequently in summer and do a poor job of dehumidification. In this example, the contractor selected a 3-ton heat pump. From capacity rating tables in a manufacturer's specification sheet, the heat pump unit's heating capacity is plotted as curve 2. These two lines intersect at approximately 32,000 Btu/hr, which establishes this heat pump unit's balance point as 35°F.

From the graph, the heating capacity of the unit alone is determined to be 19,500 Btu/hr at the 10°F design temperature. This leaves a remainder of 32,500 Btu/hr to be supplied by the supplemental electric heaters at this design temperature. Two 5.0-kW electric heating elements can be used, as they will provide 17,000 Btu/hr each, or a total of 34,000 Btu per hour of heating.

By reference to the graph, it is found that the heat pump unit plus the first 5.0-kW increment will meet the structure's heating requirements down to about 23°F (see curve 3). When the second 5.0-kW increment is brought into use, the structure's heating requirements can be met down to 9°F, or 1°F below the design temperature. Each of these 5.0-kW electric heating elements is generally equipped with an individual outdoor thermostat. In some places the utility may require 100% backup. A third 5.0-kW increment of electric heat could be added with its own separate outdoor thermostat. The resulting total of 15.0 kW is sufficient to meet the heating requirements of the structure at the 10°F design temperature. The full heating capacity of these three heaters can be brought into direct use, when needed, by operation of the bypass switch.

Even in southern areas where the balance point may be as low as the winter design point, indicating no need for supplemental heat, it is good technique to provide a small heater to temper the air during defrost. Defrosting may occur at outdoor temperatures up to 40°F.

As noted already, sizing a heat pump is a critical and complicated procedure. The typical problem in a northern location is compromising the equipment selection because of a small cooling load and a large heat loss. In southern states, sizing to a large cooling load may give more heating capacity than is really needed. An addtional problem in sizing heat pumps is the advent of very highly insulated "thermos bottle" homes. The heating and cooling loads in these homes are so small that the equipment selected to meet the sensible loads is too small to provide enough air flow to prevent stratification. If the equipment is sized larger to meet ventilation requirements, long "off" cycles and poor latent cooling can lead to user dissatisfaction from the cold, clammy atmosphere encountered.

A possible solution to these difficult sizing problems is the two-speed heat pump. The basic element of the two-speed heat pump is a compressor that operates at high and low speeds, thus delivering high and low capacities with commensurate changes in wattage. Additionally, the two-speed heat pump system contains an indoor DX coil, which has split refrigerant circuiting that lets just half the coil function during the low-speed cooling operation. The split coil configuration lets the two-speed heat pump deliver excellent latent cooling capability during low-speed, high airflow cooling operation.

Figure 3.44 shows how the two-speed heat pump can solve sizing problems. In a northern climate, the heat pump is sized so that the low-speed cooling capacity meets the design cooling load (at 90°F in this example). This ensures comfortable cooling with a minimum of compressor cycling. In heating, the low-speed operation is used until it can no longer meet the heat loss; then the system switches to high-speed operation. The two-speed heating capability lowers the system balance point significantly (from 32°F to 17°F in this example), thus providing a much more economical heating operation with a minimum of cycling. Additionally, if the extra cooling capacity is needed (e.g., during a large party) the heat pump system has the high-speed cooling backup capacity to meet the added requirement.

In a southern climate the heat pump is sized so that the design cooling load (at 100°F in this example) is met by the high-speed cooling capacity. The low-speed cooling operation is used at more moderate outdoor temperatures, ensuring good sensible and latent cooling with a minimum of compressor cycling. As the outdoor temperature increases, the high-speed cooling capacity is used to meet the load. During the winter, most of the heating hours are at more moderate outdoor temperatures (between 35 and 60°F), and the heat pump operates at low speed. This means less compressor

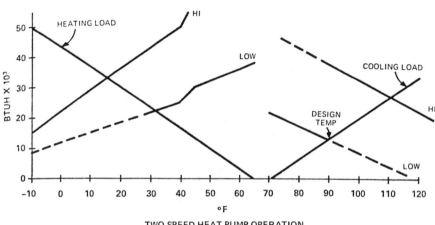

TWO SPEED HEAT PUMP OPERATION

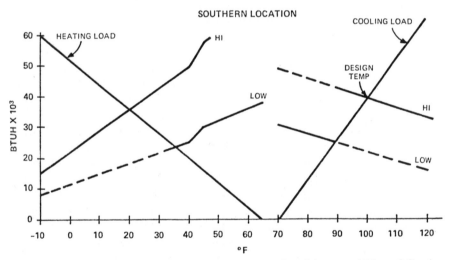

Fig. 3.44. Two-speed heat pump operation. Reproduced by permission of Carrier Copyright © Carrier Corporation 1978.

cycling for more reliable and efficient operation. If winter temperatures dip severely, the high-speed heat pump energizes to maintain the thermostat setpoint and minimize the use of costly electrical resistance heat.

In the case of a tightly insulated, low-infiltration house where low sensible heat and high latent heat loads in the summer are typical, the two-speed heat pump can meet both the sensible and latent requirements with the split evaporator coil. This coil has two separate refrigerant circuits, one of which is shut off during low-speed cooling. The result is a colder coil in the operating half, and thus a higher latent heat capacity. This coil also allows high

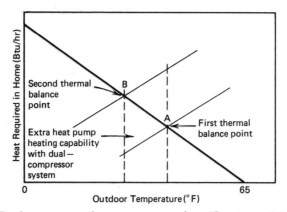

Fig. 3.45. Dual-compressor heat pump operation. (Courtesy of Carrier Corp.)

airflow to meet circulation requirements without sacrificing the latent heat capacity of the equipment.

If properly applied, the two-speed heat pump can solve many sizing problems, provide more reliable and efficient operation, and assure the homeowner comfortable conditions during all seasons.

Carrier Corporation introduced the Weathermaster III as the first commercially available dual-compressor heat pump designed for northern climates. At temperatures where conventional heat pumps switch to more costly electrical resistance heat, Weathermaster III turns on its second compressor to boost heating capacity. According to the manufacturer, both compressors work until a second, lower, thermal balance point is reached before resorting to standby electric heat. The diagram in Fig. 3.45 shows the extra heating provided by the dual-compressor heat pump. Conventional single-compressor heat pumps cannot fully satisfy indoor comfort needs at thermal balance point *A*. The second compressor extends high-efficiency heating to a second, lower thermal balance point *B*.

In climates where the need for heating exceeds the need for cooling, the dual-compressor feature also eliminates the need for oversize cooling capacity. One compressor is used for cooling, two for heating [11].

REFERENCES

1. American Society of Heating, Refrigerating and Air Conditioning Engineers, Handbook and Product Directory, 1976, Systems, ASHRAE, New York, 1976, Chap. 11.

2. Air-Conditioning and Refrigeration Institute Standard for Unitary Heat Pump Equipment, ARI Standard 240-77, 1977.

3. *Directory of Certified Unitary Heat Pumps,* Air-Conditioning and Refrigeration Institute, Arlington, VA, June 30, 1976.

4. Christian, J. E., "Unitary Air-to-Air Heat Pumps," *ICES Technol. Evaluations,* Report ANL/CES/TE 77-10, Argonne National Laboratory, July 1977.

5. Bose, J. E., "Design and Field Testing of Solar-Assisted Earth Coils," Project Report DE-AS03-79CS30210, 1 Aug. 1978–31 July 1980, Annual DOE Active Solar Heating and Cooling Contractors' Review Meeting, March 1980.

6. Comly, J. B., and Jaster, H., "Heat Pumps and Solar Energy—A Primer," General Electric Co. Rep. 78CRD251, Dec. 1978.

7. Andrew, J. W., "Optimization of Solar Assisted Heat Pump Systems Via a Simple Analytic Approach," *Proc. 2nd Ann. Systems Simulation Conf.,* BNL 27316, San Diego, CA, Jan. 23–25, 1980.

8. American Society of Heating, Refrigerating and Air Conditioning Engineers, Handbook and Product Directory, 1979, Equipment, ASHRAE, New York, 1979, Chap. 43.

9. NESCA, "Heat Pump Equipment Selection and Application," *Manual H,* National Environmental Systems Contractors Association, Washington, DC, 1977.

10. Blatt, M. H., and R. C. Erickson, "State-of-the-Art Assessment of Hybrid Heat Pumps," *EPRI Rep.* EM-1261, Res. Proj. 1201-6, December 1979.

11. Carrier Corporation, "A Guide for Residential Heat Pumps with Selection and Installation Procedures," GT 14-01, Carrier Corp., Syracuse, N.Y., 1978.

4

HEAT PUMP COMPONENTS

The basic components of heat pumps are compressors, heat exchangers, and miscellaneous refrigeration components (reversing valves, expansion devices, etc.). These components are directly related to what has been developed in the low-temperature refrigeration industry and the air-conditioning industry during the past 50 years.

Since heat pump yearly operating hours are often up to five times that of a cooling unit, and since this operation extends over a far greater range of system operating conditions, it is imperative that the total system design criteria be analyzed completely to assure maximum reliability. Improved components and protective devices contribute to better reliability, but care must be exercised in selecting components approved for the duty to which they are applied. This chapter discusses the major components and points out the characteristics or special considerations that apply to the heat pump field.

4.1. COMPRESSORS

The compressor is the heart of the refrigeration or heat pump system. The high operating temperatures and wide range of operating conditions likely to be met in space-heating heat pumps make greater demands on compressors than do air-conditioning applications, so that purpose-designed heat pump compressors are generally more reliable than units designed primarily for the smaller range of temperatures required in air-conditioning or cooling applications.

4.1.1. Compressor Types

Figure 4.1 shows schematically various types of refrigeration and heat pump compressors. *Reciprocating* compressors are used more than any other type for systems in the general range of 0.5–100 tons, and larger. Reciprocating compressors are used in unitary heat pumps and, in most cases, are either fully hermetic or accessible hermetic and should be designed and approved for this usage. The open compressor (Fig. 4.2) has no motor, base, or controls. The hermetic compressor (Fig. 4.3) has an integral motor with a separate control panel.

The internal parts of a basic hermetic compressor are shown in Fig. 4.4. This particular compressor is a Tecumseh CL compressor, which is built in the 3–5-ton range. It will have a total oil charge of 55 fluid oz. Some smaller air-conditioning compressors will have an oil charge of about 45 fluid oz. The oil level, with the compressor not running, will be approximately to the bottom of the lower cylinder (Fig. 4.5). On startup, the oil will be changed into a fine mist and mixed with refrigerant, which is then carried through the entire system. For this reason it is important that the lines be sized and bent so as not to restrict the flow of oil returning to the compressor crankcase.

Suction gas enters the compressor shell, usually at the top, through the

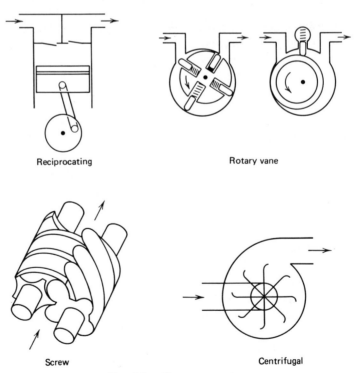

Reciprocating   Rotary vane

Screw   Centrifugal

Fig. 4.1. Compressor types.

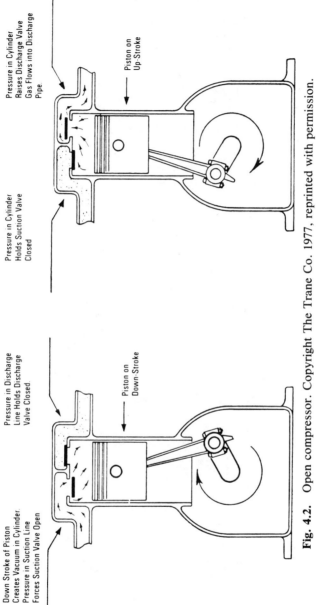

Pressure in Cylinder
Raises Discharge Valve
Gas Flows into Discharge
Pipe.

Piston on
Up-Stroke

Pressure in Cylinder
Holds Suction Valve
Closed

Pressure in Discharge
Line Holds Discharge
Valve Closed

Piston on
Down-Stroke

Down Stroke of Piston
Creates Vacuum in Cylinder.
Pressure in Suction Line
Forces Suction Valve Open

Fig. 4.2. Open compressor. Copyright The Trane Co. 1977, reprinted with permission.

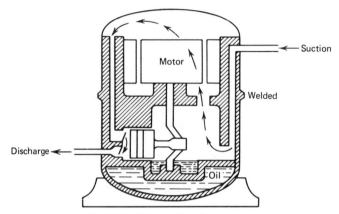

Fig. 4.3. Hermetic compressor.

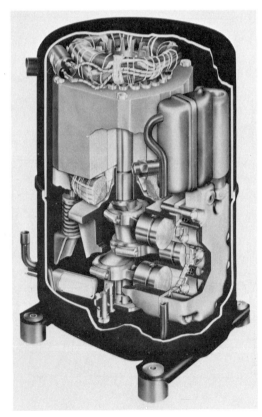

Fig. 4.4. Cutaway view of hermetic compressor. Courtesy Tecumseh Products Co.

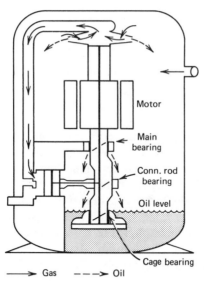

Motor

Main
bearing

Conn. rod
bearing

Oil level

Cage bearing

⟶ Gas   ---⟶ Oil

Fig. 4.5.   Oil and gas flow paths.

suction tube. The suction tube opens directly into the compressor shell, and there is no additional tubing. Thus, the compressor shell is subjected to the low side pressures of the system which, for R-22, will run between 75 and 85 psig. Suction gas totally fills the inside of the shell and removes some of the heat of the motor. It is drawn into the cylinder through the oil separator.

To control compressor capacity, *hot gas bypass* or *cylinder unloading* may be used. The preferred method, cylinder unloading, uses a mechanism which *automatically holds open some cylinder suction valves when less than full capacity is required.* Cylinder unloading provides unloaded starting, requires less power, has a wider capacity range, and is quiet in operation. *Multistep cylinder unloading* reduces compressor capacity in predetermined steps, thereby maintaining evaporator-compressor balance.

An example of a useful presentation of capacity data is given in Fig. 4.6. This is a typical set of curves for a four-cylinder semihermetic compressor, $2\frac{3}{8}$-in. bore, $1\frac{3}{4}$-in. stroke, 1720 rpm, operating with R-22. A set of power curves for the same compressor is also shown. Figure 4.7 shows the heat rejection curves for the same compressor. Compressor curves should be labeled with the following information:

1. Compressor identification.
2. Degrees subcooling, or statement that data has been corrected to zero degrees subcooling.
3. Compressor speed.
4. Type of refrigerant.

5. Suction gas superheat.

6. Compressor ambient temperature.

7. External cooling requirements (if required).

8. Maximum power or maximum operating conditions.

9. Minimum operating conditions at fully loaded and fully unloaded operation.

The *centrifugal* compressor characteristics essentially do not meet the needs of air-source heat pumps. High compression ratios or high lifts, associated in many instances with low gas volume resulting from low-load conditions, cause the centrifugal to surge. Therefore, most centrifugal applications in heat pumps have been limited to heat-transfer systems, storage systems, and hydronically cascaded systems. With these applications, the centrifugal compressor enables the heat pumps to enter the field of the large multistory buildings, and installations with double-bundle condensers up to 7500 tons have been accomplished successfully. The transfer cycles permit low com-

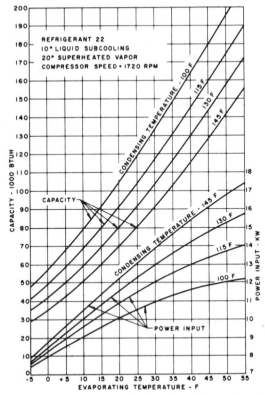

Fig. 4.6. Typical capacity and power input curves for a hermetic reciprocating compressor. Reprinted by permission from *ASHRAE 1975 Equipment Handbook*.

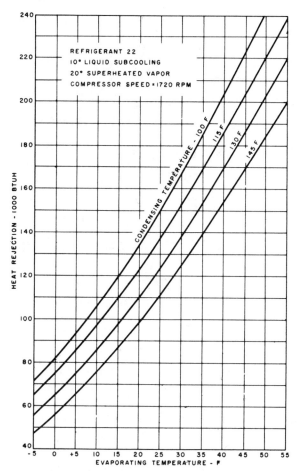

Fig. 4.7. Typical heat-rejection curves for a hermetic reciprocating compressor. Reprinted by permission from *ASHRAE 1975 Equipment Handbook*.

pression ratios, and many single- and two-stage units with R-11 and R-12 are operational. When heating water at a higher temperature than usual is desired (such as 130°F), three-stage machines have been used. A cutaway view of a centrifugal compressor is shown in Fig. 4.8.

A *rotary* compressor has characteristics similar to a reciprocating compressor except that it has low clearance and high volumetric efficiency. From this standpoint it is well suited to heat pump service, providing about 30% greater capacity at 0–110°F lift than a medium-clearance reciprocating compressor. However, this characteristic also tends to increase the power demand at the maximum cooling load conditions.

Although most manufacturers of unitary heat pumps utilize welded hermetic reciprocating motor compressors in their heat pump equipment, a

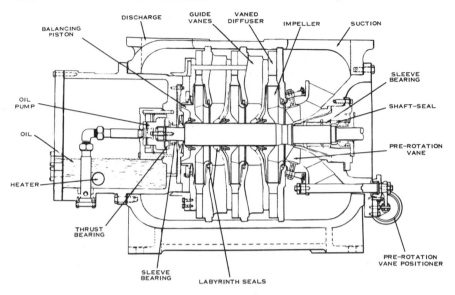

Fig. 4.8. Centrifugal refrigeration compressor. Reprinted by permission from *ASHRAE 1979 Equipment Handbook*.

split-system product with a welded hermetic rotary motor compressor has been introduced in the market. Advantages claimed for the rotary as opposed to the reciprocating positive-displacement compressor are (1) simplicity of construction (fewer moving parts, more rugged rotor, single discharge valve, etc.), (2) better ability to withstand liquid slugging, and (3) denser suction gas, since suction gas motor cooling is not used. The greater amount of refrigeration oil claimed as a benefit as far as compressor protection is concerned can serve to reduce refrigeration capacity and efficiency elsewhere in the system by decreasing evaporation rates and inhibiting heat transfer. Other disadvantages of the rotary compressor are (1) the greater possibility of leakage of high-pressure gas to the low-pressure side past rotor and vane clearances, (2) greater heat losses to ambient air due to a "hotshell" design (in the case in question), and (3) the possibility of clearance volume increases (with attendant volumetric efficiency loss) due to vane and system wear.

At present, reciprocating compressors are most widely used, with rotary compressors being restricted to the first stage of a staged-compression system and not used during the cooling cycle. Designers also should investigate the merits of *screw-type* compressors, which tend to offer high lift characteristics, even at low capacities.

In Fig. 4.9, a rotating vane, a rolling piston, and a screw-type rotary compressor are illustrated. Figure 4.10 shows a packaged heat pump with a rotary screw compressor.

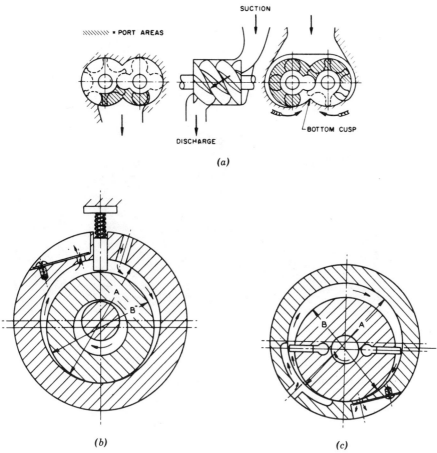

Fig. 4.9. Types of rotary compressors. (*a*) Rolling piston. (*b*) Helical rotary (screw). (*c*) Rotating vane. Reprinted by permission from *ASHRAE 1975 Equipment Handbook*.

4.1.2. Compressor Performance

One of the important thermodynamic considerations for the positive-displacement compressor is the effect of the clearance volume (i.e., the volume occupied by the refrigerant within the compressor that is not displaced by the moving member). This effect is illustrated in the case of the piston-type compressor by considering the clearance volume between the piston and the cylinder head when the piston is in a top dead-center position. The clearance gas remaining in this space after the compressed gas is discharged from the cylinder reexpands to a larger volume as the pressure falls to the inlet pressure (see Fig. 4.11). As a consequence, the mass of refrigerant discharged from the compressor is less than the mass that would occupy the volume

Fig. 4.10. Dunham-Bush DPX rotary screw compressor. Reprinted with permission—Dunham-Busch, Inc. 1982.

swept by the piston, measured at the inlet pressure and temperature. This effect is quantitatively expressed by the volumetric efficiency, η_v.

$$\eta_v = \frac{m_a}{m_i} \tag{4.1}$$

where m_a = actual mass of new gas entering the compressor per stroke

m_i = theoretical mass of gas represented by the displacement volume and determined at the pressure and temperature at the compressor inlet

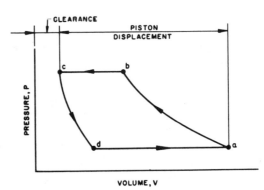

Fig. 4.11. Machine cycle for an idealized piston compressor. Reprinted by permission from *ASHRAE 1972 Handbook of Fundamentals*.

If the effect of clearance alone is considered, the resulting expression may be termed *clearance volumetric efficiency*. The expression used for grouping into one constant all the factors affecting efficiency may be termed *total volumetric efficiency*.* The clearance volumetric efficiency may be calculated with reasonable accuracy. The total volumetric efficiency is best obtained by actual laboratory tests, although a fair approximation to it may be calculated if sufficient data are available. For the simple cycle, the volumetric efficiency becomes

$$\eta_v = \frac{V_c - V_d}{V_a - V_c} = 1 + C - C\left(\frac{v_1}{v_2}\right)$$

(4.2)

where $C = V_c/(V_a - V_c)$ = clearance ratio
v_1 = specific volume of refrigerant at beginning of compression
v_2 = specific volume at end of compression

Volumetric efficiency is a measure of how well the piston displacement (size) of the compressor is utilized in moving refrigerant vapor through the cycle. The choice of refrigerant greatly affects v_1 and thus the mass flow that a given compressor displacement can deliver, since one of the most significant differences in modern refrigerants is the specific volume v_1 at a given evaporator temperature and pressure. The thermodynamic cycle for an actual single-stage system may depart significantly from the theoretical cycle. The principal departure occurs in the compressor. The most important losses are

1. Pressure drop within the compressor:
   a. Through suction, or discharge (or both), shutoff valves
   b. Across suction strainer
   c. Across motor (hermetic compressor)
   d. In manifolds, suction, and discharge
   e. Valves and valve plate ports (suction and discharge)
   f. Internal muffler
2. Heat gain to refrigerant from:
   a. Hermetic motor
   b. Friction
   c. Heat of compression—heat exchange within compressor

---

*To date various authorities are not in agreement on the best method of subdividing the volumetric efficiency. The expressions here noted are only two of those encountered. Others are *volumetric efficiency due to superheating*, usually found as an empirical equation attempting to evaluate the effect of suction-gas heat absorption from the cylinder walls; *conventional volumetric efficiency* or *theoretical volumetric efficiency*, other terms for the clearance volumetric efficiency; *compression efficiency*, a measure of the deviation of the actual compression process from the adiabatic; *volumetric efficiency without clearance*, an empirical grouping and evaluation of all efficiency factors other than the clearance volumetric efficiency; and *leakage volumetric efficiency*, an empirical evaluation of the valve and piston-ring leakage.

3. Valve inefficiencies due to imperfect mechanical action
4. Internal gas leakage
5. Oil circulation. Excessive quantity adversely effects compressor efficiency and lowers heat-transfer rates.
6. Clearance. The volume remaining upon completion of the compression stroke reexpands on the suction stroke and limits the intake. Hence, the greater the clearance volume, the lower the pumping or volumetric efficiency.
7. Deviation from isentropic compression. When considering the ideal compressor, an isentropic compression cycle is assumed. In the actual compressor, the compression cycle deviates from isentropic compression, due mainly to fluid and mechanical friction within the cylinder. The actual compression cycle and work of compression must be determined from an indicator card.

Figure 4.12 shows schematically a cycle closer to the actual. Pressure drop in compressor valves and heating of the vapor on the intake stroke are assumed. The compression process is assumed to be polytropic. Pressure drops in the condenser, evaporator, and piping are neglected. Heat addition in the compressor suction line and heat rejection in the compressor discharge line are assumed.

Because of valve pressure drop, the cylinder pressure during intake, $p_a = p_b$, may be less than the suction-line pressure p_3, and the cylinder pressure during discharge, $p_c = p_d$, may be greater than the discharge-line pressure p_4. Because of heat exchange between the vapor and the cylinder walls, the thermodynamic state of the vapor at position a may be different from that at position b, and the state at position c may be different from that at position d.

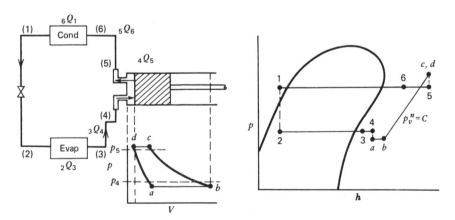

Fig. 4.12. Schematic diagrams for practical single-stage cycle. (Adapted from Threlkeld [7]).

The compressor volumetric efficiency η_v, defined as the mass of vapor actually pumped by the compressor divided by the mass of vapor that the compressor could pump if it handled a volume of vapor equal to its piston displacement and if no thermodynamic state changes occurred during the intake stroke, may be expressed by

$$\eta_v = \frac{(V_b - V_a)v_3}{(V_b - V_d)v_b} \tag{4.3}$$

But

$$V_b - V_a = (V_b - V_d) - (V_a - V_d)$$

$$V_a = V_d \left(\frac{p_d}{p_a} \right)^{1/n} = V_d \left(\frac{p_c}{p_b} \right)^{1/n}$$

If the clearance is defined as

$$C = \frac{V_d}{V_b - V_d} \tag{4.4}$$

then

$$\frac{V_b - V_a}{V_b - V_d} = 1 + C - C \left(\frac{p_c}{p_b} \right)^{1/n}$$

and

$$\eta_v = \left[1 + C - C \left(\frac{p_c}{p_b} \right)^{1/n} \right] \frac{v_3}{v_b} \tag{4.5}$$

This expression accounts for the three principal factors that affect compressor volumetric efficiency: reexpansion of clearance vapor, pressure drop in suction and discharge valves, and heating of the vapor on the intake stroke. If $C = 0$, the term in brackets is unity. When $C > 0$, drops in valve pressure also contribute to decreasing the term in brackets. The quantity v_3/v_b may be less than unity because of a pressure drop in the suction valve and because of cylinder-wall heating effects.

The total volumetric efficiency of a compressor is best obtained by actual laboratory measurements of the amount of refrigerant compressed and delivered to the condenser. It is too difficult to predict the effects of wire-drawing, cylinder-wall heating, and piston leakage to allow any degree of accuracy in most cases.

The total volumetric efficiency can be *approximated* if the pressure drop through the suction valves and the temperature of the gases at the end of the suction stroke are known and if it is assumed that there is no leakage past the pistons during compression. This approximation may be computed by

$$\eta_{tv} = \left[1 + C - C \left(\frac{p_d}{p_s} \right)^{1/n} \right] \left(\frac{p_c}{p_s} \right) \left(\frac{T_3}{T_c} \right) \tag{4.6}$$

TABLE 4.1.   Isentropic Exponent (c_p/c_v) for Various Refrigerant Vapors

Refrigerant	R-11	R-12	R-22	R-502
Vapor temperature, °F	86	50	86	50
c_p/c_v	1.11	1.13	1.16	1.135

where the subscript c refers to the compressor cylinder and s refers to the evaporator or the suction line just adjacent to the compressor. The temperature rise of the gases passing from the suction line into the cylinders is difficult to evaluate. If tests are to be made to determine this temperature rise, it is usually more practical to run complete tests and determine the total volumetric efficiency directly. Equation (4.6) applies only to calculations for the determination of the piston displacement required per ton of refrigeration and then only *when v_a and p_s are at evaporator and not cylinder conditions and p_d is at condenser pressure.*

Table 4.1 shows values of the isentropic exponent for several refrigerants. This exponent may be used as an approximation to the actual exponent *n*, where more accurate data are not known.

A compressor normally employed for comfort cooling use will have a clearance volume (ratio of gas volume remaining in cylinder after compression stroke to total swept-cylinder volume) of about 5%. A decrease in capacity at low evaporator temperatures corresponding to low heat-source temperatures is also evident, as well as a decrease in power at low evaporator and condensing temperatures. If the compressor has low clearance volume, for example, 2.5%, then it is more suitable for low-temperature operation and will provide, for example, about 15% greater refrigerating capacity at an evaporator temperature of 0°F and a condenser temperature of 110°F. However, this compressor has somewhat more power demand under maximum cooling load conditions than one of medium clearance.

It is obvious that more total heat capacity can be obtained at low outdoor temperatures by deliberately oversizing the compressor. When this is done, it may be necessary to provide some type of capacity reduction by means of two-speed motor drives, cylinder cutouts, or other methods.

The disadvantage of this arrangement is that the greater number of operating hours that occur at the higher suction temperatures must be served with the compressor in the unloaded condition, which generally causes lower efficiency. Therefore, the annual operating cost will tend to rise. It is also true that the additional initial cost of the oversized compressor must be economically justified by the gain in heating capacity. One method proposed for increasing the heating output at low temperatures involves the use of *staged compression,* in which one compressor may pump from −20°F suction temperature to 40°F condensing temperature, and a second compressor compresses the vapor from 40 to 120°F. In such an arrangement, it is possible to interconnect any two compressors so that they are in parallel, both

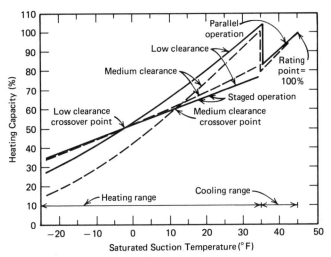

Fig. 4.13. Comparison of parallel and staged operation. Capacities of medium- and low-clearance staged compressors are based on rating point capacities for the respective compressors in parallel.

pumping from approximately 45 to 120°F at the normal cooling rating point, while at some predetermined outdoor temperature for heating they are reconnected so that they pump in staged relationship.

Figure 4.13 shows the performance of such a pair of compressors for compressors of both medium and low clearance volume. It is apparent that at low suction temperatures the reconnection into a staged relationship does provide some added capacity. It should also be understood that the motor selection involved must be based on the maximum loading conditions for summer operation, even though the low-stage compressor has a greatly reduced power requirement under the heating condition.

The COP will be approximately proportional to the ratio of the refrigeration capacity when comparing compressors that are coupled in parallel to compressors that are compound-staged, at any set of conditions, depending somewhat on the motor characteristics when lightly loaded.

EXAMPLE 4.1

Saturated R-12 vapor at 10°F enters the compressor of a single-stage system. Discharge pressure is 200 psia. Pressure drop in the compressor suction valves is 2 psi and through the discharge valves is 4 psi. The vapor is superheated by 20°F in the cylinder during the intake stroke. Compressor clearance is 4%. Determine (a) the volumetric efficiency and (b) the compressor pumping capacity for a piston displacement of 50 cfm.

Solution. From the data,

$C = 0.04$

$p_c = 200 + 4 = 204$ psia

$p_b = 29.3 - 2 = 27.3$ psia

$v_3 = 1.324$ ft^3/lb (Table A-1 at 10°F)

At $t_b = 30$°F and $p_b = 27.3$ psia,

$$v_b = 1.5 \text{ ft}^3/\text{lb (Fig. A-1)}$$

Assuming that $n = c_p/c_v = 1.13$,

$$\eta_v = \left[1 + 0.04 - (0.04)\left(\frac{204}{27.3}\right)^{1/1.13}\right]\frac{1.324}{1.5} = 0.71$$

$$m = \frac{\eta_v(\text{P.D.})}{v_3} = \frac{(0.71)(50)}{1.324} = 26.8 \text{ lb/min}$$

4.2. HEAT-TRANSFER COMPONENTS

4.2.1. Heat Exchangers

A heat exchanger is a device that permits the transfer of heat from a warm fluid to a cooler fluid through an intermediate surface and without mixing of the fluids. Simple heat-exchanger schematics are given in Fig. 4.14.

The correct choice of heat exchangers and their sizing is probably the most important factor in designing an efficient and economic heat pump. It is also in some ways the most difficult. The complex geometries of heat exchangers, together with wide variations in operating conditions, make precise calculation of sizes from basic physical principles impracticable, and various empirical factors have to be applied to heat-transfer relationships that have themselves been determined by experiment. Although such factors may be well established for available heat-exchanger geometries, details may be difficult to obtain from exchanger manufacturers anxious to maintain their competitive position.

Whether the heat exchanger is selected as an off-the-shelf item or designed especially for the application, the following factors are almost always taken into consideration:

1. Heat-transfer requirements
2. Cost
3. Physical size
4. Pressure-drop characteristics

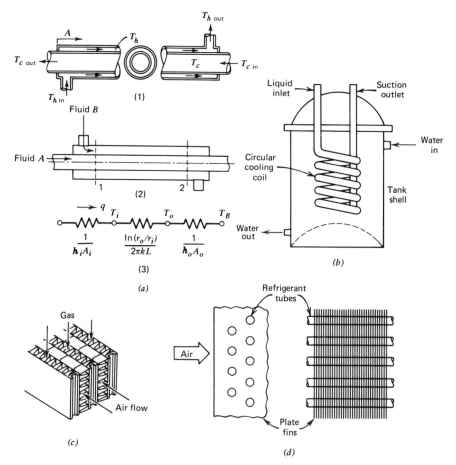

Fig. 4.14. (a) Double-pipe heat exchanger. *1*, diagram of simple counterflow exchanger; *2*, schematic of parallel-flow exchanger; *3*, thermal resistance network for either. (b) Shell-and-coil exchanger. (c) Plate-type heat exchanger. (d) Finned-coil (continuous fin) heat exchanger.

When heat is exchanged between two fluids flowing continuously through a heat exchanger, the heat transfer is usually calculated using the familiar rate equation

$$q = UA \, \Delta t_m \tag{4.7}$$

where U is the overall coefficient of heat transfer from fluid to fluid, A is an area associated with the coefficient U, and Δt_m is the mean temperature difference.

The temperature of fluids in a heat exchanger generally varies from point to point as heat flows from the hotter to the colder fluid. Thus, no single temperature difference exists as the driving force for the heat transfer. To

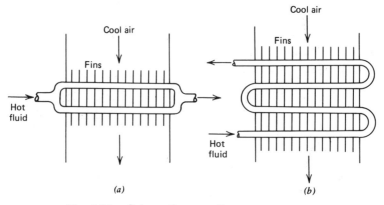

Fig. 4.15. Schematic cross-flow arrangements.

determine the rate of heat transfer with this changing temperature difference, the equation

$$dq = U \, dA \, \Delta t$$

must be integrated over the heat-transfer area A along the length of the heat exchanger. Carrying out this integration for either concurrent or countercurrent flow patterns yields

$$q = UA \frac{\Delta t_a - \Delta t_b}{\ln(\Delta t_a / \Delta t_b)} \tag{4.8}$$

where the subscripts a and b refer to the respective ends of the exchanger. It is usually convenient to use an average effective temperature difference Δt_m, called the logarithmic mean temperature difference, which must then be evaluated as

$$\Delta t_m = \frac{\Delta t_a - \Delta t_b}{\ln(\Delta t_a / \Delta t_b)} \tag{4.9}$$

When air is one of the fluids, pure counterflow is generally not practicable. The most economical heat exchanger is usually the finned-tube type employing some form of *cross-flow* arrangement as shown in Fig. 4.15.

It is convenient to express the mean temperature difference Δt_m for a cross-flow heat exchanger as

$$\Delta t_m = F \, \Delta t_{m,cf} \tag{4.10}$$

where F is a correction factor and $\Delta t_{m,cf}$ is the logarithmic mean temperature difference calculated for *pure counterflow*. Cross-flow correction factors are given in Figs. 4.16 and 4.17.

If one fluid temperature remains constant, the logarithmic mean tempera-

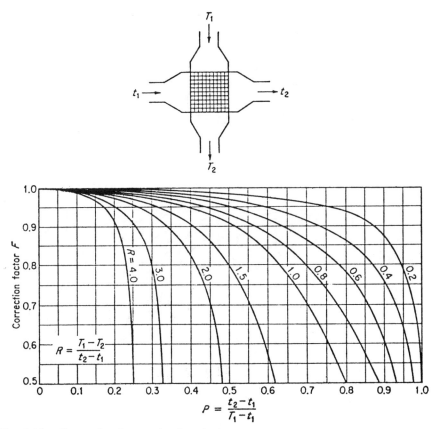

Fig. 4.16. Correction-factor plot for single-pass cross-flow exchanger, both fluids unmixed. (Extracted from "Mean Temperature Difference in Design," by R. A. Bowman, A. C. Mueller, and W. M. Nagel, published in *Trans. ASME*, **62** (1940), with permission of the publishers, the American Society of Mechanical Engineers.)

ture difference applies regardless of the flow arrangement. The overall heat-transfer coefficient may be based on either the inside or outside area of the tube at the discretion of the designer.

After a period of operation, the heat-transfer surfaces for a heat exchanger may become coated with various deposits present in the flow systems, or the surfaces may become corroded as a result of the interaction between the fluids and the material used for construction of the heat exchanger. In either event, this coating represents an additional resistance to the heat flow and thus results in decreased performance. The overall effect is usually represented by a fouling factor, or fouling resistance, R_f, which must be included along with the other thermal resistances making up the overall heat-transfer coefficient.

Resistance due to fouling depends upon the material of which the tube is

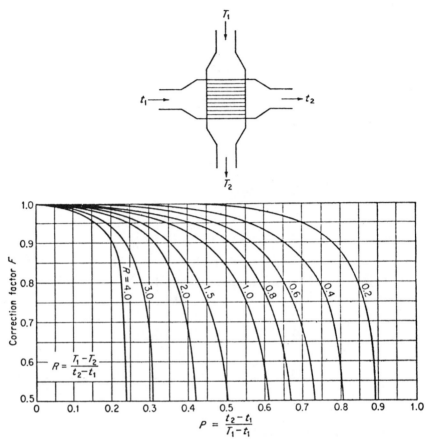

Fig. 4.17. Correction factor to counterflow Δt_m for cross-flow heat exchangers, fluid on shell side mixed, other fluid unmixed, one tube pass. (Extracted from "Mean Temperature Difference in Design," by R. A. Bowman, A. C. Mueller, and W. M. Nagel, published in *Trans. ASME*, **62** (1940), with permission of the publishers, the American Society of Mechanical Engineers.)

made, nature of the fluid, and fluid velocity. Copper tubes usually become fouled at a slower rate and to a lesser degree than steel tubes. Fluid velocities less than approximately 3 ft/s promote additional fouling by permitting static accumulation of deposits. Recommended fluid velocities for copper and steel tubes are 5–10 ft/s, provided pressure drop limitations are not exceeded. Table 4.2 lists recommended fouling factors, expressed as R_f.

Water and brine coolers operating at normal conditions do not require the addition of a fouling resistance to the refrigerant side. Caution should be exercised when designing evaporators for service below $-20°F$, because the oil that is circulating through the refrigerant circuit may congeal on the tube surface and add an appreciable resistance to heat flow. This is particularly true of the direct-expansion cooler.

TABLE 4.2. Fouling Factors

Type of Fluid	Fouling Factor, R_f (hr·ft²·°F/Btu)
Treated boiler feed water	0.001
Fuel oil	0.005
Alcohol vapors	0.0005
Steam, non-oil-bearing	0.0005
Industrial air	0.002
Refrigerating liquid	0.001

Comprehensive tables of fouling resistances are given in reference 2. The fouling factors should be applied as indicated in the following equation for the overall design heat-transfer coefficient U_o:

$$U_o = \frac{1}{1/h_o + R_o + R_k + R_i A_o/A_i + A_o/h_i A_i} \qquad (4.11)$$

where U_o = design overall coefficient of heat transfer based on a unit area of the outside tube surface

h_o = average unit-surface conductance of the fluid on the outside of tubing

h_i = average unit-surface conductance of fluid inside tubing

R_v = unit fouling resistance on outside of tubing

R_i = unit fouling resistance on inside of tubing

R_k = unit resistance of tubing

A_o/A_i = ratio of outside tube surface to inside tube surface

For preliminary estimates of heat-exchanger sizes and performance parameters, it is often sufficient to know the order of magnitude of the overall transmittance under average service conditions.

Due to the many variables that enter into the determination of overall heat-transfer rates, Table 4.3 gives only approximate minimum and maximum values of U.

The higher values of U generally apply to loadings, Q/A, corresponding to a ΔT_m of 12°F or more, at the higher velocities. The lower values of U apply to loadings corresponding to ΔT_m of 8°F and less, and at low fluid velocity.

EXAMPLE 4.2

Water is cooled from 190°F to 150°F while passing through an exchanger at the rate of 20,000 lb/hr. Water entering at 90°F flows through the outer side of the heat exchanger at 30,000 lb/hr. The overall heat transfer coefficient for the exchanger (U) is 300 Btu/(hr·ft²·°F). Determine the areas required for (a) parallel flow and (b) counterflow.

TABLE 4.3. Overall Heat-Transfer Coefficients for Liquid Coolers [1]

Type of Evaporator	Minimum	Maximum
Flooded shell and plain tube		
Water to Refrigerants 12, 22, and 717	130	190
Brine to Refrigerant 717	45	100
Brine to Refrigerant 12 or 22	30	90
Flooded shell and finned tube		
Water to Refrigerant 12 or 22	90	140
Direct expansion, shell and plain tube		
Water to Refrigerant 12, 22, or 717		
(refrigerant in tubes)	80	160
Brine to Refrigerant 12, 22, or 717		
(refrigerant in tubes)	60	140
Direct expansion, shell and internal finned tubes		
Water to Refrigerant 12 or 22 (refrigerant in tubes)	160	250
Non-salt brines to Refrigerant 12 or 22	100	170
Shell and plain tube coil (water in shell)		
Refrigerant 12, 22, or 717 in coil	10	25
Baudelot cooler		
Flooded, Refrigerant 12 or 22 to water	100	200
Direct expansion, Refrigerant 717 to water	60	150
Refrigerant 12 or 22 to water	60	120
Double-pipe cooler, Refrigerant 717 to water	50	150
Refrigerant 717 to brine	50	125
Spray type shell-and-tube water coolers		
Refrigerants 11, 12, 22, 113, or 717	150	250
Tank and agitator, coil-type water cooler, flooded		
Refrigerant 717	80	125
Refrigerant 12 or 22	60	100
Tank, ammonia, Refrigerant 717 to brine cooling,		
coils between can in ice tank	15	40
Tank, high velocity raceway type, Refrigerant 717 to		
brine	80	110

Solution

$$q = 20,000(190 - 150) = 30,000(T_o - 90) = 800,000$$

$$T_o = 116.7°F$$

Parallel flow:

$$\Delta T_m = \frac{(190 - 90)(150 - 116.7)}{\ln(100/33.3)} = 60.6°F$$

$$A = \frac{q}{U \, \Delta T_m} = \frac{800,000}{(300)(60.6)} = 44 \text{ ft}^2$$

Counterflow:

$$\Delta T_m = \frac{(190 - 116.7) - (150 - 90)}{\ln(73.3/60)} = 66.6°F$$

$$A = \frac{800,000}{(300)(66.6)} = 40 \text{ ft}^2$$

4.2.2. Extended Surface Coils

Most heat pump coils consist of tubes with fins attached to their outer surface. The purpose of the fins is to increase the area on the air side, where the convection coefficient is generally much lower than on the refrigerant or water side. Refrigerant or water flows inside the tubes, and air flows over the outside of the tubes and the fins. When a refrigerant evaporates in the tubes, the coil is called a "direct-expansion" coil. When, on the other hand, a secondary refrigerant, such as chilled water, carries away the heat, it is chilled by an evaporator in the machine room. The air-conditioning systems in many large buildings use a central water chiller and distribute chilled water throughout the building.

Several terms denoting features of coil construction should be defined first. The *face area* of the coil is the cross-sectional area of the air stream at the entrance of the coil. The *face velocity* of the air is the volume or rate of airflow divided by the face area. The *surface area* of the coil is the heat-transfer area in contact with the air. The *number of rows of tubes* is the number of rows measured in the direction of airflow.

In contrast to bare-pipe coils of the same capacity, finned coils are much more compact, have a much smaller weight, and are usually less expensive. The secondary surface area of a finned coil may be 10–30 times or more than of the bare tubes.

The primary surface is the surface of the round tubes or pipes, which are arranged in a repetitive pattern with respect to the airflow. The secondary surface (fins) consists of thin metal plates or a spiral ribbon uniformly spaced or wound along the length of the primary surface, and in intimate contact with it for good heat transfer from primary to secondary surface.

The bond between secondary and primary surface is a most important factor in the performance of extended-surface coils. This bond must be maintained permanently to assure continuation of rated performance. The heat-transfer bond between the fin and the tube may be achieved in numerous ways, among which bonding the fins to the tubes mechanically is employed most frequently. The bonding is generally accomplished by expanding the tubes into the tube holes in the fins to obtain a permanent mechanical bond. The tube holes are frequently provided with a formed fin collar, which provides the area of thermal contact and may serve as a means of spacing the fins uniformly along the length of the tubes.

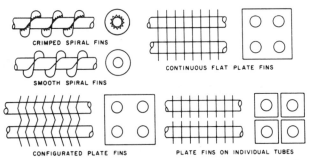

CRIMPED SPIRAL FINS

SMOOTH SPIRAL FINS

CONTINUOUS FLAT PLATE FINS

CONFIGURATED PLATE FINS

PLATE FINS ON INDIVIDUAL TUBES

Fig. 4.18. Types of fin-coil arrangements. Reprinted by permission from *ASHRAE 1975 Handbook of Fundamentals*.

The fins of spiral or ribbon-type fin coils are tension-wound onto the tubes. In addition, an alloy with a low melting point, such as solder, may be used to provide a metallic bond between fins and tube. Some types of spiral fins are knurled into a shallow groove on the exterior of the tube. Sometimes the fins are formed out of the material of the tube itself.

The fin designs most frequently used for heating coils are flat-plate fins, plate fins of special shapes, and spiral or ribbon fins, as illustrated in Fig. 4.18.

Copper and aluminum are the materials most commonly used in the fabrication of extended-surface coils. Tubing made of steel or various copper alloys is used in applications where corrosive forces might attack the coils from either inside or outside. The most common combination for low-pressure applications is aluminum fins on copper tubes.

Cooling coils for water or for volatile refrigerants most frequently have aluminum fins and copper tubes, although copper fins on copper tubes are also used, and the combination of aluminum fins on aluminum tubes is finding use. Adhesive bonding is sometimes used in making header connections, return bends, and fin-tube joints, particularly for aluminum-to-aluminum joints. There are many makes of cooling coils of the lightweight extended-surface type for both heating and cooling with tubes commonly $\frac{1}{4}$, $\frac{3}{8}$, $\frac{1}{2}$, $\frac{5}{8}$, $\frac{3}{4}$, and 1 in. outside diameter, and with fins spaced 3–14 per inch. The tube spacing generally varies from about $\frac{5}{8}$ to $2\frac{1}{2}$ in. on centers, depending upon the width of individual fins and on other considerations of performance. Fin spacing should be chosen for the duty to be performed, with special attention being paid to air friction, possibility of lint accumulation, and, especially at lower temperatures, the consideration of frost accumulation.

The addition of fins to the tubes greatly increases the outer surface area but at the expense of decreasing the mean temperature difference between the surface and the air stream. Whereas the thermal resistance of the bare tube may be negligible, the thermal resistance of the extended surface may be considerable.

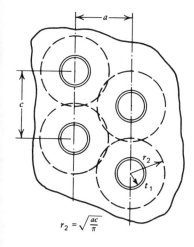

$$r_2 = \sqrt{\frac{ac}{\pi}}$$

Fig. 4.19. Approximation method for treating a rectangular-plate fin of uniform thickness in terms of a flat circular-plate fin of equal area.

A significant quantity in evaluating the thermal effectiveness of fins is the *fin efficiency* ϕ.

The rectangular-plate fin of uniform thickness is commonly used in finned coils for heating or cooling air. It is not possible to obtain an exact mathematical solution for the efficiency of such a fin. Carrier and Anderson [4] have shown that an adequate approximation is to assume that the fin area served by each tube is equivalent in performance to a flat circular-plate fin of equal area. Figure 4.19 shows the method where the equivalent outer radius of the circular fin is determined as

$$r_2 = \sqrt{\frac{ac}{\pi}} \tag{4.12}$$

The corresponding efficiency for the flat circular-plate fin can be obtained from Fig. 4.20.

Design problems with finned-tube heat exchangers involve solution of the Eq. (4.7):

$$q = U_o A_o \Delta t_m$$

where, for most cases,

$$U_o = \left[\frac{A_o}{A_{P,i} h_i} + \frac{A_o}{A_{P,i} h_{d,t}} + \frac{1 - \phi}{h_{c,o}(A_{P,o}/A_F + \phi)} + \frac{1}{h_{c,o}} \right]^{-1} \tag{4.13}$$

using the nomenclature given in Fig. 4.21.

The sensible heat-transfer capacity of a coil can also be expressed by the following basic equation:

$$q = U_o(\Delta t_m) K N A_a \tag{4.14}$$

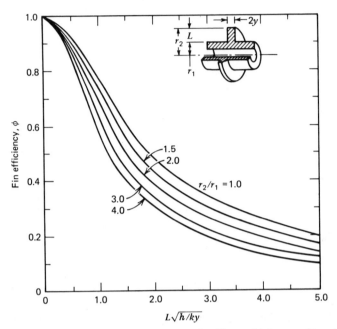

Fig. 4.20. Efficiency for a circular-plate fin of uniform thickness. [Reprinted from "Efficiency of Extended Surfaces," by K. A. Gardner, *ASME Trans.*, **67**, 625 (1945).]

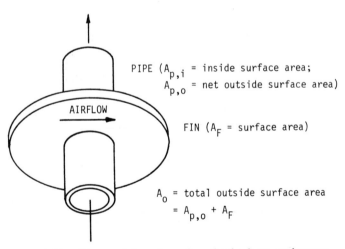

PIPE ($A_{p,i}$ = inside surface area;
$A_{p,o}$ = net outside surface area)

FIN (A_F = surface area)

A_o = total outside surface area
= $A_{p,o}$ + A_F

Fig. 4.21. Nomenclature for a finned-tube heat exchanger.

where q = total amount of heat transferred by the coil
U_o = overall heat-transfer coefficient of the coil
Δt_m = the mean temperature difference between the fluid flowing inside the coil tubes and the air passing over the coil
K = coil surface parameter, A_o/NA_a
N = number of rows of coil tubes in direction of air flow
A_a = coil face area, ft^2

In the case of dehumidifying coils, it is also important that the proper amount of surface area be installed to obtain the ratio of air-side sensible-to-total heat required for maintaining the air dry-bulb and wet-bulb temperatures in the conditioned space. The same room air conditions can be maintained with different air quantities (including outside and return air) through the coil. However, for a given total air quantity with fixed percentages of outside and return air, there is only one set of air conditions leaving the coil that will maintain the room air conditions. Once the air quantity and leaving air conditions at the coil have been selected, for a given coil surface design and arrangement, there is usually only one combination of face area, row depth, and air face velocity that will maintain the required room air conditions. Therefore, in making final coil selections, it is necessary to recheck the initial selection to assure that the leaving air conditions, as calculated by the coil selection procedure, will match those determined from the cooling-load estimate. This may involve a reselection with changes in air face velocity, coil size, and coil depth to obtain a more economical selection. The overall heat-transfer coefficient for a finned-tube coil for the sensible transfer of heat (heating or cooling) can be expressed by the following simplified basic equation:

$$\frac{1}{U_o} = \frac{B}{h_i} + \frac{BL}{k} + \frac{1}{h_a \phi_s} \tag{4.15}$$

where U_o = overall heat-transfer coefficient
h_i = film coefficient of heat transfer between the internal surface of the coil and the fluid flowing within the coil tubes
B = ratio between external and internal heat-transfer surface areas. This ratio is introduced into the equation in order to place the internal heat-transfer fluid coefficients on the basis of external surface.
h_a = heat-transfer film coefficient between the air and the external surface of the coil
k = conductivity of coil tube material
L = thickness of tube wall
ϕ_s = surface efficiency, expressed as a fraction

When the tube walls of the coil are relatively thin and made of materials having high conductivity, as is the case for all lightweight finned-tube heat transfer surfaces, the term BL/k in Eq. (4.15) becomes negligible and is usually disregarded.

For typical finned-tube coil designs, the ratio B expressing the ratio of the total external surface to the internal surface varies from approximately 10 to 35. Sometimes an allowance is made for imperfect bonding of the fins to the tubes, but this effect is difficult to evaluate, and with good construction it should be small. It is more important to include a deposit coefficient for the inside surface of the tubes.

A typical detailed method of rating and selecting dehumidification coils involves the use of certain interrelated factors:

1. Room sensible heat factor
2. Mean surface temperature
3. Coil bypass factor

Room Sensible Heat Factor (SHF).   This factor represents the ratio of the room sensible heat to the room total heat.

$$\text{SHF} = \frac{q_s}{q_s + q_l} \tag{4.16}$$

where q_s = room sensible heat, Btu/hr
q_l = room latent heat, Btu/hr

Mean Coil Surface Temperature.   For a given room condition and room sensible-heat factor, there is only one saturated air temperature that, when supplied to the conditioned space in the correct air quantity, will satisfy the room condition and will remove sensible and latent heat at the exact rates gained to the room. This is the temperature of the *effective air,* which has contacted the cooling surface and has been brought down to saturation at the mean surface temperature.

When the room sensible-heat factor is known, a surface temperature can be selected that will satisfy both the sensible and latent heat loads. This specific saturated air temperature can be determined by substituting in Eq. (4.16) for coil SHF as follows:

$$\text{SHF} = \frac{0.244(t_1 - t_2)}{0.244(t_1 - t_2) + 1076(W_1 - W_2)} \tag{4.17}$$

where W_1 = humidity ratio (moisture content) at room conditions, lb/lb
W_2 = humidity ratio (moisture content) at coil surface temperature, lb/lb
t_1 = room dry-bulb temperature, °F
t_2 = mean coil surface temperature, °F

The required mean surface temperature can also be calculated, when conditions of entering and leaving temperatures and coil bypass factor are known, by use of the equation:

$$\bar{h}_c = \bar{h}_e - \frac{\bar{h}_e - \bar{h}_i}{1 - F_b} \qquad (4.18)$$

where $\bar{h}_c$ = enthalpy of air at coil surface temperature
$\bar{h}_e$ = enthalpy of air at entering wet-bulb temperature
$\bar{h}_i$ = enthalpy of air at leaving wet-bulb temperature

Coil Bypass Factor (F_b).   In the usual application of a cooling surface, where condensation of moisture from the air is taking place, the air mixture leaving the coil has a temperature and moisture content somewhat higher than that equivalent to the surface temperature. This is due to the effect of a certain amount of air bypassing the coil surface. Bypassed air represents that portion of total air through the cooling coil that has not contacted the cooling surface and therefore has not been cooled or dehumidified to the coil surface temperature condition.

$$F_b = \frac{\text{bypassed air}}{\text{total air through coil}}$$

The bypass factor is determined from test data and is dependent on many factors, such as depth of coil, type of surface, and air velocity. Values for high-temperature conditioning coils, where 7–14 fins/in. are used, vary from 0.05 to 0.30 depending on fin spacing and number of rows deep. For applications where 3 fins per inch of tubing or prime surface pipe is used, bypass factors range from 0.25 for 8-row finned coils to 0.59 for 10-row prime surface-type coils. Stated in another way, the bypass factor is a measure of the relation between the temperature of the air mixture leaving the coil and the mean coil surface temperature (see Fig. 4.22).

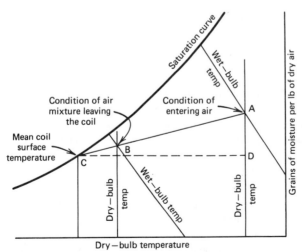

Fig. 4.22.   Psychrometric chart showing mean coil surface temperature, bypass factor, and room sensible heat factor (no ventilation air). Bypass factor $F_b = BC/AC$. Contact factor $= 1 - F_b = AB/AC$. Room sensible-heat factor $= CD/CA$.

TABLE 4.4.   Typical Bypass Factors for Finned Coils [3]

Depth of Coils (rows)	Without Sprays		With Sprays[a]	
	8 fins/in.	14 fins/in.	8 fins/in.	14 fins/in.
	Velocity (fpm)			
	300–700	300–700	300–700	300–700
2	.42–.55	.22–.38		
3	.27–.40	.10–.23		
4	.19–.30	.05–.14	.12–.22	.03–.10
5	.12–.23	.02–.09	.08–.14	.01–.08
6	.08–.18	.01–.06	.06–.11	.01–.05
	.03–.08		.02–.05	

[a]The bypass factor with spray coils is decreased because the spray provides more surface for contacting the air.

Table 4.4 contains bypass factors for a wide range of coils. This range is offered to provide sufficient latitude in selecting coils for the most economical system. Table 4.5 lists some of the more common applications with representative coil bypass factors. This table is intended only as a guide for the design engineer.

Figure 4.23 presents an example of manufacturer's data on unit sizes and bypass factors.

TABLE 4.5.   Typical Bypass Factors for Various Applications

Coil Bypass Factor	Type of Application	Example
0.30–0.50	A *small* total load or a load that is somewhat larger with a low sensible heat factor (high latent load)	Residence
0.20–0.30	Typical comfort application with a *relatively small* total load or a low SHF with a somewhat larger load	Residence, small retail shop, factory
0.10–0.20	Typical comfort application	Dept. store, bank, factory
0.05–0.10	Applications with high internal sensible loads or requiring a large amount of outdoor air for ventilation	Dept. store, restaurant, factory
0–0.10	All outdoor air applications	Hospital operating room, factory

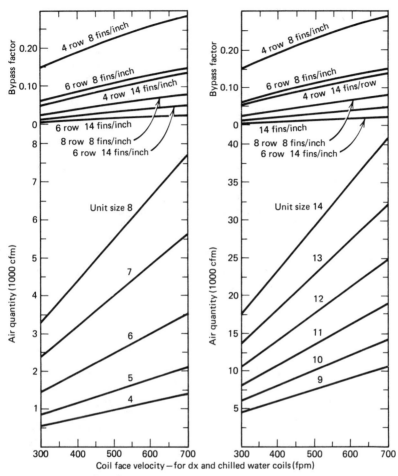

Fig. 4.23. Unit size and bypass factors.

The total refrigeration load, q_t, of a cooling and dehumidifying coil (or air washer) per pound of dry air is indicated in Fig. 4.24 and consists of the following components:

1. The sensible heat q_s removed from the dry air and moisture in cooling from entering temperature t_1 to leaving temperature t_2.
2. The latent heat q_L removed to condense the moisture at the dew-point temperature t_4 of the entering air.
3. The heat of subcooling q_w removed from the condensate in cooling it from the condensing temperature t_4 to the leaving condensate temperature t_3.

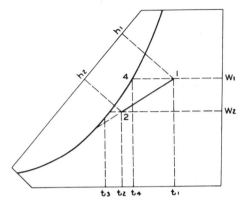

Fig. 4.24. Psychrometric performance of cooling and dehumidifying coil. Reprinted by permission from *ASHRAE 1975 Equipment Handbook.*

Items 1, 2, and 3 may be related by the equation

$$q_t = q_s + q_L + q_w \tag{4.19}$$

If only the total heat value is desired, it may be computed by

$$q_t = (h_1 - h_2) - (W_1 - W_2)h_{f3} \tag{4.20}$$

where h_1 and h_2 = enthalpy of air at points 1 and 2, respectively
W_1 and W_2 = humidity ratio at points 1 and 2, respectively
h_{f3} = enthalpy of saturated liquid at the final temperature t_3

If a breakdown into latent and sensible heat components is desired, the following relations may be used. The latent heat may be found from

$$q_L = (W_1 - W_2)h_{fg4} \tag{4.21}$$

where h_{fg4} = latent heat of water vapor at the condensing temperature t_4.

The sensible heat may be shown to be

$$q_s + q_w = (h_1 - h_2) - (W_1 - W_2)h_{g4} + (W_1 - W_2)(h_{f4} - h_{f3}) \tag{4.22}$$

or

$$q_s + q_w = (h_1 - h_2) - (W_1 - W_2)(h_{fg4} + h_{f3}) \tag{4.23}$$

where h_{g4} = $h_{fg4} + h_{f4}$ = enthalpy of saturated water vapor at the condensing temperature t_4
h_{f4} = enthalpy of saturated liquid at the condensing temperature t_4

The last term in Eq. (4.23) is the heat of subcooling the condensate from the condensing temperature t_4 to its final temperature t_3. Then,

$$q_w = (w_1 - w_2)(h_{f4} - h_{f3}) \tag{4.24}$$

The final condensate temperature t_3 leaving the system is subject to substantial variations, depending upon the method of coil installation, as affected by coil face orientation, airflow direction, air duct insulation, and so forth. In practice, t_3 is frequently the same as the leaving wet-bulb temperature. Within the normal air-conditioning range, precise values of t_3 are not necessary, since heat of the condensate, $(W_1 - W_2)h_{f3}$, removed from the air usually represents about 0.5–1.5% of the total refrigeration load.

EXAMPLE 4.3

A circumferential fin of rectangular cross section surrounds a 1-in. diameter tube. The length of the fin is $\frac{1}{4}$ in., and the thickness is $\frac{1}{8}$ in. The fin is constructed of mild steel, $k = 29$ Btu/(hr·ft²·°F) per foot. If air flows over the fin so that a convective heat-transfer coefficient of 5.2 Btu/(hr·ft²·°F) is experienced and the temperature of the base and air are 500 and 100°F, respectively, determine the heat transfer from the fin to the fluid.

Solution

$$r_i = \tfrac{1}{2} \text{ in.} = 0.0417 \text{ ft} \qquad r_o = \tfrac{3}{4} \text{ in.} = 0.0625 \text{ ft} \qquad \frac{r_o}{r_i} = 1.5$$

$$L = 0.25 \text{ in.} \qquad t = 0.125 \text{ in.}$$

$$L\left(\frac{2h}{Kt}\right)^{1/2} = \frac{0.25}{12}\left[\frac{2(5.2)}{(29)}\left(\frac{0.125}{12}\right)\right]^{1/2} = 0.122$$

$$\eta_{fin} \approx 0.98 = \frac{Q_{actual}}{Q \text{ if fin temp} = T_o}$$

$$A_{fin\ surface} = \pi[(0.0625)^2 - (0.0417)^2]2 + 2\pi(0.0625)\left(\frac{0.125}{12}\right) = 0.0177 \text{ ft}^2$$

$$Q_{actual} = 0.98[5.2(0.0177)(500 - 100)] = 36.1 \text{ Btu/hr}$$

4.2.3. Coil Ratings

Steam and hot water coils are usually *rated* within limits that may be exceeded for special applications.

Air face velocity	200–1500 fpm, based on air at standard density of 0.075 lb/ft³
Entering air temperature	−20°F to 100°F for steam coils; 0–100°F for hot water coils.
Steam pressures	2–250 psig at the coil stream supply connection (pressure drop through the steam control valve must be considered)
Hot water temperature	120–250°F
Water velocity	0.5–8 ft/s

Individual installations vary widely, but the following values can be used as a guide: The most common *air face velocities* used are between 500 and 600 fpm. *Delivered air temperatures* vary from about 72°F for ventilation only to about 150°F for complete heating. *Steam pressures* vary from 2 to 10 psig, with 5 psig most common. A minimum steam pressure of 5 psi is recommended for systems with entering air temperatures below freezing. *Water temperatures* for comfort heating are commonly between 180 and 200°F with *water velocities* between 4 and 6 fps.

Water quantity is usually based on about 20°F temperature drop through the coil. Air resistance is usually limited to between ⅜ and ⅝ in. of water for commercial buildings, and to about 1 in. of water for industrial buildings. High-temperature water systems have water temperatures commonly between 300 and 400°F, with up to 100°F drop through the coil.

Based on information in ARI Coil Standard 410-64 [5], dry surface (sensible cooling) coils and dehumidifying coils (which accomplish both cooling and dehumidification), particularly for field-assembled coil banks or factory-assembled central station type air conditioners using different combinations of coils, are usually rated within the following limits:

Entering air	65–100°F db, 60–85°F wb
Air face velocity	300–800 fpm (sometimes as low as 200 and as high as 1500)
Volatile refrigerant saturation temperature	30–55°F at coil suction outlet (refrigerant vapor superheat at coil suction outlet is 6° or higher)
Entering water temperature	35–65°F
Water quantity	1.2–6 gpm per ton (equivalent to a water temperature rise of 4–20°)
Water velocity	1–8 fps

However, for special applications, the range may be exceeded in the above design variables.

Heat-transfer capacity information given in manufacturers' catalogs is usually based on actual laboratory tests. To simplify coil selection, data are presented in the form of tables or charts.

Table 4.6 is an excerpt from a coil catalog. The complete catalog gives the coil performance at other dry- and wet-bulb temperatures of entering air and at additional face velocities. Several characteristics of coil performance can be demonstrated by the data from Table 4.6:

1. Plotting on the psychrometric chart the points representing the condition of the air at the outlet from successive rows of tubes shows the curvature to be similar to Fig. 4.25.

TABLE 4.6.   Performance of McQuay, Inc. Direct-Expansion Cooling Coils Using
Refrigerant-12. Air Enters at 90°F db and 74°F wb

	400 FPM Face Velocity			600 FPM Face Velocity	
Rows of Tubes	Final DB Temp. (°F)	Final WB Temp. (°F)	Rows of Tubes	Final DB Temp. (°F)	Final WB Temp. (°F)
35°F Refrigerant Temperature					
3	64.4	61.2	3	68.1	64.0
4	59.1	57.2	4	63.8	61.2
5	55.7	54.6	5	60.2	58.5
6	52.8	52.0	6	57.1	56.1
8	48.0	47.7	8	52.4	51.9
40°F Refrigerant Temperature					
3	66.0	62.8	3	69.4	65.3
4	61.8	59.9	4	65.5	62.9
5	58.5	57.4	5	62.3	60.6
6	56.1	55.3	6	59.6	58.6
8	52.0	51.7	8	55.6	55.1
45°F Refrigerant Temperature					
3	67.8	64.6	3	70.8	66.7
4	64.1	62.2	4	67.2	64.6
5	61.2	60.1	5	64.4	62.7
6	59.2	58.4	6	62.1	61.1
8	55.8	55.5	8	58.8	58.3

2. Each successive row of tubes removes less heat than its predecessor.
    (This fact can be shown by determining the reduction in enthalpy of
    air through each successive row of tubes.)
3. A lower refrigerant temperature with a given face velocity causes a
    greater ratio of latent to sensible heat removal.
4. An increase in the face velocity increases the capacity but also in-
    creases the dry-bulb and wet-bulb temperatures of the outlet air.

4.2.4. Condensers

The function of the condenser is to remove the heat from the refrigerant by
passing cooler air over the heat-transfer surface. Refrigerant will change
from a gas to a liquid as it passes through the condenser and at the exit will
be 100% liquid. This is piped through the liquid line to be ready for another
trip through the system.

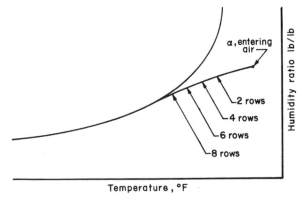

Fig. 4.25. Actual coil condition curve.

The condenser consists of copper tubing arranged in a serpentine fashion and two or three rows deep (Fig. 4.26), as the primary heat-transfer surface. Attached to the tubing will be thin aluminum fins, spaced anywhere from 6 to 14 fins per inch of tube. These act as a secondary surface to increase the effectiveness of the heat dissipation. Several rows of tubes and fins are common. The tubes will usually be staggered to get the maximum air turbulence and heat transfer from the passing air, Fig. 4.27. Outside or ambient air is forced over the surface of the condenser by a condenser fan which, in residential sizes, will move 1000–2000 cfm of air over the fins. Condensers can be mounted horizontally, with the air blown from bottom to top, or they may be arranged vertically, with the air being either blown through or sucked through.

Condensers can be simple air-cooled units (Fig. 4.28) or simple water-cooled shell and coil units (Fig. 4.29), or complex water-cooled units with

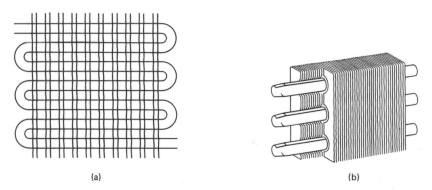

(a)   (b)

Fig. 4.26. Typical condenser arrangements. (*a*) Serpentine pattern. (*b*) Cutaway of condenser tubes with fins.

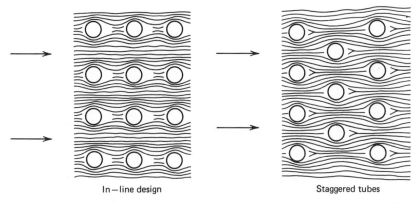

In—line design   Staggered tubes

Fig. 4.27. Condenser tubes are usually staggered to increase heat transfer.

tubes inside a shell (Fig. 4.30), or a compact evaporative type of condensing unit (Fig. 4.31).

Refrigerant vapor is condensed inside the tubes of the air-cooled condenser by giving up heat to the air circulated across the condenser surface. However, water-cooled condensers have cooling water flowing inside the tubes and refrigerant condensing inside the shell but outside the tubes.

For water-cooled condensers, two types of controls regulate condenser water flow, an automatic water regulating valve and a control circuit incorporating a water pump. For air-cooled condensers, control is obtained with shutters or floodback control.

Because a heat pump uses both coils as either a condenser or evaporator depending upon whether heating or cooling is called for, the terminology for the components of a heat pump is different than for a straight air-conditioning system. The coils are referred to as the *outside coil* and the *inside coil* rather than condenser and evaporator.

The outdoor unit will look physically just about the same as an air-conditioning condensing unit (Fig. 4.32). It will contain a coil, fan, fan motor, compressor, and control vestibule with contactor, relays, capacitors,

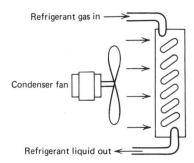

Refrigerant gas in

Condenser fan

Refrigerant liquid out

Fig. 4.28. Air-cooled condenser.

Refrigerant
gas in

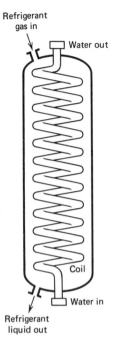

Water out

Coil

Water in

Refrigerant
liquid out

Fig. 4.29.   Water-cooled condenser.

and so on. It will also contain an expansion valve (or RFC metering device), check valve, defrost control with clock timer, and reversing valve. These last four parts would not be included in an air-conditioning condensing unit.

Figures 4.33, 4.34, and 4.35 show various ways of presenting the thermal performance of condensers.

4.2.5.   Evaporators

Direct expansion and *shell-and-tube evaporators* are the types employed in most refrigeration systems. The direct expansion evaporator is basically a coil containing refrigerant over which air is passed for cooling purposes. The

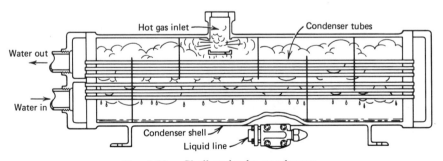

Fig. 4.30.   Shell-and-tube condenser.

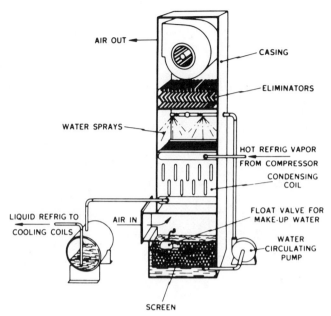

AIR OUT

CASING

ELIMINATORS

WATER SPRAYS

HOT REFRIG VAPOR
FROM COMPRESSOR

CONDENSING
COIL

LIQUID REFRIG TO
COOLING COILS

AIR IN

FLOAT VALVE FOR
MAKE-UP WATER

WATER
CIRCULATING
PUMP

SCREEN

Fig. 4.31. Evaporative condenser. Copyright The Trane Co. 1965. Reprinted with permission.

shell-and-tube evaporator has a heavy shell containing water, which passes over the outside of the refrigerant tubes within the shell.

A dry expansion coil, shown in Fig. 4.36, is used in many air-conditioning applications. The general construction is that of tubes and fins, which give a large area for heat transfer. Some of the area is used for changing the liquid to vapor, while the rest of the tube area is used for heating the vapor. The tubes and fins must be kept clean.

Another type of coil is the flooded coil (Fig. 4.37). In this case a constant level of liquid is maintained and a smaller coil is required than if it were a dry expansion coil.

Figure 4.38 shows a typical direct-expansion shell-and-tube evaporator or chilling unit. These units are used on intermediate or large water-chilling units.

There are two common types of evaporators in residential systems. One is an *A coil*, which has two evaporators connected together and mounted at an angle to each other, roughly forming the letter A (Fig. 4.39, bottom). An A coil is primarily used on an up-flow furnace but it also can be used with a horizontal or down-flow furnace. The second type is a *slab coil*, which is built as a flat rectangle. It can be mounted vertically, within horizontal furnaces, or the slab coil can be mounted at an angle for up-flow, down-flow or horizontal applications (Fig. 4.39, top).

An evaporator has two basic functions: to reduce the temperature of the

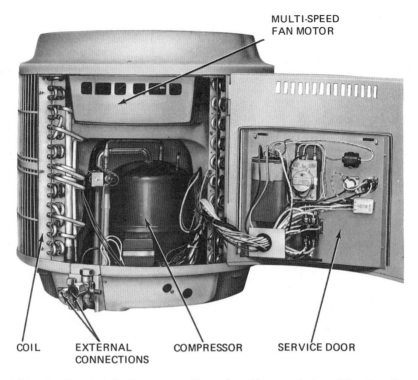

MULTI-SPEED
FAN MOTOR

COIL EXTERNAL COMPRESSOR SERVICE DOOR
CONNECTIONS

Fig. 4.32. Outdoor unit for heat pump. Reproduced by permission of Carrier, Copyright © Carrier Corporation 1980.

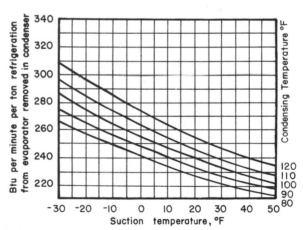

Fig. 4.33. Heat removed in Refrigerant-12 condenser.

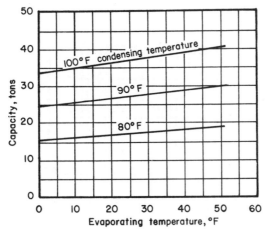

Fig. 4.34. Performance of a Refrigerant-12 condenser, 8 in. by 7 ft, four-pass, made by the Vilter Manufacturing Company. The condenser is supplied with 46 gpm of water which enters at 65°F. The condensing area is 134 ft².

indoor air passing over it, and to reduce the moisture content and de-humidify the air. In doing this, the refrigerant changes from a liquid to a gas while passing through the evaporator. Condensate is formed as part of the dehumidifying process. As the air is cooled, its ability to hold water vapor is reduced. Some of this airborne water vapor condenses out as water, which must be piped to a drain connection, usually through a plastic hose. A graphical method of presenting evaporator performance is given as Fig. 4.40.

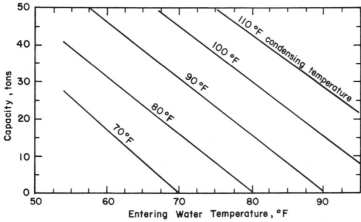

Fig. 4.35. Performance of a Worthington Corporation four-pass Refrigerant-12 con-denser. The condenser has a shell diameter of 16 in. and a length of 74¼ in. The evaporating temperature on which the capacities are based is 40°F, and the rate of flow of cooling water is 60 gpm.

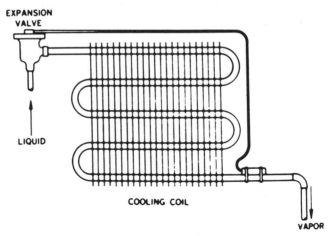

Fig. 4.36.   Refrigerant-to-air heat exchanger. Copyright The Trane Co. 1965, reprinted with permission.

4.2.6.   Heat Exchanger Effectiveness

Since thermal effectiveness is a measure of the thermodynamic availability loss in a heat exchanger, improved heat exchange is clearly one means for improving overall system performance. Heat-exchanger effectiveness for an evaporator or a condenser is an exponential function of the number of heat-transfer units UA, where U is the overall air-side based heat-transfer coefficient and A is the corresponding heat-transfer area. Hence, increasing the overall heat-transfer coefficient (e.g., by increased air flow, since the air-side

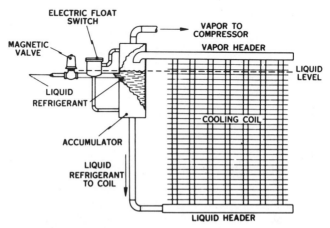

Fig. 4.37.   Flooded evaporator. Copyright The Trane Co. 1965. Reprinted with permission.

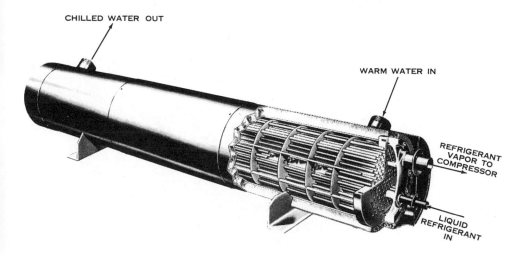

CHILLED WATER OUT

WARM WATER IN

REFRIGERANT VAPOR TO COMPRESSOR

LIQUID REFRIGERANT IN

Fig. 4.38. Shell-and-tube water chiller. Copyright The Trane Co. 1965. Reprinted with permission.

heat-transfer resistance is liable to be controlling in most cases of practical interest) or the coil area will produce the desired result.

Taking up first the question of increasing the coil area, it should be noted that this, indeed, is what many manufacturers have been doing in the last several years. Outdoor coil areas of some manufacturers' equipment nearly doubled since 1973. Comparing, for example in Table 4.7, General Electric Co.'s high-efficiency (1977) Weathertron Executive heat pump with its earlier (1976) 2-ton heat pump, nearly 60% more coil face area per Btu of nominal (cooling) capacity is provided with the newer model.

Increasing the coil area obviously increases heat-exchanger cost. Besides cost, however, there are other limitations to increasing the size of the heat exchangers. Pietsch [6] has discussed the problem of the decrease in dehumidification capacity in the cooling mode that results from a large indoor coil and the attendant higher coil temperature. Increasing the size of the outdoor coil (the condenser, in the cooling mode) increases the amount of refrigerant charge required. According to Pietsch, empirical data on system reliability show loss of reliability beyond a certain charge quantity (due in part to increased probability of liquid slugging). Hence, this approach has its restrictions.

Increased airflow (i.e., coil face velocity) for both the indoor and outdoor coils is also a means of improving heat-exchanger effectiveness through its effect on U, the overall heat-transfer coefficient. Since the refrigerant-side condensing or boiling heat-transfer coefficient is much greater than on the air side, the air-side film coefficient is controlling, that is, $U \sim h_{\text{air}}$, the latter

Fig. 4.39. Two types of evaporators. (Top) Slab coil. (Bottom) A-coil. Courtesy of York Division of Borg-Warner and Coleman Corporation.

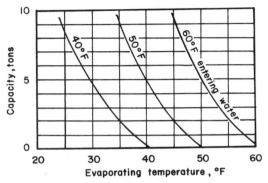

Fig. 4.40. Performance of a liquid-chiller type of evaporator, DXH 1413 M, Refrigerant-12, manufactured by Acme Industries, Inc. The surface area is 302 ft², and the water flow is 180 gpm.

varying approximately as the eight-tenths power of the face velocity. Fan power, however, increases at an even faster rate with increased velocity (varying as the velocity squared), and this method has severe restrictions also.

4.3. EXPANSION (METERING) DEVICES

The control, expansion, or metering device must match the flow of refrigerant to the load on the evaporator, the pumping capacity of the compressor, and the ability of the condenser to reject heat. It determines the capacity of the system, which is probably the most important function in the entire system.

TABLE 4.7. Trends in Electric Heat Pump Heat Exchanger Area as Contributor to Higher COP[a]

	1976 Model	1977 Model
Nominal cooling capacity, Btu/hr	24,000	26,000
Outdoor coil face area, ft²	8.5	14.5
Face area/capacity ratio, ft²/(Btu/hr) × 10⁴	3.5	5.6
Heating COP		
17°F	1.6	1.9
47°F	2.5	2.7
Cooling EER, 95°F	7.0	8.1

[a] Not all of the COP improvement shown is, of course, attributable solely to increased coil area.

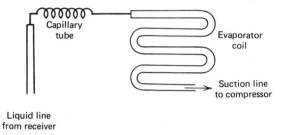

Fig. 4.41.   Capillary tube.

Several types of devices are used to lower the pressure of the liquid refrigerant before it enters the evaporator. The capillary tube (Fig. 4.41) is a small-bore tube that acts as a restriction and reduces the pressure. It is commonly used on fractional tonnage units that do not require accurate control. On an air-source heat pump that must operate over a wide range of evaporating temperatures, a capillary tube will tend to pass refrigerant at an excessive rate at low back pressures, causing liquid floodback to the compressor. In some cases, suction-line accumulators or charge-control devices are added to minimize this effect.

Expansion devices for controlling the refrigerant flow in heat pumps are normally thermostatic expansion valves. If the circuiting is arranged so that the refrigerant line upon which the control bulb is placed can become the compressor discharge line, the resulting pressure developed in the power element of the valve may be excessive, requiring a special control charge or pressure-limiting element. When a thermostatic expansion valve is applied to an outdoor air coil, a special *cross-charge* is desirable to limit the superheat at low temperatures and thereby obtain better use of the coil.

When an expansion valve is attached to a coil that is operated as a condenser, a bypass with a check valve is normally provided. The thermostatic expansion valve (Fig. 4.42) is a more complex device and also has better control capability for the evaporator. This type of device is very common on domestic and commercial air-conditioning and refrigeration units, since it will control the flow of refrigerant to the evaporator.

While a compressor is a constant-displacement pump, that is, it has a fixed cylinder capacity plus a constant operating speed, the volume of refrigerant pumped can vary with the density of the refrigerant in pounds per cubic foot. Since the refrigerant, as it goes through the compressor, is in the vapor state, its volume and density will be affected by its temperature. As temperature increases, the volume increases, and therefore the pumping rate of the compressor, in terms of pounds per hour pumped, will decrease due to decreased density. The temperature of the refrigerant gas as it reaches the compressor is dependent upon the load on the evaporator, since the greater the load the more the refrigerant temperature will increase.

The metering device must always reduce the flow of refrigerant if the outside air temperature increases or if the indoor air temperature decreases. Conversely, it must increase the flow of refrigerant if the outside air temper-

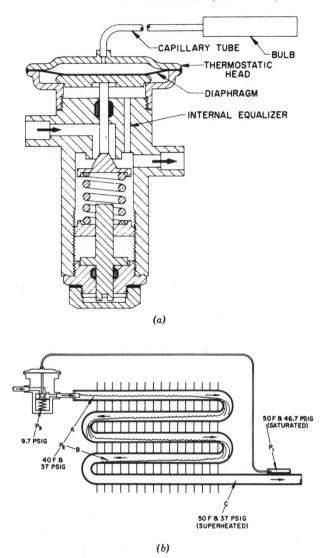

(a)

(b)

Fig. 4.42. (a) Thermostatic expansion valve and (b) bulb location. Reprinted by permission from *ASHRAE 1975 Equipment Handbook*.

ature decreases or if the indoor air temperature increases. Outdoor air temperature varies with weather conditions, and indoor air temperature varies with the load inside the house. This load can be due to infiltration, which increases as outside air temperature goes up, people, lights, cooling or anything else that increases the indoor air temperature.

An expansion valve is a metering device used on many air-conditioning systems, although it is normally not used with systems under 4 tons (Fig. 4.43). The expansion valve controls the flow of refrigerant by means of a needle valve placed in the refrigerant line. Liquid refrigerant enters the valve

Fig. 4.43. Thermostatic expansion valve. (Courtesy Sporlan Valve Co.)

from the high-pressure side of the system and passes through the needle valve to the low-pressure side. Here its pressure is reduced, causing a portion of it to vaporize immediately, cooling the balance of the refrigerant at this point.

Some expansion valves will have a series of tubes at the exit so that they may feed more than one row of an evaporator. These are called distributors and are designed to feed the several rows of a multicircuited evaporator simultaneously (Fig. 4.44).

Fig. 4.44. Distributor allows expansion valve to feed several rows of the evaporator simultaneously. (Courtesy Sporlan Valve Co.)

The thermostatic expansion valve has three major functions:

1. *Throttling Action.* A thermostatic expansion valve separates the high and low sides of the system. This pressure difference, between the condenser and evaporator, is maintained by the valve to assure the most efficient system performance.

2. *Modulating Action.* The valve feeds liquid refrigerant into the evaporator in the proper amounts at all times. If too much liquid refrigerant enters the evaporator, not all of it will change to a vapor and there is danger of passing liquid into the compressor, damaging the compressor valves. If there is too little refrigerant in the evaporator, all of the liquid will be evaporated before it makes a complete circuit, reducing the system capacity and starving the evaporator. By modulating the flow, the valve will maintain the proper amount of liquid in the system.

3. *Controlling Action.*   The valve must also respond to load changes in the system. As the load in the system increases, the valve moves to the wide-open position, lowering the pressure drop across the valve port and allowing more refrigerant to flow. Thus, the compressor capacity balances with the increased load on the system. As the load decreases, the valve closes, reducing the amount of refrigerant available to the compressor.

The opening and closing of this valve is controlled by a combination of forces that constantly monitor the temperature of the refrigeration system. An expansion valve consists of an outlet and inlet for the refrigerant and a diaphragm separating the evaporator inlet and outlet pressures.

In addition, there is a thermal bulb attached to the suction line at the exit of the evaporator. This bulb is filled with the same refrigerant as is used in the system. The bulb senses the temperature at the evaporator exit and is connected to the diaphragm by a capillary tube. Therefore, the operation of the valve is affected by three basic forces (Fig. 4.45):

1. Bulb temperature (pressure at the exit of the evaporator) acting on one side of the diaphragm, which tends to open the valve (P_1)

2. Evaporator or suction pressure acting on the opposite side of the diaphragm, which tends to close the valve (P_2)

3. Spring pressure on the evaporator side of the diaphragm, which also tends to close the valve (P_3)

When the pressures above and below the diaphragm are equal, the valve is said to be in equilibrium. That is, it will pass the correct amount of refrigerant required by the conditions of the system. The remote bulb senses the leaving temperature of the refrigerant gas as it passes out of the evaporator. This gas should contain no liquid which, being noncompressible, could damage the compressor.

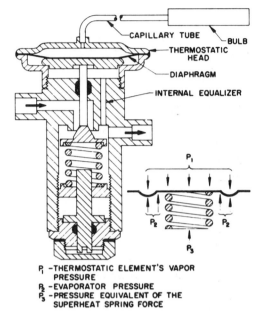

CAPILLARY TUBE
BULB
THERMOSTATIC HEAD
DIAPHRAGM
INTERNAL EQUALIZER

P_1
P_2 P_2
P_3

P_1 –THERMOSTATIC ELEMENT'S VAPOR PRESSURE
P_2 –EVAPORATOR PRESSURE
P_3 –PRESSURE EQUIVALENT OF THE SUPERHEAT SPRING FORCE

Fig. 4.45. Three pressures are used to balance expansion valve flow rate with evaporator demand. Reprinted by permission from *ASHRAE 1975 Equipment Handbook*.

The pressure exerted by the sensing bulb is offset by the adjustable spring at a pressure that assures that the leaving vapor is superheated and contains no liquid. In residential air-conditioning systems, the spring is set to assure 6–15° of superheat, with an average of about 10°. As system load changes, the suction temperature and pressure of the inlet and outlet of the evaporator will change, but the *difference* in temperature or superheat should remain the same.

If the load on the evaporator increases, the refrigerant passing through the coil will pick up more heat and will be at a higher temperature than the refrigerant at the inlet of the evaporator. The expansion valve bulb at the exit of the evaporator will sense this and exert additional pressure on the diaphragm, opening the valve and allowing more refrigerant to flow. Liquid refrigerant at the entrance to the evaporator will evaporate at a faster rate, raising the evaporator pressure and bringing the forces again in balance at the new setting. Thus, the evaporator pressure acting on one side of the diaphragm and the additional bulb pressure acting on the other will tend to offset each other, and the valve will adjust to the new load conditions. The additional pressures on both sides will tend to open the valve, since more pressure is exerted in that direction even though the superheat setting or difference in temperature remains the same.

If there is a very light load on the evaporator, the liquid admitted to the evaporator will not vaporize completely until it almost reaches the outlet of

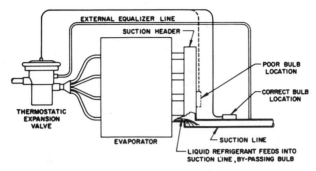

Fig. 4.46. Correct locations for the remote bulb and the equalizer line. Reprinted by permission from *ASHRAE 1975 Equipment Handbook.*

the coil. Therefore, its temperature and pressure will be very close to the temperature and pressure at which it entered. This reduction in temperature will be sensed by the feeler bulb, which will exert less pressure on the diaphragm. The valve will tend to close, reducing the flow of liquid. Although the inlet and outlet temperatures and pressures are lower, the difference in pressure will remain the same as the superheat setting, so the valve can adjust to this change in load.

In all cases, the valve adjusts to changes in load by changes in refrigerant flow. This flow rate then matches the pumping rate of the compressor so that the system is always in balance and maintains its basic capacity.

Multicircuited evaporators will also have an accumulator or manifold to collect the refrigerant gas from the several circuits and pass it into the single suction line, returning to the compressor. In this instance it is important that the feeler bulb will be strapped to the manifold or final collecting point and not one of the individual circuits (Fig. 4.46).

If there are no unusual pressure drops in the evaporator (normal design pressure drop is 2 psig), the inlet gas pressure is reasonably close to the exit gas pressure. In this case, the valve senses pressure at the evaporator inlet, point P_2 in Fig. 4.45.

If the pressure across the evaporator exceeds 2 lb, an excessive amount of inlet gas pressure offsets the pressure generated at the remote sensing bulb. In one example, 6 psi pressure drop required an additional 6° of superheat to offset the pressure on the diaphragm. While the system was attempting to achieve equilibrium at this higher superheat, the evaporator was starved for lack of sufficient refrigerant.

To correct this problem, an external equalizer is tapped into the suction line at a point downstream from the sensing bulb and connected to the underside of the valve diaphragm (Fig. 4.46). In this way, the effective pressure at the bulb is fed to the valve, and superheat is maintained at a normal level. As the result, the evaporator receives the required amount of refrigerant despite the pressure drop through the coils.

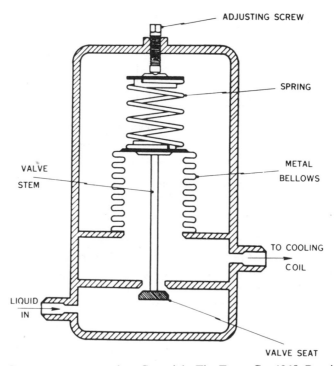

Fig. 4.47. Constant-pressure valve. Copyright The Trane Co. 1965. Reprinted with permission.

A third type of expansion device is the constant-pressure valve (Fig. 4.47), which maintains a constant temperature in the evaporator and is used in water chillers, coolers, and charge-limited systems.

4.4. REVERSING VALVES

A *reversing valve* changes the system from the cooling to the heating mode. This changeover requires the use of a valve or valves in the refrigerant circuit, except where the change is accomplished in fluid circuits external to the refrigerant circuit. Reversing valves are usually pilot-operated by means of solenoid valves, which admit head and suction pressures to move the operating elements.

Physically, a typical reversing valve looks like Fig. 4.48 on the outside. Internally (Fig. 4.49), it is composed of two pistons connected to a sliding block or cylinder with two openings. The illustration shows heating and cooling modes. The four-way valve is actuated by a solenoid valve that uses high-pressure compressor gas to move the piston left or right depending on which mode is needed. The plunger position changes the vent lines accord-

Fig. 4.48. Reversing valve. Courtesy Ranco Controls Corporation.

ingly. In the crossover action, too rapid a change in pressures could result in system shock and excessive noise.

The reversing valve has three major parts. The solenoid is a special adaptation of a magnetic relay, which, when its coil is energized, will draw up an iron plunger or stem into the magnetic field. This stem acts as a valve in the pilot circuit of the reversing valve, closing off one pilot tube and opening the other to the center pilot tube connected to the suction line of the compressor. This causes one tube to have high-pressure gas and the other low-pressure gas, and this is the second major assembly in the reversing valve. These tubes are connected to the ends of the main reversing valve body, which contains a piston that will slide to one end or the other, depending upon which end has the high pressure. As the piston slides toward the low-pressure end, it will open up the tube to one coil from the compressor discharge line and form a circuit with the tube from the other coil to the compressor suction line. This will change the direction of the refrigerant flow in the system and change over the coil functions. Note again that the routing of the refrigerant from the reversing valve to the compressor and back to the reversing valve assures that the flow of gas is always in the same direction through the compressor. Bleed ports within the pistons of the main valve body help to maintain the position of this valve until such time as the valve is reversed by the solenoid and pilot circuit.

Notice that when the system is reversed, some other things have to come about in order for it to work properly. One is that the control point for the metering device, whether it be an RFC system or a thermostatic expansion valve, now changes from the entrance of one coil to the entrance of the other. A common system will therefore have an expansion valve at the entrance of each valve, so that only the expansion valve on the coil being used as an evaporator will meter the refrigerant in the system. Some systems will have one expansion valve, two distributors, and four check valves,

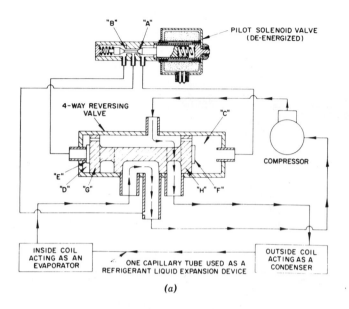

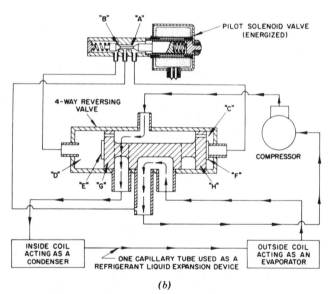

Fig. 4.49. Operation of reversing valve. (*a*) Cooling mode. (*b*) Heating mode. Reprinted by permission from *ASHRAE 1975 Equipment Handbook*.

while others may have only one capillary tube and one special capillary control check valve with a small additional capillary tube. With the check valves in the system, it is therefore possible to route the refrigerant in either direction and, as this is done, bypass one control device in the circuit.

4.5. FILTER-DRIER

On nearly every heat pump or refrigeration unit there will be a filter-drier to clean out any dirt and remove water that can clog the expansion device. A typical design is shown in Fig. 4.50. A filter-drier is designed to remove both solids and undesirable solubles, like water and acid, from the refrigerant. There are several types of filter-driers, but most combine an activated alumina desiccant (drying agent) with a molecular sieve. A metal screen will be used at the entrance to trap solids, and the design of the core assures that all of the refrigerant will come into contact with the drying agents. Filters are mounted in the liquid line, usually near the condensing unit. Whenever the compressor must be changed because of a burnout, the liquid line drier should also be changed. It is also good practice to change the drier on an older system that has been opened or is suspected of containing moisture.

A reversible filter-drier (Fig. 4.51) is for use on heat-pump systems. A blend of desiccants is used in the molded core to obtain maximum water capacity, acid removability, and the ability to remove oleo-resin and other oil breakdown products. The single core is assembled in a shell with two check valves at either end so that filtration occurs on the large exterior core surface. No dirt is released when the flow direction reverses.

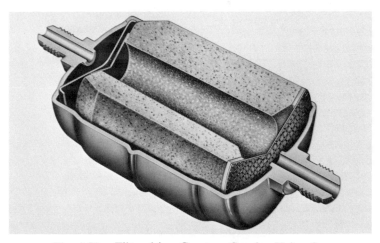

Fig. 4.50. Filter-drier. Courtesy Sporlan Valve Co.

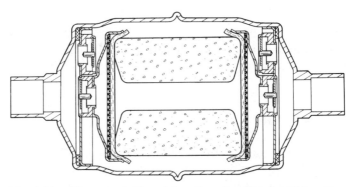

Fig. 4.51.   Heat pump filter-drier. Courtesy Sporlan Valve Co.

4.6.   ACCUMULATOR

A *refrigerant receiver* and/or *accumulator,* which is commonly used to provide a storage place for liquid refrigerant, is particularly useful in a heat pump to take care of the unequal refrigerant requirements of heating and cooling. A accumulator as applied to heat pumps is shown in Fig. 4.52.

4.7.   REFRIGERANT PIPING

Hot gas lines carry hot refrigerant gas from the compressor to the condenser, while suction lines carry cool refrigerant gas from the evaporator to the compressor. The liquid line carries liquid refrigerant from the condenser

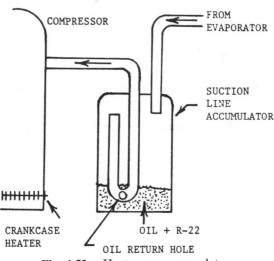

Fig. 4.52.   Heat pump accumulator.

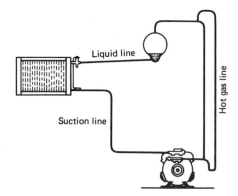

Fig. 4.53. Refrigerant piping.

to the evaporator (Fig. 4.53). Proper sizing of these lines depends on pressure drop and velocity for hot gas and suction lines and pressure drop only for liquid lines.

The *suction line* connects the evaporator to the compressor. Since this line must carry gas, which has a greater volume than liquid, it must have a larger diameter to compensate for this. This line will be insulated because it will be colder than the ambient air and will have a tendency to collect condensate (water) on its surface.

A second line will run from the compressor discharge valve to the inlet side of the condenser. Since the compressor and condenser are mounted in the same cabinet, this will be a relatively short line. It is called the *hot gas* or *discharge line*. It will be smaller than the suction line because the gas is now compressed into a smaller volume. It will be much hotter than the other lines in the system because the gas has its normal heat content plus additional heat from the compressor motor plus the heat of compression.

A third line will run from the outlet of the condenser, which is usually at its lowest point, back to the inlet side of the evaporator. This line carries liquid refrigerant, which has been condensed in the condenser. It is called the *liquid line*. If the liquid line is exposed to high ambient temperatures, such as runs on flat roofs or through kitchens or unvented attics, then it should also be insulated.

Horizontal suction lines should be pitched toward the compressor (approximately $\frac{1}{8}$ in. for each 10 ft of run) to assure good oil return to the compressor. A certain amount of oil will always circulate within the system, so deep bends or traps should be avoided in order to allow the oil to return to the compressor.

If the evaporator is 10 ft or more below the compressor, then an oil trap must be provided in the suction line near the evaporator (Fig. 4.54). The reason is that the oil returning to the compressor has only the velocity of the return gas to carry it upwards against the force of gravity. As it rises, it tends to adhere to the walls of the tube and run back down, collecting at the lowest

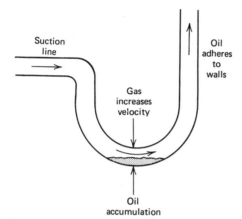

Fig. 4.54. Oil trap in the suction line helps oil return to the compressor.

point. With a trap, the effective tube area is reduced, which increases the velocity of the return gas at this point. The gas will reach sufficient velocity to force additional oil out of the trap and carry it back to the compressor. This eliminates the possibility of oil collecting in the evaporator.

Only refrigeration-grade copper tubing should be used in air-conditioning systems. This special tubing is seamless and has been deoxidized and dried prior to being coiled into 50- or 100-ft rolls, so it can be used on the job without additional dehydration. It should be noted that the system must still be dehydrated after installation is complete. This tubing is known as Type L and is available in either soft or hard copper in a number of sizes. It is recommended for use with either flared fittings or brazed connections and is the most popular tubing for refrigeration work. Note that refrigeration tubing is sized by its outside diameter (OD).

4.8. SUPPLEMENTARY ELECTRIC HEATERS

Supplementary *resistance heaters* may be incorporated either within the unit or external to it. Figure 4.55 shows one type of resistance element and its delay relay. When installed in the distribution ductwork, the heaters are frequently controlled to temper the air during defrosting operations using *reverse-cycle* defrost.

The main power for the electric heat is equipped with a separate fused disconnect switch. The heating elements are controlled by individual heating delay relays that bring on each element through a low-voltage delay circuit. A step-down transformer is connected to the high-voltage circuit and provides 24-V power for the low-voltage control circuit. The heating delay relays are controlled by a low-voltage single pole relay in series with the delay heaters.

(b)

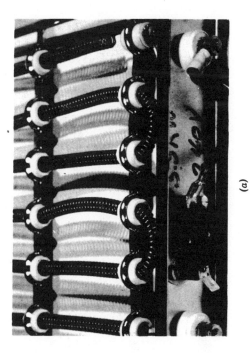

(a)

Fig. 4.55. Supplementary electric heater. (Left) Heating element and limit. (Right) Delay relay. Courtesy Lennox Industries, Inc.

235

4.9.   CONTROLS

4.9.1.   Pressure Controls

Most compressors will include a low-pressure control, and some also will have a high-pressure control, usually mounted on the compressor shell. A small tube or capillary will be connected to a port on the service valve so it can sense pressure under operating conditions. Both controls may be combined into one control called a dual pressure control (Fig. 4.56). They are wired electrically in series into the low-voltage control circuit.

The low-pressure control is in the circuit primarily to protect against loss of refrigerant charge. However, if the suction pressure goes below the control setting (about 5–8 psig) for any reason, it will trip, shutting off the compressor. It is an automatic reset control; that is, it will reset itself when the suction pressure rises above the setting.

The high-pressure control is in the circuit to protect against failure of the condenser fan motor or any other condition that causes the head pressure to increase to an unsafe point. It usually has a manual reset feature, which means that once it trips it must be reset manually before the compressor will run.

4.9.2.   Defrost Control

When the defrost cycle is initiated, several things happen simultaneously to defrost the coil. (1) The cycle is reversed through the reversing valve so that the outdoor coil becomes the condenser coil. This routes hot gas through the

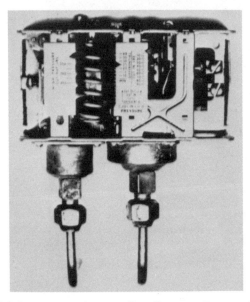

Fig. 4.56.   High-Low pressure controller. Courtesy Lennox Industries, Inc.

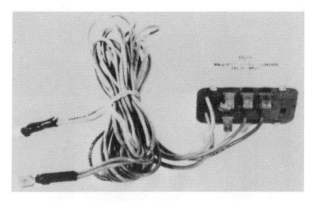

Fig. 4.57. Time/temperature defrost control. Courtesy Lennox Industries, Inc.

outdoor coil increasing its temperature, thus melting whatever ice or frost has built up on it. (2) In order to accomplish this more quickly and efficiently, the outdoor fan is turned off so that there will be no cold outdoor air passing over the coil during the defrost cycle. (3) Since the indoor coil now becomes the evaporator and as such is taking heat out of the air passing over it and going into the space, it is necessary to add suplementary heat to this airstream so that cold air will not be delivered inside the space during the defrost cycle. Therefore, during the defrost cycle the supplementary strip heaters are turned on to maintain the temperature of the conditioned air within the space.

It can be seen that with these additional operations on the heating and defrost cycle, it is necessary to add certain relays and other controls that would not normally be found on an air-conditioning system (Figs. 4.57, 4.58). Some models use a clock timer with a temperature-sensitive overload attached to a return bend or crossover on the outdoor coil, while other models use a temperature differential device to control the defrost cycle. The time-temperature or clock timer control is normally set to have a 90-min time cycle, but this cycle can be reduced manually to about 30 min. The temperature differential control will normally initiate a defrost cycle about once an hour, depending upon icing conditions. A clock timer in the clock timer control is energized and operated all the time the compressor is running. After the compressor has run for a total period of 90 min, it will close its contacts, which are in series with the temperature-sensitive overload switch. If the coil has been reduced in temperature so that this overload switch has closed, then a defrost cycle will be initiated.

Another means for defrosting involves a temperature differential control in which two temperature-sensing elements are used, one responsive to the outdoor air temperature and the other responsive to the temperature of the refrigerant in the coil (Fig. 4.59). This is sometimes accomplished electronically, using thermistors as the sensors. As frost accumulates, the differential

Fig. 4.58. Air-pressure sensor for demand defrost systems. Courtesy Lennox Industries, Inc.

between outdoor temperature and refrigerant temperature will increase, causing a defrost cycle to be initiated. The system will be restored to operation when the refrigerant temperature in the coil reaches a specified temperature indicating that defrosting has been completed. When the outdoor air temperature decreases, the temperature differential between outdoor air and refrigerant decreases, allowing greater frost buildup before the defrost cycle is initiated, unless compensation is provided.

A third method of starting the defrost cycle is to use a pressure control that reacts to the air-pressure drop across the coil. Under conditions of frost accumulation, the airflow will be reduced and the increased pressure drop across the coil will initiate the defrost cycle. Again, the preferred method of terminating the defrost cycle is to use a refrigerant temperature control that measures the temperature of the liquid refrigerant in the coil.

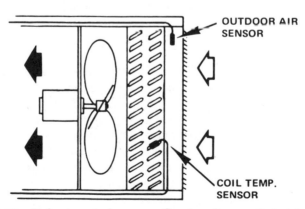

Fig. 4.59. Defrost control sensors. Courtesy Lennox Industries, Inc.

4.9.3. Thermostats

Provisions should be made in the control system to prevent energizing the supplemental electric heater until (1) the heat pump system is unable to satisfy the heating requirement when operating at full capacity and (2) the outdoor air is below a predetermined outdoor temperature. In this way the coincident electricity kilowatt demand will be minimized, with a resulting lower average power cost. In some installations, it may be necessary to energize the supplemental heat during the defrosting cycle in order to minimize any cooling effect within the structure, but during the heating cycle this normally will not exceed the maximum coincident demand.

In air-to-water and water-to-water systems, an outdoor reset control of the hot-water temperature may be desirable for improved economy. During milder weather, the warm-water temperature is rescheduled to a lower level, since less heat dissipation is required; this permits the compressor to operate at a lower condensing temperature.

There are several methods of system changeover in common use.

1. A *conditioned space thermostat* is commonly used for residences and small commercial applications.
2. An *outdoor air thermostat* (with provision made for manual override for variable solar and internal load conditions) may be used to reverse the cycle on larger installations, where it may be difficult to find a location in the conditioned space that will reflect the desired operating cycle for the total building.
3. *Manual changeover.*
4. A *sensing device* that responds to the greater load requirement, heating or cooling, is generally applied on simultaneous heating and cooling systems.

In the heat pump system it is important that space thermostats be interlocked with ventilation dampers so that both are operating on the same cycle. When on the heating cycle, the fresh-air damper should be positioned for minimum ventilation air, with the space thermostat calling for increased ventilation air only if the conditioned space becomes too warm. Fan or pump interlocks, or both, are generally provided to prevent the heat pump system from operating if accessory equipment is not available for use.

A different type of thermostat is used with a heat pump because of these special operating characteristics. Either a three- or four-bulb thermostat may be used. In a thermostat with four mercury bulbs (Fig. 4.60), one bulb controls cooling, one controls the reversing valve, and two control the heating. The reversing valve solenoid is therefore energized during the cooling cycle. A thermostat with three mercury bulbs has one bulb to control cooling and two bulbs to control heating. The reversing valve solenoid is energized in the heating cycle.

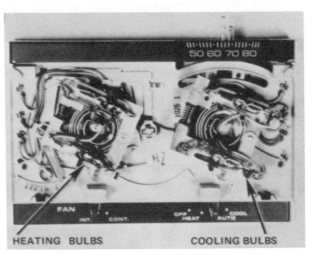

Fig. 4.60. Four-bulb thermostat. *Cooling:* First stage cooling bulb energizes the reversing valve. Second stage cooling bulb operated the compressor. *Heating:* First stage heating bulb operates the compressor. Second stage heating bulb energizes the strip heaters should the indoor temperature drop 2 F below the thermostat setting. Courtesy Lennox Industries, Inc.

Another method is to use a two-bulb heating-and-cooling thermostat to control the compressor and reversing valve and add an outside thermostat to bring on the strip heat whenever the outside temperature drops to the balance point of the unit. The balance point is the outside temperature at which the heat loss of the house exactly matches the capacity of the unit. Any drop in outside temperature below the balance point therefore requires additional heat input, which is supplied by the strip heaters.

A four-bulb heat pump thermostat (as shown in Fig. 4.61) is designed to control the compressor, reversing valve, and auxiliary resistance heat. The system switch has four positions labeled "Cool," "Auto," "Heat," and "Off." The system can function on cooling or on heating. With the switch on Auto, the unit will heat or cool automatically. There is also a fan switch marked "Cont" for continuous blower and "Int" for intermittent (blower cycles with compressor).

When the system switch is in the Cool position, the first-stage cooling bulb will energize the reversing valve through the R circuit and switch the refrigeration system to the cooling position. After the room temperature increases 2°F, the second-stage cooling bulb will energize the compressor contactor through the M circuit. If not on continuous blower, the indoor blower motor is energized through F in the thermostat. If on Cont, the indoor blower is already running.

When the system switch is in the Heat position, the first-stage heat bulb will energize the compressor contactor through the M circuit. The reversing valve circuit (R) is deenergized, which positions the refrigeration system correctly for heating. If on Int, the fan circuit will be activated automatically when the compressor or M circuit is energized, bringing on the indoor blower motor whenever the compressor is operating. The second-stage heat bulb will make if there is an additional 2° drop in the indoor temperature. The second-stage heating circuit is switched through Y and will energize the auxiliary heat relay, which will switch on the auxiliary resistance heaters. After the thermostat is satisfied and shuts down the compressor, if the fan is on Int the indoor blower delay relay will hold on the blower for an additional 90 s. The reason for the delay on the indoor blower is that the strip heaters have a delay of 45 s on shutoff. It is necessary to be sure the blower is operating when the heaters are on.

4.10. TYPICAL WIRING CIRCUITS

Figures 4.62 and 4.63 are combined wiring diagrams of all the heat pump components including the low-voltage control circuits. The heart of the unit is the thermostat, which controls the cooling and two stages of heating. Other wiring diagrams are similar but may vary by protection circuits or small changes in the defrost controls.

4.11. ESTIMATING PART-LOAD PERFORMANCE OF COMPONENTS AND SYSTEMS

Manufacturers' and developers' data almost uniformly represent equipment performance on a steady-state or full-load basis, although it has been known

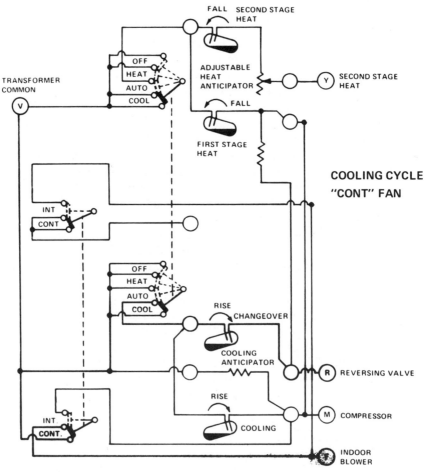

Fig. 4.61. Bulb thermostat. Courtesy Lennox Industries, Inc.

for some time that in addition to frost buildup on outside coils, equipment cycling due to part-load operation can substantially reduce heat pump and air-conditioner efficiencies. In order to present results for HVAC equipment more adequately, transient or part-load and efficiency losses must be accounted for.

Design performance data of a piece of equipment are obtainable from the manufacturer. Off-design performance data are much harder to obtain; in some cases, the manufacturer does not know the off-design performance. For this reason, equipment simulation algorithms should be designed to accept the design data as input. The algorithm can use a set of general correction functions to modify the design data for off-design performance. If the off-design performance is known, the algorithm should allow this to override the general correction functions.

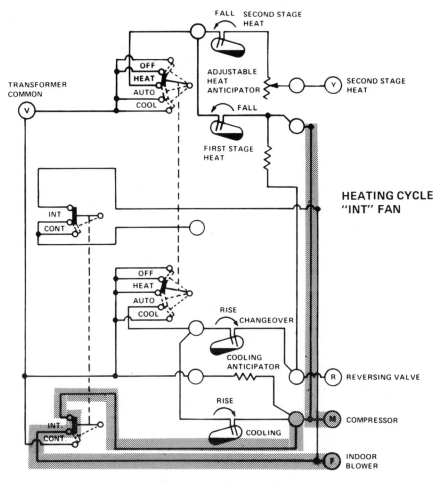

Fig. 4.61. (*Continued*)

The hourly dependence on a relevant parameter Z affected by an arbitrary variable X can be calculated as

$$Z = Z_{\text{design}} f(X)$$

If Z is dependent on two variables, then

$$Z = Z_{\text{design}} f(X)F(Y)$$

or

$$Z = Z_{\text{design}} f'(X, Y)$$

Since at the design conditions, $Z = Z_{\text{design}}$, the correction functions must be normalized so that at the design conditions,

$$f(X_{\text{design}}) = f(Y_{\text{design}}) = f(X_{\text{design}}, Y_{\text{design}}) = 1.0$$

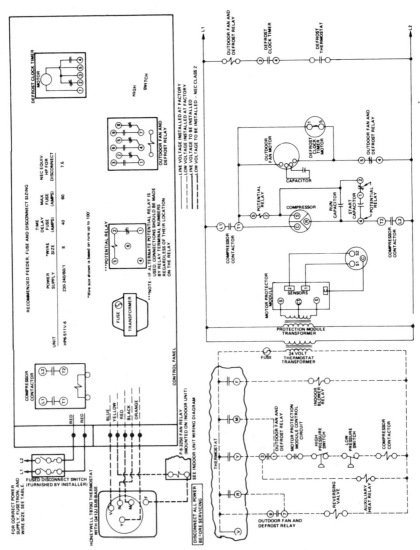

Fig. 4.62. Typical four-ton heat pump wiring diagram. Courtesy Lennox Industries, Inc.

244

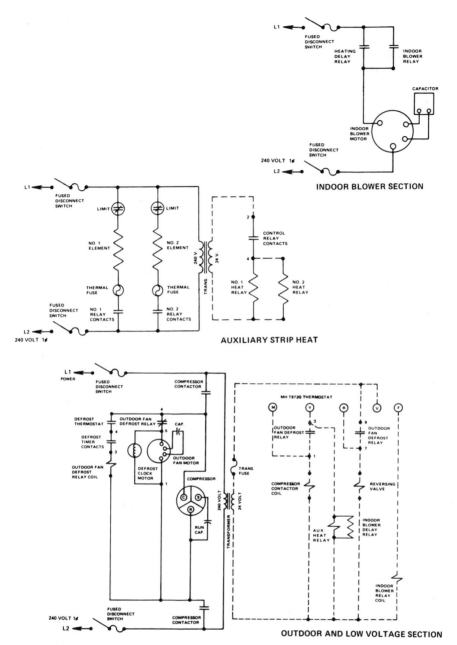

Fig. 4.63. Heat pump wiring diagram. Courtesy Lennox Industries, Inc.

The above expressions are only a few of many ways to model the performance of a piece of equipment. The functions may be of any appropriate form, including polynomials, exponentials, or Fourier series. Most computer algorithms use polynomials, usually third-order or less. The necessary degree of the polynomial depends on the behavior of the equipment. Figure 4.64 illustrates five of the polynomial forms commonly used. It is also possible to use tables of data in place of polynomial functions; however, this method may require excessive data input and computer memory.

Equation fitting is to a great extent an art. There is no methodical procedure for selecting the most applicable form of equation. It is assumed that tabular or graphical data are available from such sources as manufacturers' catalogs for the component. It is further understood that all physical laws or mathematical relations should be used to advantage. For example, in simulating the performance of heat exchangers, heat-transfer laws should be used to the greatest extent possible.

When no unique insight into the component performance is available, a polynomial representation is probably the best choice:

$$Z = a_0 + a_1 x + a_2 x^2 + \cdots + a_n x^n$$

For simplicity, the degree used should be no higher than necessary.

Figure 4.65 represents typical performance of a small (3–10-ton) unitary heat pump under summer cooling (air-conditioning) conditions as a function of the ambient dry-bulb temperature. The characteristics of equipment from different manufacturer will vary, dependent on such variables as compressor clearance, motor design, amount of heat-exchange surface used in evaporator and condenser, and types of fans and fan motors used, but some characteristic can be developed from published ratings and auxiliary data on components.

A significant point in considering such data is that it represents the stabilized running condition at some specified refrigerant suction temperature or room fan airflow and return air conditions. Such data provide only limited information on the energy consumption of systems in actual operation. This is true for a number of reasons, the most important of which is probably the method of controlling the system. Another factor that does not follow simple theory in residential and commercial systems is the dehumidification. It has been found that the moisture actually removed by the equipment is a function of the evaporator operating characteristics and the equipment operating hours rather than any calculated latent load in the space. Thus it is difficult to predict exactly what the actual space humidity ratio will be.

The last variable affecting part-load performance of unitary systems is, of course, the length of the starting transient. In some split systems with long interconnecting lines, several minutes elapse before essentially stable operation is achieved. Gas heat pump systems may require 5 min or more to reach a normal operating condition.

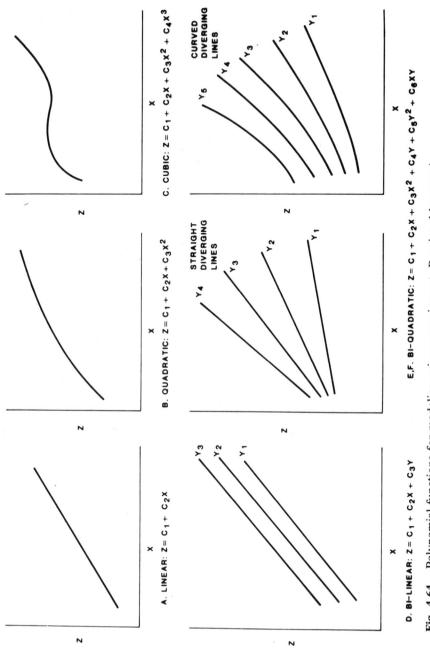

A. LINEAR: $Z = C_1 + C_2 X$

B. QUADRATIC: $Z = C_1 + C_2 X + C_3 X^2$

C. CUBIC: $Z = C_1 + C_2 X + C_3 X^2 + C_4 X^3$

D. BI-LINEAR: $Z = C_1 + C_2 X + C_3 Y$

E.F. BI-QUADRATIC: $Z = C_1 + C_2 X + C_3 X^2 + C_4 Y + C_5 Y^2 + C_6 XY$

Fig. 4.64. Polynomial functions for modeling primary equipment. Reprinted by permission from *ASHRAE 1981 Fundamentals Handbook*.

247

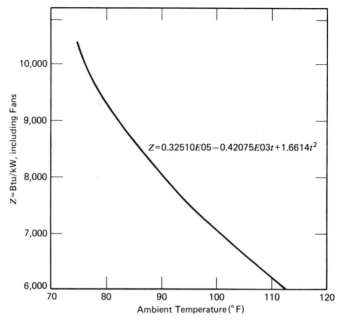

Fig. 4.65.   Performance of 3–10-ton unitary heat pump system.

4.12.   COMPONENT EFFECT ON SYSTEM PERFORMANCE

One of the principal design aspects of a heat pump system, whether factory-built or field-assembled, is the selection of the components, including the compressor, condenser, evaporator, refrigerant flow control device, fan, motor, and controls. The characteristics of each of the components are related to the other components, and the system formed by them must perform properly at both design conditions and every other condition that may be expected during operation. Most heat pump systems operate with varying conditions of both the fluid from which heat is removed and the fluid to which heat is discharged.

Selection of the best and most economical components for a particular system requires a familiarity with the effects of changing operating conditions on the performance of each component. The next step is to study the behavior of the entire system. To make this study, the individual performance characteristics of the compressor, condenser, expansion device, and evaporator must be combined. The operation of these components is interdependent. This section outlines a technique for analyzing the performance of the complete system. Essentially the operating conditions are determined by solving simultaneous equations representing the performance of each component. The mathematical representation of the performance of a component such as the compressor, for example, would be very complex. It is much simpler to solve the simultaneous equations by graphic means.

The actual performance of the component can be obtained from catalog data and plotted on a graph either directly or after suitable calculations.

In the design of a heat pump system, one of the most important considerations is that of establishing the proper relationship or "balance" between the vaporizing and condensing sections of the system. It is important to recognize that whenever an evaporator and a condensing unit are connected together in a common system, a condition of balance is established between the two such that the rate of vaporization is equal to the rate of condensation. Since all the components in a heat pump system are connected together in series, the refrigerant flow rate through all components is the same. It follows, therefore, that the capacity of all the components must be the same.

When the system components are seleted to have equal capacities at the system design conditions, the point of system equilibrium or balance will occur at the system design conditions. On the other hand, when the components selected do not have equal capacities at the system design conditions, system equilibrium will be established at operating conditions other than the system design conditions and the system will not perform satisfactorily.

The flow characteristics of the expansion valve have a decided effect on the functioning of the entire system. The combination of the flow characteristics of the expansion device and the compressor has been discussed in Section 4.3. To simplify the considerations in the present section, the expansion device is expected to deliver enough refrigerant to balance the compressor flow rate at all suction and condensing conditions. A float valve or a thermostatic expansion valve would meet this requirement.

The next step in analyzing the performance of the complete vapor-compression system will be to study the behavior of the compressor and the condenser operating together. These two components combine to form the "condensing unit," whose performance can be found by superimposing condenser curves as in Fig. 4.34 on compressor curves as in Fig. 4.6. The result appears in Fig. 4.66.

At any operating point of the condensing unit, identical values of the condensing temperature, evaporating temperature, and number of tons must prevail for both the compressor and condenser. Operating points occur at the intersections of common condensing temperatures. A line connecting the operating points represents the performance of the condensing unit. As the evaporator temperature drops, the condensing temperature also decreases. The physical explanation for the characteristic is that a lower evaporator temperature reduces the rate at which the compressor can pump refrigerant, which in turn allows condensation at a lower temperature.

Another characteristic of the condensing unit that is often important is the effect on the refrigeration capacity of changes in condensing-water temperature at a constant evaporator temperature. Figures 4.34 and 4.35 may be combined to give this characteristic in the manner shown in Fig. 4.67. The circled points on the graph on the right are the operating points where the evaporator temperatures, condenser temperatures, and capacities are identical for the compressor and the condenser.

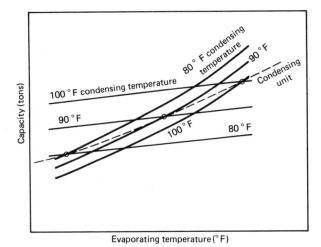

Fig. 4.66. Performance of a condensing unit as a function of the evaporating temperature.

Now that the combined performance of the compressor and the condenser has been evaluated, attention may be shifted to the evaporator. For any particular evaporator and condensing unit connected together in a common system, the relationship established between the two, that is, the point of system balance, can be evaluated graphically by plotting evaporator capacity and condensing unit capacity on a common graph. Using data taken from the manufacturers' rating tables, condensing unit capacity is plotted against suction temperature, whereas evaporator capacity is plotted against evaporator temperature differential, TD. A graphical analysis is shown in Fig. 4.68.

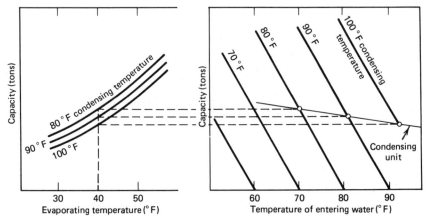

Fig. 4.67. Performance of a condensing unit as a function of the temperature of the entering condenser water.

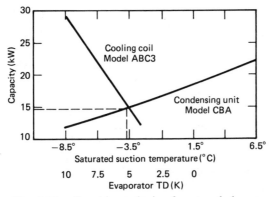

Fig. 4.68.   Graphic analysis of system balance.

With regard to Fig. 4.68, the following procedure is used in making a graphical analysis of the system equilibrium conditions:

1. Using manufacturer's catalog data, plot the capacity curve for the condensing unit. Since condensing-unit capacity is not exactly proportional to suction temperature, the condensing-unit capacity curve will ordinarily have a slight curvature.

2. From the evaporator manufacturer's catalog data, plot the evaporator capacity curve. Since evaporator capacity is assumed to be proportional to the evaporator temperature differential, TD, the evaporator capacity curve is a straight line, the position and direction of which is adequately established by plotting evaporator capacity at any two selected TDs.

More complete sets of performance data for a particular combination of evaporator and condensing unit are shown in Fig. 4.69. Here the evaporator is an air-cooling coil, but appropriate curves can be drawn for any type of evaporator compressor or condensing unit if the necessary performance data are available.

To illustrate the analysis of a system, assume that an air-cooling coil has been selected to satisfy an actual air-conditioning load in which it must remove 250,000 Btu/hr when air at 65°F wb enters and the evaporating temperature is about 48°F. Coil capacity lines for various entering wet-bulb temperatures have been plotted in Fig. 4.69. These lines are drawn about 3 psi to the left of the evaporating temperature to allow for that pressure loss between the evaporator outlet and the compressor inlet. It is required that this 250,000 Btu/hr be removed by the condensing unit when air is entering it at 95°F. The condensing-unit capacity curves in Fig. 4.69 are for a unit that meets the requirement. Note that the 95°F condensing-unit line intersects the 65°F coil line at about 260,000 Btu/hr and just under 46°F at the 3 psi

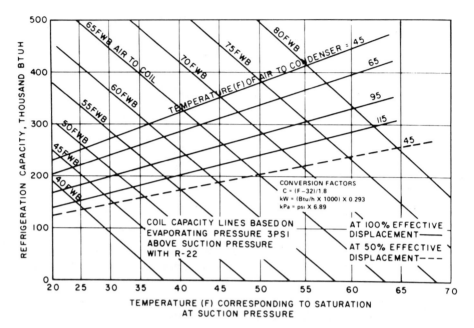

Fig. 4.69. Performance diagram for air-cooling coil and condensing unit with various temperatures of air to condenser. Reprinted by permission from *ASHRAE 1975 Equipment Handbook.*

higher evaporator pressure. At design conditions, performance is almost as desired; however, design conditions do not always occur.

If the wet-bulb temperature entering the coil is very much higher, the 75°F wb line intersects the 95°F line at over 300,000 Btu/hr and 55°F suction. This may overload the compressor motor. If air enters the condenser at 45°F and the air entering the coil is at 50°F wb, the intersection is at 250,000 Btu/hr and 26°F suction. This new balance point causes several changes.

The condensing pressure will be lower. The actual condensing temperature (and corresponding pressure) at this suction and air temperature can be obtained from the condenser manufacturer's data. The difference available for transfer of liquid to the evaporator can then be compared to the value required for the new flow rate and the actual liquid line, valves, and so on, for the particular installation. Figure 4.69 assumes that the available pressure difference is adequate. If it is not, the suction pressure (and temperature), as well as capacity, will fall to the point where the pressure difference available matches the value required.

The evaporating pressure at a little above 26°F, or lower as a result of liquid pressure loss, may result in a coil surface below freezing. The lowest surface temperature can be calculated from the coil characteristics and the temperature of the air leaving the coil. If the surface is below freezing, it may still be acceptable if it is not below the dew point of the air passing over it.

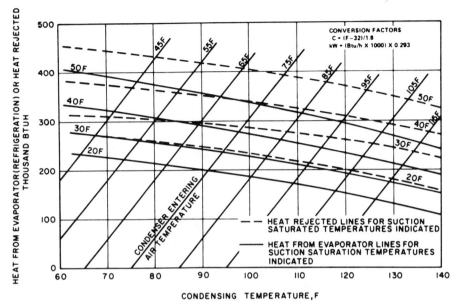

Fig. 4.70. Performance diagram for air-cooled condenser with compressor refrigeration capacity and heat rejection. Reprinted by permission from *ASHRAE 1975 Equipment Handbook.*

Reducing condenser capacity so that the condensing pressure is maintained at about the same value as if 95°F air entered it results in approximately 200,000 Btu/hr capacity at 31°F suction.

If the compressor can be controlled to operate at 50% capacity, the balance with 50°F wb air to the coil and 45°F air to the condenser will be at about 150,000 Btu/hr and 36°F suction. This latter arrangement results in less power per unit of refrigeration.

Figure 4.69 includes the condenser, but only in combination with a particular compressor. The curves show the performance of the combination without any indication of the condensing temperature at various operating conditions. If a particular combination of compressor and condenser is used, and the performance is available for the combination at all desired operating conditions, there would be no need to consider the condensing temperature except to check that there is adequate pressure difference to transfer liquid refrigerant from condenser to evaporator.

When the condenser and compressor are selected separately or the performance is not available for all the operating conditions expected or to be considered, separate compressor and condenser performance curves should be plotted. Figure 4.70 is an example of such curves (the compressor heat-rejection curves are based on a hermetic compressor with the motor waste heat also transferred to the refrigerant). Compressor performance requires two curves from each suction saturation temperature, the solid line indicating the heat from the evaporator and the broken line indicating the heat to the condenser. The difference between the two represents the energy to lift the heat from the evaporating temperature to the condensing temperature. The larger the difference in temperatures, the larger the heat energy added by the compressor for the lift. The heat added is almost directly proportional to the temperature difference in the range 0–150°F for differences up to 100°F, assuming there are no other changes.

Since the heat rejected is the amount that enters the condenser, the intersections of the broken lines with the condenser performance lines are the only intersections of any significance. If the suction saturation temperature is 40°F and the temperature of air to the condenser is 105°F, the intersection of the broken line for 40°F with the condenser line for 105°F, as shown in Fig. 4.70, is at about 129°F condensing temperature and 290,000 Btu/hr heat rejected. Moving vertically down to the solid line for 40°F, the heat removed from the evaporator at this condition is about 213,000 Btu/hr.

A chart like Fig. 4.70 facilitates consideration of the effects of various condensers and air rates over the condenser. The capacity lines for a condenser of twice the capacity would have angles with the vertical, the tangents of which would be half of those shown.

Reducing the airflow rate over the condenser increases the angle with the vertical. Of course, the capacity of the condenser does not change proportionately with a change in air rate. At 50% air rate, the capacity would be more than 50%. The actual value depends on the characteristics of the particular condenser.

If the condenser to be used is water-cooled (possibly for heating water), capacity lines can be constructed on the chart. For this purpose, temperature spacings as in Fig. 4.70 would denote cooling-water temperatures.

Lines representing condenser performance with various airflow or water flow rates can be drawn if the actual condenser performance at such conditions is known or calculated.

Many variables can influence the pressure difference required to transfer refrigerant liquid from the condenser to the evaporator. When all the factors arc known, this difference can be plotted to assist in the prediction of performance.

In Fig. 4.71 (based on R-22), the capacity of the expansion valve is plotted against pressure loss through the valve. The plot includes correction for the greater Btu/lb available with the lower temperature liquid when the condensing temperature is lower. To allow for losses in addition to those through the expansion valve, the suction and condensing temperature lines for an air-cooled condenser have been so plotted that the difference in pressure at the intersections is about 35–40 psi less than the actual difference.

The unit capacity curve for constant load assumes air entering the evaporator at a constant rate and nearly constant enthalpy. As the condensing temperature is reduced, the capacity of the combined system compo-

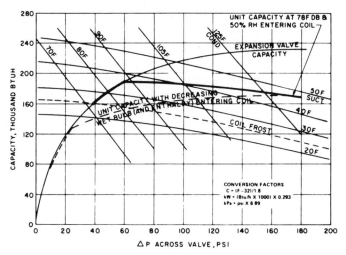

Fig. 4.71. Refrigeration capacity vs. Expansion valve capacity. (Reprinted by permission from *ASHRAE 1975 Equipment Handbook*.)

nents increases somewhat, even at the lower suction pressure, until the expansion valve capacity is reached. As condensing temperature is further reduced, the suction temperature and capacity are limited by the expansion valve capacity. The coil surface temperature falls below freezing because of reduction in pressure difference between condenser and evaporator. As this capacity curve at constant load is followed from right to left, the condensing temperature decreases.

The particular charts shown in Figs. 4.69, 4.70, and 4.71 are examples of the many possibilities for graphically illustrating the performance of each item so as to show the actual result at design conditions, as well as the effect of various changes. In the design of heat pump systems, superposition of the heat-gain or heat-loss characteristics of the conditioned space on the capacity characteristic of the cooling or heating system often will be of value in determining the degree of complexity actually required to meet the objectives of the system.

While the component-balancing process is essential to the detailed technical problems of component selection, it also helps in the development of an overall appraisal of system performance. The designer may consider the system as merely an assembly of components, each having a known sensitivity to the variables that affect its performance, thereby overlooking the fact that the sensitivity of the system as a whole to the external variables is generally of a lower order. This can result in unwarranted system and control complexity, which will have an important influence on the cost and reliability of the overall system. A study of the way in which components balance out when combined will yield valuable information on the overall response of the system to its environment. This information can be applied to the more detailed considerations of determining the extent to which specific refinements are warranted.

The power required by the compressor is indeed important, although it was not one of the operating characteristics investigated in this section. Using the values of condensing and evaporating temperatures found from methods outlined in this section, the power required can be found directly from catalog data.

REFERENCES

1. American Society of Heating, Refrigerating and Air-Conditioning Engineers, *ASHRAE Handbook and Product Directory, 1979, Equipment,* Atlanta, GA, 1979.
2. *Standards of Tubular Exchanger Manufacturers Association,* 6th ed., Tubular Exchanger Manufacturers Association, New York, NY, 1978.
3. Carrier Air Conditioning Company, *Carrier Handbook of Air Conditioning System Design,* McGraw-Hill, New York, 1965.
4. Carrier, W. H. and Anderson, S. W., "The Resistance to Heat Flow Through Finned Tubing," *ASHVE Trans.,* Vol. 50, 1944.
5. *ARI Standard* 410-72, "Forced-Circulation Air-Cooling and Air-Heating Coils," Air-Conditioning and Refrigeration Institute, Arlington, VA, 1972.
6. Pietsch, J. A., "The Unitary Heat Pump Industry—25 Years of Progress," *ASHRAE J.,* **19**(7), 15–18, (1977).
7. Threlkeld, J. L. *Thermal Environmental Engineering,* Second Edition, Prentice-Hall, Inc., Englewood Cliffs, NJ, 1970.

PROBLEMS

1. Given that a compressor using R-22, condensing at 80°F and evaporating at 20°F, find the enthalpy of the refrigerant when
    (a) It enters the compressor
    (b) It enters the condenser
    (c) It enters the evaporator
    (d) Find the horsepower required for the compressor.

2. Given that a compressor uses R-11, which condenses at 100°F and evaporates at −10°F (under regular assumptions), calculate the compressor displacement, the mass flow rate, and the horsepower required for the compressor.

3. A reference book on refrigeration indicates that a compressor using R-22 will require a displacement of 40.59 cfm/ton for evaporating at −100°F and condensing at −30°F. Is this correct?

4. An *ideal* R-11 vapor compression unit operates between 40°F evaporating and 100°F condensing temperatures in a water-chiller application. For a 2-ton unit, give:

(a)   Refrigerant flow rate, lb/min

(b)   Compressor displacement, cfm

(c)   Horsepower

5. A study is to be made of the influence on power requirements of the difference between the temperature of the refrigerant in the condenser and the temperature of the surroundings. For this purpose consider the ideal cycle with R-12 as the refrigerant. The temperature of the surroundings is 100°F, and the temperature of the refrigerant in the evaporator is 10°F. Plot a curve of horsepower per ton of refrigeration for temperature differences of 0–100°F between the refrigerant in the condenser and the surroundings.

6. An air conditioner uses R-12 and develops 3 tons when operating between 40°F evaporating and 95°F condensing temperatures. The compressor has a volumetric efficiency of 80%. R-12 leaves the condenser and immediately enters the expansion valve subcooled by 10°F. The vapor entering the compressor and leaving the evaporator is superheated by 10°F. A water-cooled condenser is used with a water temperature rise of 15°F.

(a)   Determine motor size, hp.

(b)   Determine condenser water flow rate, gpm.

(c)   Plot cycle on a pressure–enthalpy diagram.

7. An R-12 vapor-compression refrigeration system has an air-cooled condenser and a subcooling-superheating, liquid-vapor counterflow heat exchanger. Condensation, without subcooling, is at a saturation pressure corresponding to 10°F above ambient air temperature of 90°F. Subcooling is to within 20°F of the evaporation temperature of 10°F. There is no superheating in the evaporator. Assume isentropic compression, and ignore friction losses. Use chart values to calculate the COP as a refrigerator. Sketch the thermodynamic cycle on both the pressure–enthalpy and temperature–entropy coordinates to identify significant state points.

8. Air enters a coil at 95°F db and 78°F wb and leaves at 62°F db and 60°F wb. The condensate is assumed to be at a temperature of 56°F. Find the total, latent, and sensible cooling loads on the coil with air at 14.7 psia.

9. Air enters a direct-expansion coil at 85 db and 70 wb and leaves at 62 db and 90 RH.

(a)   How much sensible heat and how much latent heat is removed from the air by the coil?

(b)   What is the apparatus dew point?

(c)   How much condensate drains off the coil?

10. Air enters a direct-expansion coil at 90 db, 60 RH and leaves the coil at 60 db, 95 RH. Find:
    (a) Heat removed per pound of air
    (b) Moisture condensed per pound of air
    (c) SHR for the condition line

11. Water at 51°F is chilled in an evaporator to 40°F at 60 pounds per minute. The heat transfer area is 20 ft², and the heat exchanger has an overall heat-transfer coefficient of 60 Btu/(hr·ft²·°F). The direct expansion evaporator uses R-12 and operates at 35°F. Find the evaporator effectiveness.

12. A water-cooled R-12 condenser has a U value of 104 Btu/(hr·ft²·°F) based on the outside tube area of 175 ft² when 38 gpm of water enters at 60°F and leaves at 80°F. If the refrigeration system performance is limited by this condenser operation, determine the maximum refrigerating capacity at an evaporating temperature of 10°F for the system, Btu/hr and tons.

13. The condenser in a heat pump system has water flowing through the tubes with a convective coefficient of 185 Btu/(hr·ft²·°F). The water enters at 60°F and leaves at 80°F. Refrigerant-12 at 90°F condenses on the outside of the tubes. The tubes are ½-in. OD copper. For a rated capacity of 45,000 Btu/hr, determine:
    (a) Required water flow rate, gpm
    (b) Length of tubing required, ft

14. Design a shell-and-tube heat exchanger to cool 18 gpm of water from 48°F to 40°F using R-12 evaporating at 40 psia on the outside of the tubes. Tubes are to be of copper with 0.555 in. ID and 0.625 in. OD. Determine:
    (a) Number of tubes per pass
    (b) U value for the exchanger, Btu/(hr·ft²·°F) based on A_i.

15. A condensing unit consists of a compressor whose performance is shown in Fig. P4.15A and a condenser whose performance is shown in Fig. P4.15B. Entering water temperature to the condenser is 65°F. Evaporator temperature is constant at 40°F.
    (a) Specify the rated cooling capacity of the condensing unit, Btu/hr.
    (b) If a ⅛-in. soft scale deposit builds up on the condenser tubes, with the reduction in heat transfer shown in the table below, determine the new cooling capacity of the condensing unit, Btu/hr.

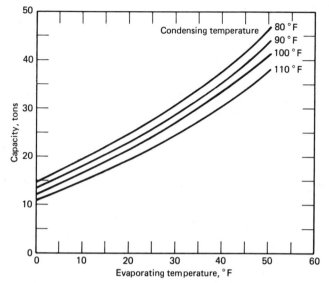

Fig. P4.15A. Performance of a Carrier Refrigerant-12 compressor.

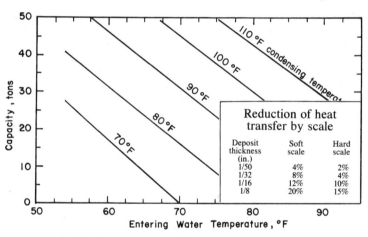

Fig. P4.15B. Performance of a Worthington Corporation four-pass Refrigerant-12 condenser. The condenser has a shell diameter of 16 in. and a length of 74¼ in. The evaporating temperature on which the capacities are based is 40°F, and the rate of flow of cooling water is 60 gpm.

16. A DXH 1413 M evaporator whose performance is given in Figs. P4.16A and P4.16B is to be used for supplying chilled water at 40°F with return water at 50°F. Capacity is to be 40 tons. Determine: (a) required water flow rate, gpm, and (b) the U value for the exchanger at this operating condition.

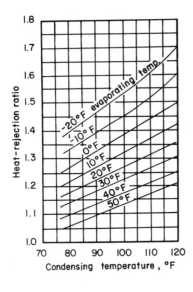

Fig. P4.16A. Ratio of heat rejected at the condenser to refrigerating effect for R-12 and R-22.

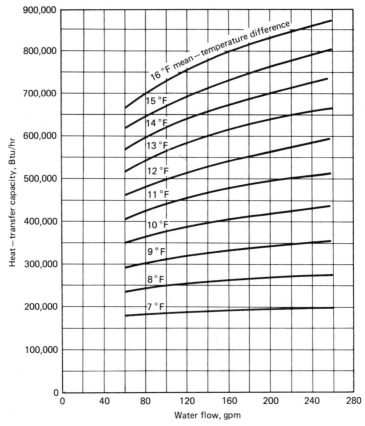

Fig. P4.16B. Performance of DXH 1413 M liquid chiller using R-12. External surface area of tubes is 302 ft^2.

260

17. An expansion device has a mass flow rate for R-12 given by

$$m = 60 + 0.25\Delta p$$

where m = flow rate, lb/min, and Δp = pressure drop across valve, psi. For an evaporator temperature of 0°F and a condenser temperature of 100°F, estimate the piston displacement required for a compressor if $C = 0.04$ and $n = 1.1$ for the compression process.

18. A heat pump is to be used for both air-conditioning and heating a home, maintaining the interior at 75°F all year long. In summer, when the outside air temperature is 96°F, the heat gain by the house is 36,000 Btu/hr. In winter, when the outside air temperature is 4°F, the heat loss by the house is 52,000 Btu/hr. If the heat pump is sized for cooling, determine the *minimum* amount of auxiliary heating (Btu/hr) needed for the 4°F winter condition.

19. A water-source heat pump uses 50°F lake water pumped to an evaporator whose performance characteristics are shown in Fig. P4.19A. The compressor's performance is shown in Fig. P4.19B. A matching heating coil is used for the condenser with a condensing temperature of 120°F. The heat rejection ratios of Fig. P4.19C are applicable. The heat pump is to be used for a large residence whose heat loss is shown in Fig. P4.19D. Determine:

(a) For an outdoor design temperature of 5°F, the amount of supplemental resistance heat to be installed, kW

(b) The total electric power for the heating system at design conditions, kW

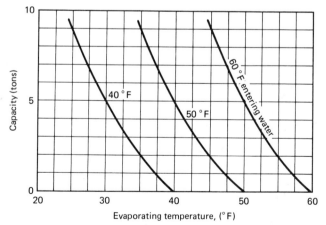

Fig. P4.19A. Evaporator performance.

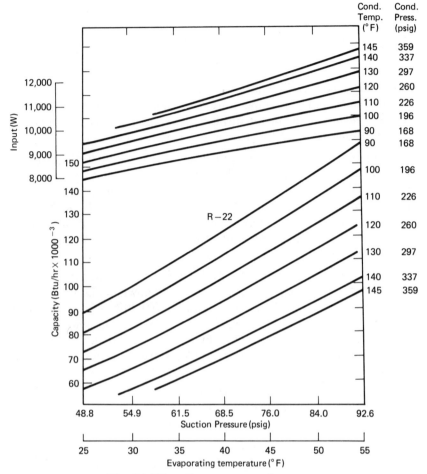

Fig. P4.19B. Compressor performance.

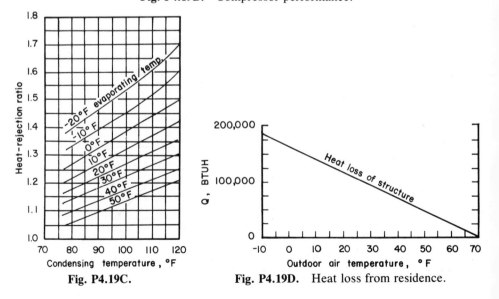

Fig. P4.19C.

Fig. P4.19D. Heat loss from residence.

5

GAS-FIRED AND HYBRID HEAT PUMPS

5.1. GAS-FIRED HEAT PUMPS [1]

Two-thirds of residential and over half of commercial gas consumption is for space-conditioning (heating and cooling) purposes. In total, more than 30% of the natural gas retailed is sensitive to heating/cooling systems technology. Any improvement in this technology would have a significant national impact on the consumption of gas and could be more than twice as great as technological advances in any other market sector.

The seasonal efficiency of a properly sized, properly installed conventional gas furnace is about 65%, and the COP of electric air conditioners is about 0.6. Using the contemporary gas furnace–electric air-conditioner system as a base, it is estimated that seasonal improvements in overall gas utilization of 2.0 to 2.5 can be achieved by using gas-fired heat pumps.

Business market analyses have indicated a cumulative total available heat pump market of 10,278,000 units during the period 1981–1990. Based on low, nominal, and high percentage market potential, the cumulative total of gas-fired heat pumps would be 208,000 (low), 419,000 (nominal), and 879,000 (high) for the projected 10-year period.

While it is recognized that the gas heat pump is not the total answer to conservation in the use of gas energy, it does have the potential for reducing the consumption level of gas for heating and cooling purposes in a reasonable period of time. To better appreciate the magnitude of these savings, the data generated to date can be expressed in some meaningful ways. For example, the average 1975 natural gas consumption per household was 124.8 Mcf. In 1990, with reduced consumption through the use of heat pumps,

263

238,000–528,000 additional households could be satisfied with the gas savings alone, without increasing supply requirements. The incentive is clear: The greater the number and the sooner gas-fired heat pump products can be introduced to the marketplace, the sooner benefits will be realized. However, there are still developmental and product-line complexities to be overcome.

Recognizing the energy conservation potential and advantages of an on-site heat-activated heat pump, the American Gas Association (AGA) initiated a gas heat pump research program long before the energy crisis became a pressing public concern. Within the overall objective of developing gas-fired heat pumps for the residential and commercial markets, this program has four specific goals:

1. To have gas-fired heat pumps available for the residential market by the early 1980s.

2. To have gas-fired heat pumps available for the commercial market by the mid-1980s.

3. To demonstrate gas-fired heat pumps with a reliability equivalent to that of a conventional gas-fired air furnace.

4. To achieve a first cost on initial gas-fired heat pumps within 15% of a combination forced-air furnace–electric air-conditioner system with a payback period of less than four years.

An interesting feature of the gas-fired heat pump systems is that their performance does not decrease with temperature as fast as the performance of conventional heat pumps.

5.2.   THERMAL ENGINE HEAT PUMPS [2]

While the electric heat pump employs an electric motor hermetically sealed into the refrigerant loop to drive the Rankine-cycle compressor, other means of providing compressor shaft power can be used as well. Indeed, using a prime mover such as a thermal engine as a power source has a considerable advantage in that engine waste heat rejected in coolant or exhaust gas can be used to supplement the refrigeration cycle output in the heating mode, thus substantially increasing the overall on-site system COP. On the other hand, the extra heat produced becomes a liability in the cooling mode, in that it has to be rejected without significantly diminishing the refrigerating capacity of the system.

This chapter discusses a number of new heat pump concepts using an on-site thermal energy to drive a refrigeration machine. In most cases of interest, the latter is of the usual Rankine-cycle type; in some cases, a cycle operating entirely in the gas phase is treated. Some generally applicable considerations in the design and operating characteristics of thermal engine heat pumps will be noted before specific heat pump concepts are discussed. Use has been made of a number of comprehensive discussions on this subject published by Wurm and Rush [3], Wurm et al. [4], and Colosimo [5], and an AGA research project report edited by Wurm [6].

Depending on the particular prime mover or refrigeration cycle contemplated, thermal engine heat pumps have a number of significant advantages in comparison to, say, electric heat pumps. The major ones are:

1. High overall efficiency (in a primary energy sense) in comparison to state-of-the-art electric heat pumps in heating mode and gas or oil furnaces
2. Ability to utilize prime mover waste heat to supplement refrigeration cycle on heating mode
3. Capability to modulate heat-pump capacity by varying fuel input rate to prime mover
4. Availability of on-site waste heat for evaporator defrost
5. Adaptability to total-energy systems or for operation independent of utility lines

Some of these characteristics are discussed further below.

The major inducement to thermal engine heat pump development is, of course, the combination of high engine efficiency and high refrigeration cycle COP obtainable in principle. Waste-heat utilization on-site, facilitated defrost, and the ability to modulate capacity all add to high seasonal performance in the heating mode. In the cooling mode, electric heat pumps have certain advantages, but it is possible to design a thermal engine heat pump to "hold its own" on cooling as well.

Prime movers, especially turbomachinery, are relatively easily speed-modulated by controlling fuel flow to the combustor or bypassing some of the power fluid around the prime mover. This allows heat pump capacity to be more easily modulated than with the alternating current induction motors presently employed in electric heat pumps.

Having a source of waste heat on-site not only makes supplemental heat available for augmenting the output but also allows its use for efficient evaporator coil defrosting in the heating mode. This may potentially eliminate what is frequently a major reliability problem with electric heat pumps as well as a cause of some efficiency loss.

There are also significant disadvantages to thermal engine heat pumps in general, in contrast to electric heat pumps or gas oil furnaces:

1. Requirement to reject prime mover waste heat in the cooling mode
2. Greater system complexity, which could result in potentially higher system cost and lower reliability

3. Difficulty in designing totally hermetic refrigeration system
4. Aggravated noise and pollution problems due to on-site prime mover operation and fuel combustion
5. Dependence on availability of scarce fossil fuels (natural gas, fuel oil)

As before, these apply in varying degrees depending on the particular system being considered.

In elaboration of some of the disadvantages listed above, it should be noted that the fact of having a fuel-burning prime mover on site, while highly advantageous in the heating mode, becomes a decided liability on cooling in that additional heat-transfer surface and pump or fan energy are required to reject engine waste heat.

Experience with small residential and commercial HVAC and refrigeration systems has shown that hermetic systems are preferred for reliability. In systems that operate above atmospheric pressure, refrigerant charge leakage is a possibility. In systems that operate at subatmospheric pressure, infiltration of air (or possibly other fluids) can occur. Noncondensible gases in condensers, especially, tend to diffuse and form a film adjacent to the condenser wall, thus increasing the inside film resistance to heat transfer.

With an on-site prime mover, design of a hermetic refrigerant loop becomes difficult although not impossible. One solution is to use a magnetic coupling, as in the Brayton-Rankine case. While eliminating a shaft, the mass of magnetic material required to transmit sufficient power at minimum slip may be significant in relation to the rest of the rotating group, thus increasing the polar moment of inertia of the shaft and lowering the critical rotational speed at the onset of shaft whirl. However, this may not be as serious a constraint as it might appear. The Brayton-Rankine system, which transmits 13 hp across a samarium-cobalt coupling, is designed to operate at speeds up to 90,000 rpm.

Another interesting solution to this problem is evident in the design of the (linear) free-piston Stirling-Rankine heat pump. This design employs an inertial compressor consisting of a cylindrical refrigerant compressor housing rigidly connected to the power piston. The cylinder contains a free inertial mass, the compressor piston, which compresses the refrigerant vapor as the housing reciprocates. The refrigerant suction and discharge lines are in the form of tubular helical springs that extend and contract with the motion of the compression cylinder. Some possible problems with this design may be (1) lack of sufficient gas pressure or "spring" in the heating mode to bounce back the inertial piston due to the lower mass density of the refrigerant, and (2) suction and discharge tube failure due to large number of flexing cycles encountered over the expected operating life of the equipment.

Rotating high-speed machinery or unbalanced engine compressors are bound to increase noise and vibration levels over hermetic electric motor compressors or gas furnaces. Much of this noise can be eliminated in a closed system with proper vibration damping and sound insulation, but again at some increased cost. On-site fuel combustion may increase emission levels of carbon monoxide, nitrogen oxides, and particulates unless an existing furnace burning the same fuel is being replaced. Again, these problems are amenable to design solutions.

Another limitation of a combination on-site prime mover and heat pump is the greater system complexity, in most cases, in comparison to an electric heat pump. This is not to say that an elegantly simple design such as the free-piston Stirling engine compressor with only three moving parts (in the engine compressor) is not possible, but, in general, practical thermal engine heat pump designs have to have:

Two different working fluids, requiring their sealing and isolation from each other

A large number of moving machined surfaces requiring machining to close tolerances and lubrication. (Most heat-pump concepts use refrigerant oil, but the engine lubrication may be dry.)

Several thrust and journal bearings

Feed pumps or blowers, and engine coolant pumps

Engine as well as heat-pump control devices

A separate starting system in some cases

Materials to withstand high static or dynamic operating stresses at high temperature

This increased complexity is likely to result in a high first cost for the equipment and potential problems of system and component reliability, especially in a product that is expected to operate essentially maintenance-free and unattended for a long time.

Now that some general advantages and disadvantages of thermal engine heat pumps have been discussed, some specific embodiments of prime movers and refrigeration machines will be described. Thermal engine heat pumps to be discussed include

Stirling-Rankine heat pump

Rankine-Rankine heat pump

Brayton-Rankine heat pump

Otto-Rankine heat pump

Ericsson-Ericsson heat pump

Table 5.1 summarizes some pertinent characteristics of prototype systems under development.

TABLE 5.1. Summary of Some Thermal Engine Heat-Pump Characteristics [2]

	Stirling-Rankine	Rankine-Rankine	Brayton-Rankine	Otto-Rankine	Ericsson-Ericsson
Engine Data					
Thermodynamic cycle	Free piston Stirling engine	Rankine steam turbine	Brayton gas turbine	Braun linear Otto engine	Ericsson
Power	3 kW	31 hp (23 kW)	13 hp (9.6 kW)		2 hp (1.5 kW)
Power fluid	Helium	Steam	Combustion gas	Combustion gas	Hydrogen
Efficiency (max.)	32–36%	23%	27%	33%	
Frequency	30 Hz	500 Hz	1170–1500 Hz	Variable	
Coolant	Glycol water	Glycol-water			
Fuel (s)	Gas (other fuel, solar)	Gas	Gas (fuel oil)	Gas	Gas (other fuel, solar)
Self-starting	Yes	Yes	No	No	Yes
Compressor Data					
Thermodynamic cycle	Rankine	Rankine	Rankine	Rankine	Ericsson
Refrigerant	R-22	R-11	R-12	R-22	Hydrogen
Type	Inertial (recip.)	Centrifugal	Centrifugal	Reciprocating	Reciprocating
Nominal cooling	3 tons	7.5 tons	10 tons	20 tons	
Hermetic	Yes	No	Yes	Yes	Yes
Other Data					
Electrical power required?	Yes	None (generates 8 kVA own power)	Yes	Yes	None

5.2.1. Stirling-Rankine Heat Pump [1, 2]

Figure 5.1 shows a theoretical Stirling engine operating a heat pump. The most significant advantage of the free-piston Stirling engine as a heat-pump prime mover is the high engine efficiency that is theoretically possible. The free-piston Stirling engine has only two moving parts, and the compressor has one, thereby providing for potentially high reliability. Also, the design of the system facilitates quick disassembly for maintenance.

The nominal performance goals for the Stirling-Rankine heat pump are a heating COP of 1.5 and a cooling COP of 0.9. The first prototype unit has been fabricated and assembled and is now being tested. This system, now in a 3-ton laboratory breadboard configuration, uses a free-piston Stirling engine developed with AGA support and a vapor-compression Rankine refrigeration loop. It is intended for residential and light commercial systems of up to 10 tons. The work is being performed by General Electric, with funding by AGA, GE, and the Department of Energy. It is a linear engine that follows the Stirling cycle by using a gas (helium) spring to provide the motion of a light displacer and a heavy piston. It is coupled with an inertia compressor that tends to oscillate out of phase with the Stirling engine to provide the pumping action. The R-22 refrigerant is transferred to the compressor through a steel tube-spring.

The refrigerant loop operates in the standard heat-pump cycle. An addi-

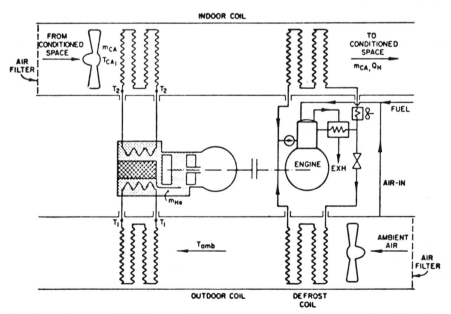

Fig. 5.1. Theoretical Stirling-cycle heat pump system using engine-rejected heat for heating purposes. Reprinted from HCP/M2121-01, Gordian Associates, Inc.

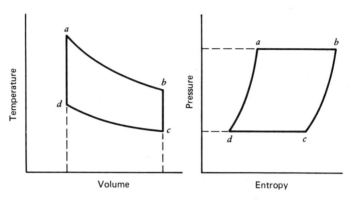

Fig. 5.2. The Stirling cycle. Cycle efficiency $\eta = (T_a - T_c)/T_a$. Source: Gordian Associates, Inc.

tional water-glycol loop is used to recover the engine's waste heat for supplementary use. In the heating mode, it transfers heat to the indoor coil; in the cooling mode it goes to the outdoor coil.

The ideal Stirling cycle, shown in Fig. 5.2, has received considerable attention over the years because of its potential for high efficiency. The constant-volume, constant-temperature cycle has the same efficiency as the Carnot cycle, but unlike a Carnot, a practical Stirling engine can be and has been built. In cost-effectiveness terms, the Stirling engine lost out to other engine cycles, particularly the Rankine-cycle steam engine, but modern technology combined with other benefits of the cycle have aroused new interest in the Stirling engine. Recent developments have tended to focus upon automotive applications because of the inherent low pollution levels of the external-combustion Stirling engine. These automotive engines have rotary shaft power output and require a high pressure differential, low leakage, and dynamic seals. However, the engine does not necessarily have to have rotary shaft power output.

The Stirling engine was originally used in the early 19th century to operate a piston pump to drain deep mines. In most refrigeration and heat pump devices, a single-piston positive-displacement pump is used to compress the vapor. This type of pump needs linear power. Fortunately, a free-piston Stirling engine has been developed that is a linear engine that could be matched with a reciprocating inertia compressor in a hermetically sealed assembly. Additionally, the engine is self-starting, does not require oil lubrication, has no gears, and has low loadings on the piston rings. Recognizing all these advantages, AGA negotiated a position on patent rights and, in 1972, initiated a project to further develop the engine for gas industry applications.

The Stirling cycle consists of a constant-temperature expansion, a constant-volume reduction in temperature and pressure, a constant-temperature compression, and a constant-volume heating. To accomplish this cycle, two pistons are required. The displacer piston, usually called the displacer, moves the working gas through the heating and cooling steps of the cycle. The power piston, usually called the piston, provides the working gas compression and takes the power out during the expansion. Obviously, the phase relationship between the displacer and piston is important if the cycle is to work.

In rotary Stirling engines, the displacer and piston are placed in separate cylinders and connected with a rather elaborate set of linkages. In the free-piston Stirling engine, the displacer and piston are placed in a single cylinder. This engine depends upon the mass differences among a light displacer, a heavy piston, and a gas spring formed by the pressure variations in the bounce space or what would typically be called the engine's crankcase.

In order to take the work out of the engine, a shaft could be brought through the engine housing and connected to the piston of a reciprocating compressor. However, this would require a sliding seal that would be effective with helium at 1000 psi in the bounce space. An alternative technique is to place an inertia compressor within the engine's crankcase and bring the refrigerant through the engine casing. This would permit the engine to be hermetically sealed. Figure 5.3 illustrates how this is accomplished. The engine's piston is coupled to the compressor-cylinder housing of the inertia compressor. The mass of the compressor aids in keeping the proper phase relationship between the engine piston and displacer. As the engine piston and compressor cylinder oscillate, the compressor piston tends to oscillate out of phase with the cylinder, providing a pumping action. The compressor assembly is hermetically sealed in the bounce space of the engine. The refrigerant is transmitted to the compressor through a steel tube spring.

The free-piston Stirling engine and inertia compressor can be used in a heat pump system as shown in Fig. 5.4. The refrigerant loop operates in the standard heat pump cycle. An additional heat-transfer medium is used to recover the engine's waste heat for supplemental heat. In the heating mode, the water-glycol loop will transfer heat to the indoor coil; in the cooling mode, to the outside coil.

A number of potential problem areas exist that require further development. For example, the decreased refrigerant pressure during heating-mode operation apparently produces insufficient "gas spring" to prevent the inertial compressor piston from banging against the ends of the compressor housing.

Engine hot end materials are required to withstand helium temperatures of 1200°F and pressures as high as 1050 psig. This places stringent requirements on materials of construction. While the engine-compressor has only three moving parts, close tolerances to minimize yaw of piston and displacer and low dead volume are critical if design efficiency is to be achieved. The ability of helical spring compressor suction and discharge tube joints to withstand an estimated 6–7 billion flexures has to be proven. It should be

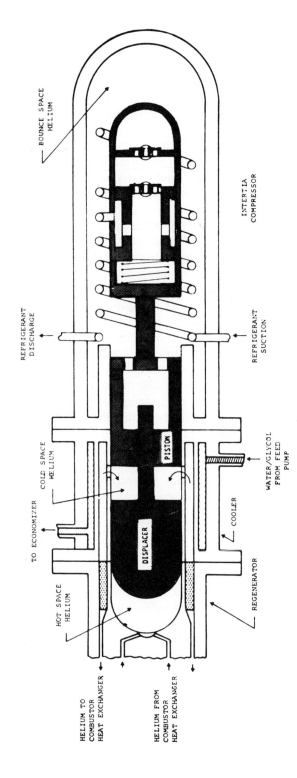

Fig. 5.3. Free-piston stirling engine with inertial compressor. Reprinted from HCP/M2121-01, Gordian Associates, Inc.

272

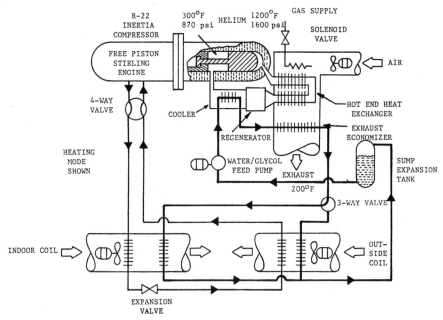

Fig. 5.4. Stirling-Rankine heat pump scheme.

noted that the control over engine operation does not appear to be "positive" in the sense of control over internal combustion engines or electric motors. The engine may continue to operate after the fuel supply is cut off, apparently until the hot end cools down sufficiently. Experimental engines have been observed to develop abnormal vibrations (shaking) in operation.

The estimated specific power of the projected engine is lower than for Rankine, Otto, or Diesel cycle engines. This usually implies a costlier engine.

Many of the problems cited are of course expected in an experimental device and should be amenable to design solution. Needless to say, critical questions that remain to be resolved are cost and operating reliability of production-type equipment. Although the free-piston Stirling engine seems attractive as a heat pump prime mover in many respects, it is not in commercial production at present.

5.2.2. Rankine-Rankine Heat Pump [2]

A Rankine-cycle heat pump powered by a steam turbine is being developed by Consolidated Natural Gas Service Company (CNG) with the participation of Mechanical Technology Inc. and the Department of Energy. The production version of the system is to be 100 tons applied or built up instead of unitary. The heat pump compressor, augmented by turbine waste-heat

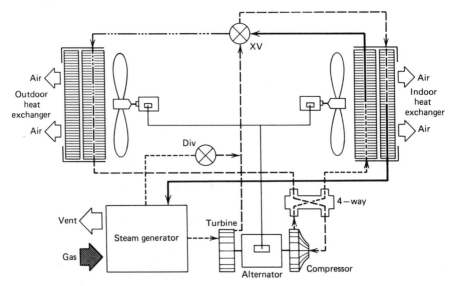

Fig. 5.5. DOE-CNG-MTI gas heat pump system schematic—heating service. Reprinted from HCP/M2121-01, Gordian Associates, Inc.

recovery, provides heat down to a temperature of $-10°F$; below that, direct steam injection into the steam coil in the supply air duct is planned.

Figure 5.5 shows the prime mover and heat pump flow schemes in the heating mode; the cooling mode is shown in Fig. 5.6. The steam turbine is a radial entry machine driving an alternator and centrifugal R-11 refrigerant compressor via a rigid shaft. The alternator generates all auxiliary power to drive a condensate-return pump and combustion air blower for the compact steam boiler and indoor and outdoor fans. The turbine is rated at 31 shaft hp and operates in the 30,000–60,000 rpm range. All turbine waste heat is recovered for maximum efficiency.

The Rankine-Rankine heat pump design exhibits a number of interesting features. First, the power fluid (water) is, of course, stable, nontoxic, and nonflammable. While the cost effectiveness of small steam turbines has not yet been universally demonstrated (except where special needs such as for reliability in emergencies or use of waste heat can be justified), the designers claim that the use of steam reduces necessary heat-exchanger area for engine heat rejection, since its film heat-transfer coefficients are higher than for air or refrigerant at the same Reynolds number, due to the higher Prandtl number of steam. A design feature of interest (and on which CNG has filed for patent protection) is the sequencing of the steam and refrigerant heat exchange in the indoor and outdoor coils (Figs. 5.5, 5.6). Steam and refrigerant are in a cross-flow configuration with respect to air but in counterflow with respect to each other. This maximizes the heat-transfer effectiveness of the heat exchangers and keeps down the refrigerant-compressor pressure ratio.

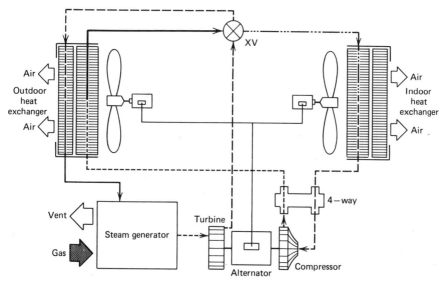

Fig. 5.6. DOE-CNG-MTI gas heat pump system schematic—cooling service. Reprinted from HCP/M2121-01, Gordian Associates, Inc.

It should be noted that turbomachinery is potentially highly reliable for long-time unattended operation, although there is admittedly little experience with small-capacity steam expanders. Other advantages are that the system is self-starting and, because the steam turbine cycle is basically a closed loop, with proper design, noise and vibration are not expected to be a problem.

The major drawbacks to this concept appear to be few but should be mentioned. First, the refrigeration loop is not hermetic. Hence, a dynamic shaft seal is necessary on the common shaft linking the steam turbine to the alternator and refrigerant compressor. Engineering a reliable dynamic shaft seal to minimize infiltration of air (both the water and the refrigerant loop have sections where the pressure is subatmospheric) without incurring a significant efficiency penalty is not a trivial problem. Air infiltration can cause corrosion on the steam side and reduce heat-exchanger overall heat-transfer rates on the refrigerant side and hence reduce the capacity.

The Rankine-Rankine heat pump is intended for commercial applications in areas where the dominant requirement is for heating. As such it would compete primarily with the "year-round" air conditioner using gas for heating.

5.2.3. Brayton-Rankine Heat Pump

Another gas-fired heat pump being considered is the Brayton power cycle (gas turbine) driving a vapor compressor in a Rankine cycle. The objective of the unit is to develop a subatmospheric gas-fired Brayton-cycle engine

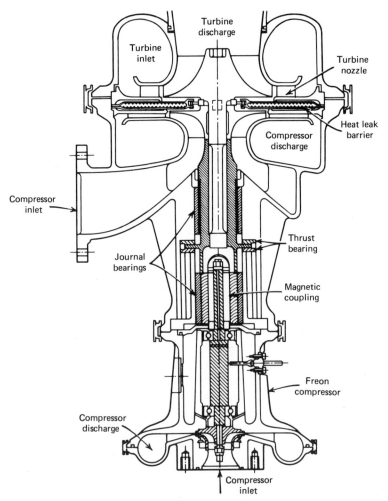

Fig. 5.7. Brayton-Rankine magnetic coupling heat pump. Reprinted from HCP/ M2121-01, Gordian Associates, Inc.

driving a centrifugal R-12 compressor via a hermetically sealed magnetic coupling. The development is tailored to a 10-ton rooftop vapor-cycle air conditioner, with the engine and compressor replacing the electric motor and positive displacement compressor. This concept is depicted in Fig. 5.7.

The work is being performed by Garrett Corp. and Dunham-Bush, Inc., on a cooperative funding basis with AGA.

In order to obtain the advantages of a closed Brayton cycle without the disadvantages of an expensive heat exchanger, a subatmospheric Brayton engine has been developed. Air enters through a recuperator, then into a combustor, then enters the turbine at 1500°F and 14.5 psia, expands to 1146°

at 5.8 psia, and goes through a heat exchanger and into a compressor, where the gas stream is brought up to slightly above atmospheric pressure and exhausted. A possibility exists of also using an economizer in parallel with the recuperator or heat exchanger. There are two heat-transfer loops. The vapor-compression loop uses R-12, and the water-glycol heat-recovery loop dumps heat to the outdoor coil during the cooling mode and, through the indoor coil, provides supplemental heat during the heating mode.

During heat-pump operation (see Fig. 5.8), the exhaust heat from the engine will be utilized in addition to the vapor-cycle unit operating in the heat-pump mode, thus providing a heating COP as high as 1.5 and a cooling COP of 1.10 as an air conditioner. When developed, this concept can provide an all-gas year-round space-conditioning unit that can reduce the gas consumer's energy expense, reduce seasonal peak-load demands, and conserve source energy.

The gas-turbine heat pump development emphasizes commercial or multifamily applications. A "breadboard" system has been operated, and development of a commercial prototype is under way. Commercial production is expected by the late 1980s, with units expected to range from 7.5 to 25 tons in capacity. The installed cost is expected to be 15–20% higher than existing systems, with payback in fuel savings in about 2 years. The Garrett Corporation has reportedly achieved a cycle efficiency of 38%, using a small regenerated gas turbine.

Some potential problem areas with the Brayton-Rankine concept are as follows: (1) Clearly, the high (90,000 rpm maximum) rotating speeds and turbine wheel temperatures of the order of 1500°F place stringent requirements on rotating group materials. (2) The turbine is not self-starting, requiring a relatively complex vacuum or a refrigerant bottoming start cycle. (3) The present plan of using reverse-cycle defrost (rather than turbine waste heat) will incur an efficiency penalty.

The Brayton-Rankine heat pump is being developed as a unitary gas heat pump for commercial applications. While the design is amenable to scale-up, according to the developers, it is unlikely that a smaller unit than the prototype 10-ton (or perhaps 7½-ton) size could be developed at a reasonable first cost. However, at the Glynwed Central Resources Unit in Solihull, England, work is under way on a miniature turbine for a residential size unit [8]. Glynwed has spent $200,000 on the world's smallest turbocompressor alone, though it looks simple enough, as shown in Fig. 5.9. On the heat-pump side there is a two-stage expansion turbine, and at the other end of the shaft, a radial compressor. It is only 3 in. long but incorporates a lot of research and development.

Know-how was needed to create a miniature turbo machine that could run efficiently at 150,000 rpm, a speed that gives the best match between blade speed and nozzle velocity. Mass-production of these precision parts is one of the many problems yet to be licked. The basic cycle is shown in Fig. 5.10.

Water-outlet temperature in the prototype is a modest 60°C (140°F),

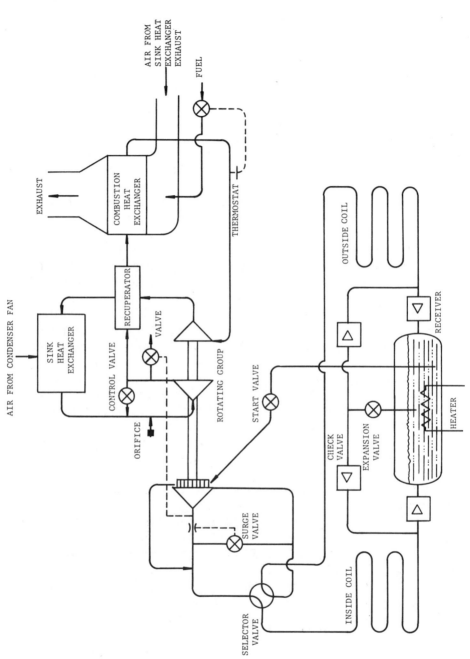

Fig. 5.8. Schematic heat pump configuration for the Brayton/Rankine concept. Reprinted from HCP/M2121-01, Gordian Associates, Inc.

Fig. 5.9. Three-inch-long turbine rotor for heat pump. Reprinted with permission from *Popular Science* © 1979, Times Mirror Magazines, Inc., Illustration by Eugene J. Thompson.

barely high enough for effective space heating with ordinary radiators. Fan convectors are seen as the most suitable units for a low-temperature heating system. They have a highly efficient surface-to-air heat transfer, and the fan motor uses no more electricity than a 30-W light bulb. They would work well in the cooling mode, too, as forced air reduces the condensation problem with panel radiators. Glynwed had already designed an unobtrusive convector, only 4 in. deep, with thermoelectronic speed control for the quiet centrifugal blower.

Heat-pump performance, using ambient air, falls off as outdoor temperature drops. To counter this, an auxiliary gas heater for the water circuit is planned as a backup in really cold weather. The future package will include microprocessor controls for economical year-round operation. Initial cost of the new heating system would be two or three times that of a conventional boiler.

5.2.4. Otto-Rankine Heat Pump

Another concept that has been advanced is the use of a spark-ignition free-piston reciprocating gas engine as a heat-pump prime mover. The engine was developed as a driver for air compressors by A. T. Braun. Participants in the proposed application of the engine to drive a commercial-scale Rankine-

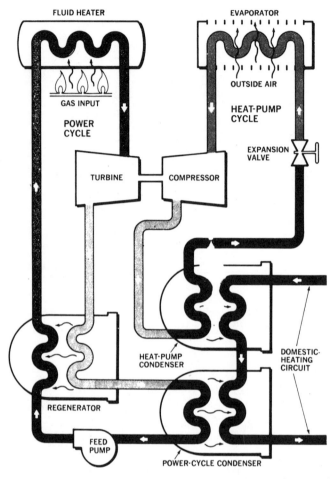

Fig. 5.10. Glynwed Brayton-Rankine heat pump. Reprinted with permission from *Popular Science* © 1979, Times Mirror Magazines, Inc., Illustration by Eugene J. Thompson.

cycle heat pump include Tectonics Research Inc. (a Braun subsidiary) and Columbia Gas System Service Co.

The Braun linear engine has been described in some detail in an article by Braun and Schweitzer [7]. Its major advantages over crank-type reciprocating engines are (1) smaller number of moving parts, (2) lower cost due to simpler construction and fewer critical tolerances, and (3) lower fuel consumption at equivalent compressor gas (air) output. The developers claim that this engine compressor is closer to commercialization that any of the other engine-driven heat pumps, in that major components of the air-compressor version of the system have been tested for over 45,000 hr of running time, and a heat pump type of test engine has been successfully run for

15,000+ hr. The developers also claim that the engine compressor has a higher COP than other engine systems, has a higher part-load efficiency (due to the ability of the prime mover to modulate speed), and is of simpler construction. The compressor can run without refrigerant oil.

However, this engine-driven heat pump has a number of apparent limitations. Although the refrigerant compressor in the design configuration is hermetic, the membrane isolating the refrigerant loop from the prime mover is subject to fatigue failure as a result of a large number of flexures over the expected 60,000-hr design life of production-type equipment. Indeed, although it has fewer moving parts than an automotive-type, crank-driven engine, the power transmission from the engine to the refrigerant compressor uses a number of mechanical components that are potentially subject to high wear in heat pump service under possibly very adverse conditions (e.g., cold-soak startup). Also, while the engine itself may have only a few moving parts, it requires carburetion or fuel injection, ignition, a startup system, and other auxiliaries. Finally, the Braun linear engine is an internal combustion engine developing relatively high combustion-chamber pressures at high temperature. Thus noise and pollutant emissions are expected to be a problem, which, although amenable to solution, would surely detract from overall system efficiency.

5.2.5. Ericsson-Ericsson Heat Pump

The heat pumps described in the previous sections of this chapter have been two-fluid heat pumps; that is, different fluids were used in the engine and in the refrigeration loop. This is because higher overall efficiencies are obtained, in practice, if the power fluid and refrigerant are each selected on the basis of offering the optimal set of thermodynamic, transport, flammability, toxicity, cost, and other properties. On the other hand, separate power and refrigeration loops make for a more complex and costly machine. Hence, in principle, there are certain advantages in a single-fluid thermal engine heat pump concept. One such concept is a gas-fired Ericsson-cycle engine driving an Ericsson-cycle refrigerating machine.

The Ericsson cycle is similar to the Stirling cycle except that the latter ideally uses constant-volume regenerative heat exchange, whereas the Ericsson cycle uses constant-pressure regenerative processes. That is, in the ideal Ericsson engine, the working gas (e.g., hydrogen, helium, carbon dioxide, or air) is compressed isothermally by the power piston, heated at constant pressure to the higher cycle temperature, expanded isothermally to produce work, and cooled at constant pressure to the lower cycle temperature. As with the Stirling cycle, the thermodynamic efficiency of the Ericsson cycle is identical to that of a Carnot machine operating between the same terminal temperatures. Again, however, machines approximating the Ericsson cycle are easier to build than Carnot machines. The Ericsson refrigerator employs the same processes as the Ericsson engine but in a re-

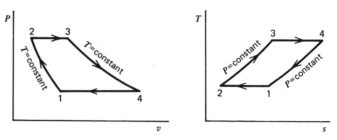

Fig. 5.11. The air-standard Ericsson cycle.

verse cyclical order (counterclockwise on a *T-s* diagram). Thus work has to be input to the refrigerator to produce a net refrigerating effect.

The Ericsson cycle is shown in the *P-v* and *T-s* diagrams of Fig. 5.11. This cycle differs from the Stirling cycle in that the constant-volume processes of the Stirling cycle are replaced by constant-pressure processes. In both cycles there is an isothermal compression and expansion. In Fig. 5.12, the Ericsson cycle is accomplished in a device that is essentially a gas turbine. If we assume an ideal heat-transfer process in the regenerator, that is, no pressure drop and an infinitesimal temperature difference between the two streams, and reversible compression and expansion processes, then this device operates on the Ericsson cycle. Since all the heat is supplied and rejected isothermally, the efficiency of this cycle will equal the efficiency of a Carnot cycle operating between the same temperatures.

The difficulties in achieving such a cycle are primarily those associated with heat transfer. It is difficult to achieve an isothermal compression or expansion in a machine operating at a reasonable speed, and there will be pressure drops in the regenerator and a temperature difference between the two streams flowing through the regenerator. However, the gas turbine with intercooling and regenerators is a practical attempt to approach the Ericsson cycle.

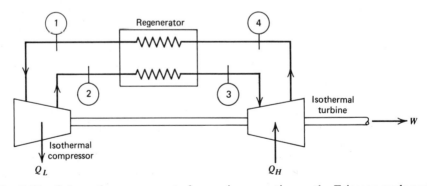

Fig. 5.12. Schematic arrangement of an engine operating on the Ericsson cycle and utilizing a regenerator.

An Ericsson-Ericsson heat pump has been proposed by Energy Research and Generation Inc. (Benson [9]). The concept utilizes a single-fluid hermetic free-piston machine with an engine displacer, heat-pump displacer, and two phaser pistons actuated by working gas pressure. The engine is valveless. Since this machine is still in the proof-of-concept stage (as far as heat pump development is concerned), the developers' claim of attaining 80% of the efficiency of an ideal Carnot engine heat pump at the same operating temperature and pressure conditions has yet to be proven.

5.3. ABSORPTION-CYCLE HEAT PUMPS

The absorption heat pump is unlike the other gas-fired heat pumps in that it does not actually involve a heat engine as prime mover or a compressor. It utilizes two fluids, which are pumped through the system. One fluid (the refrigerant) is sequentially absorbed, boiled out, condensed, and reabsorbed in the second fluid (the absorbent) to produce the heat pump action. In this system the only components with moving parts are basically the solution pump and air blowers.

Efficiency of the absorption heat pump is very much a function of the fluids used in the system. Units operating on the absorption cycle to provide space cooling and refrigeration have been commercially available for many years. The most common systems have working fluids of lithium bromide–water or ammonia-water. These systems are well known and characterized. In the lithium bromide–water systems, water is the refrigerant, thus restricting cycle operation to temperatures above the freezing point of water. This limitation makes the application of lithium bromide–water systems as heat pumps impractical.

However, ammonia-water absorption refrigeration systems can be converted to operate as heat pumps, just as mechanical refrigeration systems can be converted. Studies using ammonia and water as working fluids have predicted the performance for a gas-absorption heat pump at nominal heating and cooling conditions. These studies show potential advantages for gas-absorption heat pump systems over conventional systems, particularly for heating.

Figure 5.13 shows a generalized diagram of an absorption refrigeration cycle.

Three gas-absorption heat pump concepts have been investigated in a study conducted for AGA by General Electric. An ammonia-water system, like today's ammonia-water absorption chiller, was taken as the baseline system. The Whirlpool Corp. designed and built a prototype 3-ton ammonia-water heat pump that demonstrated the performance indicated by the studies mentioned above. Because this system required relatively little modification of currently marketed hardware, it represents a potentially near-term gas-

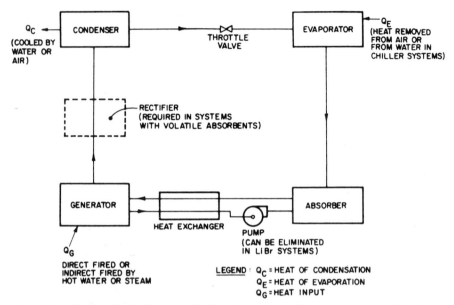

Fig. 5.13. Generalized diagram of the absorption-refrigerant process. Reprinted from HCP/M2121-01, Gordian Associates, Inc.

fired heat pump and was therefore selected as the baseline gas-absorption heat pump concept.

Two advanced concepts considered to be representative of performance improvements achievable by the early 1980s were evaluated. The first of these is an upgraded ammonia-water system that operates at boiler temperatures on the order of 350–400°F and employs a novel generator-absorber heat exchanger to reduce the gas-heat input requirements, thereby increasing the cycle COP. Refrigeration COP improvements of 75% over today's ammonia-water systems appear achievable. The second advanced system consists of a cycle employing a new absorption working fluid combination. The advanced fluid is used in a system having the same basic components as the ammonia-water system.

An absorption heat pump utilizing a proprietary fluorocarbon refrigerant and aromatic hydrocarbon absorbent is being developed for residential use by Allied Chemical Corporation and Phillips Engineering under sponsorship of the Gas Research Institute. Figure 5.14 shows this single-effect machine with absorber and generator heat exchangers, a separator heat exchanger, and a precooler. Three fluid-circulation pumps are also shown. The goals for operating COP of this unit are 1.2 for heating and 0.5 for cooling. Further improvement in heating COP could be obtained by recovering the waste heat from the generator section of the unit. It is believed that significant advances in fluids, fluid pumps, and heat exchangers have overcome problems that hindered earlier efforts to develop an absorption heat pump. This system is

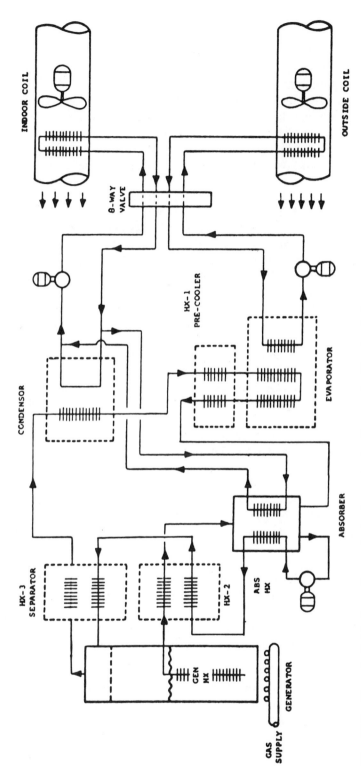

Fig. 5.14. Organic fluid absorption-cycle heat pump—heating mode. Reprinted from HCP/M2121-01, Gordian Associates, Inc.

285

expected to cost about 15% more than a gas furnace–electric air conditioner combination and will probably be most competitive in areas where heating is the major requirement.

Arkla Industries Inc., the only U.S. manufacturer of heat-powered unitary air conditioners, has been selected to conduct research, development, and demonstration of the air-source heat-powered heat pump in a project that is being sponsored by the U.S. Department of Energy (DOE) in support of the federal energy conservation program for residential and commercial systems.

The program, which is administered by Union Carbide Corp.'s Nuclear Power Division at Oak Ridge, Tennessee, involves adapting the ammonia-water absorption technology used in Arkla's gas-fired air-conditioning unit to achieve heating outputs greater than 100% of the value of fuel input. Successful completion of the project will lead to a fossil-fuel heat pump using substantially less energy than the oil- and gas-fired space-heating equipment manufactured at the beginning of the decade.

Absorption air conditioners are inherently more expensive than electric systems because (1) they require more heat exchangers and (2) a unit's cooling surface must be large enough to reflect both heat from the combustion process and heat removed from the space that was cooled. (In electric systems, the heat from generation is rejected at the electric generator site.) Absorption units had a lower operating cost than electric chillers in the era of cheap gas and are still competitive in some areas, but their use is limited by first cost.

A typical single-effect absorption chiller has a COP of 0.52; double-effect units can have COPs of 0.88. These COPs must, however, be carefully qualified, as they do not include the electricity used for fans and pumps. They cannot be compared directly with the COPs of electric chillers, since fuel costs are different and the COP of electric units does not include power-plant losses. It is reasonable to expect that improvements in current designs could lead to significant improvements in performance.

The Iron Fireman double-effect chiller, which was manufactured for a time, was able to achieve a COP of 1.2 (not including boiler losses and electric energy requirements). Some engineers believe it would be possible to increase this COP for double-effect absorption devices to the range of 1.35.

Absorption-cycle heat pumps, depending somewhat on the cycle and the particular design configuration employed, have the folllowing major *advantages:*

Inherently simple and potentially highly reliable equipment, in that the solution pump and air-moving fans (or water pumps for water-cooled systems) are the only moving parts (no compressor is needed)

High efficiencies in the heating mode due to the possibility of heat recovery from on-site combustion as supplementary heat to that provided by the absorption-cycle condenser itself

Lower parasitic energy requirements than for thermal engine heat pumps

Refrigerant-absorbent fluids can be used that are chemically nonreactive with atmospheric ozone

The generator of the absorption cycle heat pump can be energized more easily by a solar thermal source, if of sufficiently high temperature, than a thermal engine

Availability of on-site waste heat for evaporator defrost

Inherent capability of storing energy as high-pressure refrigerant vapor.

Absorption-cycle heat pumps also have a number of *disadvantages:*

Comparatively large heat-exchanger areas (at attendant high first cost) required for realization of acceptable COP

Usually lower cooling-mode COP than obtainable with Rankine refrigeration machinery (including electric heat pumps)

Toxicity of ammonia required that direct refrigerant/conditioned air contact in water-ammonia systems must be avoided (necessitating the use of either a water loop or other intermediate means of heat transfer)

Corrosiveness of water-ammonia systems requires the use of steel rather than aluminum or copper as material of construction; the organic absorption system uses aluminum

Diameter limitation of generator pressure vessel requires use of a tall unit for adquate vapor capacity, with attendant bulkiness of equipment

"Plumbing" of absorption machines, although obviously dependent on the particular design, appears to be more complex than that of a Rankine cycle machine

Dependence on availability of scarce fossil fuels (e.g., natural gas).

A problem that needs mention is the fact that initial experience with water-ammonia absorption-cycle gas air conditioners has been rather dismal, and their market share, since their highest penetration (about 4%) of the residential unitary air-conditioning market, has decreased steadily. Cost and equipment reliability problems have been the cause, so that only one of the original three unitary gas air-conditioner manufacturers remains in the market today.

5.4. CHEMICAL HEAT PUMPS

Chemical heat pumps operate with cycles that have some similarity to absorption systems. However, these chemical cycles use the latent heat of absorption, or of chemical reactions, as the primary heating mechanism, whereas absorption units use the latent heat of condensation and evaporation of refrigerant driven by absorption, for example, of ammonia in water,

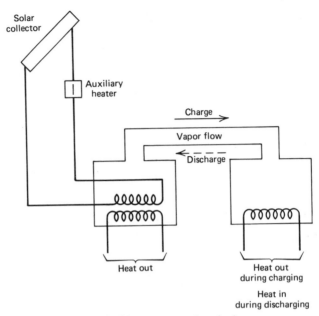

Fig. 5.15. Simple chemical heat pump–chemical energy storage system.

or water in a concentrated lithium bromide solution. An example of a chemical heat pump is the absorption of water vapor into concentrated sulfuric acid.

Chemical heat pump/chemical energy storage (CHP/CES) systems promise considerable advantages over other heating and cooling systems, both solar and nonsolar, and are being actively developed under DOE funding [10]. The advantages include considerably smaller storage volumes than any other concept for either heating or cooling, efficient long-term low-loss storage at ambient temperature, the ability to be trickle-charged during off-peak hours and used for load leveling, and the potential of reducing the required collector area in solar heating-and-cooling systems by as much as 70%.

Figure 5.15 shows a schematic of a simple CHP/CES system. It is apparent that the basic principles of operation of a CHP/CES system are similar to conventional absorption-cycle refrigeration machinery. However, the different chemicals used and the energy storage feature require the development of totally different system design and control approaches, including new heat-exchanger designs, new reactor designs, new valving and manifolding, and other components. Separation of the chemicals is achieved by heating at 60–200°C, after which the sensible and/or latent heat in both chemicals is either used for heating or is rejected. The mutual affinity of the chemicals due to their difference in chemical potential can be used during the recombination process to cause the temperature of one of them to drop below the temperature of the surroundings, allowing it to absorb heat. Both the heat of reaction and the sensible or latent heat absorbed from the surroundings are released at a temperature higher than the surroundings as the chemicals recombine, thus achieving the pumping of heat and completing the cycle. Note that either heating or cooling can be accomplished with the device. Chemical storage is achieved merely by storing the chemicals separately and allowing them to recombine only when heating or cooling is required.

Numerous chemicals and design approaches can be used in CHP/CES systems. These systems can be classified into several categories: open and closed cycles, and liquid-liquid, solid-liquid, or solid-solid systems. An *open-cycle* system is one in which one of the constituent chemicals (usually water) is discarded to and reclaimed from the environment rather than being stored. Desiccant dehumidification systems are an example of this class of system. *Closed-cycle* systems, on the other hand, store and recycle all of the constituents.

A *liquid-liquid* system is one in which both constituents are stored in liquid form, one chemical typically being boiled out of solution with the other chemical and then condensed. A *solid-liquid* system is one in which one of the chemicals is a solid and the other can be driven off from the solid and condensed to be stored as a liquid. Finally, a *solid-solid* system is one in which three chemicals are involved rather than two. In this case the chemical that acts as the working fluid for heat pumping is driven off from one solid substrate and is absorbed in another solid substrate to be stored. This approach is typically used when the working fluid is not readily condensible at the temperatures and pressures of interest. Active development programs in all of the above areas are proceeding under funding from the Energy Storage and Solar Heating and Cooling Divisions of DOE, with primary technical management provided by Sandia Laboratories in Livermore, California. Systems under study include, among others, $CaCl_2-H_2O$, $MgCl_2-H_2O$, $CaCl_2-NH_3$, $MgCl_2-NH_3$, $MnCl_2-SH_3$, $SrCl_2-NH_3$, $CaCl_2-CH_3OH$, $MgCl_2-CH_3OH$, $H_2SO_4-H_2O$, metal hydrides, and zeolites [1–6].

The required heat input temperature of 60–200°C allows CHP/CES systems to operate efficiently from a variety of heat sources, including solar, fossil fuel, electricity, and waste heat. The heat pump feature of CHP/CES systems allows them to provide both heating and cooling. The heating COP anticipated range, 1.2–1.7, allows CHP/CES systems to significantly reduce energy consumption compared to non–heat pump systems. This leads to both reduced fuel costs and reduced capital investment in the heat-input device, be it a solar collection device, fossil-fuel burner, or other energy converter. The anticipated cooling COP for CHP/CES systems ranges from .6 to .7, which is comparable with conventional absorption cycle cooling devices.

A very attractive feature of most CHP/CES systems under study is their extremely high energy storage densities compared to more conventional energy storage devices. Volumetric energy storage densities of CHP/CES systems are 3–10 times that of typical unpressurized hot water storage, 3–4 times that of pressurized hot water storage, 4–10 times that of hot-rock storage, 2–5 times that of phase-change storage, and 4–10 times that of ice storage.

The use of reactive chemicals rather than water, rocks, or some other medium requires greater safety precautions for the CHP/CES systems. No insurmountable safety hazards are anticipated for any of the CHP/CES systems, and the costs of meeting safety requirements are under investigation.

CHP/CES systems for heating applications are generally more complex than alternative systems, and studies are in progress to reduce the number of valves and other components to a minimum.

5.5. SOLAR-FIRED HEAT PUMPS

Solar-fired air-conditioning, using the heat from the sun to drive any of several possible types of refrigeration cycle, is still primarily a research subject. Those technologies that have proven to be feasible technically in commercial packages (absorption cycles) are still too costly to compete with electrically driven air conditioners. Other technologies with interesting or promising features are still in the early stages of development to assure that their size, reliability, operating characteristics, and cost will allow them to compete. Examples of these are heat engines (e.g., Rankine expanders coupled to more-or-less conventional compressors) and chemical heat pumps.

Heat engines driving vapor compressors currently appear to represent the most promising technology for high-efficiency gas or solar-fired air-conditioning. They promise high efficiency or COP (cooling effect divided by thermal input). Systems using solar-fired Rankine engines typically have performance figures in the following ranges:

COP, 0.4–0.7 depending on condensing temperature

Thermal source temperature (collector temperature), 120°C

Air flow requirements, approximately 2000 cfm per ton of cooling capacity, about twice the requirement for a conventional heat pump.

5.6. HYBRID HEAT PUMPS

The hybrid heat pump is a space-conditioning system consisting of an electric heat pump with a fossil-fuel furnace backup. The electric heat pump used in the hybrid heat pump has the same components as the electric heat pump used in a conventional heat pump employing electrical resistance heating for backup.

Hybrid heat pumps using a combustion furnace for supplemental heat, introduced a number of years ago, seem to be attracting increased market interest for principally two reasons: (1) They are applicable to the retro market as add-ons to existing furnaces, and (2) they have a better annual load factor (ratio of average power demand to peak demand) than resistance heat–supplemented heat pumps. With the exception of certain control components designed to regulate compressor and furnace operation, essentially standard split-system heat pump components are utilized: a heat pump outdoor unit with fan and heat exchanger, a compressor (a separate unit in some models), and a conditioning or indoor coil for the air handler. Duct work for heat pumps is larger than for furnaces because the lower output temperature with heat pumps requires greater airflow.

Operation of the hybrid heat pump/fossil furnace differs from that of the conventional heat pump/electric furnace. With the conventional heat pump, the heat pump indoor coil is upstream of the furnace and can operate simultaneously with the electric furnace. With the hybrid heat pump, the heat pump cannot be placed upstream of the fossil furnace because condensation could corrode the furnace heat exchanger. Also, with fossil furnaces, a flame can turn Refrigerant-22 into phosgene gas. The indoor coil must therefore be placed downstream of the furnace. With the coil in this position, the furnace and heat pump cannot operate simultaneously, because high furnace exit temperatures exceed the normal refrigerant temperature in the condenser, making normal heat pump operation impossible.

Operation of the heat pump and furnace alternately rather than simultaneously results in higher consumer energy costs than if the heat pump could be operated as much as possible above the breakeven temperature. Conversely, alternate operation reduces electrical energy requirements at low outside temperatures, providing load-factor improvement potential for utilities.

There are two basic types of heat pump/furnace control. In one, the heat pump does not operate below the balance point or breakeven temperature, whichever is higher. The balance-point temperature is the outdoor temperature at which the heat-pump heating capacity equals the building heating load. The breakeven temperature is the outdoor temperature at which the consumer is economically indifferent to using the heat pump or fossil furnace. Below this temperature, operation of the fossil furnace is more economical than the heat pump, while above this temperature, because heat pump efficiency increases with increasing outdoor temperature, it is more economical to operate the heat pump.

The other operational mode either controls the heat pump off the first stage of the thermostat and the furnace off the second stage of the thermostat or by a combination of first- and second-stage operations. Control is based on the juxtaposition of the outdoor temperature, balance-point temperature, and breakeven temperature.

Specific heat pump/furnace control scenarios are shown in Table 5.2.

TABLE 5.2. Heat Pump/Furnace Control Scenarios*

Manufacturer	Control Scenario
BDP (Bryant, Day and Night, Payne)	A. When the economic breakeven temperature is above the balance-point temperature, the heat pump will operate at all outdoor temperatures above the breakeven temperature, and the fossil fuel furnace will operate at all temperatures below the breakeven temperature.
	B. When the breakeven temperature is below the balance point and the outdoor temperature is below the breakeven temperature, the furnace operates exclusively and is energized by the first stage of the wall thermostat.
	C. When the breakeven temperature is below the balance point and the outdoor temperature is between the balance point and breakeven point, the heat pump is energized by the first-stage thermostat. Since the heat pump cannot carry the full load, the second stage of the wall thermostat will call for additional heat. The fossil furnace is turned on and the heat pump is turned off. The furnace stays on until both the second and first stages of the thermostat are satisfied. Thereafter the cycle repeats for all temperatures between the breakeven temperature and the balance-point temperature.
Carrier	Outdoor thermostat is set at breakeven temperature or balance-point temperature, whichever is higher. Heat pump operates exclusively above this temperature; furnace operates exclusively below this temperature.
Fedders (Climatrol/ Mueller)	Control scenario A when the breakeven temperature is computed to be higher than the balance-point temperature. (An outdoor thermostat is used with scenario A to activate the furnace below the breakeven temperature and the heat pump above this temperature.) When the breakeven temperature is below the balance point, no outdoor thermostat is used and, as in scenario C, the heat pump is deactivated and the furnace activated when the second stage of the thermostat calls for additional heat.
Friedrich	Outdoor thermostat is set at breakeven temperature. If outdoor temperature is below this temperature, only the furnace will operate. When the outside temperature is above the breakeven temperature, the

TABLE 5.2. (Continued)

Manufacturer	Control Scenario
	heat pump will be activated by the indoor thermostat and the furnace will operate only when the heat pump cannot handle the load. (Similar to scenarios B and C.) Below 0°F outside temperature, the heat pump will be locked out. It can come on again at 10°F. There is a 4-min delay when switching from furnace to heat pump operation.
General Electric	Can be operated in either unrestricted or restricted mode. In the unrestricted mode, the heat pump operates any time first-stage heat is called for. When second-stage heat is required, the heat pump is cut off and the furnace operates until both stages are satisfied. An outdoor thermostat is not required for unrestricted operation. (Similar to scenario C.)
	For restricted operation, scenario A is used with an outdoor thermostat.
	An emergency heat switch can convert the system to "furnace only" operation regardless of outdoor temperature.
Lennox	Normally, scenario C is utilized with no outdoor thermostat. An outdoor thermostat can be used, however, if economics justify employing scenarios A and B. After the furnace shuts off, the heat pump cannot come on until the plenum temperature has dropped to the 90–100°F range. Blower operation is continuous.
Rheem	Two kits can be used. One kit uses an outdoor thermostat and restricts heat-pump operation below a preset temperature. When the outdoor temperature warms up approximately 5°F, it will lock out the furnace and let the heat pump do the heating. The set point is recommended to be slightly higher than the balance-point temperature. (This is similar to scenario A.)
	The other kit does not use an outdoor thermostat. Heat-pump operation is unrestricted as in scenario C, with the furnace deactivated when the second-stage thermostat opens. There is a 5-min delay between call for first-stage heat and compressor start-up. The blower operates during this period. A low-temperature cutoff shuts off the compressor when the outside temperature is below 15°F.

TABLE 5.2.   (Continued)

Manufacturer	Control Scenario
York	An outdoor thermostat is set to the balance-point temperature. Heat pump operates exclusively above this outside temperature. Below this temperature the heat pump runs off the first-stage thermostat. When the second stage of the thermostat is closed, the furnace is activated and the heat pump will be deactivated. The furnace will shut off when the second-stage thermostat becomes satisfied. The furnace blower will continue to run since the first-stage thermostat is not satisfied. When the supply air thermostat closes (at approximately 98°F) the heat pump will be activated if both 5 min have passed since the heat pump was deactivated and the first-stage thermostat has not been satisfied. Outdoor sensor can be set to either $+10°F$ or $-10°F$ to prohibit heat-pump operation below this point. Below the temperature cutoff the furnace operates off the second-stage thermostat.

*Blatt, M. H., and Erickson, R. C., "State-of-the-Art Assessment of Hybrid Heat Pumps," Report EM-1261, Research Project 1201-6, prepared for EPRI, December 1979.

Conservative practice in terms of reliability is to use a scenario such as Carrier's where the heat pump does not operate at all below the balance point or breakeven temperature, whichever is higher. The scenario preferred by the utilities contacted maximizes the heat pump operation below the balance point. This latter scenario is similar to the one used by the other heat pump manufacturers and in Table 5.2. For this scenario the heat pump operation is controlled off the first stage of the thermostat and the furnace can be controlled off the second stage of the thermostat or by a combination of first- and second-stage operations.

In addition to the typical balance point for a heat pump–residence combination there is an "economic balance point" in add-on heat pumps that is determined by the cost of electrical power and the cost of the supplemental fuel on the existing furnace. The relationship between electricity costs and gas costs as related to the economic balance point and the add-on heat pump is shown in Fig. 5.16.

For illustrative purposes, the operation of one commercially available system is described in the following paragraphs. Figure 5.17 presents a schematic of the system.

This system is designed for use with air heaters burning fossil fuels and may be applied to a new or existing installation. The system comprises an

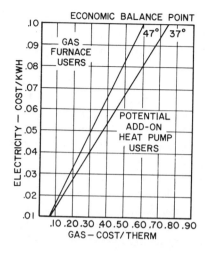

ECONOMIC BALANCE POINT

Fig. 5.16. Selection of add-on heat pumps.

outdoor heat pump unit, indoor coil, refrigerant lines, control box (supplied with a separate heat pump relay and defrost limit controls), and air heater. The heat pump system is installed as usual, with the exception that it is matched to an add-on indoor coil on an air heater rather than the customary fan-coil unit.

The two-stage heating and single-stage cooling thermostat controls the system, with heat pump operation (primary heat source) for the first stage and air-heater operation (supplementary heat source) for the second stage. In heat pump–supplementary electric heat applications, one or more out-

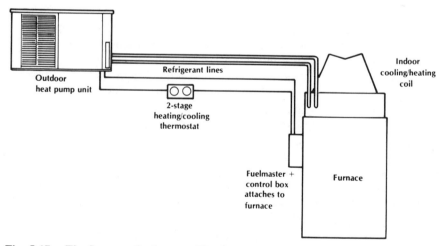

Fig. 5.17. The Lennox Fuelmaster Plus heat pump system, based on The Lennox split air-air heat pump. Courtesy Lennox Industries, Inc.

door thermostats are used to cycle the heaters as the outdoor temperature drops. However, in the Fuelmaster Plus system the indoor thermostat controls the supplementary heat (air heater) and the outdoor thermostat is not required.

Heating requirements on mild heating days will be provided on demand of the first stage of the thermostat with the Fuelmaster Plus. When the outdoor temperature reaches the "balance point" of the structure (heat loss equals heat pump heating capacity) and the first stage of the thermostat is not satisfied, the second stage will activate the air heater (secondary heat source). On heater start-up, a heat relay in the Fuelmaster Plus control box terminates the heat pump operations. When the second stage (air-heater operation) of the thermostat is satisfied and the plenum temperature reaches 32–38°C, a heat pump delay reactivates the heat pump as the heat source until conditions warrant a repeat of the cycle for air-heater operation. Fan operation is continuous during the heat pump–air heater cycle. Additionally, when the outdoor temperature drops below the low-temperature compressor monitor setting, the heat-pump operation stops and the air heater provides the heating requirements.

When a defrost cycle is initiated (heat pump system changes from heating to cooling) and the indoor coil starts to supply cold air, the Fuelmaster Plus controls will automatically start the air-heater operation to temper the air, thus avoiding the distribution of cold air in the conditioned area. When the supply air temperature reaches 43–49°C, the defrost limit control terminates the air-heater operation, preventing excessive temperatures in the supply air. If after a defrost cycle the air temperature downstream from the coil exceeds the 46°C closing point of the heat pump delay, the compressor operation will stop until the plenum temperature has dropped to between 32 and 38°C.

During the cooling season, the Fuelmaster Plus heat pump operates in reverse, with the indoor coil supplying the cooling requirements.

The Carrier Optimizer heat pump is designed for use with existing hydronic heating systems either (1) installed with a duct system to provide cooling and most heating, with the existing hydronic system left intact for cold weather backup duty only; or (2) with a Carrier MPX heat exchanger connecting the existing hydronic system to the outdoor heat-pump section. Typical piping schematics are shown in Fig. 5.18, where the interconnecting heat exchanger is labeled 09WQ.

In a discussion of the testing of Lennox hybrid heat pumps, Gilles [11] and Romancheck [12] show that the peak load for winter peaking utilities can be reduced by shifting from electric heat to fuel oil during peak periods. Hybrids also improve load factors by requiring electrical usage during off-peak periods. Benefits to the consumer include lower energy costs and the ability to satisfy heating-season requirements with a single tankful of fuel oil.

One summer-peaking utility is encouraging customers to use heat pumps

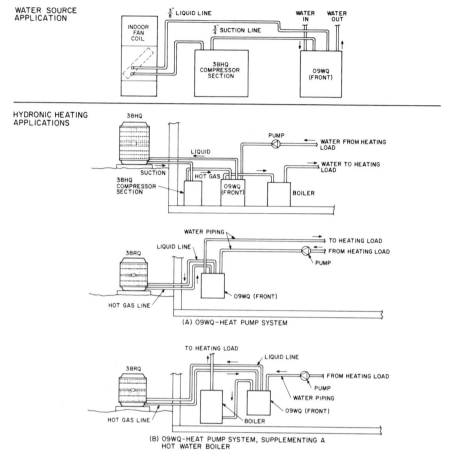

Fig. 5.18. Piping schematics for hydronic heat pump systems. Reproduced by permission of Carrier Corporation. Copyright © 1982 by Carrier Corporation.

when replacing central air conditioners on forced-air fossil-fueled furnaces. This provides additional winter load at off-peak conditions, and their winter peak occurs at a temperature at which the heat pump is off.

A combination utility with nearly equal summer and winter electrical peaks is interested in hybrid heat pumps mainly to improve the electrical load factor by shifting peak electricity consumption to gas consumption. Increases in peak gas consumption can be accommodated by the utility more easily than increases in peak electricity consumption.

A winter-peaking utility is advocating first-cost incentives to the consumer to encourage installation of hybrid heat pumps instead of electrical resistance heating. Use of hybrid heat pumps would directly relieve peak

TABLE 5.3. Typical Seasonal Heating Costs, High-Performance Heat Pumps, Oil-Furnace Backup ($\Delta T_f = 3.0°F$) [13]

Sites	Hybrid Heat Pump			Conventional Heat Pump	All Electrical Resistance	Fossil Fuel (Oil)
	Electricity	Fossil Fuel (Oil)	Total			
Minneapolis	270.78	494.77	725.54	739.92	1146.52	919.24
Chicago	397.04	285.28	682.32	927.08	1726.90	737.18
Boston	182.82	133.13	315.95	265.18	521.15	593.03
Seattle	155.93	190.69	346.62	270.29	614.66	872.46
Kansas City	273.84	125.61	399.45	458.28	851.36	455.33
Philadelphia	273.24	200.78	474.02	520.23	1068.84	677.17
Sacramento	71.75	10.72	82.47	75.82	180.37	260.16
Las Vegas[a]	—	—	—	—	—	—
Dallas	80.97	15.96	96.93	96.28	248.61	179.16
Los Angeles[a]	—	—	—	—	—	—
New Orleans[a]	—	—	—	—	—	—
Miami	10.62	0.77	1.4	11.25	33.23	20.80

[a]Oil not available.

TABLE 5.4. First-Cost Differentials for Hybrid Heat Pumps Compared to Other Heating Systems [13]

Hybrid Heat Pump Compared to:	Installation Type		Furnace Type		Cost Differential,[a] Hybrid − Alternate, $
	New	Retrofit	Oil	Natural Gas	
Air conditioner/	x		x		767
fossil furnace	x			x	760
		x	x		767
		x		x	760
Conventional heat	x		x		457
pump	x			x	200
Air conditioner/	x		x		1136
electric furnace	x			x	

[a] Best estimate (selected from a wide variance of data).

load increases, since they would not be operating at temperatures as low as those at which the system peaks.

In a recent study by Blatt and Erickson [13] for EPRI, energy costs per heating season were computed using a seasonal performance model. An example of these energy costs for a high-performance hybrid heat pump using an oil-furnace backup is shown in Table 5.3. Equipment cost differentials and a wide range of estimates were obtained by contacting manufacturers and utilities. Table 5.4 presents the data selected for use in the life-cycle cost estimates. The heat pump portion of the hybrid heat pump and conventional heat pump are assumed to be identical with regard to components, defrost control, and defrost operation.

An earlier study showed that hybrid heat pumps did not compare favorably with other types of heat pumps based on calculated first cost and annual energy cost. The poor showing of hybrids may have been due to the inefficiency of the early model chosen for examination. Life-cycle cost calculations by Blatt and Erickson [13] showed that a significant market exists for hybrid heat pumps as retrofit units (for replacement of air conditioners in systems with adequate duct work) or as new units in winter peaking areas where natural gas is not available. For the new units, utility intervention in the form of a rebate to customers for installing hybrids is required to overcome first-cost differentials. These rebates can be justified economically by utility load management savings resulting from peak shaving. Utility and manufacturer publicity is essential to making the cost advantages of the hybrid heat pump known to the consumer.

The main institutional barrier to installation of hybrid heat pumps is the objection voiced by gas utilities to being required to maintain a complete distribution system while supplying only small quantities of natural gas to users of the hybrid heat pump. If gas companies are allowed to increase rates to hybrid heat pump customers, this could have a significant effect on hybrid market penetration.

5.7. SUMMARY

Of the gas-fired heat pump concepts being investigated, the Stirling-Rankine system is by far the best performing system. The COP of the gas-absorption system does outperform the Stirling-Rankine heat pump at ambient temperatures below 0°F, but this is a result of an assumption that the Stirling-Rankine system is operated at constant speed. On the cooling-mode side, the Stirling-Rankine is the only gas-fired heat pump system that can compete with an electrically driven vapor-compression system.

In system capacity relative to the design-point cooling capacity, the Stirling engine efficiency is so high that the engine waste heat is not as significant in contributing to the overall heating capacity as the Brayton-Rankine system. As a consequence, the heating-to-cooling ratio of the Stirling-Rankine gas-fired heat pump may require considerably more make-up energy in the cooler climates, unless multiload operation proves feasible. However, Didion et al. [14] recently carried out experiments at the National Bureau of Standards with an air-source Rankine-cycle heat pump powered by a Philips I-98 gas-fired, water-cooled Stirling engine. The maximum thermal efficiency of the engine was 28% over the range of conditions tested. The experiments found that between 50 and 60% of the fuel energy input was rejected to engine coolant (in contrast to, say, 15–30% for an internal combustion engine), energy that is potentially partially reclaimable as supplemental heat for heating-mode operation. Another finding was that higher efficiency was obtained at the lower engine speed as the compressor was speed-locked in constant ratio to the engine through a pulley drive. Hence, the authors suggest that speed and capacity modulation is indicated only below low-speed heat pump balance point.

The COPs for heat-fired heat pumps cannot be compared directly with COPs of electric heat pumps. In order to obtain comparable "system efficiency" for an electric system, the electric COPs must be reduced by the efficiency of converting primary fuels to electricity and transmitting this energy to a heat pump system. The average generating efficiency of U.S. utilities is approximately 29%; the average transmission losses, approximately 9%. Under these assumptions, an electric heat pump with a heating COP of 3.0 and a cooling COP of 2.5 would have an effective "system" COP of 0.79 for heating and 0.66 for cooling.

An add-on heat pump is a remote heat pump conditioning coil (and associated hardware) fitted into the supply air duct of a forced-air gas or oil furnace. Such a system, in addition to providing cooling in the summer, would be able to utilize the high heating efficiency of the heat pump cycle in milder weather but rely on the combustion of fuel during the very coldest weather when the full capacity of the furnace can be utilized to best advantage. Most important, this system can be retrofitted to an existing home, provided the forced air ducts are large enough to accommodate air-conditioning.

REFERENCES

1. Sarkes, L. A., Nicholls, J. A., and Menzer, M. S., "Gas Fired Heat Pumps: An Emerging Technology," *ASHRAE J.,* **19**(3), 36–41 (1977).

2. _____,"Heat Pump Technology—A Survey of Technical Development Market Prospects," HCP/M2121-01, Contract No. EX-76-C-01-2121, Gordian Associates, Inc., June 1978.

3. Wurm, J., and Rush, W. F., "Evaluation of Engine-Driven Heat Pump Systems of Small Capacities," paper presented at 14th Intern. Congr. Refrigeration, Moscow, Sept. 20–30, 1975 (IGT, Chicago, 1975).

4. Wurm, J., Rush, W. F., Weil, S. A., Dufour, R. J., and Lassota, M. J., *An Assessment of Selected Heat-Pump Systems* (Annual Rept. on Proj. HC-4-20 1976).

5. Colosimo, D. D., "On-Site Heat Activated Heat Pumps," paper presented at ERDA Conference on Technical Opportunities for Energy Conservation in Appliances (Boston, May 11, 1976).

6. Wurm, J., "An Assessment of Selected Heat Pump Systems," Project HC-4-20 for the American Gas Assoc., March, 1974.

7. Braun, A. T., and Schweitzer, P. H., "The Braun Linear Engine," paper presented at SAE Intl. Automotive Engineers Congress, Detroit, Michigan, January 8–12, 1973.

8. Scott, D., "Turbine-Drive Heat Pump Doubles Heating/Cooling Efficiency," *Popular Science,* October 1979.

9. Benson, G. M., "Thermal Oscillator," U.S. Patent No. 3,928,974 (December 30, 1975).

10. Hiller, C. C., "Development of Chemical Heat Pumps/Chemical Energy Storage Systems for Heating and Cooling Applications," *Proc. 4th Annual Heat Pump Techn. Conf., April 9–10, 1979.*

11. Gilles, T. C., "Summary of Remarks Presented to Task Force VIII, Argonne National Laboratories," Lennox Industries, June 8, 1976.

12. Romancheck, R., "Possible Methods for Retarding Growth Rates," Pennsylvania Power and Light Company, March 16, 1977.

13. Blatt, M. H., and Erickson, R. C., "Residential Hybrid Heat Pump State-of-the-Art Assessment," *ASHRAE Trans.* **86**(2), 1980.

14. Didion, D., Maxwell, B., and Ward, D., "A Laboratory Investigation of a Stirling Engine Driven Heat Pump," paper presented at International Seminar on Heat and Mass Transfer in Buildings, Dubrovnik, Yugoslavia, August 29–September 3, 1977.

6

HEATING AND COOLING LOADS

6.1. LOAD COMPONENTS

One of the most important steps in the design of an environmental control or air-conditioning system is estimation of the space heat and moisture loads. The size, construction, and use of the space must be known. The schedule of occupancy and activity of occupants must be estimated. All equipment emitting heat and/or moisture in the space and schedules of operation must be determined. The space conditions to be maintained as well as the environment external to the space must be specified.

Load calculations can be used to accomplish one or more of the following objectives:

Provide information for equipment selection and HVAC system design;
Provide data for evaluation of the optimum possibilities for load reduction;
Permit analysis of partial loads as required for system design, operation, and control.

These objectives can be met by making accurate load calculations and understanding the basis for the loads.

Details on the methodology for heat-loss and heat-gain calculations can be found in the *ASHRAE Handbook and Product Directory, 1977, Fundamentals** [1] and elsewhere.

*Referred to hereafter as *1977 Fundamentals*.

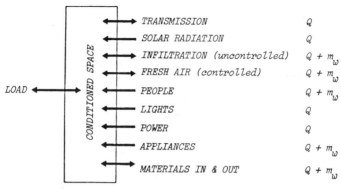

Fig. 6.1. Components of heating and cooling loads. Reprinted by permission from *ASHRAE Environmental Control Principles*.

Figure 6.1 indicates the components of heating and cooling loads.

6.2. INDOOR DESIGN CONDITIONS

The temperatures that should be maintained in the rooms of a building vary over a wide range, the exact temperatures depending primarily upon the use for which the rooms are planned.

In Figure 6.2 the most recently developed comfort envelope is plotted on the coordinates of the ASHRAE psychrometric chart. The comfort zone is the one recommended in ASHRAE Comfort Standard 55-74 [2], which applies generally for average clothing and activity. The most commonly recommended design conditions for comfort are:

Dry-bulb air temperature	76°F (24.5°C)
Relative humidity	40% (20–60%)
Air velocity	45 fpm (0.23 m/s)

Table 6.1 gives inside design temperatures commonly used for some other conditions. These temperatures are dry-bulb temperatures and do not necessarily imply comfort conditions, since comfort is also influenced by relative humidity, air motion, activity of the occupants, and radiation effects. For spaces heated by radiation methods, slightly lower dry-bulb temperatures may be acceptable.

The effect of relative humidity on all aspects of human comfort has not yet been completely established. Low relative humidity may well be undesirable for reasons other than those based on thermal comfort. Low levels will no doubt lead to increased evaporation from the membranes of the nose and throat and drying of the skin and hair. Some medical opinion attributes the increased incidence of respiratory complaints to drying out of the mu-

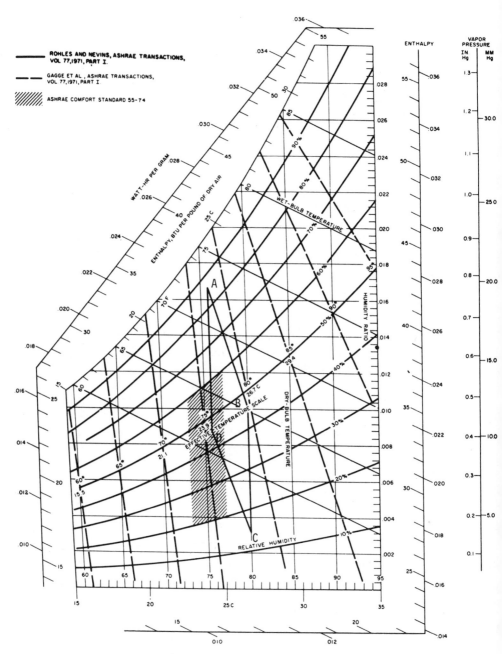

Fig. 6.2. The ASHRAE comfort chart. Reprinted by permission from *ASHRAE Handbook of Fundamentals 1977.*

TABLE 6.1.   Inside Dry-Bulb Temperatures

Type of Space	Temperature, °F	(°C)
Computer rooms	72 ± 2	(22 ± 1)
Dairy cattle barn	50–75	(10–24)
Factories		
Light work	60–70	(16–21)
Heavy work	58–68	(14–20)
Gymnasiums	60–70	(16–21)
Hospitals		
Operating rooms	70–95	(21–35)
Patients' rooms	74–76	(23–24)
Kitchens and laundries	66	(19)
Paint shops	80	(27)
Steam baths	110	(43)
Stores	70–74	(21–23)

cous membranes due to low humidity indoors in winter. In spite of lack of complete information, it seems logical to assume that extremes of humidity are undesirable and that, for human comfort, relative humidity should be kept within the broad range of 30–70%.

When inside temperatures for a given type of installation are selected for design purposes, the temperature at the *breathing line* of 5 ft (or less frequently at the 30-in. *comfort level*) is considered. However, in making calculations for heat-transfer losses, the breathing-line temperature may be far from the average (mean) room temperature because of the tendency for air to stratify as the warm air rises to the top. For ceilings not over 20 ft high it has been found that the temperature rises approximately 2% for each foot of height above the breathing line.

With types of heating systems having positive (fan) air circulation, as when the air is distributed into the space in such a manner as to oppose the tendency of warm air to rise, the temperature difference between floor and ceiling may be quite small, and the foregoing rule should not be followed. For this condition a temperature-rise factor of 1% per foot of height above the breathing-line temperature should be used. Where ceiling heights exceed 15 ft, allow 1% per foot of height up to 15 ft, and then allow 0.1°F for each foot of height in excess of 15 ft. Very often a temperature at the floor 5°F less than at the breathing line is assumed.

6.3.   OUTDOOR DESIGN CONDITIONS: WEATHER DATA

Data on outdoor design conditions are presented for over 1000 stations in the United States, Canada, and 102 other countries in the *1977 Fundamentals* volume [1]. Table 6.2 provides selected values from this source. All data

TABLE 6.2. Outdoor Air Design Conditions[a]

City[b]	Latitude, deg N	Winter Design Dry-Bulb, °F		Summer Design Dry-Bulb and Mean Coincident Wet-Bulb, °F			Mean Daily Range °F
		99%	97.5%	1%	2.5%	5%	
Albuquerque, NM	35	12	16	96/61	94/61	92/61	27
Atlanta, GA	33	17	22	94/74	92/74	90/73	19
Boston, MA	42	6	9	91/73	88/71	85/70	16
Chicago, IL (CO)	41	−3	2	94/75	91/74	88/73	15
Cleveland, OH	41	1	5	91/73	88/72	86/71	22
Dallas, TX	32	18	22	102/75	100/75	97/75	20
Denver, CO	39	−5	1	93/59	91/59	89/59	28
Detroit, MI	42	3	6	91/73	88/72	86/71	20
Fairbanks, AK	64	−51	−47	82/62	78/60	75/59	24
Honolulu, HI	21	62	63	87/73	86/73	85/72	12
Houston, TX (AP)	29	27	32	96/77	94/77	92/77	18

Indianapolis, IN	39	−2	2	92/74	90/74	87/73	22
Kansas City, MO	39	2	6	99/75	96/74	93/74	20
Las Vegas, NV	36	25	28	108/66	106/65	104/65	30
Los Angeles, CA	34	37	40	93/70	89/70	86/69	20
Memphis, TN	35	13	18	98/77	95/76	93/76	21
Miami, FL	25	44	47	91/77	90/77	89/77	15
Minneapolis, MN	44	−16	−12	92/75	89/73	86/77	22
Mobile, AL	30	25	29	95/77	93/77	91/76	16
New Orleans, LA	30	29	33	93/78	92/78	90/77	16
New York, NY	40	11	15	12/74	89/73	87/72	17
Philadelphia, PA	39	10	14	93/75	90/74	87/72	21
Phoenix, AZ	33	31	34	109/71	107/71	105/71	27
St. Louis, MO (AP)	38	2	6	97/75	94/75	91/74	21
San Diego, CA	32	42	44	83/69	80/69	78/68	12
Seattle, WA (CO)	47	22	27	86/68	82/66	78/65	19
Washington, DC	38	14	17	93/75	91/74	89/74	18

[a] Adapted by permission from ASHRAE Handbook and Product Directory, 1977 Fundamentals [1].

[b] AP = airport; CO = community.

307

used in determining design conditions are based on detailed records from official weather stations of the U.S. Weather Bureau, U.S. Air Force, U.S. Navy, Canadian Atmospheric Environment Service, and weather organizations of the various countries noted.

The 99% and 97.5% winter design values represent the temperatures which equalled or exceeded these portions of the total hours in the months of December, January, and February (a total of 2160 hr) in the northern hemisphere, and the months of June, July, and August in the southern hemisphere, with a total hour count of 2208 hr. In a normal winter, there would be approximately 22 hr at or below the 99% design value and approximately 54 hr at or below the 97.5% design value. Outdoor humidity can be assumed at 70–80% in winter for most areas.

Recommended summer design dry-bulb and coincident wet-bulb temperatures and outdoor daily temperature ranges are also provided in Table 6.2. The dry-bulb temperatures presented represent values that have been equalled or exceeded in 1, 2.5, and 5% of the total hours during the summer months of June through September (a total of 2928 hr) in the northern hemisphere. The coincident wet-bulb temperature listed with each design dry-bulb temperature is the mean of all wet-bulb temperatures occurring at the specific dry-bulb design temperature. In a normal summer, there would be approximately 30 hr at or above a 1% design value and approximately 150 hr at or above a 5% design value.

The outdoor daily range of dry-bulb temperatures gives the difference between the average daily maximum and average daily minimum temperatures during the warmest month at each station.

6.4. OTHER FACTORS AFFECTING DESIGN CONDITIONS

With increasing need to conserve energy, adjustments may be mandated by the federal government in the allowable winter and summer design conditions and possibly thermostat settings. The U.S. Congress under the Energy Policy and Conservation Act of 1975 has established federal guidelines that require that certain levels of energy efficiency be achieved in new building construction. In the interest of energy conservation, new buildings that are used primarily for human occupancy (including residences, office space, portions of factory and industrial occupancies, educational facilities, or shelter for public assembly, business, etc.) must meet certain minimum design requirements that will enable the efficient use of energy in them. Many standards have been published since 1973, but ASHRAE 90-75, "Energy Conservation in New Building Design" [3] is most widely adopted. ASHRAE Standard 90-75 sets forth requirements for the design of new buildings covering their exterior envelope and selection of their HVAC, service water heating, electrical distribution and illuminating systems, and equipment for effective use of energy. Portions of the standard relating to design conditions are excerpted in Table 6.3.

TABLE 6.3.   Design Conditions from ASHRAE Standard 90

Heating, Ventilating and Air-Conditioning (HVAC) Systems

Scope. This section covers determination of heating and cooling loads, design requirements, and control requirements for general comfort applications in new buildings where normally clothed people are engaged in sedentary or near-sedentary activities . . .

Design Parameters. The following design parameters *shall* be used for HVAC system design load calculations for general comfort applications:

Outdoor Design Conditions. Winter and summer outdoor design conditions *shall* be selected for listed locations in Chapter 23 of the *1977 ASHRAE Handbook of Fundamentals* from the columns of $97\frac{1}{2}$ percent values for winter and $2\frac{1}{2}$ percent values for summer.

Indoor Design Conditions.
Winter: The recommended heating design condition is 72°F (22°C) dry bulb. If humidification is provided, it *shall* be designed to a *maximum* relative humidity of 30 percent.
Summer: Where comfort air conditioning is required or used, the recommended indoor design condition is 78°F (25.5°C) dry bulb for cooling. The actual design relative humidity within the comfort envelope as defined in ASHRAE Standard 55-74 "Thermal Environmental Conditions for Human Occupancy" *shall* be selected for minimum total HVAC system energy use.
"Winter" conditions prevail when a heating system is installed and used, and "summer" conditions prevail when a cooling system is installed and used.
Where heating systems are not normally installed the *recommended* design condition is 78°F (25.5°C) dry bulb, and where cooling systems are not normally installed the *recommended* design condition is 72°F (22°C).
Although the *recommended* design points are established above, the HVAC system design *should* permit operating at minimum energy levels within the boundaries of the comfort envelope.
Due to internal heat gains or losses, it *may* be more energy efficient to operate at other than the design points for major portions of the year.

SOURCE: Reprinted by permission from *ASHRAE Environmental Control Principles,* Fig. I-27.

Figure 6.3 provides another example of possible changes in indoor design conditions based on energy considerations. A heating or cooling system is sized to meet design conditions but functions at partial capacity most of the time. The ability to properly control the system at all times is one of the most important factors in the design of the system. The proper use and application of controls should receive primary consideration at the time the heating or cooling system is being designed. The designer of the system and its controls should take into consideration the expected use and type (single-level, multistory, etc.) of buildings in which they are to be used, the size of system, and

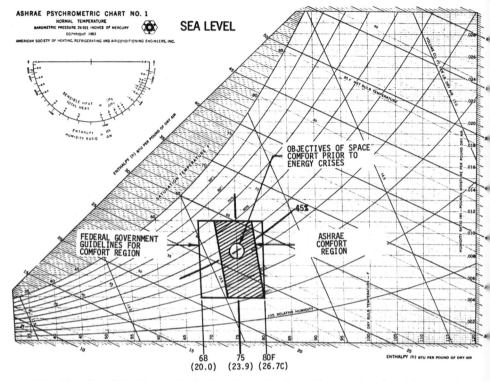

Fig. 6.3. Possible changes in indoor "comfort" conditions. Reprinted by permission from *ASHRAE Environmental Control Principles*.

the desired results, including the conservation of energy. Quality control equipment should be consistent with the accuracy of control required. Where results more precise than those for normal comfort applications are required, such as in certain manufacturing areas, constant-temperature rooms, and so on, both the air-conditioning system and the control system should be given special consideration so that the required results can be produced. Control devices can be used that produce almost any degree of control required, but it is useless to provide such controls unless the air-conditioning system is capable of properly responding to the demands of the controllers. For example, it is virtually impossible to maintain close control of temperature and humidity by starting and stopping a refrigeration compressor or by opening and closing a refrigerant valve. It is necessary instead to employ refrigeration equipment that will permit proportional control, or to use a chilled-water system with a cooling coil or spray dehumidifier that will permit proportional control. It is neither good practice nor economical to select equipment capable of producing far more precise control than the application requires, or to complicate the system to obtain special sequences or cycles of operation when they are not necessary. It is well to remember that the system must be adjusted and maintained in operation for many years, and that the simplest system that will produce the necessary results is usually the best.

EXAMPLE 6.1

Specify *completely* the indoor and outdoor design conditions for both summer and winter for an apartment building in Cleveland, Ohio.

Solution.   (Selection according to Table 6.3.)

Summer		Winter	
Outside	Inside	Outside	Inside
88°F db	78°F db	5°F db	72°F db
72°F wb	60% RH[a]	80% RH	30% RH
41° lat			
22° range			

[a] Fig. 6.2.

6.5.   INFILTRATION AND VENTILATION

Ventilation of a conditioned space may be the result of natural infiltration, alone or in combination with intentional mechanical ventilation. Natural infiltration will vary with the indoor-outdoor temperature difference, wind velocity, and the tightness of the construction.

Infiltration is the air leakage through cracks and interstices, around windows and doors and through floors and walls of a building of any type from a one-story house or commercial building to a multistory skyscraper. The magnitude of infiltration depends on the type of construction, workmanship, and condition of the building. The rate of infiltration cannot be controlled by the inhabitants of the building to any considerable extent.

Natural ventilation, on the other hand, is the intentional displacement of air through specified openings, such as windows, doors, and ventilators.

Outside air infiltration may account for a significant proportion of the heating or cooling requirements for buildings. It is therefore important to be able to make an adequate estimate of its contribution with respect to both design loads and seasonal energy requirements. Air infiltration is also an important factor in determining the relative humidity that will occur in buildings or, conversely, the amount of humidification or dehumidification required to maintain given humidities.

There are two methods of estimating air infiltration in buildings. In one case the estimate is based on measured leakage characteristics of the build-

TABLE 6.4. Air Changes Occurring under Average Conditions in Residences, Exclusive of Air Provided for Ventilation[a]

Kind of Room or Building	Number of Air Changes per Hour
Rooms with no windows or exterior doors	$\frac{1}{2}$
Rooms with windows or exterior doors on one side	1
Rooms with windows or exterior doors on two sides	$1\frac{1}{2}$
Rooms with windows or exterior doors on three sides	2
Entrance halls	2

[a]For rooms with weatherstripped windows or with storm sash, use two-thirds of these values.

SOURCE: Reprinted by permission from *ASHRAE Handbook of Fundamentals 1977*, [1].

ing components and selected pressure differences. This is known as the *crack method,* since cracks around windows and doors are usually the major source of air leakage. The other method is known as the *air-change method* and consists of assuming a certain number of air changes per hour for each room, the number of changes assumed being dependent upon the type, use, and location of the room. The crack method is generally thought to be more accurate, provided that leakage characteristics and pressure differences can be properly evaluated. Otherwise the air change method may be justified. To date, however, there is no substitute for judgment based on experience in the estimation of infiltration.

The accuracy of estimating infiltration for design load calculations by the crack method is restricted both by the limitations in information on air leakage characteristics of components and by the difficulty of estimating the pressure differences under appropriate design conditions of temperature and wind.

Table 6.4 may be used to determine an infiltration allowance for each room; the total allowance for a residence is the sum of the allowances of the individual rooms. After selecting the appropriate number of air changes (AC) per hour, the infiltration rate can be converted to cubic feet per minute (cfm) as follows:

$$(\text{cfm})_{\text{inf}} = \left(\frac{\text{AC}}{\text{hr}} \times \text{volume}\right) \div 60 \, \frac{\text{min}}{\text{hr}} \qquad (6.1)$$

Alternatively, the Achenbach-Coblentz wind speed and temperature correlation may be used for estimating the hourly infiltration rate through an average residence. The correlation is based on experimental measurements of actual homes and is of the form

$$I = A + BV + C|\Delta T| \qquad (6.2)$$

where I = infiltration rate, AC/hr
        V = wind speed, mph
      ΔT = temperature difference between the interior of the residence and the outside, °F

and A, B, and C are empirical constants characteristic of the residence being measured. Values for a typical house are A = 0.25, B = 0.02165, and C = 0.00833.

For nonresidential buildings, Table 6.5 may be used for approximate values of infiltration. In commercial buildings, infiltration may occur mostly from door openings. Infiltration through a swinging door is about 900 ft³ (25.5 m³) per person for a single-bank entrance and 550 ft³ (15.6 m³) per person for a vestibule entrance. However, it is only about 60 ft³ (1.7 m³) per person for a manually operated revolving door and 32 ft³ (0.9 m³) per person for a motor-driven one.

Infiltration should not be confused with the outdoor air requirement of a space. An air-conditioned space requires positive introduction of outdoor (fresh) air (1) for dilution of tobacco smoke and odors and (2) as make-up air for exhaust fans. Except in some residences, infiltration should never be relied upon to provide the outdoor air requirement.

Mechanical ventilation may be provided in residential air-conditioning systems. It is not a code requirement in most localities, and only general guidance can be given in determining its desirability. Field experience indicates that natural air leakage is adequate to provide sufficient ventilation in many installations. Nevertheless, the need for positive ventilation must be

TABLE 6.5.   Probable Building Air Changes per Hour by Natural Effects

Type of Building	Min	Max
Stores and shops	$\frac{1}{2}$	1
Commercial buildings	$\frac{1}{2}$	$1\frac{1}{2}$
Industrial buildings	2	3
Public or institutional buildings	$1\frac{1}{2}$	3
Offices, apartments, and hotels		
0 walls with windows	$\frac{1}{4}$	$\frac{1}{2}$
1 wall with window	$\frac{1}{2}$	1
2 walls	$\frac{3}{4}$	$1\frac{1}{2}$
3 or 4 walls with windows	$1\frac{1}{2}$	1

NOTE. In commercial buildings with doors leading to nonconditioned areas, add 100 ft³/hr for each person entering or leaving. For a 36-in. door standing open, allow 48,000 ft³/hr. Revolving doors can be calculated at 60 ft³ per person passing through, or if the door is equipped with a brake, 50 ft³ per person.

TABLE 6.6. Ventilation Requirements for Occupants[a]

	Estimated Persons per 1000 ft² Floor Area	Required Ventilation Air per Human Occupant			
		Minimum		Recommended	
		cfm	l/s	cfm	l/s
Residential					
Single-unit dwellings					
General living areas, bedrooms, utility rooms	5	5	2.5	7–10	3.5–5
Kitchens, baths, toilet rooms	—	20	10	30–50	15–25
Multiple-unit dwellings and mobile homes					
General living areas, bedrooms, utility rooms	7	5	2.5	7–10	3.5–5
Kitchens, baths, toilet rooms	—	20	10	30–50	15–25
Garages	—	1.5	7.5	2–3	10–15
Commercial					
Public rest rooms	100	15	7.5	20–25	10–12.5
General requirements—merchandising (apply to all forms unless specially noted)					
Sales floors (basement and ground floors)	30	7	3.5	10–15	5–7.5
Sales floors (upper floors)	20	7	3.5	10–15	5–7.5
Storage areas (serving sales areas and storerooms)	5	5	2.5	7–10	3.5–5
Dressing rooms	—	7	3.5	10–15	5–7.5
Malls and arcades	40	7	3.5	10–15	5–7.5
Shipping and receiving areas	10	15	7.5	15–20	7.5–10
Warehouses	5	7	3.5	10–15	5–7.5
Elevators	—	7	3.5	10–15	5–7.5
Bank vaults	—	5	2.5	5	2.5
Dining rooms	70	10	5	15–20	7.5–10
Kitchens	20	30	15	35	17.5
Cafeterias, short order; drive-ins, seating areas	100	30	15	35	17.5
Bars (predominantly stand-up)	150	30	15	40–50	20–25
Cocktail lounges	100	30	15	35–40	17.5–20

Hotels, motels, resorts

Bedrooms	5	7	3.5	10–15	5–7.5
Living rooms (suites)	20	10	5	15–20	7.5–10
Baths, toilets (attached to bedrooms)	—	20	10	30–50	15–25
Corridors	5	5	2.5	7–10	3.5–5
Lobbies	30	7	3.5	10–15	5–7.5
Conference rooms (small)	70	20	10	25–30	12.5–15
Assembly rooms (large)	140	15	7.5	20–25	10–12.5

Offices

General office space	10	15	7.5	15–25	7.5–12.5
Conference rooms	60	25	12.5	30–40	15–20
Drafting rooms, art rooms	20	7	3.5	10–15	5–7.5
Doctors' consultation rooms	—	10	5	10–15	7.5–10
Waiting rooms	30	10	5	15–20	7.5–10

Institutional

Schools

Classrooms	50	10	5	10–15	5–7.5
Multiple use rooms	70	10	5	10–15	5–7.5
Laboratories	30	10	5	10–15	5–7.5
Craft and vocational training shops	30	10	5	10–15	5–7.5
Music, rehearsal rooms	70	10	5	15–20	7.5–10
Auditoriums	150	5	2.5	5–7.5	2.5–3.8
Gymnasiums	70	20	10	25–30	12.5–15
Libraries	20	7	3.5	10–12	5–6
Common rooms, lounges	70	10	5	10–15	5–7.5
Offices	10	7	3.5	10–15	5–7.5
Lavatories	100	15	7.5	20–25	10–12.5

Industrial (Including Agricultural Processing)
(Occupational safety laws in the various states usually regulate the ventilation requirements, which are almost always far in excess of the ventilation requirements for the occupants.) ASHRAE Standard 62-73 lists requirements for occupants only. In general, 25 cfm (12.5 l/s) per occupant is recommended except for mining or metalworking, where 40 cfm (20 l/s) is recommended.

[a]Abridged by permission from ASHRAE Standard 62-73 [5].

evaluated for each installation. Considerations include owner desires, entertaining by the owner, etc. In larger homes, ventilation should be provided. Ventilation air is usually introduced at the rate of about one air change per hour. It is considered desirable to install a manually operated, locking-type damper in the outdoor air duct so the owner can have the option of ventilating positively or relying on natural air leakage.

Local codes and ordinances frequently specify outdoor air ventilation requirements for public places and industrial installations. Spaces like auditoriums, clubrooms, dance halls, and theaters usually take in enough outside air to offset infiltration.

Table 6.6, abridged from ASHRAE Standard 62-73, lists required ventilation quantities for 100% outdoor air of quality meeting the standard. If adequate temperature regulation is provided and filtration restricts particulates to 60 μg/m^3, the outdoor air may be reduced to 33% of the listed quantity. If, in addition, efficient absorption or other odor- and gas-removal equipment is provided, it may be reduced to 15%. In no case, however, should the outdoor air quantity be less than 5 cfm (2.5 liters/s) per person. Existing building codes and other legal standards may have more stringent requirements than the standard. ASHRAE Standard 90-75 cites the minimum quantity of ASHRAE Standard 62-73 as the value to be used on load computations to establish an energy budget for a building.

Thermal loads and the associated energy requirements due to the introduction of outdoor air are determined from the indoor and outdoor conditions and the quantity of incoming air. In winter this heat loss includes (1) the sensible heat loss, or the heat required to warm the entering outdoor air and (2) the latent heat loss, or the heat equivalent moisture that must be added.

The formula for the energy required to warm the outdoor air that enters a room to the temperature of the room is

$$q_s = c_p \dot{V} \rho (t_i - t_o) \qquad (6.3)$$

where q_s = heat required to raise temperature of air leaking into building from t_o to t_i, Btu/hr
  c_p = specific heat of air, Btu/lb
  $\dot{V}$ = volume of outdoor air entering building, ft^3/hr
  ρ = density of air at temperature t_o, lb/ft^3

When moisture is added to air to maintain proper winter comfort conditions, it is necessary to determine the energy required to evaporate the water added. This heat for humidification may be calculated from

$$q_l = \dot{V} \rho (W_i - W_o) h_{fg} \qquad (6.4)$$

where q_l = heat required to increase moisture content of air entering building from W_o to W_i, Btu/hr
  $\dot{V}$ = volume of outdoor air entering building, ft^3/hr

ρ = density of air at temperature t_i, lb/ft^3
W_i = humidity ratio of indoor air, lb/lb dry air
W_o = humidity ratio of outdoor air, lb/lb dry air
h_{fg} = latent heat of water, Btu/lb

In summer, wind and pressure differences cause air of higher temperature and moisture content to infiltrate through the cracks around doors and windows, resulting in localized sensible and latent heat gains. Also, some outdoor ventilation air is usually necessary in order to eliminate odors. This outdoor air imposes a cooling and dehumidifying load in the cooling coil because the heat and/or moisture must be removed. Heat gains due to infiltration and ventilation can be computed using

$$q_s = (1.10 \text{ cfm})\Delta t \qquad (6.5)$$

$$q_l = (4840 \text{ cfm})\Delta W \qquad (6.6)$$

EXAMPLE 6.2

Specify the required amount of outside air for ventilation for:
a.   A 12 ft × 12 ft private office with 8-ft high walls.
b.   A department store with 20,000 ft^2 floor area.

Solution
a.   15 cfm (Table 6.6)

b.   Table 6.6: $\dfrac{\sim 30 \text{ people}}{1000 \text{ ft}^2}$ and $\dfrac{7 \text{ cfm}}{\text{person}}$

$\dfrac{20,000}{1000} \times 30 \times 7 = 600 \times 7 = 4200$ cfm

EXAMPLE 6.3

A small factory with a 10-ft high ceiling is shown in Fig. E6.3. There are normally 22 employees in the shop area and 4 employees in the office area. On a winter day when the outside temperature is 0°F, the office is maintained at 75°F, 25% RH and the shop is kept at 68°F with no humidity control. Determine for *each* area:

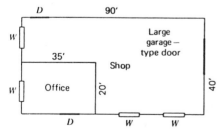

Fig. E6.3.

a.   Infiltration, cfm.
b.   Minimum required outside air, cfm.
c.   Sensible heat loss due to infiltration, Btu/hr.
d.   Latent heat loss due to infiltration, Btu/hr.

Solution
a.   Office area:

$$V = 20 \times 35 \times 10 = 7000 \text{ ft}^3; \ AC/hr \approx 1\tfrac{1}{2}$$

$$\text{Infiltration} = \frac{1.5(7000)}{60} = 175 \text{ cfm}$$

Shop area:

$$V = 90 \times 40 \times 10 - 7000 = 29,000 \text{ ft}^3; \ AC/hr = 3$$

$$\text{Infiltration} = \frac{3(29,000)}{60} = 1450 \text{ cfm}$$

b.   Office:

15 cfm/person $\times$ 4 = 60 cfm

Shop:

25 cfm/person $\times$ 22 = 550 cfm (Industrial, Table 6.6)

c.   Office:

$$Q_S = 1.10 \times 175 \times (75 - 0) = 14,400 \text{ Btu/hr}$$

Shop:

$$Q_S = 1.10 \times 1450 \times (68 - 0) = 108,500 \text{ Btu/hr}$$

d.   Office:

$$Q_L = 4840 \times 175 \times (.0046 - .070787) = 3230 \text{ Btu/hr}$$

Shop:

$$Q_L = 0, \text{ no humidifier}$$

6.6.   HEAT TRANSFER COEFFICIENTS

Whenever there is a difference in temperature between two areas (indoor-outdoor), heat will flow from the warmer area to the cooler area. The flow or transfer of heat will take place by one or more of three methods: conduction, convection, or radiation.

Conduction is the transfer of heat through a solid. When the end of a poker is left in a fire, the handle will also be warmed even though it is not in

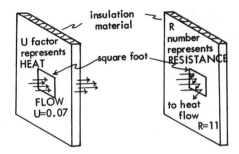

Fig. 6.4. *U* factor and *R* number.

direct contact with the flame. The heat flows along the poker by conduction. The rate of flow is influenced by the temperature difference, the area of the material, the distance through the material from warm side to cool side, and the thermal conductivity of the material. Insulating materials have low conductivity, which, combined with their thickness, provides a barrier that slows the transfer of heat by conduction.

Convection transfers heat by movement through liquids or gases. As a gas such as air is heated, it expands, becomes lighter, and rises. It is then displaced with cooler air, which follows the same cycle, carrying heat with it. The continuous cycle of rising warm air and descending cool air is a convection current. By dividing a large space into many small spaces and providing barriers that restrict convection, the flow of heat can be slowed. Among other methods, the mat of random fibers in mass-type insulation, such as mineral wood or wood fiber insulation, provides such a barrier.

Radiation is a method of heat transfer whereby a warm object can heat a cool object without the need of a solid, liquid, or gas between them. Heat from the sun passing through the vacuum of space and heating the earth is an example of radiation. In order for radiant heat to be transferred, the objects must "see" each other. Standing in front of the campfire results in the warming of those parts of the body that the fire "sees." To warm the other parts, you must reverse your position. The body can also be the radiator, and heat can be radiated to objects of cooler temperature.

Two terms are used to indicate the relative insulating value of materials and sections of walls, floors, and ceilings. One is called the *U* factor (or *U* value) and the other the *R* number. The *U* factor indicates the rate at which heat flows through a specific material or a building section like that shown in Fig. 6.4. The smaller the *U* factor, the better the insulating value of the material or group of materials making up the wall, ceiling, or floor.

The *R* number indicates the ability of one specific material, or a group of materials in a building section, to resist heat flow. Many insulating materials now have their *R* number stamped on the outside of the package, batt, or blanket. Table 6.7 shows the *R* number and relative heat-resisting values of several of these materials. The *R* number for batt, blanket, and loose fill insulation as listed in Table 6.7 is for a thickness of 1 in. The *R* number for

TABLE 6.7.   Relative Heat Resistance Value for Specific Building Materials

MATERIAL DESCRIPTION	MATERIAL DENSITY	MATERIAL THICKNESS INCHES	R NUMBER FOR THICKNESS SHOWN TO LEFT	RELATIVE VALUE OF RESISTANCE TO HEAT FLOW
Building Paper	–	–	0.06	
Gypsum Plaster, sand aggregate	105	1/2	0.09	
Structural Glass	–	–	0.10	
Air Surface, 15 m.p.h. wind (outside surface)	–	–	0.17	
Asbestos – Cement siding or shingles	–	–	0.21	
Gypsum or Plaster board	50	3/8	0.32	
Stone, lime or sand	–	4	0.32	
Concrete, sand-gravel aggregate	140	4	0.32	
Built-up roofing	70	3/8	0.33	
Brick, face	130	4	0.44	
Air surface, still air, horizontal ordinary materials, heat flow-up	–	–	0.61	
Plywood	34	1/2	0.63	
Air surface, still air, vertical, ordinary materials, heat flow, horizontal	–	–	0.68	
Wood siding, bevel, 1/2" x 8 ", lapped	–	–	0.81	
Wood shingle siding, 16 in. 7 1/2" exposure	–	–	0.87	
Oak, maple & similar hardwoods	45	1	0.91	
Air space, vertical, ordinary materials, horizontal heat flow	–	3/4-4	0.97	
Clay tile, one cell deep	–	4	1.11	
Concrete block, 3 core, sand-gravel aggregate	–	8	1.11	
Acoustical tile, wood or cane fiber	–	1/2	1.19	
Fir, pine & similar softwoods	32	1	1.25	
Insulation board, impregnated	20	1/2	1.32	
Concrete, lightweight aggregate (expanded shale, clay, slate, slag)	80	4	1.50	
Air space, vertical, bounded by reflective material	–	3/4-4	1.70	
Concrete block, 3 core, cinder aggregate	–	8	1.72	
Concrete block, 3 core, lightweight aggregate	–	8	2.00	
Vermiculite, expanded	7.0	1	2.08	
Carpet & Fibrous pad	–	–	2.08	
Cellular glass insulation board	9.0	1	2.50	
Roof insulation, preformed for above deck	–	1	2.78	
Mineral wool, loose fill, from slag, glass or rock	2.0-5.0	1	3.33	
Wood fiber, loose fill, redwood, hemlock or fir	2.0-3.5	1	3.33	
Plastic, foamed	1.62	1	3.45	
Macerated paper or pulp	2.0-3.5	1	3.57	
Corkboard, without added binder	6.5-8.0	1	3.70	
BATT & BLANKETS BOUNDED BY NON-REFLECTIVE MATERIAL				
Mineral Wool, fibrous form, (rock, slag or glass)	1.5-4.0	1	3.70	
Wood fiber, multilayer, stitched expanded	1.5-2.0	1	3.70	
Cotton fiber	0.8-2.0	1	3.85	
Wood fiber	3.2-3.6	1	4.00	

SOURCE: *Insulation and Ventilation for Missouri Homes,* University of Missouri Extension Division, C747, Reprint 12/68/5M.

greater thicknesses can be determined by multiplying the thickness desired, in inches, by the R number listed. (*Example:* The R number of 1 in. of wood fiber, the last material listed, is 4. Therefore, the R number for 3 in. is 4×3, or R equals 12.) The greater the R number, the greater the insulating value of the material and the lower the heat loss. A high R number thus means lower heating and cooling costs and less energy to maintain a comfortable temperature.

A total resistance to heat flow through a flat ceiling, floor, or wall (or a curved surface if the curvature is small) is equal numerically to the sum of the resistances in series

$$R_T = R_1 + R_2 + R_3 + \cdots + R_n \tag{6.7}$$

where R_1, R_2, etc., are the individual resistances of the components, and R_T is total resistance.

As U is the conductance of the entire building section, it is the inverse of R_T, or

$$U = \frac{1}{R_T} = \frac{1}{R_1 + R_2 + R_3 + \cdots + R_n} \tag{6.8}$$

or

$$U = \left[\frac{1}{h_i} + \Sigma\frac{x}{k} + \Sigma\frac{1}{C} + \frac{1}{h_o}\right]^{-1} = \frac{1}{\Sigma R} = \frac{1}{R_T} \tag{6.9}$$

In order to compute the U value of a construction, it is first necessary to know the conductivity and thickness of homogeneous materials, the conductance of nonhomogeneous materials (such as concrete blocks), the surface conductances of both sides of the construction, and the conductances of any air spaces or the thermal resistances of individual elements.

For a known wall, the U factor may be calculated if the surface coefficients h_i and h_o, the required thermal conductivities (k values), and the required thermal conductances (C values) are available. Fortunately, a vast amount of experimental data have been published covering a wide variety of building materials. The ASHRAE *Handbook 1977 Fundamentals* shows extensive tables for the various coefficients and also the calculated U factors for many wall constructions. Table 6.8 gives experimentally determined values of the surface coefficients h_i and h_o. These values are generally applicable to most building wall problems. They apply specifically for the case where the ambient air and surrounding surfaces are at the same temperature ($h = h_c + h_R$). Table 6.9 shows experimentally determined values of the thermal resistance for plane air spaces. Both Tables 6.8 and 6.9 show the significant influence of direction of heat flow and the effect of surface emissivities upon the coefficients.

Table 6.10 shows thermal conductivities, conductances, and resistances for selected building materials. If the conductivities or conductances of materials in a wall are dependent on temperature, the mean temperature must be known to obtain the correct value. In such cases, it is usual to use a trial-and-error procedure. First, the mean temperature of each layer is estimated, and conductivities or conductances are obtained. The total resistance is then calculated, and then the temperature at each interface can be calculated. This procedure can be repeated until the conductivities or conductances have been selected correctly for the resulting mean temperatures.

In many installations, components are arranged so that parallel heat flow paths of different conductances result. If there is no lateral heat flow be-

TABLE 6.8. Surface Conductances and Resistances for Air[a,b]

Section A.   Surface Conductances and Resistances

Position of Surface	Direction of Heat Flow	Non-reflective $\epsilon = 0.90$		Reflective $\epsilon = 0.20$		Reflective $\epsilon = 0.05$	
		h_i	R	h_i	R	h_i	R
Still Air							
Horizontal	Upward	1.63	0.61	0.91	1.10	0.76	1.32
Sloping, 45°	Upward	1.60	0.62	0.88	1.14	0.73	1.37
Vertical	Horizontal	1.46	0.68	0.74	1.35	0.59	1.70
Sloping, 45°	Downward	1.32	0.76	0.60	1.67	0.45	2.22
Horizontal	Downward	1.08	0.92	0.37	2.70	0.22	4.55
Moving Air		h_o	R	h_o	R	h_o	R
Any position							
15-mph wind (for winter)	Any	6.00	0.17				
7.5-mph wind (for summer)	Any	4.00	0.25				

Section B.   Reflectivity and Emittance Values of Various Surfaces and Effective Emittances of Air Spaces

Surface	Reflectivity, %	Average Emittance ϵ	Effective Emittance E of Air Space	
			One Surface Emittance ϵ; the Other 0.90	Both Surfaces Emittance ϵ
Aluminum foil, bright	92–97	0.05	0.05	0.03
Aluminum sheet	80–95	0.12	0.12	0.06
Aluminum coated paper, polished	75–84	0.20	0.20	0.11
Steel, galvanized, bright	70–80	0.25	0.24	0.15
Aluminum paint	30–70	0.50	0.47	0.35
Building materials: wood, paper, masonry, nonmetallic paints	5–15	0.90	0.82	0.82
Regular glass	5–15	0.84	0.77	0.72

[a] All conductance values expressed in Btu/(hr·ft²·°F).

[b] A surface cannot take credit for both an air space resistance value and a surface resistance value. No credit for an air space value can be taken for any surface facing an air space of less than 0.5 in.

SOURCE: Reprinted by permission from *ASHRAE Handbook of Fundamentals 1977.*

tween paths, each path may be considered to extend from inside to outside, and the average overall coefficient is then

$$U_{av} = a(U_a) + b(U_b) + c(U_c) + \cdots \tag{6.10}$$

where $a, b, c, \cdots$ are respective fractions of a typical basic area composed of several different paths whose overall coefficients are U_a, U_b, U_c, etc. This equation is used to correct for parallel heat flow through framing and may be rewritten as

$$U_{av} = S(U_s) + (1 - S)(U_i) \tag{6.11}$$

where U_{av} = average U value for building section
U_i = U value for area between framing members
U_s = U value for area backed by framing members
S = fraction of area backed by framing members

For 16-in. oc framing incuding multiple studs, plates, headers, sills, band joists, and so on, the framing is typically about 20%, or $S = 0.20$. For a frame wall with 24-in. oc stud space, the framing factor is estimated at 15% or $S = 0.15$.

A construction may be made up of two or more layers of unequal area, separated by an air space and arranged so that heat flows through the layers in series. The most common construction is a ceiling-and-roof combination, where the attic space is unheated and unventilated. A combined coefficient based on the most convenient area, say the ceiling area, from air inside to air outside can be calculated from

$$R_{T,c} = R_c + \frac{R_r}{n} \quad \text{and} \quad U_c = \frac{1}{R_{T,c}} \tag{6.12}$$

where U_c = combined coefficient based on ceiling area
n = ratio of roof to ceiling area, A_r/A_c

Figure 6.5 shows the insulating value and R number of a wall section with different amounts of insulation. This also shows the R number of each material in the wall. The heat loss through the wall section with $3\frac{5}{8}$ in. of insulation is 31% less than it is through the wall containing only 2 in. of insulation.

In calculating U values, exemplary conditions of components and installations are assumed (i.e., that insulating materials are uniformly of the nominal thickness and conductivity, air spaces are of uniform thickness and surface temperatures, effects due to moisture are not involved, and installation details are in accordance with design). Common variations in conditions, materials, workmanship, and so on can introduce much greater variations in U values than the variations resulting from assumed mean temperatures and temperature differences. From this, it is also clear that the use of more than two significant figures in stating a U value may assume more precision than can possibly exist.

TABLE 6.9. Thermal Resistances of Plane Air Spaces[a,b]

Position of Air Space	Direction of Heat Flow	Air Space Mean Temp (°F)	Temp Diff (°F)	1.5-in. Air Space Value of E					3.5-in. Air Space Value of E				
				0.03	0.05	0.2	0.5	0.82	0.03	0.05	0.2	0.5	0.82
Horiz.	Up	90	10	2.55	2.41	1.71	1.08	0.77	2.84	2.66	1.83	1.13	0.80
		50	30	1.87	1.81	1.45	1.04	0.80	2.09	2.01	1.58	1.10	0.84
		50	10	2.50	2.40	1.81	1.21	0.89	2.80	2.66	1.95	1.28	0.93
		0	20	2.01	1.95	1.63	1.23	0.97	2.25	2.18	1.79	1.32	1.03
		0	10	2.43	2.35	1.90	1.38	1.06	2.71	2.62	2.07	1.47	1.12
		−50	20	1.94	1.91	1.68	1.36	1.13	2.19	2.14	1.86	1.47	1.20
		−50	10	2.37	2.31	1.99	1.55	1.26	2.65	2.58	2.18	1.67	1.33
Slope, 45°	Up	90	10	2.92	2.73	1.86	1.14	0.80	3.18	2.96	1.97	1.18	0.82
		50	30	2.14	2.06	1.61	1.12	0.84	2.26	2.17	1.67	1.15	0.86
		50	10	2.88	2.74	1.99	1.29	0.94	3.12	2.95	2.10	1.34	0.96
		0	20	2.30	2.23	1.82	1.34	1.04	2.42	2.35	1.90	1.38	1.06
		0	10	2.79	2.69	2.12	1.49	1.13	2.98	2.87	2.23	1.54	1.16
		−50	20	2.22	2.17	1.88	1.49	1.21	2.34	2.29	1.97	1.54	1.25
		−50	10	2.71	2.64	2.23	1.69	1.35	2.87	2.79	2.33	1.75	1.39
		90	10	3.99	3.66	2.25	1.27	0.87	3.69	3.40	2.15	1.24	0.85
		50	30	2.58	2.46	1.84	1.23	0.90	2.67	2.55	1.89	1.25	0.91
		50	10	3.79	3.55	2.39	1.45	1.02	3.63	3.40	2.32	1.42	1.01

Position of Air Space	Direction of Heat Flow	Mean Temp. °F	Temp. Diff. °F										
Vertical	Horiz.	0	20	2.76	2.66	2.10	1.48	1.12	2.88	2.78	2.17	1.51	1.14
		0	10	3.51	3.35	2.51	1.67	1.23	3.49	3.33	2.50	1.67	1.23
		−50	20	2.64	2.58	2.18	1.66	1.33	2.82	2.75	2.30	1.73	1.37
		−50	10	3.31	3.21	2.62	1.91	1.48	3.40	3.30	2.67	1.94	1.50
Slope, 45°	Down	90	10	5.07	4.55	2.56	1.36	0.91	4.81	4.33	2.49	1.34	0.90
		50	30	3.58	3.36	2.31	1.42	1.00	3.51	3.30	2.28	1.40	1.00
		50	10	5.10	4.66	2.85	1.60	1.09	4.74	4.36	2.73	1.57	1.08
		0	20	3.85	3.66	2.68	1.74	1.27	3.81	3.63	2.66	1.74	1.27
		0	10	4.92	4.62	3.16	1.94	1.37	4.59	4.32	3.02	1.88	1.34
		−50	20	3.62	3.50	2.80	2.01	1.54	3.77	3.64	2.90	2.05	1.57
		−50	10	4.67	4.47	3.40	2.29	1.70	4.50	4.32	3.31	2.25	1.68
Horiz.	Down	90	10	6.09	5.35	2.79	1.43	0.94	10.07	8.19	3.41	1.57	1.00
		50	30	6.27	5.63	3.18	1.70	1.14	9.60	8.17	3.86	1.88	1.22
		50	10	6.61	5.90	3.27	1.73	1.15	11.15	9.27	4.09	1.93	1.24
		0	20	7.03	6.43	3.91	2.19	1.49	10.90	9.52	4.87	2.47	1.62
		0	10	7.31	6.66	4.00	2.22	1.51	11.97	10.32	5.08	2.52	1.64
		−50	20	7.73	7.20	4.77	2.85	1.99	11.64	10.49	6.02	3.25	2.18
		−50	10	8.09	7.52	4.91	2.89	2.01	12.98	11.56	6.36	3.34	2.22

[a] All resistance values expressed in (hour)(square foot)(degree Fahrenheit temperature difference) per Btu

[b] Values apply only to air spaces of uniform thickness bounded by plane, smooth, parallel surfaces with no leakage of air to or from the space. Thermal resistance values for multiple air spaces must be based on careful estimates of mean temperature differences for each air space.

SOURCE: Reprinted by permission from *ASHRAE Handbook of Fundamentals 1977.*

TABLE 6.10. Thermal Properties of Typical Building and Insulating Materials (Design Values)[a]

Description	Density (lb/ft³)	Conductivity (k)	Conductance (C)	Customary Unit Resistance (R) Per inch thickness (1/k)	Customary Unit Resistance (R) For thickness listed (1/C)	Specific Heat [Btu/(lb·°F)]	SI Unit Resistance (R) m·K/W	SI Unit Resistance (R) m²·K/W
Building Board[b]								
Asbestos-cement board	120	4.0	—	0.25	—	0.24	1.73	
Asbestos-cement board 0.125 in.	120	—	33.00	—	0.03			0.005
Asbestos-cement board 0.25 in.	120	—	16.50	—	0.06			0.01
Gypsum or plaster board 0.375 in.	50	—	3.10	—	0.32	0.26		0.06
Gypsum or plaster board 0.5 in.	50	—	2.22	—	0.45			0.08
Gypsum or plaster board 0.625 in.	50	—	1.78	—	0.56			0.10
Plywood (Douglas fir)	34	0.80	—	1.25	—	0.29	8.66	
Plywood (Douglas fir) 0.25 in.	34	—	3.20	—	0.31			0.05
Plywood (Douglas fir) 0.375 in.	34	—	2.13	—	0.47			0.08
Plywood (Douglas fir) 0.5 in.	34	—	1.60	—	0.62			0.11
Plywood (Douglas fir) 0.625 in.	34	—	1.29	—	0.77			0.19
Plywood or wood panels 0.75 in.	34	—	1.07	—	0.93	0.29		0.16
Vegetable fiber board								
Sheathing, regular density 0.5 in.	18	—	0.76	—	1.32	0.31		0.23
0.78125 in.	18	—	0.49	—	2.06			0.36
Sheathing, intermediate density 0.5 in.	22	—	0.82	—	1.22	0.31		0.21
Tile and lay-in panels, plain or acoustic 0.5 in.	18	0.40	—	2.50	—	0.14	17.33	
0.5 in.	18	—	0.80	—	1.25			0.22
0.75 in.	18	—	0.53	—	1.89			0.33

Building Membrane							
Vapor—permeable felt	—	—	16.70	—	0.06	—	0.01
Vapor—seal, 2 layers of mopped 15-lb felt	—	—	8.53	—	0.12	—	0.02
Vapor—seal, plastic film	—	—	—	—	Negl.	—	
Finish Flooring Materials							
Carpet and fibrous pad	—	—	0.48	—	2.08	0.34	0.37
Carpet and rubber pad	—	—	0.81	—	1.23	0.33	0.22
Insulating Materials							
Blanket and Batt							
Mineral fiber, fibrous form processed from rock, slag, or glass							
approx. 2–2.75 in.	0.3–2.0	—	0.143	—	7	0.17–0.23	1.23
approx. 3–3.5 in.	0.3–2.0	—	0.091	—	11		1.94
approx. 3.50–6.5 in.	0.3–2.0	—	0.053	—	19		3.35
approx. 6–7 in.	0.3–2.0		0.045		22		3.87
approx. 8.5 in.	0.3–2.0		0.033		30		5.28
Board and Slabs							
Cellular glass	8.5	0.38	—	2.63	—	0.24	18.23
Glass fiber, organic bonded	4–9	0.25	—	4.00	—	0.23	27.72
Expanded rubber (rigid)	4.5	0.22	—	4.55	—	0.40	31.53
Expanded polystyrene extruded, cut cell surface	1.8	0.25	—	4.00	—	0.29	27.72
Masonry Materials[c]							
Cement mortar	116	5.0	—	0.20	—	—	1.39
Sand and gravel or stone aggregate (not dried)	140	12.0	—	0.08	—		0.55
Stucco	116	5.0	—	0.20	—		1.39

TABLE 6.10. (Continued)

Description	Density (lb/ft³)	Conductivity (k)	Conductance (C)	Customary Unit Resistance (R) Per inch thickness (1/k)	For thickness listed (1/C)	Specific Heat [Btu/(lb·°F)]	SI Unit Resistance (R) m·K / W	m²·K / W
Masonry Units								
Brick, common	120	5.0	—	0.20	—	0.19	1.39	
Brick, face	130	9.0	—	0.11	—		0.76	
Roofing								
Asbestos-cement shingles	120	—	4.76	—	0.21	0.24		0.04
Asphalt roll roofing	70	—	6.50	—	0.15	0.36		0.03
Asphalt shingles	70	—	2.27	—	0.44	0.30		0.08
Built-up roofing 0.375 in.	70	—	3.00	—	0.33	0.35		0.06
Siding Materials (on Flat Surface)								
Asphalt insulating siding (0.5 in. bed.)	—	—	0.69	—	1.46	0.35		0.26
Wood, drop, 1 × 8 in.	—	—	1.27	—	0.79	0.28		0.14
Woods								
Maple, oak, and similar hardwoods	45	1.10	—	0.91	—	0.30	6.31	
Fir, pine, and similar softwoods	32	0.80	—	1.25	—	0.33	8.66	

[a]These constants are expressed in Btu per (hour)(square foot)(degree Fahrenheit temperature difference). Conductivities (k) are per inch thickness, and conductances (C) are for thickness or construction stated, not per inch thickness. All values are for a mean temperature of 75°F.

[b]Boards, panels, subflooring, sheathing, woodboard panel products

[c]Concretes

SOURCE: Adapted by permission from *ASHRAE Handbook of Fundamentals 1977.*

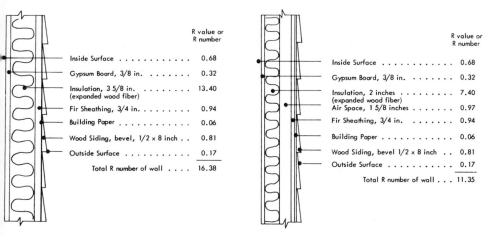

Fig. 6.5. Thermal resistance of frame wall.

When any insulation is installed, irregular areas must be given careful attention. Blanket-type insulation should be sealed by stapling or taping to the floor and ceiling plates and also to the studs. Insulating material should be carefully fitted around all plumbing, wiring, and other projections. The proper thickness should be maintained throughout the walls, ceilings, and floors. To get the most value from reflective materials, such as aluminum foil facing on batts or blankets, allow a $\frac{3}{8}$-in. air space between the foil and the wallboard.

Loose fill insulation, for either new or remodeled homes, must be blown into the walls, floors, and ceilings with special equipment and by competent operators so that the proper density of the insulation will be obtained. Insulation should be blown into each wall cavity at both top and bottom so that all spaces will be filled and variations in density will be minimized. Variations in insulation density will affect the resistance value and R number of the material. Loose fill insulation tends to settle resulting in high density insulation below and none above. Batt or blanket insulation is therefore preferable for vertical spaces in new construction.

Any insulation will resist heat flow in proportion to its R number *only* if it is installed according to these general recommendations and the manufacturer's instructions. Wall voids are frequently left open at the top of the stud space, allowing outside air to enter and materially affect the U value of the wall.

Walls insulated with fibrous materials (mineral wool, glass fibers, wood fibers, etc.) that are permeable to movement of water vapor must be provided with suitable vapor barriers. Vapor barriers are necessary for refrigerated structures in all climates and for all insulated structures, including residences, in cold climates. A typical practice is to provide a single vapor barrier suitably placed on the warm side of the insulation. The vapor barrier should be located such that its temperature will be higher than the dew-point temperature of the air on the humid side of the wall. In cold-storage wall construction, multiple vapor barriers are often applied when the insulation is installed in multiple layers.

Often a membrane type of vapor barrier is used. Membrane barriers include aluminum foil sheets, laminates of aluminum foil and treated papers, coated felts and papers, and plastic films. Rigid sheets and coatings made of asphaltic, resinous, or polymeric materials may be used as vapor barriers.

Vapor barriers do not necessarily prevent all flow of water vapor. They should be of sufficient effectiveness that vapor-flow continuity can be maintained and condensation of moisture in interior parts of walls can be minimized.

A thorough discussion of insulations, vapor-proofing of walls, and effects of moisture in building construction is given in *1977 Fundamentals* [1].

Table 6.11 gives overall U values for glass and plastic panels including the convective resistance on each side. Similar values for doors are given in Table 6.12.

Shading devices such as venetian blinds, draperies, and roller shades will reduce the U value substantially if they fit tightly to the window jambs, head, and sill, and are made of nonporous material. As a rough approximation, tight-fitting shading devices may be considered to reduce the U value of vertical exterior single glazing by 25% and of vertical exterior double glazing and glass block by 15%. These adjustments should not be considered in choosing heating equipment, but may be used for calculating design cooling loads.

EXAMPLE 6.4

Find the overall coefficient of heat transfer and the total thermal resistance for the following exterior wall exposed to a 15-mph wind: Face brick veneer, $\frac{25}{32}$ in. insulating board sheathing, 3 in. fiberglass insulation in stud space, and $\frac{1}{4}$ in. walnut veneer plywood panels for the interior.

Solution

	R
Outside air (15 mph)	0.17
Face brick, 4 in.	0.44
Ins. board sheathing	2.06
Air space, $\frac{3}{4}$ in.	1.01
Fiberglass, 3 in.	11.00
Plywood, $\frac{1}{4}$ in.	0.31
Inside air	0.68
	$\Sigma R = 15.67$

$$U = 0.0638$$

EXAMPLE 6.5

a. Compute the U value for a wall of frame construction consisting of $\frac{1}{2}$ in. $\times$ 8 in. bevel siding, permeable felt building upper, $\frac{25}{32}$ in. wood fiber sheathing, 2-by-4 studding on 16-in. centers, and $\frac{3}{4}$-in. metal lath and sand plaster. Outside wind velocity is 15 mph. Neglect the effect of the studs.

b. Determine U if the space between the studs is filled with fiberglass blanket insulation. Neglect the effect of the studs.

c. Repeat (b), including the framing effect.

Solution

a.

	R
Outside air	0.17
Siding	0.81
Felt paper	0.06
Sheathing	2.06
Air space	0.97
Lath and plaster	0.10
Inside air	0.68
	$\Sigma R = 4.85$

$$U = \frac{1}{4.85} = 0.206 \text{ Btu/(hr·ft}^2\text{·°F)}$$

b. $R_{\text{with insulation}} = 4.85 - 0.97 + 11.0 = 14.88$

$U = 0.067 \text{ Btu/(hr·ft}^2\text{·°F)}$

c. $U_i = 0.067$

$R_s = 4.85 - 0.97 + 4.35 = 8.23$

$U_s = 0.12$

$U_{\text{av}} = S(U_s) + (1 - S)(U_i) = 0.20(0.12) + (0.80)(.067)$

$U_{\text{av}} = 0.078 \text{ Btu/(hr·ft}^2\text{·°F)}$

EXAMPLE 6.6

The ceiling of a residence has a winter U value of 0.27. The roof is pitched such that the ratio of roof to ceiling area is 1.3. The roof construction is such that its U value is 0.85. Determine the combined coefficient from inside air to outdoor air.

TABLE 6.11. Coefficients of Transmission (U) of Windows, Skylights, and Light-Transmitting Partitions

Description	Exterior[a] Winter	Exterior[a] Summer	Interior
Part A Vertical Panels (Exterior Windows, Sliding Patio Doors, and Partitions)—Flat Glass, Glass Block, and Plastic Sheet			
Flat glass[b]			
Single glass	1.10	1.04	0.73
Insulating glass, doubles[c]			
0.1875-in. air space[d]	0.62	0.65	0.51
0.25-in. air space	0.58	0.61	0.49
0.5-in. air space	0.49	0.56	0.46
0.5-in. air space, low emittance coating			
$e = 0.20$	0.32	0.38	0.32
$e = 0.40$	0.38	0.45	0.38
$e = 0.60$	0.43	0.51	0.42
Insulating glass, triple[c]			
0.25-in. air spaces	0.39	0.44	0.38
0.5-in. air spaces	0.31	0.39	0.30
Storm windows			
1-in. to 4-in. air space	0.50	0.50	0.44
Plastic sheet			
Single glazed			
0.125-in. thick	1.06	0.98	—
0.25-in. thick	0.96	0.89	—
0.5-in. thick	0.81	0.76	—
Insulating unit—double[c]			
0.25-in. air space	0.55	0.56	—
0.5-in. air space[c]	0.43	0.45	—
Glass block			
6 × 6 × 4 in. thick	0.60	0.57	0.46
8 × 8 × 4 in. thick	0.56	0.54	0.44
with cavity divider	0.48	0.46	0.38
12 × 12 × 4 in. thick	0.52	0.50	0.41
with cavity divider	0.44	0.42	0.36
12 × 12 × 2 in. thick	0.60	0.57	0.46
Part B Horizontal Panels (Skylights)—Flat Glass, Glass Block, and Plastic Domes			
Flat glass			
Single glass	1.23	0.83	0.96
Insulating glass, double[c]			
0.1875-in. air space	0.70	0.57	0.62
0.25-in. air space	0.65	0.54	0.59
0.5-in. air space[c]	0.59	0.49	0.56

TABLE 6.11. **(Continued)**

Description	Exterior[a] Winter	Exterior[a] Summer	Interior
0.5-in air space, low emittance coating			
$e = 0.20$	0.48	0.36	0.39
$e = 0.40$	0.52	0.42	0.45
$e = 0.60$	0.56	0.46	0.50
Glass block			
11 × 11 × 3 in. thick with cavity divider	0.53	0.35	0.44
12 × 12 × 4 in. thick with cavity divider	0.51	0.34	0.42
Plastic domes			
Single-walled	1.15	0.80	—
Double-walled	0.70	0.46	—

Part C   Adjustment Factors for Various Window and Sliding Patio Door Types (Multiply U Values in Parts A and B by These Factors)

Description	Single Glass	Double or Triple Glass	Storm Windows
Windows			
All glass	1.00	1.00	1.00
Wood sash, 80% glass	0.90	0.95	0.90
Wood sash, 60% glass	0.80	0.85	0.80
Metal sash, 80% glass	1.00	1.20	1.20
Sliding patio doors			
Wood frame	0.95	1.00	—
Metal frame	1.00	1.10	—

[a] See Part C for adjustment for various window and sliding patio door types.

[b] Emittance of uncooled glass surface = 0.84.

[c] Double and triple refer to the number of lights of glass.

[d] These values are for heat transfer from air to air, Btu/(hr·ft²).

SOURCE: Reprinted by permission from *ASHRAE Handbook of Fundamentals 1977*.

Solution

$$R_T = \frac{1}{U_1} + \frac{1}{n_2 U_2} = \frac{1}{0.27} + \frac{1}{(1.3)(0.85)} = 4.61$$

$$U_T = 0.217 \text{ Btu/(hr·ft}^{2.\circ}\text{F)}$$

EXAMPLE 6.7

The west wall of a College Hills residence is 70 ft long by 8 ft high. The wall contains four 3 ft × 5 ft wood-sash 80% glass single-pane windows, each with a storm window; one double-glazed ($\frac{1}{2}$-in. air space) picture window $5\frac{1}{2}$

TABLE 6.12.   Coefficients of Transmission (U) for Slab Doors[a]

	Winter			
	Solid Wood, No Storm Door	Storm Door[c]		Summer, No Storm Door
Thickness[b]		Wood	Metal	
1-in.	0.64	0.30	0.39	0.61
1.25-in.	0.55	0.28	0.34	0.53
1.5-in.	0.49	0.27	0.33	0.47
2-in.	0.43	0.24	0.29	0.42

	Steel Door	
	Winter	Summer
1.75 in.		
A[d]	0.59	0.58
B[e]	0.19	0.18
C[d]	0.47	0.46

[a] Btu per (hr·ft^2·°F)

[b] Nominal thickness.

[c] Values for wood storm doors are for approximately 50% glass; for metal storm door values apply for any percent of glass.

[d] A = mineral fiber core (2 lb/ft^3).

[e] B = Solid urethane foam core with thermal break.

[f] C = solid polystyrene core with thermal break.

SOURCE: Reprinted by permission from *ASHRAE Handbook of Fundamentals 1977*.

ft × 10 ft; and one 1¾-in.-thick solid wood door, 3 ft × 7 ft, with a glass storm door. Specify the U value and corresponding area for the door and each of the various windows with normal winter air velocities.

Solution

Windows (4):

$U = 0.50 \times 0.90$ (Table 6.11)
$\quad = 0.450$
$A_w = 4 \times 3 \times 5 = 60 \text{ ft}^2$

Picture Window:

$U = 0.49 \times 1$ (Table 6.11)
$\quad = 0.49$
$A_{pw} = 5.5 \times 10 = 55 \text{ ft}^2$

Door:

$U = 0.26$ (Table 6.12)
$A_d = 3 \times 7 = 21 \text{ ft}^2$

6.7. SURFACE TEMPERATURES

It is often necessary to determine the inside surface temperature or the temperature of the surfaces within the structure. The temperature on the inside-wall surface or ceiling surface of a building cannot be considered to be the same as the air temperature inside the building. This surface temperature depends on the convection (film) conditions in the building, the insulating ability of the wall, and the outside conditions of temperature and wind. If the surface temperature is lower than the inside dew-point temperature, moisture will condense. A wet wall or ceiling results, and the moisture may cause serious damage to plaster and woodwork, besides constituting a nuisance. In winter, if the insulation effectiveness of the wall cannot be increased, this moisture formation necessitates lowering the inside relative humidity, or decreasing the inside film resistance by increasing the air circulation over the inside surfaces. Even if the dew problem does not exist, a wall of inadequate insulating capacity may chill the occupants in winter, by radiation to the cold wall, and cause discomfort in summer, by permitting too much heat inflow.

In winter, condensation of humidity from the inside air on windows often presents a serious problem. The use of double-glass windows or storm windows probably furnishes the best means of alleviating such a nuisance.

The calculation of inside surface temperature for a given wall at certain inside and outside temperatures can most easily be made by making use of the ratio of the surface-film resistance to that of the whole wall. Or, more generally, the resistance through any two paths of heat flow is proportional to the temperature drops through these paths, and can be expressed as

$$\frac{R_1}{R_2} = \frac{t_i - t_x}{t_i - t_o} \tag{6.13}$$

where R_1 = the resistance from the indoor air to any point in the structure at which the temperature is to be determined

R_2 = the overall resistance of the wall from indoor air to outdoor air

t_i = indoor air temperature

t_x = temperature to be determined

t_o = outdoor air temperature

EXAMPLE 6.8

Determine the inside surface temperature for a wall having an overall coefficient $U = 0.25$, indoor air temperature 70°F, and outdoor air temperature -20°F.

Solution

$$R_1 = \frac{1}{h_i} = 0.68 \text{ (Table 6.8)}$$

$$R_2 = \frac{1}{U} = \frac{1}{0.25} = 4.00$$

$$\frac{0.68}{4.00} = \frac{70 - t_x}{70 - (-20)}$$

$$t_x = 54.7°F$$

6.8. TEMPERATURES IN ADJACENT UNCONDITIONED SPACES

The heat loss or gain between conditioned rooms and unconditioned rooms or spaces must be based on the estimated or assumed temperature in such unconditioned spaces. This temperature will normally lie in the range between the indoor and outdoor temperatures. The temperature in the unconditioned space may be estimated as

$$t_u = \frac{\begin{aligned}t_i(A_1U_1 + A_2U_2 + A_3U_3 + \cdots) + \\ t_o(2.16V_o + A_aU_a + A_bU_b + A_cU_c + \cdots)\end{aligned}}{\begin{aligned}A_1U_1 + A_2U_2 + A_3U_3 + \cdots + \\ 2.16V_o + A_aU_a + A_bU_b + A_cU_c + \cdots\end{aligned}} \qquad (6.14)$$

where t_u = temperature in unheated space, °F

t_i = indoor design temperature of heated room, °F

t_o = outdoor design temperature, °F

$A_1, A_2, A_3, \cdots$ = areas of surface of unheated space adjacent to heated space, ft^2

$A_a, A_b, A_c, \cdots$ = areas of surface of unheated space exposed to outdoors, ft^2

$U_1, U_2, U_3, \cdots$ = overall heat-transfer coefficients of surfaces of A_1, A_2, A_3, etc.

$U_a, U_b, U_c, \cdots$ = overall heat-transfer coefficients of surfaces of A_a, A_b, A_c, etc.

V_o = rate of introduction of outside air into the unheated space by infiltration and/or ventilation, cfm

Reasonable accuracy for ordinary rooms may be attained if the following rules are used in determining the design temperatures.

1. *Cooling with Unconditioned Room Adjacent.* Select for computation a temperature equal to $t_i + (t_o - t_i) \times 0.667$. In other words, add to the room temperature two-thirds of the difference between the indoor and outdoor temperatures.

2. *Heating Season, with Adjacent Room Unheated.* Take $\Delta t = (t_i - t_o) \times 0.5$. That is, use one-half the temperature difference between inside and outside in computing the heat loss through the wall to the adjacent room.

Temperatures in unconditioned spaces having large glass areas and two or more surfaces exposed to the outdoors (such as sleeping porches or sun parlors) and/or large amounts of infiltration (such as garages with poorly fitting doors) are generally assumed to be that of the outdoors.

EXAMPLE 6.9.

Calculate the temperature in an unheated attic, assuming: $t_i = 70°F$; $t_o = 10°F$; $A_c = 1000 \text{ ft}^2$; $A_r = 1200 \text{ ft}^2$; $A_w = 100 \text{ ft}^2$; $A_g = 10 \text{ ft}^2$; $U_r = 0.50 \text{ Btu/}$ $(\text{hr·ft}^2\text{·°F})$; $U_c = 0.40 \text{ Btu/(hr·ft}^2\text{·°F})}$; $U_w = 0.30 \text{ Btu/(hr·ft}^2\text{·°F})$; $U_g = 1.13$ $\text{Btu/(hr·ft}^2\text{·°F})$; $V_o = 0.5 \text{ cfm/ft}^2$.

Solution. Substituting these values into Eq. (6.14),

$$t_a = \frac{(1000 \times 0.40 \times 70) + 10[(2.16 \times 1000 \times 0.5) + (1200 \times 0.50) + (100 \times 0.30) + (10 \times 1.13)]}{1000 \times [0.40 + (2.16 \times 0.5)] + (1200 \times 0.50) + (100 \times 0.30) + (10 \times 1.13)}$$

$t_a = 45,213/2121 = 21.3°F$

6.9. HEATING LOAD

6.9.1. General Concepts

Prior to designing a heating system, an estimate must be made of the maximum probable heat loss of each room or space to be heated, based on maintaining a selected indoor air temperature during periods of design outdoor weather conditions. The heat losses may be divided into two groups: (1) transmission losses or heat transmitted through the confining walls, floor, ceiling, glass, or other surfaces, and (2) infiltration losses or heat required to warm outdoor air that leaks in through cracks and crevices, around doors and windows, or through open doors and windows, or heat required to warm outdoor air used for ventilation.

The ideal solution to the basic problem that confronts the designer of a heating system is to design a plant that has a capacity at maximum output just equal to the heating load that develops when the most severe weather conditions for the locality occur. However, where night setback is used, some excess capacity may be needed unless the owner is aware that under some conditions of operation he may not have the ability to elevate the temperature.

In most cases, economics interferes with the attainment of this ideal. Studies of weather records show that the most severe weather conditions do not repeat themselves every year. If heating systems were designed with adequate capacity for the maximum weather conditions on record, there would be considerable excess capacity during most of the operating life of the system.

In many cases, occasional failure of a heating plant to maintain a preselected indoor design temperature during brief periods of severe weather is not critical. However, the successful completion of some industrial or commercial processes may depend upon close regulation of indoor temperatures. These are special cases and require extra study before assigning design temperatures.

Theoretically, as a basis for design, the most unfavorable combination of temperature and wind speed should be chosen. It is entirely possible that a building might require more heat on a windy day with a moderately low outdoor temperature than on a quiet day with a much lower outdoor temperature. However, the combination of wind and temperature that is the worst would differ with different buildings, because wind speed has a greater effect on buildings that have relatively high infiltration losses. It would be possible to compute the heating load for a building for several different combinations of temperature and wind speed and to select the worst combination on record, but designers generally do not feel that such a degree of refinement is justified.

Normally the heating load is estimated for the winter design temperature occurring at night; therefore, no credit is taken for the heat given off by internal sources (people, lights, etc.).

The heat supplied by persons, lights, motors, and machinery should always be ascertained in the case of theaters, assembly halls, industrial plants, and commercial buildings such as stores and office buildings, but allowances for such heat sources must be made only after careful consideration of all local conditions. In many cases, these heat sources may materially affect the size of the heating plant and may have a marked effect on the operation and control of the system. In any evaluation, however, the night, weekend, and any other unoccupied periods must be evaluated in order to ascertain if the heating system has sufficient capacity to bring the building to the stipulated indoor temperature before the audience arrives. In industrial plants, quite a different condition may exist, and heat sources, if always available during occupancy, must be substituted for a portion of the heating requirements. In no case should the actual heating installation (exclusive of heat sources) be reduced below that required to maintain at least 40°F in the building.

The allowable temperature swing should be considered in the evaluation of the heating load. Capacity requirements may be reduced when the temperature within the space is allowed to drop a few degrees during periods of design load. Although the practice of drastically lowering the temperature to 50 or 55°F when the building is unoccupied may be effective in reducing fuel consumption, additional equipment capacity is required for pickup. In the case of intermittently heated buildings, additional heat is required for raising the temperature of the air, the building materials, and the material contents of the building to the specified indoor temperature. The rate at which this additional heat must be supplied depends upon the heat capacity of the structure and its material contents, and upon the time in which these are to be heated. In some cases, the capacity provided to temper outdoor ventilation air can be diverted to warmup or pickup capacity if the outdoor intake damper is closed until the building is warmed up and occupied.

Because design outdoor temperatures are generally quite lower than outdoor temperatures typically experienced during operating hours, many engineers make no allowances for this additional pickup heat in most buildings. However, if optimum equipment sizing is used (minimum safety factor), the additional pickup heat should be computed and allowed for as conditions require. In the case of churches, auditoriums, and other intermittently heated buildings, additional capacity on the order of 10% of the sum of all other heating loads should be provided.

6.9.2. General Procedure

To calculate the design heating load for a building or space, the following steps should be followed:

1. Select outdoor design weather conditions; temperature, wind direction, and wind speed (Section 6.3).
2. Select the indoor air temperature to be maintained in each room during coldest weather (Section 6.2).
3. Estimate temperatures in adjacent unheated spaces (Section 6.8).
4. Select or compute heat-transfer coefficients for outside walls and glass; for inside walls, nonbasement floors, and ceilings, if these are next to unheated spaces; and the roof if it is next to heated spaces (Section 6.6).
5. Determine net area of outside wall, glass, and roof next to heated spaces, as well as any cold walls, floors, or ceilings next to unheated spaces. Such determinations are made from building plans or from the actual building, using inside dimensions.
6. Compute heat-transmission losses for each kind of wall, glass, floor, ceiling, and roof in the building by multiplying the heat-transfer coefficient in each case by the area of the surface and the temperature difference between indoor and outdoor air, or adjacent unheated space (Section 6.9.3).

7. Compute heat losses from basement or grade-level slab floors by the methods given in Section 6.9.4.

8. Compute the energy associated with infiltration of cold air around outside doors and windows (Section 6.5).

9. The sum of the transmission losses or heat transmitted through the confining walls, floor, ceiling, glass, and other surfaces, plus the energy associated with the cold air entering by infiltration or required to replace mechanical exhaust, represents the total heating load.

10. In buildings that have a reasonably steady internal heat release of applicable magnitude from sources other than the heating system, a computation of this heat release under design conditions should be made for deduction from the total of the heat losses computed above.

11. Additional heating capacity may be required for intermittently heated buildings to bring the temperature of the air, confining surfaces, and building contents to the design indoor temperature within a specified time.

Table 6.13 summarizes the methodology for calculating the design heating loads.

6.9.3. Heat Loss for Above-Grade Components

The heat loss by conduction and convection through walls, windows, doors, and other structural components is determined by the following equation:

$$q = U(t_i - t_o)A \qquad (6.15)$$

where q = heat transfer through the wall, roof, ceiling, floor, or glass, Btu/hr

A = area of wall, glass, roof, ceiling, floor, or other exposed surface, ft^2

U = air-to-air heat transfer coefficient, Btu/(hr·ft^2·°F)

t_i = indoor air temperature near surface involved, °F

t_o = outdoor air temperature, or temperature of adjacent unheated space, °F

6.9.4. On-Grade and Below-Grade Heat Losses

The allowance made for basement heat loss depends on whether or not the basement is heated. If the basement is heated to a specified temperature, heat loss is calculated in the usual manner, based on proper wall and floor U values and outdoor air and ground temperature. Heat loss through windows and walls above grade is based on outdoor air temperature and proper air-to-

TABLE 6.13. Design Heating Loads

Heating Load	Equation[a]
Roofs, walls, glass	$q = U \times A \times TD$
Floors over exterior space	$q = U \times A \times TD$
Floors on or below grade	$q = U \times A \times TD$
	and/or
	$q = F \times P \times TD$
Walls below grade	$q = U \times A \times TD$
Ventilation and infiltration air	
Sensible	$q_s = 1.10 \times \text{cfm} \times \Delta t$
Latent	$q_l = 4840 \times \text{cfm} \times \Delta W$
Total	$q = 4.5 \times \text{cfm} \times \Delta h$

[a]TD = temperature difference.

air U values. Heat loss through basement walls below grade is based on floor and wall U values for surfaces in contact with the soil, and on proper ground temperature.

If a basement is completely below grade and unheated, its temperature normally will range between that in the rooms above and that of the ground. Of course, basement windows will lower basement temperature when it is cold outdoors, and heat given off by the heating plant will increase the basement temperature. The exact basement temperature is indeterminate if the basement is not heated. In general, heat from the heating plant warms the air near the basement ceiling sufficiently to make an allowance unnecessary for floor heat loss from rooms located over the basement.

The heat loss through basement floors and through building-wall surfaces below ground level is difficult to compute with accuracy, since the ground temperature varies with depth and with the amount of heat flowing into it from the structure above. The ground temperature near the surface varies with the season of the year and the climate; in the United States, however, frost seldom penetrates more than 4 ft, and below 3 or 4 ft the ground temperature undergoes only moderate swings the year round. If fact, it is customary to consider the temperature of groundwater as indicative of sub-surface earth temperature. Groundwater temperature, even in the most northerly sections of the United States, is seldom below 40°F and in warm sections of the country seldom exceeds 60°F. When below-grade spaces are conditioned as living space, the walls should be furred and finished with a vapor barrier, insulating board, and some type of finish layer such as paneling. This will add thermal resistance to the wall and reduce the effect of the uncertainty of the data.

Below-grade basement-wall heat losses are given in Table 6.14 for uninsulated concrete walls as well as walls to which 1, 2, or 3 in. of insulation has been added. An average value of 9.6 (Btu/hr)(in.)/(ft^2·°F) has been assumed

TABLE 6.14.   Heat Loss below Grade in Basement Walls[a], Btu/(hr·ft²·°F)

Depth (ft)	Path Length Through Soil (ft)	Uninsulated	Insulated		
			1 in.	2 in.	3 in.
0–1 (1st)	0.68	0.410	0.152	0.093	0.067
1–2 (2nd)	2.27	0.222	0.116	0.079	0.059
2–3 (3rd)	3.88	0.155	0.094	0.068	0.053
3–4 (4th)	5.52	0.119	0.079	0.060	0.048
4–5 (5th)	7.05	0.096	0.069	0.053	0.044
5–6 (6th)	8.65	0.079	0.060	0.048	0.040
6–7 (7th)	10.28	0.069	0.054	0.044	0.037

[a] k_{soil} = 9.6 (Btu/hr)(in.)/(ft²·°F); $k_{insulation}$ = 0.24 (Btu/hr)(in.)/(ft²·°F).

SOURCE: Reprinted by permission from *ASHRAE Handbook of Fundamentals 1977*, p. 24.4, Table 1.

for conductivity of soil; and insulation is assumed to have a thermal conductivity of 0.24 (Btu/hr)(in.)/(ft²·°F).

The average rate of heat loss through the floor may be taken as equal to that from a point located one-quarter of the basement width from the side wall. The path length from this point varies with both depth of basement below grade and width of basement. Shallow, narrow basements will have higher heat loss per square foot than deep, wide basements. Typical values are given in Table 6.15.

Selecting the appropriate design temperature difference can be a problem. Although internal design temperature is given by basement air temperature, none of the usual external design air temperatures are applicable because of the heat capacity of the soil. However, ground surface temperature is known to fluctuate about a mean value by an amplitude A, which will vary with geographic location and surface cover. Thus, suitable external design temperatures can be obtained by subtracting A for the location from the mean

TABLE 6.15.   Heat Loss through Basement Floors, Btu/(hr·ft²·°F)

Depth of Foundation Wall below Grade (ft)	Width of House (ft)			
	20	24	28	32
5	0.032	0.029	0.026	0.023
6	0.030	0.027	0.025	0.022
7	0.029	0.026	0.023	0.021

SOURCE: Reprinted by permission from *ASHRAE Handbook of Fundamentals 1977*, p. 24.4, Table 2.

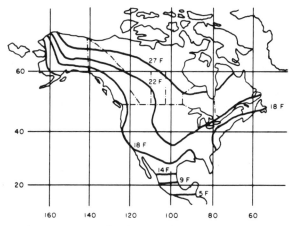

Fig. 6.6.   Lines of constant amplitude. Reprinted by permission from *ASHRAE Handbook of Fundamentals 1977.*

annual air temperature, $\bar{t}_a$. Values for $\bar{t}_a$ can be obtained as the average winter temperature from Table 7.1; and A can be estimated from the map in Fig. 6.6.

In one type of building construction that has found frequent use, the structure is placed on a concrete slab laid directly on the ground over a well-drained gravel or ash fill. For heating, in this case, the effect of loss into the ground and air near the outside edge of the slab is extremely important, and it is almost essential to provide for edge insulation.

When considering the heat losses for floor slabs at or near grade level, we must take into account two situations. The first is the unheated slab where heat is supplied to the space from above. The second situation results when the air-duct system is installed beneath the slab, with air discharged around the perimeter of the structure. In both cases, most of the heat loss is from the edge of the slab. Compared to the total heat losses of the structure, this loss may not be significant; however, from the viewpoint of comfort the loss is important. Proper insulation around the perimeter of the slab is essential in severe climates to ensure a warm floor.

Experiments with unheated floor slabs indicate that heat loss from a concrete slab floor on grade is more nearly proportional to perimeter than to area of the floor, and that the heat loss can be estimated by means of the equation

$$q = F_2 P(t_i - t_o) \tag{6.16}$$

where q = heat loss of floor, Btu/hr
    F_2 = heat loss coefficient, Btu/hr per linear foot of exposed edge per °F temperature difference between the indoor air and the outdoor air

TABLE 6.16.   Heat Loss of Concrete Floors At or Near Grade Level per Foot of Exposed Edge

Outdoor Design Temperature, °F	Heat Loss per Foot of Exposed Edge, Btu/hr	
	Recommended 2-in. Edge Insulation	1-in. Edge Insulation
−20 to −30	50	55
−10 to −20	45	50
0 to −10	40	45
	1-in. Edge Only Insulation	No Edge Insulation[a]
−20 to −30	60	75
−10 to −20	55	65
0 to −10	50	60

[a]This construction not recommended; shown for comparison only.

SOURCE: Reprinted by permission from *ASHRAE Handbook of Fundamentals 1977*, p. 24.5, Table 3.

P = perimeter or exposed edge of floor, ft
t_i = indoor air temperature, °F
t_o = outdoor air temperature, °F

The value of F_2 ranged from 0.81 for a floor with no edge insulation to 0.55 for a floor with 1 in. of edge insulation.

Values of heat loss per linear foot of exposed edge given in Table 6.16 are sufficiently accurate for design calculations. The insulation should extend under the flow horizontally for 2 ft. It can also be located along the foundation vertical wall with equal effectiveness if it extends 2 ft below floor level as shown in Fig. 6.7.

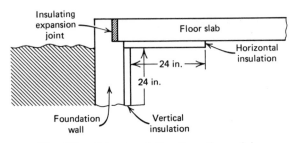

Fig. 6.7.   Edge insulation for a floor slab.

EXAMPLE 6.10

Determine the heat loss for a basement in St. Louis which is 60 ft × 40 ft × 7 ft high, of standard concrete construction, and entirely below grade.

Solution. Perimeter = 2(60 + 40) = 200 ft

Walls	Heat Loss
1st	0.41
2nd	0.222
3	0.155
4	0.119
5	0.096
6	0.079
7	0.069
	1.15 Btu/(hr·ft·°F)

Wall loss = 1.15(200) = 230

Floor (Table 6.15)
40 ft wide
q = .021 Btu/(hr·ft^2·°F)
Area = 60 × 40 = 2400 ft^2
Floor loss = 0.021(2400) = 50.4
Design $\Delta t = t_i - (\bar{t}_a - A) = 70 - (44 - 22) = 48$
Q = 48(50.4 + 230) = 13,500 Btu/hr

EXAMPLE 6.11

For a building with design conditions of 75°F indoor and 15°F outdoor, determine the heat loss through each of the following components:

a. Slab floor, 56 ft × 28 ft, on grade without perimeter insulation.
b. Single glass double-hung window, 3 ft × 5 ft, with storm window in common metal frame
c. 1½ in. thick solid wood door, 3 ft × 7 ft, with wooden storm door
d. Sliding patio door, 6 ft × 7 ft, metal frame with double insulating glass having ¼ in. air space.

Solution

a. $q = F_2 P(t_i - t_o) = 0.81[2(56 + 28)](75 - 15) = 8165$ Btu/hr
b. $q = (.50)(1.20) \times (5 \times 3) \times (75 - 15) = 540$ Btu/hr
c. $q = (.27)(3 \times 7) \times (75 - 15) = 340$ Btu/hr
d. $q = (.58)(1.10) \times (6 \times 7) \times (75 - 15) = 1610$ Btu/hr

EXAMPLE 6.12

A typical residence may have a total wall area of 1496 ft^2 (excluding windows and doors) and may be located where winter design conditions are 72°F indoor and 2°F outdoor. The basic wall construction includes 1 by 8 wood drop siding, $\frac{3}{4}$-in. plywood sheathing, 2 by 6 studs, and $\frac{3}{8}$-in. plasterboard. Determine the design heat loss through the wall (Btu/hr):

a.   If uninsulated in the stud space.

b.   If full-wall glass fiber insulation is in the stud space.

Solution

	R_s	R_{ag}	R_i
Outside air, 15 mph	0.17	0.17	0.17
1 × 8 wood drop siding	0.79	0.79	0.79
$\frac{3}{4}$ in. plywood sheathing	0.93	0.93	0.93
5$\frac{1}{2}$ in. wood @1.25	6.88		
or			
Still air		0.68	
Still air		0.68	
or			
5$\frac{1}{2}$ in. glass fiber insulation			19.00
$\frac{3}{4}$ in. plasterboard	0.32	0.32	0.32
Inside air	0.68	0.68	0.68
$\Sigma R =$	9.77	4.25	21.89
$U =$	0.1024	0.2353	0.0457

Uninsulated:
$U_{av} = 0.15(0.1024) + 0.85(0.2353) = 0.215$
$Q = UA\Delta t = 0.215(1,496)(72 - 2) = 22,500$ Btu/hr
Insulated:
$U_{av} = 0.15(0.1024) + 0.85(0.0457) = 0.054$
$Q = UA\Delta t = 0.054(1,496)(72 - 2) = 5,650$ Btu/hr

6.10.   COOLING LOAD

6.10.1.   General Concepts

The variables affecting cooling-load calculations are numerous, often difficult to define precisely, and always intricately interrelated. Many of the components of the cooling load vary in magnitude over a wide range during a

24-hr period. Since these cyclic changes in load components are often out of phase with each other, a detailed analysis is required to establish the resultant maximum cooling load for a building or zone. A zoned system (i.e., a system serving several areas that each has its own temperature control) must often handle peak loads in different hours. Also, at certain times of the day during the heating season, some perimeter zones may require heating and others may require cooling.

Economic considerations must be of particular influence in the selection of equipment for cooling-season operation in comfort air-conditioning, and this fact, coupled with present inadequacies in available data and knowledge of the air-conditioning art, places a premium on the experienced judgment essential to successful design or practice. Variations in the weather, building occupancy, and other factors affecting load necessitate careful coordination of controls to regulate simultaneously the system components and the equipment in order to maintain the desired room conditions.

The calculation procedures presented deal with the various instantaneous rates of heat gain, both sensible and latent, in a conditioned space. There may be an appreciable difference between the net instantaneous rate of heat gain and the actual cooling load at any instant. This difference is caused by the storage and subsequent release of heat by the structure and its contents. Thermal storage effects may be quite important in determining an economical cooling equipment capacity.

In air-conditioning design, it is important to differentiate among four related, but distinct, heat flow rates, each of which varies with time: (1) space heat gain, (2) space cooling load, (3) space heat extraction rate, and (4) cooling coil load.

Space heat gain (instantaneous rate of heat gain) is the rate at which heat enters into and/or is generated within a space at a given instant of time. Heat gain is classified by

1. The mode in which it enters the space.
2. Whether it is a sensible or latent gain.

The first classification is necessary because different fundamental principles and equations are used to calculate different modes of energy transfer. The heat gain occurs in the form of

1. Solar radiation through transparent surfaces
2. Heat conduction through exterior walls and roofs
3. Heat conduction through interior partitions, ceilings, and floors
4. Heat generated within the space by occupants, lights, and appliances
5. Energy transfer as a result of ventilation and infiltration of outdoor air
6. Miscellaneous heat gains

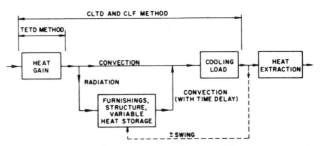

Fig. 6.8. Origin of difference between magnitudes of instantaneous heat gain and instantaneous cooling load. Reprinted by permission from *ASHRAE Handbook of Fundamentals 1977.*

The second classification, sensible or latent, is important for proper selection of cooling equipment. The heat gain is sensible when there is a direct addition of heat to the conditioned space by any or all mechanisms of conduction, convection, and radiation. The heat gain is latent when moisture is added to the space (e.g., by vapor emitted by occupants). To maintain a constant humidity ratio in the enclosure, water vapor in the cooling apparatus must condense out at a rate equal to its rate of addition into the space. The amount of energy required to do this, the latent heat gain, essentially equals the product of the rate of condensation and latent heat of condensation. The distinction between sensible and latent heat gain is necessary for cooling apparatus selection. Any cooling apparatus has a maximum sensible heat removal capacity and a maximum latent heat removal capacity for particular operating conditions.

Space cooling load is the rate at which heat must be removed from the space to maintain room air temperature at a constant value. Note that the summation of all space instantaneous heat gains at any given time does not necessarily equal the cooling load for the space at that same time. The space heat gain by radiation is partially absorbed by the surfaces and contents of the space and does not affect the room air until some time later. The radiant energy must first be absorbed by the surfaces that enclose the space (i.e., walls, floor, and ceiling) and the material in the space. As soon as these surfaces and objects become warmer than the space air, some of their heat will be transferred to the air in the room by convection. Since their heat storage capacity determines the rate at which their surface temperatures increase for a given radiant input, it governs the relationship between the radiant portion of heat gain and the corresponding part of the space cooling load (see Fig. 6.8). The thermal storage effect can be important in determining an economical cooling equipment capacity.

Space heat extraction rate is the rate at which heat is removed from the conditioned space. It equals the space cooling load only when room air temperature is kept constant, which rarely occurs. Usually, the control system, in conjunction with intermittent operation of the cooling equipment, will cause a "swing" in room temperature. Therefore, provided the control system is simulated properly, computation of the space heat-extraction rate results in a more realistic value of energy removal at the cooling equipment than merely using the values of the space cooling load.

Cooling coil load is the rate at which energy is removed at the cooling coil that serves one or more conditioned spaces in any central air-conditioning system. It is equal to the instantaneous sum of the space cooling load (or space heat-extraction rate if the space temperature is assumed to "swing") for all the spaces served by the system, plus any additional load imposed on the system external to the conditioned space. Such additional load components include heat gain into the distribution system between the individual spaces and the cooling equipment and outdoor hot and moist air introduced into the distribution system through the cooling equipment.

There are basically five different components of heat gain that must be calculated in any space cooling load computation: (1) heat gain through exterior walls and roofs, (2) heat gain through fenestrations (light-transmitting areas), (3) heat gain through interior partitions, ceilings, and floors, (4) heat sources within the conditioned space, and (5) heat gain due to ventilation and infiltration.

6.10.2. Exterior Walls and Roofs

The technique for calculating the space cooling-load component as a result of heat gain through exterior roofs and walls involves the concept of sol-air temperature. *Sol-air temperature* is that temperature of the outdoor air which, in the absence of all radiation exchanges, would give the same rate of heat entry into the surface as would exist within the actual combination of incident solar radiation, radiant energy exchange with the sky and other outdoor surroundings, and convective heat exchange with the outdoor air.

The *transfer function* method can be used to compute the one-dimensional transient heat flow through various sunlit roofs and walls. The heat gain is then converted to cooling load by using the room transfer functions. The results were then generalized to some extent by dividing the cooling load by the U factor for each roof or wall. The results thus obtained are in units of total equivalent cooling load temperature difference (CLTD). The corresponding cooling load for exterior wall and roofs is thus obtained as

$$q = UA \text{ CLTD} \qquad (6.17)$$

The heat flow through a similar roof or wall (similar in thermal mass or weight as well as U value) can be obtained by multiplying the total CLTDs listed in Tables 6.17 and 6.18 by the U value of another roof or wall, respectively. Table 6.19 provides a description of the wall construction types listed in Table 6.18.

The CLTD values in the tables were computed for an indoor air tempera-

TABLE 6.17. Cooling Load Temperature Differences for Calculating Cooling Load from Flat Roofs

Roof No	Description of Construction	Weight lb/ft²	U-value Btu/(h ft²·°F)	1	2	3	4	5	6	7	8	9	10	11	12	13	14	15	16	17	18	19	20	21	22	23	24	Hour of Maximum CLTD	Minimum CLTD	Maximum CLTD	Difference CLTD
	Without Suspended Ceiling																													
1	Steel sheet with 1-in. (or 2-in.) insulation	7 (8)	0.213 (0.124)	1	-2	-3	-3	-5	-3	-3	6	19	34	49	61	71	78	79	77	70	59	45	30	18	12	8	5	14	-5	79	84
2	1-in. wood with 1-in. insulation	8	0.170	6	3	0	-1	-3	-3	-2	4	14	27	39	52	62	70	74	74	70	62	51	38	28	20	14	9	16	-3	74	77
3	4-in. l.w. concrete	18	0.213	9	5	2	0	-2	-3	-3	-1	9	20	32	44	55	64	70	73	71	66	57	45	34	25	18	13	16	-3	73	76
4	2-in. h.w. concrete with 1-in. (or 2-in.) insulation	29	0.206 (0.122)	12	8	5	3	0	-1	-1	3	11	20	30	41	51	59	65	66	66	62	54	45	36	29	22	17	16	-1	67	68
5	1-in. wood with 2-in. insulation	19	0.109	3	0	-3	-4	-5	-7	-6	-3	5	16	27	39	49	57	63	64	62	57	48	37	26	18	11	7	16	-7	64	71
6	6-in. l.w. concrete	24	0.158	17	13	9	7	6	3	1	3	7	9	13	19	23	33	43	51	58	62	64	62	57	50	42	35	18	1	64	63
7	2.5-in. wood with 1-in. insulation	13	0.130	29	24	20	16	13	10	7	6	6	9	13	20	27	34	42	48	53	55	56	54	49	44	40	34	19	6	56	50
8	8-in. l.w. concrete	31	0.126 (0.120)	35	30	26	22	18	14	11	8	7	7	9	13	19	25	33	39	46	50	53	54	53	49	45	40	20	7	54	47
9	4-in. h.w. concrete with 1-in. (or 2-in.) insulation	52	0.200	25	22	18	15	12	9	8	8	10	14	20	26	33	40	46	50	53	53	52	48	43	38	34	30	18	8	53	45
10	2.5-in. wood with 2-in. insulation	13 (52)	0.093	30	26	23	19	16	13	10	9	8	9	13	17	23	29	36	41	46	49	51	50	47	43	39	35	19	8	51	43
11	Roof terrace system	75	0.106	34	31	28	25	22	19	16	14	13	13	15	18	22	26	31	36	40	44	45	46	45	43	40	37	20	13	46	33
12	6-in. h.w. concrete with 1-in. (or 2-in.) insulation	(75)	0.192 (0.117)	31	28	25	22	20	17	15	14	13	13	15	18	22	31	36	40	43	45	45	44	42	40	37	34	19	14	45	31
13	4-in. wood with 1-in. (or 2-in.) insulation	17 (18)	0.106 (0.078)	38	36	33	30	28	25	22	20	18	17	16	17	18	21	24	28	32	36	39	41	43	43	42	40	22	16	43	27
	With Suspended Ceiling																													
1	Steel Sheet with 1-in. (or 2-in.) insulation	9 (10)	0.134 (0.092)	2	0	-2	-3	-4	-4	-1	9	23	37	50	62	71	77	78	74	67	56	42	28	18	12	8	5	15	-4	78	82
2	1-in. wood with 1-in. insulation	10	0.115	20	15	11	8	5	3	2	3	7	13	21	30	40	48	55	60	62	61	58	51	44	37	30	25	17	2	62	60
3	4-in. l.w. concrete	20	0.134	19	14	10	7	4	2	0	0	4	10	19	29	39	48	56	62	65	64	61	54	46	38	30	24	17	0	65	65
4	2-in. h.w. concrete with 1-in. (or 2-in.) insulation	30	0.131	28	25	23	20	17	15	13	13	16	20	25	30	35	39	43	46	46	47	46	44	41	38	35	32	18	13	47	34
5	1-in. wood with 2-in. insulation	10	0.083	25	20	16	13	10	7	5	5	7	12	18	25	33	41	48	53	57	57	56	52	46	40	34	29	18	5	57	52
6	6-in. l.w. concrete	26	0.109	32	29	26	23	19	16	13	10	8	7	8	11	16	22	29	36	42	48	52	54	54	51	47	42	20	7	54	47
7	2.5-in. wood with 1-in. insulation	15	0.096	34	31	29	26	23	21	18	16	15	15	16	18	21	25	30	34	38	41	43	44	44	42	40	37	21	15	44	29
8	8-in. l.w. concrete	33	0.093 (0.090)	39	36	33	29	26	23	20	18	14	14	15	17	20	25	29	34	38	42	45	45	46	44	42	42	21	14	46	32
9	4-in. h.w. concrete with 1-in. (or 2-in.) insulation	53 (54)	0.128	30	29	27	26	24	22	21	20	20	21	22	24	29	32	34	34	36	38	38	38	37	36	34	31	19	20	38	18
10	2.5-in. wood with 2-in. insulation	15	0.072	35	33	30	28	26	24	23	23	23	22	20	22	25	28	32	35	38	40	41	41	40	39	37	37	21	18	41	23
11	Roof terrace system	77	0.082	30	29	28	27	26	25	24	23	22	22	23	23	25	28	29	31	34	31	32	33	33	33	32	32	22	22	33	11
12	6-in. h.w. concrete with 1-in. (or 2-in.) insulation	77 (77)	0.125 (0.088)	29	28	27	26	25	24	23	22	21	21	22	23	25	30	32	33	34	34	34	34	33	32	31	34	20	21	34	13
13	4-in. wood with 1-in (or 2-in) insulation	19 (20)	0.082 (0.064)	35	34	33	32	31	29	27	26	24	23	22	22	24	25	27	30	32	34	36	37	39	37	36	33	23	21	37	16

1 Application: These values may be used for all normal air-conditioning estimates; usually without correction (except as noted below) in latitude 0° to 50° North or South when the load is calculated for the hottest weather.

2 Corrections: The values in the table were calculated for an inside temperature of 78 F and an outdoor maximum temperature of 95 F, with an outdoor daily range of 21 deg F. The table remains approximately correct for other outdoor maximums (93-102 F) and other outdoor daily ranges (16-34 deg F), provided the outdoor daily average temperature remains approximately 85 F. If the outdoor air temperature is different from 78 F and/or the outdoor daily average temperature is different from 85 F, the following rules apply: (a) For room air temperature less than 78 F, add the difference between 78 F and room air temperature; if greater than 78 F, subtract the difference. (b) For outdoor daily average temperature less than 85 F and the daily average temperature; if greater than 85 F, add the difference.

3 Attics or other spaces between the roof and ceiling: If the ceiling is insulated and a fan is used for positive ventilation in the space between the ceiling and roof, the total temperature difference for calculating the room load may be decreased by 25%. If the attic space contains a return duct or other air plenum, care should be taken in determining the portion of the heat gain that reaches the ceiling.

4 Light Colors: Multiply the CLTD's in the table by 0.5. Credit should not be taken for light-colored roofs except where the permanence of light color is established by experience, as in rural areas or where there is little smoke.

5 For solar transmission in other months: The table values of temperature differences calculated for July 21* will be approximately correct for a roof in the following months:

North Latitude

Latitude	Months
0°	All Months
10°	All Months
20°	All Months except Nov., Dec., Jan.
30°	Mar., Apr., May, June, July, Aug., Sept.
40°	April, May, June, July, Aug.
50°	May, June, July

South Latitude

Latitude	Months
0°	All Months
10°	All Months
20°	All Months except May, June, July
30°	Sept., Oct., Nov., Dec., Jan., Feb., Mar.
40°	Oct., Nov., Dec., Jan., Feb.
50°	Nov., Dec., Jan.

6 For each 7 increase in R-value due to insulation added to the roof structures (Table 4), use a CLTD for a roof whose weight is approximately the same but whose CLTD has a maximum value 2 hr later. If this is not possible, due to having already selected the roof with the longest time lag, use a Δt in the load calculation equal to the difference between the 24 hr average sol-air temperature and the room air temperature.

SOURCE: Reprinted by permission from *ASHRAE Handbook of Fundamentals 1977*, p. 25.4, Table 5.

351

TABLE 6.18. Cooling Load Temperature Differences for Calculating Cooling Load from Sunlit Walls

North Latitude Wall Facing	1	2	3	4	5	6	7	8	9	10	11	12	13	14	15	16	17	18	19	20	21	22	23	24	Hr of Maximum CLTD	Minimum CLTD	Maximum CLTD	Difference CLTD
Group A Walls																												
N	14	14	14	13	13	13	12	12	11	11	10	10	10	10	10	10	11	11	12	12	13	13	14	14	24	10	14	4
NE	19	19	18	18	17	17	16	15	15	15	15	15	16	16	17	18	18	18	19	19	20	20	20	20	22	15	20	5
E	24	23	23	22	22	21	20	19	19	19	18	18	18	19	20	21	23	24	25	25	25	25	24	24	22	18	25	7
SE	24	23	23	22	21	20	20	19	18	18	18	18	18	18	20	21	22	23	23	24	24	24	24	24	22	18	24	6
S	20	20	20	19	18	18	17	16	15	15	14	14	14	14	15	15	17	19	20	21	22	22	21	21	23	14	20	6
SW	25	25	25	24	24	23	22	21	20	20	19	19	19	19	19	19	20	20	22	23	24	25	25	26	24	17	25	8
W	27	27	26	26	25	24	24	23	22	21	21	20	19	19	19	19	20	20	22	23	25	26	27	27	1	18	27	9
NW	21	21	21	20	20	19	18	18	17	16	16	15	14	14	14	14	15	15	16	17	18	19	20	21	1	14	21	7
Group B Walls																												
N	15	14	14	13	12	11	11	10	9	9	9	9	9	10	11	12	13	14	15	14	14	15	15	15	24	8	15	7
NE	19	18	17	16	15	14	13	12	12	13	14	15	16	17	18	19	20	20	20	21	21	21	21	20	21	12	21	9
E	23	22	21	20	18	17	16	15	14	15	17	19	21	22	24	25	26	26	26	27	26	26	25	24	20	15	27	12
SE	23	22	21	20	18	17	16	15	15	16	18	20	22	23	24	25	26	26	26	26	26	25	24	23	21	14	26	12
S	20	20	19	18	17	16	15	13	12	11	11	11	12	13	14	15	17	19	20	21	22	22	21	20	23	11	22	11
SW	27	26	25	24	22	21	20	18	16	15	14	14	13	14	15	17	20	23	25	27	28	28	28	27	24	13	28	15
W	29	28	27	26	24	23	21	19	18	17	16	15	14	14	14	15	17	19	22	24	27	29	30	30	22	14	30	16
NW	23	22	21	20	19	18	17	15	14	13	12	12	11	11	12	13	15	15	16	18	20	22	23	23	22	11	23	9
Group C Walls																												
N	15	14	13	12	11	10	9	8	7	7	7	8	9	10	12	13	14	14	15	16	17	17	17	16	22	7	17	10
NE	19	17	16	15	13	11	10	11	13	15	17	19	20	21	21	22	22	23	23	23	23	22	21	20	20	10	23	13
E	22	21	19	18	16	14	12	12	14	16	19	22	25	27	29	29	29	30	29	29	28	26	24	23	18	12	30	18
SE	22	21	19	18	16	14	13	12	13	15	17	19	22	24	26	28	29	29	29	29	28	27	24	23	19	12	29	17
S	21	19	18	17	15	14	13	11	9	9	10	11	12	14	17	20	22	24	25	26	26	25	24	22	20	9	26	17
SW	29	27	25	24	22	20	18	16	15	14	14	14	15	16	19	21	24	27	30	32	33	33	32	31	22	11	33	22
W	31	29	27	25	23	21	19	17	16	15	14	14	14	15	16	18	20	22	25	28	32	34	35	34	22	12	35	23
NW	25	23	21	20	18	16	14	13	11	10	10	11	12	13	15	16	18	22	24	26	27	27	26	22	22	10	27	17
Group D Walls																												
N	15	14	13	12	11	10	8	6	6	6	6	7	8	10	12	13	15	17	18	19	19	19	18	16	21	6	19	13
NE	17	15	14	13	11	10	8	7	10	14	17	20	22	23	24	24	24	25	25	24	23	22	20	18	19	7	25	18
E	19	17	15	13	11	10	8	8	11	17	22	27	30	32	33	32	32	32	31	30	28	26	24	22	16	8	33	25
SE	20	17	15	13	11	10	8	7	9	15	20	26	30	32	32	32	32	32	31	30	28	26	24	22	17	8	32	24
S	19	17	15	13	11	9	7	6	6	7	9	12	16	20	24	26	27	29	29	29	27	24	21	19	19	6	29	23
SW	28	25	22	19	16	14	11	9	8	8	10	12	14	16	20	24	27	31	35	38	38	37	34	31	21	8	38	30
W	31	27	24	21	18	15	12	10	9	9	10	12	14	15	17	21	27	32	36	38	40	41	38	34	21	9	41	32
NW	25	22	19	17	14	12	10	9	7	7	8	10	12	14	17	21	24	27	31	32	32	32	30	27	22	7	32	25

Group E Walls

Hour	1	2	3	4	5	6	7	8	9	10	11	12	13	14	15	16	17	18	19	20	21	22	23	24
N	12	10	8	7	5	3	2	1	1	2	3	5	6	7	9	11	13	14	15	17	19	20	20	19
NE	13	11	9	7	6	4	3	5	5	5	5	7	9	11	13	15	17	19	22	24	25	26	26	22
E	14	12	10	8	6	5	4	6	10	17	26	31	34	36	36	37	36	34	33	32	30	33	38	33
SE	15	12	10	8	7	5	4	6	10	17	25	31	33	37	37	36	35	33	32	31	30	31	37	32
S	22	18	15	12	9	6	4	3	5	7	11	17	19	24	29	34	43	44	45	43	38	34	31	31
SW	25	21	17	14	11	9	6	6	5	6	8	12	18	24	32	39	44	49	50	53	49	45	40	40
W	20	17	14	11	9	7	6	6	5	6	8	11	14	20	27	34	43	50	54	57	60	49	43	43
NW	16	14	11	9	7	6	5	5	5	6	7	10	13	15	20	26	32	35	42	46	44	38	33	33

Group F Walls

Hour	1	2	3	4	5	6	7	8	9	10	11	12	13	14	15	16	17	18	19	20	21	22	23	24
N	8	6	5	3	2	1	1	0	-1	2	4	6	9	11	14	17	19	19	21	22	23	23	23	23
NE	9	7	5	5	2	1	0	-1	5	14	23	28	30	29	27	27	27	26	24	24	22	22	30	29
E	10	7	6	4	3	2	1	0	6	17	28	38	44	45	43	41	36	34	32	30	27	24	45	43
SE	10	8	6	4	3	2	1	-1	3	10	18	28	36	41	42	43	41	39	36	34	31	28	25	41
S	15	10	8	5	3	1	-1	-2	3	7	13	20	27	34	38	39	38	35	39	38	38	39	39	38
SW	17	15	11	7	5	3	1	0	4	8	11	17	26	35	44	50	53	49	57	54	53	53	53	48
W	14	11	9	7	5	3	2	1	5	8	11	14	20	28	39	49	57	67	60	60	67	60	60	57
NW	11	10	8	6	4	2	1	0	3	5	8	11	13	15	20	27	35	43	46	46	44	46	46	44

Group G Walls

Hour	1	2	3	4	5	6	7	8	9	10	11	12	13	14	15	16	17	18	19	20	21	22	23	24
N	3	2	1	1	0	-1	-1	1	2	7	8	9	12	15	18	21	23	22	26	25	26	24	26	27
NE	3	2	2	1	0	-1	0	5	9	27	36	35	30	26	26	22	18	18	22	25	22	22	39	40
E	4	2	2	1	0	-1	0	11	17	31	47	50	45	40	33	30	24	21	19	27	24	30	55	56
SE	4	4	2	1	0	-1	0	5	18	32	42	49	48	42	36	30	27	24	21	20	27	31	51	52
S	4	4	2	1	0	-1	0	5	12	22	31	39	45	46	43	37	37	25	20	15	14	46	46	47
SW	5	5	3	2	1	0	1	9	16	26	38	52	59	63	61	52	52	24	17	13	16	63	63	63
W	6	5	3	2	1	0	1	8	15	27	41	67	72	67	56	48	67	29	20	15	17	72	72	71
NW	5	5	2	1	0	-1	0	8	15	18	21	41	48	55	55	41	25	27	25	17	18	55	55	55

1. Application: These values may be used for all normal air-conditioning estimates; usually without correction (except as noted below) when the load is calculated for the hottest weather.

2. Corrections: The values in the table were calculated for an inside temperature of 78 F and an outdoor daily range of 21 deg F. The table remains approximately correct for other outdoor maximums (93-102 F) and other outdoor daily ranges (16-34 deg F), provided the outdoor daily average temperature remains approximately 85 F. If the room temperature is different from 78 F and/or the outdoor daily average temperature is different from 85 F, the following rules apply: (a) For room air temperature less than 78 F, add the difference between 78 F and room air temperature; if greater than 78 F, subtract the difference. (b) For outdoor daily average temperature less than 85 F, subtract the difference between 85 F and the daily average temperature; if greater than 85 F, add the difference. The table values will be approximately correct for the east or west wall in any latitude (0° to 50° North or South) during the hottest weather.

3. Color of exterior surface of wall: For light colors, multiply the Wall-Cooling Load Temperature Difference in the tables by 0.65. Use temperature differences for light walls only when the permanence of the light wall is established by experience. For cream colors, use the light wall values. For medium colors, interpolate half-way between the dark and light values. Medium colors are medium blue and green, bright red, light brown, unpainted wood, and natural color concrete. Dark blue, red, brown, and green are considered dark colors.

4. Correction for other months and latitudes: The CLTD is calculated by adding this correction factor to the CLTD of the above table: $[(I_{DT-new}/I_{DT-Table\ 22})-1](t_{ea}-t_{oa})$ where I_{DT-new} is the sum of two appropriate half-day totals of solar heat gain in Tables 17 to 25 of Chapter 26.

5. For each 7 increase in R-value due to insulation added to the wall structures in Table 6, use the CLTD for the wall group with the next higher letter in the alphabet. When the insulation is added to the exterior of the construction rather than the interior, use the CLTD for the wall group two letters higher. If this is not possible, due to having already selected a wall in Group A, use a Δt in the load calculation equal to the difference between the 24 hr average sol-air temperature and the room air temperature.

SOURCE: Reprinted by permission from AHSRAE Handbook of Fundamentals 1977, p. 25.9, Table 7.

TABLE 6.19.   Wall Construction Group Description

Group No.	Description of Construction	Weight(lb/ft²)	U-value (Btu/(h·ft²·F)	Code Numbers of Layers (see Table 8)
4-in. Face Brick+(*Brick*)				
C	Air Space+4-in. Face Brick	83	0.358	A0, A2, B1, A2, E0
D	4-in. Common Brick	90	0.415	A0, A2, C4, E1, E0
C	1-in. Insulation or Air space+4-in. Common Brick	90	0.174-0.301	A0, A2, C4, B1/B2, E1, E0
B	2-in. Insulation+4-in. Common Brick	88	0.111	A0, A2, B3, C4, E1, E0
B	8-in. Common Brick	130	0.302	A0, A2, C9, E1, E0
A	Insulation or Air space+8-in. Common Brick	130	0.154-0.243	A0, A2, C9, B1/B2, E1, E0
4-in. Face Brick+(*H.W. Concrete*)				
C	Air Space+2-in. Concrete	94	0.350	A0, A2, B1, C5, E1, E0
B	2-in. Insulation+4-in. concrete	97	0.116	A0, A2, B3, C5, E1, E0
A	Air Space or Insulation+8-in. or more Concrete	143-190	0.110-0.112	A0, A2, B1, C10/11, E1, E0
4-in. Face Brick+(*L.W. or H.W. Concrete Block*)				
E	4-in. Block	62	0.319	A0, A2, C2, E1, E0
D	Air Space or Insulation+4-in. Block	62	0.153-0.246	A0, A2, C2, B1/B2, E1, E0
D	8-in. Block	70	0.274	A0, A2, C7, A6, E0
C	Air Space or 1-in. Insulation+6-in. or 8-in. Block	73-89	0.221-0.275	A0, A2, B1, C7/C8, E1, E0
B	2-in. Insulation+8-in. Block	89	0.096-0.107	A0, A2, B3, C7/C8, E1, E0
4-in Face Brick+(*Clay Tile*)				
D	4-in. Tile	71	0.381	A0, A2, C1, E1, E0
D	Air Space+4-in. Tile	71	0.281	A0, A2, C1, B1, E1, E0
C	Insulation+4-in. Tile	71	0.169	A0, A2, C1, B2, E1, E0
C	8-in. Tile	96	0.275	A0, A2, C6, E1, E0
B	Air Space or 1-in. Insulation+8-in. Tile	96	0.142-0.221	A0, A2, C6, B1/B2, E1, E0
A	2-in. Insulation+8-in. Tile	97	0.097	A0, A2, B3, C6, E1, E0
H.W. Concrete Wall+(*Finish*)				
E	4-in. Concrete	63	0.585	A0, A1, C5, E1, E0
D	4-in. Concrete+1-in. or 2-in. Insulation	63	0.119-0.200	A0, A1, C5, B2/B3, E1, E0
C	2-in. Insulation+4-in. Concrete	63	0.119	A0, A1, B6, C5, E1, E0
C	8-in. Concrete	109	0.490	A0, A1, C10, E1, E0
B	8-in. Concrete+1-in. or 2-in. Insulation	110	0.115-0.187	A0, A1, C10, B5/B6, E1, E0
A	2-in. Insulation+8-in. Concrete	110	0.115	A0, A1, B3, C10, E1, E0
B	12-in. Concrete	156	0.421	A0, A1, C11, E1, E0
A	12-in. Concrete+Insulation	156	0.113	A0, C11, B6, A6, E0
L.W. and H.W. Concrete Block+(*Finish*)				
F	4-in. Block+Air Space/Insulation	29	0.161-0.263	A0, A1, C2, B1/B2, E1, E0
E	2-in. Insulation+4-in. Block	29-37	0.105-0.114	A0, A1, B3, C2/C3, E1, E0
E	8-in. Block	47-51	0.294-0.402	A0, A1, C7/C8, E1, E0
D	8-in. Block+Air Space/Insulation	41-57	0.149-0.173	A0, A1, C7/C8, B1/B2, E1, E0
Clay *Tile*+(*Finish*)				
F	4-in. Tile	39	0.419	A0, A1, C1, E1, E0
F	4-in. Tile+Air space	39	0.303	A0, A1, C1, B1, E1, E0
E	4-in. Tile+1-in. Insulation	39	0.175	A0, A1, C1, B2, E1, E0
D	2-in. Insulation+4-in. Tile	40	0.110	A0, A1, B3, C1, E1, E0
D	8-in. Tile	63	0.296	A0, A1, C6, E1, E0
C	8-in. Tile+Air Space/1-in. Insulation	63	0.151-0.231	A0, A1, C6, B1/B2, E1, E0
B	2-in. Insulation+8-in. Tile	63	0.099	A0, A1, B3, C6, E1, E0
Metal Curtain Wall				
G	With/without Air Space+1-in./2-in./3-in. Insulation	5-6	0.091-0.230	A0, A3, B5/B6/B12, A3, E0
Frame Wall				
G	1-in. to 3-in. Insulation	16	0.081-0.178	A0, A1, B1, B2/B3/B4, E1, E0

SOURCE: Reprinted by permission from *ASHRAE Handbook of Fundamentals 1977*, p. 25.8, Table 6.

ture of 78°F, an outdoor maximum temperature of 95°F, and an outdoor mean temperature of 85°F, with an outdoor daily range of 21°F (and a solar radiation variation typical of 40°N latitude on July 21). If the indoor air temperature is not 78°F and/or the outdoor daily mean temperature is not 85°F, the following corrections apply:

1.   For indoor air temperatures less than 78°F, add the difference between 78°F and indoor air temperature to the CLTD; if greater than 78°F, subtract the difference.

2. For outdoor daily mean temperatures less than 85°F, subtract the difference between 85°F and the daily mean temperature from the CLTD; if greater than 85°F, add the difference.

6.10.3. Fenestrations

Fenestration is the term used here to designate any light-transmitting opening in a building wall or roof. The opening may be glazed with single- or multiple-sheet plate or float glass, pattern glass, plastic panels, or glass block. Interior or exterior shading devices are usually employed, and some glazing systems incorporate integral sunlight control devices.

Heat admission or loss through fenestration areas is affected by many factors of which the following are the most significant.

1. Solar radiation intensity and incident angle
2. Outdoor-indoor temperature difference
3. Velocity and direction of air flow across the exterior and interior fenestration surfaces
4. Low-temperature radiation exchange between the surfaces of the fenestration and the surroundings
5. Exterior and/or interior shading

The total instantaneous rate of heat gain through a glazing material can be obtained from the heat balance between a unit area of fenestration and its thermal environment:

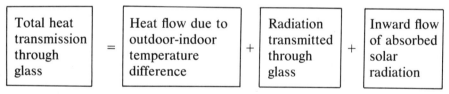

In this equation, the last two terms on the right are present only when the fenestration is irradiated and are therefore related to the incident radiation. Combining the last two terms,

$$
\boxed{\begin{array}{l}\text{Total heat}\\\text{transmission}\\\text{through}\\\text{glass}\end{array}} = \boxed{\begin{array}{l}\text{Conduction}\\\text{heat gain}\end{array}} + \boxed{\begin{array}{l}\text{Solar heat}\\\text{gain}\end{array}}
$$

In this way, the heat gain is divided into two components: (1) the conduction heat gain (or loss) due to differences between outdoor and indoor air temperature; and (2) the solar heat gain (SHG), due to transmitted and absorbed solar energy.

For the conduction heat gain, the overall heat-transfer coefficient ac-

TABLE 6.20.   Cooling Load Temperature Differences for Conduction through Glass[a]

Hour	2	4	6	8	10	12	14	16	18	20	22	24
CLTD, F	0	−2	−2	0	4	9	13	14	12	8	4	2

[a] Corrections: The values in the table were calculated for an inside temperature of 78°F and an outdoor maximum temperature of 95°F with an outdoor daily range of 21°F. The table remains approximately correct for other outdoor maximums (93–102°F) and other outdoor daily ranges (16–34°F), provided the outdoor daily average temperature remains approximately 85°F. If the room air temperature is different from 78°F, and/or the outdoor daily average temperature is different from 85°F, the following rules apply: (a) For room air temperature less than 78°F, add the difference between 78°F and room air temperature; if greater than 78°F, subtract the difference. (b) For outdoor daily average temperature less than 85°F, subtract the difference between 85°F and the daily average temperature; if greater than 85°F, add the difference.

SOURCE: Reprinted by permission from *ASHRAE Handbook of Fundamentals 1977*, p. 25.11.

counts for the heat-transfer processes of: (1) convection and long-wave radiation exchange outside and inside the conditioned space; and (2) conduction through the fenestration material. Consequently, in calculating cooling-load factors for this component, the conduction heat gain was treated in a manner similar to that through walls and roofs. The transfer function coefficients used to convert the heat gain to cooling load are the same as for walls and roofs. The resulting cooling load temperature differences are given in Table 6.20. The cooling load due to conduction and convection heat gain is then calculated as

$$q = UA \text{ CLTD} \qquad (6.18)$$

where A is the net glass area of the fenestration in square feet.

The heat gain due to the transmitted and absorbed solar energy, grouped in the term solar heat gain (SHG), is present in cooling load calculations only when fenestration is irradiated. ASHRAE has developed a method for estimating SHG through fenestrations, based on a reference glazing material of double-strength (0.125-in.) sheet glass. The solar heat gain through this reference material, called the solar heat gain factor (SHGF) is calculated for daylight hours of the 21st day of each month, for 17 principal orientations. Table 6.21 presents the table of solar heat gain factors for 40°N latitude. Values for other latitudes are available from ASHRAE [1].

To account for the different types of fenestration and shading devices used, the shading coefficient (SC), relating SHG through a glazing system under a specific set of conditions to SHG through the reference glazing material under the same conditions, is defined:

$$SC = \frac{\text{SHG of fenestration}}{\text{SHG of double-strength glass}}$$

Most fenestration has some type of internal shading to provide privacy and aesthetic effects, as well as to give varying degrees of sun control. The shading coefficients applicable to some of the more widely used types of internal shading are given in Tables 6.22, 6.23, and 6.24.

The effectiveness of any internal shading device depends on its ability to reflect incoming solar radiation back through the fenestration before it can be absorbed and converted into heat within the building.

For calculation, the cooling load due to solar radiation must be analyzed in one of two cases: (1) presence of interior shading or (2) absence of interior shading. In converting heat gain to cooling load, the time lag due to the radiant solar energy entering the space is variable; for example, it differs when energy is absorbed by interior draperies from which it is absorbed by the floor. This difference will appear in the cooling load factors used to multiply the solar heat gains.

Cooling load due to solar radiation through fenestration is calculated from the relation

$$q = \text{Area} \times \frac{\text{shading}}{\text{coefficient}} \times \frac{\text{maximum}}{\text{solar heat gain}} \times \frac{\text{cooling load}}{\text{factor}}$$

The total load through fenestration is the sum of the load due to conduction heat gain and the load due to solar heat gain:

$$q = UA \times \text{CLTD} + A \times \text{SC} \times \text{SHGF}_{\text{max}} \times \text{CLF} \qquad (6.19)$$

The *area* is the net glass area of the fenestration. The *maximum solar heat gain* is obtained for the appropriate latitude, month, and surface orientation from Table 6.25. The CLF values for the three common room thermal characteristics are listed in Table 6.26 and 6.27.

In general, external shading devices that eliminate all direct radiation and still permit free air movement at the outer fenestration surface will reduce the solar heat gain by as much as 80%. In this category are found architectural treatments such as overhangs, wings, and screen walls, as well as horizontal and vertical louvers and vented awnings.

During certain seasons of the year and for some exposures, horizontal projections can result in considerable reductions in solar heat gain by providing shade. This is particularly applicable to southern, southeastern, and southwestern exposures during the late spring, summer, and early fall. On eastern and western exposures during the entire year, and on southerly exposures during the winter, the solar altitude is generally so low that horizontal projections, in order to be effective, would have to be excessively long.

The horizontal projection, P_h, required to produce a given shadow height, S_H, on a window or wall for any time of day or year is related to the profile angle Ω (Table 6.28) by

$$P_h = S_H \cot \Omega \qquad (6.20)$$

TABLE 6.21. Solar Intensity and Solar Heat Gain Factors* for 40° North Latitude

Date	Solar Time am	Direct Normal Btuh/ft²	Solar Heat Gain Factors, Btuh/ft²																	Solar Time pm
			N	NNE	NE	ENE	E	ESE	SE	SSE	S	SSW	SW	WSW	W	WNW	NW	NNW	HOR	
Jan 21	8	142	5	5	17	71	111	132	133	114	75	22	6	5	5	5	5	5	14	4
	9	239	12	12	13	74	154	205	224	209	160	82	13	12	12	12	12	12	55	3
	10	274	16	16	16	31	124	199	241	246	213	146	51	17	16	16	16	16	96	2
	11	289	19	19	19	20	61	156	222	252	244	198	118	28	19	19	19	19	124	1
	12	294	20	20	20	20	21	90	179	234	254	234	179	90	21	20	20	20	133	12
	HALF DAY TOTALS		61	61	73	199	452	734	904	932	813	561	273	101	62	61	61	61	354	
Feb 21	7	55	2	3	2	40	51	53	47	34	14	2	2	2	2	2	2	2	4	5
	8	219	10	11	23	129	183	206	199	160	94	18	10	10	10	10	10	10	43	4
	9	271	16	16	50	107	186	234	245	218	157	66	17	16	16	16	16	16	98	3
	10	294	21	21	22	49	143	211	246	243	203	129	38	21	21	21	21	21	143	2
	11	304	23	23	23	24	71	160	219	244	231	184	103	27	23	23	23	23	171	1
	12	307	24	24	24	24	25	86	170	222	241	222	170	86	25	24	24	24	180	12
	HALF DAY TOTALS		84	86	152	361	648	916	1049	1015	821	508	250	114	85	84	84	84	548	
Mar 21	7	171	9	29	93	140	163	161	135	86	22	8	8	8	8	8	8	8	26	5
	8	250	16	18	91	169	218	232	211	157	74	17	16	16	16	16	16	16	85	4
	9	282	21	22	47	136	203	238	236	198	128	40	22	21	21	21	21	21	143	3
	10	297	25	25	27	72	153	207	229	216	171	95	29	25	25	25	25	25	186	2
	11	305	28	28	28	30	78	151	198	213	197	150	77	30	28	28	28	28	213	1
	12	307	29	29	29	29	31	75	145	191	206	191	145	75	31	29	29	29	223	12
	HALF DAY TOTALS		114	139	302	563	832	1035	1087	968	694	403	220	132	114	113	113	113	764	
Apr 21	6	89	11	46	72	87	88	76	52	18	5	5	5	5	5	5	5	5	11	6
	7	206	16	71	140	185	201	186	143	75	16	14	14	14	14	14	14	14	61	5
	8	252	22	44	128	190	224	223	188	124	41	22	21	21	21	21	21	21	123	4
	9	274	27	29	80	155	202	219	203	156	83	29	27	27	27	27	27	27	177	3
	10	286	31	31	37	92	152	187	193	170	121	56	32	31	31	31	31	31	217	2
	11	292	33	33	34	39	81	130	160	166	146	102	52	35	33	33	33	33	243	1
	12	293	34	34	34	34	36	62	108	142	154	142	108	62	36	34	34	34	252	12
	HALF DAY TOTALS		154	265	501	758	957	1051	994	782	488	296	199	157	148	147	147	147	957	
May 21	5	1	0	1	1	1	1	0	0	0	0	0	0	0	0	0	0	0	0	7
	6	144	36	90	128	145	141	115	71	18	10	10	10	10	10	10	10	11	31	6
	7	216	28	102	165	202	209	184	131	54	20	19	19	19	19	19	19	19	87	5
	8	250	27	73	149	199	220	208	164	93	29	25	25	25	25	25	25	25	146	4
	9	267	31	42	105	164	197	200	175	121	53	32	30	30	30	30	30	30	195	3
	10	277	34	36	54	105	148	168	163	133	83	40	35	34	34	34	34	34	234	2
	11	283	36	36	38	48	81	113	130	127	105	70	42	38	36	36	36	36	257	1
	12	284	37	37	37	38	40	54	82	104	113	104	82	54	40	37	37	37	265	12
	HALF DAY TOTALS		215	404	666	893	1024	1025	881	601	358	247	200	180	176	175	174	175	1083	
Jun 21	5	22	10	17	21	22	20	14	6	2	1	1	1	1	1	1	1	2	3	7
	6	155	48	104	143	159	151	121	70	17	13	13	13	13	13	13	13	14	40	6
	7	216	37	113	172	205	207	178	122	46	22	21	21	21	21	21	21	21	97	5
	8	246	30	85	156	201	216	199	152	80	29	27	27	27	27	27	27	27	153	4
	9	263	33	51	114	166	192	190	161	105	45	33	32	32	32	32	30	32	201	3
	10	272	35	38	63	109	145	158	148	116	69	39	36	35	35	35	34	35	238	2
	11	277	38	39	40	52	81	105	116	110	88	60	41	39	38	38	36	38	260	1
	12	279	38	38	38	40	41	52	72	89	95	89	72	52	41	40	38	38	267	12
	HALF DAY TOTALS		253	470	734	941	1038	999	818	523	315	236	204	191	188	187	186	188	1126	

Solar Heat Gain Factors* (Btu/h·ft²) — by month, solar time, and exposure

Date	AM	N	NNE	NE	ENE	E	ESE	SE	SSE	S	SSW	SW	WSW	W	WNW	NW	NNW	HOR	PM
Jul 21	5	2	2	2	2	2	1	1	0	0	0	0	0	0	0	0	0	0	7
	6	37	89	125	142	137	112	68	18	11	11	11	11	11	11	11	12	32	6
	7	30	102	163	198	204	179	127	53	21	20	20	20	20	20	20	20	88	5
	8	28	75	148	196	216	203	160	90	30	26	26	26	26	26	26	26	145	4
	9	32	44	106	163	193	196	170	118	52	33	31	31	31	31	31	31	194	3
	10	35	37	56	106	146	165	159	129	81	41	36	35	35	35	35	35	231	2
	11	37	38	40	50	81	111	127	123	102	69	43	39	37	37	37	37	254	1
	12	38	38	38	40	41	55	80	101	109	101	80	55	41	40	38	38	262	12
	HALF DAY TOTALS	223	411	666	885	1008	1003	858	584	352	248	204	186	181	180	180	181	1076	
Aug 21	6		44	68	81	82	71	48	17	6	5	5	5	5	5	5	5	12	6
	7		71	135	177	191	177	135	70	17	16	16	16	16	16	16	16	62	5
	8		47	126	185	216	214	180	118	41	23	23	23	23	23	23	23	122	4
	9		31	82	153	197	212	196	151	80	31	28	28	28	28	28	28	174	3
	10		33	40	93	150	182	187	165	116	56	34	32	32	32	32	32	214	2
	11		35	36	41	81	128	156	160	141	99	52	37	35	35	35	35	239	1
	12		35	35	36	38	63	106	138	149	138	106	63	38	36	35	35	247	12
	HALF DAY TOTALS		273	498	741	928	1013	956	751	474	296	205	166	157	156	156	156	946	
Sep 21	7		27	84	125	146	144	121	77	21	9	9	9	9	9	9	9	25	5
	8		19	87	160	205	218	199	148	71	18	17	17	17	17	17	17	82	4
	9		23	47	131	194	227	226	190	124	41	23	22	22	22	22	22	138	3
	10		27	28	71	148	200	221	209	165	93	30	27	27	27	27	27	180	2
	11		29	29	31	78	147	192	207	191	146	77	31	29	29	29	29	206	1
	12		30	30	30	32	75	142	185	200	185	142	75	32	32	30	30	215	12
	HALF DAY TOTALS		142	291	534	787	980	1033	925	672	396	222	137	119	118	118	118	738	
Oct 21	7		3	20	36	45	47	42	30	12	2	2	2	2	2	2	2	4	5
	8		12	49	123	173	195	188	151	89	18	11	11	11	11	11	11	43	4
	9		17	23	104	180	225	235	209	151	64	18	17	17	17	17	17	97	3
	10		21	22	50	139	205	238	235	196	125	38	22	21	21	21	21	140	2
	11		24	24	25	71	156	212	236	224	178		28	24	24	24	24	168	1
	12		25	25	25	27	85	165	216	234	216	165	85	27	27	25	25	177	12
	HALF DAY TOTALS		89	152	351	623	878	1006	974	791	493	247	117	89	89	88	88	540	
Nov 21	8		5	18	69	108	128	129	110	72		6	5	5	5	5	5	14	4
	9		12	13	73	151	201	219	204	156	80	13	12	12	12	12	12	55	3
	10		16	16	31	122	196	237	242	204	143	50	17	16	16	16	16	96	2
	11		19	19	20	61	154	218	248	209	194	116	28	19	19	19	19	123	1
	12		20	20	20	21	89	176	231	250	231	176	89	21	21	20	20	132	12
	HALF DAY TOTALS		63	75	198	445	721	887	914	798	551	269	101	63	63	63	63	354	
Dec 21	8	3	3		41	67	82	84	73	50			3	3	3	3	3	6	4
	9	10	10	11	60	135	185	205	194	151	83	13	10	10	10	10	10	39	3
	10	14	14	14	25	113	188	232	239	210	146	55	15	14	14	14	14	77	2
	11	17	17	17	17	56	151	217	249	242	198	120	28	17	17	17	17	104	1
	12	18	18	18	18	19	89	178	233	253	233	178	89	19	18	18	18	113	12
	HALF DAY TOTALS	52	52	56	146	374	649	822	867	775	557	276	94	53	52	52	52	282	

Half Day Totals computed by Simpson's Rule with time interval equal to 10 minutes.
*Total Solar Heat Gains for DS (0.125 in.) sheet glass.
Based on a ground reflectance of 0.20 and values in Tables 1 and 26.

SOURCE: Reprinted by permission from ASHRAE Handbook of Fundamentals 1977, p. 26.22, Table 22.

TABLE 6.22.   Shading Coefficients for Single Glass and Insulating Glass[a]

Type of Glass	Nominal Thickness (in.)	Solar Trans.	Shading Coefficient	
			$h_o = 4.0$	$h_o = 3.0$
Single Glass				
Clear	$\frac{1}{8}$	0.84	1.00	1.00
	$\frac{1}{4}$	0.78	0.94	0.95
	$\frac{3}{8}$	0.72	0.90	0.92
	$\frac{1}{2}$	0.67	0.87	0.88
Heat Absorbing	$\frac{1}{8}$	0.64	0.83	0.85
	$\frac{1}{4}$	0.46	0.69	0.73
	$\frac{3}{8}$	0.33	0.60	0.64
	$\frac{1}{2}$	0.24	0.53	0.58
Insulating Glass				
Clear Out, Clear In	$\frac{1}{8}$	0.71	0.88	0.88
Clear Out, Clear In	$\frac{1}{4}$	0.61	0.81	0.82
Heat Absorbing Out, Clear In	$\frac{1}{4}$	0.36	0.55	0.58

[a] Refers to factory-fabricated units with $\frac{3}{16}$, $\frac{1}{4}$, or $\frac{1}{2}$-in. air space or to prime windows plus storm sash.
SOURCE: Reprinted by permission from *ASHRAE Handbook of Fundamentals 1977*.

The *profile angle*, also called the *vertical shadow line angle*, is the angle between a line perpendicular to the plane of the window and the projection of the earth-sun line in a vertical plane that is also normal to the window.

The use of an overhang to shade glass is an excellent way of reducing the solar heat gain. For the SHGF for a shaded window, the values for north glass should be used. North glass does not have sun shining directly on it, so the SHGFs listed for it are based on only diffuse solar radiation (as on shaded glass). If a window is partially shaded, use the north value of SHGF for the shaded portion and the regular (east, west, south, etc.) SHGF for the rest of the window.

A vertical fin or projection P_v required to produce a given shadow width S_W on a window or wall for any given time of day and year is related to the vertical surface solar azimuth γ by

$$P_v = S_W \cot \gamma \qquad (6.21)$$

A window with a significant depth of reveal will generally have part of the glass area shaded by the mullions and the transom. The area that is shaded varies continuously throughout the day but it can be estimated readily by treating mullions and transoms as vertical and horizontal projections.

The width, M, of the shadow from a mullion that projects a distance P ft beyond the plane of the glass is

$$M = P \tan \gamma \qquad (6.22)$$

TABLE 6.23. Shading Coefficients for Single Glass with Indoor Shading by Venetian Blinds or Roller Shades

			Type of Shading				
			Venetian Blinds		Roller Shade		
					Opaque		Translucent
Glass	Nominal Thickness[a] (in.)	Solar Trans.	Medium	Light	Dark	White	Light
Clear	$\frac{3}{32}$–$\frac{1}{4}$	0.87–0.80					
Clear	$\frac{1}{4}$–$\frac{1}{2}$	0.80–0.71					
Clear Pattern	$\frac{1}{8}$–$\frac{1}{2}$	0.87–0.79	0.64	0.55	0.59	0.25	0.39
Heat-Absorbing Pattern		—					
Tinted	$\frac{3}{16}$, $\frac{7}{32}$	0.74, 0.71					
Heat-Absorbing	$\frac{3}{16}$, $\frac{1}{4}$	0.46					
Heat-Absorbing Pattern	$\frac{3}{16}$, $\frac{1}{4}$	0.59, 0.45	0.57	0.53	0.45	0.30	0.36
Tinted	$\frac{1}{8}$, $\frac{7}{32}$						
Heat-Absorbing or Pattern	—	0.44–0.30	0.54	0.52	0.40	0.28	0.32
Heat-Absorbing	$\frac{3}{8}$	0.34					
Heat-Absorbing or Pattern	—	0.29–0.15	0.42	0.40	0.36	0.28	0.31
		0.24					

[a] Refers to factory-fabricated units with $\frac{3}{16}$, $\frac{1}{4}$, or $\frac{1}{2}$-in. air space or to prime windows plus storm sash.

SOURCE: Adapted by permission from *ASHRAE Handbook of Fundamentals 1977*.

TABLE 6.24. Shading Coefficients for Insulating Glass[a] with Indoor Shading by Venetian Blinds or Roller Shades

Type of Glass	Nominal Thickness (in.), each light	Solar Trans. Outer Pane	Solar Trans. Inner Pane	Venetian Blinds Medium	Venetian Blinds Light	Roller Shade Opaque Dark	Roller Shade Opaque White	Roller Shade Translucent Light
Clear out	$\frac{3}{32}$, $\frac{1}{8}$	0.87	0.87	0.57	0.51	0.60	0.25	0.37
Clear in								
Clear out	$\frac{1}{4}$	0.80	0.80					
Clear in	$\frac{1}{4}$	0.80	0.80	0.39	0.36	0.40	0.22	0.30
Heat-absorbing out	$\frac{1}{4}$	0.46						
Clear in	$\frac{1}{4}$		0.80					

[a] Refers to factory-fabricated units with $\frac{3}{16}$, $\frac{1}{4}$, or $\frac{1}{2}$-in. air space, or to prime windows plus storm windows.

SOURCE: Adapted by permission from ASHRAE Handbook of Fundamentals 1977.

TABLE 6.25. Maximum Solar Heat Gain Factors [Btu/(h·ft²)]

	0°N Latitude						8°N Latitude					
	N	NE/NW	E/W	SE/SW	S	HOR	N	NE/NW	E/W	SE/SW	S	HOR
Jan.	34	88	234	235	118	296	32	71	224	242	162	275
Feb.	36	132	245	210	67	306	34	114	239	219	110	294
Mar.	38	170	242	170	38	303	37	156	241	184	55	300
Apr.	71	193	221	118	37	284	44	184	225	134	39	289
May	113	203	201	80	37	265	74	198	209	97	38	277
June	129	206	190	65	37	255	90	200	200	82	39	269
July	115	201	195	77	38	260	77	195	204	93	39	272
Aug.	75	187	212	112	38	276	47	179	216	128	41	282
Sep.	40	163	231	163	40	293	38	149	230	176	56	290
Oct.	37	129	236	202	66	299	35	112	231	211	108	288
Nov.	35	88	230	230	117	293	33	71	220	238	160	273
Dec.	34	71	226	241	138	288	31	54	215	247	180	264

	16° N Latitude						24° N Latitude					
	N	NE/NW	E/W	SE/SW	S	HOR	N	NE/NW	E/W	SE/SW	S	HOR
Jan.	30	55	210	251	199	248	27	41	190	253	227	214
Feb.	33	96	231	233	154	275	30	80	220	243	192	249
Mar.	35	140	239	197	93	291	34	124	234	214	137	275
Apr.	39	172	227	150	45	289	37	159	228	169	75	283
May	52	189	215	115	41	282	43	178	218	132	46	282
June	66	194	207	99	41	277	56	184	212	117	43	279
July	55	187	210	111	42	277	45	176	213	129	46	278
Aug.	41	168	219	143	46	282	38	156	220	162	72	277
Sep.	36	134	227	191	93	282	35	119	222	206	134	266
Oct.	33	95	223	225	150	270	31	79	211	235	187	244
Nov.	30	55	206	247	196	246	27	42	187	249	224	213
Dec.	29	41	198	254	213	234	25	29	179	252	237	199

	32° N Latitude						40° N Latitude					
	N	NE/NW	E/W	SE/SW	S	HOR	N	NE/NW	E/W	SE/SW	S	HOR
Jan.	24	29	175	249	246	176	20	20	154	241	254	133
Feb.	27	65	205	248	221	217	24	50	186	246	241	180
Mar.	32	107	227	227	176	252	29	93	218	236	206	223
Apr.	36	146	227	187	115	271	34	140	224	203	154	252
May	38	170	220	155	74	277	37	165	220	175	113	265
June	47	176	214	139	60	276	48	172	215	161	95	268
July	40	167	215	150	72	273	38	163	216	170	109	262
Aug.	37	141	219	181	111	265	35	135	216	196	149	247
Sep.	33	103	215	218	171	244	30	87	205	226	200	215
Oct.	28	63	195	239	215	213	25	49	180	238	234	177
Nov.	24	29	173	245	243	175	18	20	151	237	250	132
Dec.	22	22	162	246	252	158	18	18	135	232	253	112

	48° N Latitude						56° N Latitude					
	N	NE/NW	E/W	SE/SW	S	HOR	N	NE/NW	E/W	SE/SW	S	HOR
Jan.	15	15	118	216	245	85	10	10	74	169	205	40
Feb.	20	36	168	242	250	138	16	21	139	223	244	91
Mar.	26	80	204	239	228	188	22	65	185	238	241	149
Apr.	31	132	219	215	186	226	28	123	211	223	210	195
May	35	158	218	192	150	247	36	149	215	206	181	222
June	47	165	215	180	134	252	53	161	213	195	167	231
July	37	156	214	187	146	244	37	147	211	201	177	221
Aug.	33	128	211	208	180	223	30	119	203	215	203	193
Sep.	27	72	191	228	220	182	23	58	171	227	231	144
Oct.	21	35	161	233	242	136	16	20	132	213	234	91
Nov.	15	15	115	212	240	85	10	10	72	165	200	40
Dec.	13	13	91	195	233	64	7	7	46	135	170	23

	64° N Latitude					
	N	NE/NW	E/W	SE/SW	S	HOR
Jan.	3	3	15	67	96	8
Feb.	11	11	89	177	210	45
Mar.	18	47	159	226	239	105
Apr.	25	113	201	225	224	160
May	48	150	211	215	204	192
June	62	162	213	208	193	203
July	49	148	207	211	200	192
Aug.	27	109	193	217	217	159
Sep.	19	43	148	213	227	101
Oct.	11	11	83	167	199	46
Nov.	4	4	15	66	93	8
Dec.	0	0	1	10	14	1

SOURCE: Reprinted by permission from *ASHRAE Handbook of Fundamentals 1977*, p. 25.12, Table 10.

TABLE 6.26. Cooling Load Factors for Glass without Interior Shading (Includes Reflective and Heat-Absorbing Glass)

N. Latitude Fenestration Facing	Room Construction	Solar Time, hr																							
		1	2	3	4	5	6	7	8	9	10	11	12	13	14	15	16	17	18	19	20	21	22	23	24
N	L	0.17	0.14	0.11	0.09	0.08	0.33	0.42	0.48	0.56	0.63	0.71	0.76	0.80	0.82	0.82	0.79	0.80	0.84	0.61	0.48	0.38	0.31	0.25	0.20
	M	0.23	0.20	0.18	0.16	0.14	0.34	0.41	0.46	0.52	0.59	0.65	0.70	0.73	0.75	0.76	0.74	0.75	0.79	0.61	0.50	0.42	0.36	0.31	0.27
	H	0.25	0.23	0.21	0.20	0.19	0.38	0.45	0.50	0.55	0.60	0.65	0.69	0.72	0.73	0.72	0.70	0.70	0.74	0.57	0.46	0.39	0.34	0.31	0.28
NE	L	0.04	0.04	0.03	0.02	0.02	0.23	0.41	0.51	0.51	0.45	0.39	0.36	0.33	0.31	0.28	0.26	0.23	0.19	0.15	0.12	0.10	0.08	0.06	0.05
	M	0.07	0.06	0.06	0.05	0.04	0.21	0.36	0.44	0.45	0.40	0.36	0.33	0.31	0.30	0.28	0.26	0.23	0.21	0.17	0.15	0.13	0.11	0.09	0.08
	H	0.09	0.08	0.08	0.07	0.07	0.23	0.37	0.44	0.44	0.39	0.34	0.31	0.29	0.27	0.26	0.24	0.22	0.20	0.16	0.14	0.13	0.12	0.11	0.10
E	L	0.04	0.04	0.03	0.02	0.02	0.19	0.37	0.51	0.57	0.57	0.51	0.42	0.36	0.32	0.29	0.25	0.22	0.19	0.14	0.12	0.09	0.08	0.06	0.05
	M	0.07	0.06	0.06	0.05	0.04	0.18	0.33	0.44	0.50	0.51	0.45	0.39	0.35	0.32	0.29	0.26	0.23	0.21	0.17	0.15	0.13	0.11	0.10	0.08
	H	0.09	0.09	0.08	0.08	0.07	0.21	0.34	0.45	0.50	0.49	0.43	0.36	0.32	0.29	0.26	0.24	0.22	0.19	0.17	0.15	0.13	0.12	0.11	0.10
SE	L	0.05	0.04	0.04	0.03	0.02	0.13	0.28	0.43	0.55	0.62	0.63	0.57	0.48	0.42	0.37	0.33	0.28	0.24	0.19	0.15	0.12	0.10	0.08	0.07
	M	0.09	0.08	0.07	0.06	0.05	0.14	0.26	0.38	0.48	0.54	0.55	0.51	0.45	0.40	0.36	0.33	0.29	0.25	0.21	0.18	0.16	0.14	0.12	0.10
	H	0.11	0.10	0.10	0.09	0.08	0.17	0.28	0.40	0.49	0.53	0.53	0.48	0.41	0.36	0.33	0.30	0.27	0.24	0.20	0.18	0.16	0.14	0.13	0.12
S	L	0.08	0.07	0.05	0.04	0.04	0.06	0.09	0.14	0.22	0.34	0.48	0.59	0.65	0.65	0.59	0.50	0.43	0.36	0.28	0.22	0.18	0.15	0.12	0.10
	M	0.12	0.11	0.09	0.08	0.07	0.08	0.11	0.14	0.21	0.31	0.42	0.52	0.57	0.58	0.53	0.47	0.41	0.36	0.29	0.25	0.21	0.18	0.16	0.14
	H	0.13	0.12	0.12	0.11	0.10	0.12	0.14	0.17	0.24	0.33	0.43	0.51	0.56	0.55	0.50	0.43	0.38	0.32	0.26	0.22	0.20	0.18	0.16	0.15
SW	L	0.12	0.10	0.08	0.06	0.05	0.06	0.08	0.10	0.12	0.14	0.16	0.24	0.36	0.49	0.60	0.66	0.66	0.58	0.43	0.33	0.27	0.22	0.18	0.14
	M	0.15	0.13	0.12	0.10	0.09	0.09	0.10	0.12	0.13	0.15	0.17	0.23	0.33	0.44	0.53	0.58	0.59	0.53	0.41	0.33	0.28	0.24	0.21	0.18
	H	0.15	0.14	0.13	0.12	0.11	0.12	0.13	0.14	0.16	0.17	0.19	0.25	0.34	0.44	0.52	0.56	0.56	0.49	0.37	0.30	0.25	0.21	0.19	0.17
W	L	0.12	0.10	0.08	0.07	0.05	0.06	0.07	0.08	0.10	0.11	0.13	0.14	0.20	0.32	0.45	0.57	0.64	0.61	0.44	0.34	0.27	0.22	0.18	0.14
	M	0.15	0.13	0.11	0.10	0.09	0.09	0.09	0.10	0.11	0.12	0.13	0.14	0.19	0.29	0.40	0.50	0.56	0.55	0.41	0.33	0.27	0.23	0.20	0.17
	H	0.14	0.13	0.12	0.11	0.10	0.11	0.12	0.13	0.13	0.14	0.15	0.16	0.21	0.30	0.40	0.49	0.54	0.52	0.38	0.30	0.24	0.21	0.18	0.16
NW	L	0.11	0.09	0.08	0.06	0.05	0.06	0.08	0.10	0.12	0.14	0.16	0.17	0.19	0.23	0.33	0.47	0.59	0.60	0.43	0.33	0.26	0.21	0.17	0.14
	M	0.14	0.12	0.11	0.09	0.08	0.09	0.10	0.11	0.13	0.14	0.16	0.17	0.18	0.21	0.30	0.42	0.51	0.53	0.39	0.32	0.26	0.22	0.19	0.16
	H	0.14	0.12	0.11	0.11	0.10	0.10	0.12	0.13	0.15	0.16	0.18	0.19	0.19	0.22	0.30	0.41	0.50	0.51	0.36	0.29	0.23	0.20	0.17	0.15
HOR	L	0.11	0.09	0.07	0.06	0.05	0.07	0.14	0.24	0.36	0.48	0.58	0.66	0.72	0.74	0.73	0.67	0.59	0.47	0.37	0.30	0.24	0.19	0.16	0.13
	M	0.16	0.14	0.12	0.11	0.09	0.11	0.16	0.24	0.33	0.43	0.52	0.59	0.64	0.67	0.66	0.62	0.55	0.47	0.38	0.32	0.28	0.24	0.21	0.18
	H	0.17	0.16	0.15	0.14	0.13	0.15	0.20	0.27	0.36	0.45	0.52	0.59	0.62	0.64	0.62	0.58	0.51	0.42	0.35	0.29	0.26	0.23	0.21	0.19

L=Light construction: frame exterior wall, 2-in. concrete floor slab, approximately 30 lb of material/square feet of floor area.
M=Medium construction: 4-in. concrete exterior wall, 4-in. concrete floor slab, approximately 70 lb of building material/square feet of floor area.
H=Heavy construction: 6-in. concrete exterior wall, 6-in. concrete floor slab, approximately 130 lb of building materials/square feet of floor area.

SOURCE: Reprinted by permission from *ASHRAE Handbook of Fundamentals 1977*, p. 25.13, Table 11.

TABLE 6.27. Cooling Load Factors for Glass with Interior Shading (Includes Reflective and Heat-Absorbing Glass)

N. Latitude Fenestration Facing	Room Construction	Solar Time, hr																							
		1	2	3	4	5	6	7	8	9	10	11	12	13	14	15	16	17	18	19	20	21	22	23	24
N	L	0.07	0.05	0.04	0.04	0.05	0.70	0.65	0.65	0.74	0.81	0.87	0.91	0.91	0.88	0.84	0.77	0.80	0.92	0.27	0.19	0.15	0.12	0.10	0.08
	M	0.08	0.07	0.06	0.06	0.07	0.73	0.66	0.65	0.74	0.80	0.86	0.89	0.89	0.86	0.82	0.75	0.78	0.91	0.24	0.18	0.15	0.13	0.11	0.09
	H	0.09	0.09	0.08	0.07	0.09	0.75	0.67	0.66	0.74	0.80	0.86	0.89	0.88	0.85	0.80	0.73	0.76	0.88	0.23	0.17	0.14	0.13	0.11	0.10
NE	L	0.02	0.01	0.01	0.01	0.02	0.55	0.76	0.75	0.60	0.39	0.31	0.28	0.27	0.25	0.23	0.20	0.16	0.12	0.06	0.05	0.04	0.03	0.02	0.02
	M	0.03	0.02	0.02	0.02	0.03	0.56	0.76	0.74	0.58	0.37	0.29	0.27	0.26	0.24	0.22	0.20	0.16	0.12	0.06	0.05	0.04	0.04	0.03	0.03
	H	0.03	0.03	0.02	0.03	0.04	0.57	0.77	0.74	0.58	0.36	0.28	0.26	0.25	0.23	0.21	0.19	0.16	0.11	0.06	0.05	0.05	0.04	0.04	0.04
E	L	0.02	0.01	0.01	0.01	0.01	0.45	0.71	0.80	0.77	0.64	0.43	0.29	0.25	0.23	0.20	0.17	0.14	0.10	0.06	0.05	0.04	0.03	0.02	0.02
	M	0.03	0.02	0.02	0.02	0.02	0.47	0.72	0.80	0.76	0.62	0.41	0.27	0.24	0.22	0.20	0.17	0.14	0.11	0.06	0.05	0.04	0.04	0.03	0.03
	H	0.04	0.03	0.03	0.03	0.02	0.48	0.72	0.80	0.75	0.61	0.40	0.25	0.22	0.21	0.19	0.16	0.14	0.10	0.06	0.05	0.05	0.04	0.04	0.04
SE	L	0.02	0.02	0.01	0.01	0.01	0.29	0.56	0.74	0.82	0.81	0.70	0.52	0.35	0.30	0.26	0.22	0.18	0.13	0.08	0.06	0.05	0.04	0.03	0.03
	M	0.03	0.03	0.02	0.02	0.02	0.30	0.56	0.74	0.81	0.79	0.68	0.49	0.33	0.28	0.25	0.22	0.18	0.13	0.08	0.07	0.06	0.05	0.04	0.04
	H	0.04	0.04	0.04	0.03	0.04	0.31	0.57	0.74	0.81	0.79	0.67	0.48	0.31	0.27	0.23	0.20	0.17	0.13	0.07	0.07	0.06	0.05	0.05	0.05
S	L	0.03	0.03	0.02	0.02	0.02	0.08	0.15	0.22	0.37	0.58	0.75	0.84	0.82	0.71	0.53	0.37	0.29	0.20	0.11	0.09	0.07	0.06	0.05	0.04
	M	0.04	0.04	0.03	0.03	0.03	0.09	0.16	0.22	0.38	0.58	0.75	0.83	0.80	0.68	0.50	0.35	0.27	0.19	0.11	0.09	0.08	0.07	0.05	0.05
	H	0.05	0.05	0.04	0.04	0.04	0.11	0.17	0.24	0.39	0.59	0.75	0.82	0.79	0.67	0.49	0.33	0.26	0.18	0.10	0.08	0.07	0.06	0.06	0.05
SW	L	0.05	0.04	0.03	0.02	0.02	0.06	0.10	0.13	0.16	0.18	0.22	0.38	0.59	0.76	0.84	0.83	0.72	0.48	0.18	0.13	0.11	0.08	0.07	0.06
	M	0.06	0.05	0.04	0.04	0.03	0.07	0.11	0.14	0.16	0.19	0.22	0.38	0.59	0.75	0.83	0.81	0.69	0.45	0.15	0.12	0.10	0.08	0.07	0.06
	H	0.06	0.05	0.05	0.04	0.04	0.08	0.12	0.15	0.18	0.20	0.23	0.39	0.59	0.75	0.82	0.80	0.68	0.43	0.14	0.11	0.09	0.08	0.07	0.06
W	L	0.05	0.04	0.03	0.02	0.02	0.05	0.08	0.11	0.13	0.14	0.15	0.17	0.30	0.53	0.72	0.83	0.83	0.63	0.19	0.14	0.11	0.08	0.07	0.06
	M	0.05	0.05	0.04	0.04	0.03	0.06	0.09	0.11	0.13	0.15	0.16	0.17	0.31	0.53	0.72	0.82	0.81	0.61	0.16	0.12	0.10	0.08	0.07	0.06
	H	0.05	0.05	0.04	0.04	0.04	0.07	0.10	0.12	0.14	0.16	0.17	0.18	0.31	0.54	0.71	0.81	0.80	0.59	0.15	0.11	0.09	0.07	0.06	0.06
NW	L	0.04	0.04	0.03	0.02	0.02	0.06	0.10	0.13	0.16	0.19	0.20	0.21	0.22	0.30	0.52	0.73	0.83	0.71	0.19	0.13	0.10	0.08	0.07	0.05
	M	0.05	0.04	0.04	0.03	0.03	0.07	0.11	0.14	0.17	0.19	0.20	0.21	0.22	0.30	0.52	0.73	0.82	0.69	0.16	0.12	0.09	0.08	0.07	0.06
	H	0.05	0.04	0.04	0.04	0.04	0.08	0.12	0.15	0.18	0.20	0.21	0.22	0.23	0.30	0.52	0.73	0.81	0.67	0.15	0.11	0.08	0.07	0.06	0.05
HOR	L	0.04	0.03	0.03	0.02	0.02	0.10	0.26	0.43	0.59	0.72	0.81	0.87	0.87	0.83	0.74	0.60	0.44	0.27	0.15	0.12	0.09	0.08	0.06	0.05
	M	0.06	0.05	0.04	0.04	0.03	0.12	0.27	0.44	0.59	0.72	0.81	0.85	0.85	0.81	0.71	0.58	0.42	0.25	0.14	0.12	0.10	0.08	0.07	0.06
	H	0.06	0.06	0.06	0.05	0.05	0.13	0.29	0.45	0.60	0.72	0.81	0.85	0.84	0.79	0.70	0.56	0.40	0.23	0.13	0.11	0.09	0.08	0.08	0.07

L=Light construction: frame exterior wall, 2-in. concrete floor slab, approximately 30 lb of material/sq ft of floor area.
M=Medium construction: 4-in. concrete exterior wall, 4-in. concrete floor slab, approximately 70 lb of building material/sq ft of floor area.
H=Heavy construction: 6-in. concrete exterior wall, 6-in. concrete floor slab, approximately 130 lb of building materials/sq ft of floor area

SOURCE: Reprinted by permission from *ASHRAE Handbook of Fundamentals 1977*, p. 25.13, Table 12.

TABLE 6.28. Solar Position and Related Angles for 40° North Latitude

	Solar Time	Solar Time		Profile (Shadow Line) Angles												Angles of Incidence Vertical Surfaces													Solar Time
DATE	AM	ALT	AZ	N	NNE	NE	ENE	E	ESE	SE	SSE	S	SSW	SW	WSW	N	NNE	NE	ENE	E	ESE	SE	SSE	S	SSW	SW	WSW	HOR	PM
DEC	8	5	53			35	11	7	6	6	6	9	21					82	60	37	16	10	31	53	76			85	4
	9	14	42				37	20	15	14	15	18	30	78					71	50	29	14	24	44	65	87		76	3
	10	21	29				72	38	26	21	21	23	31	54					84	63	43	26	22	35	55	75		69	2
	11	25	15					61	37	28	25	26	31	43	75					76	56	38	26	29	44	63	83	65	1
	12	27	0					90	53	35	28	27	28	35	53					90	70	51	34	27	34	51	70	63	12
JAN + NOV	8	8	55			38	15	10	8	8	10	14	34					80	58	36	15	13	34	56	78			82	4
	9	17	44				40	24	18	17	18	23	37	87					70	48	29	17	27	46	68	89		73	3
	10	24	31				72	41	29	24	24	27	36	61					82	62	43	27	25	38	57	77		66	2
	11	28	16					63	41	32	29	29	35	48	78					76	57	40	29	32	47	65	84	62	1
	12	30	0					90	56	39	32	30	32	39	56					90	71	52	37	30	37	52	71	60	12
FEB + OCT	7	4	72		43	9	6	4	4	5	7	14					85	63	41	18	6	27	50	72				86	5
	8	15	62			43	23	17	15	15	19	29	69					74	52	32	16	22	41	63	84			75	4
	9	24	50			80	45	31	25	24	27	35	56					86	65	46	30	25	36	54	74			66	3
	10	32	35				70	47	36	32	33	38	50	75					79	61	44	33	34	46	63	82		58	2
	11	37	19					67	49	40	37	39	45	60	85					75	58	45	37	41	53	69	87	53	1
	12	39	0					90	65	49	41	39	41	49	65					90	73	57	44	39	44	57	73	51	12
MAR + SEP	7	11	80		43	13	13	12	12	14	21	50					78	56	34	15	17	37	58	80				79	5
	8	23	70		35	45	29	24	23	25	31	50					88	67	47	30	23	33	51	71				67	4
	9	33	57			72	48	37	33	33	38	50	75					80	61	45	34	35	46	63	81			57	3
	10	42	42				69	53	45	42	43	50	64	87					76	60	48	42	45	56	71	88		48	2
	11	48	23				90	71	57	50	48	50	57	71					90	75	62	52	48	52	62	75		42	1
	12	50	0					90	72	59	52	50	52	59	72					90	76	63	54	50	54	63	76	40	12
APR + AUG	6	7	99	40	14	9	8	8	9	12	29					81	59	37	15	12	32	54	77					83	6
	7	19	89		42	26	20	19	20	26	41	88					69	48	29	19	29	48	68	89				71	5
	8	30	79		71	46	35	31	31	35	47	72					80	61	44	32	32	44	62	81				60	4
	9	41	67			67	51	44	41	43	51	66	90					74	58	46	41	46	58	73	90			49	3
	10	51	51			85	69	58	52	51	55	63	77					86	72	61	53	51	57	67	80			39	2
	11	59	29				86	73	64	60	59	62	69	81					87	75	66	60	59	63	71	82		31	1
	12	62	0					90	78	69	63	62	63	69	78					90	80	70	64	62	64	70	80	28	12
MAY + JUL	5	2	115	5	3	2	2	2	3	6						65	43	20	3	25	47	70						88	7
	6	13	106	40	20	15	13	13	16	25	62					75	53	32	14	20	40	61	83					77	6
	7	24	97	75	42	30	25	24	27	36	58					84	64	44	28	25	37	55	76					66	5
	8	35	87		63	47	38	35	37	44	59	86					74	57	43	36	40	53	70	88				55	4
	9	47	76		82	64	53	48	47	51	61	77					84	69	57	48	47	54	66	80				43	3
	10	57	61			80	68	61	58	58	63	73	86					82	70	62	58	59	65	75	86			33	2
	11	66	37				84	75	69	66	67	71	77	87					84	76	70	66	67	71	78	87		24	1
	12	70	0					90	82	76	71	70	71	76	82					90	82	76	72	70	72	76	82	20	12
JUN	5	4	117	9	6	4	4	5	7	14						63	40	18	6	28	50	72						86	7
	6	15	108	40	22	16	15	16	19	31	75					72	51	30	15	23	43	64	86					75	6
	7	26	100	71	42	31	27	26	30	40	66					81	61	43	29	28	40	59	79					64	5
	8	37	91	89	63	47	39	37	40	48	64					89	72	55	42	37	43	56	73					53	4
	9	49	80		79	63	54	49	50	54	65	82					82	68	56	50	50	57	69	84				41	3
	10	60	66			78	68	62	60	61	67	77	89					80	70	63	60	62	69	78	89			30	2
	11	69	42				83	76	71	69	70	74	81	89					83	76	71	69	70	75	81	89		21	1
	12	73	0					90	84	78	75	73	75	78	84					90	84	78	75	73	75	78	84	17	12
				N	NNW	NW	WNW	W	WSW	SW	SSW	S	SSE	SE	ESE	N	NNW	NW	WNW	W	WSW	SW	SSW	S	SSE	SE	ESE	HOR	PM

Dates vary year to year within plus or minus three days of the twenty-first day of the month.

SOURCE: Adapted by permission from *ASHRAE Handbook of Fundamentals 1977*, p. 26.6, Table 7.

where γ is the wall solar azimuth. When γ is greater than 90°, the entire window is in the shade.

Similarly, the height of the shadow cast by a transom that projects P ft beyond the plane of the glass is

$$T = \frac{P \tan \beta}{\cos \gamma} \qquad (6.23)$$

where β is the solar altitude. The solar angles are shown in Fig. 6.9. Thus, the sunlit area of the window is

$$\text{Area in sun} = (\text{width} - M)(\text{height} - T) \qquad (6.24)$$

6.10.4. Interior Partitions

Whenever a conditioned space is adjacent to a space in which a different temperature prevails, transfer of heat through the separating structural section must be considered. The heat transfer rate, q, in Btu/hr, is given by

$$q = UA(t_b - t_i) \qquad (6.25)$$

where U = coefficient of overall heat transfer between the adjacent and the conditioned space, Btu/(hr·ft^2·°F)

A = area of separating section concerned, ft^2

t_b = average air temperature in adjacent space, °F

t_i = air temperature in conditioned space, °F

Temperature t_b may have any value over a considerable range according to conditions in the adjacent space. The temperature in a kitchen or boiler room may be as much as 15–50°F above the outdoor air temperature. It is recommended that actual temperatures in adjoining spaces be measured wherever practicable. Where nothing is known except that the adjacent space is of conventional construction and contains no heat sources, $t_b - t_i$ should be considered the difference between the outdoor air and conditioned-space design dry-bulb temperatures *minus* 5°F. In some cases, the air temperature in the adjacent space may always closely correspond to the outdoor air temperature.

For floors directly in contact with the ground, or over an underground basement that is neither ventilated nor warmed, heat transfer may be neglected for cooling load estimates.

6.10.5. Heat Sources within Conditioned Space

Values of energy transfer into a space due to occupants, electric motors, commercial cooking appliances, and various other miscellaneous appliances likely to be found in air-conditioned enclosures can be found in tables in the *ASHRAE Handbook* [1]. Where the sources also add moisture to the space,

$$\cos \theta = \cos \beta \cos \gamma \sin \Sigma + \sin \beta \cos \Sigma$$

where

Σ = tilt angle of the surface from the horizontal.

When the surface is horizontal, $\Sigma = 0$ deg and:

$$\cos \theta_H = \sin \beta$$

For a vertical surface, $\Sigma = 90$ deg and:

$$\cos \theta_V = \cos \beta \cos \gamma$$

Wall Orientations and Azimuths, Measured from the South

Orientation	N	NE	E	SE	S	SW	W	NW
Wall Azimuth, deg	180	135	90	45	0	45	90	135

To find the wall-solar azimuth for other orientations,

For walls facing East of South:

$\gamma = \phi - \psi$ a.m.
$\gamma = \phi + \psi$ p.m.

For walls facing West of South:

$\gamma = \phi + \psi$ a.m.
$\gamma = \phi - \psi$ p.m.

Treat negative values of γ as if they were positive. If γ is greater than 90 deg, the surface is in the shade.

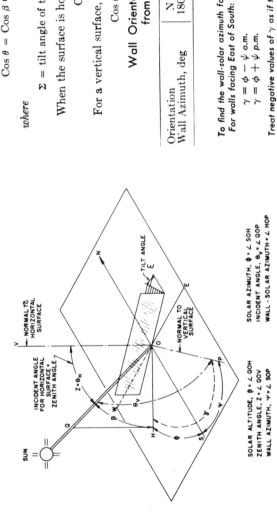

SOLAR ALTITUDE, β = ∠ OOH
ZENITH ANGLE, Z = ∠ OOV
WALL AZIMUTH, γ = ∠ SOP

SOLAR AZIMUTH, φ = ∠ SOH
INCIDENT ANGLE, θ_v = ∠ OOP
WALL - SOLAR AZIMUTH = ∠ HOP

Solar Angles for Vertical and Horizontal Surfaces

Fig. 6.9. Solar angles. Reprinted by permission from *ASHRAE Handbook of Fundamentals 1977.*

the total energy must be divided into sensible and latent parts. Generally, these internal heat gains consist of some or all of the following:

1. Lighting
2. People
3. Appliances
4. Electric motors
5. Miscellaneous sources, such as steam or hot water pipes running through the conditioned space, hot water tanks in the space, etc.

Lighting.   An accurate estimate of the space-cooling load imposed by lighting, often the major space-load component, is essential in air-conditioning system design. Calculation of this load component is not straightforward; the rate of heat gain to the air caused by lights can be quite different from the power supplied to the lights.

Some of the energy emanating from lights is in the form of radiation, which affects the air only after it has been absorbed by walls, floors, and furniture and has warmed them to a temperature higher than the air temperature. This absorbed energy, stored by the structure, contributes to the space-cooling load after a time lag and is present after the lights are switched off.

Tables 6.29 and 6.30 provide the a values and b classifications necessary to obtain the CLF from Table 6.31. At any time, the space-cooling load due to lighting is the heat gain multiplied by the CLF (Table 6.31). This CLF assumes that (1) the conditioned space temperature is maintained at a constant value; and (2) the cooling load and power input to the lights eventually become equal if lights are on long enough.

If the cooling system operates only during the occupied hours, the CLF should be considered 1.0. Where one portion of the lights are on during one operation and another portion are on another schedule, treat each separately. Where lights are on 24 hr a day, use a CLF of 1.0.

In general, the instantaneous rate of heat gain from electric lighting (in Btu/hr) can be calculated from

$$q = \frac{\text{total}}{\text{light wattage}} \times \frac{\text{use}}{\text{factor}} \times \frac{\text{special}}{\text{allowance factor}} \times 3.41 \quad (6.26)$$

The *total light wattage* is obtained from the ratings of all fixtures installed for general illumination or special use. The *use factor* is the ratio of the wattage in use, for the conditions under which the load estimate is being made, to the total installed wattage. For commercial applications such as stores, the use factor is generally unity. The *special allowance factor* is introduced for fluorescent fixtures and fixtures requiring more energy than their rated wattage. For fluorescent fixtures, the special allowance factor accounts for ballast losses, which can be as high as 2.19 for 32-W single-lamp high-output

TABLE 6.29.   Design Values of *a* Coefficient Features of Room Furnishings, Light Fixtures, and Ventilation Arrangements

a	Furnishings	Air Supply and Return	Type of Light Fixture
0.45	Heavyweight, simple furnishings, no carpet	Low rate; supply and return below ceiling $(V \leqslant 0.5)^a$	Recessed, not vented
0.55	Ordinary furniture, no carpet	Medium to high ventilation rate; supply and return below ceiling or through ceiling grill and space $(V \geqslant 0.5)^a$	Recessed, not vented
0.65	Ordinary furniture, with or without carpet	Medium to high ventilation rate or fan coil or induction type air-conditioning terminal unit; supply through ceiling or wall diffuser; return around light fixtures and through ceiling space $(V \geqslant 0.5)^a$	Vented
0.75 or greater	Any type of furniture	Ducted returns through light fixtures	Vented or free-hanging in air stream with ducted returns

^{a} V is room air supply rate in cfm per square foot of floor area.

SOURCE: Reprinted by permission from *ASHRAE Handbook of Fundamentals 1977*, p. 25.15, Table 13.

fixtures on 277 V. Rapid-start, 40-W lamp fixture allowance factors vary from a low of 1.18 for two lamps on 277 V to a high of 1.30 for one lamp on 118 V.

People.   The rates at which heat and moisture are given off by human beings depend on activity, modes of dress, and environmental conditions. However, in Table 6.32 some practical values for these rates are given for commonly encountered conditions, activities, and dress appropriate for the listed applications.

The latent heat gain caused by human beings can be considered an instantaneous cooling load, but the total sensible heat gain is not converted directly to cooling load. The radiant portion is first absorbed by the surroundings, then convected to the room at some later time, depending on the

TABLE 6.30. *b* **Classification Values Calculated for Different Envelope Constructions and Room Air Circulation Rates**

Room Envelope Construction[a] (mass of floor area, lb/ft^2)	Room Air Circulation and Type of Supply and Return[b]			
	Low	Medium	High	Very High
2-in. wood floor (10)	B	A	A	A
3-in. concrete floor (40)	B	B	B	A
6-in. concrete floor (75)	C	C	C	B
8-in. concrete floor (120)	D	D	C	C
12-in. concrete floor (160)	D	D	D	D

[a] Floor covered with carpet and rubber pad; for a floor covered only with floor tile, take next classification to the right in the same row.

[b] Low: Low ventilation rate, minimum required to cope with cooling load due to lights and occupants in interior zone. Supply through floor, wall, or ceiling diffuser. Ceiling space not vented and $h = 0.4$ Btu/(hr·ft^2·°F) (h = inside surface convection coefficient used in calculation of b classification.)

Medium: Medium ventilation rate, supply through floor, wall, or ceiling diffuser. Ceiling space not vented and $h = 0.6$ Btu/(hr·ft^2·°F).

High: Room air circulation induced by primary air of induction unit or by fan coil unit. Return through ceiling space and $h = 0.8$ Btu/(hr·ft^2·°F).

Very High: High room air circulation used to minimize temperature gradients in a room. Return through ceiling space and $h = 1.2$ Btu/(hr·ft^2·°F).

SOURCE: Reprinted by permission from *ASHRAE Handbook of Fundamentals 1977*, p. 25.15, Table 14.

thermal characteristics of the room. The radiant portion of the sensible heat loss from people is generally near 70%, varying by only a few percent. This 70% value was used to generate Table 6.33, which considers the storage effect in its results. The instantaneous sensible cooling load is the product of the sensible heat loss from people (Table 6.32) and the CLF (Table 6.33). This CLF is a function of the time people spend in the conditioned space and the time elapsed since first entering.

If the space temperature is not maintained constant during the 24-hr period (for example, if the cooling system is shut down during the night), a CLF of 1.0 should be used. This pulldown load results when the stored sensible heat in the structure is not removed and thus reappears as a cooling load when the system is started the next day.

If there is a high occupant density, as in theaters and auditoriums, the quantity of radiation to the walls and room furnishings is reduced. In this situation, a CLF of 1.0 should be used.

Appliances and Equipment. In estimating a cooling load, heat gain from all heat-producing appliances must be taken into account. The most common heat-producing appliances found in conditioned areas are those used for

TABLE 6.31. Cooling Load Factors

"a" Coef-ficient	"b" Classification	0	1	2	3	4	5	6	7	8	9	10	11	12	13	14	15	16	17	18	19	20	21	22	23
0.45	A	0.02	0.46	0.57	0.65	0.72	0.77	0.82	0.85	0.88	0.46	0.37	0.30	0.24	0.19	0.15	0.12	0.10	0.08	0.06	0.05	0.04	0.03	0.03	0.02
	B	0.07	0.51	0.56	0.61	0.65	0.68	0.71	0.74	0.77	0.34	0.31	0.28	0.25	0.22	0.20	0.18	0.16	0.15	0.13	0.12	0.11	0.10	0.09	0.08
	C	0.11	0.55	0.58	0.60	0.63	0.65	0.67	0.69	0.71	0.28	0.26	0.25	0.23	0.22	0.20	0.19	0.18	0.17	0.16	0.15	0.14	0.13	0.12	0.12
	D	0.14	0.58	0.60	0.61	0.62	0.63	0.64	0.65	0.66	0.22	0.22	0.21	0.20	0.20	0.19	0.19	0.18	0.18	0.17	0.16	0.16	0.16	0.15	0.15
0.55	A	0.01	0.56	0.65	0.72	0.77	0.82	0.85	0.88	0.90	0.37	0.30	0.24	0.19	0.16	0.13	0.10	0.08	0.07	0.05	0.04	0.03	0.03	0.02	0.02
	B	0.06	0.60	0.64	0.68	0.71	0.74	0.76	0.79	0.81	0.28	0.25	0.23	0.20	0.18	0.16	0.15	0.13	0.12	0.11	0.10	0.09	0.08	0.07	0.06
	C	0.09	0.63	0.66	0.68	0.70	0.71	0.73	0.75	0.76	0.23	0.21	0.20	0.19	0.18	0.17	0.16	0.15	0.14	0.13	0.12	0.11	0.11	0.10	0.10
	D	0.11	0.66	0.67	0.68	0.69	0.70	0.71	0.72	0.72	0.18	0.18	0.17	0.17	0.16	0.16	0.15	0.15	0.14	0.14	0.13	0.13	0.13	0.12	0.12
0.65	A	0.01	0.66	0.73	0.78	0.82	0.86	0.88	0.91	0.93	0.29	0.23	0.19	0.15	0.12	0.10	0.08	0.06	0.05	0.04	0.03	0.03	0.02	0.02	0.01
	B	0.04	0.69	0.72	0.75	0.77	0.80	0.82	0.84	0.85	0.22	0.19	0.18	0.16	0.14	0.13	0.12	0.10	0.09	0.08	0.08	0.07	0.06	0.06	0.05
	C	0.07	0.72	0.73	0.75	0.76	0.78	0.79	0.80	0.82	0.18	0.17	0.16	0.15	0.14	0.13	0.12	0.11	0.11	0.10	0.10	0.09	0.08	0.08	0.07
	D	0.09	0.73	0.74	0.75	0.76	0.77	0.77	0.78	0.79	0.14	0.14	0.13	0.13	0.12	0.12	0.12	0.11	0.11	0.11	0.10	0.10	0.10	0.10	0.09
0.75	A	0.01	0.76	0.80	0.84	0.87	0.90	0.92	0.93	0.95	0.21	0.17	0.13	0.11	0.09	0.07	0.06	0.05	0.04	0.03	0.02	0.02	0.02	0.01	0.01
	B	0.03	0.78	0.80	0.82	0.84	0.85	0.87	0.88	0.89	0.15	0.14	0.13	0.11	0.10	0.09	0.08	0.07	0.07	0.06	0.05	0.05	0.04	0.04	0.04
	C	0.05	0.80	0.81	0.82	0.83	0.84	0.85	0.86	0.87	0.13	0.12	0.11	0.10	0.10	0.09	0.09	0.08	0.08	0.07	0.07	0.06	0.06	0.06	0.05
	D	0.06	0.81	0.82	0.82	0.83	0.83	0.84	0.84	0.85	0.10	0.10	0.10	0.09	0.09	0.09	0.08	0.08	0.08	0.08	0.07	0.07	0.07	0.07	0.07

Lights on for 8 hr

"a" Coef-ficient	"b" Classification	0	1	2	3	4	5	6	7	8	9	10	11	12	13	14	15	16	17	18	19	20	21	22	23
0.45	A	0.05	0.49	0.59	0.67	0.73	0.78	0.83	0.86	0.89	0.91	0.93	0.94	0.95	0.51	0.41	0.33	0.27	0.22	0.17	0.14	0.11	0.09	0.07	0.06
	B	0.13	0.57	0.61	0.65	0.69	0.72	0.75	0.77	0.79	0.82	0.83	0.85	0.87	0.43	0.39	0.35	0.31	0.28	0.25	0.23	0.21	0.18	0.17	0.15
	C	0.19	0.63	0.65	0.67	0.69	0.71	0.73	0.74	0.76	0.77	0.79	0.80	0.81	0.37	0.35	0.33	0.31	0.29	0.27	0.26	0.24	0.23	0.21	0.20
	D	0.22	0.66	0.67	0.68	0.69	0.70	0.71	0.72	0.73	0.74	0.74	0.75	0.76	0.32	0.31	0.30	0.29	0.28	0.27	0.26	0.26	0.25	0.24	0.23
0.55	A	0.04	0.58	0.66	0.73	0.78	0.82	0.86	0.89	0.91	0.93	0.94	0.95	0.96	0.42	0.34	0.27	0.22	0.18	0.14	0.11	0.09	0.07	0.06	0.05
	B	0.11	0.65	0.68	0.72	0.74	0.77	0.79	0.81	0.83	0.85	0.86	0.88	0.89	0.35	0.32	0.28	0.26	0.23	0.21	0.19	0.17	0.15	0.14	0.12
	C	0.15	0.69	0.71	0.73	0.75	0.76	0.78	0.79	0.80	0.81	0.83	0.84	0.85	0.30	0.29	0.27	0.25	0.24	0.22	0.21	0.20	0.19	0.17	0.16
	D	0.18	0.72	0.73	0.74	0.75	0.76	0.76	0.77	0.78	0.78	0.79	0.80	0.80	0.25	0.25	0.24	0.24	0.23	0.22	0.22	0.21	0.20	0.20	0.19
0.65	A	0.03	0.67	0.74	0.79	0.83	0.86	0.89	0.91	0.93	0.94	0.95	0.96	0.97	0.33	0.26	0.21	0.17	0.14	0.11	0.09	0.07	0.06	0.05	0.04
	B	0.09	0.73	0.75	0.78	0.80	0.82	0.84	0.85	0.87	0.88	0.89	0.90	0.91	0.27	0.25	0.22	0.20	0.18	0.16	0.15	0.13	0.12	0.11	0.10
	C	0.12	0.76	0.78	0.79	0.80	0.81	0.83	0.84	0.85	0.86	0.86	0.87	0.88	0.24	0.22	0.21	0.19	0.18	0.17	0.16	0.15	0.14	0.13	0.13
	D	0.14	0.79	0.79	0.80	0.80	0.81	0.82	0.82	0.83	0.83	0.84	0.84	0.85	0.20	0.20	0.19	0.18	0.18	0.17	0.17	0.16	0.15	0.15	0.15
0.75	A	0.02	0.77	0.81	0.85	0.88	0.90	0.92	0.94	0.95	0.96	0.97	0.97	0.98	0.23	0.19	0.15	0.12	0.10	0.08	0.06	0.05	0.04	0.03	0.03
	B	0.06	0.81	0.82	0.84	0.86	0.87	0.88	0.90	0.91	0.92	0.93	0.93	0.94	0.19	0.18	0.16	0.14	0.13	0.12	0.10	0.09	0.08	0.08	0.07
	C	0.09	0.83	0.84	0.85	0.86	0.87	0.88	0.88	0.89	0.90	0.90	0.91	0.91	0.17	0.16	0.15	0.14	0.13	0.12	0.12	0.11	0.10	0.10	0.09
	D	0.10	0.85	0.85	0.86	0.86	0.86	0.87	0.87	0.88	0.88	0.88	0.89	0.89	0.14	0.14	0.14	0.13	0.13	0.12	0.12	0.11	0.11	0.11	0.11

Lights on for 12 hr

0.45	A	0.12	0.54	0.63	0.70	0.76	0.81	0.85	0.88	0.90	0.92	0.94	0.95	0.96	0.97	0.97	0.98	0.98	0.54	0.43	0.35	0.28	0.23	0.18	0.15
	B	0.23	0.66	0.69	0.72	0.75	0.78	0.80	0.82	0.84	0.85	0.87	0.88	0.89	0.90	0.91	0.92	0.93	0.49	0.44	0.39	0.35	0.32	0.29	0.26
	C	0.29	0.72	0.74	0.75	0.77	0.78	0.80	0.81	0.82	0.83	0.84	0.85	0.86	0.87	0.88	0.88	0.89	0.45	0.42	0.39	0.37	0.35	0.33	0.31
	D	0.31	0.75	0.76	0.77	0.77	0.78	0.79	0.79	0.80	0.81	0.81	0.82	0.82	0.83	0.83	0.84	0.84	0.40	0.39	0.37	0.36	0.35	0.34	0.33
0.55	A	0.10	0.63	0.70	0.76	0.81	0.84	0.87	0.90	0.92	0.93	0.95	0.96	0.97	0.97	0.98	0.98	0.99	0.44	0.35	0.28	0.23	0.18	0.15	0.12
	B	0.19	0.72	0.75	0.77	0.80	0.82	0.84	0.85	0.87	0.88	0.89	0.90	0.91	0.92	0.92	0.94	0.94	0.40	0.36	0.32	0.29	0.26	0.24	0.21
	C	0.24	0.77	0.79	0.80	0.81	0.82	0.83	0.84	0.85	0.86	0.87	0.88	0.88	0.89	0.90	0.90	0.91	0.37	0.34	0.32	0.30	0.29	0.27	0.25
	D	0.26	0.80	0.80	0.81	0.82	0.82	0.83	0.83	0.84	0.84	0.85	0.85	0.86	0.86	0.86	0.87	0.87	0.33	0.32	0.31	0.30	0.29	0.28	0.27
0.65	A	0.07	0.71	0.77	0.81	0.85	0.88	0.90	0.92	0.94	0.95	0.96	0.97	0.97	0.98	0.98	0.99	0.99	0.34	0.27	0.22	0.18	0.14	0.12	0.09
	B	0.15	0.78	0.81	0.82	0.84	0.86	0.87	0.88	0.90	0.91	0.92	0.92	0.93	0.94	0.94	0.95	0.96	0.31	0.28	0.25	0.23	0.20	0.18	0.16
	C	0.18	0.82	0.83	0.84	0.85	0.86	0.87	0.88	0.89	0.89	0.90	0.90	0.91	0.92	0.92	0.93	0.93	0.28	0.27	0.25	0.24	0.22	0.21	0.20
	D	0.20	0.84	0.85	0.85	0.86	0.86	0.87	0.87	0.88	0.88	0.88	0.89	0.89	0.89	0.90	0.90	0.90	0.25	0.25	0.24	0.23	0.22	0.22	0.21
0.75	A	0.05	0.79	0.83	0.87	0.89	0.91	0.93	0.94	0.95	0.96	0.97	0.98	0.98	0.98	0.99	0.99	0.99	0.24	0.20	0.16	0.13	0.10	0.08	0.07
	B	0.11	0.85	0.86	0.87	0.89	0.90	0.91	0.92	0.93	0.93	0.94	0.95	0.95	0.96	0.96	0.97	0.97	0.22	0.20	0.18	0.16	0.15	0.13	0.12
	C	0.13	0.87	0.88	0.89	0.89	0.90	0.91	0.91	0.92	0.92	0.93	0.93	0.94	0.94	0.94	0.95	0.95	0.20	0.19	0.18	0.17	0.16	0.15	0.14
	D	0.14	0.89	0.89	0.89	0.90	0.90	0.90	0.91	0.91	0.91	0.91	0.92	0.92	0.92	0.93	0.93	0.93	0.18	0.18	0.17	0.17	0.16	0.15	0.15

Lights on for 16 hr

SOURCE: Reprinted by permission from *ASHRAE Handbook of Fundamentals 1977*, p. 25.16, Table 15A,C,E.

TABLE 6.32. Rates of Heat Gain from Occupants of Conditioned Spaces[a]

Degree of Activity	Typical Application	Total Heat Adults, Male			Total Heat Adjusted[b]			Sensible Heat			Latent Heat		
		Watts	Btuh	kcal/hr	Watts	Btuh	kcal/hr	Watts	Btuh	kcal/hr	Watts	Btuh	kcal/hr
Seated at rest	Theater, movie	115	400	100	100	350	90	60	210	55	40	140	38
Seated, very light work writing	Offices, hotels, apts	140	480	120	120	420	105	65	230	55	55	190	50
Seated, eating	Restaurant[c]	150	520	130	170	580[c]	145	75	255	60	95	325	80
Seated, light work, typing	Offices, hotels, apts	185	640	160	150	510	130	75	255	60	75	255	65
Standing, light work or walking slowly	Retail Store, bank	235	800	200	185	640	160	90	315	80	95	325	80
Light bench work	Factory	255	880	220	230	780	195	100	345	90	130	435	110
Walking, 3 mph, light machine work	Factory	305	1040	260	305	1040	260	100	345	90	205	695	170
Bowling[d]	Bowling alley	350	1200	300	280	960	240	100	345	90	180	615	150
Moderate dancing	Dance hall	400	1360	340	375	1280	320	120	405	100	255	875	220
Heavy work, heavy machine work, lifting	Factory	470	1600	400	470	1600	400	165	565	140	300	1035	260
Heavy work, athletics	Gymnasium	585	2000	500	525	1800	450	185	635	160	340	1165	290

[a]Note: Tabulated values are based on 78 F room dry-bulb temperature. For 80 F room dry-bulb, the total heat remains the same, but the sensible heat value should be decreased by approximately 8% and the latent heat values increased accordingly.

[b]Adjusted total heat gain is based on normal percentage of men, women, and children for the application listed, with the postulate that the gain from an adult female is 85% of that for an adult male, and that the gain from a child is 75% of that for an adult male.

[c]Adjusted total heat value for eating in a restaurant, includes 60 Btuh for food per individual (30 Btu sensible and 30 Btu latent).

[d]For bowling figure one person per alley actually bowling, and all others as sitting (400 Btuh) or standing and walking slowly (790 Btuh).

Also refer to Tables 4 and 5, Chapter 8.
All values rounded to nearest 5 watts or kcal/hr or to nearest 10 Btuh.

SOURCE: Reprinted by permission from *ASHRAE Handbook of Fundamentals 1977*, p. 25.17, Table 16.

TABLE 6.33. Sensible Heat Cooling Load Factors for People

Total Hours in Space	Hours after Each Entry Into Space																							
	1	2	3	4	5	6	7	8	9	10	11	12	13	14	15	16	17	18	19	20	21	22	23	24
2	0.49	0.58	0.17	0.13	0.10	0.08	0.07	0.06	0.05	0.04	0.04	0.03	0.03	0.02	0.02	0.02	0.02	0.01	0.01	0.01	0.01	0.01	0.01	0.01
4	0.49	0.59	0.66	0.71	0.27	0.21	0.16	0.14	0.11	0.10	0.08	0.07	0.06	0.06	0.05	0.04	0.04	0.03	0.03	0.03	0.02	0.02	0.02	0.01
6	0.50	0.60	0.67	0.72	0.76	0.79	0.34	0.26	0.21	0.18	0.15	0.13	0.11	0.10	0.08	0.07	0.06	0.06	0.05	0.04	0.04	0.03	0.03	0.03
8	0.51	0.61	0.67	0.72	0.76	0.80	0.82	0.84	0.38	0.30	0.25	0.21	0.18	0.15	0.13	0.12	0.10	0.09	0.08	0.07	0.06	0.05	0.05	0.04
10	0.53	0.62	0.69	0.74	0.77	0.80	0.83	0.85	0.87	0.89	0.42	0.34	0.28	0.23	0.20	0.17	0.15	0.13	0.11	0.10	0.09	0.08	0.07	0.06
12	0.55	0.64	0.70	0.75	0.79	0.81	0.84	0.86	0.88	0.89	0.91	0.92	0.45	0.36	0.30	0.25	0.21	0.19	0.16	0.14	0.12	0.11	0.09	0.08
14	0.58	0.66	0.72	0.77	0.80	0.83	0.85	0.87	0.89	0.90	0.91	0.92	0.93	0.94	0.47	0.38	0.31	0.26	0.23	0.20	0.17	0.15	0.13	0.11
16	0.62	0.70	0.75	0.79	0.82	0.85	0.87	0.88	0.90	0.91	0.92	0.93	0.94	0.95	0.95	0.96	0.49	0.39	0.33	0.28	0.24	0.20	0.18	0.16
18	0.66	0.74	0.79	0.82	0.85	0.87	0.89	0.90	0.92	0.93	0.94	0.94	0.95	0.96	0.96	0.97	0.97	0.97	0.50	0.40	0.33	0.28	0.24	0.21

SOURCE: Reprinted by permission from *ASHRAE Handbook of Fundamentals 1977*, p. 25.17, Table 17.

TABLE 6.34. Recommended Rate of Heat Gain from Commercial Cooling Appliances Located in the Air-Conditioned Area[a]

Appliance	Capacity	Overall Dimensions, in. Width × Depth × Height	Miscellaneous Data	Manufacturer's Input Rating Btu/hr	Probable Max. Hourly Input, Btu/hr	Recommended Rate of Heat Gain, Btu/hr			With Hood,[b] All Sensible
						Without Hood			
						Sensible	Latent	Total	
Gas-Burning, Counter Type									
Broiler-griddle		31 × 20 × 18		36,000	18,000	11,700	6,300	18,000	3,600
Coffee brewer per burner			With *warm* position	5,500	2,500	1,750	750	2,500	500
Water heater burner			With storage tank	11,000	5,000	3,850	1,650	5,500	1,100
Coffee urn	3 gal	12-inch dia.		10,000	5,000	3,500	1,500	5,000	1,000
	5 gal	14-inch dia.		15,000	7,500	5,250	2,250	7,500	1,500
	8 gal twin	25-inch wide		20,000	10,000	7,000	3,000	10,000	2,000
Deep fat fryer	15 lb fat	14 × 21 × 15		30,000	15,000	7,500	7,500	15,000	3,000
Dry food warmer per sq ft of top				1,400	700	560	140	700	140
Griddle, frying per sq ft of top				15,000	7,500	4,900	2,600	7,500	1,500
Short order stove, per burner			Open grates	10,000	5,000	3,200	1,800	5,000	1,000
Steam table per sq ft of top				2,500	1,250	750	500	1,250	250
Toaster, continuous	360 slices/hr	19 × 16 × 30	2 slices wide	12,000	6,000	3,600	2,400	6,000	1,200
	720 slices/hr	24 × 16 × 30	4 slices wide	20,000	10,000	6,000	4,000	10,000	2,000

Electric, Counter Type

	Capacity	Size, in.	Miscellaneous	W	Btu/hr					
Coffee brewer										
per burner				625	2,130	1,000	770	230	1,000	340
per warmer				160	545	300	230	70	300	90
automatic	240 cups/hr	27 × 21 × 22	4-burner + water heater	5,000	17,000	8,500	6,500	2,000	8,500	1,700
Coffee urn	3 gal			2,000	6,800	3,400	2,550	850	3,400	1,000
	5 gal.			3,000	10,200	5,100	3,850	1,250	5,100	1,600
	8 gal. twin			4,000	13,600	6,800	5,200	1,600	6,800	2,100
Deep fat fryer	14 lb fat	13 × 22 × 10		5,500	18,750	9,400	2,800	6,600	9,400	3,000
	21 lb fat	16 × 22 × 10		8,000	27,300	13,700	4,100	9,600	13,700	4,300
Dry food warmer, per sq ft of top				240	820	400	320	80	400	130
Egg boiler	2 cups	10 × 13 × 25		1,100	3,750	1,900	1,140	760	1,900	600
Griddle, frying, per sq ft of top				2,700	9,200	4,600	3,000	1,600	4,600	1,500
Griddle-grill		18 × 20 × 13	Grid, 200 in.²	6,000	20,400	10,200	6,600	3,600	10,200	3,200
Hotplate		18 × 20 × 13	2 heating units	5,200	17,700	8,900	5,300	3,600	8,900	2,800
Roaster		18 × 20 × 13		1,650	5,620	2,800	1,700	1,100	2,800	900
Roll warmer		18 × 20 × 13		1,650	5,620	2,800	2,600	200	2,800	900
Toaster, continuous	360 slices/hr	15 × 15 × 28	2 slices wide	2,200	7,500	3,700	1,960	1,740	3,700	1,200
	720 slices/hr	20 × 15 × 28	4 slices wide	3,000	10,200	5,100	2,700	2,400	5,100	1,600
Toaster, pop-up	4 slice	12 × 11 × 9		2,540	8,350	4,200	2,230	1,970	4,200	1,300
Waffle iron		18 × 20 × 13	2 grids	1,650	5,620	2,800	1,680	1,120	2,800	900

TABLE 6.34. (Continued)

Appliance	Capacity	Overall Dimensions, in. Width × Depth × Height	Miscellaneous Data	Manufacturer's Input Rating hp	Manufacturer's Input Rating Btu/hr	Probable Max. Hourly Input, Btu/hr	Without Hood Sensible	Without Hood Latent	Without Hood Total	With Hood,[b] All Sensible
Steam Heated										
Coffee urn	3 gal.			0.2	6,600	3,300	2,180	1,120	3,300	1,000
	5 gal.			0.3	10,000	5,000	3,300	1,700	5,000	1,600
	8 gal. twin			0.4	13,200	6,600	4,350	2,250	6,600	2,100
Steam table per sq ft of top			With insets	0.05	1,650	825	500	325	825	260
Bain marie per sq ft of top			Open Tank	0.10	3,300	1,650	825	825	1,650	520
Oyster steamer per sq ft of top				0.5	16,500	8,250	5,000	3,250	8,250	2,600
Steam kettles per gal. capacity			Jacketed type	0.06	2,000	1,000	600	400	1,000	320
Compartment steamer per compartment		24 × 25 × 12 compartment	Floor mounted	1.2	40,000	20,000	12,000	8,000	20,000	6,400
Compartment steamer	3 pans	12 × 20 × 2.5	Single counter unit	0.5	16,500	8,250	5,000	3,250	8,250	2,600
Plate warmer per cu ft				0.05	1,650	825	550	275	825	260

[a]Data was determined by assuming the hourly heat input was 0.50 times the manufacturer's energy input rating. This is felt to be conservative on the average but could result in heat gain estimates higher or lower than actual heat gains depending on the appliance. Consult the text for additional discussion.

[b]For poorly designed or undersized exhaust systems, the heat gains in this column should be doubled and half the increase assumed as latent heat.

SOURCE: Abridged by permission from *ASHRAE Handbook of Fundamentals 1977*.

378

food preparation in commercial and industrial food service establishments such as restaurants, hospitals, schools, hotels, and in-plant cafeterias.

Table 6.34 presents recommended data for estimating the rate of heat gain from commercial cooking appliances. Table 6.35 suggests rates of heat gain for miscellaneous electric and gas appliances. The effect of the sensible heat from appliances and laboratory equipment on the cooling load is delayed in the same manner as the radiant function of the sensible heat gain of other load components already discussed. Table 6.36 provides the cooling load factors (CLF) for appliances. The sensible portion of the cooling load is obtained by multiplying the sensible heat gain by the appropriate CLF.

When equipment of any sort is operated within the conditioned space by electric motors, the heat equivalent of this operation must be considered in the heat gain. As with other cooling load components, the cooling load due to motor-driven equipment is obtained by multiplying the heat gain from the equipment by its CLF. In this case, the CLFs for appliances without hoods (Table 6.36) should be used.

6.10.6. Ventilation and Infiltration

Wind and pressure differences cause outdoor air that is higher in temperature and moisture content to infiltrate through the cracks around doors and windows, resulting in localized sensible and latent heat gains. Also, some outdoor ventilation air is usually necessary to eliminate any odors. This outdoor air imposes a cooling and dehumidifying load in the cooling coil because the heat and/or moisture must be removed. Heat gains due to infiltration and ventilation can be computed using

$$q_s = 1.10 \times \text{CFM} \times \Delta t \tag{6.27}$$

$$q_t = 4840 \times \text{CFM} \times \Delta W \tag{6.28}$$

where CFM = cubic feet per minute of outdoor air, cfm
Δt = difference between outdoor and indoor temperatures, °F
ΔW = difference between outdoor and indoor humidity ratios, lb/lb

These equations are valid for calculating the cooling load due to infiltration of outside air and also due to the positive introduction of air for ventilation, provided it is introduced directly into the space. When the ventilation air is introduced through the air-conditioning system and imposes a load directly on the cooling coil, it is handled as a coil load.

6.10.7. Cooling Load Summary

Table 6.37 summarizes the sources of the space cooling load discussed in the previous sections.

TABLE 6.35. Rate of Heat Gain from Miscellaneous Appliances

Appliance	Miscellaneous Data	Manufacturer's Rating		Recommended Rate of Heat Gain, Btuh			Appliance	Miscellaneous Data	Manufacturer's Rating		Recommended Rate of Heat Gain, Btuh		
		Watts	Btuh	Sensible	Latent	Total			Watts	Btuh	Sensible	Latent	Total
Electrical Appliances							Gas-Burning Appliances						
Hair dryer	Blower type	1580	5400	2300	400	2700	Lab burners						
Hair dryer	Helmet type	705	2400	1870	330	2200	Bunsen	0.4375-in. barrel		3000	1680	420	2100
Permanent wave machine	60 heaters @25 W, 36 in normal use	1500	5000	850	150	1000	Fishtail	1.5-in. wide		5000	2800	700	3500
							Meeker	1-in. diameter		6000	3360	840	4200
Neon sign, per linear ft of tube	0.5 in., dia			30		30	Gas light, per burner	Mantle type		2000	1800	200	2000
	0.375 in., dia			60		60	Cigar lighter	Continuous flame		2500	900	100	1000
Sterilizer, instrument		1100	3750	650	1200	1850							

SOURCE: Reprinted by permission from *ASHRAE Handbook of Fundamentals 1977*, p. 25.19, Table 19.

TABLE 6.36. Sensible Heat Cooling Load Factors for Appliances (A, Hooded, B, Unhooded)

A

Total Operational Hours	Hours after appliances are on																							
	1	2	3	4	5	6	7	8	9	10	11	12	13	14	15	16	17	18	19	20	21	22	23	24
2	0.27	0.40	0.25	0.18	0.14	0.11	0.09	0.08	0.07	0.06	0.05	0.04	0.04	0.03	0.03	0.03	0.02	0.02	0.02	0.02	0.01	0.01	0.01	0.01
4	0.28	0.41	0.51	0.59	0.39	0.30	0.24	0.19	0.16	0.14	0.12	0.10	0.09	0.08	0.07	0.06	0.05	0.05	0.04	0.04	0.03	0.03	0.02	0.02
6	0.29	0.42	0.52	0.59	0.65	0.70	0.48	0.37	0.30	0.25	0.21	0.18	0.16	0.14	0.12	0.11	0.09	0.08	0.07	0.06	0.05	0.05	0.04	0.04
8	0.31	0.44	0.54	0.61	0.66	0.71	0.75	0.78	0.55	0.43	0.35	0.30	0.25	0.22	0.19	0.16	0.14	0.13	0.11	0.10	0.08	0.07	0.06	0.06
10	0.33	0.46	0.55	0.62	0.68	0.72	0.76	0.79	0.81	0.84	0.60	0.48	0.39	0.33	0.28	0.24	0.21	0.18	0.16	0.14	0.12	0.11	0.09	0.08
12	0.36	0.49	0.58	0.64	0.69	0.74	0.77	0.80	0.82	0.85	0.87	0.88	0.64	0.51	0.42	0.36	0.31	0.26	0.23	0.20	0.18	0.15	0.13	0.12
14	0.40	0.52	0.61	0.67	0.72	0.76	0.79	0.82	0.84	0.86	0.88	0.89	0.91	0.92	0.67	0.54	0.45	0.38	0.32	0.28	0.24	0.21	0.19	0.16
16	0.45	0.57	0.65	0.70	0.75	0.78	0.81	0.84	0.86	0.87	0.89	0.90	0.92	0.93	0.94	0.94	0.69	0.56	0.46	0.39	0.34	0.29	0.25	0.22
18	0.52	0.63	0.70	0.75	0.79	0.82	0.84	0.86	0.88	0.89	0.91	0.92	0.93	0.94	0.95	0.95	0.96	0.96	0.71	0.58	0.48	0.41	0.35	0.30

B

Total Operational Hours	Hours after appliances are on																							
	1	2	3	4	5	6	7	8	9	10	11	12	13	14	15	16	17	18	19	20	21	22	23	24
2	0.56	0.64	0.15	0.11	0.08	0.07	0.06	0.05	0.04	0.04	0.03	0.03	0.02	0.02	0.02	0.02	0.01	0.01	0.02	0.01	0.01	0.01	0.01	0.01
4	0.57	0.65	0.71	0.75	0.23	0.18	0.14	0.12	0.10	0.08	0.07	0.06	0.05	0.05	0.04	0.04	0.03	0.03	0.02	0.02	0.02	0.02	0.01	0.01
6	0.57	0.65	0.71	0.76	0.79	0.82	0.29	0.22	0.18	0.15	0.13	0.11	0.10	0.08	0.07	0.06	0.06	0.05	0.04	0.04	0.03	0.03	0.03	0.02
8	0.58	0.66	0.72	0.76	0.80	0.82	0.85	0.87	0.33	0.26	0.21	0.18	0.15	0.13	0.11	0.10	0.09	0.08	0.07	0.06	0.05	0.04	0.04	0.03
10	0.60	0.68	0.73	0.77	0.81	0.83	0.86	0.87	0.89	0.90	0.36	0.29	0.24	0.20	0.17	0.15	0.13	0.11	0.10	0.08	0.07	0.07	0.06	0.05
12	0.62	0.69	0.75	0.79	0.82	0.84	0.86	0.88	0.89	0.91	0.92	0.93	0.38	0.31	0.25	0.21	0.18	0.16	0.14	0.12	0.11	0.09	0.08	0.07
14	0.64	0.71	0.76	0.80	0.83	0.85	0.87	0.89	0.90	0.92	0.93	0.93	0.94	0.95	0.40	0.32	0.27	0.23	0.19	0.17	0.15	0.13	0.11	0.10
16	0.67	0.74	0.79	0.82	0.85	0.87	0.89	0.90	0.91	0.92	0.93	0.94	0.95	0.96	0.96	0.97	0.42	0.34	0.28	0.24	0.20	0.18	0.15	0.13
18	0.71	0.78	0.82	0.85	0.87	0.89	0.90	0.92	0.93	0.94	0.94	0.95	0.96	0.96	0.97	0.97	0.97	0.98	0.43	0.35	0.29	0.24	0.21	0.18

SOURCE: Reprinted by permission from *ASHRAE Handbook of Fundamentals 1977*, p. 25.21, Tables 20 and 21.

TABLE 6.37.   Procedure for Calculating Space Design Cooling Load—Summary of Load Sources and Equations—Nonresidential Buildings

Load Source	Equation
External	
Roof	$q = U \times A \times \text{CLTD}$
Walls	$q = U \times A \times \text{CLTD}$
Glass	
Conduction	$q = U \times A \times \text{CLTD}$
Solar	$q = A \times \text{SC} \times \text{SHGF} \times \text{CLF}$
Partitions, Ceilings, Floors	$q = U \times A \times \text{TD}$
Internal	
Lights	$q = \text{Input} \times \text{CLF}$
People	
Sensible	$q_s = \text{No.} \times \text{sensible HG} \times \text{CLF}$
Latent	$q_l = \text{No.} \times \text{latent HG}$
Appliances	
Sensible	$q_s = \text{heat gain} \times \text{CLF}$
Latent	$q_l = \text{heat gain}$
Power	$q = \text{heat gain} \times \text{CLF}$
Ventilation and Infiltration Air	
Sensible	$q_s = 1.10 \times \text{CFM} \times \Delta t$
Latent	$q_l = 4840 \times \text{CFM} \times \Delta W$
Total	$q = 4.5 \times \text{CFM} \times \Delta h$

6.11.   SIMPLIFIED PROCEDURE FOR CALCULATING RESIDENTIAL COOLING LOAD

The procedures in the previous sections are applicable for calculating a cooling load for a residential building. However, as a result of research studies of five residences at the University of Illinois, taking into account the unique features of residential buildings and their systems, an all-industry residential cooling-load calculations procedure has been developed.

The cooling-load calculation consists of determining total sensible cooling load due to heat gain: (1) through structural components (walls, floors, and ceiling areas); (2) through windows; (3) caused by infiltration and ventilation air; and (4) due to occupancy. The latent portion of the cooling load can be evaluated separately but is usually computed as 1.3 times the calculated sensible load. For very dry climates, the factor can be reduced to 1.2.

An assumption of a single indoor temperature is not a sufficient condition for equipment selection and system design if the design goal is occupant comfort without requiring uneconomical equipment capacity. The procedure is therefore based upon an assumed indoor temperature swing of not more than 3°F on a design day, when the residence is conditioned 24 hr/day, and the thermostat setting is 75°F.

The sensible cooling load due to heat gains through walls, floors, and ceiling areas of each room is calculated using the equivalent temperature differences given in Table 6.38 and U values for summer conditions. For ceilings under naturally vented attics or beneath vented flat roofs, use the combined U value for the roof, vented space, and ceiling. Where the design temperature difference (outdoor design temperature minus 75°F) is not an even increment of 5°F, the equivalent temperature difference should be corrected 1°F for each 1°F difference from the tabulated values.

Direct application of instantaneous heat gains for flat glass has been found to result in unrealistically high cooling loads for residential installations. The data have therefore been interpreted in a manner similar to that used for equivalent temperature differences. Window factors for residential cooling load calculations are tabulated in Table 6.39.

In using these factors, the area of each window is multiplied by the appropriate glass factor. Factors incorporating shading devices, such as roller shades half-drawn, take into account both the shaded and unshaded portions of a window. When permanent shading devices over windows are used, their effects must be considered separately in determining cooling load due to heat gain through windows. Glass protected by permanent shading, such as wide roof overhangs, is usually considered north-facing glass. Hence, the line of shadow from the overhang onto the window, the shade line, must be located. Shade line factors are given in Table 6.40. These factors are expressed as the distance the shade line falls beneath the edge of the overhang width.

Table 6.41 shows factors for calculating natural infiltration, based on a leakage rate of $\frac{1}{2}$AC/hr. However, because of the effect of house configuration on areas exposed to wind, these factors are tabulated in terms of Btu's per square foot of gross exposed wall area.

Mechanical ventilation may be provided in residential air-conditioning systems. This ventilation air is usually introduced at the rate of about one air change per hour (1 AC/hr).

Even though occupant density is usually lower in residences than in many other structures, occupancy loads must be considered in residential cooling-load calculations. Heat release per occupant is usually assumed to be 225 Btu/hr of sensible heat gain. The number of occupants must be estimated as accurately as possible, with special attention paid to the owner's use of living or recreation rooms for entertaining large groups. When the owner does not indicate the need for special entertainment provisions, and the exact number of occupants is unknown, a convenient rule of thumb is to assume approximately twice as many occupants as the structure has bedrooms. The occupancy load should then be distributed about equally among the living rooms of the residence, because the maximum load period occurs when most of the residents occupy the living rather than sleeping areas.

In most cases, appliance loads should be limited to the kitchen. A value of 1200 Btu/hr of sensible heat gain released by kitchen appliances has been found satisfactory for residential cooling-load calculations. Although this

TABLE 6.38. Design Equivalent Temperature Differences

Design Temperature, °F	85		90			95			100		105	110
Daily Temperature Range[a]	L	M	L	M	H	L	M	H	M	H	H	H
WALLS AND DOORS												
1. Frame and veneer-on-frame	17.6	13.6	22.6	18.6	13.6	27.6	23.6	18.6	28.6	23.6	28.6	33.6
2. Masonry walls, 8-in. block or brick	10.3	6.3	15.3	11.3	6.3	20.3	16.3	11.3	21.3	16.3	21.3	26.3
3. Partitions, frame	9.0	5.0	14.0	10.0	5.0	19.0	15.0	10.0	20.0	15.0	20.0	25.0
masonry	2.5	0	7.5	3.5	0	12.5	8.5	3.5	13.5	8.5	13.5	18.5
4. Wood doors	17.6	13.6	22.6	18.6	13.6	27.6	23.6	18.6	28.6	23.6	28.6	33.6
CEILINGS AND ROOFS[b]												
1. Ceilings under naturally vented attic or vented flat roof—dark	38.0	34.0	43.0	39.0	34.0	48.0	44.0	39.0	49.0	44.0	49.0	54.0
—light	30.0	26.0	35.0	31.0	26.0	40.0	36.0	31.0	41.0	36.0	41.0	46.0
2. Built-up roof, no ceiling—dark	38.0	34.0	43.0	39.0	34.0	48.0	44.0	39.0	49.0	44.0	49.0	54.0
—light	30.0	26.0	35.0	31.0	26.0	40.0	36.0	31.0	41.0	36.0	41.0	46.0
3. Ceilings under unconditioned rooms	9.0	5.0	14.0	10.0	5.0	19.0	15.0	10.0	20.0	15.0	20.0	25.0
FLOORS												
1. Over unconditioned rooms	9.0	5.0	14.0	10.0	5.0	19.0	15.0	10.0	20.0	15.0	20.0	25.0
2. Over basement, enclosed crawl space or concrete slab on ground	0	0	0	0	0	0	0	0	0	0	0	0
3. Over open crawl space	9.0	5.0	14.0	10.0	5.0	19.0	15.0	10.0	20.0	15.0	20.0	25.0

[a] Daily Temperature Range
L (Low) Calculation Value: 12 deg F.
  Applicable Range: Less than 15 deg F.
M (Medium) Calculation Value: 20 deg F.
  Applicable Range: 15 to 25 deg F.
H (High) Calculation Value: 30 deg F.
  Applicable Range: More than 25 deg F.

[b] Ceilings and Roofs: For roofs in shade, 18-hr average = 11 deg temperature differential. At 90 F design and medium daily range, equivalent temperature differential for light-colored roof equals $11 + (0.71)(39 - 11) = 31$ deg F.

SOURCE: Reprinted by permission from *ASHRAE Handbook of Fundamentals 1977*, p. 25.40, Table 35.

TABLE 6.39. Design Transmitted Solar Energy, Btu/(hr·ft²)

Outdoor	Regular Single Glass						Regular Double Glass						Heat Absorbing Double Glass					
Design Temp.	85	90	95	100	105	110	85	90	95	100	105	110	85	90	95	100	105	110
No Awnings or Inside Shading																		
North	23	27	31	35	38	44	19	21	24	26	28	30	12	14	17	19	21	23
NE and NW	56	60	64	68	71	77	46	48	51	53	55	57	27	29	32	34	36	38
East and West	81	85	89	93	96	102	68	70	73	75	77	79	42	44	47	49	51	53
SE and SW	70	74	78	82	85	91	59	61	64	66	68	70	35	37	40	42	44	46
South	40	44	48	52	55	61	33	35	38	40	42	44	19	21	24	26	28	30
Draperies or Venetian Blinds																		
North	15	19	23	27	30	36	12	14	17	19	21	23	9	11	14	16	18	20
NE and NW	32	36	40	44	47	53	27	29	32	34	36	38	20	22	25	27	29	31
East and West	48	52	56	60	63	69	42	44	47	49	51	53	30	32	35	37	39	41
SE and SW	40	44	48	52	55	61	35	37	40	42	44	46	24	26	29	31	33	35
South	23	27	31	35	38	44	20	22	25	27	29	31	15	17	20	22	24	26
Roller Shades Half-Drawn																		
North	18	22	26	30	33	39	15	17	20	22	24	26	10	12	15	17	19	21
NE and NW	40	44	48	52	55	61	38	40	43	45	47	49	24	26	29	31	33	35
East and West	61	65	69	73	76	82	54	56	59	61	63	65	35	37	40	42	44	46
SE and SW	52	56	60	64	67	73	46	48	51	53	55	57	30	32	35	37	39	41
South	29	33	37	41	44	50	27	29	32	34	36	38	18	20	23	25	27	29
Awnings																		
North	20	24	28	32	35	41	13	15	18	20	22	24	10	12	15	17	19	21
NE and NW	21	25	29	33	36	42	14	16	18	21	23	25	11	13	16	18	20	22
East and West	22	26	30	34	37	43	14	16	19	21	23	25	12	14	17	19	21	23
SE and SW	21	25	29	33	36	42	14	16	19	21	23	25	11	13	16	18	20	22
South	21	24	28	32	35	41	13	15	18	20	22	24	11	13	16	18	20	22

SOURCE: Reprinted by permission from *ASHRAE Handbook of Fundamentals 1977*, p. 25.40, Table 36.

TABLE 6.40.   Shade Line Factors[a]

Direction Window Faces	Latitude, Degrees						
	25	30	35	40	45	50	55
E	0.8	0.8	0.8	0.8	0.8	0.8	0.8
SE	1.9	1.6	1.4	1.3	1.1	1.0	0.9
S	10.1	5.4	3.6	2.6	2.0	1.7	1.4
SW	1.9	1.6	1.4	1.3	1.1	1.0	0.9
W	0.8	0.8	0.8	0.8	0.8	0.8	0.8

[a]Distance shadow line falls below the edge of the overhang equals shade line factor multiplied by width of overhang. Values are averages for 5 hr of greatest solar intensity on August 1.
SOURCE: Reprinted by permission from *ASHRAE Handbook of Fundamentals 1977.*

does not equal the load that can be imposed by even one top burner or element of a domestic range, such factors as intermittent use of appliances, flywheel effects, and kitchen ventilating fans make this a reasonable and practical value to use.

As mentioned previously, the latent portion of a residential cooling load is usually estimated at 20–30% of the calculated sensible load. This is an approximation; moisture sources in a residence are numerous and, if taken individually, difficult to evaluate precisely. Furthermore, residential cooling equipment is controlled by one or more room thermostats, and humidity rarely affects a control device directly.

EXAMPLE 6.13

The west brick wall of a residence in Rolla, Missouri, has a net area (excluding windows and doors) of 510 ft^2 and a U value of 0.067 for both winter and summer. Determine:

a.   Heating load for the wall, Btu/hr

b.   Cooling load for the wall, Btu/hr

TABLE 6.41.   Sensible Cooling Load due to Infiltration and Ventilation

Design temperature, °F	85	90	95	100	105	110
Infiltration, Btu/hr per square foot of gross exposed wall area	0.7	1.1	1.5	1.9	2.2	2.6
Mechanical ventilation, Btu/(hr·cfm)	11.0	16.0	22.0	27.0	32.0	38.0

SOURCE: Reprinted by permission from *ASHRAE Handbook of Fundamentals 1977.*

Solution

a. $q_s = UA \, \Delta t = 0.067(510)(75 - 3) = 2460$ Btu/hr
b. $q_s = UA(\text{DETD}) = 0.067(510)(25.6) = 875$ Btu/hr   (for 97°F design db and M daily range)

EXAMPLE 6.14

A large window, essentially all glass, 10 ft × 5 ft, is located in the west wall of a residence in Rolla. To reduce energy requirements the window is heat-absorbing double glass, ½-in. air space, coating emissivity of 0.20, and has inside draperies that are closed at peak conditions. Determine:

a.   Heating load for the window, Btu/hr.
b.   Cooling load for the window, Btu/hr.

Solution

a. $q_s = UA \, \Delta t = 0.38(50)(72) = 1368$ Btu/hr
b. $q_s = 35.8$ Btu/hr·ft^2 × 50 ft^2 = 1790 Btu/hr   (for 97°F design db)

EXAMPLE 6.15

For a commercial building located in St. Louis, Missouri (Fig. E6.15), determine the sensible, latent, and total cooling load existing at 4:00 P.M. on August 21 for an inside design temperature of 75°F and relative humidity of 55%.

Wall: 1-in. dark stucco on 12 in. regular concrete; $U = 0.340$

Roof: 8-in. lightweight concrete; $U = 0.115$

Door: 2-in. solid wood; $U = 0.42$

Windows: South: ⅛-in. regular sheet glass with white roller shades drawn half-way   West: Insulating glass, regular sheet out/regular sheet in, no internal shading, ¼-in. air space.

People: 35 at moderate office activity

Lights: 5600 W, fluorescent

Infiltration: ½ AC/hr

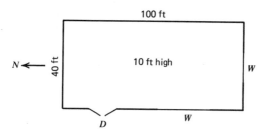

Fig. E6.15.

Solution

Outside design: 94/75, 21° range, 83.5°F av. outside, 75°F inside
Correction + 1.5°F

	U	A	CLTD	$Q_{cooling}$
East wall	0.34	1000	26.5	9,010
North wall	0.34	400	11.5	1,564
West wall	0.34	720	16.5	4,039
South wall	0.34	280	16.5	1,571
Door (west)	0.42	40	68.5	1,151
Roof	0.11	4000	40.5	17,820
			Subtotal	35,155

Glass	A	U	CLTD	Q_c	$SHGF_{max}$	CLF	SC	Q_s	$Q_{cooling}$
South	60	0.81	15.5	750	149	0.33	0.25	740	1,490
South	60	1.04	15.5	970	149	0.43	1.0	3,840	4,810
West	240	0.61	15.5	2270	216	0.49	0.81	20,580	22,850
							Subtotal		29,150

Lighting: $5600 \times 3.41 \times 1.2 \times 0.76 = 17,415$
People:
  Sensible: $35 \times 255 \times 0.8 = 7,140$
  Latent: $35 \times 255 = 8925$
Infiltration:
  Sensible: $\frac{1}{2}(4000 \times 10)\left(\frac{1}{60}\right) \times 1.10 \times (94 - 75) = 6970$

  Latent: $\frac{1}{2}(4000 \times 10)\left(\frac{1}{60}\right) \times 4840 \times (0.0144 - 0.0102) = 6780$

Sensible total	95,830 Btu/hr
Latent total	15,700 Btu/hr
Grand total	111,530 Btu/hr

EXAMPLE 6.16

For the building described in Example 6.15, the heating load is to be found.
The outside winter design temperature is 6°F and the inside winter design
temperature is 75°F. The building is a slab on grade with 2 in. of edge
insulation.

Solution.   The winter U factors are as follows:

Walls	$U = 0.35$ Btu/(hr·ft²·°F)
Roof	$U = 0.12$ Btu/(hr·ft²·°F)
Door	$U = 0.44$ Btu/(hr·ft²·°F)
Windows	$U_{south} = 1.10$ Btu/(hr·ft²·°F)
	$U_{west} = 0.58$ Btu/(hr·ft²·°F)

	U	$\times$	A	$\times$	Δt	$=$	Q
East wall	0.35		1000		69		24,150
North wall	0.35		400		69		9,660
West wall	0.35		720		69		17,388
South wall	0.35		280		69		6,762
Door	0.44		40		69		1,214
South glass	1.10		120		69		9,108
West glass	0.58		240		69		9,605
Roof	0.12		4000		69		33,210

Floor   (40 Btu/(hr·ft) $\times$ 280)   11,200

$$\text{Subtotal} = 122{,}207 \text{ Btu/hr}$$

Infiltration

$$Q_s = 0.018 \,(CFH)(t_i - t_o)$$
$$Q_s = 0.018 \times \tfrac{1}{2} \times (1000 \times 40) \times 69°F = \qquad 24{,}840$$

$$\text{Total} \qquad 147{,}047 \text{ Btu/hr}$$

REFERENCES

1. American Society of Heating, Refrigerating and Air Conditioning Engineers, *ASHRAE Handbook and Product Directory, 1977, Fundamentals,* New York, 1977.

2. "Thermal Environmental Conditions for Human Occupancy (ANSI B193.1-1976)," ASHRAE 55-74, American Society of Heating, Refrigerating and Air Conditioning Engineers, New York, 1974.

3. "Energy Conservation in New Building Design" (ASHRAE 90-75), American Society of Heating, Refrigerating and Air Conditioning Engineers, New York, 1975.

4. Shavit, G., "Energy Conservation and Fan Systems: Computer Control with Floating Space Temperature," *ASHRAE J.* Oct., 1977.

5. "Standards for Natural and Mechanical Ventilation (ANSI B194.1-1977)," ASHRAE 62-73, American Society of Heating, Refrigerating and Air Conditioning Engineers, New York, 1973.

PROBLEMS

1. Estimate the infiltration, in cfm, for each of the following:
   (a) A residence 65 ft × 24 ft × 8 ft to be constructed as a tract house.
   (b) The kitchen of a residence, 14 ft × 11 ft with two sides having windows.
   (c) An old factory building, 250 ft × 78 ft × 12 ft.

2. Estimate the sensible heat loss (Btu/hr) and the size humidifier (lbf/hr) needed due to infiltration at winter design conditions (indoor at 75°F, 20% RH) for each of the following spaces located in St. Louis, MO:
   (a) A residence, 70 ft × 28 ft × 8 ft high
   (b) An apartment, 50 ft × 40 ft × 9 ft high, two walls with windows

3. A building 20 ft × 40 ft × 9 ft has an anticipated infiltration rate of $\frac{3}{4}$ AC/hr. Indoor design conditions are 75°F, 30% RH minimum. Outdoor design temperature is 5°F.
   (a) Determine sensible, latent, and total heat loads (Btu/hr) due to infiltration.
   (b) Specify the necessary humidifier size (lb/hr).

4. Calculate the winter U value for a wall consisting of 4 in. face brick, 4 in. common brick, and $\frac{1}{2}$ in. gypsum plaster (sand aggregate).

5. Find the overall coefficient of heat transmission, U, for a wall consisting of 4 in. of face brick, $\frac{1}{2}$ in. of cement mortar, 8 in. of stone, and $\frac{3}{4}$ in. of gypsum plaster. The outside air velocity is 15 mph, and the inside air is still.

6. A concrete wall 10 in. thick is exposed to outside air at 5°F, 15 mph. Inside air temperature is 60°F. Determine the heat flow through 160 ft² of this wall.

7. A composite wall structure experiences a −10°F air temperature on the outside and a 75°F air temperature on the inside. The wall consists of a 4-in. thick outer face brick, a 2-in. batt of fiberglass insulation, and a $\frac{3}{8}$-in. sheet of gypsum board. Determine the U value and the heat flow rate, per square foot. Plot the steady-state temperature profile across the wall.

8. Find the overall heat transmission coefficient for a floor-ceiling sandwich (heat flow up) having the construction shown in Fig. P6.8.

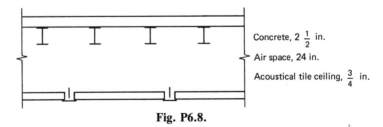

Concrete, 2 $\frac{1}{2}$ in.

Air space, 24 in.

Acoustical tile ceiling, $\frac{3}{4}$ in.

Fig. P6.8.

9. A residence has an unheated basement. Determine the U value applicable for heat transfer between the first-floor rooms and the basement if the floor joists are 2 by 10's; there is a basement ceiling of $\frac{1}{2}$-in. plasterboard, and the floor proper has a carpet and rubber pad laid on $\frac{5}{8}$-in. particleboard underlay on a $\frac{3}{4}$-in. wood subfloor. There is no insulation between the joists. Neglect the effect of the joists.

10. Determine the winter U value in W/(m^2·°C) for the wall of a building that has the following construction:

Face brick, 4 in.
Air space, $\frac{3}{4}$ in.
Concrete, 9 in.
Cellular glass board insulation, 1 in.
Plywood paneling, $\frac{1}{4}$ in.

11. An exterior wall contains a 3 ft × 7 ft wood door, $1\frac{1}{2}$ in. thick, and a 6 ft × 7 ft sliding patio door with double insulating glass having a $\frac{1}{2}$-in. air space and metal frame. Determine the summer U value for each door.

12. A prefabricated commercial building has exterior walls constructed of 2 in. expanded polyurethane bonded between $\frac{1}{4}$-in. aluminum sheet and $\frac{1}{4}$-in. veneer plywood. Design conditions include 105°F outside air temperature, 72°F indoor air, 7.5 mph wind. Determine: (a) the overall thermal resistance, (b) U, and (c) the heat gain per square foot.

13. Determine the design winter heat loss through each of the following components of a building located in Minneapolis, MN:

    (a) Wall with 648 ft^2 area, constructed of 4 in. face brick; $\frac{3}{4}$ in. plywood sheathing; $2\frac{1}{2}$ in. glass fiber insulation in 2 by 4 stud space (16 in. on center; $\frac{1}{2}$ in. plasterboard interior wall.

    (b) A 2185-ft^2 ceiling topped by a 2622 ft^2 hip roof. The ceiling consists of $\frac{1}{2}$ in. acoustical tile with R-19 insulation between the 2 by 6 (16 in. oc) ceiling joists. The roof has asphalt shingles on $\frac{3}{4}$-in. plywood sheathing on the roof rafters. Attic is unvented in winter.

    (c) Two 4 ft × 6 ft single-pane glass windows with storm windows.

14. A house has a pitched roof with an area of 1716 ft^2 and a U of 0.28. The ceiling beneath the roof has an area of 1430 ft^2 and a U of 0.075. The attic is unvented in winter, for which the design conditions are -3°F outside and 75°F inside. Determine the heat loss through the ceiling.

15. A residence has a total ceiling area of 1960 ft^2 and consists of $\frac{3}{8}$-in. gypsum board on 2 in. × 6 in. ceiling joists. Six inches of fiberglass (mineral/glass wool) insulation fills the space between the joists. The

pitched roof has asphalt shingles on $\frac{25}{32}$ in. solid wood sheathing with no insulation between the rafters. The ratio of roof area to ceiling area is 1.3. The attic is unvented in winter. The residence is located in Chicago, IL. For winter design conditions, including a 75°F db inside at the 5-ft line, determine:

(a)   Outside design temperature, °F.

(b)   Appropriate temperature difference, °F.

(c)   Appropriate overall coefficient, U, Btu/(hr·ft²·°F).

(d)   Ceiling heat loss, q, Btu/hr.

16.   An automotive repair shop is to be started in Chicago, IL. The building will be 110 ft long × 60 ft wide × 10 ft high. Walls are 12-in. concrete block (3 oval core, sand and gravel aggregate) followed by 1-in. cellular glass slab insulation and a row of 4-in. lightweight aggregate concrete block on the inside. Ceiling/roof combination is simply 4 in. of pre-formed concrete with lightweight aggregate and a density of 30 lb/ft³ covered with $\frac{3}{8}$ in. built-up roofing. Floor is 6-in. concrete slab on grade with 1-in. edge insulation. On the front side, there is one 4 ft × 6 ft single-pane window with storm window and a 30-ft wide × 8-ft high steel ($\frac{1}{8}$-in. thick) garage door. On the back side there is a single 3 ft × 7 ft solid wood (2 in. thick) door. Determine the size heating system required.

17.   The south wall of a residence located in New York City has one 3 ft × 5 ft double glass, $\frac{1}{4}$ in. air space, window with wood sash. The remainder of the 30 ft × 8 ft high wall is constructed as follows: face brick; $\frac{3}{4}$ in. air space, $\frac{25}{32}$ in. insulating board siding; full-wall glass fiber insulation in the stud space; and $\frac{1}{4}$ in. wood paneling. Neglecting the effect of the studs, determine the cooling load through the wall and window if there are closed draperies at the window.

18.   A naturally vented attic of a residence has a roof with $U = 0.23$. The ceiling beneath the attic has an area of 1960 ft² and a $U = 0.048$. The ratio of roof to ceiling area is 1.2. For a dark roof, medium daily temperature range, and outdoor design temperature of 90°F, determine the heat gain to the residence through the ceiling, Btu/hr.

19.   The southwest wall of a school building in Philadelphia, PA is 250 ft × 9 ft overall and includes 30 windows, each 6 ft × 4 ft. The wall is 12 in. concrete ($k = 12$ Btu/(hr·ft²·°F·in.) followed by 8 in. cinder aggregate concrete block. The windows are single-pane, $\frac{1}{4}$ in. regular plate glass, with light venetian blinds. Determine the design cooling load for the complete wall, assuming this wall determines the design time for the building.

20.   As a move to conserve energy, underground houses are being de-signed. One such three-bedroom house in St. Louis, MO, has walls,

floor, and roof/ceiling of 10-in.-thick concrete. The roof is covered with 6 in. of dirt having a thermal conductivity of 9.0 Btu/(hr·ft²·°F·in.). The house is overall 70 ft × 28 ft × 8 ft, interior dimensions. One air change per hour is provided. Continuous lighting of a minimum of 1120 W is required. Determine the size air conditioner required.

21. As an attempt to minimize energy requirements, a new-style residence has been constructed in Dallas, TX (100°F db; 20° daily range; $W = 0.0156$) with the following features:

No windows

Inside design conditions: 75°F, 60% RH ($W = 0.0112$)

One 3 ft × 7 ft 2 in. wood door with storm door on south side, $U = 0.26$

one 3 ft × 7 ft 2 in. wood door with storm door on east side, $U = 0.26$

Overall size: 70 ft × 28 ft × 8 ft high

Walls are frame construction with 2 × 6 in. studs and full insulation for $U = 0.043$

Attic has natural ventilation, 12 in. fiberglass insulation, and overall U of roof and ceiling of 0.021 based on ceiling area. Light color roof.

Infiltration is so small that outside air must be brought in at rate of 9 cfm/person

Estimated occupancy at design condition: 15 people

Fluorescent lights rated at 320 W will be on all the time

Floor is concrete slab on ground

Size the air-conditioning unit, Btu/hr. *NOTE:* Factor of .3 for latent is *not* applicable for this residence.

22. The office portion of a multistory commercial building is shown in Fig. P6.22. Neglecting any outside air load, determine the cooling load at 4 P.M., August 21.

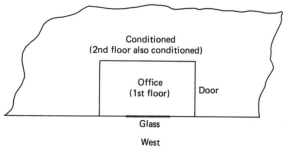

Fig. P6.22.

Inside design: 75°F, $W = 0.0102$

Outside design: 95°F db, 22° daily range, $W = .0168$, N. Lat. 40°

For office portion:

West wall: Net area = 230 ft²; $U = 0.333$
4 in. red face brick + 4 in. l.w. concrete block + $\frac{3}{4}$ in. plaster

Window: Area = 90 ft²; $U = 0.59$; double-pane, regular-sheet insulating glass

People: 4, moderately active office work

Lights: 800 W fluorescent, on continuously

23. For a commercial building located in Cleveland, OH (Fig. P6.23), determine the sensible, latent, and total cooling load existing at 4 PM on August 21 for an inside design temperature of 75°F.

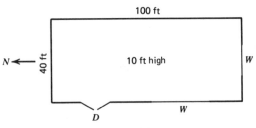

Fig. P6.23.

Wall: 1 in. dark stucco on 12 in. regular concrete; $U = 0.340$

Roof: 8 in. lightweight concrete; $U = 0.115$

Door: 2 in. solid wood; $U = 0.42$.

Windows
South: $\frac{1}{8}$ in. regular sheet glass with white roller shades drawn half-way

West: Insulating glass, regular sheet out/regular sheet in, no internal shading, $\frac{1}{4}$ in. air space.

People: 35 at moderate office activity.

Lights: 5600 W, fluorescent.

Infiltration: $\frac{1}{2}$ AC/hr

24. The basic plan for a residence in St. Louis, MO, is shown in Fig. P6.24. Construction data are as follows:

Wall: Face brick, $\frac{25}{32}$ in. insulating board sheathing, 2 × 4 studs on 16 in. centers, $\frac{3}{8}$ in. gypsum board interior.

Ceiling: 2 × 6 ceiling joists, 16 in. oc, no flooring above, $\frac{3}{8}$ in. gypsum board ceiling.

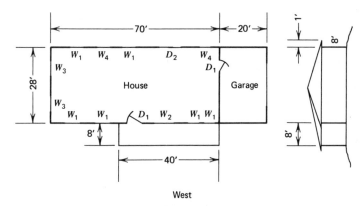

Fig. P6.24.

Roof: Asphalt shingles on solid wood sheathing, 2 × 6 rafters, no insulation between rafters, no ceiling applied to rafters, 1:4 pitch, 1-ft overhang on eaves, no overhang on gables.

Full basement: Heated, 10 in. concrete walls, all below grade; 4 in. concrete floor over 4 in. gravel.

One fireplace in living room on first floor.

Garage attached but unheated.

Windows:
   W_1 3 ft × 5 ft single glazed, double hung wood sash, weather stripped with storm window
   W_2 10 ft × 5½ ft picture window, double glazed, ½ in. air space
   W_3 5 ft × 3 ft wood sash casement, double glazed, ½ in. air space
   W_4 3 ft × 3 ft wood sash casement, double glazed, ½ in. air space

Doors:
   D_1 3 ft 0 in. × 6 ft 8 in., 1¾ in. solid with glass storm door
   D_2 Sliding patio glass door, two section, each 3 ft × 6 ft 8 in., double glazed, ½-in. air space, aluminum frame

Determine the heating and cooling loads for the residence with:
(a) Black asphalt shingles, full-wall fiberglass insulation, 6 in. fiberglass ceiling insulation, no drapes.
(b) Same as (a) except silver-white asphalt shingles.
(c) Same as (a) except lined drapes at all windows.

7

ENERGY USE DETERMINATION

7.1. ENERGY ESTIMATING

After the peak loads have been evaluated as described in previous chapters, the equipment must be selected with capacity sufficient to offset these loads. The air supplied to the space must be at the proper conditions to satisfy both the sensible and latent loads. However, peak load occurs only a few times each year, whereas partial load operation exists most of the time. Since operation is predominantly at partial load, partial-load analysis is at least as important as the selection procedure.

It is often necessary to estimate the energy requirements and fuel consumption of environmental control systems for either short or long terms of operation. These quantities can be much more difficult to calculate than design loads or required system capacity. The most direct and reliable energy-estimating procedure requires the hourly integrations of the various loads as a function of weather and use schedules and control systems. Several time-saving procedures for providing rough estimates of energy consumption are presented in this chapter.

The simplest calculation procedures use one measure to describe the annual weather conditions. The *calculated heat loss* method uses average winter temperature and seasonal efficiency. The average winter temperature for many cities is, however, close to or even above the balance point for most well-insulated residences. This, combined with the need to estimate seasonal efficiency, renders the calculated heat loss procedure invalid for most energy-estimating purposes and has therefore been omitted from this book.

7.2. THE DEGREE-DAY METHOD

The traditional degree-day procedure for estimated heating energy requirements is based on the assumptions that, on a long-term average, solar and
internal gains will offset heat loss when the mean daily outdoor temperature
is 65°F, and that fuel consumption will be proportional to the difference
between the mean daily temperature and 65°F. In other words, on a day
when the mean temperature is 20° below 65°F, twice as much fuel is consumed as on days when the mean temperature is 10° below 65°F.

Estimating the theoretical seasonal energy requirement of a conventional
heating system (gas furnace, oil furnace, electrical furnace, etc.) using the
degree-day method is relatively simple, because the efficiency of the system
is assumed to be constant regardless of outdoor temperature. The annual
theoretical heating requirement is calculated as shown below:

$$\frac{\text{Btu/hr loss} \times 24 \times \text{degree days}}{\text{Temperature difference}} = \text{Annual Btu}$$

Table 7.1 lists the average number of degree days that have occurred over
a long period of years, by months, and the yearly totals for various cities in
the United States and Canada. The number of degree days for U.S. cities
was calculated by taking the difference between 65°F and the daily mean
temperature computed as half the total of the daily maximum and the daily
minimum temperatures. The monthly averages were obtained by adding
daily degree days for each month each year, totaling the respective calendar
monthly averages for the number of years indicated, and dividing by the
number of years. The total or long-term yearly average degree-day value is
the summation of the 12 monthly averages.

Studies made by the American Gas Association up to 1932, and by the
National District Heating Association in 1932, indicate the 65°F base. However, residential insulation practices have improved from virtually none in
1930 to R-11 in walls and R-30 in ceilings today.

Internal gains have also increased dramatically. Edison Electric Institute
(EEI) reports show average residential electricity consumption of 675 kWh/
mo in 1973 versus 46 kWh in 1930. Use of a lower degree-day base than 65°F
would probably improve the procedure for annual calculations in northern
climates but substantially reduce accuracy in milder areas and seasons,
since much of the heating occurs on days with mean temperatures close to
65°F in those areas. Therefore, the use of the 65°F base has been retained,
with the addition of reduction factors to reflect actual lower heat losses.

Several organizations, including the National Bureau of Standards, the
Electric Power Research Institute, the Institute of Gas Technology, and
ASHRAE, are engaged in developing more valid simple procedures, but it is
probable that the degree-day procedure will continue to be used due to its
ease of application and suitability to quick estimates. It is important, however, to remember that the wide variations in occupant living habits and the

TABLE 7.1. Average Degree Days[a] for Cities in the United States and Canada

State and Station	Average Winter Temperature °F	°C	July	Aug.	Sept.	Oct.	Nov.	Dec.	Jan.	Feb.	Mar.	Apr.	May	June	Yearly Total
Alabama, Birmingham	54.2	12.7	0	0	6	93	363	555	592	462	363	108	9	0	2551
Alaska, Anchorage	23.0	5.0	245	291	516	930	1284	1572	1631	1316	1293	879	592	315	10864
Arizona, Tucson	58.1	14.8	0	0	0	25	231	406	471	344	242	75	6	0	1800
Arkansas, Little Rock	50.5	10.6	0	0	9	127	465	716	756	577	434	126	9	0	3219
California, San Francisco	53.4	12.2	81	78	60	143	306	462	508	395	363	279	214	126	3015
Colorado, Denver	37.6	3.44	6	9	117	428	819	1035	1132	938	887	558	288	66	6283
Connecticut, Bridgeport	39.9	4.72	0	0	66	307	615	986	1079	966	853	510	208	27	5617
Delaware, Wilmington	42.5	6.17	0	0	51	270	588	927	980	874	735	387	112	6	4930
District of Columbia, Washington	45.7	7.94	0	0	33	217	519	834	871	762	626	288	74	0	4224
Florida, Tallahassee	60.1	15.9	0	0	0	28	198	360	375	286	202	86	0	0	1485
Georgia, Atlanta	51.7	11.28	0	0	18	124	417	648	636	518	428	147	25	0	2961
Hawaii, Honolulu	74.2	23.8	0	0	0	0	0	0	0	0	0	0	0	0	0
Idaho, Boise	39.7	4.61	0	0	132	415	792	1017	1113	854	722	438	245	81	5809
Illinois, Chicago	35.8	2.44	0	12	117	381	807	1166	1265	1086	939	534	260	72	6639
Indiana, Indianapolis	39.6	4.56	0	0	90	316	723	1051	1113	949	809	432	177	39	5699
Iowa, Sioux City	43.0	1.10	0	9	108	369	867	1240	1435	1198	989	483	214	39	6951
Kansas, Wichita	44.2	7.11	0	0	33	229	618	905	1023	804	645	270	87	6	4620
Kentucky, Louisville	44.0	6.70	0	0	54	248	609	890	930	818	682	315	105	9	4660
Louisiana, Shreveport	56.2	13.8	0	0	0	47	297	477	552	426	304	81	0	0	2184
Maine, Caribou	24.4	-3.89	78	115	336	682	1044	1535	1690	1470	1308	858	468	183	9767
Maryland, Baltimore	43.7	6.83	0	0	48	264	585	905	936	820	679	327	90	0	4654
Massachusetts, Boston	40.0	4.40	0	9	60	316	603	983	1088	972	846	513	208	36	5634
Michigan, Lansing	34.8	1.89	6	22	138	431	813	1163	1262	1142	1011	579	273	69	6909
Minnesota, Minneapolis	28.3	-1.72	22	31	189	505	1014	1454	1631	1380	1166	621	288	81	8382
Mississippi, Jackson	55.7	13.5	0	0	0	65	315	502	546	414	310	87	0	0	2239
Missouri, Kansas City	43.9	6.94	0	0	39	220	612	905	1032	818	682	294	109	0	4711
Montana, Billings	34.5	1.72	6	15	186	487	897	1135	1296	1100	970	570	285	102	7049

Location	°F	°C													Total
Nebraska, Lincoln	38.8	4.11	0	6	75	301	726	1066	1237	1016	834	402	171	30	5864
Nevada, Las Vegas	53.5	12.28	0	0	0	78	387	617	688	487	335	111	6	0	2709
New Hampshire, Concord	33.0	0.60	6	50	177	505	822	1240	1358	1184	1032	636	298	75	7383
New Jersey, Atlantic City	43.2	6.56	0	0	39	251	549	880	936	848	741	420	133	15	4812
New Mexico, Albuquerque	45.0	7.20	0	0	12	229	642	868	930	703	595	288	81	0	4348
New York, Syracuse	35.2	2.11	6	28	132	415	744	1153	1271	1140	1004	570	248	45	6756
North Carolina, Charlotte	50.4	10.56	0	0	6	124	438	691	691	582	481	156	22	0	3191
North Dakota, Bismarck	26.6	-2.67	34	28	222	577	1083	1463	1708	1442	1203	645	329	117	8851
Ohio, Cleveland	37.2	3.22	9	25	105	384	738	1088	1159	1047	918	552	260	66	6351
Oklahoma, Stillwater	48.3	9.39	0	0	15	164	498	766	868	664	527	189	34	0	3725
Oregon, Pendleton	42.6	6.22	0	0	111	350	711	884	1017	773	617	396	205	63	5127
Pennsylvania, Pittsburgh	38.4	3.89	0	9	105	375	726	1063	1119	1002	874	480	195	39	5987
Rhode Island, Providence	38.8	4.11	0	16	96	372	660	1023	1110	988	868	534	236	51	5954
South Carolina, Charleston	56.4	13.9	0	0	0	59	282	471	487	389	291	54			2033
South Dakota, Rapid City	33.4	1.11	22	12	165	481	897	1172	1333	1145	1051	615	326	126	7345
Tennessee, Memphis	50.5	10.6	0	0	18	130	447	698	729	585	456	147	22	0	3232
Texas, Dallas	55.3	13.3	0	0	0	62	321	524	601	440	319	90	6	0	2363
Utah, Salt Lake City	38.4	3.89	0	0	81	419	849	1082	1172	910	763	459	233	84	6052
Vermont, Burlington	29.4	-1.11	28	65	207	539	891	1349	1513	1333	1187	714	353	90	8269
Virginia, Norfolk	49.2	9.89	0	0	0	136	408	698	738	655	533	216	37	0	3421
Washington, Spokane	36.5	2.83	9	25	168	493	879	1082	1231	980	834	531	288	135	6655
West Virginia, Charleston	44.8	7.44	0	0	63	254	591	865	880	770	648	300	96	9	4476
Wisconsin, Milwaukee	32.6	0.667	43	47	174	471	876	1252	1376	1193	1054	642	372	135	7635
Wyoming, Casper	33.4	1.11	6	16	192	524	942	1169	1290	1084	1020	657	381	129	7410
Alberta, Calgary	—	—	109	186	402	719	1110	1389	1575	1379	1268	798	477	291	9703
British Columbia, Vancouver	—	—	81	87	219	456	657	787	862	723	676	501	310	156	5515
Manitoba, Winnipeg	—	—	38	71	322	683	1251	1757	2008	1719	1465	813	405	147	10679
New Brunswick, Fredericton	—	—	78	68	234	592	915	1392	1541	1379	1172	753	406	141	8671
Nova Scotia, Halifax	—	—	58	51	180	457	710	1074	1213	1122	1030	742	487	237	7361
Ontario, Ottawa	—	—	25	81	222	567	936	1469	1624	1441	1231	708	341	90	8735
Quebec, Montreal	—	—	9	43	165	521	882	1392	1566	1381	1175	684	316	69	8203
Saskatchewan, Regina	—	—	78	93	360	741	1284	1711	1965	1687	1473	804	409	201	10806

SOURCE: Abridged by permission from *ASHRAE Handbook and Product Directory—Systems*, 1973.

[a]Based on °F, quantities may be converted to degree days based on °C by multiplying by 5/9.

assumptions inherent in this procedure may result in a variation of $\pm 20\%$ from actual fuel use on an annual basis.

7.3.   MODIFIED DEGREE-DAY PROCEDURE

The general equation for calculating the probable energy consumption by the modified degree-day method is

$$E = \left(\frac{H_L \times DD \times 24}{\Delta t \times \eta \times HV} \right) C_D \tag{7.1}$$

where E = fuel or energy consumption for the estimate period
  H_L = design heat loss, including infiltration, Btu/hr
  DD = number of 65°F degree days for the estimate period
  Δt = design temperature difference, °F
  η = rated full-load efficiency, decimal
  HV = heating value of fuel, consistent with H_L and E
  C_D = correction factor for heating effect vs degree days

Values of heating load, H_L, must be determined for the particular residence for which the estimate is being made. It must account for size, building materials, architectural features, use, and climatic conditions. Generally, the peak design heating load is used as determined in Chapter 6. Rated full-load efficiencies of fuel-fired equipment are usually in the range of 70–80% and may be obtained from the manufacturer. Figure 7.1 provides values for the correction factor C_D.

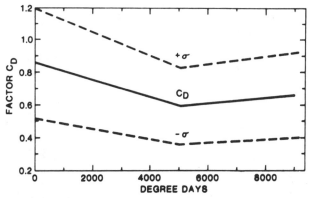

Fig. 7.1. Correction factor, C_D vs degree days. Reprinted by permission from *ASHRAE Systems Handbook 1980.*

EXAMPLE 7.1

Estimate the energy requirements to heat a residence in Atlanta, GA. The design heat loss is 35,000 Btu/hr based on 15°F outdoor and 70°F indoor.

Solution.   From Eq. 7.1,

$$E = \frac{35,000 \times 2961 \times 24}{55 \times 1.0 \times 3413} \times 0.79$$

$$= 10,468 \text{ kWh}$$

7.4.   EQUIVALENT FULL-LOAD-HOURS METHOD FOR COOLING

This procedure relies on using an estimate, based on local experience, of the ratio of annual cooling energy requirements to rated energy input of the cooling equipment. The operating cost of cooling equipment during a particular summer depends on variables such as the amounts of sunshine and rain, the number of abnormally hot or cool days, the efficiency of the equipment, and the local power rate. It is also influenced by human factors, such as operation of equipment only during the hottest weather, opening windows at night, and differences in preferred indoor temperatures. Nevertheless, it is important that lending agencies and prospective buyers of equipment be given a reasonably accurate estimate of the operating cost during normal summer weather and under usual operating conditions. Adjustments can then be made for any special conditions anticipated. The approximate electrical power inputs for the various motorized components in mechanical-cycle air conditioners are shown in Table 7.2.

TABLE 7.2.   Approximate Power Inputs

System	Compressor (kW/Design Ton)	Auxiliaries (kW/Design Ton)
Window units	1.46	0.32
Through-wall units	1.64	0.30
Dwelling unit, central air-cooled	1.49	0.14
Central, group, or bldg. cooling plants		
3–25 tons air-cooled	1.20	0.20
25–100 tons air-cooled	1.18	0.21
25–100 tons water-cooled	0.94	0.17
Over 100 tons water-cooled	0.79	0.20

SOURCE: Reprinted by permission from *ASHRAE Handbook and Product Directory—Systems*, 1973.

Energy cost per hour can be estimated by multiplying the estimated power per ton, the cooling capacity in tons, and the cost per kilowatt-hour. Thus, the estimated cost per hour for a central 3-ton air conditioner with an air-cooled condenser will be 1.63 kW × 3 tons × the electric rate. It is essential to use the correct step of the utility residential rate structure to get a good estimate. The basis of the method is the use of a table of the estimated annual hours of operation for properly sized equipment in typical cities (see Table 7.3). The values in Table 7.3 are estimated ranges based on a survey of electric utility companies and are based on an indoor temperature of 75°F. In general, residential units will be toward the lower end of the range, and light commercial toward the higher.

It must be impressed on the buyer that the energy consumption arrived at by this method is purely an estimate. It does not take into consideration any abnormally hot weather or a buyer's preference for lower indoor temperatures.

Another method has been developed using cooling degree days above 65°F as a criterion. Tabulated values of cooling degree days for various localities are available from the National Climatic Center, Asheville, North Carolina. Approximate values can be obtained from degree-day maps. The daily range of temperature for various localities is also required with this method. It is important that the method used in estimating operating cost be consistent with the method used in calculating the heat gain of the structure.

TABLE 7.3.   Estimated Equivalent Rated Full-Load Hours of Operation for Properly Sized Equipment during Normal Cooling Season

Albuquerque, NM	800–2200	Indianapolis, IN	600–1600
Atlantic City, NJ	500–800	Little Rock, AR	1400–2400
Birmingham, AL	1200–2200	Minneapolis, MN	400–800
Boston, MA	400–1200	New Orleans, LA	1400–2800
Burlington, VT	200–600	New York, NY	500–1000
Charlotte, NC	700–1100	Newark, NJ	400–900
Chicago, IL	500–1000	Oklahoma City, OK	1100–2000
Cleveland, OH	400–800	Pittsburgh, PA	900–1200
Cincinnati, OH	1000–1500	Rapid City, SD	800–1000
Columbia, SC	1200–1400	St. Joseph, MO	1000–1600
Corpus Christi, TX	2000–2500	St. Petersburg, FL	1500–2700
Dallas, TX	1200–1600	San Diego, CA	800–1700
Denver, CO	400–800	Savannah, GA	1200–1400
Des Moines, IA	600–1000	Seattle, WA	400–1200
Detroit, MI	700–1000	Syracuse, NY	200–1000
Duluth, MN	300–500	Trenton, NJ	800–1000
El Paso, TX	1000–1400	Tulsa, OK	1500–2200
Honolulu, HI	1500–3500	Washington, DC	700–1200

SOURCE: Reprinted by permission from *ASHRAE Handbook and Product Directory—Systems*, 1973.

EXAMPLE 7.2

The total design heating load on a residence in Kansas City, MO, is 112,000 Btu/hr. The design cooling load is 48,000 Btu/hr. The furnace is off during June through September. Estimate:

a.   Annual energy required for heating, Btu/yr.
b.   Annual heating cost if fuel oil with a heating value of 140,000 Btu/gal is used with a furnace having an efficiency of 75%. Fuel oil costs $1.09/gal.
c.   Maximum savings possible if the thermostat is set back to 65°F from 75°F between 10 P.M. and 6 A.M.
d.   Annual energy required for cooling, kWh/yr.

Solution

a.   Kansas City, MO:

$$t_o = 6°F \ (97\tfrac{1}{2}\% \ \text{value})$$
$$t_i = 75°F \ (\text{assumed})$$
$$DD = 4711 - 39 = 4672 \ (\text{Table 7.1})$$
$$C_D = 0.62 \ (\text{Fig. 7.1})$$

$$E = \frac{H_L}{(t_i - t_o)_{\text{design}}} \times DD \times 24 \times C_D$$

$$= \frac{112000}{(75 - 6)} \times 4672 \times 24 \times 0.62$$

$$= 112,840,000 \ \text{Btu/yr}$$

b. $$\text{Fuel} = \frac{E}{HV \times \eta} = \frac{112,840,000}{140,000 \times 0.75} = 1075 \ \text{gal}$$

$$\text{Cost} = 1075 \ \text{gal} \times \$1.09 = \$1172$$

c. $t_{o,av}$: $4672 = (365 - 122)(65 - t_{o,av}); \quad t_{o,av} = 45.8°F$

$t_{i,av}$: $t_{i,av} = \dfrac{8}{24}(65) + \dfrac{16}{24}(75) = 71.7°F$

$$\% \ \text{savings} = \frac{(75 - 45.8) - (71.7 - 45.8)}{75 - 45.8} \times 100 = 11.3\%$$

$$\text{Cost savings} = 0.113 \ (\$1172) = \$132.41$$

d.   Equivalent full-load hours ≈ 1000 hr (St. Joseph, Table 7.3)

Power input $\approx (1.49 + 0.14)$ kW/ton

$$= 1.63 \times \frac{48,000}{12,000} = 6.52 \ \text{kW}$$

Cooling energy $= 1000 \ \text{hr} \times 6.52 \ \text{kW} = 6520 \ \text{kWh}$

Even when used with applications as apparently simple as residential cooling, single-measure estimates can be grossly inaccurate if they do not consider all major factors. For example, the growth of the heat pump, the performance of which is extremely dependent on outdoor air temperature, resulted in an awareness of some of the failings of the degree-day procedure in residences.

7.5. BIN-METHOD PROCEDURE

The bin method consists of performing an instantaneous energy calculation at many different outdoor dry-bulb temperature conditions, and multiplying the result by the number of hours of occurrence of each calculation. The "bins" are usually 5°F in size and further divided into three daily 8-hr shifts. For example, a calculation is performed for 42°F outdoor (representing all occurrences of 40–44°F) and with building operation during the midnight-to-8 A.M. shift. Since there are 23 bins between − 10 and 105°F, and three shifts, 69 separate operating points would be calculated. For many applications, the number of calculations can be reduced. A residential heat pump (heating portion), for example, could be calculated for just the bins below 65°F and without the "three-shift" breakdown. The bin method may be used with or without refinements such as coincident wet-bulb conditions, depending on the anticipated impact of the additional parameters. Weather data for use with the bin method is available in *Air Force Manual* 88-8 (Chapter 6, Engineering Weather Data).* Where detailed bin data are not required, the information in Table 7.4 may be used.

The bin method of energy estimating has the advantage of allowing off-design calculations. This permits the user to accurately predict effects, such as reheat and "free cooling" that can only be guessed at with less sophisticated procedures.

7.6. HEAT PUMP RATINGS AND OPERATING CHARACTERISTICS

Heat pumps are rated, as are single-purpose air conditioners, in terms of tons of cooling. However, this nominal rating also indicates the heating capacity of the unit, bearing in mind that two temperatures are used to certify the rating. One manufacturer, for instance, lists the ARI-certified rating of a typical 2.5-ton heat pump as: 30,500 Btu/hr cooling capacity at 95° outdoor temperature and 31,000 Btu/hr heating capacity at 47° outside air

---

*Copies may be obtained from the Superintendent of Documents, Government Printing Office, Washington, DC, 26402.

temperature. At higher or lower ambient temperatures, the heating and cooling outputs diverge. Usually, the heating output at various ambients is given in 5° increments.

Most heat pumps (packaged and split) are capacity tested, rated, and certified against ARI Standard 240—on cooling, 95°F db outside air; 80°F db and 67°F wb indoor air, and on heating, 47°F db and 43°F wb outdoor air and 70°F db indoor. Heating capacity is for the compressor operation only and does not include any supplementary heating.

With a 115°F condenser, the temperature rise over the indoor coil will be considerably less than from a conventional furnace. The delivered air temperature to the room will be about 105° when it is 30° outside, or only about a 35° rise over the coil.

On the other hand, a gas or oil furnace will have an 80° rise and a delivered air temperature of 150°F. The air will feel even cooler, due to the increased volume of air, although the velocity will be about the same. This modest temperature rise will not affect the system's ability to maintain the temperature desired by the homeowner, but it will mean that the system will run for a longer period of time.

Air Conditioning and Refrigeration Institute (ARI), the industry standards association, has established that, for air-conditioning, airflow should be 400 cfm/ton over the evaporator and 700 cfm/ton over the condenser.

However, since heat pump coils must function as either condenser or evaporator, the airflow over each must be the same. This has been set at 450 cfm/ton. This is a greater volume of air over the inside coil than with a straight air-conditioning system. So, on a conversion replacement or add-on to an existing furnace, the duct system may be undersized.

Since a heat pump not only heats a home during the winter months but also cools it in the summer, how should it be sized? A heat-pump system is normally sized to handle the cooling requirements of the home (referred to in terms of Btu per hour). The selection of a heat pump for a given structure is made on the basis of its cooling capacity and matched to the calculated heat gain at a 95°F design temperature. It should not be oversized by more than 0.5 ton, or 6000 Btu/hr, even though additional capacity may be needed in the heating mode. As with air-conditioning, control of latent heat is lost with an oversized unit, because of shorter running time, and undersizing will not handle the load.

Additional heating requirements are supplied by supplemental electrical resistance heaters, in 5–10-kW increments. The number of electric heaters needed is determined by the calculated heat loss of the structure.

The *balance point* is the lowest outdoor temperature at which the heat pump can handle the heating load alone, and will vary with each installation. At temperatures below this point, supplementary heat is supplied by electric resistance heaters or, in the case of an add-on heat pump, by the furnace. Figure 7.2 illustrates the balance point.

TABLE 7.4. Hourly Weather Occurrences

Location	Outdoor Temperature, °F																		
	72	67	62	57	52	47	42	37	32	27	22	17	12	7	2	-3	-8	-13	-18
Albany, NY	588	733	740	708	652	625	647	769	793	574	404	278	184	110	63	32	10	5	4
Albuquerque, NM	767	831	719	651	687	734	741	689	552	346	154	66	21	4	1	1			
Atlanta, GA	1185	926	823	784	735	676	598	468	271	112	44	19	8	2					
Bismarck, ND	454	566	614	606	563	520	518	604	653	550	474	371	338	292	278	208	131	77	80
Burlington, VT	573	670	703	694	655	603	637	716	752	561	491	336	272	216	135	81	39	17	8
Casper, WY	423	532	592	642	606	670	782	831	806	683	495	324	200	116	73	45	30	15	5
Charleston, SC	1267	1090	889	787	651	576	434	321	192	79	27	5							
Chicago, IL	762	769	653	592	569	543	591	800	822	551	335	196	117	85	59	25	12	3	
Cincinnati, OH	879	843	726	639	611	599	627	698	711	460	249	131	68	44	18	8	2		
Cleveland, OH	763	831	732	641	638	607	620	754	806	578	355	201	111	47	22	11	2		
Dallas, TX	831	795	693	656	629	576	504	371	231	91	34	17	4	1					
Denver, CO	549	684	783	731	678	704	692	717	721	553	359	216	119	78	36	22	6	1	1
Des Moines, IA	707	751	681	600	585	512	510	627	747	557	405	281	211	152	104	59	23	8	1
Detroit, MI	721	783	695	633	592	566	595	808	884	618	377	248	131	61	17	4	1		
Houston, TX	1172	980	772	681	570	452	291	141	64	18	4	2							
Indianapolis, IN	821	815	722	585	586	579	605	712	791	551	293	152	97	60	35	13	3	2	
Jackson, MS	1169	922	790	677	618	605	484	367	224	103	41	6	2	2	1				
Jacksonville, FL	1334	975	879	692	530	355	288	154	83	24	2								
Kansas City, MO	761	723	601	572	553	562	628	625	591	407	265	175	99	51	21	4			
Las Vegas, NV	651	644	699	786	769	716	591	396	194	44	7	1	5						
Little Rock, AR	940	803	725	672	638	669	605	509	363	172	50	25	5	1					

City																		
Los Angeles, CA	881	1654	2193	1904	1054	428	107	10	631	332	169	97	45	25	8	3	1	
Louisville, KY	869	758	693	654	619	634	649	703	374	196	74	25	10	4				
Memphis, TN	977	798	715	690	618	633	614	532										
Miami, FL	1705	810	452	277	147	71	26	4										
Milwaukee, WI	597	753	749	634	585	591	611	774	913	659	421	285	176	116	83	47	18	4	3
Minneapolis, MN	621	690	695	602	588	482	500	560	632	609	514	383	311	246	186	119	62	31	16
Nashville, TN	933	838	738	697	637	619	627	565	463	263	132	67	28	9	3	1	1	
New Orleans, LA	1189	987	850	692	621	449	282	128	47	9	2							
New York, NY	926	877	754	745	722	796	838	858	603	330	188	2	26	10	1			
Oklahoma City, OK	881	769	717	643	645	611	641	570	468	287	173	77	36	12	3	1		
Omaha, NB	726	721	606	558	539	543	543	655	663	511	390	287	189	135	93	40	15	
Philadelphia, PA	863	809	735	710	663	701	758	818	654	335	189	100	32	9				
Phoenix, AZ	762	776	767	769	659	540	391	182	57	8								
Pittsburgh, PA	722	910	799	678	637	587	631	688	774	569	360	233	159	60	30	7	1	
Portland, OR	373	581	1001	1316	1274	1271	1238	772	343	123	40	10	4	1	1			
Reno, NV	418	477	572	690	845	909	890	829	733	530	387	227	101	37	4			
Salt Lake City, UT	569	615	614	635	682	685	755	831	798	564	328	158	80	41	15	4		
San Antonio, TX	1086	943	789	669	569	445	387	190	94	31	11	4	1	1	16	2		
San Francisco, CA	285	665	1264	2341	2341	1153	449	99	10									
Seattle, WA	258	448	750	1272	1462	1445	1408	914	427	104	39	20	3					
St. Louis, MO	823	728	646	575	585	578	620	671	650	411	219	134	77	40	15	7	1	
Syracuse, NY	627	735	723	717	656	641	651	720	830	547	392	282	190	102	55	23	5	
Tampa, FL	1387	1187	877	570	345	216	137	48	10	1								
Washington, DC	960	766	740	673	690	684	790	744	542	254	138	54	17	2	2			
Wichita, KS	758	709	641	603	589	592	611	584	607	426	273	161	85	45	14	3	1	2

SOURCE: Abridged by permission from *ASHRAE Handbook and Product Directory—Systems*, 1973.

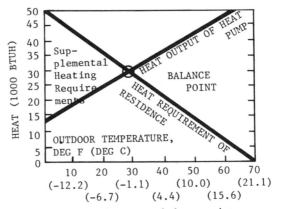

Fig. 7.2. Heat pump balance point.

Figure 7.3 shows the performance characteristics of a typical heat pump. Also shown are the heating and cooling loads for a typical building. If the balance point is above the heating design temperature t_d, supplemental heat will be required as denoted by the shaded area. Capacity, power input, and COP characteristics can vary appreciably with equipment size and design.

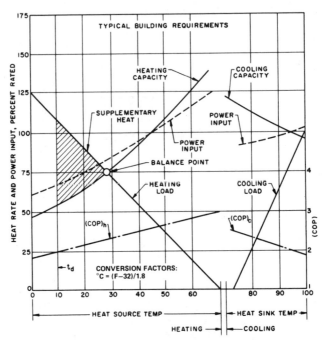

Fig. 7.3. Operating characteristics of single-stage unmodulated heat pump—air source. Reprinted by permission from *ASHRAE 1975 Equipment Handbook*.

7.7. HEAT PUMP PERFORMANCE PARAMETERS

Heat pump performance for heating at a given set of operating conditions is usually described by a coefficient of performance defined by the equation

$$\text{COP} \equiv \frac{\text{output, Btu/hr}}{\text{input, Btu/hr}} \tag{7.2}$$

$$\equiv \frac{\text{output, kW}}{\text{input, kW}}$$

However, heat pump performance at a given set of operating conditions may be described by an energy efficiency ratio defined by the equation

$$\text{EER} \equiv \frac{\text{heat output, Btu/hr}}{\text{electrical input, W}} \tag{7.3}$$

The COP and EER are useful in comparing heat pump sizes and manufacturers, but they do not provide the proper data required to compute seasonal energy savings. Table 7.5 presents a listing of the various heat pump performance parameters and their definitions.

Present steady-state ratings are not exactly adaptable to seasonable performance factors (SPF) or annual operating costs. There are too many variables in operating conditions, particularly in the heating mode (changing temperature and humidity, wind chill, frequency and duration of defrost cycles, etc.).

Another consideration is that no heat pump manufacturer presently publishes ratings for heating-cycle performance at any relative humidities but ARI conditions. The variance in heating capacity with a change in relative humidity is therefore virtually unknown.

Wind chill effect can be quite severe, not only with respect to defrosting time, but also to the loss of heating capacity, by dropping the coil temperature far below the design level for a given ambient.

Figure 7.4 shows the probable number of defrost cycles per 24 hours of compressor operation at various outdoor air relative humidities versus outdoor coil temperatures.

The introduction of greatly oversized outdoor coils as a means of improving the EER on the cooling cycle has made the problem of defrost and its effect on efficiency even more severe. These large-face-area, thin coils are fine condensers but poor low-temperature evaporators, and their increased mass requires considerably more time to bring them up to defrosting temperature with a given amount of available defrost heat, particularly if there is any wind velocity.

These difficulties have caused some heat pump producers to go to full-time cycle control systems, with 10 min termination limit. And because the defrost cycles are now definitely of long enough duration to chill the oc-

TABLE 7.5. Heat Pump Performance Parameters

Annual Performance Factor (APF)   The total heating and cooling done by a heat pump in a particular region in one year divided by the total electric power used in one year.

Coefficient of Performance, Heating (COP_H)   Ratio of the rate of heat delivery to the conditioned space to the rate of energy input, in consistent units, for a complete operating heat pump plant or some specific portion of that plant, under designated operating conditions.

Coefficient of Performance, Cooling (COP_C)   Ratio of the rate of heat removal from the conditioned space to the rate of energy input, in consistent units, for a complete heat pump or refrigerating plant or some specific portion of that plant, under designated operating conditions.

Cooling Load Factor (CLF)   Ratio of the total cooling of a complete cycle for a specific period consisting of an "on" time and "off" time to the steady-state cooling done over the same period at constant ambient conditions.

Degradation Coefficient (C_D)   Measure of the efficiency loss due to the cycling of the unit. The measure of the reduction in performance under cyclic operation.

Energy Efficiency Ratio, Heating (EER_H)   Ratio of the rate of heat delivery to the conditioned space, in Btu/hr, to the rate of energy input, in watts, for a complete operating heat pump plant or some specific portion of that plant, under designated operating conditions. (Same as COP_H except for units.)

Energy Efficiency Ratio, Cooling (EER_C)   Ratio of the rate of heat removal from the conditioned space, in Btu/hr, to the rate of energy input, in watts, for a complete heat pump or refrigerating plant or some specific portion of that plant, under designated operating conditions. (Same as COP_C except for units.)

Heating Load Factor (HLF)   Ratio of the total heating of a complete cycle for a specified period consisting of an "on" time and "off" time to the steady state heating done over the same period at constant ambient conditions which correspond to the average indoor and outdoor conditions which existed in the time period. The numerator includes the output from supplemental resistance heaters when such heaters operated during a defrost period to offset the cooling effect of the indoor coil during this period.

Heating Seasonal Performance Factor, Heating (HSPF)   Ratio of the total heat delivered over the heating season to the total energy input over the heating season, in consistent units.

Part Load Factor (PLF)   Ratio of the cyclic coefficient of performance to the steady state coefficient of performance.

Seasonal Performance Factor, Heating (SPF_H)   Ratio of the total heat delivered over the heating season to the total energy input over the heating season, in consistent units.

Seasonal Performance Factor, Cooling (SPF_C)   Ratio of the total heat removed over the cooling season to the total energy input over the cooling season, in consistent units.

Seasonal Performance Factor, Heating, Total System ($SPF_{H,S}$)   Total energy delivered over the heating season divided by total energy delivered to the heat pump system (compressor energy, indoor and outdoor fan energies, supplementary heater energy), in consistent units.

410

TABLE 7.5.   (Continued)

Seasonal Performance Factor, Cooling, Total System (SPF$_{C,S}$)   Total energy removed over the cooling season divided by total energy delivered to the complete heat pump or refrigerating plant (compressor energy, indoor and outdoor fan energies), in consistent units.

Seasonal Energy Efficiency Ratio, Heating (SEER$_H$)   Ratio of the total heat delivered over the heating season, in Btu, to the total energy input over the heating season, in watt-hours. (Same as SPF$_H$ except for units).

Seasonal Energy Efficiency Ratio, Cooling (SEER$_C$)   Ratio of the total heat removed, in Btu, during the normal usage period for cooling (not to exceed 12 months) to the total energy input, in watt-hours, during the same period. (Same as SPF$_C$ except for units).

cupied area, the initiation of defrost also energizes the first stage of supplementary heat (5 or 10 kW).

The projection of supplementary electric heat into the total watts picture may virtually wipe out any COP advantage over straight resistance heating in severe weather and neutralize the basic operating efficiency of the heat pump.

For ambient conditions under which frosting will occur, an average COP

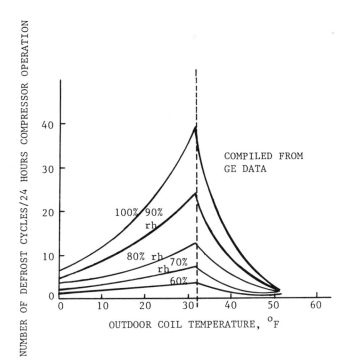

Fig. 7.4.   Defrost cycles [1].

TABLE 7.6.   Average COP over Frosting-Defrosting Cycle [3]

Outdoor Air Temperature (°C)	Frosting Interval (min)	Defrost Interval (min)	COP No Frost	Average COP Frost-Defrost Cycle	Average COP with Supplementary Electric Heat
2.5	30	3.7	2.05	1.73	1.60
2.5	90	7.3	2.05	1.69	1.48
−3	30	3.7	2.00	1.56	1.42
−3	60	5.1	2.00	1.47	1.33

over the frosting-defrosting cycle is a significant measure of system efficiency. A proper method of evaluating COP for the frosting-defrosting cycle would seem to depend upon whether the system was operating above or below the balance point. If the system is operating at ambient temperatures below the balance point, the loss in capacity during frosting must be replaced by resistance heat. Therefore, constant heating output with supplementary resistance heat appears to be a fair basis for determining average COP under these conditions.

For operation above the balance point, no additional resistance heat would be needed; therefore, an average COP without supplemental heat seems an appropriate measure for this case. Calculations were performed to determine an average effective COP under both sets of conditions. Results are shown in Table 7.6.

Average values are computed assuming that compressor power input during defrost is lost from the system. Average COP values were computed using the following expressions. For the case where the system is penalized for the use of supplementary resistance heat:

$$\text{COP} = \frac{(\dot{q}_{\text{heat pump}} + \dot{E}_{\text{res. heat}})_{\text{av } f}\Delta T_f + (\dot{E}_{\text{res. heat}})_{\text{av } d}\Delta T_d}{(\dot{E}_{\text{heat pump}} + \dot{E}_{\text{res. heat}})_{\text{av } f}\Delta T_f + (\dot{E}_{\text{res. heat}})_{\text{av } d}\Delta T_d} \qquad (7.4)$$

where ΔT_f = time duration of frosting interval
ΔT_d = time duration of defrost period
av f = average over frosting interval
av d = average during defrost period

For the case in which resistance heat is not used,

$$\text{COP} = \frac{(\dot{q}_{\text{heat pump}})_{\text{av } f}\Delta T_f}{(\dot{E}_{\text{heat pump}})_{\text{av } f}\Delta T_f + (\dot{E}_{\text{heat pump}})_{\text{av } d}\Delta T_d} \qquad (7.5)$$

In Eqs. (7.4) and (7.5), $\dot{q}$ = heating capacity and $\dot{E}$ = energy input.

7.8. SEASONAL PEFORMANCE FACTOR

A seasonal performance factor for heating (SPF$_H$) may be defined as follows:

$$SPF_H \equiv \frac{\text{total heat delivered over heating season}}{\text{total energy consumed over heating season}} \qquad (7.6)$$

$$SPF_H \equiv \frac{HPE + AE}{3413\,[HPP + AP]} \qquad (7.7)$$

where HPE = energy supplied by the heat pump during the heating season, Btu

   AE = energy supplied by auxiliary electric heat during the heating season, Btu

  HPP = energy required by the heat pump during the heating season, kWh

   AP = energy required by auxiliary electric heat during the heating season, kWh

  1 kW = 3413 Btu/hr

Seasonal performance factors make it possible to estimate seasonal energy saving assuming that the space would normally have been electrically heated. The estimate is made by dividing the energy required to heat the space using electrical resistance heating only during the heating season by the heat pump SPF. The result is the energy required to heat the same space using a heat pump system. A comparison of the two heating requirements will yield the seasonal energy savings. For example, a heat pump with an SPF of 2 would yield a 50% energy savings over electrical resistance heating. Table 7.7 compares the heating SPFs of various systems for various geographical locations. A seasonal performance factor for cooling (SPF$_C$) may be defined as

$$SPF_C \equiv \frac{HR}{HPC}$$

where HR = energy removed by the heat pump during the cooling season, Btu

  HPC = energy required by the heat pump during the cooling season

The SPF$_C$ for a heat pump system should be approximately the same as for a regular air-conditioning unit in that both operate on the same principle. However, the heat pump has the advantage that it can provide both heating and cooling.

TABLE 7.7. Heating Season Performance Factors for Heat Pumps and Other Heating Systems [2]

City	Heat Pump		Gas Furnace	Oil Furnace	Electric Furnace	Baseboard Electric
	System[a]	Compressor and Fan[b]				
Houston, TX	2.64	2.66	0.56	0.53	0.92	1.0
Birmingham, AL	2.14	2.40	0.56	0.52	0.92	1.0
Atlanta, GA	2.14	2.37	0.57	0.53	0.93	1.0
Tulsa, OK	1.83	2.39	0.59	0.55	0.93	1.0
Philadelphia, PA	1.69	2.13	0.60	0.56	0.91	1.0
Seattle, WA	1.80	1.97	0.53	0.50	0.89	1.0
Columbus, OH	1.69	2.40	0.63	0.59	0.91	1.0
Cleveland, OH	1.60	2.51	0.64	0.59	0.91	1.0
Concord, MA	1.47	2.73	0.61	0.57	0.91	1.0

[a] Heat pump system SPF: Total energy delivered to the residence divided by total energy delivered to the heat pump system (compressor energy, indoor and outdoor fan energies, supplementary heater energy). Energy delivered to the residence is just equal to thermal load in the residence.

[b] Compressor and fan SPF: Same as above except that energy delivered to the pump system is for the compressor and fans and does not include supplementary heaters.

7.9. EFFECT OF COMPONENT INEFFICIENCIES [3]

Figure 7.5 shows the manner in which the observed component inefficiencies affect the system efficiency or COP for the nonfrosting conditions of 21°C indoor and 8°C outdoor air temperatures. An ideal vapor-compression cycle using R-22 as the working fluid and operating between the temperature limits of 21 and 8°C has a COP of 21.45. Of course, a real system cannot operate without a temperature difference across the heat exchangers.

For an ideal R-22 system using the observed heat-exchanger temperatures (53°C condensing, −3°C evaporating, 27°C subcooling, and 11°C superheat), the computed COP is 5.41. Thus a 75% loss in efficiency from the ideal, reversible cycle results from the temperature differences across the heat exchangers. A compressor efficiency of 0.42 reduces the COP to 2.85, an additional 47% loss.

Heat loss from the compressor shell as measured in these tests reduces the computed COP 14% to 2.45. The ideal fan power would reduce the COP 3% to 2.38, and the observed fan and fan-motor inefficiencies further reduce the computed COP an additional 12% to 2.09, which is about equal to the observed COP. An incorrect charge of refrigerant can substantially reduce the performance even further, as shown in Fig. 7.6.

The high sensitivity of this unit to the amount of refrigerant charge is

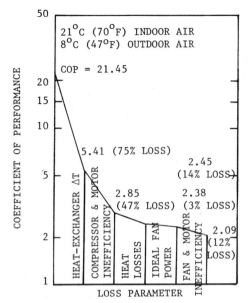

Fig. 7.5. Effect of component inefficiencies on system efficiency under steady-state, nonfrosting conditions.

probably due to the use of a capillary-tube type of refrigerant-metering device and the absence of a suction-line accumulator in the refrigerant system. During the overcharge test, the sudden decrease in the mass flow rate of refrigerant may have been caused by the refrigerant being accumulated in the indoor coil due to the capillary tube's incapacity to pass enough refrigerant to the low side of the unit.

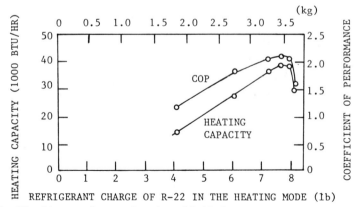

Fig. 7.6. Heating capacity as a function of R-22 charge; outdoor air in, 10°C, indoor air in, 21°C.

7.10. THERMOSTAT SETBACK WITH HEAT PUMPS

In a residential heating system, the further a thermostat setting is reduced, the greater the fuel savings. The longer the time of this setback, the greater the savings. And in dual setback (both night and day), fuel savings are usually approximately twice that of night-only setback. There is, however, an exception. While gas, oil, and electric furnaces have fixed capacities, heat pumps do not. Capacity varies with outside temperature, and this presents a problem.

Setback with a heat pump is a good idea during periods when temperatures are significantly below the system's balance point, since auxiliary resistance heating is required anyway. Similar economic advantages are inherent when temperatures remain sufficiently above the balance point. But the key to energy savings is to limit the use of auxiliary resistance heating. According to a study by Bullock [4], when the thermostat setback precipitates the otherwise unnecessary or increased use of resistance heat for morning pickup, fuel cost is actually increased.

EXAMPLE 7.3

Estimate the energy requirements assuming a 3-ton heat pump with characteristics given below to heat a residence in Washington, DC. The design heat loss is 65,000 Btu/hr based on 10°F outdoors and 70°F indoors.

	Outdoor Coil Entering Air Temp, (°F)							
	10	20	30	40	45	50	60	70
Heating Capacity, Btu/hr	19,000	24,500	29,500	34,000	36,000	38,000	40,800	42,200
Unit Input, kW	3.17	3.45	3.73	3.99	4.11	4.23	4.47	4.67
COP	1.75	2.07	2.32	2.51	2.58	2.63	2.68	2.62

Solution

The heat loss per degree is first adjusted by the use of C_D.

$$\frac{H_L}{\Delta t}(C_D) = \frac{65,000}{60}(0.79) = 856 \text{ Btu/(hr} \cdot \text{°F)}$$

| | | | | Heat Pump Alone | | | | | Supplementary Heat | | |
Outdoor Temperature (5° Increments)	Btuh Loss per 1°F (Heat Loss/TD)	Outdoor Temperature Below 65°F (65 − Column A)	Heat Loss (Btuh) B × C	Heat Pump Heating Capacity (Btuh) Mfr. Data	Heat Pump Running Time (%) D/E	Heat Pump Input (kW) Mfr. Data	Seasonal Heating Hours (Table 7.4)	Seasonal Heat Pump Input (kWh) F × G × H	Resistance Heat Input (Btuh) D − E	Resistance Heat Input (kW) J/3413	Seasonal Resistance Heat Input (kWh) H × K
A	B	C	D	E	F	G	H	I	J	K	L
62	856	3	2568	41 000	0.06	4.51	740	200	0	0	0
57	↑	18	6848	40000	0.17	4.40	673	503	0	0	0
52		13	11128	38 500	0.29	4.28	690	856	0	0	0
47		18	15428	36 500	0.42	4.14	684	1189	0	0	0
42		23	19688	35 000	0.56	4.04	790	1789	0	0	0
37		28	23928	32500	0.74	3.90	744	2147	0	0	0
32		33	28248	30500	0.93	3.78	542	1905	0	0	0
27		38	32528	28000	1.0	3.65	254	927	4528	1.33	338
22		43	36808	25000	1.0	3.51	138	484	11828	3.46	477
17		48	41,098	23500	1.0	3.37	54	182	17588	5.15	278
12		53	45368	OFF	0	0	17	—	45368	13.29	226
7		58	49678	OFF	0	0	2	—	49648	14.55	29
2		63									
− 3		68									
− 8		73									
− 13		78									
− 18 & below	↓	83									
								10182	← Totals →		1348

Fig. E7.3.

Form I, shown in Fig. E7.3, is then completed, resulting in a total of 11,530 kWh.

EXAMPLE 7.4

The four-bedroom residence described below, located between Rolla and St. Louis, MO, is to be equipped with a heat pump whose performance is given in the following table:

	Outdoor Coil Entering Air Temp, °F							
	10	20	30	40	45	50	60	70
Heating Capacity, Btu/hr	19,000	24,500	29,500	34,000	36,000	38,000	40,800	42,200
Unit Input, kW	3.17	3.45	3.73	3.99	4.11	4.23	4.47	4.67
COP	1.75	2.07	2.32	2.51	2.58	2.63	2.68	2.62

Design conditions are 75°F db indoor with humidification to 30% in winter; winter outdoor temperature of 7°F db; and summer outdoor temperatures of 95°F db and 78°F wb. For this residence, determine:

a. Heating and cooling loads

b. Annual heating energy requirements, kWh, estimated with regular "bin" method

c. Annual heating energy requirements, kWh, estimated with "bin" method but including cycling losses for this heat pump with a capillary tube with part-load degradation of performance determined as a cyclic modifier expressed by

$$C_M = 0.4761 \times RT^{0.161}$$

The basic plan of the residence is given in Fig. E7.4A.

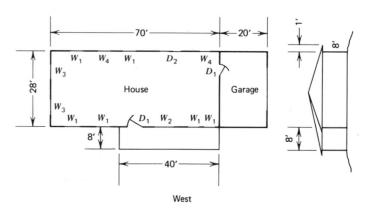

West

Fig. E7.4A.

Wall construction: Face brick, $\frac{25}{32}$ in. insulating board sheathing, 2 × 4 studs on 16 in. centers, $\frac{3}{8}$ in. gypsum board interior. Full-wall fiber-glass insulation.

Ceiling: 2 × 6 ceiling joists, 16 in. oc, no flooring above, $\frac{3}{8}$ in. gypsum board ceiling. 4 in. insulation.

Roof: Asphalt shingles on solid wood sheathing, 2 × 6 rafters, no insulation between rafters, no ceiling applied to rafters, 1:4 pitch, 1 ft overhang on eaves, no overhang on gables. Dark color.

Full basement: Heated, 10 in. concrete walls, all below grade; 4 in. concrete floor over 4 in. gravel. One fireplace in living room on first floor.

Garage: Attached but unheated.

Windows:

W_1 3 ft × 5 ft single glazed, double hung wood sash, weather stripped with storm window

W_2 10 ft × $5\frac{1}{2}$ ft picture window, double glazed, $\frac{1}{2}$ in. air space

W_3 5 ft × 3 ft wood sash casement, double glazed, $\frac{1}{2}$ in. air space

W_4 3 ft × 3 ft wood sash casement, double glazed, $\frac{1}{2}$ in. air space
    No drapes.

Doors:

D_1 3 ft 0 in. × 6 ft 8 in., $1\frac{3}{4}$ in. solid with glass storm door

D_2 Sliding patio glass door, two section, each 3 ft × 6 ft 8 in., double glazed, $\frac{1}{2}$ in. air space, aluminum frame

Solution

a.   See Figure E7.4B

b.   See Figure E7.4C

c.   See Figure E7.4D

7.11. SEASONAL PERFORMANCE OF SEVERAL HEAT PUMPS

Seasonal performances for typical air-source and water-source heat pumps have been evaluated and presented by Briggs and Shaffer [5], Commonwealth Edison [6], National Bureau of Standards [7], and Westinghouse/EPRI [8]. Results of these studies are summarized in the following sections.

7.11.1.   Air-Source Heat Pump Performance Data

In one study [5], the performance data of four air-source heat pumps of different sizes (2-, 3-, 4-, and 5-tons) from the same manufacturer were selected. The units chosen appear to be representative of what is available in the market at that time. Detailed performance data were used to determine the following results. In Fig. 7.7 the heating capacities of the four pumps are plotted as a function of outdoor air temperature. In Fig. 7.8 the cooling capacities of the same units are given as a function of the outdoor air temperature during the cooling season.

Heating performance is strongly dependent on outside temperature and is not a linear function of the outdoor air temperature. The cooling performance is relatively independent of the outdoor air temperature.

RESIDENTIAL HEATING AND COOLING EQUIPMENT SELECTION

LOCATION: *Rolla / St. Louis*
AVERAGE WINTER TEMPERATURE *44 F*
HEATING DEGREE DAYS *4900*
LATITUDE *38.50*   *D*

AVERAGE DESIGN CONDITIONS:
WINTER INDOOR db   *75* F & RELATIVE HUMIDITY *30* %
WINTER OUTDOOR db   *7* F & RELATIVE HUMIDITY _____ %
SUMMER INDOOR db   *75* F & RELATIVE HUMIDITY *-* %
SUMMER OUTDOOR db   *95* F & WET BULB   *78* F

WORKSHEET—HEAT LOSS & HEAT GAIN CALCULATIONS

	Construc-tion	U-value or Q/A HTG.	CLG.	ΔT HTG.	CLG.	AREA	LOAD, Btuh HTG.	CLG.
1. Windows (Heating)	W_1	0.504		68		90	3084	
	W_2	0.580		68		55	2169	
	W_3	0.550		68		30	1122	
	W_4	0.490		68		18	600	
2. Doors (Heating)	D_1 (W) D_2 (S)	0.31 0.27		68 68		20 20	422 367	
	D_2	0.64		68		40	1741	
3. Windows (Cooling)	W_1		W73 W(S)24 73			30 20 70		2140 920 2198
	W_2		W(S)(24)			55		1320
	W_3		N(24)			30		720
	W_4		E(73)			18		1314
4. Doors (Cooling)	D_1 (W) D_1 (S)		0.29 0.27		23.6 15.0	20 20		137 81
	D_2		E(73)			40		2920
5. Net exposed Walls and Partitions	W, N, E	0.07	0.07	68	23.6	1091	5193	1802
	S	0.07	0.07	68	15.0	204	971	214
	Basement	(3.0)	0			1568	4704	0
6. Ceiling &	Roof	0.067	0.067	68	44	1960	8930	5778
7. Floor	Basement	(1.5)				1960	2940	0

	HTG.	CLG.
8. Sub-total Btuh Loss: Total Lines 1, 2, 5, 6 & 7	32243	
9. (a) Infiltration [*1* AC/hr V/60] (1.10) ($t_1 - t_o$)	19548	
(b) Humidification (if avail.) [ ] (4840) (ΔW)	5527	
10. Total Btuh Loss (Total Lines 8 & 9)	57318	*Furnace*
11. Sub-Total (Total Lines 3, 4, 5, 6, 7)		19386
12. Appliances		1200
13. People @ 300 per person × *8* people		1800
14. Infiltration [____AC/hr V/60] (1.10) ($t_o - t_1$)		5749
15. Total of Lines 11, 12, 13 & 14		28135
16. Total Btuh Gain (Lines 15 × 1.3)		36576 AC

Fig. E7.4B

420

Outdoor Temperature (5° Increments)	Btuh Loss per °F (Heat Loss/TD)	Outdoor Temperature Below 65°F (65 − Column A)	Heat Loss (Btuh) B × C	Heat Pump Heating Capacity (Btuh) (Mfr. Data)	Heat Pump Running Time (%/100) D/E	Heat Pump Input (kW) (Mfr. Data)	Seasonal Heating Hours (Table 7.4)	Seasonal Heat Pump Input (kWh) F × G × H	Resistance Heat Input (Btuh) D − E	Resistance Heat Input (kW) J/3413	Seasonal Resistance Heat Input (kWh) H × K	Resistance Heat Input (kWh) (for Total Electric Resistance)
A	B	C	D	E	F	G	H	I	J	K	L	M
62	843	3	2529	41000	0.062	4.51	646	180	0	0	0	479
57		8	6744	40000	0.169	4.40	575	428	0	0	0	1136
52		13	10959	38500	0.285	4.28	585	714	0	0	0	1878
47		18	15174	36500	0.416	4.14	578	995	0	0	0	2570
42		23	19389	35000	0.554	4.04	620	1388	0	0	0	3522
37	C	28	23604	32500	0.726	3.90	671	1900	0	0	0	4641
32	O	33	27819	30500	0.912	3.78	650	2241	0	0	0	5298
27	N	38	32034	28000	1.00	3.65	411	1500	4034	1.2	493	3858
22	S	43	36249	25000	1.00	3.51	219	769	11249	3.3	723	2326
17	T	48	40464	23500	1.00	3.37	134	452	16964	5.0	670	1589
12	A	53	44679	OFF	0	0	77	0	44679	13.1	1009	1009
7	N	58	48894	OFF	0	0	40	0	48894	14.3	572	572
2	T	63	53109	OFF	0	0	15	0	53109	15.6	234	234
−3		68	57324	OFF	0	0	7	0	57324	16.8	118	118
−8		73	61539	OFF	0	0	1	0	61539	18.0	18	18
−13		78										
−18 & below		83										
								10567		Totals	3837	27817

$$\text{Seasonal Performance Factor} = \frac{\Sigma M}{\Sigma I + \Sigma L} = \frac{27817}{14404} = \boxed{1.93} \text{ S.P.F.}$$

Fig. E7.4C. Energy calculation form for heat pumps.

421

A	B	C	D	E	F	C_M	F'	G	H	I	J	K	L	M
Outdoor Temperature (5° Increments)	Btuh Loss per 1°F (Heat Loss/TD)	Outdoor Temperature Below 65°F (65 − Column A)	Heat Loss (Btuh) B × C	Heat Pump Heating Capacity (Btuh) Mfr. Data	Heat Pump Running Time (%) D/E	Cycling Modifier	Modified F = $\frac{F}{C_M}$ % RT	Heat Pump Input (kW) Mfr. Data	Seasonal Heating Hours (Table 7.4)	Seasonal Heat Pump Input (kWh) F × G × H	Resistance Heat Input (Btuh) D − E	Resistance Heat Input (kW) J/3413	Seasonal Resistance Heat Input (kWh) H × K	Degree Hours C × H
62	843	3	2529	41000	6.2	.639	9.7	4.51	646	283	0	0	0	1938
57	↑	8	6744	40000	16.9	.751	22.5	4.40	575	569	0	0	0	4600
52		13	10959	38000	28.5	.816	34.9	4.28	585	874	0	0	0	7605
47		18	15174	36500	41.6	.868	47.9	4.14	578	1146	0	0	0	10404
42		23	19389	35000	55.4	.909	60.9	4.04	620	1525	0	0	0	14260
37	C	28	23604	32500	72.6	.949	76.5	3.90	671	2002	0	0	0	18788
32	N	33	27819	30500	91.2	.985	92.6	3.78	650	2275	0	0	0	21450
27	A	38	32034	28000	100	1	100	3.65	411	1500	4034	1.2	493	15618
22	T	43	36249	25000	100	1	100	3.51	219	769	11249	3.3	723	9417
17	S	48	40464	23500	100	1	100	3.37	134	452	16964	5.0	670	6432
12	N	53	44679	OFF	0			0	77	0	44679	13.1	1009	4081
7	O	58	48894	OFF	0			0	40	0	48894	14.3	572	2320
2	C	63	53109	OFF	0			0	15	0	53109	15.6	234	945
−3		68	57324	OFF	0			0	7	0	57324	16.8	118	476
−8	↓	73	61539	OFF	0			0	1	0	61539	18.0	18	73
−13		78												
−18 & below		83												
										11395 ◄─	Totals ─►		**3837**	**118407**

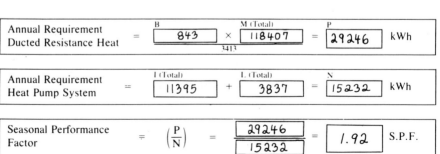

Annual Requirement Ducted Resistance Heat = $\dfrac{843 \times 118407}{3413}$ = 29246 kWh

(B = 843, M (Total) = 118407, P = 29246)

Annual Requirement Heat Pump System = 11395 + 3837 = 15232 kWh

(I (Total) = 11395, L (Total) = 3837, N = 15232)

Seasonal Performance Factor = $\left(\dfrac{P}{N}\right)$ = $\dfrac{29246}{15232}$ = 1.92 S.P.F.

Fig. E7.4D. Energy calculation form for heat pumps.

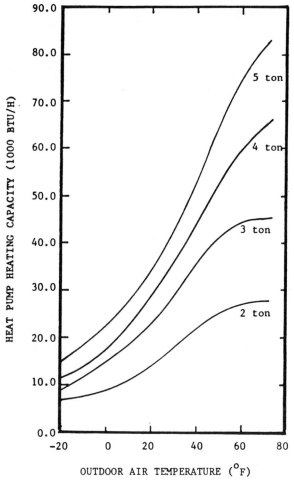

Fig. 7.7.   Heating capacities of typical air-source heat pumps [5].

The COPs for the same four units are given in Fig. 7.9, and their EERs for cooling, in Fig. 7.10. Both are strongly dependent on the outdoor air temperature. For these units there is no apparent size effect on the COP and EER; in both cases there are crossings of the various performance curves. The heat pump power requirements for these units can be found by using the definitions of COP and EER and Figs. 7.7 and 7.9 for heating and Figs. 7.7 and 7.10 for cooling.

The SPFs for the four air-source heat pumps are given in Fig. 7.11. This was done for various heat-loss factors between 250 and 1500 Btu/hr·°F. The seasonal performance factor is defined as

$$SPF = \frac{HPE + AE}{3413[HPP + AP]} \qquad (7.10)$$

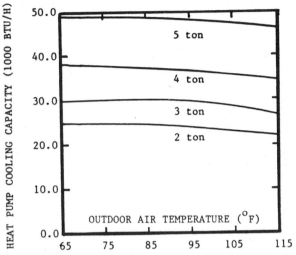

Fig. 7.8. Cooling capacities of typical air-source heat pumps [5].

where HPE = energy supplied by the heat pump during the heating season, Btu

   AE = energy supplied by auxiliary electric heat during the heating season, Btu

  HPP = energy required by the heat pump during the heating season, kWh

   AP = energy required by the auxiliary electric heat during the heating season, kWh

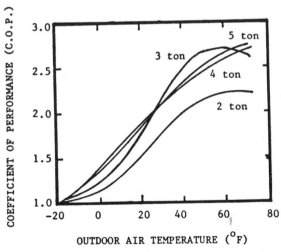

Fig. 7.9. Coefficients of performance of typical air-source heat pumps [5].

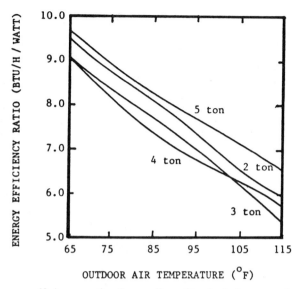

Fig. 7.10. Energy efficiency ratios for cooling of typical air-source heat pumps [5].

The results shown in Fig. 7.11 apply for the entire heating season. A heat pump system having an SPF of 1.75 would require approximately 43% less energy $(1 - 1/1.75)$ than an all-electrical heating system. The low SPF of the 2-ton heat pump is related to the low COP curve shown in Fig. 7.9, which is inherent in the particular design of that heat pump and is not typical of 2-ton heat pumps in general. These results illustrate the nonstandardization of commercially available heat pumps.

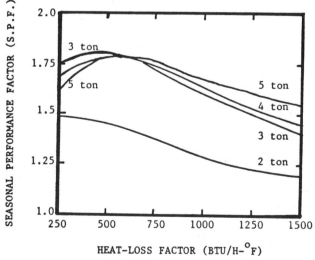

Fig. 7.11. Air-source heat pump seasonal heating performance [5].

TABLE 7.8.   Seasonal Cooling Performance (SCF) for
Air-Source Heat Pumps [5]

Heat Pump Size (tons)	Heat-Loss Factor [Btu/(hr·°F)]			
	250	500	1000	1500
2	8.31	8.14	8.11	8.11
3	7.90	7.72	7.68	7.68
4	7.64	7.45	7.40	7.39
5	8.47	8.31	8.26	8.26

A seasonal performance factor for cooling (SCF) may be defined by the following:

$$SCF = \frac{HR}{3413(HPC)} \tag{7.11}$$

where   HR = energy removed by the heat pump during the cooling season, Btu

HPC = electrical energy required by the heat pump during the cooling season, kWh

The seasonal cooling performance for the previously described air source heat pumps is shown in Table 7.8. The SCF values vary slightly from size to size but vary very little with the heat loss factor.

A joint EPRI-Association of Edison Illuminating Companies (AEIC) project "Load and Use Characteristics of Electric Heat Pumps in Single Family Residential Housing Units" [8] conducted by Westinghouse was initiated to obtain information on heat pump heating system operation based on field tests of existing installations. The field testing was performed between October 1975 and March 1977. The project was designed to record heat pump and total house electric energy usage every 15 min using magnetic tape recording equipment. A total of 120 existing installations were instrumented. They were located in 12 different geographic regions, covering a range of 300–8000 heating degree days. The heat pumps had been installed between 1968 and 1975 and ranged in size from $1\frac{1}{2}$ to 5 tons. The houses ranged from 1100 to 4000 ft^2 in size. The typical installation was a 1972 heat pump of 3-ton rating installed in a 2000-ft^2 house. In this project, 43 of the heat pump systems had timers connected to bypass the heat pump on alternate days, resulting in all heat required on those days to be supplied by the electrical resistance heaters. By comparing the energy used operating with resistance heat only with the energy used by the heat pump system during normal operation, the SPF was estimated for those houses for each of the two winter seasons. These estimated SPFs show a wide range, but when they are grouped by region the average SPF in each region was found to be relatable to the heating degree days in the 6-month period October through March in

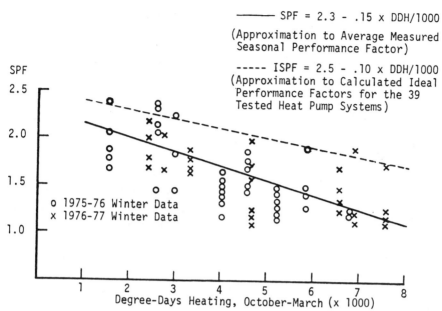

——— SPF = 2.3 - .15 x DDH/1000

(Approximation to Average Measured
Seasonal Performance Factor)

----- ISPF = 2.5 - .10 x DDH/1000
(Approximation to Calculated Ideal
Performance Factors for the 39
Tested Heat Pump Systems)

o 1975-76 Winter Data
x 1976-77 Winter Data

Fig. 7.12. Performance of installed heat pump systems [8]. Courtesy of P.J. Blake and W.C. Gernert and Electric Power Research Institute.

each region. The individual calculated SPFs are shown in Fig. 7.12 plotted against heating degree days.

Data points in Fig. 7.12 show the actual performance of the installations measured over two winters. The solid line represents the average measured SPF as a function of the heating degree days. Manufacturers' performance specifications were used together with the measured heating demands of each house to calculate the theoretical SPF for each installation, and the results were averaged to obtain the broken line shown in the figure. The horizontal dotted line represents the performance of an electrical furnace. The figure shows that the average installation achieves 88% of the expected electricity savings in a 2000-degree-day climate, but only 22% of the expected savings in an 8000-degree-day climate.

The energy used by each of the 120 heat pump heating systems was measured along with total house energy use. The houses were grouped by region and the data averaged, with the results for the first winter shown in Table 7.9. Results from the second winter were similar. These data confirm the conventional assumption that energy use is roughly proportional to the heating degree days, although there is a wide variation evident from the maximum and minimum values listed. The concept that energy use per square foot of house floor area per degree day would be less variable than the energy per degree day was tested. In general, there was little difference in variability.

7.11.2.   Water-Source Heat Pump Performance Data

The performance data of four water-source heat pumps (2-, 3-, 4-, and 5-tons from the same manufacturer) were chosen as typical data. The manufacturer of these heat pumps was not the same as for the air-source heat pumps. The performance data are shown in Table 7.10. The seasonal performance of each heat pump was evaluated for a range of heat-loss factors between 250 and 1500 Btu/hr·°F). The seasonal heating results are shown in Fig. 7.13. These figures point out that the higher source-water temperature results in a significantly higher SPF.

The seasonal heating performance as a function of the source-water temperature is shown in Fig. 7.14 for a heat-loss factor of 500 Btu/hr·°F.

The SCFs for these water-source heat pumps are identical to the EER values given in Table 7.10. This is true as long as the water temperature remains constant.

The heat pump seasonal performance of a water-source heat pump is dependent upon the water flow rate through the heat exchanger. This relationship for the four heat pumps is shown in Fig. 7.15. The SPF values are shown to increase slightly (this analysis assumes no pumping power). If significant pumping power is required for a given application, then pumping power would increase with an increase in water flow rate. This could eliminate any increase in SPF.

The following general conclusions can be formed from this study:

1.   The coefficients of performance for the air-source heat pumps tend to bunch in the general neighborhoods of 2.0–2.5 and 1.2–1.8 for outside air temperatures of 47 and 17°F, respectively.

2.   For water-source heat pumps, the COP is in the vicinity of 2.3–3.0 for 60°F water.

3.   The EER values for air-source heat pumps are in the vicinity of 6–7 Btu/(W·hr). For water-source heat pumps the EER values are near 7.5–9.0 Btu/(W·hr).

4.   The seasonal performance results for typical commercially available air-source heat pumps indicate an energy savings over electrical-resistance heating of approximately 45%. For a water-source heat pump using water between 60 and 80°F, the savings is of the order of 65%.

7.11.3.   Field Test Results—Commonwealth Edison

According to Commonwealth Edison's field tests, there is a poor correlation between COP and SPF [6]. According to test data (although labeled by the utility as still "preliminary, limited in number of observations, and strictly qualitative"), efficiencies are found to decrease on both sides of the heat

TABLE 7.9. Heat Pump Energy Use per House by Region, First Winter, October 1975 through March 1976 [8]

Degree-Days[a] Heating (DDH)	Utility	kWh(HP),[b] Avg	kWh(HP)/DDH Min	kWh(HP)/DDH Avg	kWh(HP)/DDH Max	kWh(HP)/(DDH × ft²) × 10³	kWh(TOT)/DDH Avg[c]	kWh(HP)/kWh(TOT) Avg	kWh(HP)/kWh(TOT) Min	kWh(HP)/kWh(TOT) Max
326	Florida P&L	1187	0.86	3.64	6.61	2.22	8.45[d]	.15	.03	.31
1566	Houston P&L	2537	0.78	1.62	2.62	0.88	6.02	.27	.18	.37
2576	Arkansas P&L	7255	1.51	1.90	2.42	1.01	5.06	.36	.30	.46
2962	Pacific G&E	7775	1.71	2.61	3.66	1.62	5.94	.44	.22	.58
3196	So. Cal Ed	5561	1.15	1.74	3.27	1.12	4.57	.38	.27	.67
4014	P.S. New Mex	11159	1.40	2.78	4.08	1.22	5.06	.55	.32	.75
4576	Philadelphia Elec	12218	1.57	2.67	4.04	1.09	4.87	.55	.39	.68
4602	Con Ed NY	12794	0.62	2.78	4.48	1.04	6.20	.51	.39	.62
4603	PSEG (NJ)	12428	1.78	2.70	4.78	1.25	4.60	.59	.43	.78
5202	Rochester G&E	18322	2.71	3.52	4.44	1.66	5.48	.64	.60	.77
5817	Consumers Pwr	12322	1.22	2.12	3.11	0.81	3.40	.62	.46	.73
6666	Northern States	20665	1.75	3.10	4.02	1.02	4.76	.65	.46	.85

[a]DDH is the 6-month total degree-days heating.

[b]kWh(HP) is the energy used by the heat pump heating system, including indoor fan.

[c]kWh(TOT) is the total electric energy input to the house.

[d]Because of the few hours below 65°F in this region, kWh(TOT) was based only on those hours where the temperature was below 65°. This is equivalent to a shorter heating season.

TABLE 7.10.   Typical Water-Source Heat Pump Performance Data [5]

Nominal Size Rating (tons)[a]	Water Flow Rate (gpm)[b]	Source-Water Temp. (°F)	Heating Performance Capacity (1000 Btu/hr)	COP	Cooling Performance Capacity (1000 Btu/hr)	EER [(Btu/hr)/W]
2	3.0	60	24.60	2.32	25.96	9.76
		70	27.85	2.41	24.49	8.65
		80	31.50	2.54	22.53	7.38
3	5.5	60	33.60	2.38	42.85	9.97
		70	36.90	2.40	40.05	8.84
		80	43.05	2.61	37.20	7.62
4	7.5	60	49.00	2.76	52.50	10.44
		70	55.45	2.87	49.50	9.25
		80	62.70	3.02	45.60	8.00
5	8.5	60	55.00	2.68	60.40	9.71
		70	62.05	2.78	57.00	8.61
		80	70.40	2.94	53.00	7.52

[a] One ton of cooling is equal to 12,000 Btu/hr.

[b] The lowest source-water flow rate for which data were available is listed.

pump system's balance point in the heating mode, and the same type dropoff occurs in the cooling mode.

In Fig. 7.16a, for example, the operating efficiency curve for the heating mode is seen to run significantly under the COP curve below the balance point. But the operating efficiency is also seen to drop off sharply above the balance point. It is not generally realized that this higher temperature operating efficiency dropoff is the result of cycling. The same sort of divergence in the cooling mode, which occurs for the same reason, is shown in Fig. 7.16b.

While seasonal efficiencies provide better judgment criteria than steady-state efficiencies, heat pump performance must be considered on a regional basis rather than by one national rating standard.

Commonwealth Edison chose, for purposes of comparison, 25 geographically diverse locations in the United States where the cooling design temperature is the same. Among them are cities in Alabama, New Hampshire, Florida, Minnesota, Texas, and Wisconsin. As might be expected, operating hours in these widely scattered areas vary. For example, the cooling operating hours in Corpus Christi, Texas (approximately 1175) could be five times those of Chicago, Illinois, and ten times those of Pittsburgh, Pennsylvania. Nevertheless, for sizing by cooling capacity in comparable houses using single rating point criteria, they would all take the same size heat pump. In these 25 locations, however, heating design conditions range from −15°F (Sioux Falls) to 31°F (Corpus Christi).

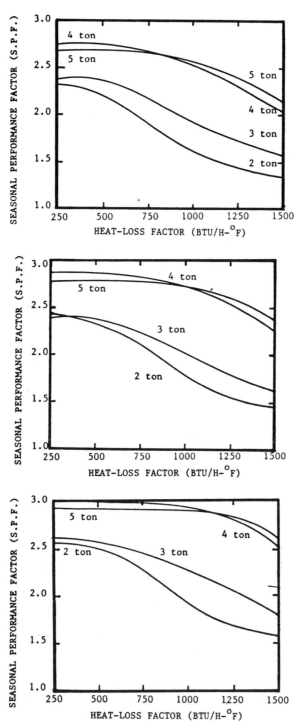

Fig. 7.13. Seasonal performance factors for a water-source heat pump at (a) 60°F, (b) 70°F, (c) 80°F.

431

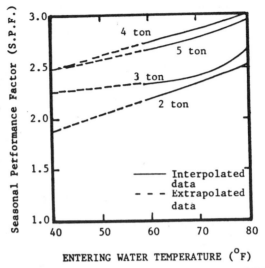

Fig. 7.14. Seasonal performance factor for a heat-loss factor of 500 Btu/(hr · °F) [5].

The ratio of heating operating hours to cooling operating hours in the various areas can be as much as 20 to 1. Some relative regional values are shown in Table 7.11.

7.11.4. Field Test Results—NBS (Bowman House)

In an attempt to obtain quantitative information on the dynamic performance of air-to-air heat pumps, a 5-ton heat pump was installed in a 20-year-

WATER FLOW RATES

Heat pump size (tons)	Low flow (gpm)	High flow (gpm)
2	3	6
3	5.5	11
4	7.5	15
5	8.5	17

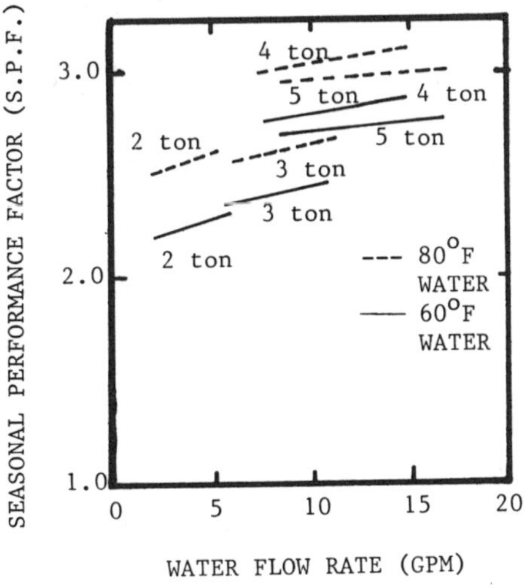

Fig. 7.15. Seasonal performance factor vs. water flow rate. Heat-loss factor 500 Btu/hr · °F) [5].

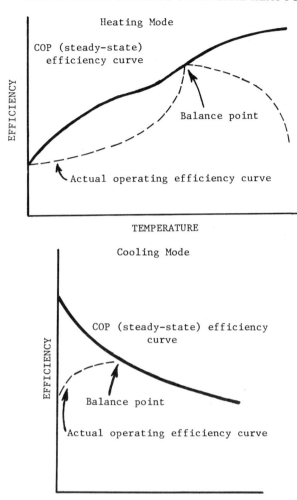

Fig. 7.16. Actual operating efficiency of heat pumps (a) during heating and (b) during cooling.

old house (known as the Bowman house) in the Washington, DC area [7]. This heat pump experiment was part of a larger program to measure experimentally the energy savings that could be achieved through retrofitting. The house studied was a single-story frame ranch-type residence that was built in the early 1950s and had a floor area of approximately 2500 ft² (232.25 m²).

The 5-ton heat pump installed in the Bowman house was a commercially available split system employing thermostatic expansion values. It was sized to meet the calculated preretrofit cooling requirements of the house, which were determined using ASHRAE procedures. Figure 7.17 is a schematic of

TABLE 7.11. Regional Variation in Heat Pump
Operating Hours

Location	Cooling Hours	Heating Hours	Total
New Orleans	1484	843	2327
Memphis	1259	2031	3290
Las Vegas	2431	1569	3910
St. Louis	1012	2988	4009
Chicago	577	3621	4198

the indoor section, which contained the indoor coil, indoor blower, supplemental resistance heaters, and the expansion valve employed when the heat
pump was used to cool the house. The outdoor section (Fig. 7.18) contained
the outdoor coil, outdoor fan, compressor, accumulator, switchover valve,
and the expansion valve used during the heating process.

The heating duct system is shown schematically in Fig. 7.19. Half the
house was over a basement and the other half over a crawl space, and there
were separate supply and return ducts for each half of the house. The return
air was fed into a return air plenum and then into the indoor air handler,
where it was either heated or cooled. Upon leaving the indoor air handler,
the air entered the two supply ducts, where airflow monitoring devices were
used to measure the mass flow rate of air delivered to each half of the house.

Since there is no perfect correlation between the hourly cooling or heating
requirement of a house and the difference between the indoor and outdoor
dry-bulb temperatures, it was necessary to determine the actual cooling and

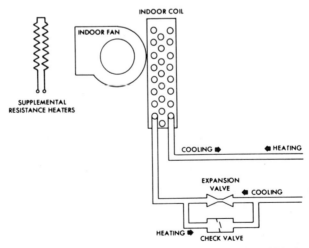

Fig. 7.17. Schematic of heat pump indoor unit [7]. Courtesy of National Bureau of
Standards.

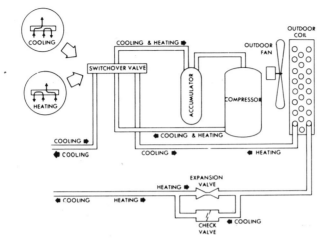

Fig. 7.18. Schematic of heat pump outdoor unit [7]. Courtesy of National Bureau of Standards.

heating load factors in order to properly evaluate the dynamic performance of the heat pump under test. The equations used to define these load factors were:

$$\text{Cooling load factor} = \frac{\text{cooling done by the heat pump in time } \tau}{\text{manufacturer's steady-state cooling capacity } (\tau)}$$

$$\text{Heating load factor} = \frac{\text{heating done by the heat pump system in time } \tau}{\text{manufacturer's steady-state heating capacity } (\tau)}$$

where τ is a time period or cycle over which the cooling or heating COP was measured and it is to be understood that the manufacturer's steady-state cooling or heating capacities correspond to the average indoor and outdoor conditions that existed in the time period τ. The numerator for the heating load factor includes the output from the first step of supplemental resistance heaters when this step operated during a defrost period to offset the cooling effect of the indoor coil during this period.

Figure 7.20 shows the heating performance of the Bowman house heat pump as a function of the heating load factor.

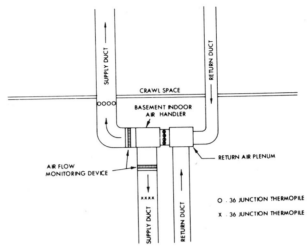

Fig. 7.19. Schematic of duct system in the Bowman house [7]. Courtesy of National Bureau of Standards.

An average hourly heating COP was determined as

$$\text{Average hourly heating COP}_{\text{measured}} \equiv \frac{1}{N} \sum_{i=1}^{N} (\text{COP})_i \qquad (7.11)$$

where $(\text{COP})_i$ = COP measured for hour i and N = the total number of hours of data. The average hourly COP was then calculated for the same

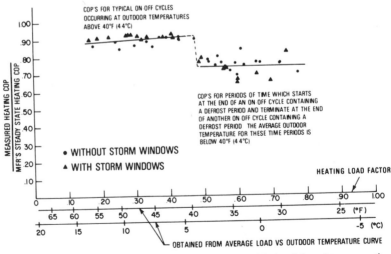

Fig. 7.20. Variation in heat pump performance with load heating operating [7]. Courtesy of National Bureau of Standards.

period by determining the manufacturer's steady-state COP for each hour, multiplying by a correction factor from Fig. 7.20:

$$\begin{matrix} \text{Average hourly} \\ \text{heating COP} \\ \text{(from Fig. 7.20)} \end{matrix} \equiv \frac{1}{N} \sum_{j=1}^{N} \left[\left(\frac{\text{Mfr's steady-state}}{\text{heating COP}} \right)_j \left(\frac{\text{correction factor}}{T_j} \right)_j \right]$$

(7.12)

where the manufacturer's steady-state heating COP is the heating COP obtained from the manufacturer's performance data for hour j, and the correction factor is the ordinate of the curve in Fig. 7.20 corresponding to average outdoor temperature T_j for hour j. The measured and calculated average hourly heating COPs were found to agree within 3%.

The SPF of the Bowman house heat pump was estimated using the bin method and weather data for Andrews Air Force Base, which is near Washington, DC. Using the manufacturer's steady-state data, the SPF was predicted to be 2.16. Employing the results in Fig. 7.20 to correct for the effect of cycling, frost buildup, and defrost, the SPF of the Bowman house heat pump turned out to be 1.74, or some 19% lower.

When the estimated seasonal heating performance of the Bowman house heat pump was traced back to the power plant, it was found that only about a half-unit of heat energy was actually delivered to the interior living space for every unit of energy consumed at the power plant. This is comparable to the probable performance of many gas- or oil-fired residential heating units. If this estimated seasonal performance of the Bowman house heat pump is representative of residential heat pumps in areas having heating seasons similar to Washington, DC, there does not appear to be any clear-cut advantage to choosing either a heat pump or a fossil-fuel heating system in these areas for the purpose of saving primary source energy.

7.11.5. Laboratory Test Results—NBS

To supplement the results of the Bowman house field study, a 3-ton heat pump was installed in a laboratory, and its full-load and part-load performance evaluated over a broad range of operating conditions [9]. A split-system air-to-air unitary heat pump was utilized in the experimental investigation. The ARI rating of the heat pump was 37,000 Btu/hr in the heating mode at outdoor dry-bulb and wet-bulb temperatures of 45°F (7.2°C) and 43°F (6.1°C), respectively, and at 70°F (21.1°C) indoor temperature. The ARI rating in the cooling mode was 36,000 Btu/hr at indoor dry-bulb and wet-bulb temperatures of 80°F (26.7°C) and 67°F (19.4°C), respectively, and at 95°F (35.0°C) outdoor temperature.

Figure 7.21 may be used to compare the measured full-load coefficient of performance and the heating capacity, integrated over a complete heating-defrosting cycle, at different outdoor temperatures with the integrated data

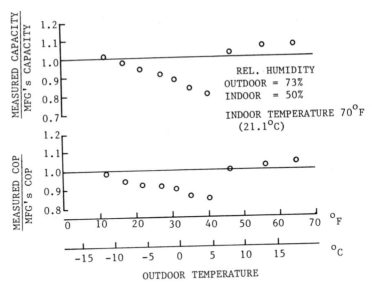

Fig. 7.21. Comparison of heating performance results with the manufacturer's reported results.

supplied by the manufacturer. The tests were conducted at an outdoor relative humidity of 73% and include the ARI rating points at outdoor temperatures of 47°F (8.3°C) and 17°F (−8.3°C).

The results presented indicate a larger capacity by approximately 5% than the manufacturer's application ratings for this model for outdoor temperatures ranging from 65°F (18.3°C) to 47°F (8.3°C). The capacity falls below the manufacturer's results by approximately 20% in the temperature range of 42°F (5.5°C) to 35°F (1.7°C). As the outdoor temperature continues to decrease, the results indicate better agreement. A similar trend is evident in the full-load coefficient of performance.

7.12. HEAT PUMP TEST STANDARD PROPOSED BY DOE [10]

Based on the recommendations from the National Bureau of Standards, the Department of Energy, in 1979, proposed a standard test method for single-phase air- and water-source heat pumps of less than 65,000 Btu/hr capacity that have electrical resistance supplementary heat. The proposed method provides for heat pumps equipped with single- and two-speed compressors, twin compressors, and for cylinder unloading.

The tests for heat pumps in the cooling mode are essentially the same as for central air conditioners (steady-state tests at 95 and 82°F coupled with optional cycling and dry-coil tests at 82°F or a degradation factor of 0.25%).

For the heating mode, four tests are used to determine steady-state efficiency, energy lost through defrosting, degradation of efficiency because of cycling, and efficiency at low ambients. The tests for heat pumps in the heating mode include steady-state tests at either 47 or 62°F and at 17°F plus a cycling test at 47°F and a defrost test at 35°F. The tests for modulating capacity units, however, are more involved. A curve is generated that is applicable to all outdoor temperatures that may be encountered over a heating season. A mathematical technique is used to incorporate different weather data for different regions to calculate the performance factors.

Similar to the air-conditioner tests, manufacturers may use a degradation factor of 0.25 to account for cycling. Also designed to reduce testing costs, demand-defrost systems may show a 5% improvement in defrost capacity compared to the more conventional time-temperature defrost method.

The proposed test standard incorporates parts of ARI's Standards 210-77 (unitary air conditioners), 240-77 (unitary heat pumps), 320-76 (water-source

TABLE 7.12. Distribution of Fractional Hours in Temperature Bin, Heating Load Hours, and Outdoor Design Temperatures for Major Climatic Regions in the Continental United States

Region:		I	II	III	IV	V	VI
Heating load hours, HLH:		750	1250	1750	2250	2750	2750
Outdoor design temperature, T_{OD} for the Region (°F)		37	27	17	5	−10	30

Fractional Hours

Bin No. (j)	T_j (°F)	Fractional Hours in Temperature Bins					
1	62	.291	.215	.153	.132	.106	.113
2	57	.239	.189	.142	.111	.092	.206
3	52	.194	.163	.138	.103	.086	.215
4	47	.129	.143	.137	.093	.076	.204
5	42	.081	.112	.135	.100	.078	.141
6	37	.041	.088	.118	.109	.087	.076
7	32	.019	.056	.092	.126	.102	.034
8	27	.005	.024	.047	.087	.094	.008
9	22	.001	.008	.021	.055	.074	.003
10	17	0	.002	.009	.036	.055	0
11	12	0	0	.005	.026	.047	0
12	7	0	0	.002	.013	.038	0
13	2	0	0	.001	.006	.029	0
14	−3	0	0	0	.002	.018	0
15	−8	0	0	0	.001	.010	0
16	−13	0	0	0	0	.005	0
17	−18	0	0	0	0	.002	0
18	−12	0	0	0	0	.001	0

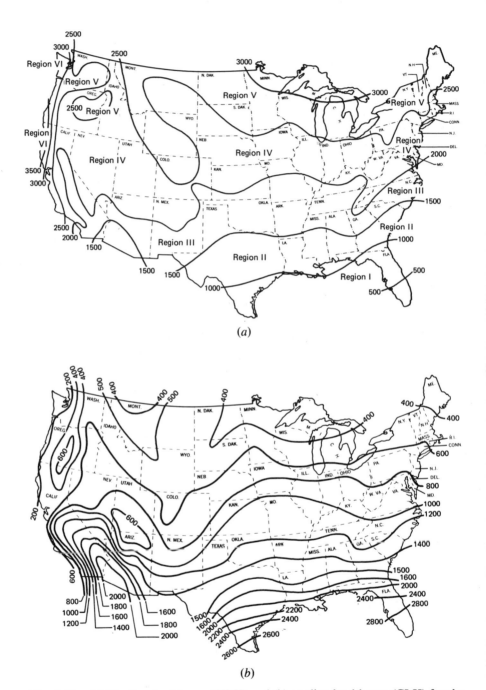

Fig. 7.22. (a) Heating load hours (HLH) and (b) cooling load hours (CLH) for the United States.

heat pumps), and refers to the ASHRAE Standard 37-69, "Methods for Testing or Rating Unitary Air-Conditioning or Heat Pump Equipment."

According to the Department of Energy, present industry standards deal only with steady-state conditions and make no allowances for performance degradation because of cycling and defrosting. It adapted the standards so that estimated annual operating costs and seasonal performance factors could be determined in both the heating and cooling modes.

For air-source central air conditioners, the annual electricity consumption for cooling is estimated by dividing the output (the cooling capacity times the representative annual average cooling load hours) by the cooling SPF (CSPF).

The annual electrical energy input for heating is estimated by dividing the output (a representative design heating requirement times the representative average heating load hours) by a representative heating seasonal performance factor (HSPF). For water-source units, the annual electrical energy input is estimated by multiplying the test-measured electrical input by the representative annual average cooling or heating load hours.

The proposed DOE standards for heat pumps define tests for the HSPF in each of six different broadly defined climatic regions of the country. Cooling seasonal performance may be specified by a CSPF or a seasonal energy efficiency ratio (SEER). In addition, an annual SPF is defined as a weighted average of the HSPF and the CSPF based on the number of heating and cooling hours in different parts of the country. Table 7.12 gives the climatic conditions to be used for each of the six regions. Figure 7.22 shows the heating load hours (HLH) and cooling load hours (CLH) throughout the continental United States. Figure 7.22a may be used to locate the six regions.

7.13. VARIATION IN HEAT PUMP PERFORMANCE

The heat pump is a refrigeration system; therefore its performance can be analyzed as such. The trends in capacity of heat pumps using various sources are the same except that the air-source heat pump has the widest extremes of performance.

The primary influence on the performance of the heat pump is the temperature of the outdoor air. During heating operation the air entering the indoor coil remains approximately constant. The outdoor air, on the other hand, may vary over a wide range of temperature. The compressor experiences wide changes in suction pressure but little change in the discharge pressure. Considering the heat pump first as a refrigeration system, the refrigeration capacity at a constant condensing temperature of 105°F is shown in Fig. 7.23. The compressor capacity diminishes at low evaporating temperatures because at low suction pressures the volumetric efficiency of the compressor is low and also the refrigerant vapor is less dense than at high suction

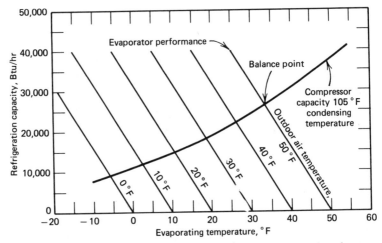

Fig. 7.23. Balance points of condensing unit and evaporator in a heat pump.

pressures. The evaporator curves show that with a given temperature of entering air, more heat can be transferred from the air to refrigerant at lower evaporating temperatures. Successive balance points are reached at various outdoor air temperatures, with the refrigeration capacity decreasing as the outdoor temperature drops.

During heating operation the heat rejection at the condenser is of main interest, because the rate of heat rejection is the heating capacity of the heat pump. The balance points on Fig. 7.23 can be transferred to the coordinates of Fig. 7.24, which are the heat transfer in Btu's per hour vs. the outdoor air temperature. For example, the refrigeration capacity from Fig. 7.23 is 15,000 Btu/hr when the outdoor air temperature is 20°F. To determine the quantity of heat rejection, the heat rejection ratio from Fig. P4.16A must be applied.

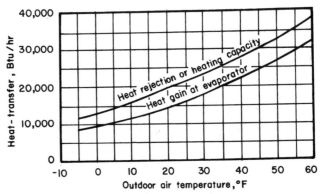

Fig. 7.24. Heating capacity of an air-source heat pump.

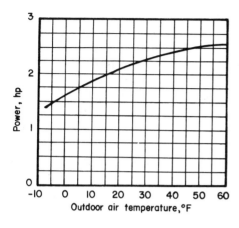

Fig. 7.25. Power required by the compressor as the outdoor air temperature changes.

At an outdoor air temperature of 20°F the evaporating temperature is 11°F. From Fig. P4.16A, with an evaporating temperature of 11°F and a condensing temperature of 105°F, the heat rejection ratio is 1.33. Multiplying 1.33 by the refrigerating capacity of 15,000 Btu/hr at an outdoor air temperature of 20°F, the heat rejection rate is found to be 20,000 Btu/hr, as shown by Fig. 7.24. The difference between the quantities of heat transfer shown by the two curves in Fig. 7.24 is the energy input at the compressor.

The power required by the compressor through a range of outdoor temperatures should·also be evaluated. For each of the balance points in Fig. 7.23, catalog data of the compressor will provide the power required. The power requirement may then be plotted as a function of the outdoor temperature as in Fig. 7.25.

When the purpose of the heat pump is to cool a structure, the outdoor temperature again exerts a crucial influence. The outdoor temperature now,

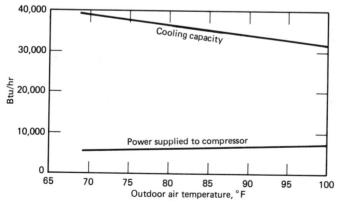

Fig. 7.26. Cooling capacity and power required by a heat pump as the outdoor temperature changes.

however, controls the condensing temperature, and the evaporator temperature changes only slightly. An increase in condensing temperature decreases the capacity of the compressor. This relationship appears in the heat pump as is shown in Fig. 7.26. When the outdoor temperature rises, the capacity decreases while the power required increases slightly.

REFERENCES

1. Healy, J. H., "The Heat Pump in a Cold Climate," 49th Annual Convention of the National Warm Air Heating and Air Conditioning Association, Jacksonville, Florida, Nov. 1962.

2. Gordian Associates, "Evaluation of the Air-to-Air Heat Pump for Residential Space Conditioning," Final Report FEA-10-04-50171-00, Federal Energy Administration, Washington, DC, 1976.

3. Domingorena, A. A., "Performance Evaluation of a Low-First-Cost, Three-Ton, Air-to-Air Heat Pump in the Heating Mode," *Proc. Third Ann. Conf. Heat Pump Technol., Oklahoma State Univ., April 10–11, 1978.*

4. Bullock, C. E., "Energy Savings Through Thermostat Setback with Residential Heat Pumps," *Trans. Am. Soc. Heating, Refrigerating Air Conditioning Eng.,* Vol. 84, Part 2, 1978.

5. Briggs, J. B., and Shaffer, C. J., "Seasonal Heat Pump Performance for a Typical Northern United States Environment," *EG&G Idaho, U.S. Dep. Energy Rep.,* HTPUMP, H001151A, October 1977.

6. *Air-Conditioning News,* May 29, 1978.

7. Kelley, G. E., and Bean, J., "Dynamic Performance of a Residential Air-to-Air Heat Pump," *NBS Building Sci. Ser.,* Vol. 93, March 1977.

8. Blake, P. J., and Gernert, W. C., "Load and Use Characteristics of Electric Heat Pumps in Single-Family Residences," prepared by Westinghouse Electric Corporation for EPRI, EPRI EA-793, Project 432-1 Final Report, Vol. 1, June 1978, pp. 2.1–2.12, 13.

9. Parken, W. H., Jr., Beausoliel, R. W., and Kelly, G. E., "Factors Affecting the Performance of a Residential Air-to-Air Heat Pump," Symposium Paper CH-77-14 #1, *ASHRAE Trans.,* **83,** Part 1, 1977.

10. "Test Procedures for Central Air Conditioners, Including Heat Pumps," *U.S. Fed. Reg.,* **44**(249), 76,700–76,723 (Dec. 27, 1979).

PROBLEMS

1. Determine the cost per 1000 Btu of supplying heat in your territory with (a) oil, (b) gas, (c) direct electric heating, and (d) a heat pump.

2. The total design heating load on a residence in Kansas City, MO, is 112,000 Btu/hr. The furnace is off during June through September. Estimate:
   (a) Annual energy requirement for heating, Btu/hr.
   (b) Annual heating cost if electric heat is used with a cost of 5¢/kWh.

(c) Maximum savings effected if the thermostat is set back to 65°F between 10 P.M. and 6 A.M., $/yr.

3. For a residence located in New Orleans, LA, the design cooling load is 40,000 Btu/hr. Determine:
   (a) Annual energy requirements for cooling, kWh.
   (b) Cost of this energy if the electricity rate is 4.5¢/kWh.

4. The total design heating load on a residence in Kansas City, MO, is 112,000 Btu/hr. Estimate the annual energy requirements for heating if the 4-ton heat pump whose performance is given in Figs. 7.7–7.10 is selected.

5. Estimate the annual energy costs for heating and cooling a residence located in Cleveland, OH, having design loads of 65,000 Btu/hr (heating) and 36,000 Btu/hr (cooling) based on a 75°F indoor temperature. The furnace is on from October 1 through May 31. Electric baseboard heat is used. Air conditioner has an EER of 7.3 Btu/(hr·W). Electricity costs 3¼¢/kWh year-round.

6. Rework Problem 6 using the heat pump of Example 7.3.

7. As a move to conserve energy, some people are designing and building underground houses. One such house has walls, floor, and roof/ceiling all of 10-in.-thick concrete. The roof is covered with 6 in. of dirt having a thermal conductivity of 9.0 Btu/(hr·ft²·°F·in.). Outdoor air design temperature is 0°F. The house is overall 70 ft × 28 ft × 8 ft, interior dimensions. Two air changes per hour are provided. Determine the furnace size required and the energy requirements for a 5900-degree-day heating season. Groundwater is at 50°F.

8. Repeat Problem 7 using a water-source heat pump.

9. Size the heat pump and estimate the annual energy requirements for heating and cooling the building shown in Fig. P7.9, which is located in Rochester, NY. There is no humidifier. Other data are:

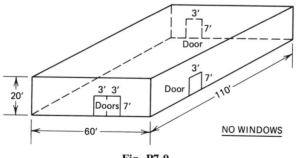

Fig. P7.9.

Walls and roof   3 in. of glass fiberboard, organically bonded, sand-
wiched between metal sheets

Floor   concrete slab on ground with edge-only insulation, R
= 2.50

Doors   2-in. thick, solid wood

10.   The residence shown in Fig. P7.10 is located in Indianapolis, IN. Size
the furnace and the air conditioner and estimate the energy uses and
costs for heating and cooling, if an all-electric system is used. Electric
rate is flat $4\frac{1}{2}$¢/kWh.

Walls   8 ft high; $U = 0.07$ (frame)

Door $U = 0.27$

Window   Single glass

Floor   Slab, 1-in., edge-only insulation

Ceiling/roof $U_{o,c} = 0.05$; dark roof; no overhang; naturally vented
attic

Infiltration Estimated at $\frac{3}{4}$ AC/hr

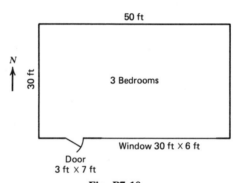

Fig. P7.10.

11.   Rework Problem 10, using the heat pump of Example 7.3.

8
COMPARATIVE ENERGY USE STUDIES

There have been a number of studies dealing with the energy use of heat pumps and other heating and/or cooling systems. These include both computer simulations and actual field measurements for both residential and commercial systems. This chapter briefly describes some of these investigations and their results.

8.1. HOWELL AND SAUER STUDY: RESIDENTIAL (1976)

A simulation study was carried out by Howell and Sauer [1] to look at the comparative energy requirements and costs of heating a typical midwestern residence with either oil, LP gas, electrical resistance heating, or air-source heat pumps.

In order to expeditiously study the relative cost of operation of residential heat pumps and furnaces, the AXCESS Energy Analysis Computer Program was used. The AXCESS (for Alternate Choice Comparison for Energy System Selection) program was designed to provide accurate economic comparisons of the different energy systems that may be used in all types of buildings. The AXCESS program simulates the operation of various types of terminal and primary systems planned for installation in buildings by making hourly calculations using weather data and building specifications. The program may also be used to evaluate various energy conservation techniques for HVAC systems in existing buildings. The program, with some limitations, can be applied with accuracy to residences as well.

447

The AXCESS program evaluates building energy requirements on an hour-by-hour basis for a full year using typical local weather data (dry-bulb temperature, relative humidity, cloud cover), building operating profiles, and load usage profiles (lighting, appliances, people). The weather data used by AXCESS comes from U.S. Weather Bureau hourly data. The user can select any station and any year of interest. For this study the weather data for the years 1970–1973 at St. Louis, Missouri, were used for all calculations.

Heating system comparisons were made with an existing house, which allowed comparison of the actual fuel utilization with the simulated fuel simulation for the installed heating unit. This comparison made it possible to validate the simulation technique. The residential structure that was simulated was a 2000-ft^2 ranch style house with a heated full basement. The structure had brick veneer with full wall insulation, 4-in. ceiling insulation, and 14% double-pane glass. The first-floor temperature was set at 75°F, while the basement was maintained at 70°F. The heating of the residence was simulated using each of the following: oil furnace with 60, 70, or 80% efficiency; LP gas furnace with 60, 70, or 80% efficiency; electric furnace with 90 or 100% efficiency; five commercially available air-to-air heat pumps with electric supplementary heat.

The house was built in 1964 and has good insulation; however, it was not designed or insulated to meet recommendations for electric heat. For the calculations, it was assumed that the infiltration rate would be 1 AC/hr, which is compatible with the life style of the occupants. For the simulation the fuel-fired furnaces were sized at an output of 64,000 Btu/hr, and the heat pumps were sized for the cooling requirements of the house, 36,000 Btu/hr. The heating capacity and input power for each heat pump were obtained from the manufacturer's data and are given in Table 8.1.

TABLE 8.1. Heat Pump Performance (from Manufacturers' Data)

Outside Air Temperature (°F)	Heating Capacity^a (Btu/hr)/Input power (kW)				
	A	B	C	D	E
−10	11,040/2.82	13,400/2.84	10,200/3.06	11,300/3.18	7,520/3.51
0	14,440/3.06	16,900/3.11	14,200/3.16	13,400/3.28	10,940/3.71
10	18,780/3.36	21,200/3.39	18,200/3.34	18,200/3.46	15,900/3.90
20	23,620/3.66	25,200/3.69	22,200/3.55	21,900/3.66	20,880/4.12
30	29,300/3.96	29,200/3.99	26,200/3.75	24,400/3.86	25,580/4.39
40	35,240/4.22	33,200/4.34	31,400/4.06	30,000/4.06	34,960/4.69
50	40,380/4.52	38,200/4.67	40,600/4.56	38,200/4.23	43,020/4.99
60	43,620/4.82	43,200/5.02	48,060/5.08	42,200/4.33	50,920/5.30
70	45,120/5.06	47,200/5.38	55,520/5.53	46,200/4.42	59,460/5.68

^{a}Including indoor fan.

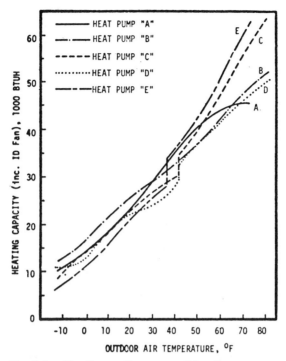

Fig. 8.1. Heating performance of five heat pumps.

The heating performance of the five selected heat pumps is shown in Fig. 8.1. The balance point for these heat pumps is generally between 25 and 30°F.

The results from the simulations are presented in Table 8.2. The fuel-fired furnaces and heat pumps were simulated for 4 yr (1970–1973) using weather data for St. Louis, Missouri. The fuel and electric power requirement, degree days, and annual heating required are given for each year and as a 4-yr average. The oil, gas, and electric furnaces were operated with continuous air circulation (CAC), and this power requirement is noted.

A limitation of the AXCESS program prevents the simulation of a typical residential system with temperature-controlled on-off cycling of the fan. For this reason CAC was used for fuel-fired furnaces, while the heat pump performance data included appropriate fan operation.

The simulated operation of the oil furnace in this house compares favorably with measured oil consumption. In the 1972–1973 heating season the actual oil used was 1041 gal. In subsequent years the oil consumption was 714, 715, and 532 gal. During the three heating seasons, fall 1973 to spring 1976, the thermostat in the house had been lowered to 72°F, the air registers adjusted to give lower bedroom temperatures, fan set to operate longer, oil

TABLE 8.2. Annual Fuel Use for Heating

Year	1970	1971	1972	1973	4 YR Av.
Degree days	4961	4510	5087	4472	4758
Heating req'd 10^6 Btu	128	120	129	118	124
Oil furnace fuel use, gal					
80% efficiency	1016	951	1030	923	980
70% efficiency	1161	1089	1177	1055	1121
60% efficiency	1355	1269	1373	1231	1307
			+ 4261 kWh fan (CAC)		
LP gas furnace fuel use, gal					
80% efficiency	1569	1471	1591	1426	1514
70% efficiency	1793	1682	1818	1629	1731
60% efficiency	2092	1951	2121	1901	2016
			+ 4261 kWh fan (CAC)		
Electric furnace, kWh					
100% efficiency	33103	31017	33559	30073	31938
90% efficiency	36781	34464	37287	33414	35487
			+ 4261 kWh fan (CAC)		
Heat pump A, kWh	19857	18171	20005	17671	18926
Heat pump B, kWh	20138	18484	20289	17977	19222
Heat pump C, kWh	21100	19400	21333	18707	20135
Heat pump D, kWh	21933	20224	22222	19401	20945
Heat pump E, kWh	22664	20835	22947	20074	21630

Building Description:
   Size: 2000 ft², ranch, with full basement (heated)
   Construction: Brick veneer, full wall insulation, 4 in. ceiling insulation, 14%
            glass (all double pane)
   Indoor temperatures: 75°F, first floor; 70°F, basement
Location: St. Louis, Missouri
Heating Values for Fuel:
   No. 2 fuel oil, 139,000 Btu/gal
   LP gas, 90,000 Btu/gal

nozzle replaced with a smaller orifice, and the garage door closed when possible. These changes contributed to the reduced oil consumption.

Several interesting results can be noted in Table 8.2. The yearly variation in fuel and power requirements was only about ±6% for 4 years in St. Louis. Continuous air circulation requires a significant quantity of power (13% of electric heat requirements). Various heat pumps can deviate in power re-

TABLE 8.3. Comparative Annual Operating Cost Estimates

	Fuel Use[a]	Unit Fuel Costs[b]	Annual Operating Cost
Oil furnace (with CAC)			
80% efficiency	980 gal	36.9¢/gal	$434
	4261 kWh	1.7¢/kWh	($362 w/o fan)
70% efficiency	1121 gal	36.9¢/gal	$486
	4261 kWh	1.7¢/kWh	($414 w/o fan)
60% efficiency	1307 gal	36.9¢/gal	$555
	4261 kWh	1.7¢/kWh	($482 w/o fan)
LP gas furnace (with CAC)			
80% efficiency	1514 gal	29.2¢/gal	$514
	4261 kWh	1.7¢/kWh	($442 w/o fan)
70% efficiency	1731 gal	29.2¢/gal	$578
	4261 kWh	1.7¢/kWh	($505 w/o fan)
60% efficiency	2016 gal	29.2¢/gal	$661
	4261 kWh	1.7¢/kWh	($588 w/o fan)
Electric furnace (with CAC)			
100% efficiency	36,199 kWh	1.7¢/kWh	$615
			($543 w/o fan)
90% efficiency	39,748 kWh	1.7¢/kWh	$676
			($603 w/o fan)
Heat pump A	18,926 kWh	1.7¢/kWh	$322
Heat pump E	21,630 kWh	1.7¢/kWh	$368

[a] Fuel use based on 4 yr average weather data.

[b] Fuel costs are for September 1976 in Rolla, MO.

Electric rate assume a minimum of 1000 kWh/month of electricity used for other than heating.

quirements by up to 14% while delivering identical heating requirements. Air-to-air heat pumps can significantly reduce energy requirements for the heating of a residence. Although not shown in the table, supplemental electrical resistance heating accounted for 27–35% of the heating energy requirements.

Comparative annual costs were calculated for the oil, LP gas, and electric furnace, and for two of the heat pumps, based on the 4-year average data and the September 1976 fuel costs in Rolla, Missouri. These results are presented in Table 8.3. These results have been translated to a bar graph for other energy costs, shown in Fig. 8.2. Annual energy costs for this residence at various oil, gas, and electric rates are given.

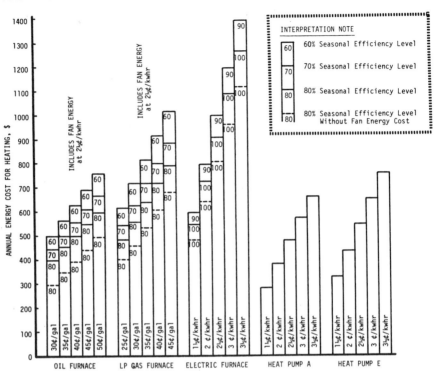

Fig. 8.2.   Annual energy costs at various oil, gas, and electric rates.

Three significant conclusions can be drawn from the cost estimates in Table 8.3. First, an oil-fired furnace (at 70% efficiency) was the least expensive to operate of the three typical residential heating units. The cost of LP gas heating (at 70% efficiency) was 19% higher, while electric resistance heating (at 100% efficiency) was 26% higher than oil. Second, both heat pumps were less expensive to operate than the oil furnace and thus less also than gas and electric furnaces. The average of the two heat pumps gives a 29% reduction in operating cost. Even if credit was given for the CAC operation on the oil furnace, the air-to-air heat pumps would consume about 17% fewer operating dollars. Third, the average operating cost of the two heat pumps is about 44% less than the operating cost of the electric resistance furnace. Giving credit for the CAC operation on the electric furnace would lower this difference in operating costs to approximately 36%.

These cost comparisons show that significant differences do exist in the operating costs for various types of residential heating systems. These differences should be taken into consideration along with initial costs and maintenance costs when selecting a residential heating unit.

It is important to note that the operating costs from this study were based on relative fuel prices existing in 1976 and have since changed.

8.2. GORDIAN STUDY: RESIDENTIAL (1976)

8.2.1. Scope

In a comprehensive study in 1976 [2], alternative heating and cooling systems were tested to determine their *energy effectiveness,* the relative
amounts of energy required to maintain comfort conditions in single-family
detached houses in the United States. In particular, the energy required by
the electric heat pump, in nine widely varying climatic and geographic regions in the United States, was estimated and compared with estimated
energy requirements for electrical-resistance and fossil-fuel space-heating
systems.

The study simulated the load on a single residential structure synthetically
located (by means of the computer) in each region and then estimated the
energy required to balance the load by simulating the performance of different heating and cooling systems. The comparison of energy effectiveness for
residential heating was based upon simultaneous computer simulations of
the electric heat pump, central gas-fired furnace, central electric furnace,
and electric baseboard. An estimate of energy effectiveness for the electric
heat pump compared with an oil-fired furnace was determined by scaling the
computer gas-furnace energy data.

After design-load calculations were completed for each of the nine cities,
equipment to be simulated was selected. Using manufacturers' performance
data, heat pump, split air-conditioning, and electric furnace operations were
simulated. A partial-load efficiency curve for the gas furnace was used in its
simulation. Because it was assumed that there are no system losses in electric baseboard heating, all energy consumed was modeled as a heat gain to
the space. In addition, duct losses, where applicable, were taken into account.

The output from the systems simulation consisted in part of hourly values
of heat pump compressors, strip heater, blower, and fan energy consumption, gas furnace gas consumption and blower energy consumption, electric
furnace energy consumption, baseboard energy consumption, and air-conditioning compressor, fan, and blower energy consumption.

8.2.2. Selection of Test Cities

Cities in which to locate the test house were selected to represent as completely as possible the major climatic regions of the continental United
States on a heating degree-day basis. The United States spans a range of
degree days from about 100 in southern Florida to more than 10,000 in the
northern areas of Minnesota and North Dakota. On the basis of population
distribution, however, the relevant degree-day range is smaller and was

taken for this study to be 500–8000 degree days. This degree-day range was then subdivided into degree-day zones, and nine test cities were selected.

The first selection consideration took the form of restricting the number of computer simulations, one to each city selected, to a practical number. The computer simulation is complicated and time-consuming and was necessarily limited to only those test cities minimally required to characterize the effect of climatic variations on energy effectiveness. Of the cities to be selected, it was mandatory that one be Columbus, Ohio, the location of the model test house that was to be simulated. Another had to be selected from the coastal portion of Washington or Oregon. The northwest part of the country represents a sufficient population in a climatic region of roughly 4000–5000 degree days that is considerably different from other parts of the country with the same number of degree days (Philadelphia, for example). The winters there are not very intense, and the annual extremes in temperature are less than in other regions with the same number of degree days. However, the average temperature is sufficiently below 65°F to make the heating season relatively long. In addition, there are long periods of rain and generally high humidity. It was anticipated that these climatic conditions would produce an anomalous residential heating situation that required special attention, therefore demanding that Portland, Oregon, or Seattle, Washington, be selected as one of the nine test cities.

It was also considered desirable to select test cities that had a history of heat pump activity. The purpose here was to make available field-performance data in areas of high heat pump saturation that could be used to compare the computer simulation resulting from the present study. There had also been sufficient independent energy effectiveness studies performed by electrical utilities in regions of high heat pump concentration to attempt to include cities in those regions as test sites for this study. On this basis, particular attention was focused on Birmingham, Alabama; Atlanta, Georgia; Cleveland, Ohio; and Concord, New Hampshire as highly desirable locations for energy effectiveness simulations.

The final selection of test cities is shown in Table 8.4 along with the heating degree days (and in one case, cooling hours) characterizing each city from a number of sources. The variation among the data serves to clearly illustrate the difficulty associated with using degree days as a measure of heating requirements. Although a precisely defined variable, degree days for any location strongly depends on the time interval over which the measured degree days are averaged to produce an annual estimate.

The procedure for determining the "average" year was adapted from a procedure proposed by the ASHRAE Task Group on Energy Requirements for Heating and Cooling. The typical year weather data selected for each city is shown in Table 8.4 along with the heating degree days for that year which now no longer necessarily agrees exactly with the number of degree days for which the cities were originally selected.

TABLE 8.4. Weather Data for Test Cities

City	ASHRAE[a] Heating Degree Days	NAHB[b] Heating Degree Days	NAHB[b] Cooling Hours	NOAA[c] Heating Degree Days	Weather Data Used in This Study Heating Degree Days	Weather Data Used in This Study Year of Weather Data	Weather Data Used in This Study Deviation Level
Houston, TX	1278	1400	1900	1375	1290	1955	3%
Birmingham, AL	2551	2600	1350	2475	2844	1955	7%
Atlanta, GA	2961	3000	1000	2913	2821	1959	6%
Tulsa, OK	3860	3800	1600	3655	3504	1950	6%
Philadelphia, PA	4486	4400	700	4866	4508	1951	8%
Seattle, WA	4424	5200	100	4372	4407	1960	2%
Columbus, OH	5660	5600	600	5458	5467	1964	10%
Cleveland, OH	6351	6400	500	5972	6097	1964	10%
Concord, NH	7383	7400	400	7448	7377	1964	10%

[a]Heating degree days from *ASHRAE 1973 Systems Handbook*, Chapter 43, "Energy Estimating Method." Data is credited to *Monthly Normals of Temperature, Precipitation, and Heating Degree Days, 1962* published by the United States Weather Bureau for the period 1931 to 1960 inclusive with revisions from the 1963 Revisions to the U.S. Weather Bureau publication.

[b]*Insulation Manual—House/Apartments*, published by National Association of Home Builders (NAHB), September 1971, Rockville, Maryland. Heating degree days are credited to ASHRAE as in (a) above. Discrepancy between ASHRAE degree day value for Seattle (4424) and NAHB's value (5200) is due to the Seattle-Tacoma Airport weather station location for NAHB and a city weather station for ASHRAE. NAHB values are rounded off to the nearest 200 degree days. Cooling hour data is credited to the *Engineering Weather Data Manual* rounded to the nearest 50 cooling hours.

[c]Degree day values for each city shown are averaged from the *1934–1973 Annual Summary with Comparative Data*, National Oceanic and Atmospheric Administration (NOAA), Environmental Data Service, U.S. Department of Commerce.

SOURCE: Gordian Associates [2].

455

8.2.3. Selection and Description of Test House

The particular structure selected for simulation is an existing single-family detached residence located in the northeast area of Columbus, Ohio. The selection was based primarily on the fact that a large amount of field test data had been gathered by Ohio State University researchers for this particular house under occupied conditions for a period of about 2 years. Information, such as internally generated heat loads and measured energy consumption for different seasons and for heating and cooling systems identical with the systems to be simulated, could then be advantageously used in the present study. In particular, computer simulation of energy requirements for the test house in the Columbus, Ohio, location could be compared with measured energy requirements and thereby serve to validate the computer models.

Another factor in its selection is that the Columbus test house is relatively new, incorporating modern construction techniques, and as such is representative of new single-family residential units currently being built in the country. Thus, the absolute amounts of energy computed by the simulation for the various heating and cooling systems takes on a valuable degree of relevance.

General physical characteristics of the test house are reported in Fig. 8.3.

8.2.4. Heating and Cooling Systems Selection

The simulation of energy required to balance thermal loads for the test house in each city studied was performed for four different combinations of heating and cooling systems. The systems simulated were:

1. Air-to-air split heat pump
2. Gas furnace and central air-conditioning
3. Electric furnace and central air-conditioning
4. Electric baseboard and central air-conditioning

The basis for the equipment selection was a combination of design heat loads for heating equipment and cooling loads (both sensible and latent) for the air-conditioning equipment. These heating and cooling loads for the nine cities are tabulated in Table 8.5 along with the $97\frac{1}{2}\%$ dry-bulb temperature (the lowest winter temperature achieved $97\frac{1}{2}\%$ of the time) used to estimate design heat load and the $2\frac{1}{2}\%$ dry-bulb and $2\frac{1}{2}\%$ wet-bulb temperatures (the highest wet- and dry-bulb temperatures reached $2\frac{1}{2}\%$ of the time) used to estimate sensible and latent heat cooling loads.

One of the contributions to the heat balance on the residential structure is the internal generation of heat. This important term reduces the energy input from a heating system during the heating season to maintain residential comfort and adds to the burden of a cooling system during warm weather

Heat Transfer Surface	Insulation Specified[a]	Total Resistance to Heat Transfer[b] [(hr·ft²·°F)/Btu]	Overall Heat Transfer Coefficient[c] [Btu/(hr·ft²·°F)]	FHA Minimum Overall Heat Transfer Coefficient [Btu/(hr·ft²·°F)]
House style		Two stories, wood frame, 2-car garage		
Living area		1850 ft² (excluding basement)		

<table>
House style Two stories, wood frame, 2-car garage
</table>

House style   Two stories, wood frame, 2-car garage
Living area   1850 ft² (excluding basement)
Room complement
  First floor   Foyer, living room, dining room, family room, kitchen, powder room
  Second floor   Master bedroom, large bedroom, two small bedrooms, two baths
Exposure   Front of house faces north
Exterior surface   Aluminum siding
Doors   One, wood panel construction
Roof   Black asphalt shingles over building paper and $\frac{5}{8}$-in. plywood deck
Glass areas   16 glass areas, including patio sliding door. All windows are aluminum cased, double-hung, weatherstripped, double pane insulating glass. Basement windows (2) are single glazed, $\frac{1}{8}$-in. sheet glass
Basement
  Walls   11-course, 8-in. concrete block
  Floor   4-in. concrete slab over 4-in. gravel bed, 662 ft²
  Crawl space   4-in. gravel bed over plastic vapor barrier

Insulation properties

Heat Transfer Surface	Insulation Specified[a]	Total Resistance to Heat Transfer[b] [(hr·ft²·°F)/Btu]	Overall Heat Transfer Coefficient[c] [Btu/(hr·ft²·°F)]	FHA Minimum Overall Heat Transfer Coefficient [Btu/(hr·ft²·°F)]
Exterior walls	R-9	13.47	0.074	0.08
Ceilings (room to attic)	R-19	20.77	0.048	0.05
Overhang	R-19	23.22	0.043	0.05
Garage wall to house interior	R-9	12.32	0.081	0.08
Garage ceiling to house interior	None	4.64	0.216	Optional— not required
	R-7	11.64	0.086	0.10
	R-9	13.64	0.073	0.08
Floor section over basement	None			Not required
Basement foundation walls	None			Not required

[a] R values refer to resistance to heat transfer (in hr·ft²·°F/Btu)

[b] Total resistance to heat transfer and heat transfer coefficient for each surface including specified insulation.

[c] Reciprocal of total resistance to heat transfer, in previous column.

Fig. 8.3. General characteristcs of test house [2].

TABLE 8.5. Design Temperatures and Heat Load for Heating and Cooling Equipment [2]

| | Heating | | | Cooling | | |
	$97\frac{1}{2}\%$ Design db Temperature	Design Heat Load (Btu/hr)	$2\frac{1}{2}\%$ Design db Temperature	$2\frac{1}{2}\%$ Design wb Temperature	Design Sensible Cooling Load (Btu/hr)	Design Total Cooling Load (Btu/hr)
Houston	32	27,030	94	80	24,069	36,844
Birmingham	22	33,915	94	78	25,335	35,803
Atlanta	23	33,288	92	77	23,126	32,994
Tulsa	16	37,671	99	78	29,383	38,351
Philadelphia	15	38,242	90	77	24,880	35,302
Seattle	32	27,621	79	65	18,757	20,365
Columbus	7	42,913	88	76	23,468	33,090
Cleveland	7	42,913	89	75	24,454	32,845
Concord	−7	52,088	88	73	23,837	30,428

months. These internally generated loads are due to lights, appliances, and the activities of people within the residence. The heat produced is in the form of sensible heat, which increases the dry-bulb temperature in the house, and latent heat, which raises the humidity. Both forms are generally beneficial during the heating season, while both add to the burden of maintaining comfort level during the cooling season.

The hourly heat load contributions from human activity were modeled according to standard engineering practice. Some direct measurements and modeling techniques were used to simulate the internal heat loads due to appliances and lights.

8.2.5. Results of Simulations

A summary of computer-simulated energy-consumption data for the different heating and cooling systems studied is presented in Table 8.6. The seasonal and annual energy totals for the test house are assembled for each of the nine cities selected to represent U.S. climatic regions. Heating degree days for each city are those reported by NOAA (see Table 8.4), while the cooling hours and heat pump performance factors at each location were calculated from the computer simulation as explained in the notes to Table 8.4. The energy data shown are for consumption at the point of application, that is by the residential homeowner, for the different heating and cooling systems that are most often encountered. For all-electric systems (heat pump, electric furnace, and baseboard) energies consumed are expressed in single units of kilowatt-hours. For gas- and oil-fired heating systems combined with all-electric air-conditioning, energy consumption is reported as the quantity of fossil fuel required (cubic feet of gas or gallon of oil) and kilowatt-hours of electric energy. Most of the electric energy required for the gas and oil systems combined with air-conditioning is for the cooling-season operation of the air conditioner. However, some small amount of electric energy is reported for the heating season and represents energy needed for fans that distribute heated air.

The reason for the low value of power required for the cooling season with window-air conditioning units is that not all of the upstairs was cooled. Only two units were installed on the second floor, so some bedrooms were not air-conditioned.

A direct comparison of the energies required by different systems to comfort-air condition the test house is difficult with the results as reported in Table 8.6. To assist in the comparison, all energies were converted to an equivalent basis and uniformly reported as therms (1 therm = 100,000 Btu) in Table 8.7. The conversion of kWh to therms is simple and direct and uses the equivalency factor 3412 Btu/kWh. The gas- and oil-fired systems are slightly more complicated and required knowledge of the heating value of each fuel in each of the test cities. Average heating values for gas and oil reported in 1973 on a state basis were used.

TABLE 8.6. Fuels and Electric Energy Consumed by Test House in Selected Cities [2]

Cities	Length of Season (mo)	Degree Days or Cooling Hours	Heat Pump kWh	Heat Pump SPF	Gas Furnace and Central A/C Gas (ft³)	Gas Furnace and Central A/C Electric (kWh)	Oil Furnace and Central A/C Oil (gal)	Oil Furnace and Central A/C Electric (kWh)	Electric Furnace and Central A/C (kWh)	Baseboard Electric and Central A/C (kWh)	Baseboard Electric and Window A/C (kWh)
Houston											
Heating season	4	1290	1,416	2.64	18,137	60	134	64	3,372	3,092	3,092
Cooling season	8	2022	9,283	2.47	—	8019	—	8019	8,019	8,019	4,818
Total annual			10,699		18,137	8079	134	8083	11,391	11,111	7,910
Birmingham											
Heating season	6	2844	3,832	2.14	43,224	144	347	154	8,107	7,432	7,432
Cooling season	6	1480	6,790	2.47	—	5867	—	5867	5,867	5,867	4,455
Total annual			10,622		43,224	6011	347	6021	13,974	13,299	11,887
Atlanta											
Heating season	5	2821	3,837	2.14	43,965	147	340	157	8,286	7,596	7,596
Cooling season	7	1393	6,333	2.51	—	5485	—	5485	5,485	5,485	4,235
Total annual			10,170		43,965	5632	340	5642	13,771	13,081	11,831
Tulsa											
Heating season	6	3504	7,653	1.83	73,358	211	583	226	14,364	13,367	13,369
Cooling season	6	876	4,731	2.53	—	4365	—	4365	4,365	4,365	3,501
Total annual			12,384		73,358	4576	583	4591	18,729	17,734	16,870

Philadelphia											
Heating season	7	4508	8,013	1.69	80,161	279	610	299	15,729	14,386	14,386
Cooling season	5	917	4,123	2.54	—	3583	—	3583	3,583	3,583	2,847
Total annual			12,136		80,161	3862	610	3882	19,312	17,969	17,283
Seattle											
Heating season	9	4407	7,682	1.80	76,134	294	579	315	14,363	12,818	12,818
Cooling season	3	532	1,697	2.38	—	1382	—	1382	1,392	1,382	1,101
Total annual			9,379		76,134	1676	579	1697	15,745	14,201	13,919
Columbus											
Heating season	7	5407	10,796	1.69	99,738	340	706	364	19,199	17,561	17,561
Cooling season	5	861	3,896	2.53	—	3275	—	3275	3,275	3,275	2,804
Total annual			14,692		99,738	3615	706	3539	22,474	20,836	20,365
Cleveland											
Heating season	7	6097	13,242	1.60	116,600	403	826	431	22,739	20,798	20,798
Cooling season	5	670	2,981	2.59	—	2597	—	2597	2,597	2,597	2,304
Total annual			16,223		116,600	2999	826	3028	25,336	23,395	23,102
Concord											
Heating season	9	7377	16,631	1.47	130,587	461	948	493	26,061	23,784	23,784
Cooling season	3	435	1,938	2.58	—	1813	—	1813	1,813	1,813	1,029
Total annual			18,569		130,587	2274	948	2306	27,874	25,597	24,813

TABLE 8.7. Equivalent Energy Consumed by Test House in Selected Cities (Therms) [2]

Cities	Length of Season (mo)	Degree Days or Cooling Hours	Heat Pump SPF	Heat Pump therms	Gas Furnace and Central A/C (therms)	Oil Furnace and Central A/C (therms)	Electric Furnace and Central A/C (therms)	Electric Baseboard and Central A/C (therms)	Electric Baseboard and Window A/C (therms)
Houston									
Heating season	4	1290	2.64	48	188	200	115	105	105
Cooling season	8	2022	2.47	317	274	274	274	274	164
Total annual				365	462	474	389	379	269
Birmingham									
Heating season	6	2844	2.14	131	454	485	276	253	263
Cooling season	6	1480	2.47	232	200	200	200	200	152
Total annual				363	654	685	476	453	405
Atlanta									
Heating season	5	2821	2.14	131	458	490	283	259	259
Cooling season	7	1393	2.51	216	187	187	187	187	144
Total annual				347	645	677	470	446	463
Tulsa									
Heating season	6	3504	1.83	261	770	824	490	456	456
Cooling season	6	876	2.53	161	149	149	149	149	119
Total annual				422	919	973	639	605	575

Philadelphia									
Heating season	7	4508	1.69	273	822	880	537	490	490
Cooling season	5	917	2.54	141	122	122	122	122	98
Total annual				414	944	1002	659	612	588
Seattle									
Heating season	9	4407	1.80	262	818	875	490	437	437
Cooling season	3	532	2.38	58	47	47	47	47	38
Total annual				320	865	922	537	484	475
Columbus									
Heating season	7	5407	1.69	368	956	1023	655	599	599
Cooling season	5	861	2.53	133	111	111	111	111	95
Total annual				501	1067	1134	766	710	694
Cleveland									
Heating season	7	6097	1.60	452	1118	1197	776	710	710
Cooling season	5	670	2.59	102	88	88	88	88	78
Total annual				554	1206	1285	864	798	788
Concord									
Heating season	9	7377	1.47	568	1322	1414	889	811	811
Cooling season	3	435	2.58	66	62	62	62	62	35
Total annual				634	1384	1476	951	873	846

The results of the computer simulation as reexpressed in Table 8.7 clearly show that less total energy is expended by the heat pump to satisfy seasonal heating demand in a particular residence than for any other heating system. The low total energy demand of the heat pump relative to other heating systems is sustained throughout the degree-day range of the nine cities examined. The closest competitor of the heat pump in energy efficiency is electric-baseboard resistance heating followed closely by the central electric furnace. The fossil-fuel systems, gas and oil, appear to require the largest amount of energy input to accomplish the same heating task as the heat pump.

To quantify these generalized findings further, heating SPFs were calculated for each of the heating systems examined. Seasonal performance factors provide a direct index of system efficiency with respect to energy effectiveness, as shown in Table 8.8. For the nine test cities, the heat pump SPF for an entire heating season increases from Concord, New Hampshire, the coldest region with the greatest heating degree days, to Birmingham, Alabama, and Tulsa, Oklahoma, the warmest regions with lower heating degree days. This inverse relationship between heating SPF and degree days for a heat pump is qualitatively correct and is similar to the inverse relationship between heat pump efficiency and the temperatures of cold outside air from which heat is abstracted to heat the resistance.

The SPF for electric baseboard heating is shown uniformly to be 1.0. That is, every kilowatt-hour of electric energy used by a baseboard heater is converted directly into heat in the residence, and there is no mechanism for energy loss. Some small energy loss is associated with the electric furnace, which requires more energy input to heat the residence than does baseboard heating. This is due to inherent inefficiencies in hot-air-heated systems, such as distribution losses in ducts and fans which circulate heated air through the house but run at less than 100% efficiency. A comparison of the heating-season energy requirements shows the electric furnace is about 91–93% as efficient as baseboard heating, with 92–93% being more representative of warmer climates.

Fossil-fuel combustion systems show a severe reduction in energy efficiency due to losses that are characteristic only of these types of systems. In particular, venting hot combustion gases through the chimney, incomplete combustion (in the case of oil), and the need to preheat cold outside air contribute heavily to energy inefficiency. Gas-furnace efficiencies of 56–64%, relative to electric baseboard electric systems, are shown in Table 8.8. There is a noticeable tendency for the efficiency of the gas furnace to decrease in warmer regions, to less than 60%, due to more rapid on-off cycling than in colder regions. The furnace needs to be operated for shorter periods in warmer regions to balance heat losses from the residence. This means that the furnace is run lower down on the partial-load curve, with a commensurate reduction in efficiency compared to full-load efficiencies reported as 80% by most manufacturers and distributors of gas-furnace equipment. The

TABLE 8.8.   Heating Season Performance Factors for Heat Pumps and Other Heating Systems [2]

City	Heat Pump		Gas Furnace	Oil Furnace	Electric Furnace	Baseboard Electric
	System[a]	Compressor and Fan[b]				
Houston	2.64	2.66	0.56	0.53	0.92	1.0
Birmingham	2.14	2.40	0.56	0.52	0.92	1.0
Atlanta	2.14	2.37	0.57	0.53	0.93	1.0
Tulsa	1.83	2.39	0.59	0.55	0.93	1.0
Philadelphia	1.69	2.13	0.60	0.56	0.91	1.0
Seattle	1.80	1.97	0.53	0.50	0.89	1.0
Columbus	1.69	2.40	0.63	0.59	0.91	1.0
Cleveland	1.60	2.51	0.64	0.59	0.91	1.0
Concord	1.47	2.73	0.61	0.57	0.91	1.0

[a] Season performance factors (heat pump system): Total energy delivered to the residence divided by total energy delivered to the heat pump system (compressor energy, indoor and outdoor fan energies, supplementary heater energy). Energy delivered to the residence is just equal to thermal load in the residence.

[b] Season performance factor (compressor and fan): Same as above except that energy delivered to the pump system is for the compressor and fans, but does not include supplementary heaters.

oil furnace shares these same energy-inefficient characteristics and, as explained previously, requires about 7% more energy than the gas furnace to accomplish the same heating task. The net oil-furnace efficiency is therefore shown as about 50–59% relative to electric baseboard heating.

The efficiency of the heat pump for space-conditioning in comparison to central air-conditioning and air-conditioning using window units is shown in Table 8.7. The heat pump is seen to be less energy efficient than central air-conditioning in cooling. Energy efficiency ratios (EER) of the heat pump for the entire cooling season vary from 8.1 (Btu/hr)/W in Seattle to 8.8 (Btu/hr)/W in Cleveland and Concord. Central air-conditioner EERs are 15–20% greater than for the heat pump.

The tendency for both heat pump and central air-conditioning cooling efficiency to increase in cooler climates is an inherent characteristic of vapor-compression refrigeration cycles. The lowest total energy consumption for summer air-conditioning is for cooling provided by window air conditioners. Part of the reason for the window air-conditioner's advantage over both heat pumps and central air conditioners is obviously the fact that only part of the heat load on the residence is being satisfied, about 57–89% for the cases examined here. A countereffect is that the energy efficiency of the window air conditioners used was generally lower than that of either the heat pump or central air conditioners. Bedroom air conditioners had reported EERs of 5.6 (Btu/hr)/W.

TABLE 8.9. Comparison of Gordian and Delene Studies

	Atlanta		Philadelphia		Seattle	
	Delene	Gordian	Delene	Gordian	Delene	Gordian
Heat pump SPF[a]						
(heating only)	2.09	2.14	1.92	1.69	2.24	1.80
Gas consumed, 10^3 ft^{3b}	55.1	44.0	106.7	80.2	108.9	76.1
Oil consumed, gal[c]	424	340	820	610	837	579
Electric energy for baseboard heating,						
kWh	9,700	7,590	18,760	14,386	19,150	12,818

[a]Delene does not report heat pump SPF. Instead they were estimated as in this report by forming the ratio: energy consumed by electric baseboard heaters/energy consumed by heat pump.

[b]Delene used a constant gas-furnace efficiency of 60% for all locations. For this study, gas-furnace efficiencies were 57% for Atlanta, 60% for Philadelphia, 53% for Seattle.

[c]Delene used a constant oil-furnace efficiency of 55% for all locations. For this study, oil furnace efficiencies were 53% for Atlanta, 56% for Philadelphia, 50% for Seattle.

SOURCE: Gordian Associates [2]

Despite the slightly lower efficiency of the heat pump compared with central air-conditioning, the net annual energy consumption of the heat pump is consistently less than any combination of electrical-resistance or fossil-fuel heating equipment with central air-conditioning. The reason for the heat pump's superiority is the high energy utilization efficiency it enjoys during winter heating.

8.3. DELENE STUDY: RESIDENTIAL (1974)

In a 1974 study, Delene [3] reported costs and energy requirements of space-conditioning equipment similar to the Gordian work for a single-family residence in nine cities, each city representing weather conditions for the nine census divisions of the United States. The test house modeled by Delene appears to be quite similar to the Columbus, Ohio, test house the Gordian study is based on, although insufficient details are available to thoroughly compare them. Some of the cities in the Delene study were also test cities in the Gordian report, and a comparison of some of the energy utilization results is shown in Table 8.9.

8.4. UNION ELECTRIC STUDY: RESIDENTIAL (1980)

Coinciding with a general interest in energy conservation, the electric heat pump has been installed in increasing numbers of homes in the Missouri

TABLE 8.10. General Characteristics and Calculated Heat Losses of Seven St. Louis Area Heat Pump Test Homes

Home	Estimated Age (yr)	Living Area (ft^2)	Basement Area (ft^2)	Infiltration Rate (AC/hr)	Calculated Heat Loss in W (80°F T.D.)
1	37	837	837	0.60	14,620
2	19	1697	1697	0.75	26,600
3	16	1755*	1098	0.60	20,600
4	18	1515	1287	0.60	19,380
5	45	1041	1041	0.75	18,240
6	8	2780*	1486	0.60	26,770
7	19	1189	1151	0.65	13,990

[a]Total square footage of both floors of 1½ or 2-story homes.

SOURCE: Raab [4].

areas served by Union Electric Company. About 50% of home owners in the U.E. region who install electric space heating systems are now choosing a heat pump to provide both cooling and heating to their residences. More than 3000 residences were equipped with a heat pump in Union Electric's service area during 1978. Comparing this number with an estimate of 50 installed during 1972 gives a good indication of the rapid escalation of heat pump installations during recent years.

Late in 1976, Union Electric Company decided to meter heat pump installations in the homes of some of their customers to determine actual electricity consumption compared to that predicted in our calculations.

Data were gathered during the 1977–1978 heating season on seven homes located throughout St. Louis County, Missouri. The homes ranged in age from 8 to 45 years, and in every instance the heat pump was installed as a replacement of an existing forced warm-air heating system. Only one of the homes was originally built for and equipped with an electric heating system. Table 8.10 gives explanatory data on the general characteristics of the test homes involved. Variation in infiltration rates in terms of air changes per hour is due to a judgment factor related to the generally observed tightness of the window and door areas of the home. Calculations include basement areas maintained at comfort conditions for the entire heating season.

No changes were made in the construction or thermal characteristics of the homes with the installation of the air-to-air heat pumps. No major changes in ductwork or supply-and-return registers were made. The test homes were representative of a large number of existing dwelling units in the St. Louis and St. Louis County areas and of the heat pump installation skills of heating and air-conditioning contractors doing business in the area.

Heat pumps used were "off-the-shelf" models manufactured by Amana, Carrier, General Electric, Lennox, Singer, and Westinghouse. None of them

TABLE 8.11. Comparison of Actual to Estimated Energy Consumption (1977–1978 Heating Season) of Seven St. Louis Area Heat Pump Test Homes [4][a]

Home	Estimated kWh per 4900 DD	Actual kWh	Metered DD	Adjusted Actual per 4900 DD	Deviation from Estimate
1	11,066	11,422	5667	9,876	−11%
2	24,271	12,528	5667	10,831	−55%
3	17,498	14,344	5667	12,401	−29%
4	15,223	14,621	5667	12,641	−17%
5	17,800	8,712	3739	11,417	−36%
6	21,656	12,755	5348	11,687	−46%
7	10,193	7,663	5667	6,625	−35%

[a]D.D. = degree days.

were of the more recent higher-efficiency type. No special supervision of the heat pump installations was provided.

The method employed by Union Electric Company to estimate seasonal energy consumption of the heat pumps installed in the test homes coincided with that generally used by heat pump manufacturers, with one exception. Manufacturers generally use the same design conditions, 70°F indoor temperature and −5°F outdoor temperature for the heat pump as they do for their electric furnaces. Union Electric theorized that because of cooler supply-air temperatures, customers living with heat pumps would maintain higher indoor temperatures throughout the heating season than they would with an electric or gas-fired furnace. For that reason, a 5°F indoor temperature increment was added to that which Union Electric used in its electric resistance consumption estimate.

Table 8.11 gives the results of the metered 1977–78 heating season and a comparison with the estimated energy consumption for each home. In every instance, the heat pumps consumed less energy than that which was predicted by the calculating method.

In 1960, a study of 23 heat pump homes was conducted by Union Electric Company. The result of that study in terms of energy consumption was an average of 0.0581 kWh/(DD·1000 Btu/hr). Also, a wide deviation of 50% below and above the average figure was noted in this 1960 study.

The present heat pump test (see Table 8.12) using the same basis of comparison, develops an average of 0.0331 kWh/(DD·1000 Btu/hr) energy consumption, with a deviation from the average of only 26%. Thus, a substantial improvement in terms of operating efficiency and consistency of performance is indicated for the pumps generally available in the mid-1970s compared to those available in the mid-1950s.

Since the metering segregated many of the components of the heating system, the usage of each of these components by winter months and for the season, shown in Table 8.13, is of interest. Defrosting does not appear to add a major number of kilowatt-hours to heating consumption over a winter

TABLE 8.12. Heat Pump Heating Energy Consumption for Seven St. Louis Area Homes [4]

Home	Actual[a] kWh/(DD·1000 Btu/hr)	Deviation From Average of .0331
1	0.0403	+ 22%
2	0.0244	− 26%
3	0.0360	+ 9%
4	0.0390	+ 18%
5	0.0374	+ 13%
6	0.0261	− 21%
7	0.0283	− 15%

[a] Includes all energy used for crankcase heat, indoor blower, defrost cycles, and heating cycles of compressor and resistance heat.

season. However, the consumption of the compressor in the defrost mode is not the total cost of defrosting. Since the resulting cooling by the refrigeration cycle inside the home must be offset through the operation of resistance heating, the total effect of operating in the defrost mode is estimated to be at least three times the consumption recorded in the defrost mode. Even then, this would appear to be less than 4% of the heating kilowatt-hour consumption of an average heat pump over a winter season of operation.

Another item of interest is the percentage of resistance heating consumption to the total required for heating. In the test, this amounted to 36.79% as an average of all pumps. A theoretical calculation of heat pump performance using manufacturers' data (of units having the efficiencies of those installed),

TABLE 8.13. Integrated Heating Component Energy Consumption of Seven St. Louis Area Heat Pump Homes, 1977–1978 Heating Season [4]

	Number of Homes Metered	Compressor kWh (Including Sump Heater)		Indoor Blower and Res. (kWh)	Total Heating (kWh)	Adjusted Degree Days Metered
		Defrost Mode	Heating Mode			
Oct. 1977	5	14	1,130	225	1,369	282
Nov.	6	86	4,675	1,606	6,367	601
Dec.	6	154	9,122	5,068	14,344	910
Jan. 1978	7	268	13,049	11,092	24,409	1330
Feb.	7	216	12,114	6,888	19,218	1223
Mar.	7	167	8,164	3,704	12,035	835
Apr.	7	34	1,935	1,176	3,145	284
Total season		939	50,189	29,759	80,887	
Percent of total kWh		1.16%	62.05%	36.79%	100%	

TABLE 8.14. Monthly Kilowatt-Hours per Degree Day per 1000 Btu's per Hour of Seven St. Louis Area Heat Pump Homes during the 1977–1978 Heating Season [4]

Homes	Oct. (282 DD)	Nov. (601 DD)	Dec. (910 DD)	Jan. (1330 DD)	Feb. (1223 DD)	Mar. (835 DD)	Apr. (284 DD)
1	0.034	0.029	0.041	0.043	0.040	0.039	0.038
2	0.015	0.021	0.026	0.031	0.027	0.022	0.016
3	0.020	0.029	0.041	0.043	0.037	-0.034	0.016
4	0.011	0.029	0.044	0.048	0.047	0.046	0.020
5	NA	NA	NA	0.040	0.040	0.035	0.049
6	NA	0.021	0.028	0.025	0.027	0.025	0.021
7	0.008	0.024	0.031	0.034	0.032	0.026	0.013
Average	0.0176	0.026	0.035	0.038	0.036	0.032	0.022

[a]Includes compressor operation in defrost and heating modes, crankcase heater, indoor blower and supplemental resistance heaters.

and based upon normal St. Louis winter temperatures, will project electricity consumption for resistance heating in the neighborhood of 30% of the total. Since the 1977–1978 heating season was colder than normal, a higher use of resistance heat would be expected, and the measured amount of 36.79% would seem to indicate a reasonable relationship to the theoretical projections. The data on kilowatt-hours per degree day per 1000 Btu's per hour (Mbh) in Table 8.14 shows the effect of heavier reliance on resistance heating and reduced efficiency of the refrigeration cycle during colder weather.

As degree days per month increase, the energy consumption per degree day also increases. The fact that heat pumps appear to be more efficient (when measured against degree days) in the fall of the year than they are in the spring may be due to the heat sink that the earth provides as storage from warm summer temperatures. These data, indicating an average energy consumption greater in April than in October for nearly equal degree days, suggests that any performance test conducted with heat pumps comparing the first half of a winter season with the last half (adjusted for degree days) may give incorrect results.

The method generally employed to describe the energy savings capability of a heat pump is to compare its kilowatt-hour consumption over a heating season, including the use of the supplementary electric resistance heating, to that which would be required using straight electric resistance heating alone.

$$\text{Seasonal performance factor (SPF)} = \frac{\text{resistance heating kWh required}}{\text{heat pump system kWh required}}$$

The calculated SPF of each of the heat pump installations in the test is shown in Table 8.15. The average calculated SPF for the seven heat pump

TABLE 8.15.   Calculated Seasonal
Performance Factors of Seven St. Louis Area
Heat Pump Homes during 1977–1978

Home	Metered kWh/(DD·Mbh)	Calculated SPF[a]
1	.0403	1.69
2	.0244	2.79
3	.0360	1.89
4	.0390	1.74
5	.0374	1.81
6	.0261	2.60
7	.0283	2.40

[a] $\dfrac{0.068/\text{metered kWh}}{\text{DD} \times \text{Mbh})}$

SOURCE: Raab [4].

installations over the 1977–1978 winter metered period of 5465 degree days was 2.1 : 1.

The performance of the pumps in this test, which covered a winter heating season 11.5% colder than normal, indicated a real potential for improved energy efficiency through the use of air-to-air heat pumps in both existing and new St. Louis area homes.

8.5.   GORDIAN STUDY: RESIDENTIAL (1978)

8.5.1.   Methodology

The purpose of the energy consumption simulation study carried out by Gordian Associates in 1978 [5] was to compare on an equal basis the electric and/or fossil-fuel energy required by the various heat pumps and other systems under consideration in meeting the same seasonal heating or cooling load in identical selected test buildings. For the residential-size HVAC equipment, the simulation determined the hourly heating and cooling energy requirements of a single-family two-story frame dwelling with a heated area of 1850 ft^2.

The residential building HVAC system computer simulations were accomplished in two parts: a load simulation and a system simulation. The load program is a modification of the Post Office Program developed by General American Transportation Corporation for the U.S. Postal Service.

Simulation of specific HVAC equipment operating parameters was computerized using manufacturers' performance data or prototype data for advanced heat pumps as supplied by developers. The input energy to each system required to maintain comfort conditions (i.e., satisfy heating and cooling loads) in each building was thus calculated. In broad outline, the computer simulation procedure is similar to that used in an earlier heat pump study for the FEA (Gordian Associates, [2]). However, considerable refinements of the methods used in the earlier study were incorporated in the present work. A general description of some important features is offered below.

The residential building used represented a real existing structure for which detailed architectural information was available. Important characteristics of the building, such as the composite construction of walls and roofs, window areas, and foundation construction, were thus readily available for the accurate estimation of hourly heating and cooling loads by the computerized load programs.

The residential structure load calculation program accepted ASHRAE load-calculation algorithms. However, these have been modified by the latest information currently available. In addition, both the load program and the system simulation method have been validated in previous investigations by comparing computer results with measured data for existing buildings [5].

A number of important factors, generally neglected in modeling HVAC systems, but having important effects on energy consumption, were included in the 1978 study:

Supplemental resistance heat required to temper cold air admitted to a residential structure during the defrosting of the outdoor heat exchanger or coil during winter heating was included in the simulation of conventional electric heat pumps. Heat pump manufacturers' performance data generally include compressor energy used during the defrost period, when the heat pump cycle is reversed and hot condensing refrigerant is used to melt ice on the outdoor coil instead of being delivered to the indoor coil for heating. The heat pump acts as an air conditioner during defrost periods, and air circulated through the residence is cooled. The amount of supplemental energy used to partially heat the cooled circulating air to the comfort level is dependent upon too many variables to be included in manufacturers' system performance data, but the effect was modeled for each conventional electric heat pump used in this study.

At outdoor dry-bulb temperatures below 40°F and at sufficiently high humidities, some residual frost is always present on the outdoor heat-transfer coil. An ice layer of sufficient thickness will eventually reduce the heating capacity of a heat pump, due to reduced heat-transfer rates in the outdoor refrigerant evaporator. The effect of outdoor coil ice film on heat pump operating efficiency was modeled with data reported in the literature.

The on-off cyclic operation of heating and cooling systems has the effect of reducing operating efficiency below the steady-state performance reported by equipment manufacturers for heating and cooling. Published studies quantifying dynamic efficiency losses for conventional electric heat pumps and natural gas and oil furnaces were used to model cyclic effects. The gas and oil furnace efficiency data also include stack combustion and infiltration losses.

Current installation practices for HVAC systems generally require insulating basement supply and return air ductwork for electric furnaces and electric heat pumps. No ductwork insulation is generally required for fossil-fuel-fired furnaces even if installed in combination with an electric air conditioner, except in isolated cases to prevent condensation inside supply ducts. The effect of heat losses for uninsulated ductwork was examined and found to be sufficiently high for the residential building to significantly affect seasonal performance for heating and cooling. A uniform modeling procedure, including ductwork insulation for all systems, was adopted so that the energy efficiency of all systems could be compared on the same basis.

Performance data for advanced heat pump systems were generally available from developers for only a single size machine currently under development, usually a 3-ton cooling capacity heat pump for residential service. Performance data for other size heat pumps required for some cities were scaled from prototype data by a uniform procedure confirmed in discussions with several manufacturers. In only one case did a heat pump developer undertake the task of scaling performance data for a prototype machine to other sizes [5].

Computerized residential load calculations and systems energy simulations were performed on the test house in nine different cities within the continental United States. These were the same cities used in the earlier study by Gordian and Associates described in Section 8.2.

8.5.2. Residential Building Selection and Design-Load Calculation

The test house selected for computer simulation of residential heating and cooling loads was a single-family detached two-story residence. This residence is an existing structure located in the suburbs of Columbus, Ohio. It was chosen for this study principally because it is serving in an ongoing study of residential load and system simulation modeling under the guidance of research staff at the Ohio State University. Research results obtained with this house have directly contributed to the load and systems simulation programs used in this study. In fact, the results have served to validate the modeling procedures used.

The test house is occupied by a family of four, two adults and two children. Detailed data on internally generated loads due to human activity and appliance usage have been measured over a period of several years and incorporated in this study.

Front elevation and floor plans of the test house are shown in Fig. 8.4, and general construction features given in Fig. 8.3. Insulation used in the test house thermal-load program and heat-transfer characteristics of opaque surfaces are shown in Table 8.16 along with a comparison to ASHRAE 90-75 standards. Air infiltration rates around exterior doors, windows, and sliding doors are shown in Table 8.17, where they are compared with ASHRAE 90-75 standards.

Design heating and cooling loads to be used for heating and cooling equipment selection were determined using the load-simulation program, with design weather data in lieu of NOAA weather data. Design heating and cooling loads estimated in this manner are shown in Table 8.5.

8.5.3. Heating and Cooling System Selection

A total of 16 heating and cooling systems were selected for the examination of energy usage for residential comfort conditioning. These systems were

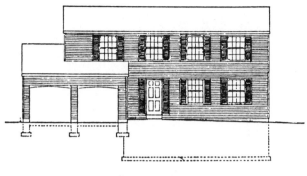

front elevation

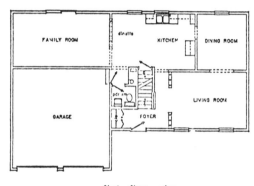

first floor plan

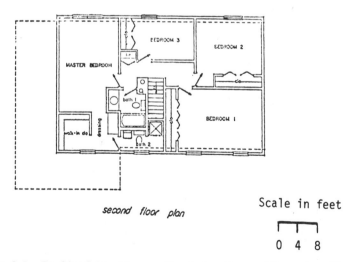

second floor plan

Scale in feet

0   4   8

Fig. 8.4. Residential test house. Front elevation and floor plans [2].

475

TABLE 8.16. Test House Insulation Used and Comparison of Heat-Transfer Coefficients with ASHRAE 90-75 Standards

	EXTERIOR WALLS			CEILINGS			FLOORS OVER UNHEATED SPACES		
	Insulation Specified	Overall Thermal Transmittance (BTU/Hr-Ft²°F) ASHRAE 90-75	Overall Thermal Transmittance (BTU/Hr-Ft²°F) Test House	Insulation Specified	Overall Thermal Transmittance (BTU/Hr-Ft²°F) ASHRAE 90-75	Overall Thermal Transmittance (BTU/Hr-Ft²°F) Test House	Insulation Specified	Overall Thermal Transmittance (BTU/Hr-Ft²°F) ASHRAE 90-75	Overall Thermal Transmittance (BTU/Hr-Ft²°F) Test House
Houston	R-9	0.29	0.132	R-19	0.05	0.048	R-7	0.305	0.086
Birmingham	R-9	0.27	0.132	R-19	0.05	0.048	R-7	0.220	0.086
Atlanta	R-9	0.27	0.132	R-19	0.05	0.048	R-7	0.200	0.086
Tulsa	R-9	0.25	0.132	R-19	0.05	0.048	R-7	0.135	0.086
Philadelphia	R-9	0.24	0.132	R-19	0.05	0.048	R-9	0.080	0.073
Seattle	R-9	0.24	0.132	R-19	0.05	0.043	R-9	0.060	0.073
Columbus	R-9	0.23	0.132	R-19	0.05	0.048	R-9	0.080	0.073
Cleveland	R-9	0.22	0.132	R-19	0.05	0.048	R-9	0.080	0.073
Concord	R-9	0.20	0.132	R-19	0.05	0.048	R-9	0.080	0.073

SOURCE: Gordian and Associates [5].

TABLE 8.17. Comparison of Test House Air Infiltration Rates with ASHRAE 90-75 Standards[a]

City	Windows (ft³/min per ft of sash crack)		Sliding Doors (ft³/min per ft² of door area)		Entrance Doors (ft³/min per ft² of door area)	
	ASHRAE 90-75	Test House	ASHRAE 90-75	Test House	ASHRAE 90-75	Test House
Houston	.5	.25	.5	.25	1.25	1.25
Brimingham	.5	.25	.5	.25	1.25	1.25
Atlanta	.5	.25	.5	.25	1.25	1.25
Tulsa	.5	.25	.5	.25	1.25	1.25
Philadelphia	.5	.25	.5	.25	1.25	1.25
Seattle	.5	.25	.5	.25	1.25	1.25
Columbus	.5	.25	.5	.25	1.25	1.25
Cleveland	.5	.25	.5	.25	1.25	1.25
Concord	.5	.25	.5	.25	1.25	1.25

(a) Air infiltration data for windows and sliding doors from "Technical Data - Computing Heat Loss for Anderson Windows and Sliding Doors", Anderson Corporation, Bayport, Minnesota (Dec. 1971)

SOURCE: Gordian and Associates [5].

divided into two broad classifications: conventional heating and cooling systems representing equipment more or less extensively used in residential applications, and new technology advanced heating and cooling systems currently being developed.

Conventional Heating and Cooling Systems. Within the classfication of conventional heating and cooling systems, six represent electric heat pumps, while four are fossil-fuel-fired warm-air furnaces or electric resistance heaters in combination with electric air conditioners. A more detailed categorization of conventional equipment examined is shown below. Each designation indicates a single representative system unless otherwise indicated.

Split-system electric heat pumps
　Standard electric heat pump (moderate COP/EER)
　High-efficiency electric heat pumps (three different systems) (high COP/EER)
Add-on split-system electric heat pump
　Hybrid electric heat pump added onto gas- or oil-fired central warm-air furnaces.
　Heat-only electric heat pump added onto a central electric furnace
Fossil-fuel-fired warm-air furnaces
　Gas-fired
　Oil-fired

Electric resistance heating
Central electric furnace
Baseboard convectors

Mechanical air conditioners used in conjunction with the heat-only heat pump, fossil-fuel furnaces, and electric resistance heating

Some operating characteristics of these systems will be found in Table 8.18.

The basic component of the split-system electric heat pump is an electrically driven reciprocating Rankine-cycle compressor using a fluorocarbon refrigerant as working fluid. The standard for this group is a split-system heat pump with a heating COP of 2.3–2.6 (47°F rating) and an EER of 6.8–7.1 for cooling (95°F rating). This heat pump, with minor modifications, has been commercially available from several manufacturers for about 20 years. The high-efficiency split-system heat pumps represent rather new improvement of the standard system (more efficient compressors, larger heat exchangers, lower balance point, new defrost control) and have become available only within the last few years. Three different manufacturers' models of high-efficiency split electric heat pump systems were examined to determine if any significant performance differences for these new systems could be discerned. As a group, the high-efficiency systems exhibit increased heating COPs to 2.7–3.1 (47°F) and cooling EERs to 7.7–8.4 (95°F), or approximately 15–25% increased efficiency for heating and 15–20% increased efficiency for cooling at rating points compared with the standard electric heat pump.

The hybrid electric heat pump has the basic configuration of the standard split-system electric heat pump and provides both heating and cooling capability. In very cold weather, when the load on the residence exceeds the capacity of the heat pump, the heat pump is shut down and an oil- or gas-fired warm-air furnace is used to satisfy the load. Thus, during very cold weather, the fossil-fuel furnace supplies not only supplementary heat but also the baseload. When the heating load for the residence is satisfied, the furnace is shut down. The control system is designed so that the heat pump is always started up first, even below the balance point, and furnace heat is called for only when the heat pump cannot meet the load on the residence. The gas or oil furnace is sized to be capable of satisfying the design heating load without the support of the heat pump.

The heat-only electric heat pump is an add-on system that can be used in conjunction with fossil-fuel furnaces or, more frequently, with central electric warm-air furnaces. It is identical in operation to the split-system electric heat pump except that there is no reversing valve to allow the system to operate in the air-conditioning mode. Because the heat-only electric heat pump has no air-conditioning capability, it uses hot compressor discharge gas for defrost control rather than hot liquid as do conventional electric heat pumps. In this way there is some capacity for residential heating during defrost. Supplemental heat is provided by the central electric furnace below

TABLE 8.18. General Characteristics of HVAC Systems[f]

	Heating COP (a) 17°F	Heating COP (a) 47°F	Cooling EER[a] 95°F (COP)[g]	Heating Capacity (Btuh)(a) 17°F	Heating Capacity (Btuh)(a) 47°F	Cooling Capacity (Btuh)(a) 95°F	Auxiliaries (w/ton)
ELECTRIC HEAT PUMPS							
Electric Heat Pump - Standard	1.8	2.6	6.9	22,400	39,000	36,000	277
Electric Heat Pump - High Efficiency I	2.1	3.1	7.9	22,600	38,900	34,100	267
- High Efficiency II	2.1	2.8	7.8	21,000	37,500	33,500	-
- High Efficiency III	2.0	2.8	8.2	22,000	39,000	36,000	275
Hybrid Electric Heat Pump	1.4	2.1	6.4	22,000	37,000	36,000	314
Electric Heat Pump - Heat Only	2.3	3.2	-	21,500	36,500	-	-
Advanced Electric Heat Pump I	2.2	2.7	7.5	38,100	36,600	39,500	0
THERMAL ENGINE AND ABSORPTION CYCLE HEAT PUMPS			(COP)				
Free Piston Stirling-Rankine Gas Heat Pump	1.4	1.7	(0.9)[g]	55,600	68,100	35,000	517
V-Type Single Cylinder Stirling-Rankine Gas Heat Pump	1.3	1.5	(1.1)[g]	41,000	58,000	36,000	-
Organic Fluid Absorption Gas Heat Pump	1.1	1.2	(0.5)[g]	76,000	87,000	36,000	268
OTHER SYSTEMS							
Gas Warm Air Furnace	80%(b)	80%(b)	-	64,000(c)	64,000(c)	-	-
Oil Warm Air Furnace	85%(b)	85%(b)	-	85,000(c)	85,000(c)	-	-
Central Electric Furnace	100%(b)	100%(b)	-	(d)	(d)	-	-
Baseboard Convectors	100%(b)	100%(b)	-	(e)	(e)	-	-
Central Air Conditioner - High Efficiency	-	-	7.7	-	-	35,500	238
- Standard	-	-	6.5	-	-	35,000	280

(a) Rating conditions are 47/43-70 for high temperature heating, 17/15-70 for low temperature heating, and 80/67-95 for cooling. (b) Manufacturer's full load system efficiency. (c) Constant capacity at all outdoor temperatures. (d) Total heating capacity varied by location from 9.6 to 15.36 kW. (e) Total heating capacity varied by location from 9.75 to 16.0 kW. (f) Characteristics of heat pumps are for 3 ton (cooling) models of each type. (g) COP for compressor only; auxiliaries not included.

SOURCE: Gordian and Associates [5].

the balance point. The heat-only heat pump has an oversized compressor relative to compressor heating capacity for standard electric heat pumps sized to meet an air-conditioning design load. Compared to the high-efficiency electric heat pumps, the heat-only heat pump is similar in rated heating capacity at 47°F but somewhat higher in heating COP.

With the exception of the heat-only electric heat pump, heat pump equipment was selected based on the design total cooling load for the test house in each of the test cities, with proper consideration given to the design latent load. Manufacturers' published performance data were used to select the closest available off-the-shelf equipment for each case. Supplemental electrical-resistance heaters were sized for 80% of design heating load in all cases. Resistance elements for heat-only heat pumps are sized for 100% design heating load.

Other conventional heating systems are represented by oil and gas furnaces and electrical-resistance heating. In each case, manufacturers' off-the-shelf equipment were selected with maximum heating capacities equal to or greater than the design heating load for the test house. In order to properly compare annual energy consumption for residential heating and cooling, electric central air-conditioning systems were added to the oil, gas, and central electric furnaces. Two central air conditioners were examined: a high-efficiency unit (COP 7.1–7.8 at 95°F) generally selected by individual homeowners, and a standard unit (COP 6.2–6.8 at 95°F), most usually selected by builders as original equipment for developments. All of the final results of energy use are reported for the high-efficiency central air conditioner.

For the test house heated with baseboard convectors, room air conditioners were selected as cooling equipment. The entire first floor was assigned two large air conditioners, one in the family room and another in the living room. The two largest bedrooms were assigned two smaller room air conditioners. Capacities of the room air conditioners were estimated by apportioning the design total cooling load among the rooms to be conditioned on the basis of relative floor area. The air conditioners selected were of the high-EER type.

Advanced Heating and Cooling Systems. Six different advanced systems were examined, four by computer simulation and two by scaling calculated energy consumption data for similar systems on a city-by-city basis. The systems were classified as follows:

> Advanced electric heat pumps (two systems)
> Thermally activated heat pumps
>> Free-piston Stirling-Rankine gas heat pump
>> V-type single-cylinder Stirling-Rankine gas heat pump
>> Organic liquid absorption gas heat pump
> Pulse combustion gas furnace

One of the advanced electric heat pumps examined is based on a proprietary design for which full details were not reported. This system was designed for increased heating performance in cold climates (COP of 2.7 at 47°F) with a lower balance point than conventional electric heat pumps, and required no defrosting or supplemental heating. Cooling performance was reported to be about equal to standard electric heat pumps.

The advanced electric heat pump briefly described immediately above was the only electric system for which adequate performance data were available for simulation. In order to explore the performance anticipated for advanced technology electric heat pumps, energy-consumption data were estimated for a hypothetical advanced electric system. This heat pump was assumed to have a high performance capability based upon a number of modifications being explored at that time by equipment manufacturers. Examples of these modifications include the use of fossil-fuel-independent energy sources (such as solar energy) and increased low-temperature heating capacity brought about by series or parallel compression or continuous capacity modulation.

Three heat-activated heat pumps were examined within this group of advanced systems. Two represent Stirling-Rankine-cycle gas heat pumps, and the third a gas-fired absorption-system heat pump. One of the Stirling-Rankine systems employs a free-piston Stirling-cycle engine driving a Rankine-cycle heat pump. In its present configuration the system operates cyclically, starting up and shutting down in response to an indoor thermostat. The second Stirling-Rankine system consists of a V-type single-cylinder Stirling engine driving a reciprocating Rankine-cycle heat pump compressor. This prime mover can be modulated, however, and the capacity of the heat pump can therefore be modulated to respond to load changes. Little is known about the configuration of the gas-fired absorption heat pump except that the working fluid in the heat pump cycle is an organic fluid of proprietary composition. Characteristic of all three heat-activated heat pumps is their high heating capacity rating compared with conventional electric heat pumps of identical cooling rating.

One additional heating system included in this group was a pulse combustion furnace. Energy consumption for residential heating with this system was not simulated, however, but scaled from conventional gas-furnace energy-consumption calculations assuming a 92.5% combustion efficiency (Hollowell, Ref. 6).

8.5.4. General Findings

The results of the computerized simulation of heating and cooling systems are shown in three different ways in Tables 8.19 through 8.26. Fuels and electric energy consumed by conventional and advanced HVAC systems to provide comfort conditioning for the test house in nine cities representing U.S. climatic conditions are shown in Tables 8.19 through 8.22. Seasonal

TABLE 8.19.   Fuels and Electric Energy Consumed for Space-Conditioning by Test Houses in Selected Cities—Conventional Electric Heat Pumps

	Length of Season (months)	Degree Days or Cooling Hours	Electric Heat Pump-Standard Elec. (kWh)	Electric Heat Pump-High Efficiency I Elec. (kWh)	Electric Heat Pump-High Efficiency II Elec. (kWh)	Electric Heat Pump-High Efficiency III Elec. (kWh)
Houston						
Total Heating Season	4	1,290	1,402 (1,261)*	1,173 (1,056)	1,254 (1,125)	1,331 (1,196)
Total Cooling Season	8	2,022	7,215 (6,273)	5,930 (5,170)	6,536 (5,607)	6,286 (5,478)
Total Annual			8,617 (7,534)	7,103 (6,234)	7,790 (6,732)	7,617 (6,674)
Birmingham						
Total Heating Season	6	2,844	3,787 (3,414)	3,250 (2,940)	3,379 (3,053)	3,603 (3,256)
Total Cooling Season	6	1,480	5,010 (4,301)	4,236 (3,657)	4,561 (3,868)	4,369 (3,754)
Total Annual			8,797 (7,715)	7,486 (6,597)	7,940 (6,921)	7,972 (7,010)
Atlanta						
Total Heating Season	5	2,821	3,789 (3,437)	3,235 (2,935)	3,562 (3,231)	3,612 (3,284)
Total Cooling Season	7	1,393	4,700 (4,001)	3,980 (3,406)	4,381 (3,778)	4,100 (3,497)
Total Annual			8,489 (7,438)	7,215 (6,341)	7,943 (7,009)	7,712 (6,781)
Tulsa						
Total Heating Season	6	3,504	7,878 (7,252)	6,855 (6,304)	6,741 (6,198)	7,687 (7,069)
Total Cooling Season	6	876	3,873 (3,261)	3,158 (2,669)	3,449 (2,931)	3,889 (3,277)
Total Annual			11,751 (10,513)	10,013 (8,973)	10,190 (9,129)	11,576 (10,346)
Philadelphia						
Total Heating Season	7	4,508	7,613 (6,958)	6,615 (6,048)	6,798 (6,215)	7,283 (6,663)
Total Cooling Season	5	917	3,006 (2,549)	2,527 (2,155)	2,725 (2,283)	2,622 (2,227)
Total Annual			10,619 (9,507)	9,142 (8,203)	9,523 (8,498)	9,905 (8,890)
Seattle						
Total Heating Season	9	4,407	6,861 (6,101)	6,238 (5,499)	5,982 (5,333)	6,101 (5,431)
Total Cooling Season	3	532	981 (875)	801 (719)	905 (795)	837 (739)
Total Annual			7,842 (6,976)	7,039 (6,218)	6,887 (6,128)	6,938 (6,170)
Columbus						
Total Heating Season	7	5,407	10,338 (9,488)	9,212 (8,434)	9,283 (8,513)	9,968 (9,146)
Total Cooling Season	5	861	2,801 (2,396)	2,351 (2,023)	2,543 (2,148)	2,441 (2,092)
Total Annual			13,139 (11,884)	11,563 (10,457)	11,826 (10,661)	12,409 (11,238)
Cleveland						
Total Heating Season	7	6,097	12,364 (11,362)	11,077 (10,145)	11,112 (10,205)	11,928 (10,963)
Total Cooling Season	5	670	2,153 (1,831)	1,792 (1,532)	1,945 (1,634)	1,878 (1,600)
Total Annual			14,517 (13,193)	12,869 (11,677)	13,057 (11,839)	13,806 (12,563)
Concord						
Total Heating Season	9	7,377	16,796 (15,743)	15,613 (14,611)	14,225 (13,306)	14,891 (13,943)
Total Cooling Season	3	435	1,366 (1,184)	1,128 (978)	1,263 (1,077)	1,098 (942)
Total Annual			18,162 (16,927)	16,741 (15,589)	15,488 (14,383)	15,989 (14,885)

* Numbers in parentheses are on-site fuels and electric energy consumption calculated on steady state or full load performance basis, i.e., no dynamic efficiency losses are included.

SOURCE: Gordian and Associates [5].

performance factors are shown in Tables 8.23 and 8.24. The annual SPFs for each system shown in these tables are combined heating and cooling loads for an entire year divided by the annual energy consumption of the system. Finally, the total thermal equivalent of the fuels and electric energy consumed by each heating and cooling system is shown in Tables 8.25 and 8.26. The data in these last tables were produced by multiplying each form of energy consumed by its gross heating value or, in the case of electric energy, the thermal equivalent of a kilowatt-hour.

The information listed in Tables 8.25 to 8.26 is presented graphically in Fig. 8.5–8.8, changed, however to show energy per square foot of residence. The graphical correlations shown are related to a well-known method recommended by ASHRAE (1976) for the estimation of seasonal heating energy. The correlating parameter shown as the abcissa of the graphs, if multiplied by the identify of 24 hr/day, is the equivalent of seasonal load, for

TABLE 8.20. Fuels and Electric Energy Consumed for Space-Conditioning by Test Houses in Selected Cities—Add-On Heat Pumps

	Electric Heat Pump-Heat Only With Central Air Conditioner High Efficiency	Hybrid Electric Heat Pump - Gas Furnace		Hybrid Electric Heat Pump - Oil Furnace	
	Elec. (kWh)	Gas (cft)	Elec. (kWh)	Oil (Gal)	Elec. (kWh)
Houston					
Total Heating Season	1,172 (1,049)*	943 (847)	1,572 (1,416)	7 (7)	1,572 (1,416)
Total Cooling Season	6,510 (5,664)	--	7,570 (6,506)	--	7,570 (6,506)
Total Annual	7,682 (6,713)	943 (847)	9,142 (7,922)	7 (7)	9,142 (7,922)
Birmingham					
Total Heating Season	3,138 (2,831)	6,141 (5,157)	3,719 (3,450)	45 (38)	3,719 (3,450)
Total Cooling Season	4,524 (3,887)	--	5,273 (4,476)	--	5,273 (4,476)
Total Annual	7,662 (6,718)	6,141 (5,157)	8,992 (7,926)	45 (38)	8,992 (7,926)
Atlanta					
Total Heating Season	3,185 (2,899)	5,903 (4,701)	3,777 (3,520)	44 (35)	3,777 (3,520)
Total Cooling Season	4,242 (3,614)	--	4,959 (4,178)	--	4,959 (4,178)
Total Annual	7,427 (6,513)	5,903 (4,701)	8,736 (7,698)	44 (35)	8,736 (7,698)
Tulsa					
Total Heating Season	6,201 (5,694)	15,802 (14,008)	5,742 (5,373)	114 (101)	5,743 (5,374)
Total Cooling Season	3,788 (3,194)	--	4,024 (3,329)	--	4,024 (3,329)
Total Annual	9,989 (8,888)	15,802 (14,008)	9,766 (8,702)	114 (101)	9,767 (8,703)
Philadelphia					
Total Heating Season	6,176 (5,645)	17,377 (15,173)	6,738 (6,313)	127 (111)	6,740 (6,314)
Total Cooling Season	2,711 (2,300)	--	3,171 (2,658)	--	3,171 (2,658)
Total Annual	8,887 (7,945)	17,377 (15,173)	9,909 (8,971)	127 (111)	9,911 (8,972)
Seattle					
Total Heating Season	5,445 (4,855)	20,249 (11,424)	5,792 (6,270)	147 (85)	5,794 (6,271)
Total Cooling Season	889 (786)	--	1,048 (914)	--	1,048 (914)
Total Annual	6,334 (5,641)	20,249 (11,424)	6,840 (7,184)	147 (85)	6,842 (7,185)
Columbus					
Total Heating Season	8,509 (7,775)	36,344 (30,892)	6,784 (6,698)	262 (223)	6,786 (6,700)
Total Cooling Season	2,527 (2,163)	--	2,957 (2,500)	--	2,957 (2,500)
Total Annual	11,036 (9,938)	36,344 (30,892)	9,741 (9,198)	262 (223)	9,743 (9,200)
Cleveland					
Total Heating Season	10,221 (9,354)	47,207 (38,902)	7,568 (7,728)	340 (281)	7,572 (7,732)
Total Cooling Season	1,940 (1,652)	--	2,278 (1,916)	--	2,278 (1,916)
Total Annual	12,161 (11,006)	47,207 (38,902)	9,846 (9,644)	340 (281)	9,850 (9,648)
Concord					
Total Heating Season	13,028 (12,191)	58,538 (52,890)	8,179 (8,018)	420 (380)	8,186 (8,024)
Total Cooling Season	1,184 (1,017)	--	1,431 (1,191)	--	1,431 (1,191)
Total Annual	14,212 (13,208	58,538 (52,890)	9,610 (9,209)	420 (380)	9,617 (9,215)

* Numbers in parentheses are on-site fuels and electric energy consumptions calculated on steady state or full load performance basis, i.e. no dynamic efficiency losses are included.

SOURCE: Gordian and Associates [5].

either heating or cooling seasons, based upon design conditions and average heating or cooling degree days at a particular geographical location.

As would be expected, the energy consumption reflects the climatic conditions in each city; as the number of heating degree days increases, energy consumption for heating increases. The converse is true for cooling systems. It can also be noted that SPFs for heating and cooling have in general an inverse relationship to the number of heating or cooling degree days in each particular city. Of course, this is due to the operating characteristics of the system simulated, particularly those of heat pumps and air conditioners. (Fossil-fuel warm-air furnace and electrical-resistance heating systems are less sensitive to climatic conditions.) Heat pump COPs for heating decrease with decreasing outdoor temperature, and vapor-compression air-conditioner efficiency decreases with increasing outdoor temperature. Hence, it

TABLE 8.21. Fuels and Electric Energy Consumed for Space-Conditioning by Test Houses in Selected Cities—Other Conventional Heating and Cooling Systems*

	Gas Warm Air Furnace With Central Air Conditioner-High Efficiency		Oil Warm Air Furnace With Central Air Conditioner-High Efficiency		Central Electric Furnace With Central Air Conditioner-High Efficiency	Baseboard Convectors With Window Air Conditioners
	Gas (cft)	Elec. (kWh)	Oil (Gal)	Elec. (kWh)	Elec. (kWh)	Elec. (kWh)
Houston						
Total Heating Season	17,112	98	119	60	3,039	2,974
Total Cooling Season	-	6,510	-	6,510	6,510	3,402
Total Annual	17,112	6,608	119	6,570	9,549	6,376
Birmingham						
Total Heating Season	41,176	236	289	161	7,450	7,295
Total Cooling Season	-	4,524	-	4,524	4,524	2,386
Total Annual	41,176	4,760	289	4,685	11,974	9,681
Atlanta						
Total Heating Season	42,078	241	296	163	7,603	7,441
Total Cooling Season	-	4,242	-	4,242	4,242	2,289
Total Annual	42,078	4,483	296	4,405	11,845	9,730
Tulsa						
Total Heating Season	73,451	397	510	286	13,342	13,142
Total Cooling Season	-	3,788	-	3,788	3,788	1,909
Total Annual	73,451	4,185	510	4,074	17,130	15,051
Philadelphia						
Total Heating Season	78,406	448	554	325	14,454	14,243
Total Cooling Season	-	2,711	-	2,711	2,711	1,578
Total Annual	78,406	3,159	554	3,036	17,165	15,821
Seattle						
Total Heating Season	69,235	348	492	190	12,906	12,629
Total Cooling Season	-	889	-	889	889	768
Total Annual	69,235	1,237	492	1,079	13,795	13,397
Columbus						
Total Heating Season	94,396	540	668	407	17,677	17,418
Total Cooling Season	-	2,527	-	2,527	2,527	1,507
Total Annual	94,396	3,067	668	2,934	20,204	18,925
Cleveland						
Total Heating Season	110,649	633	785	487	21,015	20,708
Total Cooling Season	-	1,940	-	1,940	1,940	1,247
Total Annual	110,649	2,573	785	2,427	22,955	21,955
Concord						
Total Heating Season	127,258	603	900	483	24,013	23,622
Total Cooling Season	-	1,184	-	1,184	1,184	804
Total Annual	127,258	1,787	900	1,667	25,197	24,426

* All fuels and electric energy reported represents part load results, i.e. dynamic efficiency losses are included.

TABLE 8.22. Fuels and Electric Energy Consumed by Test Houses in Selected Cities: Advanced Heating and Cooling Systems*

	Advanced Electric Heat Pump I Electric (kWh)	Advanced Electric Heat Pump II Electric (kWh)	Free Piston Stirling - Rankine Gas Heat Pump Gas (cft)	Elec (kWh)	V-Type Single Cylinder Stirling-Rankine Gas HP Gas (cft)	Elec (kWh)	Pulse Combustion Gas Furnace with Central Air Conditioner-High Efficiency Gas (cft)	Elec (kWh)	Organic Fluid Absorption Gas Heat Pump Gas (cft)	Elec (kWh)
Houston										
Total Heating Season	**	722	6,109	242	5,604	247	11040	104	8,477	150
Total Cooling Season	**	3,622	45,369	1,587	51,835	1,750	-	5664	76,446	1,546
Total Annual	**	4,344	51,478	1,829	57,439	1,997	11040	5768	84,923	1,696
Birmingham										
Total Heating Season	2,977	1,994	15,373	618	14,264	639	26447	250	21,226	378
Total Cooling Season	4,034	2,510	33,204	1,092	38,608	1,261	-	3887	55,927	1,066
Total Annual	7,011	4,504	48,577	1,710	52,872	1,900	26447	4137	77,153	1,444
Atlanta										
Total Heating Season	3,015	2,001	15,625	626	14,499	640	27058	256	21,587	384
Total Cooling Season	3,760	2,335	30,944	1,020	34,979	1,196	-	3614	52,096	998
Total Annual	6,775	4,336	46,569	1,646	49,478	1,836	27058	3870	73,683	1,382
Tulsa										
Total Heating Season	5,489	4,458	28,366	1,130	27,498	1,317	47433	423	38,967	689
Total Cooling Season	3,082	1,903	24,299	834	29,829	1,173	-	3194	40,893	817
Total Annual	8,571	6,341	52,665	1,964	57,327	2,490	47433	3617	79,860	1,506
Philadelphia										
Total Heating Season	5,896	4,110	30,321	1,225	28,301	1,248	50920	481	41,774	745
Total Cooling Season	2,402	1,483	19,968	654	22,125	781	-	2300	33,590	641
Total Annual	8,298	5,593	50,289	1,879	50,426	2,029	50920	2781	75,364	1,386
Seattle										
Total Heating Season	**	3,441	26,093	1,062	17,266	1,900	45721	379	36,180	651
Total Cooling Season	**	511	7,134	216	7,449	298	-	786	11,950	216
Total Annual	**	3,952	33,227	1,278	24,715	2,198	45721	1165	48,130	867
Columbus										
Total Heating Season	7,391	5,515	37,899	1,543	35,618	1,549	61645	582	51,956	930
Total Cooling Season	2,254	1,398	18,662	612	21,272	713	-	2163	31,394	600
Total Annual	9,645	6,913	56,561	2,155	56,890	2,262	61645	2745	83,350	1,530
Cleveland										
Total Heating Season	8,787	6,598	45,012	1,833	42,400	1,822	72867	688	61,713	1,104
Total Cooling Season	1,729	1,069	14,259	472	15,683	556	-	1652	23,954	464
Total Annual	10,516	7,667	59,271	2,305	58,083	2,378	72367	2340	85,667	1,568
Concord										
Total Heating Season	**	9,116	52,361	2,236	49,749	2,823	83325	656	71,851	1,317
Total Cooling Season	**	691	9,186	294	9,208	467	-	1017	15,432	288
Total Annual	**	9,807	61,547	2,530	58,957	3,290	83325	1673	87,283	1,605

* The energies presented may be underestimated for the Advanced Electric Heat Pump I, free Piston Stirling-Rankine gas heat pump and the Organic Fluid Absorption Gas Heat Pump due to possible underestimation of cycling and defrost losses. If the dynamic efficiency losses of the advanced heat pumps were similar to those of the conventional electric (Rankine cycle) heat pump, then adjusted estimates of energies which include these effects can be obtained by multiplying the values in this table by a factor of 1.097 for the heating season, 1.164 for the cooling season and 1.118 for annual energy values.

** System is not sized for application in these cities.

SOURCE: Gordian and Associates [5].

TABLE 8.23. Seasonal Performance Factors On-Site—Conventional Electric Heat Pumps and Other Systems*

Heating	Houston	Birmingham	Atlanta	Tulsa	Philadelphia	Seattle	Columbus	Cleveland	Concord
Electric Heat Pump - Standard	2.14 (2.38)*	1.93 (2.14)	1.97 (2.17)	1.68 (1.82)	1.87 (2.05)	1.84 (2.07)	1.69 (1.84)	1.68 (1.82)	1.41 (1.50)
Electric Heat Pump - High Efficiency I	2.56 (2.84)	2.25 (2.49)	2.31 (2.55)	1.93 (2.10)	2.16 (2.36)	2.03 (2.30)	1.89 (2.07)	1.87 (2.04)	1.51 (1.62)
Electric Heat Pump - High Efficiency II	2.39 (2.67)	2.17 (2.40)	2.10 (2.31)	1.96 (2.13)	2.10 (2.30)	2.12 (2.37)	1.88 (2.05)	1.86 (2.03)	1.66 (1.78)
Electric Heat Pump - High Efficiency III	2.25 (2.51)	2.03 (2.25)	2.07 (2.27)	1.72 (1.87)	1.96 (2.14)	2.07 (2.33)	1.75 (1.91)	1.74 (1.89)	1.59 (1.70)
Hybrid Electric Heat Pump - Gas Furnace	1.62 (1.80)	1.33 (1.47)	1.36 (1.53)	1.27 (1.38)	1.21 (1.33)	1.03 (1.32)	1.00 (1.11)	0.97 (1.08)	0.93 (1.01)
Hybrid Electric Heat Pump - Oil Furnace	1.61 (1.78)	1.32 (1.46)	1.34 (1.51)	1.27 (1.39)	1.20 (1.32)	1.08 (1.30)	1.00 (1.11)	0.97 (1.08)	0.93 (1.01)
Electric Heat Pump - Heat Only	2.56 (2.86)	2.33 (2.58)	2.35 (2.58)	2.13 (2.32)	2.31 (2.53)	2.32 (2.61)	2.05 (2.24)	2.03 (2.21)	1.81 (1.94)
Gas Warm Air Furnace	0.59	0.59	0.59	0.60	0.61	0.61	0.62	0.63	0.62
Oil Warm Air Furnace	0.61	0.61	0.61	0.63	0.62	0.62	0.63	0.64	0.64
Central Electric Furnace	0.98	0.98	0.98	0.98	0.98	0.98	0.98	0.98	0.98
Baseboard Convectors	1.00	1.00	1.00	1.00	1.00	1.00	1.00	1.00	1.00
Cooling									
Electric Heat Pump - Standard	1.97 (2.26)	1.94 (2.26)	1.95 (2.29)	1.92 (2.29)	1.97 (2.32)	2.09 (2.34)	1.96 (2.30)	1.99 (2.34)	1.98 (2.28)
Electric Heat Pump - High Efficiency I	2.31 (2.64)	2.28 (2.56)	2.30 (2.68)	2.30 (2.72)	2.32 (2.73)	2.56 (2.85)	2.32 (2.69)	2.37 (2.77)	2.39 (2.75)
Electric Heat Pump - High Efficiency II	2.14 (2.50)	2.12 (2.50)	2.10 (2.44)	2.17 (2.55)	2.16 (2.57)	2.27 (2.59)	2.15 (2.54)	2.19 (2.61)	2.13 (2.50)
Electric Heat Pump - High Efficiency III	2.26 (2.59)	2.23 (2.59)	2.24 (2.62)	1.91 (2.27)	2.26 (2.66)	2.46 (2.79)	2.25 (2.63)	2.29 (2.69)	2.46 (2.86)
Hybrid Electric Heat Pump	1.85 (2.15)	1.83 (2.16)	1.83 (2.18)	1.82 (2.20)	1.84 (2.20)	1.94 (2.22)	1.84 (2.18)	1.87 (2.22)	1.84 (2.22)
Central Air Conditioner - High Efficiency	2.18 (2.50)	2.15 (2.51)	2.16 (2.54)	1.96 (2.33)	2.18 (2.57)	2.32 (2.62)	2.18 (2.54)	2.21 (2.60)	2.23 (2.59)
Central Air Conditioner - Standard	1.82 (2.09)	1.80 (2.09)	1.81 (2.12)	1.87 (2.21)	1.82 (2.15)	1.97 (2.20)	1.82 (2.12)	1.85 (2.17)	1.97 (2.28)
Window Air Conditioners	2.19 (2.52)	2.32 (2.69)	2.30 (2.69)	2.28 (2.69)	2.28 (2.69)	2.10 (2.34)	2.31 (2.69)	2.29 (2.69)	2.32 (2.69)
Annual Heating and Cooling									
Electric Heat Pump - Standard	2.00 (2.28)	1.94 (2.21)	1.94 (2.23)	1.76 (1.96)	1.90 (2.12)	1.87 (2.10)	1.75 (1.96)	1.73 (1.89)	1.45 (1.55)
Electric Heat Pump - High Efficiency I	2.35 (2.67)	2.26 (2.56)	2.31 (2.61)	2.05 (2.29)	2.30 (2.46)	2.09 (2.36)	1.48 (2.19)	1.24 (2.14)	1.57 (1.69)
Electric Heat Pump - High Efficiency II	2.18 (2.53)	2.15 (2.44)	2.10 (2.37)	2.03 (2.26)	2.14 (2.37)	2.14 (2.40)	1.44 (2.15)	1.91 (2.11)	1.70 (1.83)
Electric Heat Pump - High Efficiency III	2.26 (2.58)	2.10 (2.41)	2.15 (2.45)	1.78 (2.02)	2.04 (2.27)	2.12 (2.39)	1.85 (2.04)	1.81 (1.99)	1.65 (1.77)
Hybrid Electric Heat Pump - Gas Furnace	1.81 (2.08)	1.57 (1.79)	1.58 (1.82)	1.42 (1.60)	1.34 (1.50)	1.15 (1.38)	1.12 (1.25)	1.05 (1.19)	0.98 (1.06)
Hybrid Electric Heat Pump - Oil Furnace	1.80 (2.08)	1.57 (1.79)	1.58 (1.82)	1.42 (1.60)	1.34 (1.49)	1.15 (1.38)	1.12 (1.25)	1.05 (1.18)	0.98 (1.06)
Gas Warm Air Furnace/Central Air Conditioner - High Efficiency	1.48	1.01	0.99	0.80	0.77	0.68	0.75	0.72	0.67
Oil Warm Air Furnace/Central Air Conditioner - High Efficiency	1.50	1.03	1.01	0.83	0.79	0.69	0.76	0.73	0.69
Central Electric Furnace/Central Air Conditioner - High Efficiency	1.73	1.42	1.40	1.20	1.17	1.07	1.13	1.08	1.04
Baseboard Convectors/Window Air Conditioners	1.63	1.33	1.31	1.16	1.13	1.06	1.10	1.07	1.04
Electric Heat Pump - Heat Only/Central Air Conditioner - High Efficiency	2.24 (2.56)	2.22 (2.54)	2.24 (2.56)	2.07 (2.28)	2.27 (2.54)	2.32 (2.61)	2.08 (2.31)	2.06 (2.27)	1.84 (1.99)

* Numbers in parentheses represent total on-site seasonal performance factors assuming steady state or full load performance, dynamic efficiency losses are not included.

SOURCE: Gordian and Associates [5].

TABLE 8.24. Seasonal Performance Factors On-Site—Advanced Heating and Cooling Systems*

	Houston	Birmingham	Atlanta	Tulsa	Philadelphia	Seattle	Columbus	Cleveland	Concord
HEATING									
Advanced Electric Heat Pump I	**	2.46	2.48	2.37	2.42	**	2.35	2.35	**
Advanced Electric Heat Pump II	3.43	2.81	2.93	2.28	2.64	2.98	2.40	2.38	2.02
Free Piston Stirling Rankine Gas Heat Pump	1.48	1.43	1.44	1.40	1.41	1.45	1.38	1.38	1.35
V-Type Single Cylinder Stirling Rankine Gas HP	1.59	1.52	1.53	1.41	1.50	1.82	1.45	1.45	1.34
Organic Fluid Absorption Gas Heat Pump	1.14	1.11	1.11	1.09	1.10	1.13	1.08	1.08	1.06
Pulse Combustion Furnace	0.90	0.91	0.91	0.93	0.93	0.93	0.94	0.94	0.94
COOLING									
Advanced Electric Heat Pump I	**	2.41	2.43	2.44	2.45	**	2.43	2.47	**
Advanced Electric Heat Pump II	3.29	3.29	3.29	3.22	3.33	3.57	3.33	3.33	3.36
Free Piston Stirling Rankine Gas Heat Pump	0.95	0.90	0.91	0.93	0.91	0.90	0.90	0.92	0.90
V-Type Single Cylinder Stirling Rankine Gas HP	1.06	1.04	1.07	0.98	1.10	1.16	1.07	1.12	1.17
Organic Fluid Absorption Gas Heat Pump	0.59	0.56	0.56	0.58	0.56	0.56	0.56	0.57	0.56
ANNUAL HEATING AND COOLING									
Advanced Electric Heat Pump I	**	2.43	2.45	2.40	2.43	**	2.37	2.37	**
Advanced Electric Heat Pump II	3.31	3.08	3.12	2.56	2.82	3.06	2.59	2.52	2.11
Free Piston Stirling Rankine Gas Heat Pump	1.01	1.07	1.09	1.19	1.21	1.33	1.22	1.27	1.28
V-Type Single Cylinder Stirling Rankine Gas HP	1.11	1.17	1.21	1.19	1.33	1.65	1.31	1.36	1.31
Organic Fluid Absorption Gas Heat Pump	0.64	0.71	0.72	0.83	0.86	0.99	0.88	0.94	0.97
Pulse Combustion Furnace/Central Air Conditioner-High Efficiency	1.91	1.43	1.41	1.19	1.14	1.02	1.11	1.06	1.01

* The performance factors presented may be underestimated for the advanced electric heat pump I, free piston Stirling-Rankine Gas heat pump and organic fluid absorption gas heat pump due to possible underestimation of cycling and defrost losses. If the dynamic efficiency losses of the advanced heat pumps were similar to those of the conventional electric (Rankine cycle) heat pump, then adjusted estimates of performance factors which include these effects can be obtained by multiplying the values in this table by a factor of 1.097 for the heating season, 1.164 for the cooling season and 1.118 for annual performance factor values.

** System is not sized for application in these cities

SOURCE: Gordian and Associates [5].

TABLE 8.25. Total Fuel and Electric Energy Consumption On-Site—Conventional Heat Pumps and Other Systems* (MBtu)

Heating	Houston	Birmingham	Atlanta	Tulsa	Philadelphia	Seattle	Columbus	Cleveland	Concord
Electric Heat Pump - Standard	4.79 (4.30)*	12.92 (11.65)	12.93 (11.73)	26.88 (24.75)	25.98 (23.74)	23.42 (20.82)	35.28 (32.38)	42.20 (38.78)	57.33 (53.73)
Electric Heat Pump - High Efficiency I	4.00 (3.60)	11.09 (10.04)	11.04 (10.02)	23.40 (21.52)	22.58 (20.64)	21.29 (18.77)	31.44 (28.79)	37.81 (34.63)	53.29 (49.87)
Electric Heat Pump - High Efficiency II	4.28 (3.84)	11.53 (10.42)	12.16 (11.03)	23.01 (21.16)	23.20 (21.21)	20.42 (18.20)	31.68 (29.06)	37.93 (34.83)	48.55 (45.41)
Electric Heat Pump - High Efficiency III	4.54 (4.08)	12.30 (11.12)	12.33 (11.21)	26.24 (24.13)	24.86 (22.74)	20.82 (18.54)	34.02 (31.22)	40.71 (37.42)	50.82 (47.59)
Hybrid Electric Heat Pump - Gas Furnace	6.31 (5.68)	18.84 (16.93)	18.79 (16.72)	35.40 (32.35)	40.38 (36.72)	40.02 (32.82)	59.50 (53.75)	73.04 (65.28)	86.45 (80.25)
Hybrid Electric Heat Pump - Oil Furnace	6.37 (5.74)	18.98 (17.07)	18.96 (16.87)	35.45 (32.42)	40.59 (36.93)	40.16 (33.28)	59.54 (53.86)	73.11 (65.49)	86.38 (80.23)
Electric Heat Pump - Heat Only	4.00 (3.58)	10.71 (9.66)	10.87 (9.89)	21.17 (19.43)	21.08 (19.27)	18.58 (16.57)	29.04 (26.53)	34.89 (31.93)	44.47 (41.61)
Gas Warm Air Furnace	17.45	41.98	42.90	74.81	79.94	70.42	96.24	112.81	129.32
Oil Warm Air Furnace	16.81	40.73	41.68	71.82	78.10	69.15	94.27	110.78	126.86
Central Electric Furnace	10.37	25.43	25.95	45.54	49.33	44.05	60.33	71.72	81.96
Baseboard Convectors	10.15	24.90	25.40	44.85	48.61	43.10	59.45	70.68	80.62
Cooling									
Electric Heat Pump - Standard	24.63 (21.41)	17.10 (14.68)	16.04 (13.66)	13.22 (11.13)	10.26 (8.70)	3.35 (2.99)	9.56 (8.18)	7.35 (6.25)	4.66 (4.04)
Electric Heat Pump - High Efficiency I	20.24 (17.67)	14.46 (12.48)	13.58 (11.63)	10.78 (9.11)	8.63 (7.36)	2.74 (2.45)	8.03 (6.90)	6.12 (5.23)	3.85 (3.34)
Electric Heat Pump - High Efficiency II	22.31 (19.14)	15.57 (13.51)	14.95 (12.89)	11.77 (10.00)	9.30 (7.79)	3.09 (2.71)	8.68 (7.33)	6.64 (5.57)	4.31 (3.68)
Electric Heat Pump - High Efficiency III	21.46 (18.70)	14.91 (12.83)	13.99 (11.94)	13.28 (11.19)	8.95 (7.60)	2.86 (2.52)	8.33 (7.14)	6.41 (5.46)	3.75 (3.21)
Hybrid Electric Heat Pump	25.84 (22.01)	18.00 (15.27)	16.92 (14.26)	13.73 (11.61)	10.82 (9.07)	3.58 (3.12)	10.09 (8.53)	7.78 (6.54)	4.89 (4.06)
Central Air Conditioner - High Efficiency	22.22 (19.33)	15.44 (13.27)	14.48 (12.34)	12.93 (10.90)	9.26 (7.85)	3.04 (2.68)	8.62 (7.38)	6.62 (5.64)	4.04 (3.47)
Central Air Conditioner - Standard	26.50 (23.10)	18.45 (15.88)	17.31 (14.77)	13.60 (11.47)	11.07 (9.41)	3.54 (3.18)	10.31 (8.84)	7.92 (6.76)	4.68 (4.04)
Window Air Conditioners	11.61	8.14	7.81	6.52	5.39	2.62	5.14	4.25	2.74
Annual Heating and Cooling									
Electric Heat Pump - Standard	29.41 (25.71)	30.02 (26.33)	28.97 (25.38)	40.11 (35.88)	36.24 (32.45)	26.76 (23.81)	44.84 (40.56)	49.55 (45.03)	61.99 (57.78)
Electric Heat Pump - High Efficiency I	24.24 (21.28)	25.55 (22.52)	24.62 (21.64)	34.18 (30.62)	31.20 (28.00)	24.03 (21.22)	39.47 (35.70)	43.92 (39.86)	57.14 (53.21)
Electric Heat Pump - High Efficiency II	26.59 (22.98)	27.10 (23.62)	27.11 (23.92)	34.78 (31.16)	32.51 (29.00)	23.51 (20.92)	40.36 (36.39)	44.57 (40.40)	52.86 (49.09)
Electric Heat Pump - High Efficiency III	26.00 (22.78)	27.21 (23.95)	26.32 (23.15)	39.51 (35.31)	33.80 (30.35)	23.68 (21.06)	42.36 (38.36)	47.12 (42.88)	54.57 (50.80)
Hybrid Electric Heat Pump - Gas Furnace	32.15 (27.89)	36.83 (32.21)	35.72 (30.98)	49.14 (43.71)	51.20 (45.79)	43.59 (35.94)	69.59 (62.29)	80.81 (71.82)	91.34 (84.32)
Hybrid Electric Heat Pump - Oil Furnace	32.22 (27.95)	36.97 (32.35)	35.89 (31.13)	49.19 (43.78)	51.41 (46.00)	43.74 (36.40)	69.63 (62.40)	80.89 (72.02)	91.24 (84.30)
Gas Warm Air Furnace/Central Air Conditioner - High Efficiency	39.64	57.42	57.38	87.74	89.30	73.46	104.86	119.43	133.36
Oil Warm Air Furnace/Central Air Conditioner - High Efficiency	39.03	56.17	56.16	84.74	87.36	72.18	102.89	117.42	130.91
Central Electric Furnace/Central Air Conditioner - High Efficiency	32.59	40.87	40.43	58.47	58.59	47.09	68.95	78.34	86.00
Baseboard Convectors/Window Air Conditioners	21.76	33.04	33.21	51.37	54.00	45.72	64.59	74.93	83.36
Electric Heat Pump - Heat Only/Central Air Conditioner - High Efficiency	26.22 (22.91)	26.15 (22.93)	25.35 (22.23)	34.10 (30.33)	30.34 (27.12)	21.62 (19.25)	37.66 (33.91)	41.51 (37.57)	48.51 (45.08)

* Numbers in parentheses represents total on-site energy consumption assuming steady state or full-load performance dynamic efficiency losses are not included.

SOURCE: Gordian and Associates [5].

TABLE 8.26. Total Fuel and Electric Energy Consumption On-Site—Advanced Heating and Cooling Systems* (MBtu)

	Houston	Birmingham	Atlanta	Tulsa	Philadelphia	Seattle	Columbus	Cleveland	Concord
Heating									
Advanced Electric Heat Pump I	**	10.16	10.29	19.07	20.12	**	25.35	30.12	**
Advanced Electric Heat Pump II	2.46	6.81	6.83	15.52	14.03	11.74	18.82	22.52	31.11
Free Piston Stirling-Rankine Gas Heat Pump	6.93	17.48	17.76	32.22	34.50	29.72	43.17	51.27	60.00
V-Type Single Cylinder Stirling-Rankine Gas HP	6.45	16.44	16.68	31.99	32.56	23.75	40.90	48.62	60.06
Organic Fluid Absorption Gas Heat Pump	8.99	22.52	22.90	41.32	44.32	38.40	55.13	65.48	76.35
Pulse Combustion Furnace	11.39	27.30	27.93	48.88	52.56	47.01	63.63	75.22	86.06
Cooling									
Advanced Electric Heat Pump I	**	13.77	12.83	10.52	8.20	**	7.69	5.90	**
Advanced Electric Heat Pump II	12.50	8.57	7.97	6.49	5.06	1.74	4.77	3.65	2.36
Free Piston Stirling-Rankine Gas Heat Pump	50.79	36.94	34.43	27.15	22.20	7.87	20.75	15.87	10.19
V-Type Single Cylinder Stirling-Rankine Gas HP	57.81	42.91	39.06	33.83	24.79	8.47	23.71	17.58	10.80
Organic Fluid Absorption Gas Heat Pump	81.72	59.57	55.50	43.68	35.78	12.68	33.44	25.54	16.41
Annual Heating and Cooling									
Advanced Electric Heat Pump I	**	23.93	23.12	29.59	28.32	**	33.04	36.02	**
Advanced Electric Heat Pump II	14.96	15.37	14.80	22.02	19.09	13.49	23.59	26.17	33.47
Free Piston Stirling-Rankine Gas Heat Pump	57.72	54.42	52.19	59.37	56.70	37.59	63.91	67.14	70.19
V-Type Single Cylinder Stirling-Rankine Gas HP	64.25	59.36	55.74	65.83	57.35	32.22	64.61	66.20	70.86
Organic Fluid Absorption Gas Heat Pump	90.71	82.08	78.40	85.00	80.10	51.09	88.57	91.02	92.76
Pulse Combustion Furnace/Central Air Conditioner - High Efficiency	30.83	40.63	40.33	53.83	60.45	49.71	71.05	80.88	89.65

* All data in this table are based on steady state or full-load performance operation,
  dynamic efficiency losses are not included. If dynamic losses of the advanced heat pumps were similar to those of the conventional
  electric heat pumps, then multiplying the numbers in the table above by the following factors, 1.097 for heating season values, 1.164
  for cooling season values, and 1.118 for annual values will result in approximate total energy consumption corrected for dynamic
  efficiency effects. Exceptions include the modulated V-type Single Cylinder Stirling-Rankine Gas Heat Pump for which the tabular data
  already include dynamic effects and the Pulse Combustion Furnace for which data was scaled from Gas Warm Air Furnace values.

SOURCE: Gordian and Associates [5].

** System is not sized for application in these cities.

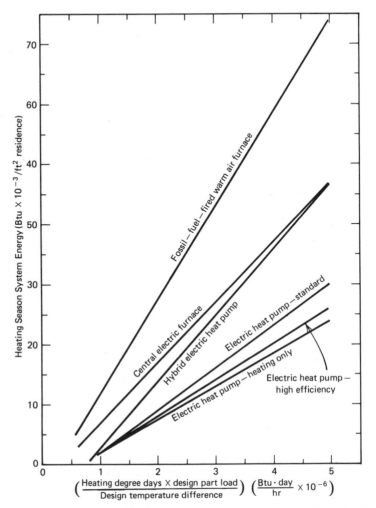

Fig. 8.5. Correlation of residential heating energy for conventional heating and cooling systems (data points not included for clarity).

would be expected that high heat pump efficiencies on heating are experienced in the warm cities (such as Atlanta and Birmingham), while cooler cities (such as Cleveland and Concord) would enjoy high cooling system efficiencies.

The simulations of commercially available heating and cooling systems were performed on a full-load or steady-state basis, taking into account dynamic efficiency losses. These latter results are shown in parentheses in Tables 8.19, 8.20, 8.23, and 8.25. As can be seen in these tables, dynamic efficiency losses, due to such factors as equipment cycling and defrosting, substantially increase energy requirements for space-conditioning, and

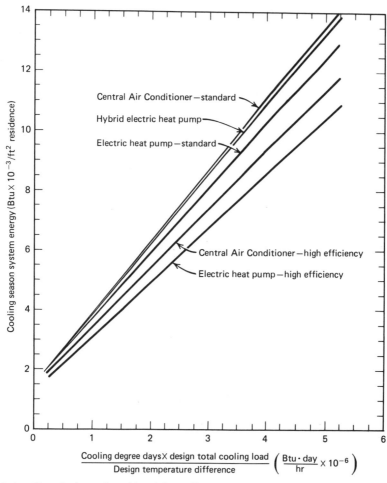

Fig. 8.6. Correlation of residential cooling energy for conventional heating and cooling systems (data points not included for clarity).

hence lower both heating and cooling SPFs. Indeed, a reduction in energy consumption of about 6–10% for the heating season and 13–16% for the cooling season was observed when the simulation was performed on the traditional basis of assuming steady-state operation. In that sense, then, failure to account for losses due to defrost and equipment dynamics, as has been the practice in the past, results in an overestimate of the SPF and an underestimate of the energy consumption. Unfortunately, however, dynamic efficiency data are not available for the nonmodulating advanced heat pumps (for which they are specifically needed). Therefore, the results presented in Tables 8.22, 8.24, and 8.26 for advanced heating and cooling systems could be determined only on a full-load or steady-state basis (except for

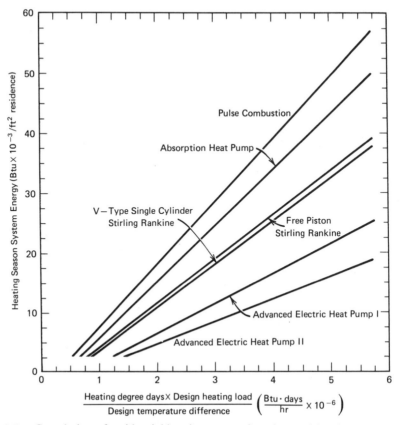

Fig. 8.7. Correlation of residential heating energy for advanced heating and cooling systems (data points not included for clarity).

those systems indicated) from system developers' data. It would be expected that in actual operation, dynamic efficiency losses would reduce system performance to some extent. If these systems experienced dynamic losses similar to those of conventional electric heat pumps, average increases in energy consumption of 6–10% for heating and 13–16% for cooling would be expected.

In the simulation of currently available electric heat pumps (standard, high-efficiency, hybrid, and heat-only), the effect on system performance of outside-coil defrosting during the heating season was included. A significant portion of the seasonal energy consumption can be assigned to this process, as can be seen in Table 8.27. These results are specific for the standard electric heat pump but are similar to those of other electric heat pumps simulated. As can be seen, up to almost 8% of seasonal energy consumption is due to resistance heat used in tempering of air into the conditioned space as the heat pump operates in the defrost mode.

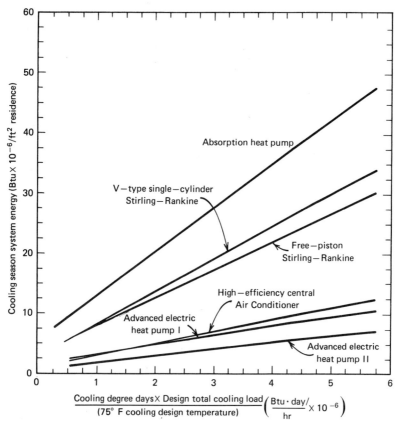

Fig. 8.8. Correlation of residential cooling energy for advanced heating and cooling systems (data points not included for clarity).

TABLE 8.27. Defrost Energy

City	Heating Degree Days	Number of Defrost Cycles	Resistance Heat During Defrost (% Total Energy)
Houston	1290	118	3.8
Birmingham	2844	397	5.5
Atlanta	2821	370	4.4
Tulsa	3504	657	4.7
Philadelphia	4508	699	5.1
Seattle	4407	1208	7.9
Columbus	5407	1262	7.5
Cleveland	6097	1523	7.6
Concord	7377	1727	7.5

The results from the simulation of heat pump defrost-cycle energy consumption may be compared to experimentally measured values presented by Groff and Bullock [7] for a residence located in Syracuse, New York (6463 heating degree days). A total of 1222 defrost cycles were measured over the entire heating season, which is less than would be expected based on the data in Table 8.27. However, the average defrost cycle was about 3.3 min, lending credence to the $3\frac{1}{2}$-min cycle assumed in this simulation. Groff and Bullock also present results for a number of residences in Minneapolis (8382 heating degree days). Here, the number of defrost cycles varies from 1653 to 2164, in an inverse relationship to the capacity of the heat pump. Again, these figures would appear to be less than the number of cycles expected by extrapolation of the data in Table 8.27. In reality, initiation of a defrost cycle is dependent on a number of factors, such as outdoor humidity, temperature, wind speed, and heat exchanger design, and is a most difficult effect to estimate. However, the simple approach used in this study reasonably approximates measured defrost occurrences. As an approximation to the defrost energy consumption, the results presented here also appear satisfactory.

8.5.5. Specific Findings: Conventional Systems

In Fig. 8.5, heating season energy consumption per square foot of residential building heated area is shown as a function of the heating correlation parameter. (The greater the value of the parameter, the colder the location or the greater the house heat loss.) The curve for the central electric furnace and baseboard convectors separates all-electric HVAC systems from systems consuming both electric energy and fossil fuels. Compared with the central electric furnace and baseboard convectors, which have heating SPFs of 0.98 and 1.0, respectively, electric heat pumps have heating SPFs greater than 1; systems firing fossil fuels on-site have SPFs less than 1. Thus, gas or oil furnaces consume considerably more total energy, about 62% more, than do all-electric resistance heating systems. The hybrid heat pump, which uses fossil fuel only below the residence balance point, consumes less total energy than systems operating most exclusively on fossil fuels (and only a relatively small amount of electric energy for circulating fans). The hybrid heat pump saves about 33–64% of the on-site energy consumed by a gas or oil furnace, the larger amount representing warmer climates and the smaller percentage saving representing colder climates. The reason for the variation in energy saved is mainly the improved performance achieved by an electric heat pump in warmer regions than in colder climates.

Figure 8.5 also shows that all three types of all-electric heat pumps considerably reduce heating-season energy consumption compared with electrical resistance heating and fossil-fuel heating systems, the greatest savings occurring in warmer climates. The all-electric heat pump with the highest energy consumption, the standard electric heat pump, uses only about 29–53% of the energy consumed by electrical resistance heating and about 56–

73% of the energy consumed by fossil-fuel warm-air furnaces. Thus, in general, quite significant savings in total energy consumed on-site for heating can be realized by the use of conventional electric heat pumps.

Upon comparing the different electric heat pumps, the data show that the new high-efficiency heat pumps use an average of 10% less energy for heating than the standard electric heat pump. The heat pump with the lowest energy consumption, the heat-only heat pump, reduces heating-season energy consumption by an average of 19% compared with the standard electric heat pump.

It must be pointed out that while a comparison of total energy consumption at the residential site is quite valid as a measure of the thermal equivalent of all forms of energy consumed by the HVAC system, comparing electric energy consumption with the consumption of fossil fuel burned on-site on this basis does not account for fuels consumed at the electric generating plant. To do so would require the energy balance envelope to be extended to include the utility as well.

Turning now to the conventional electrically operated cooling systems, Fig. 8.6 graphically displays the energy consumed in the operation of heat pumps in the cooling mode and central air conditioners. High-efficiency electric heat pumps show the lowest seasonal energy consumption for cooling. Compared with the standard electric heat pump, the high-efficiency units reduce cooling energy consumption by about 13%. The hybrid heat pump appears to have the largest energy consumption of all the heat pumps, but this should be considered a temporary finding. The heat pump portion of the hybrid system is an early standard model and is less efficient for either heating or cooling (see Table 8.18 where comparative ratings are tabulated) than the new model which has since been developed by the manufacturer.

The standard central air conditioner usually installed in tract housing developments consumes the greatest amount of energy for cooling of all the systems shown, about 17% more than the high-efficiency central air conditioner. Energy consumption by the high-efficiency central air conditioner is less than that of the standard electric heat pump in all regions of the country examined and about equal to the average for the high-efficiency electric heat pump. Cooling-season energy consumption for room air conditioners in the test house is quite low. It must be remembered, though, that the entire house is not cooled in this case, and therefore one would expect this method of cooling to be the lowest energy user of all the cooling systems examined.

On an annual basis, the combination of baseboard convectors for heating and room air conditioners for cooling consumes the smallest total amount of on-site energy, one-third less than the combination of warm-air fossil-fuel furnace and high-efficiency central air conditioner. The heat-only electric heat pump plus the high-efficiency central air conditioner and the high-efficiency electric heat pumps consume about the same amount of annual total energy, which is about 36–70% of the annual total energy consumed by the combination of warm-air furnaces and high-efficiency central air conditioners. Total annual energy savings for the standard electric heat pump relative to warm-air furnaces with central electric air conditioners is about 26–64% and is not as great as for the high-efficiency electric heat pumps.

8.5.6. Specific Findings: Advanced Systems

Figures 8.7 and 8.8 graphically display heating and cooling energy usage per square foot of test house area for advanced heating and cooling systems. Considering heat-activated heat pumps first, the data show Stirling-Rankine-cycle heat pumps to have smaller energy consumption for heating and cooling than the absorption heat pump, some 25% less seasonal energy for heating and 34% less seasonal energy for cooling.

It would appear from Figs. 8.7 and 8.8 that the V-type single-cylinder Stirling-Rankine heat pump is only marginally better for heating than the free-piston Stirling-Rankine heat pump, while for cooling, the free-piston system is more energy efficient than the V-type system. However, this may have as much to do with the assumption of capacity modulation for the latter system in the determination of the basic COP data on which the calculations of this report are based as with inherent system differences. Hence, for the V-type single-cylinder Stirling-Rankine heat pump, dynamic efficiency losses should be relatively small, and the data reported here should reasonably approximate energy consumption for the fully assembled field-operated machine. In its current design concept, the free-piston Stirling-Rankine machine will cycle. Based on this assumption, it would experience dynamic efficiency losses due to cycling not accounted for in the steady-state calculation. Both would also, of course, experience additional losses due to defrost. These cannot be specified on the basis of data available.

Without fully and properly measuring performance data for the advanced heat pump equipment, there is no completely satisfactory way of determining its energy consumption in actual practice. However, an approximation can be made if it is assumed that the ratio of energy consumed including dynamic efficiency losses to the energy consumed based on manufacturers' data is the same as that for conventional electric heat pumps. In that case the data of Table 8.19 for the standard electric heat pump can be used to predict these ratios. When this is done for heating-season, cooling-season, and annual energy usage averaged for all nine test cities, the following factors are produced:

	Energy consumption including dynamic efficiency losses	÷	Energy consumption based on manufacturers' steady-state performance
Heating season	1.097		
Cooling season	1.164		
Annual	1.118		

These factors can now be used to estimate energy consumption for equipment whose performance data are based on steady-state values. It should be understood that these factors provide only crude approximations to reality. It is not quite correct to assume, for example, that the free-piston Stirling engine will have the same dynamic efficiency losses as the electric motor driving the reciprocating compressor of the standard electric heat pump. It might be expected that such loses will be collectively greater for the free-piston Stirling-Rankine heat pump than for a conventional electric heat pump, since in the former case the prime mover will exhibit its own dynamics.

The factors given in the table can be applied to all the advanced heating and cooling systems to produce energy-consumption data on the same basis as for conventional systems with the exception of the following systems: advanced electric heat pump II, pulse combustion furnace, and the V-type single-cylinder Stirling-Rankine heat pump. The first and last are modulated systems for which dynamic efficiency losses should be quite small. The pulsed combustion furnace efficiency used already represents an integrated value over the heating season.

Application of the factors developed above to the energy data for the advanced electric heat pump I indicates that in the heating mode the system can offer energy savings over commercially available electric heat pump systems. However, in the cooling mode this system appears to perform only slightly better than the electric heat pump standard.

All gas heat pumps under consideration in this study consume substantially more energy on-site for both heating and cooling than commercially available HVAC systems, regardless of whether the comparison is made with or without consideration of dynamic efficiency losses.

The advanced electric heat pump II represents a hypothetical system employing realizable modifications to conventional electric heat pumps but having a significantly better operating performance. Energy-consumption data for this hypothetical heat pump were not separately calculated, but rather were scaled on energy-consumption data produced in the simulation of the standard electric heat pump. The SPF for this high-efficiency heat pump was taken as 2.4 in Columbus, an increase of 71% over an SPF of 1.4 for the standard electric heat pump in the same city. This increase is a reasonable assumption of the higher performance characteristics of a newer system employing advanced techniques (Comly et al. [8]). All other SPFs and energy consumed for heating and cooling the test house in each of the nine test cities with this heat pump were then produced by scaling values for the standard electric heat pump using the factor 2.4/1.4.

A similar technique was used for the pulse combustion furnace. Based on information that this type of system would have a seasonal efficiency of about 90–95% (Hollowell [6]), the gas energies calculated for the conventional gas warm-air furnace were adjusted to reflect an overall efficiency of 92.5%. Electric energy to drive the indoor circulating fan was assumed to be slightly higher for the pulse combustion furnace than for the gas warm-air furnace. Comparing the pulse combustion furnace at a 92.5% efficiency with a gas warm-air furnace at a 60% efficiency suggests that a 32% saving in total energy would result with the more efficient system.

8.6. GORDIAN STUDY: COMMERCIAL (1978)

8.6.1. Methodology

In addition, the Gordian study [5] compared energy effectiveness of advanced heat pump systems with available conventional heating and cooling systems for commercial types of buildings. This study parallels the residential study described in Section 8.5. Steps in the overall procedure are quoted from Ref. [5] as follows:

Commercial building selection. A single commercial building was selected and thoroughly characterized with respect to construction details for the purpose of estimating heating and cooling design loads for the structure in each of nine cities located in different climatological regions of the United States. This characterization included structural features of the building, including insulation details needed for the subsequent calculation of hourly heating and cooling loads and systems energy requirements in each climatological location.

Weather data and design load calculations. Design heating and cooling loads were estimated for the selected building. Weather data representing long-term average weather conditions at each of the nine cities were also used for the examination of energy effectiveness of alternative heating and cooling systems for commercial building applications.

Systems selection. Conventional heating and cooling systems were selected from among available representative equipment generally used for the commercial structure chosen. Advanced heat pumps were chosen from among those currently being developed for similar commercial building applications. Equipment sizes were then specified for the commercial building in each of the nine cities based upon estimated heating and cooling design loads.

Computer model. A computerized calculation was made to estimate monthly and annual heating and cooling loads and systems energy input. Performance data for the equipment selected were acquired from manufacturers (heating and cooling capacities and operating COP as a function of outdoor temperature, airflows and parasitic energy requirements, special characteristics of each equipment type). Performance data were modeled in a manner compatible with the computerized calculation of the commercial building heating and cooling loads and system energy requirement.

The final computer load and systems energy calculation were worked up in consistent form suitable for comparative analyses. These include a comparison of on-site energy consumption for conventional systems and advanced heat pumps, seasonal and annual performance factors.

8.6.2. Commercial Building Selection and Design-Load Calculation

The selection of a single commercial building that is then artificially moved to different climatic regions of the United States is purely a device through which a reliable comparison of energy effectiveness for alternative heating and cooling systems may be made. The building is used essentially to determine the heating and cooling loads that must be satisfied by the competing systems. To some extent, then, an almost arbitrary selection of building type could be made. However, the type of advanced technology heat pumps currently under development should obviously be a determinant in building selection. Several market studies commissioned by organizations involved in advanced heat pump development have indicated that light commercial structures (e.g., banks, community or professional services buildings, light manufacturing plants, and small office buildings) requiring unitary equipment of less than 25 tons cooling capacity represented the largest single potential. Typical are the 10-ton cooling capacity Brayton-Rankine and the 7.5-ton Rankine-Rankine gas heat pumps under current development for light commercial building applications.

Several small commercial buildings were examined, and a small office building under construction in Mountainside, New Jersey, was selected. The Mountainside building is a two-story cement-block structure having 5580 ft^2 of conditioned space. Architectural plans were made available as well as details of HVAC equipment selected for the building in its New Jersey location. Architectural views of the office building are shown in Fig. 8.9 and general features outlined in Table 8.28. The insulation and heat-transfer properties of the walls, roof, foundation, and glass areas of the building are compared with ASHRAE 90-75 thermal design standards for nonresidential buildings (ASHRAE Standard 90-75) in Table 8.29. Design loads were calculated for each location using the NESCA Manual J Supplement procedure [9] and are shown in Table 8.30.

8.6.3. Heating and Cooling System Selection

Five different heating and cooling systems were selected for study. Three represent conventional "baseline" systems frequently installed in the light commercial building application used here, and two are advanced heat-activated heat pumps that were being developed for similar installations. The specific systems selected were:

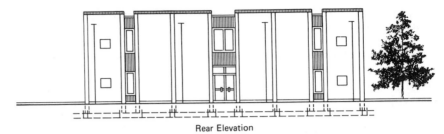

Rear Elevation

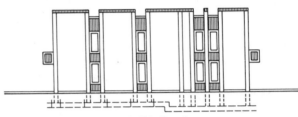

Right Side Elevation

Front Elevation

Fig. 8.9. Commercial building—architectural sketches.

Electric heat pump, standard
Electric resistance heating/electric air-conditioning
Rooftop gas furnace/electric air-conditioning
Brayton-Rankine gas heat pump
Rankine-Rankine gas heat pump

Because the advanced heat pump systems being developed are in the $7\frac{1}{2}$–10-ton nominal cooling capacity range, and also because conventional systems are available within this size range, it was desirable to subdivide the building so that multiples of the basic $7\frac{1}{2}$–10-ton sizes could be used. This was readily done by dividing the building into two zones, each zone having half the total floor area and half the design heating and cooling loads shown for the nine cities in Table 8.30. When this is done the nine cities fall into two groups.

TABLE 8.28.   Mountainside Commercial Building General Characteristics

Building Style	Two stories, masonry construction
Gross Floor Area	5580 square feet
Exposure	Front of building faces north
Occupancy	40 persons (maximum)
Lighting and Equipment Baseload	14 KW (maximum)
Construction Walls	Stucco surface over 12" concrete blocks (38# light, filled), 1-1/4" furring (16" on centers) and 1/2" styrofoam. Inside surface is 1/2" gypsum board. Net wall area: 4724 square feet.
Roof	3/8" asphalt built up roof over 1/2" plyscore sheathing, 2" x 10" rafters (16" on centers), fiberglass insulation (6") above 1/2" acoustical tile. Roof area: 2790 square feet.
Windows and Doors	Safety glass, 1/2" double glazed. Total glass area: 732 square feet.
Floor Slab	Concrete slab, on grade, 24" insulation
Ceilings	Hung ceiling between floors carries ducts and electric and water services. Ceiling height is 11 feet for each floor
Stairwell	Center staircase located just inside front entrance, double back design with turnaround landing.

SOURCE: Gordian and Associates [5].

The first group is represented by "warmer" locations, Houston through Philadelphia, for which 10 tons of cooling capacity is required. Within this group the largest design total cooling load is about 103,000 Btu/hr (Houston) and the smallest is 94,000 Btu/hr (Philadelphia) just above the $7\frac{1}{2}$-ton size. The second group consists of "colder" regions, Seattle through Concord, within which the largest design cooling load is about 89,000 Btu/hr (Columbus) and the smallest is about 63,000 Btu/hr (Seattle). For cities in this group, $7\frac{1}{2}$ tons of cooling capacity is required. It might possibly have been better to select a lower cooling capacity system for Seattle, but this was not done for two reasons. Commercially available rooftop single-package gas furnace/electric air conditioners are not made in less than $7\frac{1}{2}$-ton cooling capacities, and the Brayton-Rankine gas heat pump is not scheduled for commercialization below $7\frac{1}{2}$ tons of cooling capacity because of technical and cost factors. The system sizes selected for Seattle are therefore all

TABLE 8.29. Mountainside Commercial Building: Thermal Insulation Properties Compared with ASHRAE Standard 90-75

City	Heating Degree Days	Construction Wall Overall Heat Transfer Coeffient (Btu/Hr/ft2/°F) ASHRAE	Building	Roof Overall Heat Transfer Coeffient (Btu/Hr/ft2/°F) ASHRAE	Building	Slab on Grade Heat Transfer Resistance Factor (°F-Hr-ft2/Btu) ASHRAE	Building	Overall Thermal Transfer Factor (Btu/Hr/ft2) ASHRAE	Building
Houston	1278	0.366	0.121	0.100	0.034	1.672	5.263	30.498	12.091
Birmingham	2551	0.341	0.121	0.100	0.034	2.527	5.263	31.589	12.316
Atlanta	2961	0.333	0.121	0.100	0.034	2.802	5.263	31.593	12.214
Tulsa	3360	0.316	0.121	0.094	0.034	3.406	5.263	32.405	12.646
Philadelphia	4486	0.304	0.121	0.089	0.034	3.826	5.263	33.248	12.338
Seattle	4424	0.305	0.121	0.090	0.034	3.785	5.263	35.443	12.453
Columbus	5660	0.282	0.121	0.080	0.034	4.615	5.263	33.500	12.308
Cleveland	6351	0.268	0.121	0.074	0.034	5.079	5.263	33.784	12.437
Concord	7303	0.249	0.121	0.066	0.034	5.771	5.263	34.330	12.533

(a) ASHRAE refers to heat transfer coefficients and thermal resistance factors recommended for commercial buildings in ASHRAE Standard 90-75.

(b) Building refers to heat transfer coeffient and thermal resistance factors calculated for the Mountainside commercial office building.

SOURCE: Gordian and Associates [5].

TABLE 8.30. **Mountainside Commercial Building Heating and Cooling Design Loads**

	Heating		Cooling			
	97½% Design Dry Bulb Temperature (°F)	Design Heat Load (Btuh)	2½% Design Dry Bulb Temperature (°F)	2½% Design Wet Bulb Temperature (°F)	Design Sensible Cooling Load (Btuh)	Design Total Cooling Load (Btuh)
Houston	32	86,151	94	80	143,447	206,119
Birmingham	22	104,613	94	78	143,447	195,667
Atlanta	23	106,664	92	77	139,380	189,992
Tulsa	16	118,971	99	78	155,656	203,852
Philadelphia	15	121,022	90	77	135,315	187,535
Seattle	32	86,152	79	65	112,937	124,957
Columbus	7	139,484	88	76	131,244	178,640
Cleveland	7	139,484	89	75	133,277	176,653
Concord	-7	166,150	88	73	131,244	164,972

SOURCE: Gordian and Associates [5].

oversized with respect to cooling capacity but represent realistic sizes that would be installed in that region.

A short description of each heating and cooling system selected follows.

Electric heat pump, standard.   Single package; two unit sizes employed based on design cooling load with a nominal cooling capacity of 90,000 and 120,000. Lower capacity unit has a 90,000 Btu/hr cooling capacity at 95°F ODBT* (EER 7.8) and heating capacities of 90,000 Btu/hr (COP 2.6) and 50,000 Btu/hr at 14°F ODBT (COP 1.9); airflow is 3000 cfm for heating and cooling. Higher capacity unit has a 117,000 Btu/hr cooling capacity at 95°F ODBT (EER 8.2) and heating capacities of 120,000 Btu/hr at 47°F (COP 2.1) and 67,000 Btu/hr at 17°F; airflow is 4000 cfm for heating and cooling. Supplementary heaters sized for full design heating load for each city. Systems are equipped with outdoor temperature-controlled economizer dampers. Performance data for both heating and cooling were based on manufacturer's reported steady-state data corrected for dynamic efficiency effects with data of Kelly and Bean [10].

Electric resistance heating/electric air conditioner.   Heating provided by electrical resistance heaters sized for full-mounted single-package unit with two-stage modulated capacity (50% cooling capacity for each stage) and outdoor temperature-controlled economizer dampers. Two unit sizes were required: 90,000 and 120,000 Btu/hr nominal cooling capacity, depending on design cooling load. For the 90,000 Btu/hr air conditioner (45,000 Btu/hr each stage), first stage EER 6.62 at 95°F ODBT, and combined stages EER 7.69, airflow was 3000 cfm (heating and cooling). For the 120,000 Btu/hr air conditioner (60,000 Btu/hr each stage), airflow was 4000 cfm (heating and cooling). Both systems are equipped with outdoor temperature-controlled economizer dampers. Electrical resistance heating was assumed 100% efficient. Performance data for the electric air conditioner were based on manufacturer's reported steady-state data corrected for dynamic efficiency effects with the data of Kelly and Bean [10] for electric heat pumps in the cooling mode.

Rooftop gas furnace/electric air conditioning.   Single-package; unit sizes selected on the basis of cooling design load requirements. Each unit has two compressors for two-stage cooling capacity modulation operation. The smaller cooling capacity unit was rated at 90,000 Btu/hr cooling with two-stage operation (EER 7.26 at 95°F ODBT) and 45,000 Btu/hr cooling with one-stage operation (EER 5.85 at 95°F ODBT). The accompanying two-stage gas furnace capacity is rated at 180,000 Btu/hr output for two-stage and 135,000 Btu/hr output for one-stage operation. These outputs are AGA rat-

---

*ODBT stands for "outdoor dry-bulb temperature" and refers to the air temperature at the outdoor unit heat exchanger. Heating and cooling capacity and COPs are directly dependent upon this temperature.

ings for full-load combustion efficiencies of 75% for each stage. The larger cooling capacity unit was rated at 120,000 Btu/hr cooling with two-stage operation (EER 7.19 at 95°F ODBT) and 60,000 Btu/hr with one-stage operation (EER 5.69 at 95°F ODBT). Gas furnace capacity for this larger cooling unit was 225,000 Btu/hr output for two-stage operation and 168,000 Btu/hr output for each stage. Air flows for the two units were 3000 cfm (90,000 Btu/hr nominal cooling capacity) and 4000 cfm (120,000 Btu/hr nominal cooling capacity). Steady-state furnace output capacities were reduced using dynamic performance data for residential gas furnaces. Performance data for the electric air conditioner were based on manufacturer's reported steady-state data corrected for dynamic efficiency effects with data of Kelly and Bean.

Brayton-Rankine gas heat pump.   This system is capable of modulated operation over a major portion of its heating and cooling operating range. For small outdoor dry-bulb temperature deviation ($\pm 5°F$) about 65°F, the system ventilates only. As progressively greater temperature deviations occur (and the heating and cooling loads increase), the system first operates in an on/off cycling mode during which heating and cooling capacities are constant (45–65°F ODBT for heating and 70–75°F ODBT for cooling). At ODBT below 45°F for heating and above 75°F for cooling, the system is fully modulated. Performance data provided by the manufacturer for a nominal 10-ton system showed a cooling capacity of 119,000 Btu/hr at 95°F ODBT (COP 1.06) and a heating capacity of 100,000 Btu/hr at 47°F ODBT (COP 1.32). Parasitics (indoor and outdoor coil fans, sink heat exchanger fan, and systems control) for the 10-ton system were assumed to be 2.4 kW of electric energy (0.24 kW per ton of nominal cooling capacity). Performance data for a nominal 7½-ton cooling capacity system were scaled from data for the 10-ton system, assuming that COP vs. ODBT remains constant. Estimated performance data for the 7½-ton system at ARI rating points are 89,000 Btu/hr of cooling capacity at 95°F ODBT (COP 1.6) and a heating capacity of 75,000 Btu/hr at 47°F ODBT.

Rankine-Rankine gas heat pump.   This system is a fully modulated gas-fired steam-turbine-driven heat pump. Parasitic electrical energy requirements are produced by the turbine with a high-frequency alternator. In addition to heat pump operation, steam turbine exhaust is used to heat air for space-conditioning, and at very high heating loads (low outdoor temperatures) supplemental heat is provided by direct injection of generated steam. Data provided by the developer consisted of heating and cooling capacity ratios as percentages of full-load cooling capacity and corresponding COPs related to outdoor dry-bulb temperatures. For heating service, the heat pump is cut off at 0°F and all heating energy is provided by steam injection below this temperature. Two sizes of heat pumps were modeled for the commercial building application, 120,000 Btu/hr and 90,000 Btu/hr of cooling capacity at 95°F ODBT (COP 0.73 for both sizes). On heating service, high temperature capacities (47°F ODBT) of the heat pumps were taken as 120,000 Btu/hr and 90,000 Btu/hr, respectively (representing 100% of cooling load capacity) for which both heat pumps have COPs of 1.5. The developer specifically stated that reported operational heating COPs were substantially lower than design COPs due to off-cycle thermal losses, startup losses at low loads, and modulation losses at intermediate loads. Computed system energy requirements are therefore considered to include dynamic efficiency effects.

8.6.4. Systems Simulation and Computer Modeling

The computer program used for the determination of energy effectiveness of commercial building heating and cooling systems was AXCESS (Alternate Cost Comparison for Energy Systems Simulation), a program developed for the Edison Electric Institute (EEI) by several contractors, largely for the purpose of comparing energy requirements for alternative heating and cooling systems applied to a single structure. AXCESS has a broad flexibility with respect to the buildings and HVAC equipment it can examine, as well as a number of internal variations that can be programmed.

Although AXCESS calculates heating and cooling loads and then uses input data on the HVAC systems programmed to calculate energy needed to satisfy these loads, the program does not directly output calculated loads. It was important, though, for this study that the loads which were related to calculated energy consumption be known so that performance factors for the equipment examined could be estimated. This shortcoming was surmounted by running this program with a "synthetic" HVAC system, one that had a virtually infinite capacity for heating and cooling and an energy input exactly equal to its output. In this way the printed output was exactly equal to the load on the building.

The final outcome of the computer modeling with AXCESS for the five heating and cooling systems examined is a set of monthly heating and cooling loads for the commercial building in nine climatically different regions of the country, monthly on-site energy inputs to each space-conditioning system, and 1-hr peak electric energy demands. These results are shown in Tables 8.31 to 8.33; in terms of annual fuels and energy consumed on-site, Table 8.31; SPFs in Table 8.32, thermal equivalent of annual on-site energy consumption, Table 8.33.

8.6.5. Analysis of Heat Pump Performance

Of the two all-electric systems, the electric heat pump consistently uses less electric energy than the electrical resistance heater/electric air conditioner in all nine cities. Savings in annual electric energy by the heat pump range from 11% in Houston to 47% in Seattle and are generally larger for colder climate

TABLE 8.31. Annual On-Site Fuel and Electric Energy Consumption for Combined Heating and Cooling—Light Commercial Buildings (MBtu)

Electric Heat Pump and Conventional Systems		Houston	Birmingham	Atlanta	Tulsa	Philadelphia	Seattle	Columbus	Cleveland	Concord
Electric Heat Pump - Standard	Elec (kWh)	32,364	27,036	24,768	26,382	19,635	15,435	21,594	21,156	24,108
Electric Resistance Heating/ Electric Air Conditioning	Elec (kWh)	36,357	36,528	34,107	39,549	34,317	29,286	36,885	37,881	42,138
Rooftop Gas Furnace/Electric Air Conditioning	Gas (cft)	53,817	103,479	108,930	141,489	139,422	164,826	183,675	200,079	219,330
	Elec (kWh)	33,024	23,658	21,006	20,010	13,590	6,489	10,047	8,394	8,940
Advanced Heat Pump Systems										
Brayton-Rankine Gas Heat Pump	Gas (cft)	228,438	193,050	177,384	190,635	135,753	104,769	147,654	142,914	163,317
	Elec (kWh)	5,097	4,305	3,954	4,266	3,168	2,676	3,363	3,288	4,530
Rankine-Rankine Gas Heat Pump	Gas (cft)	253,917	199,497	181,443	184,425	137,460	96,048	136,401	128,985	142,839

SOURCE: Gordian and Associates [5].

TABLE 8.32. Average On-Site Heating, Cooling, and Annual Performance Factors—Light Commercial Buildings

Electric Heat Pump and Conventional Systems	Houston	Birmingham	Atlanta	Tulsa	Philadelphia	Seattle	Columbus	Cleveland	Concord
Electric Heat Pump – Standard									
Heating	2.67	2.48	2.40	2.35	2.44	2.33	2.14	2.13	2.04
Cooling	2.14	2.13	2.13	2.13	2.11	2.18	2.19	2.20	2.18
Annual	2.17	2.20	2.20	2.20	2.24	2.28	2.16	2.16	2.09
Heating/Electric Air Conditioning									
Heating	0.95	0.95	0.94	0.95	0.95	0.97	0.98	0.97	0.98
Cooling	2.13	2.13	2.13	2.13	1.78	2.21	2.20	2.22	2.13
Annual	1.93	1.63	1.60	1.47	1.28	1.20	1.26	1.20	1.19
Rooftop Gas Furnace/Electric Air Conditioning									
Heating	0.36	0.47	0.44	0.49	0.46	0.47	0.51	0.51	0.52
Cooling	1.97	1.97	1.97	1.97	1.95	2.06	2.05	2.07	2.06
Annual	1.44	1.10	1.03	0.95	0.81	0.64	0.73	0.68	0.69
Advanced Heat Pump Systems									
Brayton-Rankine Gas Heat Pump									
Heating	1.22	1.16	1.10	1.09	1.15	1.17	1.05	1.05	1.00
Cooling	0.96	0.93	0.93	0.91	0.95	0.88	0.94	0.94	0.89
Annual	0.98	0.98	0.97	0.97	1.03	1.05	1.00	1.01	0.96
Rankine-Rankine Gas Heat Pump									
Heating	1.54	1.46	1.42	1.41	1.45	1.46	1.38	1.38	1.38
Cooling	0.91	0.93	0.93	0.95	0.92	0.98	0.95	0.98	0.99
Annual	0.95	1.02	1.02	1.08	1.09	1.24	1.17	1.21	1.20

SOURCE: Gordian and Associates [5].

TABLE 8.33. Total On-Site Energy Consumption for Combined Heating and Cooling—Light Commercial Buildings (MBtu)

Electric Heat Pump and Conventional Systems	Houston	Birmingham	Atlanta	Tulsa	Philadelphia	Seattle	Columbus	Cleveland	Concord
Electric Heat Pump - Standard	110.46	92.27	84.53	90.04	67.01	52.68	73.70	72.21	82.28
Electric Resistance Heating/Electric Air Conditioning	124.09	124.67	116.41	134.98	117.12	99.95	125.89	129.29	143.82
Rooftop Gas Furnace/Electric Air Conditioning	166.53	184.22	180.62	209.78	185.80	186.97	217.97	228.73	249.84
Advanced Heat Pump Systems									
Brayton-Rankine Gas Heat Pump	245.83	207.74	190.87	205.19	146.57	113.90	159.13	154.14	178.78
Rankine-Rankine Gas Heat Pump	253.92	199.50	181.44	184.43	137.46	96.05	136.40	128.99	142.84

SOURCE: Gordian and Associates [5].

cities than for cities in warmer climates. This is reflected also by a superior heating SPF for the heat pump (2.0–2.7 for all cities) compared with a uniform heating performance factor less than 1 for electric resistance heating, as is shown in Table 8.32. In contrast, the cooling SPFs are approximately the same for the electric heat pump and the rooftop air conditioner. Therefore, in Houston, where the greatest energy use is for cooling, there is less of an advantage for the heat pump than in Concord, New Hampshire, where the major energy requirement is for heating.

As expected in energy use comparisons on an on-site basis, the rooftop combination of gas furnace/electric air conditioner uses more energy annually than either of the conventional all-electric systems, since electric generation transmission and distribution losses are not included in this comparison. However, a part of the increased energy consumption of the gas furnace/electric air conditioner is also due to a less efficient air-conditioning operation than the all-electric system. The rated EER for the electric air conditioner used with the gas furnace is 7.2 (120,000-Btu/hr capacity, two stages) and 7.3 (90,000-Btu/hr capacity, two stages). These are lower, for example, than the EER for the standard electric heat (8.2 for the 120,000-Btu/hr cooling capacity unit and 7.8 for the 90,000-Btu/hr unit). The higher EER electric heat pumps are clearly the reason for the better cooling SPFs for the heat pump shown in Table 8.32.

Gas furnace SPFs for residential heating were previously shown to be about 0.62–0.65, whereas the rooftop gas furnace for light commercial building applications shows heating SPFs of 0.36–0.52, both results including dynamic losses. The less efficient gas furnace performance in commercial buildings is due partly to a lower full-load (steady-state) efficiency for the large rooftop units (75%) compared with residential furnaces (80%), both certified ratings by the AGA. In addition, the rooftop units are grossly oversized compared with design heating loads. All rooftop gas furnaces are rated at 180,000 Btu/hr output heating capacity (first stage) compared with a maximum heating design load of 83,000 Btu/hr for Concord and 43,000 Btu/hr for Houston. As a consequence of oversizing, the furnaces cycle more frequently, each combustion cycle being of shorter duration compared with a gas furnace whose rated output is more nearly matched to design load. Dynamic efficiency losses due to greater cycling will considerably reduce heating SPFs.

It is difficult to compare the advanced gas heat pumps with the conventional systems because of the different forms of energy utilized on-site. It is obvious that since the on-site efficiency of the electric HVAC systems is higher, the all-electric conventional systems use more on-site energy than the advanced systems. Both advanced systems show high heating SPFs, (greater than 1) compared with a heating SPF of less than 0.5 for the gas furnace. Thus, with respect to the same energy source, the advanced heat pumps show superior performance.

Upon comparing the two advanced gas heat pumps, it appears that the

Rankine-Rankine system uses more total on-site energy in all cities except Concord than does the Brayton-Rankine system. However, the latter system uses purchased electricity for its auxiliary equipment, whereas the Rankine-Rankine machine generates all of its own electricity.

REFERENCES

1. Howell, R. H., and Sauer, H. J., Jr., "Comparative Residential Energy Consumption and Fuel Costs with Various Types of Systems: Oil-, Gas-, Electric Furnaces and Heat Pumps," *Proc. Third Annual UMR-MEC Conf. Energy, Oct. 1976, Rolla, MO.*

2. Gordian Associates Inc., *Evaluation of the Air-to-Air Heat Pump for Residential Space Conditioning* (report prepared for FEA, NTIS No. PB-255-652, April 1976).

3. Delene, J. G., *A Regional Comparison of Energy Resource Use and Cost to Consumer of Alternate Residential Heating Systems* (Oak Ridge, TN, Oak Ridge National Laboratory, Nov. 1974).

4. Raab, N. L., "Metered Air-to-Air Heat Pump Energy Consumption in Seven St. Louis Area Homes," *Proc. Sixth UMR/DNR Conf. Energy, Rolla, MO, 1979.*

5. Gordian & Associates, Inc., *Heat Pump Technology, A Survey of Technical Development Market Prospects*, Report prepared for DOE, HCP/M2121-01, June 1978.

6. Hollowell, G. T., "Pulsed Combustion—An Efficient, Forced Air, Space Heating System," *Proc. Conf. Improving Efficiency and Performance HVAC Equipment and Systems*, Purdue Research Foundation 1976.

7. Groff, G. E. and Bullock, C. E., "A Computer Simulation Model for Air-Source Heat Pump System Seasonal Performance Studies," *Prod. 2nd Annual Heat Pump Technol. Conf., Oklahoma State University, Stillwater, OK, Oct. 18–19, 1976.*

8. Comly, J. B., Jaster, H., and Quaile, J. P., "Heat Pumps—Limitations and Potential" (General Electric Co. Report No. 75 CRD185, Sept. 1975).

9. Manual, J., National Environmental Systems Contractors Association, Arlington, VA, 1975.

10. Kelly, G. E., and Bean, J., "Dynamic Performance of a Residential Air-to-Air Heat Pump," *Proc. 2nd Annual Heat Pump Technol. Conf., Oklahoma State University, Stillwater, OK, Oct. 18–19, 1976.*

9

ECONOMIC ANALYSIS FUNDAMENTALS

9.1. INTRODUCTION

The selection of any heating system should be based on a compromise between its performance and economic merits. The system selected is usually determined by the user's needs, the designer's experience, local building codes, first costs, most efficient use of source energy, and the projected operating costs. Any one of these factors may alter the choice of the system. However, when the choice is between alternatives that give apparently equal results, the system with the lowest life-cycle costs should be chosen. This does not mean the system with the lowest initial cost; instead, it directs a decision toward the system with the lowest long-term costs.

To evaluate the desirability of an investment, measures are needed to compare costs with benefits. It is important that only those costs and benefits that are attributable to an investment be included in the analysis of that investment. For example, if a plant is required by law to add a pollution-control apparatus, the decision on whether or not to include an energy-recovery system should not be influenced by the costs of the basic required pollution-control system. As a further example, when equipment is replaced or repaired, costs should be allocated to each system when they are undertaken jointly for convenience.

Once a capital investment is made, its consequences are irrevocable. In fact, once a capital budget is approved and future funding is committed, it becomes particularly difficult to divert the expenditure. Because of this permanence, the analysis of a capital investment decision must be sound and rigorous, leaving little doubt that the information presented is as reasonable and as accurate as the facts allow.

512

Not only is the permanence of the decision important, the trend in increasing costs, especially energy costs, requires that careful thought be given to all aspects of capital investment decisions. This is particularly true for decisions involving buildings and building systems, for which five components stand out as requiring careful consideration:

1. Initial capital investment cost
2. Annual operating and routine maintenance costs
3. Major repairs and component replacements
4. Complete item or system replacement
5. Residual values

A sixth consideration is time. The timing factor is used to judge when costs or benefits occur and when replacements are needed. Combining the elements results in a "life cycle" for an investment decision.

Life-cycle costing is a method of expenditure evaluation that recognizes the total costs associated with the asset during the time it is in use. It is an evaluation technique, an input for decision-making.

Since the 1950s, life-cycle costing, in conjunction with present value techniques and value engineering, has been commonly applied in capital expenditure evaluation. However, for many well-intended reasons, government investment decisions have been based upon the lowest capital costs, with little attention given to the impact of life-cycle costs. Today's rapidly accelerating costs for personnel, material, energy, equipment, and facilities have motivated both government agencies and the private sector to reassess this practice. The result is reflected in the promotion of life-cycle costing analysis techniques, especially in energy management programs.

Life-cycle costing, a term that describes a broad economic philosophy, considers costs that will be experienced over an extended period of time. Many economic procedures can be used within life-cycle costing; they all produce long-term costs so that comparisons can be made between systems having different initial and long-term operating costs. To be accurate, a procedure must include all cost factors relevant to the specific investment, such as initial costs, service life, interest, energy costs, taxes, operating expenses, and cost escalation.

The owner or engineer can influence decisions within a life-cycle costing procedure by varying economic assumptions. For instance, the use of long amortization periods, low interest rates, or high annual energy cost escalation favors the use of more efficient equipment, even if a substantially higher first cost is to be incurred. System alternatives can thus be selected that best fit the financial aspects of the project and the needs of the owner.

Techniques or procedures most commonly used depend on evaluating (1) present worth, (2) uniform annual owning and operating costs, (3) rate of return, (4) rate of return on investment, (5) benefits vs. costs, (6) years to payback, and (7) cash flow. These techniques should meet most needs. Sophisticated systems, such as cash flow, are used by owners and investors to meet their particular objectives. Simpler techniques, such as cost-benefit analysis and years to payback, meet some needs but can easily lead to improper decisions, since important financial information is often neglected.

The factors used to calculate life-cycle costs are important to the final results. The most commonly used factors are defined in this chapter. The factors can be categorized broadly as (1) costs of ownership and (2) operating costs.

9.2. COSTS OF OWNERSHIP

There are costs of ownership even if there is no activity in the facility. They can be expressed as annual costs that are distributed over an extended period of time or as an equivalent total value, which is called present worth or present value. Costs of ownership include initial costs, plus property taxes, equipment rental, and insurance.

9.2.1. Initial Costs

The first significant step in analyzing a system is to determine the first or initial costs. Initial costs should include construction costs of the system and additional building costs that are attributable to the system. Costs to design the system, to administer construction, and to raise capital can be included.

9.2.2. Interest

Money has a true value since it must be borrowed, obtained from equity or investors, or diverted from other uses. Interest, the cost of borrowed capital, is a proper and important part of the life-cycle cost of an installation. The owner should establish the true rate of interest or rate of return on his investment. The minimum value assigned to money should reflect the prevailing interest rate on borrowed money. A speculative investment bears more risk, so a higher rate may be justified. Interest is also referred to as *cost of capital* or *discount rate*. For simple interest the entire amount S to be repaid at the future time is

$$S = P + Pni = P(1 + ni) \qquad (9.1)$$

where P = principal
i = rate of interest per period
n = number of periods

Simple interest does not allow for any change in the principal amount during the time period nor does it allow unpaid interest charges to be included in the principal over the time period. With compound interest, the principal is increased at the end of each period by the interest accumulated during that period. The formula for computing compound interest is

$$S = P(1 + i)^n = P(\text{CAF}) \tag{9.2}$$

where CAF is called the compound-amount factor and is equal to $(1 + i)^n$.

9.2.3. Time Period

The time period used by owners and engineers to analyze the system can be defined using several terms. *Useful life* is a term used by the Internal Revenue Service. The useful life of any item depends upon such things as the frequency with which you use it; its age when you acquired it; your policy as to repairs, renewals, and replacements; the climate in which it is used; normal progress of the arts; economic changes; inventions; and other developments within the industry.

Depreciation period is a time period estimated as the useful life of the asset over which to allocate the first cost of the asset. It forms the basis for a deduction against income in calculating income taxes. Various methods for calculating depreciation allowed by the IRS include the straight-line method and more accelerated procedures, such as declining-balance or sum-of-the-years. Straight-line depreciation uses a constant deduction D_s in each of n depreciation periods. Thus, in any period, for initial costs C and salvage S, the depreciation is

$$D_s = \frac{C - S}{n} \tag{9.3}$$

Accelerated depreciation methods use a higher depreciation during the initial years, with amounts declining as the asset ages.

Amortization period is the number of periods n selected to perform economic comparisons. It is a time over which periodic payments of monies are made to discharge a debt.

Service life is a time value that reflects the expected life of a specific component. Equipment life is highly variable due to diverse equipment applications, the preventive maintenance given, the environment, technical advancements of new equipment, and personal opinions. The values of Table 9.1 describe the replacement time of the components. Service life can be used to establish an amortization period, or, if an amortization period is given, service life can give an insight into adjusting the maintenance costs of components.

Capital recovery factor (CRF) is calculated using an interest rate i and an amortization period n that determines the uniform periodic cost needed to repay a debt or initial cost. The factor is determined by

$$\text{CRF} = \frac{i(1 + i)^n}{(1 + i)^n - 1} \tag{9.4}$$

TABLE 9.1. Equipment Service Life[a]

Equipment Item	Years	Equipment Item	Years
Air conditioners		Coils	
Window unit	10	DX, water, or steam	20
Residential single or split		Electric	15
package	15	Heat exchangers	
Commercial through-the-wall	15	Shell-and-tube	24
Water-cooled package	15	Reciprocating compressors	20
Computer room	15	Package chillers	
Heat pumps		Reciprocating	20
Residential air-to-air	10	Centrifugal	23
Commercial air-to-air	15	Absorption	23
Commercial water-to-air	19	Cooling towers	
Roof-top air conditioners		Galvanized metal	20
Single-zone	15	Wood	20
Multizone	15	Ceramic	34
Boilers, hot water (steam)		Air-cooled condensers	20
Steel water-tube	24(30)	Evaporative condensers	20
Steel fire-tube	25(25)	Insulation	
Cast iron	35(30)	Molded	20
Electric	15	Blanket	24
Burners	21	Pumps	
Furnaces		Base-mounted	20
Gas- or oil-fired	18	Pipe-mounted	10
Unit heaters		Sump and well	10
Gas or electric	13	Condensate	15
Hot water or steam	20	Reciprocating engines	20
Radiant heaters		Steam turbines	30
Electric	10	Electric motors	18
Hot water or steam	25	Motor starters	17
Air terminals		Electric transformers	30
Diffusers, grilles, and		Controls	
registers	27	Pneumatic	20
Induction and fan-coil units	20	Electric	16
VAV and double-duct boxes	20	Electronic	15
Air washers	17	Valve actuators	
Duct work	30	Hydraulic	15
Dampers	20	Pneumatic	20
Fans		Self-contained	10
Centrifugal	25		
Axial	20		
Propeller	15		
Ventilating roof-mounted	20		

[a]Obtained from a nationwide survey conducted in 1977 by ASHRAE TC 1.8.
Reprinted by permission from *ASHRAE Systems Handbook 1980*.

Table 9.2 gives CRF values.

9.2.4. Present Worth

Present worth is the current value of monies that are to be spent over the selected amortization period. In essence, it is the money necessary today for initial investment plus all the monies necessary for all future costs remaining. The term is widely used and the concept is the basis for the most commonly used life-cycle cost techniques.

Present worth factor, uniform annual series (PW_{uas}) is given in standard discount tables and is used to calculate the present worth when given a uniform annual cost. $PW_{uas} = 1/CRF$. The term does not consider escalation nor costs that are not uniform or annual. PW is obtained by multiplying annual cost by PW_{uas} or by $1/CRF$.

Compound amount factor (CAF), uniform annual series, is used to determine a future sum of money when a uniform annual payment is made. The factor is determined by

$$CAF = \frac{(1 + i)^n - 1}{i} \tag{9.5}$$

The actual sum is determined by multiplying the annual cost by CAF. The reciprocal of CAF, the sinking fund factor (SFF), is used to calculate a uniform annual cost from a given value to the end of an amortization period.

Present worth factor, single payment (PWF_{sp}), is used to calculate the present worth of a future one-time cost. The PWF_{sp} is determined by

$$PWF_{sp} = \frac{1}{(1 + i)^n} \tag{9.6}$$

This factor multiplied by the actual costs gives the present worth of a one-time cost, such as a major overhaul or equipment replacement. The reciprocal is called the compound amount factor, single payment (CAF_{sp}).

Stated alternatively, a future sum of money F has a present worth P given by

$$P = \frac{F}{(1 + i)^n} \tag{9.7}$$

Equation (9.7) indicates that the present worth of a given amount of money in the future is discounted, in constant dollars, by a factor $(1 + i)^{-1}$ for each year in the future. In this context, i is generally called the discount rate. Table 9.3 is a tabulation of present worth factors.

9.2.5. Property Tax

Property tax is a cost of ownership. The engineer must evaluate the effects of appreciation or depreciation of property values with time and the varia-

TABLE 9.2. Capital Recovery Factors

Years					Rate of Return or Interest Rate (%)					
	3½	4½	6	8	10	12	15	20	25	30
2	0.52640	0.53400	0.54544	0.56077	0.57619	0.59170	0.61512	0.65455	0.69444	0.73478
4	.27225	.27874	.28859	.30192	.31547	.32923	.35027	.38629	.42344	.46163
6	.18767	.19388	.20336	.21632	.22961	.24323	.26424	.30071	.33882	.37839
8	.14548	.15161	.16104	.17401	.18744	.20130	.22285	.26061	.30040	.34192
10	.12024	.12638	.13587	.14903	.16275	.17698	.19925	.23852	.28007	.32346
12	0.10348	0.10967	0.11928	0.13270	0.14676	0.16144	0.18448	0.22526	0.26845	0.31345
14	.09157	.09782	.10758	.12130	.13575	.15087	.17469	.21689	.26150	.30782
16	.08268	.08902	.09895	.11298	.12782	.14339	.16795	.21144	.25724	.30458
18	.07582	.08224	.09236	.10670	.12193	.13794	.16319	.20781	.25459	.30269
20	.07036	.07688	.08718	.10185	.11746	.13388	.15976	.20536	.25292	.30159
25	0.06067	0.06744	0.07823	0.09368	0.11017	0.12750	0.15470	0.20212	0.25095	0.30043
30	.05437	.06139	.07265	.08883	.10608	.12414	.15230	.20085	.25031	.30011
35	.05000	.05727	.06897	.08580	.10369	.12232	.15113	.20034	.25010	.30003
40	.04683	.05434	.06646	.08386	.10226	.12130	.15056	.20014	.25006	.30001

TABLE 9.3. Present Worth Factors

n \ i	0%	2%	4%	6%	8%	10%	12%	15%	20%	25%
1	1.0000	0.9804	0.9615	0.9434	0.9259	0.9091	0.8929	0.8696	0.8333	0.8000
2	1.0000	0.9612	0.9246	0.8900	0.8173	0.8264	0.7972	0.7561	0.6944	0.6400
3	1.0000	0.9423	0.8890	0.8396	0.7938	0.7513	0.7118	0.6575	0.5787	0.5120
4	1.0000	0.9238	0.8548	0.7921	0.7350	0.6830	0.6355	0.5718	0.4823	0.4096
5	1.0000	0.9057	0.8219	0.7473	0.6806	0.6209	0.5674	0.4972	0.4019	0.3277
6	1.0000	0.8880	0.7903	0.7050	0.6302	0.5645	0.5066	0.4323	0.3349	0.2621
7	1.0000	0.8706	0.7599	0.6651	0.5835	0.5132	0.4523	0.3759	0.2791	0.2097
8	1.0000	0.8535	0.7307	0.6274	0.5403	0.4665	0.4039	0.3269	0.2326	0.1678
9	1.0000	0.8368	0.7026	0.5919	0.5002	0.4241	0.3606	0.2843	0.1938	0.1342
10	1.0000	0.8203	0.6756	0.5584	0.4632	0.3855	0.3220	0.2472	0.1615	0.1074
11	1.0000	0.8043	0.6496	0.5268	0.4289	0.3505	0.2875	0.2149	0.1346	0.0859
12	1.0000	0.7885	0.6246	0.4970	0.3971	0.3186	0.2567	0.1869	0.1122	0.0687
13	1.0000	0.7730	0.6006	0.4688	0.3677	0.2897	0.2292	0.1625	0.0935	0.0550
14	1.0000	0.7579	0.5775	0.4423	0.3405	0.2633	0.2046	0.1413	0.0779	0.0440
15	1.0000	0.7430	0.5553	0.4173	0.3152	0.2394	0.1827	0.1229	0.0649	0.0352
16	1.0000	0.7284	0.5339	0.3936	0.2919	0.2176	0.1631	0.1069	0.0541	0.0281
17	1.0000	0.7142	0.5134	0.3714	0.2703	0.1978	0.1456	0.0929	0.0451	0.0225
18	1.0000	0.7002	0.4936	0.3503	0.2502	0.1799	0.1300	0.0808	0.0376	0.0180
19	1.0000	0.6864	0.4746	0.3305	0.2317	0.1635	0.1161	0.0703	0.0313	0.0144
20	1.0000	0.6730	0.4564	0.3118	0.2145	0.1486	0.1037	0.0611	0.0261	0.0115
25	1.0000	0.6095	0.2751	0.2330	0.1460	0.0923	0.0588	0.0304	0.0105	0.0038
30	1.0000	0.5521	0.3083	0.1741	0.0994	0.0573	0.0334	0.0151	0.0042	0.0012
40	1.0000	0.4529	0.2083	0.0972	0.0460	0.0221	0.0107	0.0037	0.0007	0.0001
50	1.0000	0.3715	0.1407	0.0543	0.0213	0.0085	0.0035	0.0009	0.0001	—
100	1.0000	0.1380	0.0198	0.0029	0.0005	0.0601	—	—	—	—

tion of the property tax rate. Tax incentives for energy conservation items may be an important consideration in determining justifiable long-term investment.

9.2.6. Insurance

Insurance is the means through which a property owner can be reimbursed for a financial loss arising out of damage to the property that necessitates its repair or its replacement. Financial loss may also include the loss of income, rents, or profits resulting from the property damage.

9.2.7. Salvage Value

Terminal values of equipment occur at the end of life or the end of the final amortization period. These values are often assumed to be zero for many building components, as replacement takes place only when the equipment is practically worthless or will cost as much to remove as its market value. Where such an assumption cannot be made, a salvage value should be included in the calculation.

9.3.   OPERATING COSTS

Operating costs are annual expenses resulting from the actual use of the system. They include costs for energy, wages, supplies, water, material, maintenance, parts, and services.

9.3.1.   Energy

Utility costs usually require a monthly calculation considering energy consumption and peak demands. Utilities then apply demand cost and step rate schedules to establish actual costs. The engineer should review these costs and service policies so that the most advantageous selection of energy type, systems, and system operation can be made. The most reliable procedure for estimating energy costs requires hourly integration for a year of calculated energy use of each building component as a function of weather, internal loads, building heat gains and losses, and ventilation. The peak and total of these hourly consumption rates should reveal the comparative cost features of alternative control systems, distribution systems, energy sources, and other design options.

The explosive rise of energy costs has clearly indicated the need to incorporate the influence of cost escalation in life-cycle costs. Fuel escalation is expected to exceed the normal rate of inflation for the foreseeable future. Although cost escalation is difficult to establish, it can be applied to most annual costs. In the long run, investors expect to compensate for declines in purchasing power caused by inflation plus a little extra. The difference between interest and inflation, called the *real return,* is historically in the 1–5% range. Escalation values affect life-cycle costing and thereby influence the selection of equipment to be used even when initial costs are greater.

The present worth of an annual cost over a selected number of time periods n, using an interest rate of money i, and a cost escalation rate j, is called present worth escalation factor (PWEF) and is determined by

$$\text{PWEF} = \sum_{x=1}^{n} x \left(\frac{1 + j}{1 + i} \right)^x \tag{9.8}$$

A simplified, one-step calculation will work when the interest and fuel cost escalation rates are not equal.

$$\text{PWEF} = \frac{[(1 + j)/(1 + i)]^n - 1}{1 - (1 + i)/(1 + j)} \tag{9.9}$$

9.3.2. Maintenance

Expenses for labor and material necessary to make repairs as well as cleaning, painting, inspection, testing, and so on, are important in determining total operating costs. Generally, routine maintenance requirements will be met by an operating engineer or staff. Extraordinary repairs are often handled by maintenance divisions, and the expense is charged back to the system's maintenance costs. In other cases, some or all maintenance is handled by outside service firms. These costs vary considerably with the type of system and the proficiency of the servicing organization. The annual maintenance allowance should be based on the entire amortization period under study rather than on the early years with their associated shakedown problems or later years when higher costs are usually experienced. If annual costs are expected to rise or fall at a rate different from interest, present worth should be recalculated using an escalation factor.

Since maintenance policies of HVAC equipment users vary, and since varying maintenance policies may affect the lives of equipment, maintenance requirements are categorized as follows:

Type I. Maintenance with emphasis on a planned basis. It includes frequent inspections, adjustments, lubrication, and parts replacement according to a planned maintenance schedule. It includes emergency repairs, repairs in anticipation of failure, recommended start-up and shutdown procedures, and scheduled major and minor overhauls.

Type II. Maintenance structured loosely on an occasional or as-needed basis. It incudes occasional inspections, adjustment, lubrication, and parts replacement, but not necessarily on a fixed schedule. It also includes emergency repairs, start-up and shutdown procedures, planned major and unplanned minor overhauls [1].

A survey conducted in 1977 by ASHRAE indicates the cost of maintenance to be in accordance with Table 9.4. Different maintenance procedures will result in different costs, which go up with increasing maintenance.

TABLE 9.4. Maintenance Costs[a]

Maintenance Policy	$/(yr·ft^2)			$/(yr·m^2)
Type I maintenance	25th percentile	=	$0.07	
	Median	=	$0.14	$1.53
	75th percentile	=	$0.24	
Type II maintenance	25th percentile	=	$0.07	
	Median	=	$0.08	$0.87
	75th percentile	=	$0.17	

[a] 1976 dollars.

SOURCE: Reprinted by permission from *ASHRAE Systems Handbook 1980*.

9.3.3. Labor

Building engineers or operators are often necessary, due to the scope of the facility. Building personnel may have functions unrelated to the system, and thus charges should be properly allocated.

9.3.4. Water Costs

Heat from a cooling system may be rejected either in favor of water from a utility company or by means of cooling towers. Water conservation equipment must often be used regardless of economics because of local regulations intended to conserve the existing supply. Water treatment costs represent a significant operating expense. Lack of proper treatment can cause high energy consumption and operating problems.

9.3.5. Income Tax

Income taxes imposed by federal, state, and local governments can be considered an operating cost. The tax rate that should be used in the calculation is the *marginal rate,* the tax to be paid on the next dollar to be earned rather than the average tax rate. The allowances for depreciation and other deductions used in calculating income taxes strongly influence investments by reducing life-cycle costs. In the private sector this should not be ignored. Significant tax credits are often available for investments and should be included as cost offsets in life-cycle costs.

9.4. SUMMARY OF INTEREST, DISCOUNT, AND ANNUITY FORMULAS

Tables 9.5 and 9.6 summarize the significant relationships between the value of money, interest rates, and time.

9.5. LIFE-CYCLE COSTING (LCC)

9.5.1. An Overview of the LCC Concept

All too often, first cost has preoccupied the minds of both the owner and designer, causing them to neglect giving proper consideration to system life and operating cost. A system that is inexpensive to buy may be expensive to operate and maintain.

   With inflation, construction costs have escalated. The cost of money and energy continue to increase dramatically, but not always in the same proportion. These factors have given rise to a more rational and factual approach to

TABLE 9.5. Summary of Interest and Annuity Formulas [3]

Symbol	Factor	Formula	Equation of Use
CAF	Compound amount	$(1 + i)^n$	Future amount $= (CAF)(\text{present amount})$
PWF	Present worth	$\dfrac{1}{(1 + i)^n}$	Present amount $= (PWF)(\text{future amount})$
SCAF	Series compound amount	$\dfrac{(1 + i)^n - 1}{i}$	Future amount $= (SCAF)(\text{regular payment})$
SFF	Sinking fund	$\dfrac{i}{(1 + i)^n - 1}$	Regular payment $= (SFF)(\text{future amount})$
SPWF	Series present worth	$\dfrac{(1 + i)^n - 1}{i(1 + i)^n}$	Present amount $= (SPWF)(\text{regular payment})$
CRF	Capital recovery	$\dfrac{i(1 + i)^n}{(1 + i)^n - 1}$	Regular payment $= (CRF)(\text{present amount})$

TABLE 9.6. Discount Formulas[a]

Nomenclature	Use	Algebraic Form
Single compound amount formula	To find F, given P To find P, given T	$F = P(1 + i)^n$ $P = T(1 + i)^n$
Single present value formula	To find P, given F	$P = F\dfrac{1}{(1 + i)^n}$
Uniform compound amount formula	To find F, given A	$F = A\dfrac{(1 + i)^n - 1}{i}$
Uniform sinking fund formula	To find A, given F	$A = F\dfrac{i}{(1 + i)^n - 1}$
Uniform capital recovery formula	To find A, given P	$A = P\dfrac{i(1 + i)^n}{(1 + i)^n - 1}$
Uniform present value formula	To find P, given A	$P = A\dfrac{(1 + i)^n - 1}{i(1 + i)^n}$

[a] P = a present sum of money
T = a past sum of money
F = a future sum of money
i = an interest rate
n = number of interest periods
A = an end-of-period payment (or receipt) in a uniform series of payments (or receipts), usually annually.

the real costs of a system, by analyzing both owning and operating costs over a fixed time period (life-cycle costs).

Life-cycle costing (LCC) is defined as the total costs of owning, operating, and maintaining a building over its economic life, including its fuel and energy costs, determined on the basis of a systematic evaluation and comparison of alternative building systems.

Over more than two decades, life-cycle costing has become a generally accepted means, in both the public and private sectors, of recognizing the sum total of all costs (and benefits) associated with a project during its estimated lifetime. As experience has grown, the application of LCC techniques has become increasingly sophisticated, evolving from the use of simple manual calculations to complex computer programs that require vast data bases. Many government agencies are currently using LCC or other economic evaluation techniques, but differences exist in applications and in technical criteria. Thus, while the LCC technique is not new, there is a lack of uniformity and consistency in its use.

In order to facilitate a uniform LCC approach, this section provides basic ground rules, assumptions, definitions, and requirements for using the LCC methodology. It may be used in conjunction with existing calculation techniques or models for estimating specific LCC parameters such as initial investment costs, future energy costs, or maintenance costs, provided that these techniques or models satisfy the criteria included here.

As applied to energy conservation projects in buildings, LCC analysis provides an evaluation of the net effect, over time, of reducing fuel costs by purchasing, installing, maintaining, operating, repairing, and replacing energy-conserving features.

LCC analysis is primarily suited for the economic comparison of alternatives. Its emphasis is on determining how to allocate a given budget among competing projects so as to maximize the overall net return from that budget. The LCC method is used to select energy conservation projects for which budget estimates must be made; however, the LCC cost estimates are not appropriate as budget estimates, because they are expressed in constant dollars (excluding inflation) and all dollar cash flows are converted to a common point in time. Hence, LCC estimates are not necessarily equivalent to the obligated amounts required in the funding years.

The results of LCC analysis are usually expressed in either *present worth dollars, uniform annual value dollars, savings to investment ratio* (SIR) (the ratio of present or annual value dollar savings to present or annual value dollar costs), or as a percentage *rate of return on the investment.*

Expressing LCC estimates in present worth dollars was discussed in Section 9.2.4. Expressing LCC estimates in uniform annual value dollars means converting all past, present, and future cash flows to their equivalent value in terms of a series of level, annual amounts, taking into account the time value of money. For example, mortgage loan payments are usually calculated using the uniform annual value method, except that the year is generally divided into 12 interest periods.

Although it is not in a strict sense an LCC measure, the time until the initial investment is recouped (*payback*) is another concept that is sometimes used to report the results of an LCC analysis. To derive any of these measures, it is important to adjust for differences in the timing of expenditures and cost savings. This time adjustment can be accomplished by a technique called *discounting*.

The major steps for performing an LCC analysis of energy conservation investments are the following:

Identify the alternative approaches to achieve the objective.

Establish a time frame for the analysis.

Identify the cost parameters to be considered in the analysis.

Convert costs and savings occurring at different times to a common time.

Determine the cost-effectiveness of the alternatives.

Analyze the results for sensitivity to the initial assumptions.

9.5.2. LCC Organization [2]

A more detailed description of the basic LCC procedure is given in the following section. However, the basic elements needed to make a thorough LCC analysis are summarized below:

Annual Cost of Ownership

1. Initial costs.   The amortization period must be determined in which the initial costs are to be recovered and converted by use of a capital recovery factor (CRF) into an equivalent annual cost. In Table 9.2, data are given for CRFs based on years of useful life and the rate of return or interest rate. The table gives a factor which, when multiplied by the initial cost of a system or component thereof, will result in an equivalent uniform annual cost for the period of time chosen.
2. Taxes
   a. Property or real estate taxes
   b. Personnel payroll taxes
   c. Building management personal property taxes
   d. Other building taxes
3. Insurance

Annual Operating Cost

1. Annual energy costs
   a. Energy and fuel costs
   b. Water charges
   c. Sewer charges
2. Annual maintenance costs
   a. Maintenance contracts
   b. General housekeeping costs
   c. Labor and material for replacing worn parts and filters
   d. Chemicals and cleaning compounds
   e. Costs of refrigerant, oil, and grease
   f. Cleaning and painting
   g. Testing
   h. Waste disposal
3. Operators.   The annual wages of building engineers and/or operators should not be included as part of maintenance but entered as a separate cost item.

9.5.3.   Time Considerations

To perform LCC analysis, it is necessary to establish a base time so that all past, present, and future costs can be converted to a common dollar measure. If LCC estimates are to be expressed in present value dollars, the base time is the present (the time at which the LCC analysis is being conducted). If LCC estimates are to be expressed in annual value dollars, the base time is actually a series of time periods of equal intervals (e.g., years) extending over the period of the analysis.

An LCC analysis requires the estimation of the economic life expectancies of the principal assets associated with each investment alternative. The economic life is that period over which the asset is expected to be retained in use as the lowest cost alternative for satisfying its intended purpose. The economic life of the building, equipment, systems, or components is often difficult to determine. Generally, the facility engineer will determine life based on available technical manuals, information from manufacturers and distributors, expectations for obsolescence, and information on the average lives of generic types of plants and equipment.

It is also necessary for the analyst to specify the length of time, or study period, over which an investment is to be evaluated. In specifying the study period, it is important that (1) all mutually exclusive alternatives be evaluated on the basis of the same study period, (2) the study period not exceed the period of intended use of the facility in which the energy conservation investment is to be made, and (3) if alternatives are evaluated for a period shorter or longer than the estimated lives of the principal assets, any significant salvage values or replacement costs should be taken into account.

One of the following four approaches is usually taken to establish the study period:

1. If it is assumed that the facility is to be used indefinitely, the study period can also be assumed to be infinite, and costs can be evaluated in annual value dollars based on the economic life of each alternative investment. For example, the annual cost of a 10-year life investment is calculated on the basis of 10 years, and the annual cost of an alternative 15-year life investment is based on 15 years. Then it is assumed that either would be used indefinitely through a series of replacements. This approach can in some cases simplify calculations because it eliminates the need to consider replacements and salvage values.

2. The study period can be set equal to a period of time that allows coincidence of the expiration of alternative investments. For example, in comparing an investment having a 10-year life to one with a 15-year life, the study period would be set equal to 30 years, with three renewals of the first investment and two renewals of the second. This approach is often taken to evaluate alternatives over an equal period of time when results are to be measured in present value dollars.

3. The study period may be a finite period of time set to reflect the period of intended use of the investment or of the facility in which the investment is to be made.

4. Alternatively, the study period may be set equal to some other finite period to reflect other constraints, such as the time over which costs and benefits can be estimated with some degree of accuracy. Both the third and fourth approaches require the inclusion of any relevant replacements or salvage values when the study period does not coincide with the expected lives of the various alternatives. It is also important in both approaches that mutually exclusive alternatives to accomplish a given objective (e.g., solar screens of type A versus solar screens of type B) be evaluated for the same finite study period.

It is unnecessary, however, to evaluate retrofit projects that are not mutually exclusive on the basis of a common study period to compare and rank them. For example, alternative solar energy systems for application to building A may be evaluated over a study period of 15 years, while alternative new plant control systems for building B may be evaluated over a period of 10 years. The economic ranking measure for each of the alternatives selected, each based on its respective study period, can then be compared without the need to convert all of the projects to the same study period.

Due to uncertainties in forecasting energy prices and in order to promote consistency, an upper limit of 25 years is imposed on the study period for analyzing energy conservation projects in existing buildings in the Federal Energy Management Program.

9.5.4. Identifying the LCC Parameters

The costs of owning, operating, and maintaining an asset over a period of time are traditionally separated into initial costs (investment) and future costs (operation, maintenance, repair, and replacement). The investment costs include all first costs that arise directly from the project, including special site-specific studies, design, and installation or construction costs, that is, all costs necessary to provide the finished project ready for use. All investment costs should be taken into account in evaluating alternatives. Sunk costs (costs incurred prior to making the LCC analysis) should not be included. Costs for studies, analyses, and so on, that are not due directly to a specific project, such as costs for preliminary energy audits or energy surveys, should not be included as an investment cost in evaluating a given project.

Future costs can be divided into energy and non-energy costs. *Energy costs* are defined here as the dollar cost of delivered energy at the building or facility boundary. Estimates of energy costs are a critical data input to the LCC evaluation. For an existing building, estimates will be needed of the building's energy requirements before it is retrofitted, and projections will be needed of its expected energy requirements after specific retrofit actions have been taken. For new buildings, it will be necessary to estimate the expected energy requirements of alternative building and system designs.

Energy requirements may be estimated at varying levels of analytical detail, utilizing past records of energy usage, walk-through surveys of facilities, reviews of specifications and drawings, engineering test data and computer analysis of energy flows. Once the impact of a given energy conservation investment has been estimated, future dollar energy savings can be projected by first determining the value of the expected yearly energy savings in today's prices and then adjusting yearly dollar savings to reflect expected increases in energy prices over the study period.

Non-energy costs are maintenance, repair, replacement, and future non-energy operating costs such as operating personnel costs. For new buildings, where LCC analysis is used to determine the basic building design, future non-energy costs may also include functional-use costs, such as nonmaintenance costs associated with performing the intended function of the building. For example, the shape of a building may affect not only its energy requirements, but also its ability to serve its intended purpose. Multistory buildings will generally use less energy per floor than single story buildings, however multistory buildings may not be satisfactory for their intended purposes.

The implementation of some energy conservation improvements may have little or no effect on maintenance, repair, or replacement costs (e.g., installing insulation in the roof of a building). If these costs are not significantly affected, they may be excluded from the analysis. Where non-energy cost changes are significant, they should be included in the analysis.

Differences in benefits from alternative investments in energy conservation should also be taken into account wherever they are significant. For example, an energy-conserving lighting system may adversely affect the quality of lighting, and thereby affect worker productivity in a significant way. A comparison of alternative energy conservation investments based solely on their energy savings and direct costs is valid only if the investments have no other important consequences.

9.5.5. Converting to Common Time and Dollar Measure

Investment in heating systems, like many capital investments, will generally require a number of expenditures spread over a period of time and will result in cost savings (or revenue receipts) also spread over time. It is necessary to convert costs and savings to a common time and a common dollar measure to account for the time value (or opportunity cost) of money. The phrase "time value of money" reflects the difference between the value of a dollar today and its value at some future time. The time dependency of value reflects not only inflation, which may erode the buying power of the dollar, but also the fact that money currently in hand can be invested to earn a real return, i.e., it has a real opportunity cost.

Inflation.   The adjustment of costs and savings to account for inflation and for the real opportunity cost of money can be accomplished in several ways. If future estimates of costs and savings include an inflation factor (to account for expected price changes), it is necessary to remove the inflation factor so that all values are expressed in *constant dollars,* which are generally *today's dollars.* This is important, because an economic evaluation makes no sense if it is made in variable-value dollars.

Inflation may be eliminated from the evaluation in any of three ways:

1. Estimates of future prices may be stated in constant dollars, by assuming that inflationary effects will cancel out, leaving base-year prices as good indicators of future constant dollar prices. Using this approach, any future prices that are expected to increase differently from general price inflation must be adjusted to include the amounts of the expected differential rates of change. For example, it may be assumed that the price of labor to perform a given maintenance service will remain at today's dollar value, but that energy prices will rise above today's level in constant dollars, i.e., they will increase faster than general price inflation, say, 3 percent faster for purposes of illustration. Today's prices for labor could then be used for estimating future maintenance costs, but today's prices for energy

would be escalated at a rate of 3 percent per year for measuring future energy costs. Future amounts that are fixed in base year dollars—for example, level mortgage payments—do not inflate with other costs and savings. Because they do not inflate, fixed payments decline in constant dollars as inflation occurs. To convert fixed amounts in future years to constant dollars requires the use of a constant dollar price deflator. If future prices are given in constant dollars, the real opportunity cost of money can subsequently be taken into account by using a technique called *discounting*. The technique will in this case employ a real discount rate that also excludes inflation.

2. A second way to eliminate inflation, used when the estimates of future costs and savings are not in constant dollars, is to apply a constant dollar price deflator to the estimates of all future costs and savings. The deflator would be applied to fixed as well as nonfixed future amounts. In this case, the subsequent adjustment for the real opportunity cost of money is performed as above, employing a real discount rate (one that excludes inflation).

3. A third way of dealing with inflation, also used when the estimates of future cash flows are not in constant dollars, is to combine the adjustment for inflation with the adjustment for the real opportunity cost of capital. This can be done by discounting future costs and savings stated in current dollars with a nominal discount rate, a rate that includes both the real opportunity cost of capital and the expected rate of inflation.

Discounting.   Discounting is performed by applying interest (discount) formulas, or corresponding discount factors calculated from the formulas, to the estimated costs and savings resulting from a given investment. The application of the appropriate formula or factor to a cash flow will convert that cost or saving to its equivalent value at the selected point in time.

The commonly used discount formulas and corresponding tables of discount factors, calculated for specific time periods and interest (or discount) rates, are provided in Tables 9.6 through 9.10.

Cash flows occurring at different times can be converted to a common basis, that is, *discounted,* by means of discounting equations (also known as interest equations). The six basic discounting equations are shown in Table 9.6 along with an abbreviated guide for their use.

To find the comparable value of a given sum of money if it were received (or disbursed) at a different time, the appropriate equation from Table 9.7 may be used, or the corresponding discount factor may be multiplied by the given sum.

The appropriate formula, or factor, to use depends on the timing of the cost or savings and on the time basis selected by the analyst for the economic evaluation. It is often necessary to use several different discounting formulas or factors to evaluate a given investment.

TABLE 9.7. Computing the Present Value of Cost and Savings Occurring at Different Times[a]

Type of Cash Flow	Description	Appropriate Discount Formula[b]	Computation by Discount Formula	Computation by Discount Factor	Source of Discount Factor[c]
Past cost (design)	$100 cost incurred 2 yr past	Single compound amount $$P = T(1 + i)^n$$	$$P = \$100(1 + .10)^2$$ $$= \$121$$	$$P = \$100(1.2100)$$ $$= \$121$$	Table 9.10; Year 1, $i = 10\%$
Future recurring cost (maintenance)	$100 cost per year over 20 yr	Uniform present value $$P = A\left[\frac{(1+i)^n - 1}{i(1+i)^n}\right]$$	$$P = \$100\,\frac{(1 + .10)^{20} - 1}{.10(1 + .10)^{20}}$$ $$= \$851.35$$	$$P = \$100(8.5135)$$ $$= \$851.35$$	Table 9.9; Year 1, $i = 20\%$, $c = 0\%$
Future nonrecurring cost (repair and replacement)	$100 replacement cost incurred in the 10th year	Single present value $$P = F\,\frac{1}{(1+i)^n}$$	$$P = \$100\,\frac{1}{(1 + .10)^{10}}$$ $$= \$38.55$$	$$P = \$100(0.3855)$$ $$= \$38.55$$	Table 9.8; Year 10, $i = 10\%$ $c = 0\%$
Future savings (energy)	$100 energy savings priced in base period dollars, escalated at 5% yearly, over 20 yr	Uniform present value, modified $$P = A\sum_{j=1}^{n}\left(\frac{1+e}{1+i}\right)^j$$ $$= A\left(\frac{1+e}{i-e}\right)\cdot$$ $$\left[1 - \left(\frac{1+e}{1+i}\right)^n\right]$$	$$P = \$100\left(\frac{1 + .05}{.10 - .05}\right)\cdot$$ $$\left[1 - \left(\frac{1 + .05}{1 + .10}\right)^{20}\right]$$ $$= \$1271.77$$	$$P = \$100\,(12.7178)$$ $$= \$1271.78$$	Table 9.9; Year 20, $i = 10\%$, $c = 5\%$

[a] A 10% real discount rate, constant dollars, and end-of-period cash flows are assumed throughout.

[b] e = price escalation rate or inflation value.

[c] Note that in both Tables 9.8 and 9.9, the "0%" column reflects a 10% discount rate without any offsetting price escalation. This column is used in conjunction with the Federal Energy Management Program and the Federal Grant Program to determine the present value factor for all non-energy items.

Table 9.7 illustrates the use of four different discount formulas and factors to convert four different types of costs and savings to a common time. A past cost, a future recurring cost, a future nonrecurring cost, and future energy savings are all expressed as though they were to be incurred now. The result obtained is called a present value.

The discounting of costs can be greatly simplified by the use of tables of discount factors such as those in Tables 9.8 through 9.10. The discount factors are calculated from the discount equations for various time periods and opportunity costs (expressed as a rate of discount or interest).

Discount Rates.   The discount rate reflects the fact that money in hand can command resources that earn a return; that is, it reflects the opportunity cost of money. In both the public and private sectors, a wide range of rates are used to discount cash flows. Discount rates typically range from rates as low as 2–3% to rates higher than 20%. The choice of rates can significantly affect the outcome of an evaluation. The higher the rate, the lower the value of future cash flows.

Timing of Cash Flows.   To discount, it is also necessary to make an assumption about the timing of cash flows within the year of occurrence. In practice, cash flows usually occur throughout the year and may not be well described by any of the following four alternative assumptions that are usually made to simplify the discounting of cash flows: (1) lump-sum, end-of-year cash flows; (2) lump-sum, beginning-of-year cash flows; (3) lump-sum, middle-of-year cash flows; and (4) continuous cash flows throughout the year. However, to describe the timing of cash flows more accurately would generally require more effort than is warranted by the resulting improvement in the economic measures; therefore, one of the above four assumptions is usually adopted.

Energy Price Escalation.   In escalating future energy savings, there may be differences in the price escalation of alternative energy sources and in the periods of time over which various escalation rates are assumed to prevail. The prices of coal, fuel oil, electricity, and natural gas are expected to rise at different rates, both relative to one another and over time. While energy prices are widely expected to increase faster than most other prices, it is not clear that very high price escalation rates will be sustained indefinitely. As was indicated earlier, one approach to dealing with the increasing uncertainty of energy prices over time in an economic evaluation is to impose a cutoff time on the study period. Another approach is to reduce the energy price escalation rate to zero or to a low level at some future point in time.

9.5.6.   Partial vs Comprehensive LCC Methods

The simplest procedures that are used by firms to evaluate alternative kinds and amounts of investments are visual inspection, *payback period,* and *re-*

turn on investment approaches, which are termed "partial" here because they do not fully assess the economic desirability of alternatives. These partial methods may be contrasted with the more complete techniques, discussed later in the section, which take into account factors such as timing of cash flows, risk, and taxation effects—factors that are required for full economic assessment of investments.

Despite their shortcomings, the partial techniques of analysis may serve a useful purpose. They can provide a first-level measure of profitability that is, relatively speaking, quick, simple, and inexpensive to calculate. They may therefore be useful as initial screening devices for eliminating the more obviously uneconomical investments. These partial techniques (particularly the payback method) may also provide needed information concerning certain sensitive features of an investment. But where partial methods are used, the more comprehensive techniques may also be needed to verify the outcome of the evaluations and to rank alternative projects as to their relative efficiency.

Other methods of financial analysis exist that avoid the problems of the partial methods by taking into account total costs and benefits over the life of the investment and the time of cash flows by discounting. Methods of this type are the present value of net benefits method, the annual net benefits method, the benefit/cost ratio method, and the internal rate of return method. The discounting of costs is an element common to all of them.

9.6. LIFE-CYCLE COST TECHNIQUES

9.6.1. Payback Method

One of the most commonly used terms to economically evaluate systems and their projected savings is years to payback. It does not look for system cost comparison at a specific life. As long as both engineer and owner agree that years to payback is reasonable and that life expectancy is not too different, it is a term that can be used in design concepts.

The payback (also known as the payout or payoff) method determines the number of years required for the invested capital to be offset by resulting benefits. The required number of years is termed the payback, recovery, or break-even period.

The measure is popularly calculated on a before-tax basis and without discounting, i.e., neglecting the opportunity costs of capital. Investment costs are usually defined as first costs, often neglecting salvage value. Benefits are usually defined as the resulting net change in incoming cash flow, or, in the case of a cost-reducing investment like energy recovery, as the reduction in net outgoing cash flow.

The payback period is usually calculated as follows:

$$\text{Payback period (PP)} = \frac{\text{first cost}}{\text{yearly benefits} - \text{yearly costs}}$$

TABLE 9.8. Single Present Value Discount Factors for Energy Price Escalation Rates from 0 to 10% (Based upon a 10% Discount Rate)[a-c]

Year	0%[d]	1%	2%	3%	4%	5%	6%	7%	8%	9%	10%
1	0.9091	0.9182	0.9273	0.9364	0.9455	0.9546	0.9636	0.9727	0.9818	0.9909	1.0000
2	0.8264	0.8430	0.8598	0.8767	0.8938	0.9111	0.9285	0.9461	0.9639	0.9818	1.0000
3	0.7513	0.7741	0.7973	0.8209	0.8451	0.8697	0.8948	0.9203	0.9464	0.9729	1.0000
4	0.6830	0.7107	0.7393	0.7687	0.7990	0.8302	0.8623	0.8953	0.9292	0.9641	1.0000
5	0.6209	0.6526	0.6855	0.7198	0.7554	0.7925	0.8309	0.8709	0.9123	0.9553	1.0000
6	0.5645	0.5992	0.6357	0.6741	0.7143	0.7565	0.8007	0.8471	0.8958	0.9467	1.0000
7	0.5132	0.5502	0.5895	0.6312	0.6753	0.7221	0.7716	0.8241	0.8795	0.9381	1.0000
8	0.4665	0.5041	0.5466	0.5910	0.6385	0.6893	0.7435	0.8015	0.8634	0.9295	1.0000
9	0.4241	0.4638	0.5068	0.5534	0.6036	0.6579	0.7165	0.7797	0.8478	0.9211	1.0000
10	0.3855	0.4258	0.4699	0.5181	0.5706	0.6279	0.6904	0.7584	0.8323	0.9126	1.0000
11	0.3505	0.3911	0.4358	0.4852	0.5396	0.5995	0.6654	0.7378	0.8172	0.9044	1.0000
12	0.3187	0.3591	0.4042	0.4544	0.5102	0.5724	0.6431	0.7178.	0.8026	0.8964	1.0000
13	0.2897	0.3297	0.3748	0.4254	0.4824	0.5463	0.6179	0.6981	0.7879	0.8882	1.0000
14	0.2633	0.3027	0.3474	0.3983	0.4560	0.5213	0.5953	0.6789	0.7734	0.8799	1.0000
15	0.2394	0.2779	0.3222	0.3730	0.4311	0.4977	0.5737	0.6605	0.7594	0.8720	1.0000

16	0.2176	0.2552	0.2987	0.3492	0.4076	0.4750	0.5528	0.6424	0.7455	0.8639	1.0000
17	0.1979	0.2344	0.2771	0.3271	0.3855	0.4536	0.5329	0.6251	0.7322	0.8564	1.0000
18	0.1798	0.2151	0.2568	0.3061	0.3642	0.4327	0.5132	0.6077	0.7185	0.8481	1.0000
19	0.1635	0.1975	0.2382	0.2867	0.3445	0.4132	0.4947	0.5913	0.7056	0.8407	1.0000
20	0.1486	0.1813	0.2208	0.2684	0.3256	0.3943	0.4766	0.5750	0.6926	0.8328	1.0000
21	0.1351	0.1665	0.2048	0.2513	0.3079	0.3764	0.4593	0.5594	0.6801	0.8253	1.0000
22	0.1229	0.1530	0.1900	0.2355	0.2913	0.3595	0.4429	0.5445	0.6681	0.8183	1.0000
23	0.1117	0.1404	0.1761	0.2205	0.2753	0.3431	0.4267	0.5295	0.6558	0.8107	1.0000
24	0.1015	0.1289	0.1633	0.2063	0.2602	0.3273	0.4110	0.5148	0.6436	0.8030	1.0000
25	0.0923	0.1184	0.1514	0.1933	0.2461	0.3126	0.3961	0.5009	0.6321	0.7959	1.0000

[a] Factors for intermediate values of energy price escalation rates, e.g., 7.5%, may be obtained by interpolation.

[b] If the 10% discount rate on which the factors are based is defined as a real rate, excluding inflation, then the energy escalation rates should also be defined as real rates, i.e., as differential rates in excess of general price inflation.

[c] Inclusion of energy escalation rates ranging from 0 to 10% is not intended to indicate anything about the appropriate projections of energy prices.

[d] The discount factors in the 0% column of this table can be converted to a midyear discounting basis by multiplying them by the factor 1.0488.

TABLE 9.9. Uniform Present Value Discount Factors for Energy Price Escalation Rates from 0 to 10% (Based upon a 10% Discount Rate)

Year	0%	1%	2%	3%	4%	5%	6%	7%	8%	9%	10%
1	0.9091	0.9182	0.9273	0.9364	0.9455	0.9546	0.9636	0.9727	0.9818	0.9909	1.0000
2	1.7355	1.7612	1.7871	1.8131	1.8393	1.8657	1.8921	1.9188	1.9457	1.9727	2.0000
3	2.4868	2.5353	2.5844	2.6340	2.6844	2.7354	2.7869	2.8391	2.8921	2.9456	3.0000
4	3.1698	3.2460	3.3237	3.4027	3.4834	3.5656	3.6492	3.7344	3.8213	3.9097	4.0000
5	3.7907	3.8986	4.0092	4.1225	4.2388	4.3581	4.4801	4.6053	4.7336	4.8650	5.0000
6	4.3552	4.4978	4.6449	4.7966	4.9531	5.1146	5.2808	5.4524	5.6294	5.8117	6.0000
7	4.8684	5.0480	5.2344	5.4278	5.6284	5.8367	6.0524	6.2765	6.5089	6.7498	7.0000
8	5.3449	5.5521	5.7810	6.0188	6.2669	6.5260	6.7959	7.0780	7.3723	7.6793	8.0000
9	5.7590	6.0159	6.2878	6.5722	6.8705	7.1839	7.5124	7.8577	8.2201	8.6004	9.0000
10	6.1445	6.4417	6.7577	7.0903	7.4411	7.8118	8.2028	8.6161	9.0524	9.5130	10.0000
11	6.4930	6.8328	7.1935	7.5755	7.9807	8.4113	8.8682	9.3539	9.8696	10.4174	11.0000
12	6.8137	7.1919	7.5977	8.0299	8.4909	8.9837	9.5095	10.0717	10.6722	11.3138	12.0000
13	7.1034	7.5216	7.9725	8.4553	8.9733	9.5300	10.1274	10.7698	11.4601	12.2020	13.0000
14	7.3667	7.8243	8.3199	8.8536	9.4293	10.0513	10.7227	11.4487	12.2335	13.0819	14.0000
15	7.6061	8.1022	8.6421	9.2266	9.8604	10.5490	11.2964	12.1092	12.9929	13.9539	15.0000
16	7.8238	8.3574	8.9408	9.5758	10.2680	11.0240	11.8492	12.7516	13.7384	14.8178	16.0000
17	8.0216	8.5918	9.2179	9.9029	10.6535	11.4776	12.3821	13.3767	14.4706	15.6742	17.0000
18	8.2014	8.8069	9.4747	10.2090	11.0177	11.9103	12.8953	13.9844	15.1891	16.5223	18.0000
19	8.3649	9.0044	9.7129	10.4957	11.3622	12.3235	13.3900	14.5757	15.8947	17.3630	19.0000
20	8.5135	9.1857	9.9337	10.7641	11.6878	12.7178	13.8666	15.1507	16.5873	18.1958	20.0000
21	8.6486	9.3512	10.1385	11.0154	11.9957	13.0942	14.3259	15.7101	17.2674	19.0211	21.0000
22	8.7715	9.5042	10.3285	11.2509	12.2870	13.4537	14.7688	16.2546	17.9355	19.8394	22.0000
23	8.8832	9.6446	10.5046	11.4714	12.5623	13.7968	15.1955	16.7841	18.5913	20.6501	23.0000
24	8.9847	9.7735	10.6679	11.6777	12.8225	14.1241	15.6065	17.2989	19.2349	21.4531	24.0000
25	9.0770	9.8919	10.8193	11.8710	13.0686	14.4367	16.0026	17.7998	19.8670	22.2490	25.0000

Disadvantages.   There are two main disadvantages of the payback method that recommend against its use as a sole criterion for investment decisions:

1. The method does not give consideration to cash flows beyond the payback period, and thus does not measure the efficiency of an investment over its entire life.

2. The neglect of the opportunity cost of capital, that is, failing to discount costs occurring at different times to a common base for comparison, results in the use of inaccurate measures of benefits and costs to calculate the payback period and therefore yields an incorrect payback period.

In short, the payback method gives attention to only one attribute of an investment, i.e., the number of years to recover costs, and, as often calculated, does not even provide an accurate measure of this. It is a measure that many firms appear to overemphasize, tending toward shorter and shorter payback requirements. Firms' preference for very short payback to enable them to reinvest in other investment opportunities may in fact lead to a succession of less efficient, short-lived projects.

Advantages.   Despite its limitations, the payback period has advantages in that it may provide useful information for evaluating an investment. There are several situations in which the payback method might be particularly appropriate:

1. A rapid payback may be a prime criterion for judging an investment when financial resources are available to the investor for only a short period of time.

2. The speculative investor who has a very limited time horizon will usually desire rapid recovery of the initial investment.

3. Where the expected life of the assets is highly uncertain, determination of the break-even life, i.e., payback period, is helpful in assessing the likelihood of achieving a successful investment.

More Accurate Method.   The shortcomings that result from failure to discount costs and the omission of important cost items can be overcome simply by using a more accurate calculation of payback. Essentially what is desired is to find the number of years, R, for which the value of the following expression is equal to zero:

$$C = \sum_{j=1}^{R} \frac{B_i - P_j}{(1 + i)^j} \qquad (9.10)$$

where C = initial investment cost

TABLE 9.10. Single Compound Amount Factors for Alternative Discount Rates

Year	1%	2%	3%	4%	5%	6%	7%	8%	9%	10%
1	1.0100	1.0200	1.0300	1.0400	1.0500	1.0600	1.0700	1.0800	1.0900	1.1000
2	1.0201	1.0404	1.0609	1.0816	1.1025	1.1236	1.1449	1.1664	1.1881	1.2100
3	1.0303	1.0612	1.0927	1.1249	1.1576	1.1910	1.2250	1.2597	1.2950	1.3310
4	1.0406	1.0824	1.1255	1.1699	1.2155	1.2625	1.3108	1.3605	1.4115	1.4641
5	1.0510	1.1041	1.1593	1.2167	1.2763	1.3382	1.4026	1.4693	1.5386	1.6105
6	1.0615	1.1262	1.1941	1.2653	1.3401	1.4185	1.5007	1.5869	1.6771	1.7716
7	1.0721	1.1487	1.2299	1.3159	1.4071	1.5036	1.6058	1.7138	1.8280	1.9487
8	1.0829	1.1717	1.2668	1.3686	1.4775	1.5938	1.7182	1.8509	1.9925	2.1436
9	1.0937	1.1951	1.3048	1.4233	1.5513	1.6895	1.8385	1.9990	2.1718	2.3579
10	1.1046	1.2190	1.3439	1.4802	1.6289	1.7908	1.9672	2.1589	2.3673	2.5937
11	1.1157	1.2434	1.3842	1.5395	1.7103	1.8983	2.1049	2.3316	2.5804	2.8531
12	1.1268	1.2682	1.4258	1.6010	1.7959	2.0122	2.2522	2.5182	2.8126	3.1384
13	1.1381	1.2936	1.4685	1.6651	1.8856	2.1329	2.4098	2.7196	3.0658	3.4523
14	1.1495	1.3195	1.5126	1.7317	1.9800	2.2609	2.5785	2.9372	3.3417	3.7975
15	1.1610	1.3459	1.5580	1.8009	2.0789	2.3966	2.7590	3.1722	3.6424	4.1772
16	1.1726	1.3728	1.6047	1.8730	2.1829	2.5404	2.9522	3.4259	3.9703	4.5950
17	1.1843	1.4002	1.6528	1.9479	2.2920	2.6928	3.1588	3.7000	4.3276	5.0545

18	1.1961	1.4282	1.7024	2.0258	2.4066	2.8543	3.3799	3.9960	4.7171	5.5599
19	1.2081	1.4568	1.7335	2.1068	2.5270	3.0256	3.6165	4.3157	5.1416	6.1159
20	1.2202	1.4859	1.8061	2.1911	2.6533	3.2071	3.8697	4.6610	5.6044	6.7275
21	1.2324	1.5157	1.8603	2.2788	2.7860	3.3996	4.1406	5.0338	6.1088	7.4002
22	1.2447	1.5460	1.9161	2.3699	2.9253	3.6035	4.4304	5.4365	6.6586	8.1403
23	1.2572	1.5769	1.9736	2.4647	3.0715	3.8197	4.7405	5.8715	7.2578	8.9543
24	1.2697	1.6084	2.0328	2.5633	3.2251	4.0489	5.0724	6.3412	7.9110	9.8497
25	1.2824	1.6406	2.0938	2.6658	3.3864	4.2919	5.4274	6.8485	8.6230	10.8347
26	1.2953	1.6734	2.1566	2.7725	3.5557	4.5494	5.8074	7.3964	9.3991	11.9182
27	1.3082	1.7069	2.2213	2.8834	3.7335	4.8223	6.2139	7.9881	10.2450	13.1100
28	1.3213	1.7410	2.2879	2.9987	3.9201	5.1117	6.6488	8.6271	11.1671	14.4210
29	1.3345	1.7758	2.3566	3.1187	4.1161	5.4184	7.1143	9.3173	12.1721	15.8631
30	1.3478	1.8114	2.4273	3.2434	4.3219	5.7435	7.6123	10.0627	13.2676	17.4494
31	1.3613	1.8476	2.5001	3.3731	4.5380	6.0881	8.1451	10.8677	14.4617	19.1943
32	1.3749	1.8845	2.5751	3.5081	4.7649	6.4534	8.7153	11.7371	15.7633	21.1138
33	1.3887	1.9222	2.6523	3.6484	5.0032	6.8406	9.3253	12.6760	17.1820	23.2252
34	1.4026	1.9607	2.7319	3.7943	5.2533	7.2510	9.9781	13.6901	18.7284	25.5477
35	1.4166	1.9999	2.8139	3.9461	5.5160	7.6861	10.6766	14.7853	20.4139	28.1024
40	1.4889	2.2080	3.2620	4.8010	7.0400	10.2857	14.9745	21.7245	31.4094	45.2593

B_j = benefits in year j
P_j = costs in year j
R = break-even number of years
i = discount rate

When yearly net benefits are uneven, an iterative process can be used to determine the solution. If, on the other hand, yearly net benefits are expected to be fairly uniform, the following equation can be used to facilitate the calculation:

$$R = \frac{\log\,(1\,+\,iC/M)}{\log\,(1\,+\,i)} \qquad (9.11)$$

where R = break-even number years
M = yearly net benefits
C = initial investment cost
i = discount rate

9.6.2. Return-on-Investment Method

Rate of return on investment is simply the rate of return less the value of money. The return on investment (ROI) or return on assets method calculates average annual benefits, net of yearly costs such as depreciation, as a percentage of the original book value of the investment.

The ROI is calculated as follows:

$$\text{Return on investment (ROI)} = \frac{\text{average annual net benefits}}{\text{original book value}} \times 100$$

Disadvantages. The return on investment method is subject to the following principal disadvantages, and therefore is not recommended as a sole criterion for investment decisions:

1. Like the payback method, this method does not take into consideration the timing of cash flows, and thereby may incorrectly state the economic efficiency of projects.
2. The calculation is based on the accounting concept of original book value, which is subject to the peculiarities of the firm's accounting practices and generally does not include all costs. This method therefore results in only a rough approximation of an investment's value.

Advantages. The advantages of the return on investment are that it is simple to compute and a familiar concept in the business community.

9.6.3. Net Present Value (Present Worth) Method

The most commonly used LCC technique is present worth. This technique compares the equivalent cash needed on hand to own and operate a system over an entire selected time period. It produces a single investment value in dollars.

This method calculates the difference between the present value of the benefits and the costs resulting from an investment. The difference between benefits and costs is the net present value of the investment. A positive net present value means that the finance position of the investor will be improved by undertaking the investment; a negative net present value means that the investment will result in a financial loss.

To use this method in the evaluation of energy-recovery investments, the benefits would be defined as positive cash flows (as, for example, would result from sales of surplus recovered waste heat), and/or reductions in cash outflows (as would result from substitution of recovered waste heat for newly generated heat).

The equation for calculating the net present value, or net benefits, is

$$\text{NPV} = \sum_{j+1}^{n} \frac{(S_j + R_j) - (I_j - V_j + M_j)}{(1 + i)^j} \tag{9.12}$$

where NPV = net present value benefits
     n = number of time intervals over which the investment is analyzed
     S_j = energy cost savings in year j
     R_j = revenue from sale of excess energy received in year j
     I_j = investment costs in year j
     V_j = salvage value in year j
     M_j = maintenance and repair costs in year j

and $i/(1 + i)^j$ = single present value discount formula

There are two acceptance criteria of a project, as evaluated with the net present value method. (1) Only those investments having positive net benefits will be accepted (unless the project is mandatory). (2) When selecting among mutually exclusive investments, the one with the highest positive net benefits will be chosen (or the one with the lowest negative net benefits if none of the alternatives has positive net benefits and the project is mandatory).

In using the net present value method to compare alternative investments, it is important to evaluate the costs and benefits of each alternative over an equal number of years. This may be done in any of the following ways, depending upon the nature of the investment.

1. The costs and benefits can be measured over a time period that is a common multiple of the economic lives of the alternatives. For example, to compare heat exchanger A with a life of 5 yr with heat exchanger B with a life of 10 yr, alternative A could be evaluated on the basis of one replacement, and alternative B on the basis of no

replacements, such that benefits and costs of both systems would be computed for 10 yr.

2. Alternatively, benefits and costs can be calculated in annual cost terms, based on 5 yr for A and 10 yr for B, and the annual benefits and costs can then be used to calculate the present value of benefits and costs for the desired number of years of service (for example, 12 yr of service for both systems). This avoids the need to find a common multiple of system life.

3. If either system will be used for only a limited period of time that is less than a common multiple of the economic lives of the alternatives, the estimated cash flows associated with each system over the period of analysis can be discounted to present value, making sure to take into account the expected remaining value of each system at the time of terminated use. For example, the problem might be to chose between heat exchanger A, with its economic life of 5 yr, and heat exchanger B, with its economic life of 10 yr, where intended use of either would be only 7 yr, due to expected closure of the plant at the beginning of the eighth year. System A would require one replacement because its expected life is only 5 yr and it is needed for 7 yr. The remaining value of the replacement after 2 yr of use would be discounted to present value and deducted from the present value cost of the system. Similarly, an estimate would be required of the value of the remaining 3 yr of life of system B. If removal costs are prohibitive, or if there is no good resale market for the equipment, the remaining value of both systems at the end of 7 yr should be evaluated at zero dollars. This holds true even though the equipment could provide additional years of service if the existing operation were continued.

4. An investment can also be evaluated on the basis of a perpetual life, or an indefinite period of use by "capitalizing" renewal costs and expected benefits. Present value benefits of an investment in perpetuity are calculated by dividing the expected annual benefits by the discount rate. Present value costs in perpetuity are calculated by converting all costs other than first costs to an annual equivalent, dividing by the discount rate, and adding the amount to the first cost. Thus, if first costs were $5000, operation costs $1000 yearly, and renewal costs $4000 every 5 yr, the present value capitalized cost of heat exchanger A in perpetual service would be equal to $5000 + ($1000/i) + [$4000(USF, i, 5 yr)/i].

The choice among these four approaches to measuring present value is often not critical to the outcome; the particular nature of the investment will generally determine which approach is used. For example, in the case of a short-lived investment with unrecoverable salvage, the third approach explained above would be preferred.

In the case where alternative investments are expected to provide the same level of benefits, the net present value method becomes equivalent to a net present cost method. In this case, the most efficient alternative may be identified as the one with the least present value of costs alone. This approach is also often used when the benefit levels cannot be quantified.

Disadvantage. A feature of the net present value method that may be a disadvantage in some applications is that, in focusing only on net benefits, it does not distinguish between a project involving relatively large benefits and costs and one involving much smaller benefits and costs, as long as the two projects result in equal net benefits. A way to avoid this problem is to compute benefit-to-cost ratios for further evaluation of the projects. Choices will be most efficient if independent projects are chosen in the order of their benefit-cost ratios, starting with the highest and working down until the budget is exhausted.

Another possible disadvantage of the net present value method is that the results are quite sensitive to the discount rate, and failure to select the appropriate rate may alter or even reverse the efficiency ranking of the alternatives. For example, with too low a rate an alternative with benefits spread far into the future may unjustifiably appear more profitable than an alternative whose benefits are more quickly realized but of lower amount in undiscounted terms. Since changing the discount rate can change the outcome of the evaluation, the rate used should be considered carefully.

As was explained earlier, the discount rate that a firm should use to discount cash flows of an investment is the firm's opportunity cost of capital, expressed as an interest rate, the rate of return that will be forgone by using the funds (resources) for the investment under consideration instead of the next best investment opportunity available.

If the firm is uncertain as to the appropriate discount rate to use, it may wish to compute the net benefits of an investment based on several alternative discount rates to test for sensitivity of the outcome to the choice of rates.

Advantage. The net present value method has the advantage of measuring the net effect of an investment over its life, taking into account the opportunity cost of capital. The method is particularly useful for determining the efficient scale or size of an investment project.

9.6.4. Net Annual Value (Benefits) Method

This technique will compare the cost of both investment and annual costs on an annual base. Thus, a cost of owning the building or system is spread over the full amortization period, as with most mortgages. In some cases an economic analysis must be considered where a component has additional cost at some particular points in its life, as with replacements or major overhauls. This cost can be included in either owning or operating costs by multiplying the one-time expense by the single-payment present worth factor.

This method takes essentially the same form as the net present value method. The difference is that all costs and benefits of the net annual benefits method are converted to a uniform annual basis instead of to present value. The equation to be used is

$$A = \sum_{j=1}^{n} \frac{(S_j + R_j) - (I_j - V_j + M_j)}{(1 + i)^j} \left[\frac{i(1 + i)^n}{(1 + i)^n - 1} \right] \quad (9.13)$$

where A = annual value of net benefits
       n = number of time intervals over which the investment is analyzed
       S_j = energy cost savings in year j
       R_j = revenue received in year j
       I_j = investment costs in year j
       V_j = salvage value in year j
       M_j = maintenance and repair costs in year j

The term $1/(1 + i)^j$, is the single present value discount formula, and $i(1 + i)^n/[(1 + i)^n - 1]$ is the uniform compound amount formula.

If alternative investments have different life expectancies, either of two approaches may be taken to compare the alternatives.

1. It may be assumed that whichever alternative is chosen will be needed for an indefinite period of time and, hence, will be renewed as needed. In this case, the annual cost of each system may be calculated simply for its expected economic life, regardless of the fact that the lives of the alternatives may be different.

2. If the use of the investment alternatives is required for only a limited time, it is necessary to calculate costs based on the planned investment period, estimating the salvage value of each alternative at the planned time of terminated use.

Disadvantage.   This method, like the present value method, has the disadvantage of failing to distinguish between projects of unequal magnitudes that yield equal net benefits. However, analysis of benefit-cost ratios can be used to overcome this problem.

Advantage.   A possible advantage of the net annual value method, as compared with the net present value method, is that the concept of an equivalent annual amount may be easier to understand than the concept of a present equivalent of all cash flows over the period of analysis.

9.6.5. Benefit/Cost Ratio Method

This technique is a comparative procedure that provides the engineer and owner a ratio of cost to savings. The benefit/cost ratio method expresses benefits as a proportion of costs, where benefits and costs are discounted to either a present value or an annual value equivalent. The equation (with benefits and costs discounted to present value) is

$$
\text{B/C} = \sum_{j=1}^{n} \left[\frac{S_j + R_j}{(1 + i)^j} \right] \sum_{j=1}^{n} \left[\frac{I_j - V_j + M_j}{(1 + i)^j} \right]
\tag{9.14}
$$

where B/C = benefit/cost ratio

n = number of time intervals over which the investment is analyzed

S_j = energy cost savings in year j

R_j = revenue received in year j

I_j = investment costs in year j

V_j = salvage value in year j

M_j = maintenance and repair costs in year j

While the net present value and the net annual value methods require that discounted benefits minus discounted costs be positive in order for an investment to be worthwhile, the benefit/cost ratio method requires that the ratio of discounted benefits to costs be greater than 1.

Disadvantages. A disadvantage of the benefit/cost ratio method is that the ratio is influenced by the decision as to whether an item is classified as a cost or as a disbenefit, that is, whether it appears in the numerator or denominator of the ratio. For many cost or benefit items, this is an arbitrary decision, but one that can lead to confusion as to the real efficiency of a project.

   Another problem with the benefit/cost ratio method is that it is subject to misapplication in determining the efficient scale of a given project. It pays to expand a project up to the point that the ratio for the last increment of the investment is equal to 1.0, assuming no alternative investment is available with a higher ratio. Because the benefit/cost ratio for the overall investment declines as the investment is expanded toward the most efficient level, a smaller, less efficient project may have a higher ratio for the total investment than a larger, more efficient project. This problem can be avoided by applying the benefit/cost ratio method to evaluate the efficiency of increments of an investment, rather than the total investment.

9.6.6. Internal Rate of Return (Cash Flow) Method

This method (not to be confused with the ROI method evaluated earlier) calculates the rate of return an investment is expected to yield. This may be contrasted with the net present value, the net annual value, and the benefit/cost ratio methods that calculate the net dollar value of the investment based on a predetermined required rate of return. The internal rate of return method expresses each investment alternative in terms of a rate of return (a compound interest rate). The expected rate of return is the interest rate for which total discounted benefits become just equal to total discounted costs (net present benefits or net annual benefits are equal to zero) or for which the benefit/cost ratio equals one. The criterion for selection among alternatives is to choose the investment with the highest rate of return.

This "discounted cash flow" approach provides the owner with a technique to incorporate variable annual outlays and taxes, with the amount and year of the cash income. To evaluate the effect of interest and time, the net cash flow must be multiplied by the single-payment present worth factor. The interest at which the summation of present worth of net cash flow is zero gives the rate of return. If this rate offers an acceptable rate of return to the investor, the proposal should be approved; otherwise, it should be rejected.

Another approach would be to obtain an investment value at a given rate of return. This is accomplished by adding the present worth of the net cash flows, but not including the investment cost.

The rate of return is usually calculated by a process of trial and error, whereby the net cash flow is computed for various discount rates until its value is reduced to zero.

Disadvantages.   There are several possible disadvantages. For one thing, under cerain circumstances there may be either no rate of return solution or multiple solutions. Second, confusion may arise when this method is used to choose among mutually exclusive alternatives. For example, if the compared alternatives are different sizes of the same project (e.g., different capacity heat pumps, the rate of return on the larger project may be lower than on the smaller project, causing the larger to appear less efficient than the smaller. However, additional investment in the larger project may nonetheless yield a positive rate of return in excess of the minimum attractive rate of return. This problem, which is comparable to the one described for the benefit/cost method, can be avoided by analyzing incremental changes in an investment. A third problem is that the rate of return may be somewhat more cumbersome to calculate than the other methods.

9.7. SUMMARY

Each of the investment evaluation methods has its particular advantages and disadvantages and will be a useful decision criterion in certain cases. For most decision problems, the net present value or the net annual value method, supplemented by benefit/cost ratios or internal rates of return, will provide adequate measures for economically efficient investment decisions.

For additional details on economic analysis and life-cycle costs, the reader is referred to References 1–4. Compound interest tables for the various economic factors can be found in Appendix C.

EXAMPLE 9.1

A student borrowed $2000 from his brother for school expenses. Repayment would be in 3 years with an interest rate of 5%. How much must the student repay in 3 years?

Student's Solution.   Student assumed simple interest. Thus,

$$S = P(1 + ni)$$

where S = amount repaid
P = amount borrowed, $2000
n = number of years in loan, 3 yr
i = interest rate, 5%

$$S = 2000[1 + 3(.05)] = \$2300$$

Brother's Solution.   Brother had intended the interest to be compounded semiannually. Thus,

$$S = P(1 + i)^n$$

where n = number of compounding periods in loan = 6

$$S = 2000(1 + .05)^6 = \$2680.19$$

EXAMPLE 9.2

The ex-student, now an engineer, had a savings plan with payroll deductions. He had 6% of his $1700/month added to the plan, which has 5.5% interest, compounded annually. How much did he have in 5 years?

Solution

$$S = \frac{R[(1 + i)^n - 1]}{i}$$

R = value added periodically
= $(.06)(1700)(12)$$/comp. period
= $1224/comp. period

$S = 1224[(1 + .055)^5 - 1]/.055$
= $6831

EXAMPLE 9.3

The young engineer bought himself a home. It cost $80,000, had a 30-year mortgage with 10⅜% annual interest. What were his monthly payments?

Solution

$$P = \frac{R[(1 + i)^n - 1]}{i(1 + i)^n}$$

where P = \$80,000
 n = 30
 i = 0.10375
 R = yearly payment

$$R = \$8752.93/\text{yr} \text{ or } \$729.41/\text{mo}$$

EXAMPLE 9.4

Determine the payback period for a furnace air heater that costs \$10,000 to purchase and install and \$3000 per year on average to operate and maintain. By preheating combustion air it is expected to save an average of 10,000 therms of gas per year at \$0.32/therm.

Solution

$$PP = \frac{\$10,000}{\$3200 - \$300} = 3.4 \text{ yr}$$

EXAMPLE 9.5

For an investment difference of \$800, a first-year savings of \$100, an interest (i) of 6%, and an escalation rate (j) of 8%, determine the years to payback.

Solution

$$800 = \frac{100[(1.08/1.06)^n - 1]}{1 - (1.06/1.08)} = 100\left(\frac{1.0189^n - 1}{0.0185}\right)$$

$$1.1480 = 1.0189^n$$

$$n = \frac{\ln 1.1480}{\ln 1.0189} = \frac{0.1380}{0.0187} = 7.37 \text{ yr to PB}$$

EXAMPLE 9.6

Find the ROI for an investment in a heat pump system:
Original book value = \$15,000
Expected life = 10 yr
Annual depreciation, using a straight-line method = $\dfrac{\$15,000}{10}$ = \$1500

Yearly operation, maintenance and repair cost = \$200
Expected annual energy savings = \$5000

Solution

$$\text{ROI} = \frac{\$5000 - (\$1500 + \$200)}{\$15,000} \times 100$$

$$= 0.22 \times 100 = 22\%$$

EXAMPLE 9.7

$800 of additional initial cost is invested to save $100 per year. Assume 20-year amortization, 6% interest, and exclude escalation. Determine benefit/cost ratio.

Solution.   Present worth is calculated by:

$$\text{PW}_{\text{savings}} = \text{Benefit} \times 1/\text{CRF}$$
$$= 100/0.8718 = 1147.05$$

The technique is completed by:

$$\text{B/C ratio} = \frac{\text{PW benefit}}{\text{PW cost}} = \frac{1147.05}{800} = 1.43$$

EXAMPLE 9.8

Calculate the internal rate of return for a heat exchanger that will cost $10,000 to install, will last 10 years, and will result in fuel savings of $3000 each year. Find which *i* will equate the following: $10,000 = $3,000(UPW, *i* = ?, 10 yr).

Solution.   To do this, the net present value (NPV) for various *i* values (selected by visual inspection) is calculated:

$$\text{NPV } 25\% = (\$3,000)(3.571) - \$10,000$$
$$= \$10,713 - \$10,000$$
$$= \$713$$
$$\text{NPV } 30\% = (\$3,000)(3.092) - \$10,000$$
$$= \$9,276 - \$10,000$$
$$= \$724$$

For *i* = 25%, the net present value is positive; for *i* = 30%, the net present value is negative. Thus, for some discount rate between 25 and 30%, present value benefits are equated to present value costs. To find the rate more exactly, without the benefit of a complete set of discount tables, one may interpolate between the two rates as follows:

$$i = 0.25 + 0.05 \frac{\$713}{\$703 - (-\$724)}$$

$$= 0.275, \text{ or } 27.5\%$$

To decide whether or not to undertake this investment, it would be necessary for the firm to compare the expected rate of return of 27.5% with its minimum attractive rate of return.

EXAMPLE 9.9

(a) The costs of heat pump systems A and B are $1000 and $1200 and the annual operating costs $110 and $100, respectively. The value of money is established at 8% and the time period (amortization) is set at 20 yr. Compare the systems on the basis of present worth.

Solution. The capital recovery factor (CRF) for 8% interest and 20-yr amortization is 0.10185. Operating cost is calculated by multiplying the annual cost by the present worth factor, or the reciprocal of the CRF:

Present Worth	System A	System B
Initial cost	$1000.00	$1200.00
Operating Cost		
110 × 1/0.10185	1080.02	
100 × 1/0.10185		981.84
Present worth	$2080.02	$2181.84

Thus, over a 20-yr period, system A is less costly to own and operate because of the cost of money. Note in the example, that if the interest had not been considered, the present worth would be the same for the two systems, or $3200.

(b) Compare the above systems on the basis of annual benefits.

Solution. The uniform annual owning cost is calculated by multiplying the initial cost by the CRF:

Costs	System A	System B
Owning cost		
(initial cost × CRF)		
$1000 × 0.10185	$101.85	
$1200 × 0.10185		$122.22
Operating cost	110.00	100.00
Uniform annual O&O cost	$211.85	$222.22

EXAMPLE 9.10

Assume that two projects are being evaluated and that A has an economic life of 8 years and B has an economic life of 12 years. Further assume that

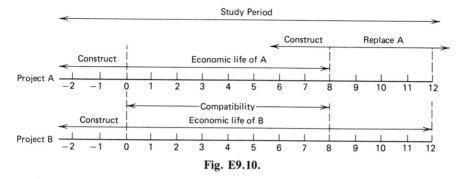

Fig. E9.10.

alternative A costs $1000 and alternative B costs $1300 and that each takes 2 years to construct with 50% of the investment cost due each year.

Solution. Figure E9.10 shows that each project will take 2 years to construct and that in year 8 a second A will have been constructed to replace the original A, which will have run its economic life. The study period ends in year 12, which is the end of the economic life of B. Consequently, in year 12, the residual value of A must be credited against the cost of replacing A.

The next step in the analysis is to discount the investment cost streams using present value factors and the method described in the previous section. This is accomplished as shown below.

Year	Investment Cost A	Investment Cost B	Annual Present Value factor	Present Value of Investment A	Present Value of Investment B
−2	500	650	1.2100	605.00	786.50
−1	500	650	1.1000	550.00	715.00
7	500	—	0.5132	256.60	—
8	500	—	0.4665	233.25	—
12	(500)	—	0.3187	(159.35)	—
Total	1500	1300		1485.50	1501.50

REFERENCES

1. *ASHRAE 1979 Systems;* see Chapter 1, ref. 15.
2. Ruegg, R. T., et al., "Life Cycle Costing: A Guide for Selecting Energy Conservation Projects for Public Buildings," *NBS Building Sci. Ser.* **113**, U.S. Dept. Commerce, 1978.
3. Stoecker, W. F., *Design of Thermal Systems,* McGraw-Hill, New York, 1980.
4. Coad, W. J., "Investment Optimization: A Methodology for Life Cycle Cost Analysis," *ASHRAE J.,* January 1977.

PROBLEMS

1. $1000 is invested at 8% interest. Find the value of this money in 10 yr.

2. Find the present worth of a sum of money that will have a value of $15,600 in 3 yr with an interest rate of 9%.

3. $100 is invested at the end of every year for 10 yr. $i = 11\%$. Find the amount accumulated.

4. It is desired to have $8000 in 5 yr. How much money should be set aside each year if the interest rate is 9%?

5. $90,000 is invested at 8.5% interest. Find the yearly withdrawal that will use up the money in 20 yr.

6. $10,000 is to be withdrawn at the end of every year for 20 yr. Find the original amount if $i = 5\%$.

7. The cost of a new heat pump system is $2000 with an expected lifetime of 20 yr. Neglect power and maintenance costs. Find the annual cost if the salvage value is $0. Assume an interest rate of 8%.

8. A new heating system has a cost of $15,000 and a salvage value of $5000, which is independent of age. The new system saves $1500 per year in fuel cost. Calculate the break-even point if $i = 8\%$. Neglect maintenance costs.

9. A new heating system costs $6000 and saves $900 in fuel costs per year. The system has a salvage value of $1000 in 20 yr. Compute the rate of return. Neglect maintenance costs.

10

OWNING AND OPERATING COSTS

10.1. INTRODUCTION

The purpose of this chapter is to provide data on the life-cycle costs of heat pump systems. In order to place these costs in the proper perspective, similar cost information is also presented for the conventional residential and light commercial space-conditioning systems. Unfortunately, in our inflationary economic environment, cost factors are increasing daily, and a cost estimate prepared yesterday is outdated today. Thus, the results of the previous (although all recent) studies reported herein must be interpreted with care.

Even though the numerical results from these investigations need updating, a study of the methodology itself is worth pursuing.

The most comprehensive evaluation of the relative costs of owning and operating air-to-air heat pumps compared with more conventional HVAC systems was conducted by Gordian Associates under FEA sponsorship and funding. The original Gordian study [1] was updated in 1978 [2]. Major emphasis in this chapter will be placed on the 1978 Gordian study, on the 1979 study by W. W. Smith [3] on hydronic heat pump economics, and on a 1980 report from the U.S. Department of Energy [4]. Before detailing these studies, some other recent investigations will be briefly reviewed.

10.2. SOME RECENT INVESTIGATIONS

10.2.1. Original Gordian (FEA) Study (1976)

Energy use and systems costs for five alternative residential space-conditioning systems were simulated for a two-story frame house occupied

TABLE 10.1. Annual Owning and Operating Costs (August 1975)

City	Heating Degree Days	Cooling Degree Days	Operating Heat Pump	Gas Furnace/ Central A/C	Oil Furnace/ Central A/C	Electric Furnace/ Central A/C	Electric Baseboards/ Window A/C Units
Houston, TX	1290	2339	$689	$545	$686	$574	$410
Birmingham, AL	2483	1928	$698	$573	$767	$685	$500
Atlanta, GA	2821	1589	$701	$561	$744	$657	$483
Tulsa, OK	3504	1949	$744	$598	$812	$736	$539
Philadelphia, PA	4508	1104	$886	$743	$935	$945	$728
Seattle, WA	4407	183	$477	$518	$685	$473	$322
Columbus, OH	5476	809	$876	$702	$965	$946	$758
Cleveland, OH	6097	613	$943	$638	$955	$1072	$824
Concord, NH	7377	349	$1045	$668	$928	$1220	$1025

by a middle class family of two adults and two children [1]. The house is actually located in Columbus, Ohio. However, for energy calculations, the house was hypothetically located in eight other cities selected to represent the other major climatic regions in which most Americans live.

The five space-conditioning systems simulated were the heat pump, gas forced-air furnace/central air conditioner, oil forced-air furnace/central air conditioner, electric forced-air furnace/central air conditioner, and baseboard resistance heaters with room air conditioners. In each city, the heating and cooling systems were sized appropriately according to established commercial practice.

The cost calculations reported in this section were based on monthly summaries of energy consumption data for each system in each city under typical weather conditions. All heating and cooling was accounted for. For example, the gas furnace energy cost includes both the cost of gas used in heating and the cost of electricity for blower operation.

The consulting firm first gathered figures on the actual house in Ohio. Then, using those figures as a basis, it ran a computer analysis calculating the comparative costs of heating and cooling in other parts of the country. Programmers input data on local fuel and labor costs and temperature ranges, got back information on how much it would cost to run a home equipped with a heat pump as against one using electric, oil, or gas heating and standard electric air conditioning. A summary of these calculations is shown in Table 10.1. Sometimes, as the chart shows, the heat pump is most economical; sometimes the economy race is won by a combination of some of the other systems.

At current energy prices, the gas furnace/central air conditioning system remains the economic choice vis-à-vis the heat pump, in most climatic regions of the United States.

The study found that a several fold increase in the price of gas in all areas studied (except Seattle, with its unusually low price for electricity) would be necessary to make the heat pump cost competitive with the gas system.

A point worth noting is that in its cooling phase, the heat pump uses somewhat more energy than a standard central air conditioner in all climatic regions of the United States. Yet despite its relative inefficiency compared with a central air conditioner, a heat pump's winter heating efficiency is so high that the total on-site energy required annually by the heat pump was less than for any combination of a heating system and air-conditioning.

10.2.2.   Hiller-Glicksman (MIT) Study (1976)

The results of a study by C. C. Hiller and L. R. Glicksman of Massachusetts Institute of Technology [5] include an economic comparison between conventional and capacity-controlled heat pumps and conventional gas and electric resistance heat, with and without air-conditioning. Various gas and electricity prices were compared for six cities in the country: San Francisco, California; Charleston, South Carolina; New York, New York; Boston, Massachusetts; Omaha, Nebraska; and Minneapolis, Minnesota. All energy use estimates were made using the bin method, with the time duration at 5°F temperature levels obtained from U.S. Weather Bureau data averaged over a 10-year period. All locations were assumed to have a constant 85% relative humidity during the heating season.

Conventional gas furnaces were sized to meet the maximum load at the coldest expected temperature at a given location. In a forced-hot-air furnace, the airflow rate is determined by the desired maximum temperature rise of the air through the furnace. A typical high-efficiency gas furnace is designed to have about an 80°F air temperature rise at maximum output, and would have a seasonal average efficiency of about 75%. Gas burner prices vary depending on the manufacturer, but reasonable estimates for typical high-efficiency models were made. Figure 10.1 shows typical purchase price as a function of heating capacity for high-efficiency gas furnaces. All comparisons in the study assumed that the heat pumps, regardless of size, used the same ducts as the gas furnace at each location.

It was assumed that the airflow rates of non-capacity-controlled units remain unchanged. However, for capacity-controlled units with balance-point temperatures lower than the conventional heat pump at a given location, two assumptions were made:

1. Outdoor airflow rates are half the full airflow rates of a non-capacity-controlled unit of comparable size.

2. Indoor airflow rates are either half the full airflow rates of a non-capacity-controlled unit of comparable size or are equal to the conventional gas-furnace airflow rate at a given location, whichever is greater.

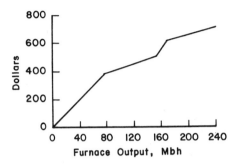

Fig. 10.1.   Gas furnace burner prices.

All heat pump installations were assumed to have 100% backup electrical resistance heaters, and all use electrical resistance auxiliary heat. All heat pumps were assumed to be turned off and 100% electrical resistance heat used at temperatures below $-10°F$, because of excessive compressor discharge gas temperature.

Total energy consumption over the heating season at each location was calculated with the heating season quantized into 5°F temperature bands, and the heat pump performance taken at the mean temperature of the band. Total seasonal energy consumption of gas, electrical resistance, and heat pump heating is summarized in Table 10.2 for each location, including SPFs.

Total seasonal energy consumption of heat pumps was calculated. Total seasonal energy consumption of gas furnaces included a portion of electrical energy due to furnace fans and was calculated in a manner similar to that for the heat pumps, using a seasonal average efficiency of 75% as compared to 65% for most existing gas or oil furnaces. Straight electrical resistance heating is assumed to use forced hot air also, with the same fans as the gas furnace.

Finally there is the question of total yearly cost for the various heating systems. The maintenance costs for all systems were neglected. This is a good approximation provided there are no major failures, such as compressor failure, since the cost of normal maintenance is much smaller than the total yearly cost of energy. Installation costs were assumed equal for all types of systems. Although heat pump installation costs are currently higher than for gas furnaces, wider use of heat pumps should reduce the cost differential as greater numbers of properly trained service personnel and improved installation practices become available. Installation costs for forced hot air electrical resistance heat would be somewhat less than for gas or heat pump heating, but the error in assuming equal installation costs is far outweighed by the high operating cost of pure electrical resistance heat. It was assumed that the various heat pump sizes and pure electrical resistance heat have the same indoor ducting size as the gas furnace at each location. The installation cost of the ducting, therefore, need not be included in the analysis. Total yearly cost comparisons hence include only two major costs: yearly energy cost and amortization of capital. Total yearly energy costs for

TABLE 10.2. Total Seasonal Energy Consumption and Seasonal Performance Factor (SPF) Data[a,b]

Location	Total Delivered Energy (Btu)	Gas ($\eta = .75$) + Electricity (Btu)		Seasonal Performance Factor					
				46°bp[c]	39°bp	37°bp	32°bp	21°bp	14°bp
San Francisco	1.57×10^8	Gas	2.06×10^8	2.56	2.60	—	2.82	—	—
		Elec.	2.14×10^6						
Charleston	1.05×10^8	Gas	1.37×10^8	2.00	2.32	—	2.66	2.85	—
		Elec.	1.76×10^6						
New York	2.25×10^8	Gas	2.96×10^8	—	—	2.19	2.37	2.73	2.67
		Elec.	2.82×10^6						
Boston	2.55×10^8	Gas	3.36×10^8	—	—	2.05	2.23	2.63	2.61
		Elec.	3.01×10^6						
Omaha	2.84×10^8	Gas	3.74×10^8	—	—	1.72	1.94	2.31	2.42
		Elec.	3.47×10^6						
Minneapolis	3.58×10^8	Gas	4.71×10^8	—	—	1.55	1.74	2.06	2.19
		Elec.	4.84×10^6						

[a] $SPF \equiv \dfrac{\text{total energy delivered}}{\text{total energy input}}$

[b] Electrical resistance SPF = 1.

[c] bp = balance point.

557

TABLE 10.3. Gas and Electrical Resistance Furnace and Air Conditioner Costs[a,b]

Location	Gas ($)	Elec. res. ($)	Air Conditioners (Fans incl.) ($)
San Francisco	350	227	1020
Charleston	400	311	1020
New York	500	424	1950
Boston	520	452	1950
Omaha	600	481	1950
Minneapolis	650	565	1950

[a] 1976 $.

[b] Carrier 50 DA 004 for San Francisco and New York and Carrier 50 DA 006 for other cities.

the various heating systems were computed for various prices of electricity and natural gas using the data of Table 10.2.

The yearly cost of capital was computed from:

$$\text{Capital cost per year} = \frac{\text{initial capital cost}}{n} \left[1 + \frac{i(n + 1)}{2} \right]$$

where i = interest rate, %/yr

n = number of years over which the cost is amortized (expected life)

A 20-year amortization period was used for the gas furnace and electrical resistance heater systems. The current life expectancy of an air-conditioning or heat pump system is on the order of 10 years and thus a 10-year amortization period for both heat pumps and air conditioners was used in most of the comparisons.

The heat pumps in this study were designed to provide both heating and air-conditioning. It is fitting, therefore, to compare the heat pumps to gas and electrical resistance heating systems with air conditioners added. Most of the economic comparisons given are between heat pumps and gas or electrical resistance heat plus air-conditioning. Comparisons of heat pumps to gas and electrical resistance heating without air-conditioning have been included for the Minneapolis and Omaha areas, although in reality, if the air-conditioning feature were not needed, a more efficient and less expensive heat pump could be built.

Initial capital costs (in 1976 $) for gas furnaces, electrical resistance heaters, and the various heat pump sizes, along with necessary air-conditioner costs are given in Tables 10.3, 10.4, and 10.5. Total yearly cost plots, comparing gas furnaces and pure electrical resistance heat with air-conditioning to various capacity-controlled heat pumps and to the conventional heat pump for a given location, are shown as a function of gas and electricity

TABLE 10.4. Conventional Heat Pump Costs and
Air Flows (Carrier)

Unit	Cost ($)	Conventional Air Flows Indoor (cfm)	Outdoor (cfm)
50 DQ 004	1200	1200	1750
50 DQ 006	2418	2100	3700
50 DQ 008	3304	3220	5200
50 DQ FICT	5600	4500	7500
50 DQ 016	6616	6330	10,000

prices for all six locations of the country in Fig. 10.2. They include an interest rate of 10%/yr, 20-year amortization of furnace and electrical resistance heaters, and 10-year amortization of heat-pump and air-conditioner costs. Figure 10.3 shows cost comparisons for the Omaha and Minneapolis areas without air-conditioning.

The first observation is that pure electrical resistance heating is extremely expensive compared to all other forms of heating considered, under all conditions studied. For example, in the New York area, at about 5¢/kWh, electrical resistance heat costs almost twice as much as the closest heat pump, about $1500/yr more for the assumed load! It is very difficult to keep accurate information on gas and electricity prices in this era of rapidly increasing prices, but in all locations, electricity is probably less than 5¢/kWh, and gas, when available, is probably around $3 per million Btu. A reasonable

TABLE 10.5. Capacity-Controlled Heat Pump Costs[a]

Location	Size[b]	Cost Breakdown	Total Cost $
San Francisco	46°	1200	1200
and Charleston	39°	2418 + 4(5) + 30 − 23 − 15 − 13	2417
	32°	3304 + 4(5) + 30 − 15 − 17 − 27	3295
	21°	5600 + 6(5) + 30 − 80 − 50 − 49	5481
Others	37°	2418	2418
	32°	3304 + 4(5) + 30 − 15 − 17 − 27	3295
	21°	5600 + 6(5) + 30 − 80 − 50 − 49	5481
	14°	6616 + 6(5) + 30 − 45 − 75 − 78	6478

[a] Cost = cost conv.

$$+ (N_{cyl})\left(\frac{\$5.00}{cyl}\right) + \$30\text{ controls} - \text{Indoor fan \& motor credit} - \text{Outdoor fan \& motor credit} - \text{Compressor motor credit}$$

[b] Shown as bp.

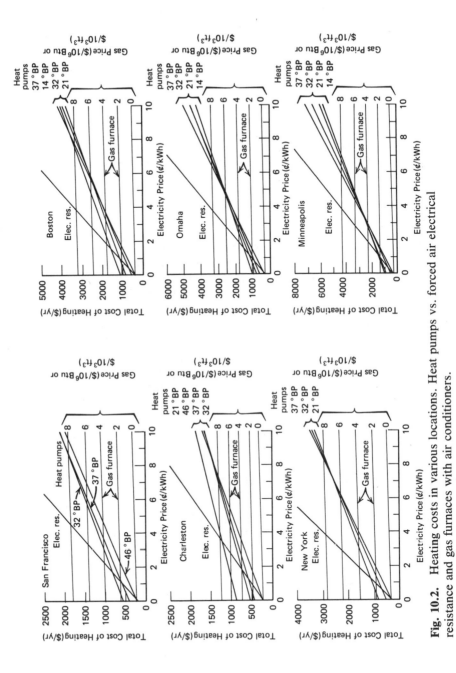

Fig. 10.2. Heating costs in various locations. Heat pumps vs. forced air electrical resistance and gas furnaces with air conditioners.

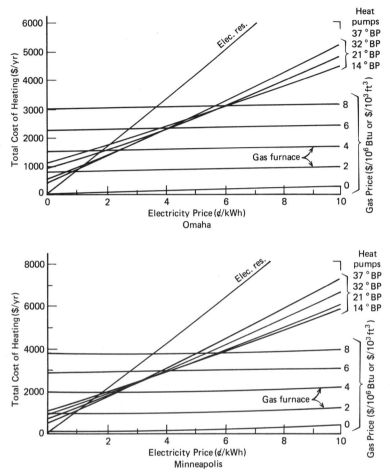

Fig. 10.3. Heating costs in Omaha and Minneapolis. Heat pumps vs. forced air electrical resistance and gas furnaces without air conditioners.

estimate for relatively near term (5 years or less) delivered price of natural gas is $5 per million Btu because of inflation, shortages, and impending federal deregulation of the interstate price of natural gas. The future prices of gas and electricity are difficult, if not impossible, to predict accurately. Most estimates by experts, however, fall within the limits of $10 per million Btu for gas, and 10¢/kWh for electricity in the next 10–15 years.

Inspecting Figs. 10.3a and b to Figs. 10.2e and f for Omaha and Minneapolis, we see that comparing heat pumps (which are also air conditioners) to gas heating without air-conditioning makes the heat pumps appreciably less competitive with gas heating but still not out of the realm of viability if gas prices rise. As mentioned earlier, if the air-conditioning feature is not desired, more efficient and less costly types of heat pumps could be selected.

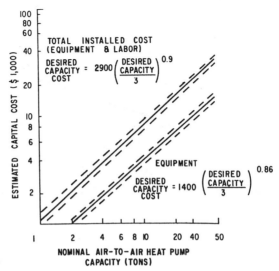

Fig. 10.4. Total air-to-air heat pump cost (mid-1976 dollars) [6].

It can be concluded that capacity-controlled heat pumps hold the potential for being economically competitive with gas or oil heating in colder climates if gas or oil prices rise faster than electricity prices. Furthermore, the colder the climate, the more desirable a low-balance-point, capacity-controlled heat pump becomes compared to conventional heat pumps.

10.2.3. Christian (Argonne) Study (1977)

In the work done at Argonne National Laboratory [6], some estimates for life costs were made. The total installed cost in mid-1976 dollars of air-to-air heat pump units was estimated by the following equation:

$$\text{Total installed cost (\$)} = 2900 \left(\frac{\text{desired capacity}}{3} \right)^{0.9}$$

Figure 10.4 shows cost values obtained from HVAC equipment distributors and contractors along with Means' 1976 Building Construction Cost Data. The equipment costs shown in Fig. 10.4 are applicable in most parts of the country; however, the installed cost will vary according to the local wage rate. The wage rate assumed in Fig. 10.4 is about $13/hr.

For single-package units, the costs shown in Fig. 10.4 can be reduced by 10–15% of the split-unit heat pump system cost. The economic life of a unitary air-to-air heat pump is estimated at about 10 years.

Typical worker-hour requirements to install a split-system heat pump are

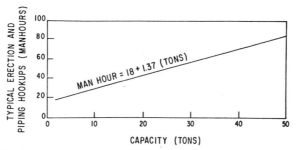

Fig. 10.5. Manhour estimates of typical split-system heat pump installation [6].

shown in Fig. 10.5. These estimates, based on a "typical" installation, include erection and piping hookups.

Some equipment cost variability results from different manufacturers' FOB prices and different distributors' assumptions as to what are considered "common accessories." One 4.8-kW electric heater coil was included for every ARI-rated ton of capacity. An additional cost of $80 was assumed for each extra 4.8-kW resistance heater. In addition to the supplementary heaters, the installation costs included external power wiring, control wiring, and condensate piping. For split systems, it was also necessary to connect the refrigerant piping between the indoor and outdoor sections; this was included in the installation costs shown in Fig. 10.4. For single-package units, the costs shown in Fig. 10.4 can be reduced by 10–15% of the split-unit heat pump system cost. The economic life of a unitary air-to-air heat pump is estimated at about 10 yr.

The annual operating and maintenance (O&M) costs (in mid-1976 dollars) can be estimated by the equation

$$\text{O \& M cost (\$/yr)} = 165 \left(\frac{\text{desired capacity}}{3} \right)^{0.5}$$

The O&M costs include a maintenance program consisting of periodic filter changes and lubrication as well as abnormal repairs. The cost estimate for electrical power is not included in this evaluation.

Figure 10.6 (top solid line) shows the estimated O&M costs for air-to-air heat pumps for years 2 through 5 in the life of the equipment. The initial cost of most heat pumps includes the first year's abnormal O&M expenses. Beyond the fifth year in the life of the equipment, the O&M costs can be expected to increase at a real rate of about 7% per year. The dashed lines indicate the variance in the cost for a preventive maintenance program. The lower solid line in Fig. 10.6 represents the cost for a service contract from the few manufacturers who have them available. Service contract costs are less than the preventive maintenance cost, because periodic filter changes, lubrication, and inspection are not included.

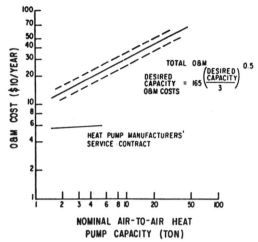

Fig. 10.6. Operation and maintenance costs (1976 $/yr) [6].

10.2.4. Nielsen (NWWA) Study (1977) [7]

Dr. Carl E. Nielsen, a professor in the Department of Physics at the Ohio State University, may have been the first person to utilize a groundwater source heat pump for residential heating and cooling. In 1948 Neilsen installed a 12,000-Btu/hr unit to condition a 500-ft^2 vacation cottage. He became convinced of the cleanliness, economy, and reliability of the groundwater heat pump system, and when the Nielsens began building a home in 1955, he installed a unit in the 2000-ft^2 two-story dwelling.

Of utmost concern to the energy consumer is the bottom line. The groundwater heat pump system must be economical to be feasible as an alternative to present heating and cooling methods.

A table prepared by Dr. Nielsen to compare 1977 annual heating costs of popular heating systems to air and water heat pumps is presented in Table 10.6.

The annual cost was obtained by adding the cost of the system amortized over an assumed lifetime to the cost of fuel used. No allowance was made for maintenance and repair bills, which were probably equivalent for the various systems. Nielsen assumed that the annual heating requirement is 100 million Btu in all cases.

The prices of systems and fuels used in computing Table 10.6 were based on data for Columbus, Ohio. These values are probably typical of prices throughout the Midwest. The energy value of the fuels assumed 65% efficiency for all fuel-burning furnaces and 100% efficiency for an electric furnace. However, the 100% electric efficiency is high, and a value of about 90% would be more typical.

The air-source heat pump is shown as having a COP of 1.9, equivalent to 190% efficiency. The groundwater-source heat pump has a COP of 3.2. This

TABLE 10.6. Annual Heating Cost (January 1977) for 100 MBtu [7]

Heating System	Initial Cost,[a] $	Annual Amortization,[b] $	Annual Fuel Cost,[c] $	Annual Heating Cost, $ (nearest $25)
Gas furnace	700	85	308	400
Oil furnace	1400	170	475	650
Coal furnace	(1400)	(170)	425	600
Water heat pump (a)[d]	3000	456	275	725
Performance 3.2 (b)	800	122	275	400
Electric furnace	900	110	880	1000
Air heat pump (a)[d]	3000	456	463	925
Performance 1.9 (b)	800	122	463	575

[a] For furnace or heat source only, excluding heat distribution system. Figures in parentheses are conjectural, quotations being unavailable.

[b] Computed at 9.0% interest and the following lifetimes: all furnaces 15 yr, heat pumps 10 yr, solar installations 20 yr.

[c] Prices of fuels: gas $2.00/1000 ft³, #2 fuel oil $0.42/gal, coal $80/ton (a 14,600 Btu/lb coal suitable for home stoker use), and electricity $0.030/kWh.

[d] Line (a) is heating cost if entire heat pump investment is charged to heating. Line (b) is additional cost for heat pump above that for summer air-conditioning, the same unit being used for both. Presumably (b) is the relevant figure in most cases. The water unit will give correspondingly lower summer cooling costs also. Data in line (a) for the water heat pump assumes that water is available without additional investment.

means that the groundwater-source heat pump is over 50% more efficient than an air-source heat pump, over three times as efficient as an electric furnace, and over four times as efficient as furnaces burning fossil fuels.

The table indicates that a gas furnace is the cheapest form of heating at $400 per year. Oil and coal furnace operations cost about the same, and minor increases in fuel costs can be expected yearly. Operation of an electric furnace costs approximately $1000/yr. An air-source heat pump has an annual operating cost of $925, while a groundwater-source heat pump has an operating cost of $725/yr. (Operating costs and cost values given for the groundwater-source heat pump do not include costs of drilling a well or wells, since this factor is highly variable and dependent on hydrological conditions. The system can utilize an existing well.) The groundwater-source heat pump becomes a most attractive investment if one considers that the system can also be used for cooling, dropping the annual operating cost for heating to $400.

10.3. THE COMPREHENSIVE GORDIAN (FEA) STUDY (1978)

10.3.1. Methodology

A section of the 1978 Gordian study [2] deals with the economic viability of heat pumps, conventional and advanced, in residential and light commercial building applications. To ascertain potential market acceptance of the heat pump, detailed installed costs and life-cycle costs were calculated. Through life-cycle cost analyses (with and without inflation), the effects of increased energy efficiency could be compared with the general trend of higher costs of installation for a heat pump system.

Energy costs were determined for all systems for which a meaningful hour-by-hour computer simulation of energy consumption could be performed. Nine cities, representing the climatic regions of the United States, were chosen for simulation. In this manner, the findings could be used broadly throughout the country. The results of this study were presented in Sections 8.2, 8.5, and 8.6.

Cost information was obtained for equipment currently available, either residential or light commercial. Most of the advanced heat pumps are still in early stages of development; hence, equipment costs for these are not generally available. However, the developers of the organic fluid absorption gas heat pump projected their wholesale price. This, together with costs of labor and materials for installation, was used in the life-cycle cost analyses.

Cost data were in general current as of the first quarter of 1977 or, in the case of the commercial systems, the second quarter of 1977. Since it was intended to have the cost analyses of conventional heat pumps and other HVAC equipment represent current commercial practice, regional representatives of the manufacturers whose equipment is featured in this study were consulted to verify compatibility of components and equipment sizing procedures. Dealers or the manufacturers themselves were the source of equipment and installation cost data used. The exception was the costs of room air conditioners, which were obtained from the Sears-Roebuck Co. *Summer 1977 Catalog*. The installed cost obtained consisted of the following major elements:

Material component prices and installation man-hours required for HVAC equipment.

Ductwork (insulated and noninsulated), material price, and installation time.

Chimney, where required.

Oil tank, where required.

Electrical service change, where required.

Labor costs for installation.

Freight costs.

Applicable state and local sales and/or use taxes.

Installing contractor's fee.

Labor costs (sheetmetal workers and other HVAC laborers) were taken as representative of the Philadelphia area, since most of the installed costs were obtained from a central New Jersey contractor. For other cities, appro-

TABLE 10.7. Annual Maintenance Cost

System	Annual Maintenance Cost ($)
Residential systems[a]	
Heat pump	149.60
Gas furnace and central air conditioner	104.50
Electric furnace and central air conditioner	109.70
Oil furnace and central air conditioner	127.90
Electric baseboard resistance heaters and room air conditioner units	110.00
Window air conditioner[b]	21.00
Commercial systems[c]	
All 15-ton heat pump systems	840.00
All 20-ton heat pump systems	1120.00

[a] National Environmental System Contractors Association (NESCA), mid-1976 prices.

[b] Sears-Roebuck Co. average price per air-conditioner unit.

[c] Tri-Boro Environmental Systems.

priate wage rates were taken from data published by the Bureau of Labor Statistics for winter 1976. Regional variations in first costs thus reflect differences in labor rates, shipping costs, and local taxes as well as equipment size differences, where appropriate.

Maintenance cost data were provided by the National Environmental Systems Contractors Association (NESCA) and represent conditions as of the third quarter of 1976. They represent the average retail price of a full-service contract to the consumer. A maintenance fee for the baseboard heaters, as recommended by NESCA, has also been included. The room air conditioner maintenance fee was obtained from Sears. For purposes of these calculations, maintenance costs were assumed not to vary regionally. An estimate of the cost of a full-service maintenance contract was obtained from the New Jersey HVAC contractor, Tri-Boro Environmental Systems, and represents prices as of the second quarter of 1977. Maintenance costs for commercial and residential systems are included as Table 10.7.

Depreciation periods for the various components of each system (e.g., the HVAC equipment, ductwork, chimney, etc.) were taken from the *ASHRAE Systems Handbook and Product Directory* [8] or represent current Internal Revenue Service recommended practices. Depreciation periods for commercial and residential systems are shown in Table 10.8.

Current state and local taxes were obtained from the sales tax departments of the individual states in which the facilities were located. A contractor's fee was also applied to determine an installed cost. Where possible,

TABLE 10.8. Depreciation Period of Heating and Cooling Systems

Component	Capital Recovery Factor, at 9% Interest	Depreciation Period (yr)
Residential systems		
Heat pump	0.1668	9[a]
Central air conditioner	0.1560	10[b]
Gas furnace	0.1203	16[b]
Oil furnace	0.1560	10[b]
Oil tank	0.1095	20[b]
Electric warm air furnace	0.1240	15[b]
Electric baseboard heaters	0.0929	40[b]
Window air conditioners	0.1560	10[b]
Ductwork, chimney	0.0929	40[c]
Commercial Systems		
Rooftop gas heating/electric cooling package	0.1560	10[d]
Single-package, two-stage central air conditioning with strip heating	0.1560	10[d]
Single-package standard electric heat pump	0.1668	9[d]
Brayton-Rankine gas heat pump	0.1668	9[d]
Rankine-Rankine gas heat pump	0.1668	9[d]

[a] Industry spokespersons give the life as 8–10 yr.

[b] ASHRAE Handbook and Product Directory, 1975 Systems.

[c] Internal Revenue Service.

[d] Tri-Boro Environmental Systems.

equipment costs and contractor's fees were verified by discussions with other contractors, with manufacturers, and by comparison with data reported in the *Means 1977 Building Construction Cost Data Book*.

The cost data reported in the study features equipment manufactured by General Electric Co.; York Division, Borg Warner Corp.; Carrier Corp.; Lennox Industries, Inc.; and Janitol Division, Tappan Company. These companies provided cost information, equipment design specifications, and recommendations concerning installation procedures.

The current rate information used in determining all energy costs was obtained by contacting the electric and gas utilities servicing each city involved in the simulation study. Oil prices were usually obtained from two private oil companies in the city. Table 10.9 summarizes the effective unit energy prices used in the study.

The cost of money was taken as 9%, the cost of a home mortgage loan at the time. (The portion of the purchase price of the system that is paid in the

TABLE 10.9. Effective Unit Energy Price for Heating and Cooling

Residential Rates	Houston $/unit	Houston $/10⁶Btu	Birmingham $/unit	Birmingham $/10⁶Btu	Atlanta $/unit	Atlanta $/10⁶Btu	Tulsa $/unit	Tulsa $/10⁶Btu	Philadelphia $/unit	Philadelphia $/10⁶Btu	Seattle $/unit	Seattle $/10⁶Btu	Columbus $/unit	Columbus $/10⁶Btu	Cleveland $/unit	Cleveland $/10⁶Btu	Concord $/unit	Concord $/10⁶Btu
Electricity Heating (kWh)	0.0271	7.94	0.0409	11.98	0.0311	9.12	0.0251	7.35	0.0479	14.04	0.0092	2.70	0.0402	11.78	0.0469	13.75	0.0387	11.34
Electricity Heating, All-Electric (kWh)	0.0217	6.36	0.0356	10.43	0.0269	7.88	0.0239	7.00	0.0285	8.35	0.0091	2.67	0.0282	8.26	0.0353	10.34	0.0387	11.34
Electricity Cooling (kWh)	0.0264	7.74	0.0409	11.98	0.0358	10.49	0.0348	10.20	0.0558	16.35	0.0092	2.70	0.0402	11.78	0.0472	13.83	0.0387	11.34
Electricity Cooling, All-Electric (kWh)	0.0264	7.74	0.0408	11.95	0.0357	10.46	0.0348	10.20	0.0542	15.88	0.0090	2.64	0.0398	11.66	0.0464	13.60	0.0387	11.34
Natural Gas Heating (10³ cft)	2.83	2.83	1.69	1.69	1.79	1.79	1.79	1.79	2.51	2.51	2.89	2.89	1.92	1.92	1.96	1.96	3.25	3.25
Natural Gas Cooling (10³ cft)	2.43	2.43	1.47	1.47	1.74	1.74	1.67	1.67	2.52	2.52	2.93	2.93	1.92	1.92	1.98	1.98	3.93	3.93
No. 2 Fuel Oil (gal*)	**	**	0.422	3.03	0.440	3.16	0.435	3.13	0.469	3.37	0.459	3.30	0.419	3.01	0.451	3.24	0.469	3.37
Commercial Rates																		
Electricity (kWh)	0.0182	5.34	0.0672	19.70	0.0529	15.51	0.0409	12.00	0.0436	12.78	0.0132	3.87	0.0633	18.55	0.0681	19.97	0.0348	10.20
Electricity All-Electric (kWh)	0.0182	5.34	0.0617	18.08	0.0484	14.19	0.399	11.70	0.0313	9.18	0.0135	3.96	0.0463	13.57	0.0479	14.04	0.0399	11.69
Natural Gas (10³ cft)	3.55	3.55	2.57	2.57	2.13	2.13	1.93	1.93	2.93	2.93	3.72	3.72	1.97	1.97	2.11	2.11	3.20	3.20

* Price quotations for first quarter, 1977.
** Not sold in Houston, assume Tulsa prices.

SOURCE: Gordian Associates Inc.

down payment on the house is assumed to have the same opportunity value to the consumer as that paid over the span on the mortgage. Hence the full installed cost of each system was amortized.)

Since the various heating and cooling systems had unequal service lives, it was necessary to compare life-cycle costs on a "perpetual replacement" basis. Two basic approaches were possible: The systems could be compared in terms of their capitalized costs or their annualized costs. The latter method was chosen because the result represents the yearly cash outflow resulting from the cost of operation and maintenance of each system (residential or commercial) as well as the annual payments required to equal the present value of the initial outlay for the installation. Where practical, costs associated with particular pieces of equipment, such as ductwork for the heat pump, were consigned to that equipment to allow the effect of different depreciation periods to be fully illustrated. The following formula (simplified for illustration purposes here) was used for the calculations:

$$
\begin{array}{c}
\text{Annualized} \\
\text{cost}
\end{array}
=
\left(
\begin{array}{c}
\text{installed} \\
\text{cost}
\end{array}
\times
\begin{array}{c}
\text{capital} \\
\text{recovery} \\
\text{factor}
\end{array}
\right)
+
\left(
\begin{array}{c}
\text{annual} \\
\text{energy} \\
\text{use}
\end{array}
\times
\begin{array}{c}
\text{unit} \\
\text{price of} \\
\text{energy}
\end{array}
\right)
+
\begin{array}{c}
\text{annual} \\
\text{maintenance} \\
\text{cost}
\end{array}
$$

Although this procedure would indeed give the yearly cost outlay necessary for a system, inflationary increases in energy and maintenance charges could not be represented.

In order to determine inflationary effects on annualized costs, a method of analysis suggested in a paper by F. R. S. Dressler [9] was utilized. The effective annual cost of owning and operating the HVAC equipment over its expected life cycle is computed, allowing for inflation in operating costs. Again, the cost of money was taken as 9%/yr. The economic life of the system was considered to be the smallest depreciation period of any major component of the system. An overall energy inflation of 8%/yr was used in the calculation. This represents current thinking as to the forecast electricity price increase in the case of total decontrol of natural gas and oil over a 9-year heat pump depreciated life. Maintenance costs, composed of both labor and material fees, were escalated at 6%/yr, which follows relevant consumer price index trends.

The determination of life-cycle costs for the alternative residential and commercial heating and cooling systems utilized the techniques outlined in the previous sections. Systems were treated in a number of different ways to represent either prevailing conditions or the most likely near future conditions. The cost of insulation of ductwork was considered. All systems are shown with insulated ducts which, according to Gordian, is the practice with electric heating systems but not with combustion systems. Costs for combustion systems with uninsulated ducts were also determined, this representing current practice. For comparative purposes, all life-cycle costs were calculated using insulated duct systems.

In a number of cases, equipment was priced as either an add-on (replacement) installation or as a new installation. The basis of a new installation involves placing an entire system into a home being built. It may include a central electric furnace, central air conditioner, and ductwork. An add-on installation cost would assume that all auxiliary equipment, such as insulated ducts or air handler, is already in place, and therefore only the cost of the HVAC equipment itself would be considered. Add-on installations, therefore, have substantially smaller first costs.

All five commercial HVAC systems were considered as new installations and were priced out including insulated ductwork, wiring, plumbing, electrical service change, etc. The single-package electric heat pump was priced including an economizer package, which comprises additional dampers, controls, and the economizer. The economizer package gives greater flexibility to the heat pump system, since outside air may be used for makeup requirements. This is also an energy-saving device, in that, on a cooler day, the outside air (which is cooler than the inside room air) can reduce the air-conditioning requirements. Seasonal and annual energy costs for conventional and advanced heat pumps and other HVAC systems account for all of heating and cooling system energy consumption. For example, the gas furnace energy cost includes both the cost of gas and the cost of electricity for indoor blower operation. In determining the residential energy cost, the test building was assumed to have a certain base load of lighting and electric or gas appliances, based on the usage reported in the Department of Commerce *1970 Census of Housing*. Energy use was simulated in the commercial building by assuming an occupancy schedule and a certain base load (including lighting, office machines). The base load was the actual maximum design load demand of 14 kW adjusted by occupancy schedule. If the electric utility offered an "all-electric" rate, a household with electric heating and cooling was assumed to have all electric appliances as well. A house with a gas- or oil-fired heating system was assumed to have appliances typical of that area (as per census data).

10.3.2. Residential HVAC System Costs

A comparison of installed cost of conventional electric heat pumps and other systems is shown in Table 10.10. The lowest cost system for any of the locations simulated is the baseboard convector with window air conditioners. This minimum cost is achieved, however, by not installing a central air distribution system and using window air conditioners which do not condition the total building space as is the case with systems that centrally distribute the comfort cooling. Since installation of a central air distribution system in an existing building has been stated to be prohibitively expensive (Hittman [10]), it should be recognized that installation of a baseboard convector/window air conditioner system would effectively prevent future upgrading with a more energy-efficient system that would require central air distribu-

TABLE 10.10. Conventional Electric Heat Pumps and Other Systems: Comparison of Installed Costs (First Quarter 1977 Dollars)

	Houston	Birmingham	Atlanta	Tulsa	Philadelphia	Seattle	Columbus	Cleveland	Concord
Electric Heat Pump-Standard	3255.24	3119.14	3079.57	3397.40	3304.03	2671.29	3234.22	3406.22	2749.34
Electric Heat Pump-High Efficiency I	3598.18	3458.84	3417.67	3913.75	3646.99	3025.20	3634.96	3816.80	3165.80
Electric Heat Pump-High Efficiency II	4170.41	4043.84	3660.18	4120.73	4232.01	3257.34	4144.46	4329.58	3558.78
Electric Heat Pump-High Efficiency III	3369.13	3234.12	3197.19	3515.02	3422.28	2853.96	3347.02	3520.43	2932.41
Hybrid Electric Heat Pump-Gas Furnace	3645.17	3491.91	3453.25	4051.71	3719.87	3260.69	3628.84	3841.92	3200.87
Hybrid Electric Heat Pump-Oil Furnace	4489.03	4367.51	4292.81	5069.55	4564.95	4034.35	4468.40	4685.62	4039.59
Gas Warm Air Furnace/Central Air Conditioner-High Efficiency	2745.00	2602.43	2552.58	2947.48	2668.18	2345.43	2717.19	2881.25	2365.95
Oil Warm Air Furnace/Central Air Conditioner-High Efficiency	3614.87	3560.62	3437.51	3832.08	3545.02	3148.33	3580.08	3754.58	3211.11
Central Electric Furnace/Central Air Conditioner-High Efficiency	2882.93	2746.14	2717.75	3090.18	2850.39	2429.62	2899.34	3060.90	2508.79
Baseboard Convectors/Window Air Conditioners	2378.14	2264.37	2165.29	2283.81	2452.54	2251.78	2367.93	2562.84	2305.43
System with Uninsulated Ducts*									
Gas Warm Air Furnace/Central Air Conditioner-High Efficiency	2558.83	2546.26	2496.41	2918.82	2612.01	2234.26	2634.26	2825.08	2282.18
Oil Warm Air Furnace/Central Air Conditioner-High Efficiency	3458.70	3504.45	3381.34	3803.42	3488.85	3037.16	3496.93	3698.41	3127.34

* Current Practice

Source: Gordian Associates, Inc.

tion. Since all of the advanced systems studied in this report use central distribution, the absence of a central distribution system would tend to restrict the owner of such a building from enjoying any possible benefits from future HVAC improvements in energy efficiency.

Of the central air distribution systems studied, the gas warm-air furnace with high-efficiency electric air conditioner was found to have the lowest first cost for all locations considered.

The first cost for electric heat pumps not unexpectedly shows a trend toward higher prices for the high-efficiency models. However, there are differences among the models priced such that the high-efficiency III is only about 5% more expensive than the standard heat pump, while high efficiency I and II models are 10–20% more expensive than the standard heat pump.

The hybrid electric heat pump with oil furnace has a much higher first cost than the other conventional heat pump systems because of higher costs for the oil furnace and auxiliary equipment. The gas furnace hybrid heat pump, while more expensive to install than the standard electric heat pump, is competitive in first cost with the high-efficiency heat pumps.

For the advanced systems, only the developer of the organic fluid absorption gas heat pump was able to provide an estimate of the equipment cost. This information was used as the basis for developing a projected equipment installed cost for a new installation and for an add-on installation as shown in Table 10.11.

TABLE 10.11. Organic Fluid Absorption Gas Heat Pump: Installed[a] (Dollars)

	New[b]	Add-On[c]
Houston	4044.09	2665.85
Birmingham	3910.89	2662.68
Atlanta	3857.44	2616.05
Tulsa	4350.97	2972.24
Philadelphia	4099.04	2745.90
Seattle	3170.65	2040.93
Columbus	4007.06	2670.30
Cleveland	4200.78	2785.15
Concord	3360.73	2226.67

[a] First quarter 1977 costs.

[b] Cost based on developer's wholesale equipment cost estimate for a 3-ton unit, scaled for certain locations as follows: Seattle, 2-ton; Concord, $2\frac{1}{2}$-ton; Tulsa, $3\frac{1}{2}$-ton.

[c] Assumes air handler and ducts in place.

SOURCE: Allied Chemical Co. (1976); Gordian Associates, Inc.

When viewed as an alternative to a conventional high-efficiency heat pump, the first cost for a new organic fluid absorption gas heat pump installation is not much greater. However, the product most likely to be the competitive option to the organic fluid absorption gas heat pump is expected to be the gas warm air furnace with a high-efficiency central air conditioner. The sizable first-cost differences in favor of the latter system clearly puts the organic fluid absorption gas heat pump at a cost disadvantage. Although the organic fluid absorption gas heat pump equipment costs used to develop the installed costs are not firm and are therefore subject to change, it is not expected that the gas heat pump would be competitive on a first-cost basis to a gas furnace with central air conditioner system.

Comparisons of total annual energy costs for conventional and advanced heat pumps are presented in Tables 10.12 and 10.13, respectively. Tables 10.14 and 10.15 present cost comparisons for currently available conventional and advanced systems on a seasonal basis.

On an annual energy cost basis, the gas warm-air furnace with high-efficiency air conditioner has a lower energy cost than the standard or even the high-efficiency electric heat pumps in a number of locations. This, of course, is not entirely related to energy efficiency but is also an artifact of the rate structure. (The variations in rates for the locations used in the simulation studies may be seen in Table 10.9, which presents effective unit energy prices for residential heating and cooling, and for commercial comfort-conditioning on an annual basis.)

Not unexpectedly, the heat pumps show the best advantage in the warmer climates, while the gas and oil furnace systems continue to be competitive in the cooler locations. Seattle, with its low electric rates, represents an anomaly in this regard. In general, the increasing costs of gas and oil have improved the economic advantage of the heat pump over those results reported in the previous Gordian study [1].

Energy costs of the highly efficient advanced systems are, of course, lower than those of conventional systems. However, it should be noted that dynamic efficiency losses due to cycling and defrost may be understated for the advanced heat pump I, free-piston Stirling-Rankine gas heat pump and the organic fluid absorption gas heat pump. If the dynamic efficiency losses of the advanced heat pumps were similar to those of the conventional electric (Rankine-cycle) heat pumps, the annual energy costs shown would be about 12% higher. If this were the case, part of the advanced system's advantage would be negated.

The rationale of life-cycle analysis is that by use of appropriate capital recovery factors, the effective owning and operating costs for each system can be determined over the life of that system, thereby putting each system into a directly comparable position.

Life-cycle costs of the residential space-conditioning systems simulated (without and with the effects of inflation on energy and maintenance costs) are shown in Tables 10.16 and 10.17, respectively. The results show that the

TABLE 10.12. Conventional Electric Heat Pumps and Other Systems: Comparison of Total Annual Energy Cost ($)*

System	Houston	Birmingham	Atlanta	Tulsa	Philadelphia	Seattle	Columbus	Cleveland	Concord
Electric Heat Pump-Standard	221.13(193.39)**	339.46(299.24)	270.09(237.42)	323.36(287.25)	380.00(336.69)	71.27(63.15)	403.4?(365.18)	536.35(490.92)	703.72(655.88)
Electric Heat Pump-High Efficiency I	182.36(160.14)	291.51(257.93)	231.69(204.62)	273.90(243.79)	325.59(289.37)	63.72(56.23)	355.98(322.26)	480.17(439.14)	648.63(603.99)
Electric Heat Pump-High Efficiency II	200.14(173.00)	308.81(270.16)	253.54(224.70)	281.45(245.80)	341.60(301.09)	62.33(55.42)	365.86(329.66)	487.51(444.84)	600.07(557.24)
Electric Heat Pump-High Efficiency III	195.14(171.06)	307.99(272.20)	244.82(216.10)	319.34(283.43)	349.80(310.87)	62.79(55.81)	379.53(344.02)	511.25(468.62)	620.62(576.71)
Hybrid Electric Heat Pump-Gas Furnace	236.80	360.23	290.95	305.88	408.39	121.37	390.96	480.55	580.53
Hybrid Electric Heat Pump-Oil Furnace	237.20	367.23	298.14	327.20	423.46	129.24	430.78	540.83	569.70
Electric Heat Pump-Heat Only/Central Air Conditioner-High Efficiency	197.67(172.90)	299.73(263.82)	239.77(211.08)	280.28(247.63)	323.09(285.84)	57.30(51.01)	344.53(310.56)	458.39(418.66)	550.64(511.75)
Gas Warm Air Furnace/Central Air Conditioner-High Efficiency	223.21	264.31	236.78	273.09	369.75	211.31	304.65	338.24	482.54
Oil Warm Air Furnace/Central Air Conditioner-High Efficiency	225.70	313.63	287.16	360.40	426.74	236.12	397.79	468.49	487.50
Central Electric Furnace/Central Air Conditioner-High Efficiency	236.97	428.82	343.92	451.16	559.10	136.18	584.93	809.13	976.26
Baseboard Convectors/Window Air Conditioners	153.17	341.39	270.16	379.50	490.12	131.91	537.40	770.05	946.35

* First quarter 1977 costs

** Numbers in parenthesis are energy costs calculated on steady state or full-load performance basis, dynamic efficiency losses are not included.

Source: Gordian Associates Inc.

TABLE 10.13. Advanced Heat Pumps and Other Systems: Comparison of Total Annual Energy Costs, ($)** (Dynamic efficiency not accounted for)

System	Houston	Birmingham	Atlanta	Tulsa	Philadelphia	Seattle	Columbus	Cleveland	Concord
Advanced Electric Heat Pump I*	***	274.41	218.95	241.16	298.29	***	305.10	403.71	***
Advanced Electric Heat Pump II	132.93	211.61	167.24	219.05	250.01	43.39	268.04	361.26	482.31
Free Piston Stirling Rankine Gas HP*	180.25	149.25	144.10	149.36	222.57	109.57	194.16	220.85	332.67
V-Type Single Cylinder Stirling Rankine HP	186.81	162.11	155.00	172.44	231.13	93.05	199.04	227.78	353.48
Organic Fluid Absorption Gas Heat Pump*	255.87	178.77	179.77	183.22	261.66	148.67	221.66	244.25	374.48
Pulse Gas Furnace/Central Air Conditioner- High Efficiency	206.56	238.79	210.20	227.74	303.60	145.92	244.55	268.19	347.00

* The energy costs presented may be understated for the advanced electric heat pump I, free piston Stirling Rankine gas heat pump and the organic fluid absorption gas heat pump due to possible understating of cycling and defrost losses. If the dynamic efficiency losses of the advanced heat pumps were similar to those of the conventional electric (Rankine cycle) heat pump, the energy costs above should be multiplied by the correction factor of 1.12 for annual energy values.

** First quarter 1977 costs

*** System is not sized for application in these cities

SOURCE: Gordian and Associates [2].

TABLE 10.14. Conventional Electric Heat Pumps and Other Systems: Comparison of Heating and Cooling Season Energy Costs, ($)*

HEATING	Houston	Birmingham	Atlanta	Tulsa	Philadelphia	Seattle	Columbus	Cleveland	Concord
Electric Heat Pump-Standard	30.44(27.46)**	134.88(123.28)	102.08(93.81)	188.54(173.56)	217.05(198.33)	62.43(55.26)	292.02(269.83)	436.35(403.80)	650.77(609.98)
Electric Heat Pump-High Efficiency I	25.61(23.14)	118.19(108.30)	89.03(81.93)	164.06(150.86)	188.59(172.42)	56.51(49.79)	262.39(241.73)	394.53(364.28)	604.93(566.10)
Electric Heat Pump-High Efficiency II	27.33(24.61)	122.21(111.89)	96.73(88.93)	161.34(148.34)	193.83(177.19)	54.18(48.28)	264.66(244.14)	395.80(365.66)	551.16(515.52)
Electric Heat Pump-High Efficiency III	28.95(26.10)	129.21(118.38)	97.93(90.23)	183.96(169.18)	207.66(189.98)	55.26(49.17)	282.37(260.72)	422.22(390.83)	578.10(540.22)
Hybrid Electric Heat Pump-Gas Furnace	36.79	144.66	113.83	165.74	236.42	111.92	273.31	375.58	525.08
Hybrid Electric Heat Pump-Oil Furnace	37.19	152.06	121.02	187.06	251.48	119.79	313.13	435.86	514.25
Electric Heat Pump-Heat Only	25.58(23.02)	114.67(104.80)	87.86(81.09)	148.41(136.25)	176.09(160.97)	49.29(43.95)	243.98(224.47)	366.83(338.67)	504.76(472.33)
Gas Warm Air Furnace	51.12	79.25	84.87	141.22	218.33	203.13	203.20	246.69	436.66
Oil Warm Air Furnace	53.61	128.57	135.25	228.53	275.32	227.94	296.34	376.94	441.62
Central Electric Furnace	64.88	248.56	192.01	319.29	412.10	128.17	484.38	717.58	930.38
Baseboard Convectors	63.54	243.76	188.21	314.28	406.07	125.01	477.61	707.61	915.21
COOLING									
Electric Heat Pump-Standard	190.69(165.93)	204.58(175.96)	168.01(143.61)	134.82(113.69)	162.95(138.31)	8.84(7.88)	111.45(95.35)	100.00(87.12)	52.95(45.90)
Electric Heat Pump-High Efficiency I	156.75(137.00)	173.32(149.63)	142.66(122.69)	109.84(92.93)	137.00(116.95)	7.21(6.46)	93.59(80.53)	85.64(74.86)	43.70(37.89)
Electric Heat Pump-High Efficiency II	172.81(148.39)	186.58(158.67)	156.81(135.77)	120.11(97.46)	147.77(123.90)	8.15(7.14)	101.20(85.82)	91.71(79.18)	48.91(41.72)
Electric Heat Pump-High Efficiency III	166.19(144.96)	178.78(153.82)	146.89(125.87)	135.38(114.25)	142.14(120.89)	7.53(6.64)	97.16(83.30)	89.03(77.79)	42.52(36.49)
Hybrid Electric Heat Pump	200.01	215.58	177.12	140.14	171.98	9.45	117.65	104.97	55.54
Central Air Conditioner-High Efficiency	172.09(149.88)	185.06(159.02)	151.91(129.98)	131.87(111.38)	151.42(128.52)	8.18(7.23)	101.45(86.90)	91.55(79.99)	45.88(39.42)
Central Air Conditioner-High Efficiency (All electric rate)	172.09(149.88)	185.06(159.02)	151.91(129.98)	131.87(111.38)	147.00(124.87)	8.01(7.06)	100.55(86.09)	91.65(79.99)	45.88(39.42)
Central Air Conditioner-Standard	205.12(178.98)	220.43(190.34)	181.11(155.19)	138.73(117.21)	181.11(154.01)	9.55(8.57)	121.06(104.34)	106.69(93.01)	53.17(45.85)
Central Air Conditioner-Standard (all Electric Rate)	205.12(178.98)	220.43(190.34)	181.11(155.19)	138.73(117.21)	175.82(149.62)	9.36(8.39)	120.15(103.43)	106.69(93.01)	53.17(45.85)
Window Air Conditioners	89.63(78.13)	97.63(84.10)	81.95(69.98)	65.22(55.18)	84.05(71.17)	6.90(6.17)	59.79(51.32)	62.44(53.29)	31.14(26.91)

* First quarter 1977 costs.

** Numbers in parenthesis are energy costs calculated on steady state or full-load performance basis, dynamic efficiency losses are not included.

Source: Gordian Associates Inc.

TABLE 10.15. Advanced Heat Pumps and Other Systems: Comparison of Heating and Cooling Season Energy Costs, ($)** (Dynamic efficiency not accounted for)

HEATING	Houston	Birmingham	Atlanta	Tulsa	Philadelphia	Seattle	Columbus	Cleveland	Concord
Advanced Electric Heat Pump I*	***	109.39	83.90	133.68	168.04	***	216.38	320.66	***
Advanced Electric Heat Pump II	18.98	90.62	67.71	138.72	153.60	38.19	202.16	302.24	451.06
Free Piston Stirling Rankine Gas HP*	23.95	53.10	49.78	78.20	135.42	86.56	133.65	168.48	281.29
V-Type Single Cylinder Stirling Rankine HP	22.53	52.02	48.19	80.81	131.48	68.39	129.53	168.36	295.33
Organic Fluid Absorption Gas Heat Pump*	28.71	52.90	52.90	86.31	141.13	111.71	137.23	173.33	302.71
Pulse Gas Furnace	34.47	53.73	58.29	95.87	152.18	137.74	143.10	176.64	301.12
COOLING									
Advanced Electric Heat Pump I*	***	165.01	135.05	107.48	130.25	***	89.72	83.05	***
Advanced Electric Heat Pump II	113.95	120.99	99.53	80.33	96.41	5.20	65.88	59.02	31.25
Free Piston Stirling Rankine Gas HP*	156.28	96.16	94.31	71.15	87.15	23.02	60.52	52.37	51.37
V-Type Single Cylinder Stirling Rankine HP	164.22	110.10	106.82	91.63	99.64	24.67	69.51	59.42	58.15
Organic Fluid Absorption Gas Heat Pump*	227.17	125.86	126.87	96.91	120.53	36.96	84.43	70.92	71.77

* The energy costs presented may be understated for the advanced electric heat pump I, free piston Stirling Rankine gas heat pump and the organic fluid absorption gas heat pump due to possible understating of cycling and defrost losses. If the dynamic efficiency losses of the advanced heat pumps were similar to those of the conventional electric (Rankine cycle) heat pump, the energy costs above should be multiplied by a correction factor of 1.09 for heating season values and by 1.18 for cooling season values.

** First quarter 1977 costs

*** System is not sized for application in these cities

SOURCE: Gordian and Associates [2].

TABLE 10.16. Life-Cycle Costs for Alternative Residential Space-Conditioning Systems—No Inflation (First Quarter, 1977 Dollars)

System as New Installation	Houston	Birmingham	Atlanta	Tulsa	Philadelphia	Seattle	Columbus	Cleveland	Concord
Electric Heat Pump-Standard	826.01	931.48	855.46	954.97	995.10	590.29	1007.59	1163.75	1236.63
Electric Heat Pump-High Efficiency I	844.39	940.19	873.46	991.63	997.90	641.77	1027.61	1176.05	1251.01
Electric Heat Pump-High Efficiency II	957.62	1055.07	935.76	1033.71	1111.49	679.10	1121.81	1268.92	1268.00
Electric Heat Pump-High Efficiency III	819.02	919.19	849.81	954.28	984.63	612.28	1002.46	1157.70	1184.07
Hybrid Electric Heat Pump-Gas Furnace	934.85	1043.42	968.04	1074.72	1119.39	755.07	1088.57	1202.94	1217.91
Hybrid Electric Heat Pump-Oil Furnace	1077.08	1194.78	1115.49	1262.58	1277.33	908.12	1269.75	1412.43	1346.72
Gas Warm Air Furnace/Central Air Conditioner-High Efficiency	671.30	699.03	664.72	753.25	806.39	605.53	750.33	803.07	882.75
Oil Warm Air Furnace/Central Air Conditioner-High Efficiency	820.34	906.55	862.76	989.05	1011.84	771.73	989.16	1081.80	1029.14
Central Electric Furnace/Central Air Conditioner-High Efficiency	705.94	885.00	795.88	952.98	1022.42	544.97	1057.11	1300.58	1450.78
Baseboard Convector/Window Air Conditioners	561.41	730.26	648.55	768.90	896.47	508.88	934.62	1186.33	1335.21
Electric Heat Pump-Heat Only	700.40	762.28	727.11	795.27	853.21	691.32	908.34	1055.34	1128.77
System as Add-On Installation									
Electric Heat Pump-Heat Only	533.65	605.74	573.43	636.75	681.04	554.38	738.60	877.61	983.82

Source: Gordian Associates Inc.

579

TABLE 10.17. Life-Cycle Costs for Alternative Residential Space-Conditioning Systems—Effects of Inflation on Energy and Maintenance Costs Included (Dollars)*

System as New Installation	Houston	Birmingham	Atlanta	Tulsa	Philadelphia	Seattle	Columbus	Cleveland	Concord
Electric Heat Pump-Standard	968.52	1125.43	1019.25	1141.92	1206.67	667.67	1229.36	1443.27	1588.90
Electric Heat Pump-High Efficiency I	970.05	1113.29	1020.56	1157.08	1185.82	688.17	1228.74	1431.15	1579.33
Electric Heat Pump-High Efficiency II	1091.01	1235.69	1092.36	1202.44	1306.37	752.59	1327.23	1527.21	1575.22
Electric Heat Pump-High Efficiency III	950.24	1099.46	1002.62	1139.48	1183.07	685.97	1213.82	1426.31	1500.22
Hybrid Electric Heat Pump-Gas Furnace	1102.79	1265.01	1159.52	1272.69	1361.92	882.84	1323.52	1479.83	1535.26
Hybrid Electric Heat Pump-Oil Furnace	1251.41	1425.63	1316.41	1476.03	1532.62	1035.52	1528.22	1715.73	1665.57
Gas Warm Air Furnace/Central Air Conditioner-High Efficiency	815.13	862.76	815.12	921.23	1021.16	743.60	933.59	1002.59	1152.21
Oil Warm Air Furnace/Central Air Conditioner-High Efficiency	973.49	1102.17	1129.80	1207.31	1262.21	929.82	1225.52	1352.39	1308.93
Central Electric Furnace/Central Air Conditioner-High Efficiency	858.21	1130.15	999.93	1208.95	1330.64	648.45	1377.84	1729.85	1908.08
Backboard Convectors/Window Air Conditioners	673.21	933.18	816.99	990.27	1171.39	610.39	1232.43	1596.77	1830.99
Electric Heat Pump-Heat Only	757.92	858.52	811.70	906.18	976.15	759.14	1060.79	1261.18	1394.56
System as Add-On Installation									
Electric Heat Pump-Heat Only	591.17	701.98	658.02	747.66	803.98	622.20	891.05	1083.45	1249.61

* Annual inflation: energy cost, 8 percent; maintenance cost, 6 percent.

Source: Gordian Associates Inc.

electric heat pump termed "standard" (which represents 1975–1976 heat pump technology in terms of COP and EER ratings) is slightly more cost effective in general than the more recent high-efficiency models under the conditions of the analysis. An exception to this generalization appears to be the high-efficiency electric heat pump III, which provides a slight annual cost advantage compared to the standard electric heat pump in all locations but Seattle, where the difference is fairly minor. For the other high-efficiency models, only at an assumed 8% annual rate of inflation in energy cost (and 6% in maintenance cost) are the high-efficiency models comparable to the standard model in terms of life-cycle owning and operating costs.

On the other hand, gas and oil combustion furnaces with electric air-conditioning still represent a low life-cycle cost system, especially in colder regions. However, the basis of the cost comparison is the furnace with insulated ducts and a minimum of oversizing. This is not quite the current practice. With uninsulated ducts, an oil furnace is attractive in comparison to the electric heat pump only in colder climates—even less so with inflation in energy costs.

As noted previously, the only advanced system for which installed costs could be developed is the organic fluid absorption gas heat pump. The life-cycle cost for this system, both new and as an add-on and with and without energy inflation, is presented in Table 10.18.

Reflecting its higher efficiency in the cooler climates, the organic gas heat

TABLE 10.18. Organic Fluid Absorption Gas Heat Pump: Life-Cycle Costs (Dollars)[a,b]

	Annualized Cost		Effective Annualized Cost with Inflation[d]	
	New	Add-On[c]	New	Add-On[c]
Houston	1007.06	850.14	1164.35	1007.43
Birmingham	917.63	772.51	1041.46	896.34
Atlanta	909.67	763.73	1033.93	887.99
Tulsa	987.66	827.59	1114.42	954.35
Philadelphia	1024.15	869.28	1184.28	1029.41
Seattle	765.77	638.70	876.54	749.47
Columbus	969.47	816.67	1111.92	959.12
Cleveland	1018.96	858.41	1171.21	1010.66
Concord	1024.15	895.49	1232.92	1104.26

[a] First quarter 1977 costs.

[b] No dynamic efficiency losses accounted for.

[c] Assumes air handler and ducts in place.

[d] Calculation on following assumptions: energy cost inflation, 8%, maintenance cost inflation, 6%, cost of money, 9%, project life, 9 yr.

SOURCE: Gordian Associates Inc.

pump life-cycle cost compares favorably with the high-efficiency electric heat pump life-cycle costs for the cooler locations simulated beginning with Tulsa. However, when compared to what may well be its strongest competitor, the gas warm-air furnace with central air conditioner, only the Concord location appears to show a slight cost advantage toward the gas heat pump. Energy cost inflation does not result in a reversal of this cost comparison situation. Also, as previously noted, the energy costs for the organic absorption gas heat pump are possibly understated by as much as 12%, which, if corrected by that quantity, would place the gas heat pump at a still lesser advantage against either the gas furnace or the electric heat pump on the basis of life-cycle costs. The organic fluid absorption gas heat pump system does appear to have a life-cycle cost advantage compared to the hybrid heat pump with gas furnace, which could also be viewed as competing for the same gas heating system customer.

When considered as an add-on system, the organic absorption system offers a life-cycle cost in line with the conventional systems. This tends to confirm the developer's belief that the organic absorption gas heat pump will be more readily able to penetrate the retrofit market.

10.3.3. Commercial HVAC Systems Costs

The previous section presented data on installed cost, energy cost, and life-cycle cost for residential building HVAC systems. In this section, a similar discussion is presented of light commercial HVAC systems for use in nonresidential buildings.

A comparison of installed cost for conventional and advanced commercial systems is presented in Table 10.19. Because of its heating system simplicity, the single-package, two-stage central air conditioner with strip heating (i.e., rooftop warm-air electric furnace) has the lowest first cost of the systems simulated. The rooftop gas furnace with electric air conditioner system is next lowest, with an installed cost 5–10% greater than the electric strip heat system. The third conventional system costed, the standard electric heat pump, is about 15–20% more expensive in first cost compared to the electric strip heat systems. In general, system installed costs are higher in the locations Houston through Philadelphia and lower in Seattle through Concord due primarily to differences in the equipment sizes. (A system capacity of 20 tons of cooling was used for the Houston through Philadelphia locations, while a capacity of 15 tons of cooling was used for the other locations.) On a cost-per-ton installed cooling capacity basis, the 20-ton installations actually have a lower unit cost than the 15-ton installations.

This same cost-per-ton effect is also seen with the two light commercial gas heat pumps studied (i.e., Brayton-Rankine and Rankine-Rankine rooftop gas heat pumps). The installed costs developed for the two advanced systems show them to be cost-competitive with the standard electric heat pump in most locations. However, they are 10–15% more costly than a gas or

TABLE 10.19. Light Commercial Heating and Cooling Systems: Comparison of Installed Costs (Second Quarter 1977 Dollars)

SYSTEM	HOUSTON	BIRMINGHAM	ATLANTA	TULSA	PHILADELPHIA	SEATTLE	COLUMBUS	CLEVELAND	CONCORD
Rooftop Gas Heating/Electric Cooling Package	19009.66	18811.27	18600.28	18790.57	19769.50	16877.56	16303.05	17346.40	15650.77
Single Package, Two Stage Central Air Conditioning with Strip Heating	17427.14	17427.20	17032.83	17223.11	18167.92	15791.99	15442.20	16473.13	14823.03
Single Package Standard Electric Heat Pump	20517.16	20528.98	20077.64	20278.64	21322.66	18583.57	18172.88	19292.55	17437.17
Brayton-Rankine Gas Heat Pump	21242.50	21298.75	20792.80	20983.08	22038.66	17812.72	17409.50	18497.66	16637.74
Rankine-Rankine Gas Heat Pump	21317.50	21374.47	20867.09	21057.37	22114.37	17888.00	17483.79	18573.01	16709.17

SOURCE: Gordian Associates Inc.

583

electric furnace with an air-conditioner system. Also, since the advanced systems costs are based on estimates of the developers and the systems are not yet commercially available, it is possible that some variation from these estimates may be found when they are put into commercial production.

The total annual energy costs for the light commercial systems simulated are presented in Table 10.20. The data indicate that the gas heat pumps, while offering lower energy costs for most locations, are more effective in the cooler climates. The exceptions for Houston and Seattle are due to the low electric rates available in these cities, along with gas rates that are higher than in the other cities studied in this report.

A further observation is that the Rankine-Rankine gas heat pump offers the best advantage in the colder locations, while the Brayton-Rankine gas heat pump performs better (relative to the Rankine-Rankine gas heat pump) in the warmer climates.

The net energy-cost-saving advantage in most locations (excluding Houston and Seattle for the reasons stated above) between a gas heat pump and the gas warm air furnace with air conditioner (a likely competitive product) is $400–600. While this appears insufficient to offer a 2-year payback of the added investment needed to install a gas heat pump (as compared to a gas warm air furnace with air conditioner), a 5-year payout appears feasible in several locations.

The life-cycle costs for the light commercial HVAC systems are presented in Table 10.21 without inflation of operating costs, and in Table 10.22 with inflation of the energy and maintenance costs. Generally, the gas heat pumps are more cost-effective in the cooler climates. However, since the gas rates per unit energy are generally lower even in those areas where its efficiency is less (i.e., warmer climates), the gas heat pump presents itself as a viable alternative for all locations except Houston with its high cooling load, and Seattle with its low electric rates.

When operating costs are inflated over the life of the project (8% for energy, 6% for maintenance), the life-cycle cost comparison offers still further advantage to the gas heat pumps because of their greater energy efficiency. (Again Houston and Seattle are outliers.) An increase in the gas inflation rate to 14% was also tested for the Rankine-Rankine gas heat pump, with results similar to those found under the 8% gas inflation rate calculation.

10.3.4. Conclusions of 1978 Gordian Study

Although the advanced residential HVAC systems offer lower energy costs, the added system first cost may not be justifiable. For example, the organic fluid absorption heat pump does not appear at this time to be life-cycle cost competitive with most of the conventional systems and, in particular, with the gas warm air furnace with central air conditioner system, which is viewed as the logical competitive product. This would be expected to have a

TABLE 10.20. Light Commercial Heat Pumps and Other Systems: Comparison of Total Annual Energy Costs (First Quarter 1977 Dollars)

System	Houston	Birmingham	Atlanta	Tulsa	Philadelphia	Seattle	Columbus	Cleveland	Concord
Gas Warm Air Furnace/ Central Air Conditioner- High Efficiency	790.74	1856.01	1342.49	1091.23	1002.03	699.61	998.00	993.07	1012.23
Central Electric Furnace Central Air Conditioner- High Efficiency	660.24	2010.78	1528.44	1513.91	1059.50	397.12	1557.39	1706.17	1727.70
Electric Heat Pump-Standard	587.73	1668.23	1198.56	1052.31	614.73	208.21	1000.78	1012.98	962.67
Brayton-Rankine Gas Heat Pump	797.34	692.16	422.60	533.05	495.18	450.37	450.00	462.73	736.62
Rankine-Rankine Gas Heat Pump	774.88	421.12	347.76	345.42	420.25	383.43	272.38	282.40	479.39

SOURCE: Gordian Associates Inc.

TABLE 10.21. Life-Cycle Costs for Light Commercial Heating and Cooling Systems (Second Quarter 1977 Dollars)

System	Houston	Birmingham	Atlanta	Tulsa	Philadelphia	Seattle	Columbus	Cleveland	Concord
Rooftop Gas Heating/Electric Cooling Package	4458.20	5505.61	4958.93	4729.24	4760.41	3771.10	3999.12	4118.06	3924.64
Single Package, Two Stage Central Air Conditioning with Strip Heating	4080.83	5444.47	4900.36	4907.39	4568.03	3299.26	4424.21	4618.77	4510.78
Single Package Standard Electric Heat Pump	4640.40	5738.19	5192.95	5070.72	4769.40	3677.83	4424.45	4577.86	4278.67
Brayton-Rankine Gas Heat Pump	5118.79	5038.32	4684.08	4816.76	4917.08	3939.21	3894.14	4042.83	4070.07
Rankine-Rankine Gas Heat Pump	5108.84	4779.91	4621.63	4641.52	4854.78	3884.83	3728.91	3875.07	3821.76

SOURCE: Gordian Associates Inc.

TABLE 10.22. Life-Cycle Costs for Light Commercial Heating and Cooling Systems Including Inflationary Effects on Energy and Maintenance Costs (Second Quarter 1977 Dollars)

System	Houston	Birmingham	Atlanta	Tulsa	Philadelphia	Seattle	Columbus	Cleveland	Concord
Rooftop Gas Heating/Electric Cooling Package	5224.39	6787.51	5992.23	5640.90	5628.89	4397.33	4769.80	4886.36	4702.01
Single Package Two Stage Central Air Conditioning with Strip Heating	4783.84	6801.29	6008.41	6023.67	5464.33	3779.05	5465.69	5732.28	5634.71
Single Package Standard Electric Heat Pump	5243.29	6810.73	6030.87	5875.54	5384.03	4028.90	5120.02	5278.73	4957.68
Brayton-Rankine Gas Heat Pump	5812.79	5686.61	5215.21	5395.89	5479.76	4395.54	4350.31	4504.54	4650.82
Rankine-Rankine Gas Heat Pump	5793.08	5310.39	5120.22	5139.10	5384.88	4312.07	3717.11	4258.40	4290.71

SOURCE: Gordian Associates Inc.

Calculation based on the following assumptions: energy inflation - 8 percent, maintenance cost inflation - 6 percent, money - 9 percent.

negative influence on the market prospects of the absorption machine for new installations. The refit market appears to offer a better potential for the organic absorption gas heat pump.

The lowest cost residential space-conditioning system (with the exception of baseboard convector/window air conditioners) continues to be the gas warm-air furnace with high-efficiency air conditioner. High-efficiency heat pumps, because of their 5–20% higher first cost over the standard heat pump, do not appear to offer any life-cycle cost advantage over the standard heat pump at 1977 energy prices. In general, the higher cost for greater energy efficiency does not appear to pay. However, the more energy-efficient systems will be the beneficiaries of future higher energy costs, which will of course improve their economics.

A potential new development, the pulse gas furnace with high-efficiency central air conditioner, offers system energy costs equal to the advanced heat pumps (and much better than for the conventional systems studied). Since the pulse gas furnace is expected to be marketed at a price near the present gas warm-air furnace, it would appear to make this system an obvious minimum-cost choice, assuming the gas hookup would be permitted.

The hybrid electric heat pumps (with gas or oil furnace) are substantially more expensive to own and operate than any of the other heat pumps studied.

The organic fluid absorption gas heat pump is not cost-effective compared to the gas warm-air furnace with central electric air conditioner. Additionally, transient losses not accounted for in the energy modeling of this system may result in energy costs of up to 12% higher.

The more attractive gas heat pump market possibility at this time is the one for light commercial service. Under present cost assumptions, the light commercial gas heat pumps are cost-effective compared to the electric heat pump for most all locations, and to the electric and gas furnace with air conditioner in the cooler climates. When operating costs are inflated, the gas heat pumps are cost-effective for all locations used in the simulation study with the exception of Houston and Seattle, both of which have low electric rates available for commercial service.

10.4.   THE SMITH (TRANE) HYDRONIC HEAT PUMP STUDY [3]

10.4.1.   System Analysis

Water-cooled heat pumps have a water-cooled condenser substituted for the air-cooled condenser of an air-to-air unit. Thus, heat is rejected to a water loop on the cooling cycle, and removed from it on the heating cycle. The source of recovered heat is the water loop rather than outside air. If the water loop were of unlimited size (e.g., connected to a lake) and of moderate temperature (safely above 32°F), then indeed the heat source could be considered free just as heat removed from the atmosphere by an air-to-air unit is considered free.

Unfortunately, most water-cooled heat pump systems do not employ water loops of unlimited size. Typically, the water-cooled heat pump installation employs a closed water loop whose water volume is increased by water storage tanks. The temperature of this loop must be held between a minimum and maximum (typically 60–90°F) to be compatible with the operation of the individual heat pumps. If most of the individual heat pumps are cooling—rejecting heat to the water loop—loop temperature increases to a maximum of 90°F; when the loop temperature is above 90°F, a closed-loop cooling tower is operated to maintain the temperature at 90°F. If, on the other hand, most of the units are on heating—removing heat from the water loop—the temperature of the water will fall to a minimum of 60°F, that level being maintained by a supplementary heater. The larger the loop's water volume, the greater is the thermal inertia and the longer the system will coast. Thus, water storage tanks are attractive in these systems.

In this study by W. W. Smith, the TRACE program, which simulated building HVAC system operation hour by hour, using U.S. Weather Bureau data for a typical year in a building's location, was used to compare a water-cooled or hydronic heat pump system with five alternative systems for a typical school in Atlanta:

1. Hydronic heat pumps.
2. Air-to-air heat pumps.
3. Two-pipe fan-coil units or unit ventilaors, electric heat.
4. Four-pipe fan-coil units or unit ventilators, oil heat.
5. Variable air volume rooftop units.

The hydronic heat pump units are single-packaged, one per room, with water-cooled condensers and have reverse cycle or heat pump capabilities. The system rejects heat through a cooling tower in summer and supplies heat with a boiler in winter. When simultaneous heating and cooling loads exist, heat is pumped from one room to another, and neither cooling tower nor boiler will operate if an energy balance of one-third cooling and two-thirds heating exists.

To improve the efficiency of this heat pump system, a 1000-gal water storage tank was added. This builds thermal inertia into the water loop and delays operation of the boiler or cooling tower whenever the system changes from predominantly cooling load to heating load, or vice versa. Essentially, the tank stores some of the daytime heat for use at night.

The second system analyzed was an air-to-air heat pump system using self-contained rooftop units. Auxiliary heat in this system was electric.

The third system chosen for comparison was a two-pipe fan-coil system with electric heat. Cooling is supplied by a central water chiller and cooling tower. Chilled water is pumped to floor-mounted fan-coil units, located in the perimeter rooms, which also have electric coils to provide heating. An air-handling unit handles the interior areas of the building using central chilled water as an electric heating coil; its operation is identical to that of a very large fan-coil unit.

The fourth alternative analyzed was a four-pipe fan-coil system. Cooling is supplied just as in the two-pipe system, except that an air-cooled water chiller was used. Heat is supplied by an oil-fired boiler, with hot water being pumped to an auxiliary water coil in each fan-coil unit and in the air handler.

The last system simulated comprised self-contained rooftop units feeding cool conditioned air to variable-volume terminal units in the individual rooms. This is the only all-air system considered. In summer, the system is controlled by varying the amounts of cool air supplied to each room in response to that room's thermostat. In winter, electric baseboard radiation and unit heaters in the ceiling plenum supply the heat.

Outside air economizers were applied to all five systems as follows: In the air-to-air heat pump system and the rooftop VAV system, the economizers were applied to both interior and perimeter; in the two- and four-pipe fan-coil systems and the hydronic heat pump system, the economizers were applied to the interior only. In addition, because there was some question as to the value of an economizer on the hydronic heat pump system, this system was run a second time with no economizer applied.

To compare the merits of a hydronic heat pump system with those of more conventional nonrecovery systems, a building with simultaneous load was needed, one representing a good heat recovery candidate. The four-story, 32,000-ft^2 vocational school building with a 24-hr computer facility in Atlanta, Georgia, shown in Fig. 10.7, provided the model.

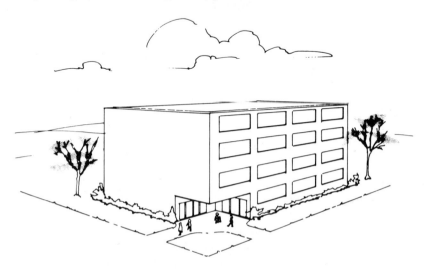

Fig. 10.7. Four-story vocational school. Reprinted by permission from *Heating, Piping, and Air Conditioning*, March 1979.

10.4.2. Economic Comparison

After determining the building's and systems' operation, the next step was to examine the economics of purchase, operation, and maintenance of the alternative systems in the vocational school and to compare life-cycle costs. The first-cost parameters used were:

1. Water-cooled heat pump system, $1400/ton
2. Air-to-air heat pump system, $1300/ton
3. Two-pipe fan-coil system with electric heat and water-cooled reciprocating chiller, $1750/ton
4. Four-pipe fan-coil system with oil heat and air-cooled reciprocating chiller, $1850/ton
5. Rooftop VAV system, $1400/ton

The computer program calculated the cooling and heating capacities required for the vocational school. The fan-coil and rooftop systems required 78 and 80 tons of cooling capacity, respectively, which are substantially below the tonnages required for the hydronic heat pump system, 98 tons, and the air-to-air heat pump system, 90.5 tons. The other systems are able to take advantage of building diversity, whereas the heat pump units must be sized for the peak load in each of the individual zones. The larger zones of the air-to-air heat pump system afford some diversity.

Based on the typical first-cost per-ton values, first costs of the five systems in the capacities required were

1. Water-cooled heat pump system, 98 tons @ $1400 per ton = $137,200
2. Air-to-air heat pump system, 90.5 tons @ $1300 per ton = $117,650
3. Two-pipe fan-coil system, 78 tons @ $1750 per ton = $136,500
4. Four-pipe fan-coil system, 78 tons @ $1850 per ton = $144,300
5. Rooftop VAV system, 80 tons @ $1400 per ton = $112,000

The diversity in this building favors the alternate systems. For the water-cooled heat pump system's first cost to equal that of the rooftop VAV system, the system would have to sell, installed, for only $1400 per ton. To equal the first cost of the air-to-air heat pump system, it would have to be $1200 per ton.

The utility costs were assumed to be as follows:

Electricity	$0.35/kWh
Oil	$0.49/gal
Water	$0.12/1000 gal

These values are representative of what one might expect for Atlanta in 1979.

Total utility bills for the 32,000-ft^2 vocational school with the various systems are compared in Table 10.23. The differences from one system to another are explained below.

Lights.   No matter what the HVAC system, the building lights react the same.

Fans.   A comparison of catalogued fan horsepowers for the most popular brand fan coils and water-cooled heat pumps reveals that fan-coil supply fans use less power. This is because of reduced internal static pressure and the use of more efficient motors.

Auxiliary Heat.   Four observations are of interest in comparing the auxiliary heat operating costs of these systems:

1.   The double pumping phenomenon of the water-cooled heat pump and the input of recovered or stored heat reduce the auxiliary heat demand compared to other systems.
2.   The air-to-air heat pump system reduces demand for electrical resistance by recovering heat from outside air and rejecting the heat of compression to the occupied space during reverse-cycle compressor operation.
3.   Although the two-pipe and four-pipe fan-coil systems meet heating demands without heat recovery, the difference in their operating costs reflects relative costs of oil and electric heat.
4.   Rooftop VAV electric heating cost is less than the electric heating cost with the two-pipe fan-coil system because the heat of lights contributes to meeting heating loads. (Floor-mounted perimeter fan-coil units do not recover this heat discharged to the ceiling plenum.)

Pumping Cost.   Even though the central systems have three separate pumps each (chilled water, hot water, and condenser water), individually they move less water against a lower pressure drop than the hydronic heat pump system pump. Also, they can be scheduled to run on demand, whereas the system pump of the hydronic heat pump system runs continuously. Naturally, pumps are not used with the unitary rooftop and air-to-air heat pump systems.

Cooling Tower (Fan and Water).   There is an operating cost penalty for the water-cooled heat pump system over the central systems because:

1.   The water-cooled heat pump system is a closed system with a closed circuit evaporative cooling tower, which requires more energy and much more water for a given duty than a central system open tower.

TABLE 10.23. Operating Costs for Various Systems Applied to Vocational School Located in Atlanta

Energy User	Hydronic Heat Pumps		Air-to-Air Heat Pumps	Two-Pipe Fan-Coil	Four-Pipe Fan-Coil	Rooftop VAV
	With Economizer	Without Economizer				
Lights	$10,494	$10,494	$10,494	$10,494	$10,494	$10,494
Fans	5,220	5,220	4,645	2,894	2,894	2,470
Auxiliary heat	361	—	670	3,215	891	2,096
Pumps	2,120	2,120	—	1,520	937	—
Cooling tower	1,460	1,722	—	900	—	—
Condenser fan	—	—	733	—	1,540	1,161
Compressors	9,745	12,102	4,310	4,414	5,400	4,029
Totals	$29,400	$31,568	$20,852	$23,437	$22,156	$20,250

SOURCE: W. W. Smith [3].

2. More heat is rejected, and a larger cooling tower capacity is required for the water-cooled heat pump system than for the central systems, since the heat of compression, and thus the total heat rejected at peak periods, is considerably greater. This is partially a function of compressor efficiency; e.g., at full cooling load, a water-cooled reciprocating chiller draws 1.0 kW/ton whereas a water-cooled heat pump draws 1.4 kW/ton. The rooftop and air-to-air heat pump systems use air-cooled condensers instead of cooling towers.

Compressors.   The compressor energy costs vary from $4029 per year for the rooftop VAV system to $12,012 for the water-cooled heat pump system. These differences are not explained solely by the differences in compressor efficiencies. For the air-to-air heat pump system, reverse-cycle compressor operation on the heating mode accounts for roughly 20% of the total. The water-source heat pump is the highest because all heat, regardless of source, involves the refrigeration system of a heating unit. Some heat is effected by two-compressor operations (sometimes called double pumping), as described earlier.

Further, the heat pump compressor efficiency is even worse on heating than on cooling. As indicated above, at full cooling load with a 90°F loop temperature, a water-cooled heat pump uses 1.4 kW/ton (EER = 8.6); at full heating load with a 60°F loop temperature, however, the heat pump uses 2.0 kW/ton (COP = 2.8).

The significantly greater number of operating hours associated with both heating and cooling operations, coupled with the lower efficiency of water-cooled heat pump compressors, result in compressor operating costs more than double those of conventional systems.

Total Utility Costs.   The cumulative totals show that the lowest energy cost system of these alternatives, on this particular building at this particular location, is the rooftop VAV system, followed closely by the air-to-air heat pump system. The most costly system to operate is the water cooled heat pump system. This shows dramatically the energy cost penalty often paid for a so-called energy saving system when it is misapplied and points to the importance of careful energy analyses before selecting a system.

Maintenance Costs.   The last variable in the life cycle cost equation is maintenance costs. In an impartial survey we found that opinions on this subject varied tremendusly, so choosing a representative figure for each of the systems compared was difficult. Maintenance costs were assumed to be equal at $22/ton annually.

The annual owning and operating costs for these systems vary by nearly $15,000 for the first year, as shown in Table 10.24.

TABLE 10.24. First-Year Owning and Operating Costs for Various Systems Applied to Vocational School Located in Atlanta

| | Hydronic Heat Pumps | | | | | |
Cost Item	With Economizer	Without Economizer	Air-to-Air Heat Pumps	Two-Pipe Fan-Coil	Four-Pipe Fan-Coil	Rooftop VAV
Installed cost	$137,200	$137,200	$117,650	$136,500	$144,300	$112,000
Principal, interest, taxes, insurance[a]	17,378	17,378	14,901	17,289	18,278	14,185
Utility cost	29,400	31,568	20,852	23,437	22,156	20,250
Maintenance	2,150	2,150	1,991	1,716	1,716	1,760
Totals	$48,928	$51,096	$37,744	$42,442	$42,150	$36,195

[a] Assumes 8% mortgage interest, 10% down payment, 3% taxes, and 0.5% insurance.

SOURCE: W. W. Smith.

10.4.3. Summary of Smith Study

The hydronic heat pump, like any system, can waste energy when it is misapplied—that is, when factors tend to favor other systems. When is it right to select a hydronic heat pump system? According to Smith,

> There are many situations where this system could be chosen for a variety of reasons. But when life cycle cost is the basis of system selection, a water cooled heat pump system will be justified only when the building exhibits all of the following characteristics:
>
> Low sustained cooling load
> No alternative to electric heat
> Small system capacity
> Significant operating time with simultaneous heating and cooling loads
>
> The design of a building and selection of an air conditioning system with energy conservation in mind should not be done by rule of thumb. A heat pump system should not be selected on the strength of its name. Rather, careful and thorough analyses must be conducted to determine the lowest life cycle cost system.

10.5. DEPARTMENT OF ENERGY STUDY [4]

In this study, published in 1980, the Department of Energy (DOE) identified 28 market areas where increased use of heat pumps in homes might reduce oil imports. These are given in Table 10.25. The 28 areas are served by electric utilities which have excess winter non-oil or gas capacity, making them suitable markets for electric heating during the winter months.

The report singles out three areas—Athens, Georgia; Detroit, Michigan; and St. Louis, Missouri—as having electric utilities that are summer-peaking and that do not rely upon scarce fuels like oil or natural gas to generate their electricity. Furthermore, the utilities have "significant" unused generation capacity during the fall, winter, and spring seasons, and their marginal cost of producing electricity in these seasons does not include the cost of building generating capacity to meet the higher summer load.

"Assuming homeowners receive the proper price signals for this off-peak electricity and for natural gas, it would be in their economic interest to replace their existing oil or gas furnace with electricity-based alternatives as the primary source of heat for their homes," the study says.

To test its thesis, DOE compared a variety of retrofit or replacement options for the three cities, which represent a full range of winter weather conditions. For each city, a 1500-ft^2 home design with insulation and

TABLE 10.25. Twenty-Eight Utilities with Excess Winter Capacity

1.	Alabama Power Co.	Alabama	
2.	Carolina Power & Light Co.	North Carolina	
3.	Central Illinois Light Co.	Illinois	
4.	Central Illinois Public Service Co.	Illinois	
5.	Cleveland Electric Illum. Co.	Ohio	
6.	Columbus & Southern Ohio Electric Co.	Ohio	
7.	Commonwealth Edison Co.	Illinois	
8.	Detroit Edison Co.	Michigan	
9.	Georgia Power Co.	Georgia	
10.	Idaho Power Co.	Idaho	
11.	Illinois Power Co.	Illinois	
12.	Indianapolis Power & Light Co.	Indiana	
13.	Interstate Power Co.	Iowa	
14.	Iowa-Illinois Gas & Electric Co.	Iowa	
15.	Iowa Public Service Co.	Iowa	
16.	Iowa Southern Utilities Co.	Iowa	
17.	Kansas City Power & Light Co.	Missouri	
18.	Louisville Gas & Electric Co.	Kentucky	
19.	Madison Gas & Electric Co.	Wisconsin	
20.	Missouri Public Service Co.	Missouri	
21.	Nebraska Public Power District	Nebraska	
22.	Northern Indiana Public Service Co.	Indiana	
23.	Northern States Power Co.	Minnesota	
24.	Omaha Public Power District	Nebraska	
25.	South Carolina Electric & Gas Co.	South Carolina	
26.	Southern Indiana Gas & Electric Co.	Indiana	
27.	Union Electric Co.	Missouri	
28.	Wisconsin Power & Light Co.	Wisconsin	

SOURCE: U.S. Dept. of Energy [4].

**TABLE 10.26.   Forecast Delivered
Energy Prices (1980 Dollars per 10^6 Btu)**

Year	Natural Gas	Electricity
1980	4.08	5.88
1981	4.35	5.92
1982	6.94	5.98
1983	6.98	6.04
1984	6.91	6.10
1985	6.99	6.16
1986	6.45	6.22
1987	7.02	6.28
1988	7.20	6.35
1989	7.37	6.41
1990	7.54	6.47
1991	7.71	6.54
1992	7.87	6.60
1993	8.05	6.67
1994	8.22	6.74
1995	8.39	6.80
1996	8.59	6.87
1997	8.78	6.94
1998	8.97	7.01
1999	9.17	7.08
2000	9.37	7.15

SOURCE: U.S. Dept. of Energy, 1980.

infiltration characteristics for its region was chosen. In addition, a weatherization package reducing each home's heating needs by 35% was costed. The base case assumed an existing central gas furnace with 10 yr of remaining life. The heating requirement for the house was used to calculate how much gas would be needed to heat the home each year, given the furnace's seasonal efficiency. This annual fuel requirement was then multiplied by the unit cost of the fuel assumed for each year from 1981 to 2000. To this was added the assumed maintenance cost for the furnace to get the total annual operating cost for the 20-yr period.

Central to DOE's analysis is its forecast delivered energy prices, shown in Table 10.26. This shows the cost of natural gas exceeding that of electricity by 1982, reflecting a quicker decontrol of natural gas prices.

The payoff is in Table 10.27, which shows the net present cost of home retrofit or replacement options for 20 years, at a 10% discount rate. The most cost-effective option is the installation of a high-efficiency heat pump with the existing furnace as a backup, along with a building conservation package.

DOE cautions, "These results are only approximations of the expected

TABLE 10.27. Net Present Cost of Residential Retrofit/Replacement Options for 1981–2000, Using 10% Discount Rate, in 1980 Dollars

Options	Detroit, MI (% of base)	Athens, GA (% of base)	St. Louis, MO (% of base)
1. Base case: Unretrofitted building and unretrofitted furnace	9028.42 (100%)	6288.63 (100%)	7426.14 (100%)
2. Building conservation package	7488.68 (83%)	5289.50 (84%)	6254.33 (84%)
3. Furnace retrofit, without building conservation package	8445.74 (94%)	6459.70 (103%)	7486.30 (101%)
with building conservation package	7211.56 (83%)	5682.01 (90%)	6476.47 (87%)
4. Replace existing furnace with high-efficiency gas furnace—without building conservation package	8793.75 (97%)	6381.39 (101%)	7428.96 (100%)
with building conservation package	7793.47 (86%)	5867.55 (93%)	6692.34 (90%)
5. Install standard heat pump with existing furnace as backup—without building conservation package	6352.31 (70%)	4160.62 (66%)	4917.65 (66%)
with building conservation package	6045.31 (67%)	3779.58 (60%)	4553.65 (61%)
6. Install high-efficiency heat pump with existing furnace as backup—without building conservation package	5984.84 (66%)	3926.02 (62%)	4504.74 (61%)
with building conservation package	6115.76 (68%)	3828.41 (61%)	4457.61 (60%)
7. Install standard heat pump with central electric resistance backup—without building conservation package	7919.94 (88%)	6028.93 (96%)	6209.13 (84%)
with building conservation package	7603.22 (84%)	5689.21 (90%)	5888.59 (79%)
8. Install high-efficiency heat pump with central resistance backup—without building conservation package	7399.03 (82%)	5775.88 (92%)	5178.45 (70%)
with building conservation package	7423.82 (82%)	5683.65 (90%)	5816.88 (78%)
9. Replace existing furnace with central electric resistance furnace—without building conservation package	6738.55 (75%)	5027.26 (80%)	5619.30 (76%)
with building conservation package	6343.20 (70%)	4795.32 (76%)	5476.91 (74%)

SOURCE: U.S. Department of Energy [4], 1980.

costs of retrofitting or replacing central heating equipment and should be used for relative comparisons." Further, it says, "One should be careful not to extrapolate the 28 utility service areas of interest here, particularly in areas where oil or natural gas is expected to be a primary fuel on the utility system in the future." As an example, it cites Boston, where much higher electricity rates "showed the heat pump to be noneconomic."

10.6.   EPRI (WESTINGHOUSE) HEAT PUMP STUDY

In 1977, the Electric Power Research Institute published a report titled "An Investigation of Methods to Improve Heat Pump Performance and Reliability in a Northern Climate" [11]. The objectives of this study were to evaluate the improvement of heat pump acceptability to both consumer and utility in the North, to configure the most promising improved system(s), and then to evaluate the impact of these upon utilities and the manufacturing industry. The conclusions from this study were as follows.

1.   The optimum seasonal performance factor as a function of electric rates has an apparent asymptote of 2.7 for single capacity machines and 3.0 for dual capacity machines. For constant-dollar electric rates of 3.6¢/kWh, the optimum seasonal performance factor for single capacity machines is approximately 2.15 and for dual capacity machines it is approximately 2.4. According to EPRI (Westinghouse) [11]:

"More efficient performance will come only at the expense of higher first cost. When the incremental increase in annual charge for all fixed charges such as capital amortization is balanced against the incremental decrease in annual energy cost as performance is improved, an optimum level of performance can be found at which the total of all annual charges is a minimum. Improving the performance of a heat pump by altering the relative proportions of its constituent components reaches a point of diminishing returns at which performance cannot be improved regardless of the cost of energy and, hence, the cost of the system. This conclusion is based upon a simplified life cycle costing model."

2.   Optimization of an improved heat pump with respect to ownership cost in the North will yield a system whose annual cost is $730/yr, a benefit of approximately $120/yr relative to a state-of-the-art unit applied in a house insulated to 1974 HUD/FHA minimum standards. An additional reduction in annual cost of approximately $50/yr is available should the reliability of the average heat pump be improved to the level of the

best manufacturers today, whose in-warranty maintenance costs average approximately $30/yr. A reasonable bound to the possible reduction in ownership cost available through system optimization and reliability improvement is then $170/yr. Should this be realized, then the heat pump's annual cost performance would be better than that of a natural gas-fired furnace with electric air-conditioning in Albany.

3. Improvement in the level of building insulation and construction technology used for a single-family dwelling, to a value of 64% of the thermal conductance specified in the 1974 revision of HUD/FHA minimum standards, will reduce the annual cost of ownership of a heat pump system by approximately $200/yr from a level of $850/yr when the annual charge for the incremental insulation cost is charged as part of the heat pump system annual cost. According to EPRI (Westinghouse) [11]:

"Reduction in ownership cost is due to both a reduced heat burden and a reduced capital cost burden since the heat pump required is of lesser capacity. Note that when an optimized, improved system is used along with better insulation, the savings are additive. The annual ownership cost can be reduced to $516/yr even with historical maintenance costs. This is approximately $90/yr less expensive than a gas furnace with electric air-conditioning, and $175/yr less expensive than an electric furnace system. Application of heat pumps to well-insulated *multi*family dwellings is more difficult to justify because the much smaller energy use, $70/yr for a heat-pump application instead of $370/yr for a single-family residence presents little opportunity for savings at current electric rates."

4. The annual cost optimization of northern climate heat pumps under the constraint of flat electric rates evolves heat pump configurations of much higher efficiency but with insignificantly reduced cold-weather electrical demand. The already adverse impact of heat pumps instead of resistance heat upon utilities is increased by optimization, since revenue will be reduced, load factor will be reduced, and the relative requirement for peaking generation will be increased. According to EPRI (Westinghouse) [11]:

"Annual cost optimization trades off annual energy cost against capital-dependent annual costs. With flat electric rates, the cost of energy is directly proportional to the quantity of energy consumed, i.e., there is no cost incurred that is a function of demand. Therefore, any incremental increase in system cost must be reflected in a compensating incremental decrease in energy use. Note that the histogram of hourly temperature occurrence for a northern city such as Albany has its peak approximately at 0°C (32°F). Note further that roughly ¾ of the hours in a heating season occur at temperatures greater than $-6.7°C$ ($20°F$). An examination of the energy use profile as a function of ambient temperature for an optimized heat pump shows that roughly 30% of the energy is consumed at temperatures below $-6.7°C$ ($20°F$). Below $-17.8°C$ ($0°F$) there is an insignificant amount of energy consumed. Since the optimization procedure used strives to reduce energy consumption, the performance of the system will be maximized for that region of ambient temperature where the most energy is consumed. The very cold ambient temperatures will not influence the system configuration. Thus, on the very cold days, the optimized system has minimum heat pump capacity and the heating load must be carried largely by electrical resistance heat. The cold-weather electrical demand for an optimized heat pump system is therefore virtually no less than it would be for a resistance heat system. Only if electrical demand is penalized will optimization reduce demand in its search for minimum ownership cost.''

In another portion of the EPRI report, analyses were carried out to identify economical and practical heat pump sizing methods for northern climates and to establish the seasonal and annual operating characteristics of state-of-the-art conventional and heat pump HVAC systems. The major conclusions from this portion of the study are as follows:

"In single-family residences state-of-the-art heat pump HVAC systems offer significantly lower annual energy costs than electric furnace and oil furnace systems with conventional air-conditioning. Gas furnace systems with conventional air-conditioning give the lowest annual energy costs in all locations except where gas rates are unusually high.''

"Energy cost patterns in multifamily residences are similar to single-family residences. However, the total annual load on the HVAC system is relatively small, particularly for the better construction level, which limits the potential for cost savings with energy-efficient systems.''

"Because of the large internal heat generation in office buildings, cooling rather than heating is the dominant load on the HVAC system, independent of the climate. Thus, energy-efficient heating systems, such as heat pumps, have little potential for reducing the total annual energy consumption and cost.''

"Conventional heat pump sizing practices (size for the design cooling load—balance point must equal or exceed the peak of the winter temperature histogram) result in total annual ownership costs near the minimum for state-of-the-art equipment over the range of energy costs considered.''

"The energy efficiency of state-of-the-art equipment in the capacity range required for the smaller, better-insulated residences of the future is relatively poor. Manufacturers should concentrate on improving the performance of units with 24,000 Btu/hr cooling capacity or less.''

"Maintenance costs for state-of-the-art heat pumps represent a significant portion of the total annual cost of ownership. The wide variation between maintenance costs of various brands indicates the need for an industry-wide quality control program."

Figure 10.8 illustrates the importance of sizing the heat pump appropriately with respect to the balance point at all northern locations. The results follow the expected trends. As electric rates and heating loads increase, the optimum heat pump size increases. As the heat pump size increases, a larger portion of the total heating load is carried by the heat pump and a smaller portion by resistance heat. Thus, the seasonal COP increases, and the annual energy consumption decreases. As electric rates and building loads increase, the energy cost becomes a larger portion of the total cost of ownership, and minimum cost of ownership is achieved with systems with lower energy costs and higher capital and maintenance costs. Table 10.28 compares the systems with minimum total cost of ownership with systems sized according to standard practice.

At an electric rate of $0.025/kWh, a 7-kW (24,000-Btu/hr) unit [0°C (32°F) balance point] is nearly optimum for all three cities. This size gives minimum total annual cost of ownership in Boston and Denver and is within $0.50/yr of the minimum in Albany. Clearly, at this electric rate, standard sizing practice gives excellent results.

At an electric rate of $0.050/kWh, the minimum total annual cost of ownership is obtained with a 8.8-kW (30,000-Btu/hr) unit [−2.5°C (27.5°F) balance point] in all three cities. With a 7-kW (24,000-Btu/hr) unit, the total annual cost of ownership is 1.1–2.1% higher. Thus, standard sizing practice is not optimum, but the cost of deviation from optimum sizing is relatively small.

At an energy cost of 0.075/kWh, the 8.8-kW (30,000-Btu/hr) unit [−2.5°C (27.5°F) balance point] is optimum for Boston and Denver, and the 10.5-kW (36,000-Btu/hr) unit [−5°C (23°F) balance point] in Albany. Since Albany has the highest annual heating load of the three cities considered, the optimum capacity is expected to be higher. The total annual cost of ownership of a conventionally sized unit [7-kW (24,000-Btu/hr)] is 3.6% higher than a 10.5-kW (36,000-Btu/hr) unit in Albany. In Boston and Denver the total annual cost for the conventionally sized unit is higher than the optimum unit by 2.2 and 2.8%, respectively.

At an electric rate of $0.10/kWh, the 14.1-kW (48,000-Btu/hr) [−10°C (14°F) balance point] unit gives the lowest total annual cost of ownership in Albany, the 10.5-kW (36,000-Btu/hr) unit has minimum cost in Denver and the 8.8-kW (30,000-Btu/hr) unit is optimum in Boston. A conventionally sized [7-kW (24,000-Btu/hr)] unit would have total annual ownership costs 5.2% higher than the minimum in Albany, 2.8% higher in Boston and 3.7% higher in Denver.

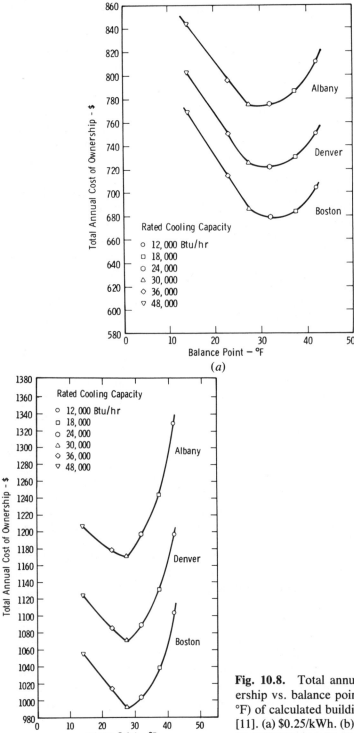

Fig. 10.8. Total annual cost of ownership vs. balance point (610 Btu/(hr · °F) of calculated building heating load) [11]. (a) $0.25/kWh. (b) $0.050/kWh. (c) $0.075/kWh. (d) $0.10/kWh. Courtesy of Electric Power Research Institute.

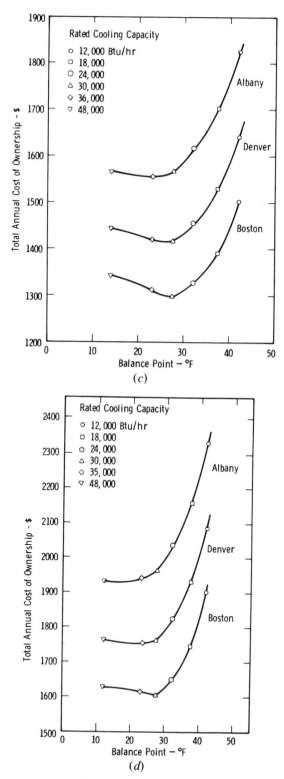

Fig. 10.8. (Continued)

TABLE 10.28.   Total Cost of Ownership of Optimally Sized Systems vs. Standard Size Systems for a Level 2 Single-Family Residence

Electric Rate ($/kWh)	City	Minimum System Cost Balance Point (°F)	Standard Total Cost of Ownership ($)	Minimum Total Cost of Ownership ($)	Net Savings ($)	Percent Savings (%)
0.025	Albany	27.5	776.50	776.04	0.46	0.06
	Boston	32.0	679.72	679.72	—	—
	Denver	32.0	723.00	723.00	—	—
0.050	Albany	27.5	1196.67	1171.59	25.08	2.10
	Boston	27.5	1003.12	991.89	11.23	1.12
	Denver	27.5	1089.67	1070.94	18.73	1.72
0.075	Albany	23.0	1616.84	1559.40	57.44	3.55
	Boston	27.5	1326.52	1297.59	28.93	2.18
	Denver	27.5	1456.34	1416.16	40.18	2.76
0.10	Albany	14.0	2037.02	1930.75	106.27	5.22
	Boston	27.5	1649.92	1603.29	46.63	2.83
	Denver	23.0	1823.02	1755.97	67.05	3.68

SOURCE: EPRI (Westinghouse) [11].

The minimum total cost of ownership system is optimum only from a total cost point of view. There are disadvantages to larger systems. In the cooling mode and during mild-weather heating the larger units will cycle more, causing more rapid fluctuations in temperature, and (in the cooling mode) humidity. Reliability may be reduced by increased cycling. The larger units occupy more space and require larger ductwork, which reduces the usable space in the residence.

Larger units may be advantageous from the point of view of the electric utility. Since a larger unit will carry a larger portion of the heating load at low ambients, less resistance heat can be installed and the peak electric demand can be reduced. Since residential customers rarely pay a demand charge, the reduced peak demand is of no direct benefit to the consumer. However, some utilities are considering instituting a demand charge for residential customers, and such a move should alter sizing philosophy.

On balance, it appears that the cost advantages of installing a heat pump sized for minimum total cost of ownership are relatively small and are partially offset by certain operating disadvantages. Thus, it is difficult to justify a sizing practice more complex than existing practice. If electric utilities institute demand charges for residential customers, the picture may be altered.

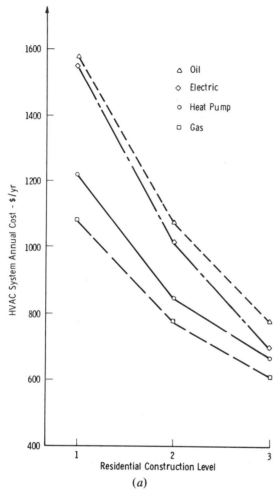

Fig. 10.9. HVAC system annual cost as a function of construction level in (a) Albany, NY; (b) Boston, MA; (c) Denver, CO [11]. Courtesy of Electric Power Research Institute.

607

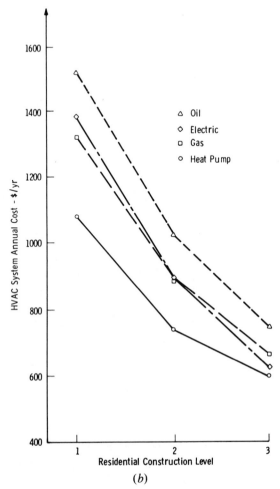

(b)

Fig. 10.9. (Continued)

608

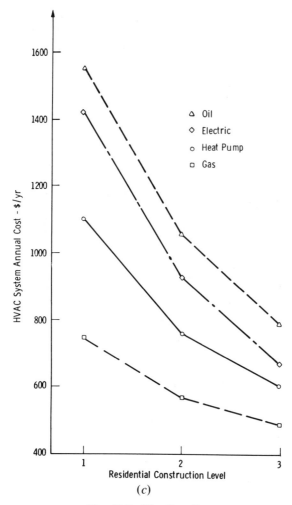

Fig. 10.9. (Continued)

Some comparative annual costs were developed for the three northern cities for oil, electric, gas, and heat pump systems for three levels of home insulation. The calculated building load constant for level 1 was 910 Btu/(hr·°F), for level 2 it was 610 Btu/(hr·°F) and for level 3 it was 390 Btu/(hr·°F). Figure 10.9 illustrates the results from these calculations.

REFERENCES

1. Gordian Associates Inc., "Evaluation of the Air-to-Air Heat Pump for Residential Space Conditioning," Report prepared for FEA, NTIS No. PB-255-652, April 1976.

2. Gordian Associates, Inc., "Heat Pump Technology, A Survey of Technical Development Market Prospects," Report prepared for DOE, HCP/M2121-01, June 1978.

3. Smith, W. W., "Water Cooled Heat Pumps: Energy Savers?" *Heating/Piping/Air Conditioning,* March 1979.

4. ———, "Reducing U.S. Oil Vulnerability," U.S. Dept. Energy, 1980.

5. Hiller, C. C., and Glicksman, L. R., *Improving Heat Pump Performance via Compressor Capacity Control—Analysis and Test,* Vols. I and II, M.I.T. Energy Laboratory, Cambridge, Mass.; NTIS Publ. No. PB-250-592, January 1976.

6. Christian, J. E., "Unitary Air-to-Air Heat Pumps," Argonne National Laboratory, ANL/CES/TE 77-10, July 1977.

7. May, J. A., and Gass, T. E., "Heat Pump Pioneer: Carl Nielsen," *Water Well J.,* April 1977.

8. American Society of Heating, Refrigerating and Air Conditioning Engineers, *Handbook and Product Directory, 1975, Systems,* ASHRAE, New York, 1975.

9. Dressler, F. R. S., "Life-Cycle Evaluation of Small/Medium HVAC Systems to Promote Energy Conservation," *Proc. Conf. Improving Efficiency in HVAC Equipment and Components,* Purdue Research Foundation, 1974.

10. Hittman Associates, Inc., "Barriers Connected with Certifying or Listing of Energy Conserving Products Used in Buildings," Report to ERDA, Nov. 1976.

11. Westinghouse Electric Corporation, *An Investigation of Methods to Improve Heat Pump Performance and Reliability in a Northern Climate,* Vols. 1, 2, 3. Report prepared for Electric Power Research Institute, EPRI EM-319, Project 544-1, January 1977.

11

RELIABILITY, MAINTENANCE, AND SERVICE

11.1. HEAT PUMP RELIABILITY

11.1.1. History of Reliability

Shortly after World War II, there were only 300–400 heat pumps installed throughout the United States for heating and cooling private homes and commercial buildings. This initial slow growth was due in part to high first cost, design problems, and insufficient incentives to overcome the problems of developing components and systems. It was not until 1953 that residential heating and cooling heat pumps experienced a significant increase in demand. Sales continued to increase between 1953 and 1959, even though failures began showing up in epidemic proportions shortly after their introduction. These machines were designed as "reversed air conditioners" without attention to the influence of specific working conditions, and the causes of the failures were not statistically analyzed. A change came about in 1959 when the U.S. Air Force stopped buying new equipment and started to analyze the performance of approximately 9000 heat pumps that were operating in several Air Force housing projects.

As a result of this study, which ended in 1965, a list of problems and potential solutions was compiled. Because they documented the specific design considerations that distinguish the reliable electrically driven heat pump from the respective air conditioners, these problems and solutions are abstracted in Table 11.1.

TABLE 11.1. Problems and Solutions in Designing a Reliable Electrically Driven Heat Pump

Problem	Solution
1. Flexing of the expansion-valve diaphragm by higher pressure differences.	Design of expansion valves for heat pump application. Mainly, the diaphragm should flex within the elastic limits.
2. Wider range of refrigerant mass flow rate corresponding to wider variation of load.	Properly sized (separate) liquid-refrigerant receiver.
3. Motor winding temperatures reaching levels dangerous to oil and refrigerant.	Internal protective relay that cuts off the motor.
4. High discharge temperatures reached when operating at very cold conditions.	Use of R-502 refrigerant, which operates at lower temperature than R-22, to reduce the danger of decomposition. The system can also be modified to ensure better subcooling and eliminate excessive superheating.
5. Short-cycling (start/stop cycle less than a minute) of a motor when switching for defrosting, e.g., overheats winding.	Installation of time-delay switch.
6. Moisture in the refrigerant increases the decomposition rate.	Necessity of detecting the moisture level below 25 ppm.
7. Activity of refrigerant driers.	Maintenance is emphasized.
8. Reduced air flow across the indoor coil, causing icing of the evaporator.	Frequent inspection and maintenance of air filters.
9. Refrigerant-circuit wiring becoming grounded to the unit, and corrosion of an electrical component, causing it to energize other equipment, making the unit unsafe.	Periodic checking of equipment grounds to control the hazard.
10. Training, service procedures, and specialized tools.	Mechanics should receive special training as well as special equipment and tools. Proficiency in ordinary refrigeration systems is inadequate.
11. Susceptibility of systems to contamination by air, moisture, and the products of oil and refrigerant breakdown.	Flushing the system with R-11 or using a filter-drier in the suction line.
12. Obtaining suction and discharge pressures with minimum loss of refrigerant.	Use of the Schraeder valve when testing the system's functioning.
13. Vibration at start-up, and running and stopping, especially when various components are out-of-tune put additional strain on piping connections.	Proper design of mounting springs.

SOURCE: Christian [11].

Although there has been much talk of heat pump failures, reliable industrywide records have not been available until recently. Alabama Power Co., which services most of that state, has been monitoring heat pump problems in their territory for the past 10 years. In January 1977 some of the findings of this 10-year program were published, and in a summary article the Supervisor of Residential Heating for the Alabama Power Company highlighted the fact that compressor failure (under a 5-year manufacturer's warranty) was not the only problem. He pointed out that a majority of the units covered under the program had withstood the test of time, but then called for improvements in fan motors [4]. It was pointed out that it now costs as much to change a fan motor in the second year from date of installation as it does a compressor—approximately $150. An overall defect frequency rate of three times the compressor failure rate led Alabama Power Company to believe improvements could be made.

In Alabama a report card is provided to the serviceman to check off the problems for each service call on a heat pump covered by this warranty program. There are 214 potential defects listed, including 17 compressor defects, 9 compressor motor defects, 36 refrigeration components, 17 refrigerant leak locations, 7 "other" components, 4 accessories, 4 miscellaneous service problems (charge, oil), and 23 installation faults.

The recap of service problems published in January 1977 for the 8521 heat pumps under the Alabama Power program shows that minor components accounted for a very small part of the trouble. For each manufacturer a percentage is listed, showing the frequency of each type of defect in ratio to that brand's total number of units, and a weighted average for defects in each category is given for all units covered. For the minor problems the weighted averages *per year* are

0.2%	General maintenance, dirty coil, etc.
0.6	Refrigerant and oil charge
1.5	Transformers, housing, drains
2.5	Thermostats, indoor and outdoor
4.4	Fan belts

Defects occurring in more than 5% of the units annually include

5.3%	Supplemental resistance heaters
5.9	Compressor failures
6.0	Internal compressor problems
8.0	Refrigerant components
8.1	Refrigerant leaks
9.9	Defrost circuit
15.5	Compressor motor
19.0	General fan problems

Not counting the compressor failures, equipment defects total 81.0% for the weighted averages. Those involving expensive components such as resistance heaters, internal compressors, compressor motors, and fans total 45.8, more than half the equipment problems during the early life of the heat pumps. The maximum range of serious problems is naturally greater than the weighted average, with individual brands showing such levels as 28.7% for fan problems, 14.7% for compressor failures, and 29.1% for compressor motors, per year.

The seriousness of this situation after 20 years of product development, field testing, and manufacturers' assurances certainly calls for a practical and realistic analysis of the underlying causes. The real problem is that the heat pump industry (manufacturers and utilities) in their anxiety to capture the residential heating trade have too frequently rushed products to market, with "make-do" remedies for the inherent conflicts with basic chemical and physical laws.

11.1.2. Compressor Failures

The compressor is the heart of the heat pump. The hermetic heat pump compressor consists of a cylindrical steel case containing a two-pole or multipole alternating current motor, whose rotor through a crankshaft and cam arrangement drives a one- to four-cylinder reciprocating compressor. Low-pressure cold refrigerant vapor at a temperature of -20 to $+50°F$ enters the welded hermetic case through the gas suction port, flows over the electric motor winding (thereby cooling it), and enters the cylinder through the intake valve at the base of the case. The vapor is adiabatically compressed from the average suction pressure of about 25–100 psia to 150–350 psia, thereby raising its temperature to about 75–140°F, and is discharged via a muffler through the discharge port. These data represent typical extreme values of the range of normal compressor operation. About one drop of lubricating oil circulates with each stroke of the piston to provide lubrication for the moving parts.

Since the cylinder clearance at the top of the compression stroke is, typically, only a few volume percent, entry of liquid refrigerant into the compression cylinder would subject the crankshaft, bearing, and valves of the compressor to violent mechanical shock. Compressor flooding or liquid slugging would, in addition, tend to dilute or completely wash away a lubricant (since the oil is soluble in the liquid) from the piston, crankshaft, bearings, and other parts subject to friction. Loss of refrigerant charge can thus be the cause of piston "freezing" or bearing failure, since lubricant circulation through the compressor and reversing valve depends on sufficient refrigerant flow.

Cole and Pietsch [1] identify three external operating conditions that place high stress on the mechanical and electrical parts of the compressor:

External Condition	Compressor Operation Condition	Effect
Heating at high outdoor temperature (above 65°F)	High suction, high discharge pressure; high refrigerant flow	High electrical input, high mechanical stress in moving parts
Cooling at low outdoor temperature, low charge, dirty filters, or otherwise restricted indoor air flow	Low suction, high discharge pressure; low refrigerant flow	High winding temperature, oil degradation
Operation at 20–40°F, freezing rain	Frequent cycling due to defrost; stops and starts at high head pressure	High mechanical wear and load on electric components

These conditions place higher stress on the equipment than operation at more normal conditions, that is, at the Air Conditioning and Refrigeration Institute (ARI) capacity rating points. ARI Standard 240 compressor capacity rating points are as follows:

Cooling (entering air): 80°F indoor db, 67°F indoor wb, 95°F outdoor db
Heating (entering air): 70°F indoor db, 45°F outdoor db, 43°F outdoor wb
Heating application rating: determined at 70°F indoor db, 20°F outdoor db, 19°F outdoor wb

Additional problem conditions include cold start, improper refrigerant charge, high or low voltage, power interruption, and system contamination. For example, when starting the compressor from cold, such as after a breakdown or long power outage, the compressor is usually the coldest point in the refrigerant circuit. Liquid refrigerant will therefore collect in the compressor base and flood it unless a crankcase heater is provided to keep the refrigerant vaporized when the compressor is not running. Operation with excessive refrigerant charge can mechanically and electrically overload the compressor; low charge produces inadequate compressor motor cooling and inadequate oil circulation. After field assembly of split-system heat pumps, contamination of the refrigerant circuit by water or air (due to incomplete evacuation before charging) sometimes occurs. It is difficult to keep dirt particles or metal shavings from entering the long runs of tubing during installation. At best, these conditions impose extra wear on the compressor, reversing valve, and other components; at worst, they can cause failure to function. Voltage variations due to surges, brownouts, or power interruptions can damage windings and other electrical components unless the system is properly designed or protected. It is clear, as one author has observed, that the hermetic heat pump compressor is given "ample cause to fail," unless adequately engineered to protect it (as by shutting down) against entering into abnormal operating modes that it is not designed to withstand.

These are some of the design obstacles: Compressors are made to compress gases; if liquid refrigerant enters, it creates problems. Compressors need oil for lubrication; refrigerants in the compressor mix readily with lubricating oil, carrying it off through the tubing and into the coils. In cold temperatures (20–40°F) the refrigerant separates out from the lubricating oil when it is in the outside coil, and the cold increases the viscosity of the oil. The oil tends to remain in the coil or the suction accumulator, rather than return to the compressor where it is needed.

Summer air-conditioning requires substantially more refrigerant than winter heating. The colder the weather, the less refrigerant is evaporated outside, creating greater risk of liquid floodback to the compressor. Liquid in the compressor increases the mechanical stresses on the internal components and tends to wash out the oil.

Compressors experience severe stresses when operating under high head and high suction pressures and also when high head pressure occurs with low suction pressure. Summer air conditioners operate well within the limits of low stress, but heat pumps in winter encounter severe stress when outside temperatures of about 25–40°F causes frosting of the outdoor coil, and in the range of 0° or below. Another stress zone occurs in mild weather when outside temperature is about 65°F, and the unit is on the heating cycle, such as in motels or apartments.

Alabama Power Co. has published various studies gathered from data collected under the company's Assured Service Heat Pump Program, which provides a service maintenance contract for up to 10 years.

A summary of data, reported by Lovvorn [2] appears in Table 11.2. Data on the equipment of three major manufacturers who carry a full line of heat pumps (i.e., packaged and split units) are combined under group I. Group II represents data on three manufacturers whose equipment is qualified packaged units only. The manufacturers whose equipment is shown in group III no longer qualify for the Alabama Power Service contract, a result not altogether surprising considering the high rate of compressor failure and the excessive service cost.

A comparison of compressor failures in split heat pump systems is shown in Fig. 11.1. Group I represents those manufacturers who, in September of 1974, had a minimum of 5 years of uninterrupted participation in the Alabama Power Company heat pump program and had at least 50 heat pumps installed. The manufacturers' equipment in group III had at one time qualified for the assured heat pump service program; however, because of

TABLE 11.2. Cost Summary and Compressor Failure Rate Data from the Alabama Power Service Program (May 1967 to September 30, 1974)

| Group | Type | Compressor Failure (%/Unit-Yr) | | Cost ($/yr) |
		In-Warranty	Out-of-Warranty	
I[a]	Packaged	4.0	5.8	22.41
	Split	4.1	3.6	26.03
II[b]	Packaged	3.8	5.3	26.75
III[c]	Packaged	9.6	12.5	39.97
	Split	10.5	10.1	50.27

[a] Grouped data for Carrier, General Electric, and Westinghouse equipment.
[b] Grouped data for Bard, Day & Night, and Fedders equipment.
[c] All others.

poor service experienced with the equipment, it could not continue to meet the program's qualifying standards.

Group I is more representative of those manufacturers who had attempted to identify the causes of the poor performance record of heat pumps in the past and had incorporated design considerations to improve the reliability.

Improvements that have been made in heat pump reliability have not been uniform. Failure rates of the compressor (the most important component of the system) have varied, according to manufacturer, from 1.6 to 23.1%, resulting in an average compressor failure of 8.5% for all nine manufactur-

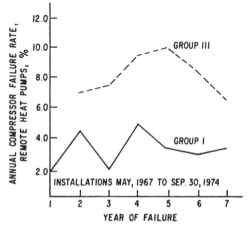

Fig. 11.1. Compressor failure rates for split heat pump system.

TABLE 11.3.   American Electric Power Maintenance and Service Costs and Compressor Failure Rate of In-Warranty Units

	1973		Five-Year Average	
Manufacturer	Cost per Ton-Year (1976 $)	Compressor Failure Rate per Unit-Year (%)	Cost per Ton-Year (1976 $)	Compressor Failure Rate per Unit-Year (%)
A	11.0	5.2	11.7	4.1
B	11.9	1.6	13.1	3.9
C	16.0	7.2	18.0	7.5
D	31.2	15.4	33.6	14.3
E	24.8	23.1	32.5	20.2
F	22.9	7.2	17.9	7.3
G	36.6	18.2	30.6	12.1
H	37.8	13.9	30.5	18.8
I	20.4	4.4	30.6	12.3
Average	18.5	8.4	20.3	8.6

SOURCE: Christian [11].

ers. Table 11.3 compares total in-warranty maintenance and service costs and compressor failure rates of the nine manufacturers with 100 or more units covered under American Electric Power Service's Protected Maintenance and Service Plan, for both 1973 and the 5-year average for January 1, 1969 through December 31, 1973. The AEP service area includes parts of Michigan, Indiana, Ohio, Kentucky, Tennessee, and West Virginia. Although a compressor failure rate of below 5%/yr may be considered acceptable, only a few manufacturers' equipment seems to meet that standard. An out-of-warranty compressor can cost up to $500 to replace.

Compressor failure, which constitutes on the average 28% of the total cost of in-warranty heat pump corrective maintenance, is of major importance in evaluating heat pump equipment reliability. By far the most costly single component of the heat pump, compressors that were not properly designed and tested for conditions under which they had to operate were, to a large degree, the early disappointing experience with heat pumps.

It is not difficult to see why the first heat pumps, which in most cases were built from standard air-conditioning equipment components, failed many times more frequently than air conditioners. Table 11.4 shows the estimated hours of cooling and heating operation for five cities of the United States from a paper by Cole [3]. The heat pump compressor, in most regions of the country, is clearly required to operate from two to eight times as many hours per year as a comparable central air conditioner.

R-22 is commonly used for heat pumps and, compared with R-502, is known to be a poor low-temperature refrigerant, involving higher compres-

TABLE 11.4. Regional Variation in Heat Pump Operating Hours

Location	Cooling Hours	Heating Hours	Total
New Orleans	1484	843	2327
Memphis	1259	2031	3290
Las Vegas	2431	1569	3910
St. Louis	1012	2988	4009
Chicago	577	3621	4198

sion ratios, higher discharge temperatures, higher motor winding temperatures, and lower capacity in low-temperature applications. Tests conducted in 1967 and 1968 on 100 heat pumps showed a substantial decrease in maintenance costs with R-502 (although there was some disagreement among the participating investigators). However, R-22 has better efficiency characteristics for the summer cooling cycle, and since many heat pumps are inferior to air conditioners in this respect this factor is an important competitive consideration, especially since R-22 also costs less.

11.1.3. Fan Motor Defect Rate

A report released in 1977 [4] by Alabama Power Co. indicated a fan motor defect rate three times that of compressors. The key words were "defect rate." The fan defect category included replacement, repair, or adjustment to either the indoor or outdoor motor and associated fan blade, blower wheel, capacitor, or relay. The heat pump units in this sample were covered by Alabama Power Company's Assured Heat Pump Service Program and installed between 1964 and 1978. This means that some current models are mixed with the older ones. The 13,000 heat pumps represented have accumulated 62,800 contract (unit) years, and thus we are confident that it is a representative sample.

The summary lists the manufacturers who have had more than 250 units covered under the program. Table 11.5 is arranged in ascending order according to the outdoor motor failure rate. It indicates a wide range of motor failure rates, particularly in the outdoor category. Hard questions must, therefore, be asked, because a fan motor failure represents the single largest cost for parts the customer must bear in the second through fifth years.

What is an acceptable failure rate?

Why are some rates three times as great as others?

Have reliability considerations become secondary to efficiency and/or cost considerations?

Have environmental conditions under which motors operate been adequately tested and/or prescribed?

TABLE 11.5.   Annual Motor Failure Rate (%)[a]

Manufacturer	Outdoor Motors	Indoor Motors
A	1	1.7
B	1.4	1.7
C	3	3.5
D	3.2	1
E	5.6	2.2
F	7.2	3.6
G	7.6	3.9
H	7.6	2
Total actual failures	5.2	2.6
	(3278)	(1623)

$$^a\% = \frac{\text{No. failures}}{\text{Contract (unit) yr}} \times 100$$

SOURCE: *Air Conditioning, Heating & Refrigeration News*, January 24, 1977.

The fan motor failure rates beyond the first year may be likened to compressor failure rates beyond the fifth year. Very few records are available for study on either, to our knowledge, outside of Alabama Power Company's program because of the inherent warranties of each.

Industry figures have indicated fan motor failure rates should be in the 1–2% range with no distinction between the indoor and outdoor. If this assumption were increased to include the 3% range, half the manufacturers included in this study would fall within the acceptable range on outdoor fan motors and all would be within the acceptable range for indoor fan motors.

Moisture and temperature protection are major factors that have been addressed by some manufacturers. They may use different types of bearings in addition to varying degrees of enclosure of the motor. Other factors in this complex problem include cost considerations that are evident in every business entity, not to mention the effort required to meet the higher efficiency requirements that have been mandated by the government.

AEP data indicated that equipment with direct-drive indoor blowers (most package units, newer split units) tended to be more reliable than equipment with belt drive blowers (earlier split models).

11.1.4.   Capillary Tubes, Expansion Valves, or Constant-Pressure Valves

There is continuing debate concerning the risks with expansion valves versus the lack of supervised control with capillary tubes. It is reported that superheat required with the former tends to cause oil breakdown and acid formation in the compressor.

The capillary tube, it is claimed, results in compressor flooding with less superheat at the lower ambient temperatures and provides better cooling of the compressor, although the valve manufacturers insist their product can do a better job if correctly applied.

Results of the AEP program indicated that expansion valves (which are mostly used in split units) required more servicing than capillary tubes (which are used in package units).

Another proposition is to use an evaporator pressure-regulating valve connected in parallel with a capillary tube feeding the indoor coil. Valves of this type are used as automatic expansion valves and are sometimes referred to as constant-pressure (CP) valves. In heat pumps using R-22, the valve is set to open in defrost cycles on a pressure drop to 40 psi. At this pressure and lower, the valve is open to provide free flow of liquid from the outdoor coil to the indoor coil immediately as it is condensed during the defrosting process. Proponents of this system view the use of an air-conditioner expansion valve as a device to meter the feed of liquid into the indoor coil during defrosting cycles as an abnormal and detrimental restriction and a hazard to the compressor.

As an example, in a heating cycle during 25°F outdoor ambient, the outdoor coil evaporating temperature is 12°F at 35 psi and the indoor coil is condensing at 100°F, 195 psi. When a defrost is triggered, the hot gas from the indoor coil surges through the reversing valve into the 12°F accumulator (where a portion is condensed), and the remainder is cooled on its way to the outdoor coil, where it gives up its remaining latent heat, melts some frost, and returns to liquid at its condensing pressure of 35 psi. This liquid is retained in the outdoor coil until the compressor reduces the indoor coil pressure sufficiently below 35 psi to cause the retained liquid to flow into it through the high resistance of the indoor coil expansion device designed for metering liquid to the indoor coil under 178 psi pressure during summer air-conditioning at 85°F outdoor ambient. The typical result is a sudden drop of pressure in the indoor coil, and in the compressor crankcase, to a 12-in. vacuum. A trickle of liquid is released into the hot indoor coil at −60°F. Within approximately 5 seconds after the defrost cycle is triggered, the crankcase oil is converted to a whipped cream–like foam, and it is pumped completely out of the compressor crankcase into the outdoor coil in approximately 23 seconds. At the same time all the liquid refrigerant in the system is pumped into the outdoor coil and retained there until it is slowly released in a trickle through the restrictive air-conditioning expansion device at the indoor coil. The defrost cycle time is abnormally extended by this undesirable flow restriction.

11.1.5. Defrost

When the heat pump is operating on the heating cycle so that the refrigerant is evaporating in the outdoor coil, and the temperature of the coil surface falls below 32°F, frost will begin to appear on the coil. If frosting is allowed to continue, the deposit of ice will gradually build up until the flow of air through the coil is restricted. This will decrease heat transfer and seriously affect the efficiency of the system. So periodic defrosting is needed to remove frost.

Defrosting is accomplished simply by reversing the cycle and directing hot gas to the outdoor coil for a period long enough to melt the ice. The outdoor fan is "off" during this action. No heat is being produced inside (actually the indoor coil is refrigerating), so it is necessary to provide supplementary heat from resistance heaters in the indoor section. Starting and stopping the defrost cycle is accomplished by one of several techniques. One method of automatically starting the defrost cycle is to measure the air pressure across the outside coil. As frost builds on the coil, it becomes more difficult for the air to find its way through the coil, so upstream pressure becomes higher than downstream pressure. When this difference in pressure reaches a predetermined point on the control, the reversing valve will be activated.

Another popular method of starting the defrost cycle uses a combination of time and temperature. A clock mechanism is set to reverse the refrigerant flow at predetermined intervals. However, if the coil temperature is above 32°F the defrost cycle will not start. The clock continues, and at the next scheduled time the temperature is again checked and defrost is started only if outside temperature conditions are sufficient to cause frost to form.

Several different methods have been used to stop the defrost cycle. One is based on time, another on temperature, and a third on pressure.

A *timing mechanism* may be activated when the defrost cycle starts, so that defrost will continue for a preset period. At the end of the time period the reversing valve again turns back to the heating cycle. The time period can be adjusted for winter conditions and an estimate of the maximum amount of frost that might be formed. It is sometimes difficult to set the timer to cover all frosting conditions and still keep the defrost cycle within reasonable limits.

A more positive method of terminating the defrost cycle is by *temperature*. While the frost is melting, the temperature in the vicinity of the outside coil will remain fairly constant, in the neighborhood of 32°F. When all the frost is gone, the temperature will begin to rise. This temperature change can be used to reverse the flow of refrigerant. If a temperature device is used to start the defrosting, the same thermocouple or thermal bulb can be used to stop the defrost cycle.

The *pressure* of the refrigerant in the outside coil can also be used to end the defrost cycle. As with temperature, the pressure of the refrigerant will stay nearly constant while the defrosting operation is going on. When the frost is all gone, the pressure of the refrigerant in the coil will rise, and this change can be used to signal the four-way valve to reverse.

The *air pressure differential* means of terminating the defrost cycle is, of course, used in conjunction with the air-pressure control that initiated the action. When the restriction is removed, the control will signal the reversing valve to return to heating.

11.1.6.   Accumulator Pros and Cons

At the termination of each defrost cycle, most of the system liquid refrigerant and the compressor lubricating oil is in the outdoor coil. At the beginning of the heating cycle immediately following, this liquid mixture is drawn in a surge into the intake port of the compressor. If a trap-type accumulator is not provided to intercept the liquid and meter its return into the compressor at a rate it can digest safely, there is a risk of severe compressor damage. In some instances, long-delayed oil return from the accumulator to the crankcase may result in compressor failure from lack of lubrication.

The use of accumulators has been, and now is, supported by compressor manufacturers as an expedient for preventing damage by liquid floodback into compressors at the beginning of heating cycles and during cold weather startups.

11.1.7.   Some Recommendations for Improved Reliability

Allan Trask in his July 1977 *ASHRAE Journal* article, "10 Design Principles for Heat Pumps" [5] presents a basic heat pump circuit *without* an accumulator, including the addition of an evaporator pressure-regulating valve connected in parallel with the capillary tube feeding the indoor coil. His ten design principles may be summarized as follows:

1. During defrost cycles, liquid condensed in the outdoor coil is immediately returned to the indoor coil. A floodback problem at the beginning of the heating cycles is eliminated, and the need for a trap-type accumulator is eliminated. Functioning under this principle may be effected by a CP valve connected in parallel with the indoor coil expansion device and set to open at 40 psi on a pressure drop for R-22.

2. The system refrigerant charge is maintained in active circulation at all times. Surplus refrigerant resulting from reduced evaporation in the outdoor coil during heating cycles is retained in the indoor coil, which then has a proportionately reduced need for condensing surface. A thermostatic expansion valve for the outdoor coil, set at 5°F superheat, will hold back the surplus in the indoor coil to accelerate defrosting.

3. At the beginning of defrost cycles, superheated flash gas flows directly through the reversing valve and the compressor to the outdoor coil. Construction under this principle ensures immediate delivery of maximum available heat for defrosting. It eliminates the detrimental

effect of the conventional trap-type accumulator on the defrosting process.

4. The compressor runs continuously below the balance point. This principle uses the full-system potential for extracting solar heat from the outdoor air. It eliminates startup problems after cold weather lockout. The continuously warm compressor eliminates the need for a crankcase heater. Excessive crankcase oil dilution is prevented by distillation as oil flows through the hot compressor crankshaft.

5. Heat pump compressors with piston rings are used instead of the type with selective piston-cylinder fits at a tolerance less than 0.001 in. Microscopic particles of core sand and copper sawdust are hidden hazards in compressors that have piston-cylinder fits around 0.0005 in. A particle wedging in the piston clearance when the motor torque is weak can lock the compressor permanently. Strainers for entrapment of particles smaller than the minimum piston fit tolerance are not practical, because the resistance to refrigerant flow would be excessive or the size too big.

6. Both coils have stacked circuits in a counterflow configuration. In defrost cycles, all the hot gas is introduced into the outdoor coil at the outside coil tube row upstream of the incoming air where most of the frost is collected. In vertical coils, adverse resistance by gravity to liquid flow will be limited to a harmless amount, and oil return under low gas flow density at low outdoor temperatures will be assured.

7. A heat-exchange manifold is used on the outdoor coil. An axial liquid line including a strainer evaporates spillover liquid. Suction gas from the coil enters the manifold at its lower end and flows upward through any liquid accumulation to atomize it for entrainment in suction gas flowing to the compressor. In cold startups, liquid slugging into the compressor is prevented. During high compression ratios, discharge temperatures are held down.

8. A surge tank is provided at the cooling cycle outlet of the indoor coil. Suction gas enters at its lower end, as in the outdoor coil manifold, to atomize spillover liquid in the same way in the event of a clogged filter or reduced blower efficiency.

9. Two tubes at the bottom front row of the outdoor coil are connected in series between the expansion devices of each coil. During heating cycles the typical problem of frost collection at the base of the outdoor coil is permanently eliminated. Its subcooling during cooling cycles is an advantage.

10. Defrost cycles are initiated and terminated by one pressure switch coordinated with an electric clock timer determining the intervals through a relay. This is the simplest of all defrost systems. It uses

only three standard components concealed from and unaffected by all external conditions. Defrost cycles are completely and reliably controlled by the pressure switch at the intervals set on the timer.

11.1.8.   Conclusions on Reliability from the Gordian Study

Analysis of heat pump reliability in the Gordian Study [6] was based on data principally from the AEP and Alabama Power Co. service program records and to a lesser degree on discussions with manufacturers and heating and air-conditioning contractors. The study was confined mainly to unitary single-phase equipment with cooling capacity of 5 tons and below, since that is the kind commonly installed in single-family residences. Based on these investigations, the conclusions were as follows:

1. If a 5% or lower yearly compressor failure rate is taken as acceptable, only a few manufacturers' unitary heat pump equipment commercially available today [note dates of Alabama Power and AEP data] will consistently meet that standard. Failure rates of other manufacturers' equipment are still intolerably high.

2. The available heat pump reliability data from two utility companies' service records show no evidence of a significant difference in compressor failure rate between warmer and colder climatic regions, although, in theory, such a difference may be presumed to exist. (Equipment in colder climates appears to require more frequent service, however.)

3. Next to quality equipment, proper application and installation have been observed to play a significant role in heat pump reliability. In the past, local utilities have exercised control over the quality of installation in only a few areas, although more are doing so today.

4. If all heat pump manufacturers would stand by the quality of their equipment by offering a 5-year full materials and labor service contract (renewable to 10 years) at competitive prices through their distributors or dealers, it is believed that inferior equipment would eventually eliminate itself from the market.

5. As far as is known, in no state are heating and air-conditioning contractors licensed. Thus there exists not even minimal control over the level of proficiency of personnel installing and servicing residential comfort-conditioning systems.

6. While sufficient instructions and guidelines are available from heat pump manufacturers and professional and trade organizations serving the industry to ensure proper application and installation of heat pumps, there appears to be no industry-accepted code of standards covering heat pump installations for use as a model by the various states.

In summary, evaluation of the data available on heat pump reliability and discussions with people from various sectors of the comfort-conditioning industry suggest that, while there has been substantial improvement in the reliability of heat pump equipment marketed, not all manufacturers have been equally successful in this regard. A concern which has been voiced repeatedly on all sides is that increased consumer and utility interest in this form of space heating may induce manufacturers with no real long-term stake in the future of the heat pump industry to reenter the market. This process could well result in a repeat of the disappointing heat pump experiences of the 1960s. Manufacturers have learned that producing a reliable heat pump requires a thoroughly researched and engineered design, tight control over the quality of manufacturing, and the ability to back equipment with proper and responsible service.

11.1.9. Recommendations on Reliability from the Westinghouse/EPRI Study

Air-to-air heat pumps were studied for three utility areas [7]. The impact of storage and solar augmentation were investigated. The proposed units for the late 1980's were configured conceptually and analyzed for their benefits and impacts on consumers, electric utilities, and the manufacturing industry. COP improvements of 30–40% over 1975 models were projected as well as life-cycle cost reductions of 9–14%. Table 11.6 summarizes the recommended features for reliability and performance enhancement.

In the EPRI Report [7] a detailed failure mode and failure cost analysis for heat pump systems was also carried out using information from the American Electric Power warranty data. This analysis led to the establishment of the rank-ordered list of heat pump problems shown in Table 11.7. Service cost, which is the product of two data elements, service calls per year and cost per service call, serves as the basis for the ranking. Note that these data are in average dollars for the period 1969–1973. Further, note that the costs reflect only those borne by the utility during the manufacturer's warranty period.

Available failure data contain no cause information nor can service calls obtain such information. A study of failure mode scenarios tends to blame the reverse-cycle defrost technique, which can lead to compressor failure through inadequate lubrication at the initiation and termination of each defrost cycle. The high frequency of refrigerant leaks indicates that inadequate reliability assurance is imposed by the industry in general. It can be reasonably inferred from the reliability analysis that a somewhat higher manufacturer cost devoted to system reliability in design and to quality assurance will lower consumer life-cycle cost.

In this study an analysis was performed for conventional systems also. The results are given in Table 11.8.

TABLE 11.6. Recommended Features for Reliability and Performance Enhancement

Feature	Function	Reliability Impact	Performance Impact	Cost/Benefit Comment
1. Suction Liquid Accumulator	1. Store liquid inventory not provided for elsewhere in the system. 2. Intercept liquid flow before ingress to compressor shell.	1. Prevent ingestion of liquid into compressor cylinder (mechanical damage). 2. Prevent dilution of compressor sump oil with refrigerant to inadequate viscosity.	Supplies saturated suction vapor rather than superheated vapor to the compressor, a volumetric efficiency advantage.	Absolute requirement but cost nonetheless is tolerable. Minimum effort needed to apply.
2. Oil-Stripping Heat Exchanger in Accumulator (subcooling high side liquid)	Provide return of high viscosity oil to compressor in most profitable thermodynamic way.	Preserves oil viscosity maximization by stripping when suction gas cooling of motor and oil is eliminated for performance improvement.	Use of liquid subcooling heat for stripping is thermodynamically superior to use of any other heat source in system.	Cost justified by COP improvement over stripping with motor heat. Minimum effort required.
3. Compressor, all Contactors and Reversing Valve Installed in Return Air Plenum (preferred indoors under roof)	Prevent distillation of refrigerant to compressor oil sump in off cycle. Prevent moisture or frost deposit in contactors. Permit comfortable service of major part of system in bad weather.	Insurance against adverse impact rather than demonstrated positive impact is to be expected. Cold start allowable.	Any heat dissipation from compressor shell will aid in supplying load and therefore improve COPH.	The magnitude of the benefits expected is not clear because compressor failures cannot be analyzed beyond "lubrication failure" level. Cost is amortization of design and manufacture change plus some additional (possibly) acoustic treatment of the compressor can. Overall a small cost feature.

627

TABLE 11.6. (Continued)

Feature	Function	Reliability Impact	Performance Impact	Cost/Benefit Comment
4a. Defrost by Cycle Reversal which Bypasses Refrigerant Flow Control (Alternate)	By removing liquid throttle from circuit, permits indoor coil to be fed without violent pump-down of low side. Bypass may be a solenoid or a (pressure) Automatic Expansion Valve or a Thermostatic (superheat) Expansion Valve.	Suppresses exposure of compressor to rapid pump-down and low mass-flow high pressure ratio operation. Shortens defrost cycle because indoor coil is now efficiently fed and loaded.	Since defrost cycle time will be less than with conventional system, it should improve seasonal performance factor.	Probably cheaper than I^2R alternate but neither is expensive. Appears to be as free of compressor stress as I^2R alternate but still requires 4-way valve automation. Automatic Expansion Valve bypass preferred over solenoid valve since the outdoor coil temperature can then be raised above indoor temperature.
4b. Defrost with Resistance Heat (Alternate)	Defrost a horizontal face coil with gravity warm air from a resistance heater below.	Suppresses cycle reversal transient effects on compressor. Insures against ice residual buildup. Permits manual cycle reversal (seasonal). Removes reliance on solenoid 4-way valve repeated functioning.	Since I^2R heaters supply defrosting energy on either this or cycle reversal system, performance penalty should be negligible provided defrosting cycle time is short and electrical demand is not penalized.	Efficiency is comparable to cycle reversal. Cost trades off cost of air heater against manual cycle reversal saving. Coil and housing costs unchanged.

5. Short Cycle Protection	Alternatives: a. Minimum time interval between automatic resets plus visual/audible alarm. b. Manual reset interlock plus visual/audible alarm when a faulted condition causes shutdown.	a. Minimize risk to windings. b. Force user intervention plus (a).	Minimize unwanted I^2R usage.	Modest cost increase
6. Hi-Pressure Cut-Out Switch	Prevent compressor operation at excessive discharge pressure.	Prevent bearing failures, piston scuffing, valve distortion. Faster acting than relying upon motor thermal protection for shutdown.	None	None
7. Thermal Tagged Coil Bypass Flow Sensor for Defrost Initiation	Senses temperature rise of constant power tagged flow in outdoor coil air bypass tube.	Provides high gain sensing—as much as 25°F temperature change at critical frost loading. Stable to gusts.	Since unnecessary defrosts will be eliminated, seasonal performance factor will be improved.	In developed production will cost no more than current barometric switch-type demand defrost sensor.
8. Pressure Switch Defrost Terminator	Senses high side pressure; terminates defrosting at saturation pressure of desired outdoor coil temperature, say 60°F.	Pressure is more reliably and precisely sensed than coil surface temperature.	Since defrost cycle will not be prematurely terminated, the number of defrost cycles required should be reduced, and therefore the seasonal performance should be improved.	No cost effect.

TABLE 11.6. (Continued)

	Feature	Function	Reliability Impact	Performance Impact	Cost/Benefit Comment
9.	Refrigerant Flow Control Comprising Two Capillaries in Series, Check Valves and Refrigerant Charge at Critical Level	Heating capillary cut to obtain saturated inlet at say −20°F evaporating. Charge set so that all liquid is in high side at say 30°F ambient and subcooling about 25°. Cooling capillary cut to yield 20° subcooled inlet at 40° evaporating.	Capillaries and check valves are intrinsically more reliable than control valves.	Heating performance is optimized by increasing subcooling with ambient temperature up to about 25–30°. The charge limit is designed to prevent further loading of high side with liquid after this level is attained.	Cost is lower than valve systems.
10.	Outside Fan Motor Off-Cycle Heated with Auxiliary Winding	Auxiliary winding and common lead are wired to lines while main winding is wired through main motor contactor. Defrost switch fan contact is in common lead. Auxiliary winding remains energized on off-cycle drawing ~10–20 watts warming power but is opened on defrost.	Outdoor fan motor is protected from off-cycle moisture condensation and grease freeze-up most likely causes of motor failure.	Electrical consumption increase is negligible.	No cost but fuse(s). May eliminate need for ball bearings.
11.	Totally enclosed outdoor fan motor	Ultimate environmental protection.	Keeps moisture and dirt out of motor.	None	Modest cost increase.

12. Inside Blower Motor. Belted, adjustable pulley.	System air flow continuously adjustable by pulley adjustment.	Easily replaceable standard motor. Belt drive is common in hot air furnaces and reliability is not a problem.	Air flow can be more precisely tailored for COP maximization.	Not Applicable.
13. Capacity Modulation via Gear Set Speed Change	To increase compressor displacement in cold weather and thereby permit reduction in I^2R usage.	Fewer contacts required. Gear reliability must be demonstrated post-design.	Better seasonal performance winter and summer.	Should be cheaper than either two compressors or two-speed motor. Gear can be speed increaser in high, reducing compressor size and cost.
14. Volumetric Efficiency Improvement	Reduce compressor system cost and improve capacity at low temperature evaporating.	Properly designed there will be no negative impact.	Improvement	Clearance ratio reduction, discharge valve and flow coefficient improvement, wall ports are best measures.

SOURCE: Kirschbaum and Veyo [7].

TABLE 11.7. Rank Ordering of Current Heat Pump Problems Based upon AEP In-Warranty Cumulative Data[a]

Component Problem	Annual Cost	% Total Cost
1. Compressor failures	$12.27	26.9
2. Refrigerant leaks	5.98	13.1
3. Outdoor fan motor	3.88	8.5
4. Indoor fan motor	2.00	4.4
5. Defrost	1.87	4.1
6. Capacitors, compressor	1.85	4.1
7. Reversing valves	1.35	3.0
8. Expansion valves	1.09	2.4
9. Others	15.35	33.5
	$45.64	100.0%

SOURCE: Kirschbaum and Veyo [7].

TABLE 11.8. Reliability and Service Cost Data for Conventional Residential Heating and Cooling Systems (Combined Heating and Cooling Components)

System Type	Service Call Rate (calls per unit per year)	Avg Annual Service Costs ($/year, 1973 $)
1. Gas furnace/gas absorption A/C	1.26	$54[a]
2. Gas furnace/vapor compression A/C	1.48	64[b]
3. Oil furnace/gas absorption A/C	2.34	81
4. Oil furnace/vapor compression A/C	2.56	95
5. Heat pump	1.36	73[c]
Heating-Only Components		
1. Gas furnace	0.40	$8.00
2. Oil furnace	1.48	35.50
3. Heat pump	1.36	72.89[c]

[a]Based on service records from a heating/cooling contractor for 48 systems with 1.5–2 yr service.

[b]Based upon detailed utility service records covering over 10,000 units for six major manufacturers for 1972 and 1973.

[c]Based upon 1973 Duquesne Light Data for combined in-warranty and out-of-warranty units.

SOURCE: Kirschbaum and Veyo [7].

11.1.10.   Results on Acceptance of Heat Pumps

In 1975 the Electric Power Research Institute and the Association of Edison Illuminating Companies began work on a jointly sponsored project, RP-432, designed to investigate the load and energy use characteristics of the electric heat pumps in single-family housing units [8].

The main objective of the study was to obtain information on the heat pump system pertinent to single-family residential housing units, which will allow for the identification of major contributors to energy use and their relationship to that use. Westinghouse Electric Corporation was the prime contractor for this project.

The sample consists of 118 single-family housing units with heat pumps drawn uniformly from 12 participating utilities. The 12 utilities represent a heating degree day spectrum ranging from 250 to 8250 degree days. Data were collected from approximately November 1975 to June 1977.

In conjunction with the monitoring and collection of physical data, each participant was required to be a respondent in an attitude and opinion survey conducted at the end of the study. The purpose of the survey was to identify and categorize the more subjective factors that contribute to heat pump acceptability and use.

In the sample, 80.2% of the participants found the heat pump system to be satisfactory. Of those, 78.2% preferred the heat pump to their previous system, 8.9% judged the heat pump and their previous system to be the same, and 6.4%, although satisfied with the heat pump, felt it was inferior to their previous system. For the remaining 6.5% of satisfied participants, the present heat pump was the first system owned.

In Table 11.9 the heat pump likes of the participants are tabulated. In Table 11.10 the heat pump dislikes of the participants have been tabulated. In addition, the test participants were asked what changes were needed to enhance heat pump acceptability; the results are given in Table 11.11. When commenting on what could be done to increase the acceptability of the heat pump system, 30.8% of the participants felt that no changes were necessary. When modifications were thought to be warranted, 14.8% mentioned the reducing of operating costs; compressor noise and equipment quality were each concerns of 10.9% of the participants.

There did not appear to be any significant correlation between attitudes toward system operation and the number of degree days in their location.

In this same study by EPRI [8], maintenance and repair data were collected during the test.

The 120 heat pump installations involved 47 different models of heat pumps from 12 different manufacturers. A summary of maintenance and repair records kept to assist in review of the data is presented in Table 11.12 to provide additional perspective to the characterization of the set of heat pumps tested.

In most cases service was performed by private contractor, and the

TABLE 11.9.   Heat Pump Likes.

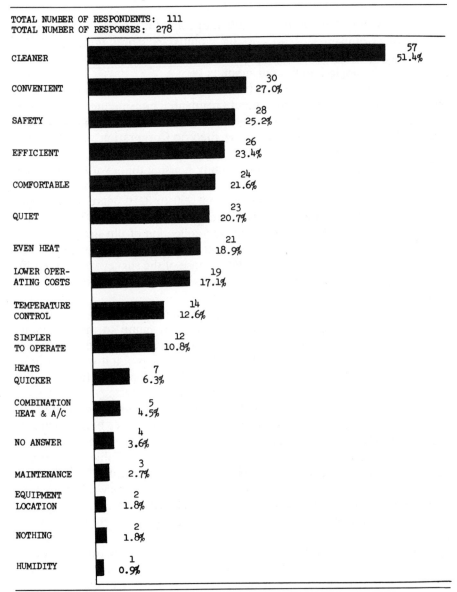

TOTAL NUMBER OF RESPONDENTS: 111
TOTAL NUMBER OF RESPONSES: 278

CLEANER	57	51.4%
CONVENIENT	30	27.0%
SAFETY	28	25.2%
EFFICIENT	26	23.4%
COMFORTABLE	24	21.6%
QUIET	23	20.7%
EVEN HEAT	21	18.9%
LOWER OPER-ATING COSTS	19	17.1%
TEMPERATURE CONTROL	14	12.6%
SIMPLER TO OPERATE	12	10.8%
HEATS QUICKER	7	6.3%
COMBINATION HEAT & A/C	5	4.5%
NO ANSWER	4	3.6%
MAINTENANCE	3	2.7%
EQUIPMENT LOCATION	2	1.8%
NOTHING	2	1.8%
HUMIDITY	1	0.9%

SOURCE: EPRI Report [8].

634

TABLE 11.10.   Heat Pump Dislikes.

TOTAL NUMBER OF RESPONDENTS: 111
TOTAL NUMBER OF RESPONSES: 160

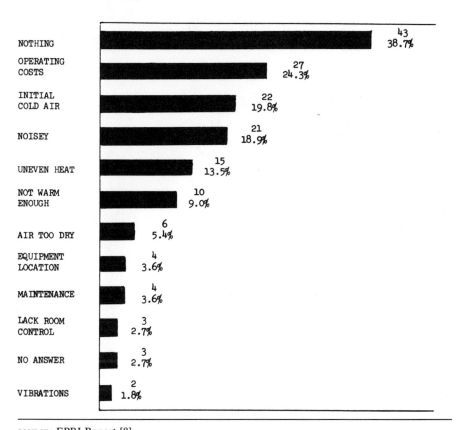

SOURCE: EPRI Report [8].

identification of the problem and description of repair performed reflect the specific experience level of the contractor. In most cases these service contractors had not installed the heat pumps at the particular houses involved in the test.

The test procedure called for an inspection of each heat pump installation prior to the start of the test and performance of any necessary maintenance at that time. Subsequent discovery of problems that must have existed at the start of the test indicates that inspections were done with different levels of thoroughness.

Some of the problems that arose during the test may have been associated with the test itself. In particular, failures of the supplementary resistance heaters or controls were more easily identified in houses that bypassed the heat pump on alternate days. No clear causal relation between the test

TABLE 11.11.   Changes to Enhance Heat Pump Acceptability.

TOTAL NUMBER OF RESPONDENTS: 111
TOTAL NUMBER OF RESPONSES: 118

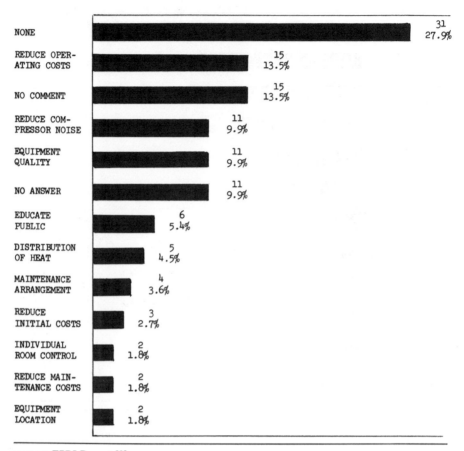

SOURCE: EPRI Report [8].

procedures and any particular equipment problems was identified, but the closer attention to performance did result in earlier identification of problems.

The data summarized in Table 11.12 represent the repair information covering the 18 months over which most of the equipment was under test. The data indicate that 54 of the 99 heat pumps for which information was received had some problem requiring attention. Of these, 21 had periodic maintenance, where many of the problems were identified and fixed. In the other 33, repair calls apparently were initiated by the customer. Not counting the periodic maintenance calls, there were approximately 1.3 service

TABLE 11.12. Repair Summary Based on Records from 99 Heat Pumps, October 1975 through March 1977[a]

Type of Problem		Number of Calls	No. of Heat Pumps Involved
Supplemental resistance heating			
Wiring and controls	19		
Outdoor thermostat	6	25	19
Heat pump controls and wiring		22	18
Compressor problems: Failures	7		
Others	12	19	17
Fans (indoor and outdoor)		16	11
Defrost controls or performance		10	8
Refrigerant leaks		7	7
Indoor thermostat		4	4
Valves (reversing, expansion, check)		4	4
Repairs total		104	54[b]
Periodic maintenance, no repairs		35	13
No maintenance, no repairs		0	32
Repairs and maintenance total		139	99

[a] Repair records were not available for 21 of the 120 test units.

[b] Different units.

SOURCE: EPRI Report [8].

calls per heat pump per 12-month period, with those requiring service averaging about two repairs each.

Because the test altered the normal repair patterns and personnel, heat pump operation, and customer awareness, it is difficult to assess typical heat pump experience from these data. Different data based on utility surveys indicated approximately 1.4 service calls per heat pump per year and 1.1 service calls per air conditioner per year. It appears the test sample was not grossly different from what one might expect from a larger sample, with respect to service calls.

The number of maintenance calls was larger than normal, primarily because several utilities established a quarterly maintenance schedule for this test. While good for keeping the equipment running throughout the test, this makes it difficult to extrapolate the number of maintenance calls to the heat pumps in general. For this test there were at least 78 periodic maintenance calls. Of these, 43 were at 21 houses that also recorded repairs; 35 were at 13 houses that did not record any repairs. Thirty-two houses recorded neither repairs nor periodic maintenance inspections.

Of the participants in the study, 71.0% felt that maintenance was not a problem, while 9.7% said that maintenance posed a minor problem. Only

19.3% felt that maintenance was a major problem. Two factors were strong contributors toward this response:

1. Of the sample, 89.0% received adequate service when maintenance was needed.
2. The expected number of annual service calls is approximately 3, while the equilibrium point between a feeling of no problem with maintenance and a problem with maintenance, irrespective of its magnitude, is approximately 5 service calls per year.

Routine maintenance of the system is performed by 85.6% of the sample. All maintenance is performed by 3.6%, while 10.8% do no maintenance at all. These maintenance statistics account for the fact that when the participants were asked what amount per month they would be willing to pay for a maintenance contract, 40.1% showed no interest and 28.0% did not have enough information to evaluate in terms of dollars. Of the remaining 31.9% willing to pay for a maintenance contract, 35.3% would be willing to pay $5–9 per month, while 35.3% would be willing to pay $10–14 per month. Only 5.8% said that they would be willing to pay more than $14 per month.

There did not appear to be any significant correlation between expected system maintenance and the number of degree days at that location.

11.2. MAINTENANCE AND SERVICE

A good operating heat pump which can be expected to have minimum maintenance costs should have the following five features:

1. Every heat pump should have an all-temperature compressor. Conventional air-conditioning systems normally operate down to an ambient temperature of about 45–50°F. However, because a heat pump must operate in temperatures below zero, it requires a more rugged and better-built compressor than those used for conventional air-conditioning.

2. During the heating season, a heat pump compressor located outdoors is subject to very low ambient temperatures, thereby causing rapid refrigerant migration, which can load the compressor with liquid refrigerant and subsequently destroy it. To vaporize this liquid refrigerant, the crankcase heater should be energized whenever the unit is not operating.

3. "Highside and lowside cutout" is a safety feature used to cut the compressor off if the highside pressure becomes excessive or in the event that lowside suction pressure reaches the critical point.

4. The suction-line accumulator stores excess liquid refrigerant that is not needed during the heating cycle; that is, the indoor coil is being used as a condenser and this coil will contain liquid refrigerant. If the unit is

suddenly switched to the cooling cycle, the outdoor coil becomes the condenser and the liquid in the indoor coil will start flowing toward the compressor suction. If there is more liquid in the coil than can be evaporated before the liquid reaches the suction line, there is a good possibility of some liquid flooding to the compressor if such a changeover occurs during defrost cycles. The suction-line accumulator catches any liquid that might escape from either coil on changeover or startup. The importance of the accumulator is revealed by the American Electric Power study of heat pump failures, which found that more heat pump compressors fail in March, April, and May than in other months of the year, indicating that stress is placed on the compressor during switchover.

5. The auxiliary heater should be controlled by a two-stage thermostat that switches on the next supplemental heater increment only if the compressor alone cannot actually maintain the preselected inside temperature.

An adequate air supply over the indoor and outdoor coils is mandatory. Ducts must be large enough to provide a minimum of 400 cfm/ton. Any restriction of the indoor air flow will cause high operating costs and may damage the equipment. In more northerly climates it is advisable to elevate the outdoor unit above the ground to prevent snow buildup in front of the outdoor coil.

While the result of misapplied furnace equipment is often no more than a lack of comfort in the conditioned space and inefficient operation, improper heat pump application and installation may precipitate equipment failure.

Since the output of the heat pump, unlike that of a gas or oil furnace or electric resistance heat, decreases as the outdoor temperature drops, proper sizing of the heat pump requires more sophisticated load calculation and duct design than the other systems do. This presents somewhat of a problem, as not all manufacturers of heat pump equipment provide the training to dealer personnel and the explicit and detailed guidelines that such application requires.

To the extent that some manufacturers fail to provide adequate training and technical guidelines and instructions to dealers that install and service their equipment, and to the extent that some distributors have not exercised proper care in assuring themselves of the qualifications of their local dealers, equipment problems will continue to be compounded by poor field practices.

A list of typical equipment problems caused by faulty applications is given in Table 11.13. Two cases of faulty installation of residential heat pumps have been described in a report by Hise [9]. While Hise does not claim that these cases are indicative of installation practices in general, they do document how failure to follow manufacturer's specifications contributed to repeated compressor and other component failures and lowered the operating efficiency of the equipment. It is of interest to note that, according to Hise,

TABLE 11.13. Typical Results of Heat Pump Misapplication

Type of Misapplication	Result
1. Oversized equipment	1. Frequent cycling, high wear, loss of temperature and humidity control
2. Inadequately sized ductwork	2. Low indoor air flow, high compressor suction and discharge pressure, high pumping rates and compressor failure
3. Undersized power wiring, especially for strip heat	3. Frequent fuse failure, fire safety hazard
4. Oversized liquid refrigerant tubing	4. Liquid slugging, compressor mechanical failure

SOURCE: Hise [9].

gas or oil forced air heating systems, while not immune from similar installation problems, would not necessarily have failed in the same circumstances, because of basic differences in these systems.

These and other case histories imply that (1) proper heat pump application and installation requires a higher degree of technical sophistication and know-how than the installation of other heating systems, (2) a heat pump is less forgiving of installation mistakes than other comfort conditioning equipment, and (3) a mechanism for ensuring that a heat pump installation conforms to good practice is necessary, but now largely nonexistent. At best, improper application can result in operation at lower-than-optimal efficiency; at worst, in repeated and expensive heating system failures.

Periodic checks of all heat pumps should include

1. Equipment grounds to control the hazard of refrigerant-circuit wiring becoming grounded to the unit, and corrosion of an electrical component, causing it to energize other equipment and make the unit unsafe.

2. Performance of the refrigerant circuit—by measuring the air temperature differential across the indoor coil and the current being drawn by the compressor.

3. Filters for cleaning.

Heat pumps are quite different from air conditioners and perform different and additional functions. Each manufacturer has his own theories and solutions, necessitating great versatility on the part of the service personnel. For example, such simple service problems as dirty filters and caked-up coils sharply reduce the air flow and cause problems. If the filter is not changed often enough the CFM will decrease and the air will feel warmer.

Troubleshooting on the job requires many deviations from standard practice. Recharging, for instance, is normally done with pressure gauges, but this is dangerous for heat pumps as it is not sufficiently accurate. The charge must be weighed instead, to avoid early compressor failure.

Moisture in the heat pump cycle can be damaging during winter operation, and recommendations have been made to service personnel to use "supersensitive" moisture detectors instead of common ones, as the latter only read down to 25 parts per million. After servicing, a new filter-drier should be installed, and the service person urged to return a few weeks later to put in a fresh one.

It appears to be the practice in the industry for heat pump manufacturers to warrant their compressors for 5 years and other heat pump components for one year from the date of purchase. The warranty covers defective materials, but not labor, refrigerant, or air filters. In many cases, the installing dealer or distributor will assume the cost of the first year's labor for preventive and corrective maintenance; however, it is not known if this practice is followed everywhere, since marketing and service policies and organization vary substantially from one location or manufacturer to another.

The American Electric Power System was the first to offer the heat pump purchaser a 5-year service and maintenance plan in 1962. The concepts and operational experience of the program have been described in a number of articles and will not be repeated here, except to indicate certain similarities to and divergences from the Alabama Power Co.'s service program initiated a few years later.

Table 11.14 summarizes the major contract provisions of the two programs, which are quite similar [10]. The contracts were sold through the utility company, which entered into the service agreement with the homeowner. The dealer performed the services as required by the contract and submitted vouchers for materials and labor expenses to the utility company for reimbursement. The servicer's diagnosis of the defect(s) found and corrective action(s) taken, if any, were appropriately coded and stored for computer retrieval and for the compilation of failure statistics. In this manner, both programs retained a detailed history of each heat pump under contract.

Certain important differences in the operation of the two programs will now be pointed out. First, when the AEP program was initiated, certain older equipment in the operating companies' service areas was accepted for the program. The AEP program accepted equipment of all manufacturers, as did the Alabama Power program initially. However, through its certification of dealers and equipment, the latter began removing from its program equipment that consistently proved to be unreliable, due to either poor quality of

TABLE 11.14   Summary of the Provisions of a Typical Heat Pump Service Contract

Items and Services Covered by Contract
1. Corrective maintenance parts and labor, including refrigerant, oil and supplies not specifically excluded, and service to keep equipment in good operating condition
2. User instruction including necessary intervals for changing air filters
3. Service during normal working hours, unless the standby electrical resistance heating equipment fails to operate during the heating season

Items and Services Excluded from Contract
1. Work required because of improper operation, negligence, or misuse of equipment; nuisance items such as improper thermostat settings or dirty air filters
2. Repairs or maintenance or responsibility for design of ducts, registers, external wiring other than unit control or power wiring foundations or supporting structures
3. Deterioration or corrosion of cabinets or compartments housing the unit, or of the condensate drain pan
4. Equipment relocation
5. Fire, water, storm damage, acts of God

Other Provisions
1. Contract and contract price renewable annually (AEP) or annually after the fifth year and thereafter (Alabama Power)
2. Contract may be terminated by owner or servicers as of any anniversary of the original installation date. Servicers must give 60 days written notice to the owner.

SOURCE: AEP [10].

the product itself or inadequate installation and service backup by the manufacturer's authorized dealers. The AEP program had no provisions for removing poor quality equipment from its service program. It is of interest to note that the Alabama Power Co. now allows the purchaser of a heat pump to enter the program within one year of purchase. In fact, they discouraged the purchase of the service contract immediately after equipment installation, thus avoiding the higher cost of correcting early failures.

Although the AEP contract was renewable annually, they would allow units that dropped out of the program to reenter it. This had the effect of increasing the number of poorer units under contract, whereas the units that operated without problems for the first few years tended to drop out of the program. The Alabama Power program did not readmit "dropouts." While both programs had specifications and guidelines governing equipment application, installation, service, and operation, Alabama Power, perhaps benefiting from the experience of AEP, was, it appears, more diligent in inspecting installations before accepting them under contract.

REFERENCES

1. Cole, M. H. and Pietsch, J. A., "Qualification of Heat Pump Design," in *Heat Pumps— Application and Reliability* (*ASHRAE Symp. Papers*, June 25–29, 1972, Nassau, Bahamas), p. 12.

2. Lovvorn, N. C., "Heat Pump Compressor Reliability," *Air Conditioning, Heating Refrigeration News,* Jan. 27, 1975.

3. Cole, M. H., "Heat Pumps Today—Part 1," in *Heat Pumps—Improved Design and Performance (ASHRAE Symp. Papers, Jan 19–22, 1970, San Francisco, CA)*, p. 41.

4. *Air Conditioning News,* Jan. 24, 1977.

5. Trask, A., "10 Design Principles for Heat Pumps," *ASHRAE J.* **19,** No. 7, July 1977.

6. Gordian Associates, Inc., "Evaluation of the Air-to-Air Heat Pump for Residential Space Conditioning," prepared for the Federal Energy Administration, April 1976.

7. Kirschbaum, H. S., and Veyo, S. E., "An Investigation of Methods to Improve Heat Pump Performance and Reliability in a Northern Climate," Westinghouse Corp., EPRI Report NU EM-319 RP 544-1, Vols. I, II, and III, January 1977.

8. Blake, P. S., and Gernert, W. C., "Load and Use Characteristics of Electric Heat Pumps in Single Family Residences," Westinghouse Corp., EPRI Report EA-793, RP 432-1, Vols. I and II, June 1978.

9. Hise, E. G., "Seasonal Fuel Utilization Efficiency of Residential Heating Systems," Oak Ridge National Laboratory Report ORNL-NSF-EP-62, April 1975.

10. American Electric Power Service Corp., "Heat Pump and Comfort Cooling Units, Protected Maintenance and Service Plan—Owner-Servicer Agreement," from ACD-14, revised August 1969.

11. Christian, J. E., "Unitary Air-to-Air Heat Pumps," ANL/CES/TE 77-10, Oak Ridge National Laboratory for Argonne National Laboratory, July 1977.

12

ADVANCES IN
HEAT PUMPS

12.1. HEAT PUMP IMPROVEMENTS

At each of the annual ASHRAE-ARI expositions since 1977, heat pumps have been strong in presence—including air-source, water-source ("hydronic" or "geothermal"), and solar-assisted models for both residential and commercial applications. Each show presents some new wrinkles in heat pump design and utilization. Many of these changes were made toward the goal of a closer approach to the ideal performance of the heat pump. Table 12.1 compares the actual integrated COP to the theoretical COP for air-to-air heat pumps at different outdoor air temperatures. There is significant room for improvement with the application of innovative design concepts. For the air-to-air heat pumps, opportunities are present for improvement in compressor efficiency at part-load operation of the heat pump, defrost initiation and termination, interface between heat pump and solar space-heating systems, and storage-augmented air-to-air heat pumps. For the water-to-air heat pump, similar opportunities exist.

12.1.1. Multispeed Compressors

In order to accommodate the variable load required by heat pumps, some method other than on-off cycling could be used. Infinitely variable speed compressors would be ideal for heat pumps; however, they are not economically feasible at this time. Two-stage or two-speed compressors are new in the heat pump market and are able to better handle the variable loads required of heat pumps. Compressor manufacturers estimate a 13–18% advan-

**TABLE 12.1. Theoretical and Actual
Coefficients of Performance [1]**

Outdoor Air Temperature (°F)	Theoretical Maximum COP	Actual Integrated COP
60	12.1	3.48
50	9.95	2.93
40	8.46	2.61
30	7.37	2.46
20	6.51	2.32
10	5.85	2.03

tage in seasonal ratings for heat pumps with a two-speed compressor over those of a single-speed model. The low-speed compressor is used for both cooling and heating when the outside temperature is above the balance point, and the high-speed compressor is used when the outside temperature is below the balance point. With this type of operation, there is less cycling of the heat pump and thus an improvement in the seasonal performance factor. This improvement in seasonal ratings will have to be weighed against the increased size and cost of the improved compressor.

12.1.2. Defrost Control

Defrost control, which is generally considered a major problem in air-to-air heat pumps, is being given widespread attention, since it has experienced difficulty in operation reliability and efficiency losses. It is not unusual for a heat pump to go through 400–500 defrost cycles per year. In common use today are defrost control operations based upon time/temperature, temperature differential, air pressure differential, and other combinations. A new approach, which shows promise of being more reliable in terms of initiation and termination of defrost, is to sense the electrical loading of the heat pump's condenser fan motor. Some manufacturers are also investigating electric heater defrost rather than hot gas defrost. Another innovation being studied is to coat (Teflon) the surfaces on the frost-forming coil to accelerate the removal of the frost.

12.1.3. Component Modifications

Some proposed modifications to the heat pump system components show promise of giving excellent performance characteristics with reduced defrost time. These modifications consist of redesigning the accumulator/receiver and the use of capillary tubes for expansion of the refrigerant. Other investigators have suggested modifications such as parallel compression processes for increased capacity, cascade cycles that use two different refrigerants over the large temperature range, multistage compression for reduced overall pressure ratios, and two-stage compression with intercooling and flash gas removal.

12.2. HEAT PUMP INNOVATIONS

Innovative heat pump systems range from ice-makers to hot water heaters. The following sections describe five such innovations:

    Ice-maker heat pumps (ACES)
    Heat pump water heaters
    Dual-source heat pumps
    Heat-only heat pumps
    Air-cycle heat pump (ROVAC)

12.2.1. Ice-Maker Heat Pumps (ACES) [2]

What about a residential, commercial, or industrial heating/cooling system that has a high coefficient of performance (COP), that could utilize solar assist and heat recovery, utilize off-peak electricity for cooling, and provide domestic hot water, and that does not require reverse-cycle refrigeration defrost? Those requirements may well be filled by the ice-maker heat pump. Although most ACES systems have been installed and evaluated under government auspices, a handful of manufacturers are now offering such systems commercially.

This type of system, also referred to as an ACES (Annual Cycle Energy System), uses water as both its heat source and heat sink. It makes ice while heating a building, and then uses the melting ice to cool the building when required.

A 2000-ft^2 ACES demonstration house sponsored by ERDA and HUD was built just outside Knoxville on property owned by the University of Tennessee. The ACES house is a two-story structure with gray vertical siding and nearly windowless walls (only 11% fenestration) (Fig. 12.1). A 250-ft^2 solar collector/radiator panel stands between the two different roof levels. In the basement is the water tank: 2500 ft^3 with 2500 ft of coils in it.

The Annual Cycle Energy System (ACES) is an integrated system for space heating and cooling and domestic water heating. Its major elements are

1. A highly efficient unidirectional heat pump with refrigerant-to-brine heat exchangers on both the evaporating and condensing side
2. Thermal storage on the low-temperature (evaporator) side

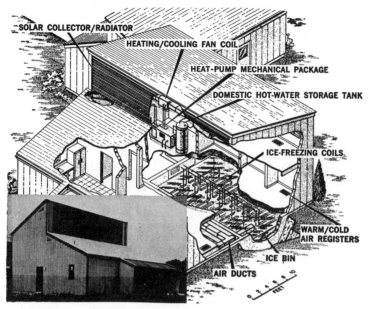

Fig. 12.1. ACES house. Reprinted with permission from *Popular Science* © 1977, Times Mirror Magazines, Inc.

3. An auxiliary heat source (e.g., solar collectors) and sink
4. A forced-air circulating system with a fan coil for space heating and cooling
5. A refrigerant-to-water heat exchanger for heating, and a tank for storing domestic hot water

Flow schematics are shown in Figures 12.2 and 12.3. In Fig. 12.3, the piping drawn in solid lines is all contained within the mechanical package. The dashed lines represent brine piping to components located externally to the mechanical package. The heat pump (compressor) always extracts heat from the evaporator (20% methanol-brine cooler), which in turn extracts heat from the ice storage bank via the circulating brine system. The heat pump always discharges its heat to one or more of the three high-temperature heat exchangers: the desuperheater, the hot water condenser, and the space-heating condenser. Thus, the refrigerant cycle is completely contained within the mechanical package and is a nonreversing cycle.

Space heating is accomplished by circulating brine from the heating condenser through the fan coil and transferring the heat to the living space by means of the circulating air system. Space cooling is accomplished by circulating brine from the ice storage bank through the fan coil and transferring heat to it from the circulating air system.

The selector valve chooses the brine circuit for heating or cooling as

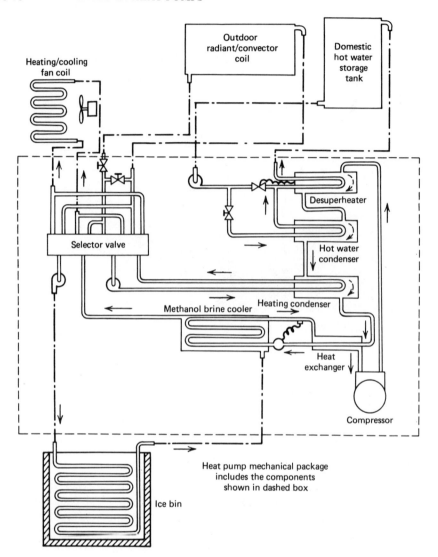

Fig. 12.2.   ACES flow schematic. (Courtesy of McQuay Perfex, Inc.)

required. Domestic water is heated by recirculation through the desuperheater during the heating season and through the desuperheater and the hot water condenser during the cooling season.

The radiant/convector panel is circuited to be able to accept or reject heat as needed. During the heating season, the panel can accept solar and convective heat and deposit it in the ice bank via the circulating brine system. This mode of operation can be selected automatically when needed (for instance, if the ice inventory exceeds 80% of the bank capacity). Ice forma-

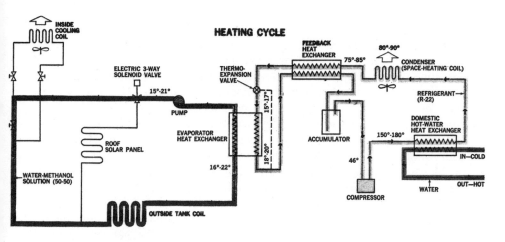

HEATING CYCLE

INSIDE COOLING COIL

ELECTRIC 3-WAY SOLENOID VALVE

15°-21°

PUMP

THERMO-EXPANSION VALVE

15°-17°

EVAPORATOR HEAT EXCHANGER

18°-20°

ROOF SOLAR PANEL

WATER-METHANOL SOLUTION (50-50)

16°-22°

OUTSIDE TANK COIL

FEEDBACK HEAT EXCHANGER

75°-85°

80°-90°

CONDENSER (SPACE-HEATING COIL)

REFRIGERANT (R-22)

ACCUMULATOR

150°-180°

DOMESTIC HOT-WATER HEAT EXCHANGER

IN—COLD

46°

OUT—HOT

WATER

COMPRESSOR

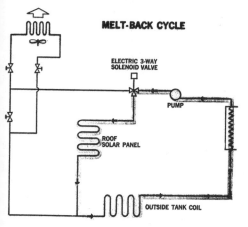

MELT-BACK CYCLE

INSIDE COOLING COIL

ELECTRIC 3-WAY SOLENOID VALVE

PUMP

ROOF SOLAR PANEL

OUTSIDE TANK COIL

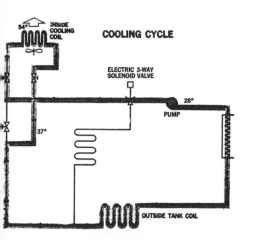

COOLING CYCLE

54°

INSIDE COOLING COIL

ELECTRIC 3-WAY SOLENOID VALVE

28°

PUMP

37°

OUTSIDE TANK COIL

ACES heating cycle (above) consists of a methanol-water loop, a refrigerant loop, and a loop for domestic hot water. Here's how it works: The methanol-water solution is pumped through the finned tubing in the water tank, where it picks up heat from the freezing water (see text). Then it enters the evaporator heat exchanger, where it transfers heat to the refrigerant, a fluid that boils at a low temperature. The refrigerant is at a very low pressure, which further reduces its boiling point. Although its temperature at this point is only 15° to 20° F, the refrigerant is boiled by the heat extracted from the methanol solution. The methanol then flows back to the water tank to pick up more heat. Meanwhile, the refrigerant goes into a feedback heat exchanger where it absorbs more heat. (The accumulator provides for expansion and contraction of the system.) The refrigerant next enters the compressor, where both its temperature and pressure are greatly elevated. This hot gas is sent to a heat exchanger to provide the domestic hot water. The refrigerant then flows through the condenser (the space-heating coil). A fan blows air over the condenser; the hot gas gives up some of its heat and condenses into a liquid. The air blowing over the condenser is thus warmed, and circulates through the ducts to heat the house. The still-warm liquid refrigerant travels through the feedback heat exchanger to help heat the incoming refrigerant about to enter the compressor. Then the returning refrigerant passes through an expansion valve, its pressure and temperature fall sharply, and it begins to boil. The cycle repeats. In the melt-back cycle (center), the methanol-water solution is sent through solar collectors on the roof, then back through the water tank to melt some of the ice. It's used where winter heating needs exceed summer cooling needs. By melting some of the ice, you keep the tank from freezing solid, and thus avoid a larger tank. For summer cooling (right), the methanol solution circulates through the coils in the ice-filled tank, then through the inside cooling coil, where a fan blows air across the cool coils and circulates it through the ducts to cool the house.

Fig. 12.3. The ACES flow diagrams. Reprinted with permission from *Popular Science* © 1977, Times Mirror Magazines, Inc.

tion is restricted to 80% of bank capacity to avoid expansion damage to the container. During the cooling season, the panel can reject heat convectively from the space-heating condenser via the circulating brine. This mode of operation can be automatically selected when needed (for instance, if the ice inventory is exhausted, the bank temperature exceeds 34°F, and outdoor ambient is below 80°F). This mode permits compressor operation under favorable conditions for heat rejection and off the usual utility peak load.

In the demonstration house the heating-cooling duct system is equipped with a motor-operated damper and is so arranged that in the normal position the blower circulates indoor air through the coil and in the economizer cycle the blower draws outside air into the house, discharges indoor air outside, and bypasses the coil. The control system places the damper in the economizer cycle position whenever the room thermostat requires cooling and the outdoor thermostat recognizes that the cooling can be accomplished by ventilation.

ACES systems have a constant capacity and constant performance because the heat source is a constant 32°F. In a MAXACES system, the heat pump makes ice during winter months. The accumulated ice is then used to cool the building during summer. A MINACES system is used in areas where more ice is made during the winter months than can be used in summer. Here a smaller tank is used, and the ice is melted by heat provided by a solar collector. In some systems, well water and even city water is used to melt the ice, which eliminates the capital expense for a solar collector.

The efficiency of the system is higher in all modes of operation than the average efficiency of conventional systems. The "heat source" for the heat pump is always near 32°F so it is never necessary to provide supplemental resistance heating, and the ACES operates with a measured COP of 2.77 as shown in Table 12.2. When providing hot water, the system has a COP slightly greater than 3.

TABLE 12.2. Full-Load Performance of the ACES System

Function	Full-Load COP
Space heating with water heating	2.77
Water heating only	3.09
Space cooling with stored ice	12.70
Space cooling with the storage > 32°F and < 45°F	10.60
Night heat rejection with water heating[a]	0.50
	(2.50)

[a] In the strict account used here, only the water heating is calculated as a useful output at the time of night heat rejection, because credit is taken for the chilling when it is later used for space cooling. This procedure results in a COP of 0.5. If the chilling credit and the water-heating credit are taken at the time of operation, then a COP of 2.5 results.

But as with all alternative-energy systems, it costs a lot more to install an ACES than it does to put in a conventional system. Just how much more is hard to say, because ACES is still experimental. However, an air-source heat pump might cost $3000, while an ACES in comparison might cost $10,000.

When compared on a life-cycle basis, however, information showed that the ACES system is competitive with the air-source system at electric rates of 4 cents/kWh.

Other systems now being installed incorporate outdoor coils and ground coils as a heat source. With the latter, as much as 1200 ft of PVC pipe is required for a typical residence.

Two methods are commonly used to make ice—either an "ice-maker" or an "ice-builder." Direct expansion or brine can be used with either system. A third system uses latent heat water modules immersed in a refrigerated brine.

Commercial buildings that have a long cooling season except for a short period each day can utilize an ACES system for cooling only. Many utilities have, or are planning to have, off-peak reduced rates, which make an ACES system attractive.

Either ice or 45° phase-change material can be used. Also, modular storage equipment can be used, which consists of cylindrical plastic tanks with internal plastic heat exchangers. Thermal losses are claimed to be low because of the poor conductivity of ice. Currently, there are seven experimental ACES systems in operation and another six under construction.

12.2.2. Water-Heating Heat Pumps

With water heating the second largest use of home energy in most locations, there has been renewed interest in the development and use of energy-conserving domestic water heaters since the 1973–1974 oil embargo. Backed by the federal government, electric utilities, and a growing number of manufacturers, air-to-water heat pumps are now a mass-market product. There are two different concepts involving the use of heat pumps for water heating. The newer development is the heat pump used only for the production of domestic hot water. The other approach makes use of all or part of the heat rejected from the condenser of a heat pump (or air conditioner) used for space-conditioning and is usually called a desuperheater.

Heat Pump Water Heaters.   Heat pumps designed to provide hot water are currently available, and proponents expect they may be able to operate with an annual water heating COP of 2 to 3. Because a heat pump removes heat from the air around it, a typical heat pump water heater will also provide space cooling about equal to that of a typical small window air-conditioner ($\frac{1}{2}$ ton) in summer.

Different from "desuperheater-type" recovery devices (where heat normally rejected outside by the cooling system is recovered first for domestic water heating), these heat pumps have water heating as their primary mission.

When installed outdoors, the heater can tap an unlimited heat source—the air around it—until the temperature drops below about 45°F, the point at which its evaporator ices over. In areas where it seldom gets this cold, the unit can switch to resistance heating until the weather warms up.

In colder climates, the heat pump water heater would be installed indoors and would provide air-conditioning and dehumidification for whatever room it was in, a bonus in summer. Even in midwinter, a unit in a utility room or furnace room could operate on heat from uninsulated ducts and furnace-jacket losses. The relatively small deficit of heat taken from the living space must be made up by the home's normal heating system. This is labeled a good tradeoff by the manufacturers, because the COP of the heat pumps is higher than that of straight resistance heaters that would otherwise be required to heat the water.

Because they promise 50–65% operating cost savings compared to heating domestic hot water with electric resistance heat or oil, areas of the country where electric rates are particularly high or where natural gas is not available will be expected areas of application. A retrofit unit on a conventional electric water heater might cost $400, and a new unit $600, not including installation. A heat pump water heater at an installed cost of about $800 is just as energy efficient and a lot less expensive in first cost than a $2000 or $3000 solar system for domestic hot water heating.

A typical American family of four that heats its water electrically might spend about $300 annually for electricity. Assuming the unit cost of electricity is not higher for families that use less electricity, a water-heating pump can cut the cost to between $100 and $150.

The concept, introduced in the 1950s, did not catch on then because electricity was cheap and heat pump technology not well developed. Now, as electricity costs escalate and fuel supplies become scarce, the idea has been revived by several manufacturers. The U.S. Department of Energy sponsored the development of the heat pump water heaters, and 225 of the units have been field tested by electric utilities across the country. The results showed that the average consumer would save about 50% of the energy used by a conventional electric resistance water heater, as well as the operating costs of such a system. Sales were projected to be about 100,000 by 1985.

A heat pump water heater works like any other heat pump. It uses standard principles of refrigeration to draw heat from the air surrounding the evaporator coil, which usually is located on top of the water tank. Then it releases that heat plus the work of compression in the condenser coil immersed in the water tank, thereby heating the water. Figure 12.4 illustrates such a hot water heating system which is termed an integral heat pump water heater.

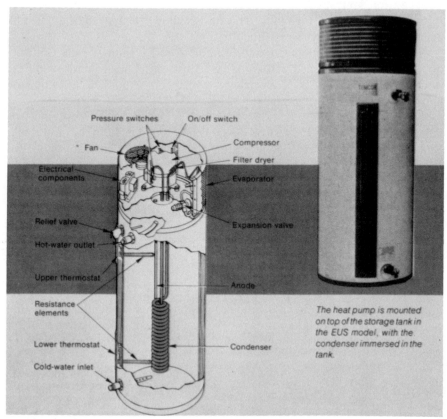

Fig. 12.4. EUS heat pump unit. Reprinted with permission from *Popular Science* ©
1978, Times Mirror Magazines, Inc.

Assuming space is available reasonably close to the water-storage tank,
retrofit heat pump water heaters can be added to any existing system. Like
the integral unit, the retrofits provide cooling and dehumidification while
efficiently heating the water. But they have more capacity than the integral
units. The retrofits deliver some 13,000 Btu/hr to the water tank while operating at about the same coefficients of performance as the integral heater.
They draw more power, produce more water heating and air-conditioning,
and cost somewhat more. Unlike the integral unit, the retrofits do not use
resistance heating for recovery. Nor is the design of the refrigerant-to-water
heat exchanger (condenser coil) constrained by the tank and flow paths
within the tank. The disadvantages of the retrofit units are that they require
more space for installation and also need an extra component, a water pump
to circulate water from the tank to the heat pump unit.

Northrup, Inc., which introduced one of the first air-to-water heat pumps designed for domestic hot water heating, have a plug-in unit that does not include a hot water storage tank. This design allows the unit to be applied to existing installations without difficulty.

Both Addison Products Co. and Comfort-Aire Division of Heat Controller, Inc., have developed self-contained air-to-water heat pump modules, powered by 115-V or 220-V electricity, which can be hooked up to new or existing hot water storage tanks.

The Efficiency II heat pump water heater, made by E-Tech, Inc., of Atlanta, heats water at the rate of 13,000 Btu/hr, equivalent to the output of a 3800-W high-recovery resistance heating element. But it consumes only 1230 W of electricity, less than one-third as much as a resistance element would use. Figure 12.5 shows the E-Tech system with do-it-yourself installation. In the water-flow circuit, a small pump circulates water from the bottom of the storage tank, through the heat exchanger in the Efficiency II, and returns it to the top of the storage tank.

The operation of the unit is controlled by a thermostat, which causes the compressor and pump to run when temperatures in the storage tank fall below the thermostat setting.

A variable-flow valve regulates the water flow rate to keep output temperatures at approximately 120°F. The positive-displacement recirculating pump tends to keep temperatures relatively consistent from top to bottom in the storage tank; a variation of 30–40°F can occur with conventional water heaters.

The Fedders-Airtemp-Climatrol unit is somewhat similar in concept except that it contains a rotary compressor. It also is available with an installation kit and enables the user to set storage temperatures at 120, 140, or 160°F. Recovery rate is about the same as an 80-gal tank with electrical resistance heating. Figure 12.6 shows the Fedders unit. It can be placed on the floor near the existing water tank, set on a counter, mounted on a nearby wall, or suspended from ceiling joists.

For nonresidential use, there are the Westinghouse Templifier (Fig. 12.7) heat pump water-heating packages for most commercial or industrial water-heating applications. They are completely factory assembled and are shipped ready to operate when connected to power and water. These complete systems include multiple hermetic motor compressors, operating and safety controls, internal power and control wiring, and a factory-installed refrigerant charge. Each compressor operates with its own condenser on a separate circuit. Unit capacity variation is accomplished by cycling compressors in response to variations in the temperature of return hot water.

Using the nonreversible heat pump principle, the Templifier recovers low-grade waste heat in the temperature range of 60–120°F (16–49°C) and amplifies it to higher, usable temperature levels. The waste heat from the source water is absorbed in the heat pump evaporator by the unit's working fluid, which is then increased in temperature and pressure by the compressor.

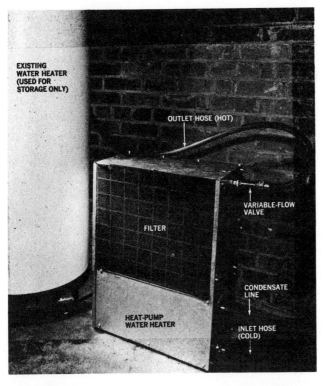

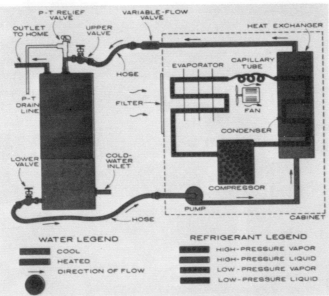

Fig. 12.5. E-Tech heat-pump water heater. Reprinted with permission from Popular Science © 1980, Times Mirror Magazines, Inc. Drawing by C. DeGroote.

Fig. 12.6. Retrofit heat pump from Fedders fits under a countertop. Reprinted with permission from Popular Science © 1982, Times Mirror Magazines, Inc.

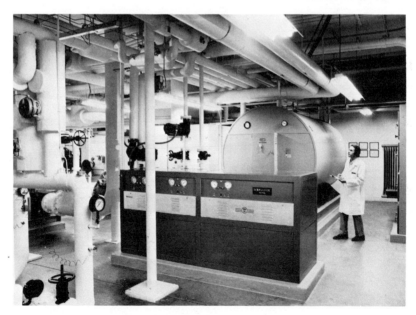

Fig. 12.7. Westinghouse Templifier heat pump water heater. Courtesy of McQuay—Perfex Inc. Templifier ®.

656

From here it goes to the condenser where it is transferred to a delivery fluid to provide useful heat at temperatures up to 220°F (104°C).

Westinghouse provides the following application suggestions:*

Hotels/Motels. Heat service hot water with waste heat from air-conditioning system; often enables shutdown of large heating boiler operating at low load and efficiency during summer months.

Office Buildings. Heat service hot water at low cost, using waste heat from computer room air-conditioning systems run year-round or from other waste heat sources. Use waste condenser water heat in summer to economically provide hot water for reheating.

Hospitals. Reclaim waste heat to provide hot water for laundries, kitchen, and washup uses.

Industrial Plants. Waste heat can be effectively reclaimed from numerous in-plant sources to provide low-cost process hot water; boiler and other feedwater heat; or space heat.

Solar Heating. By combining low-cost solar collectors (which can perform with high efficiency at low temperatures) with a Templifier, a more cost effective combination can be achieved than by direct use of a high temperature solar collection system.

Corresponding schematic diagrams are given as Fig. 12.8. Table 12.3 presents the condensed performance of the Templifier.

Desuperheater Water Heater. The second type of water-heating heat pump system is the waste-heat or desuperheating type of unit. The desuperheater water heater is a heat exchanger that uses heat that would otherwise be discharged to the atmosphere by a central air conditioner. It also can be teamed with a heat pump during the winter, but at the expense of energy available for space heating. This heat exchanger does not change the heat pump cycle, but it does allow better use of refrigerant heat. Figure 12.9 is a schematic of such a system. The discharge line from the compressor is plumbed into the heat exchanger, and the superheated refrigerant gas transfers its heat to water. Technically, the heat exchanger is a refrigerant desuperheater. The heat picked up from the refrigerant is transferred to potable water circulated from the storage tank.

The typical desuperheater water heater takes cold water from the bottom of the standard water heater and pumps it through a counterflow heat exchanger, where it picks up heat from hot (about 200°F) refrigerant gas coming off the air-conditioner compressor. Refrigerant, at a lower temperature,

*Westinghouse Technical Literature.

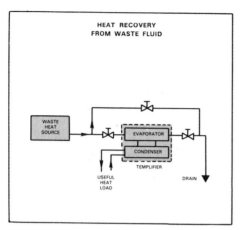

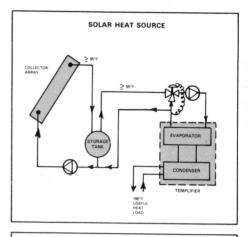

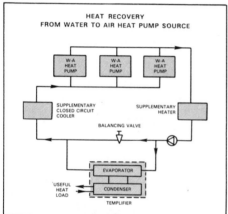

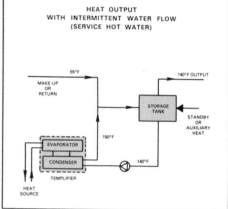

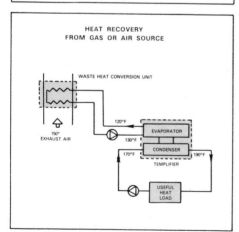

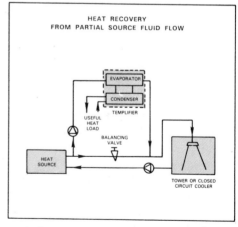

Fig. 12.8. Typical piping diagrams. (Valves, drains, vents, expansion tanks, and instrumentation must be added in accordance with good piping practice. Temperatures, where shown, are for illustration only.) Courtesy of McQuay—Perfex Inc. Templifier ®.

TABLE 12.3. Condensed Performance for Templifier on 60-Cycle Power Supply

Unit Model Number	Leaving Source Water Temp. (°F)	Maximum Leaving Hot Water Temperature											
		Series A, 135°F			Series B, 150°F			Series C, 180°F			Series D, 220°F		
		Heating Capacity (Mbh)	Input (kW)	COP	Heating Capacity (Mbh)	Input (kW)	COP	Heating Capacity (Mbh)	Input (kW)	COP	Heating Capacity (Mbh)	Input (kW)	COP
TPB020	95	—	—	—	281	21.1	3.9	168	15.7	3.1	70	8.9	2.3
	85	304	20.9	4.3	242	19.4	3.6	143	14.3	2.9	55	7.7	2.1
TPB025	95	—	—	—	368	27.6	3.9	220	20.6	3.1	91	11.6	2.3
	85	398	27.4	4.3	318	25.5	3.6	187	18.7	2.9	72	10.1	2.1
TPB030	95	—	—	—	468	35.1	3.9	280	26.2	3.1	116	14.8	2.3
	85	506	34.8	4.3	404	32.4	3.6	238	23.8	2.9	91	12.8	2.1
TPB045	95	—	—	—	704	52.6	3.9	419	39.2	3.1	173	22.2	2.3
	85	759	52.1	4.3	606	48.7	3.6	356	35.6	2.9	136	19.2	2.1
TPB055	95	—	—	—	838	62.7	3.9	499	46.7	3.1	207	26.4	2.3
	85	904	62.0	4.3	722	57.9	3.6	424	42.4	2.9	163	22.8	2.1
TPB060	95	—	—	—	938	70.2	3.9	559	52.3	3.1	231	29.6	2.3
	85	1012	69.4	4.3	808	64.8	3.6	475	47.5	2.9	182	25.6	2.1

SOURCE: Westinghouse Technical Literature.

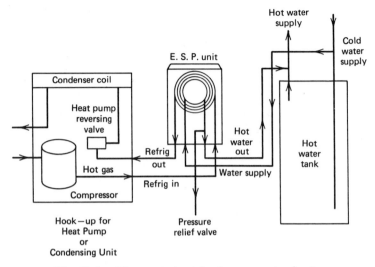

Fig. 12.9.   The waste-heat heat pump water heater.

then goes to the condenser coil, where more heat is removed and gas condenses. Next, liquid refrigerant passes through the expansion valve (not shown in Fig. 12.9) and vaporizes in the evaporator coil inside the house, absorbing heat. Hot gas returns to the compressor, and the cycle repeats. In the water loop, water passes from the heat exchanger into the top of the standard water heater. The cycle continues until water in the tank reaches a preset maximum (120–150°F). When the waste-heat water heater is hooked to a heat pump, it must be between the compressor and the reversing valve. With the heat pump in the heating mode, the refrigerant cycle is reversed: The waste-heat device takes some heat that would have gone for space heating.

Many residences, mostly in the South, have been using desuperheater water heaters for a number of years. Now, more and more companies are marketing them, including such major air-conditioner manufacturers as Carrier, Friedrich, and General Electric. Table 12.4 lists the addresses of a number of these manufacturers. Fig. 12.10 shows such a unit. Most manufacturers claim their systems can provide about 10 gal of hot water per hour per ton of air-conditioning capacity [but claims range from 3 to 18 gal/(ton·hr)]. At the rate of 10 gal/(ton·hr), a 3-ton air conditioner could provide 30 gal of hot water for every hour of compressor operation. If the compressor is on 25% of the time, for example, it would make 30 gal of hot water in 4 hours. Table 12.5 gives savings in electricity consumption for seven cities, as calculated by General Electric. Dollar savings depends on how much you would have paid to heat the water in the standard way. The annual energy savings is based on using GE's Hot Water Bank with 3-ton air conditioner.

TABLE 12.4.   Desuperheater Water Heater Manufacturers

Carrier Air Conditioning (The Hot Shot), Carrier Pkwy., Syracuse, NY 13221.
Energy Conservation Unlimited, Inc. (ECU, Hot Tao), Box 585, 311 E. Georgia
    Ave., Longwood, FL 32750.
Friedrich Air Conditioning and Refrigeration Co. (Hot Water Generator), 4200 N.
    Pan Am. Expy., San Antonio, TX 78295.
General Electric Co. (Hot Water Bank), 6200 Troup Hwy., Tyler, TX 75711.
Halstead & Mitchell (Heat Recovery Unit), Hwy. 72 West, Scottsboro, AL
    35768.
Marvair Co. (Marvair), Box 400, U.S. Hwy. 41 N., Cordele, GA 31015.
Weatherking, Inc. (Weatherking), Box 20434, Orlando, FL 32814.

The data assume a family uses 75 gal of hot water per day, that the water temperature increases from 70°F to 140°F, and that 30% losses occur in pipes through water heater jackets. In warmer climates, more waste heat is available than can be used, due to the 75-gal ceiling on hot water needs. Thus, savings per kilowatt-hour are less than in cooler climates, although total savings are more.

If the system is properly sized and installed, a desuperheater water heater can reduce the head pressure on the compressor. Thus the air conditioner will use less energy, probably 6–8% less, according to most manufacturers.

Fig. 12.10.   Waste-heat water heater (small box) made by ECU can be put outside; most must go inside for weather protection. Reprinted with permission from *Popular Science* © 1980, Times Mirror Magazines, Inc.

TABLE 12.5. Annual Energy Savings with
Waste-Heat Recovery

City	Compressor On (hr)	Electricity Saved (kWh)
Atlanta, GA	1146	2274
Chicago, IL	782	1862
Columbus, OH	758	1777
Dallas, TX	1595	2562
Jacksonville, FL	1675	2603
Louisville, KY	998	2175
Miami, FL	2448	2770

12.2.3. Dual-Source Heat Pumps

In solar heating and cooling installations, water-source heat pumps suffer from a limitation in that condenser heat on the cooling cycle must be rejected to water or, in the absence of a water-based heat sink, to air via an air-cooled heat exchanger.* This latter requirement may involve additional fan or pumping power, diminishing the COP.

A *dual-source* (water and air) heat pump circumvents these disadvantages in that the heating cycle evaporator heat exchanger can be immersed in water or air. This enables the water loop to be used during at least part of the heating cycle when the water temperature exceeds that of the ambient air, and the air coil to be used for heating under small temperature differentials or for cooling. Several major manufacturers are believed to be developing dual-source heat pumps; however, none are on the market at the present time.

12.2.4. Heat-Only Heat Pumps

Another development of interest is the heat-only heat pump. In Europe, the heat-only heat pump, as an alternative to electrical resistance heat, is basically the only type of heat pump under consideration, since air-conditioning is not widespread there. In the United States, Janitrol (a division of Tappan Co.) introduced a single size (37,000-Btu/hr nominal cooling capacity) heat-only heat pump, the Wattsaver, designed mainly as an add-on to a resistance or combustion furnace. Features of interest are the use of compressed or discharged gas for evaporator defrost (instead of hot condensing liquid) and the fact that heating-mode operation only has enabled the use of certain

---

*In some cases, nocturnal heat rejection by evaporation or thermal radiation is also a possibility. Again however, auxiliary power consumption is usually involved.

TABLE 12.6.   Comparison of Heat-Only Heat Pump with Conventional Heat Pumps

	Nominal Heating Capacity at 47°F (Btu/hr)	COP	
		17°F	47°F
Heat-only	37,000	2.3	3.2
Conventional A	37,500	2.1	2.8
Conventional B	39,000	2.0	2.8
Conventional C	37,500	2.1	3.1

SOURCES: Carrier Corp., Janitrol, General Electric, and York technical bulletins.

approaches to achieve high heating COP that might be detrimental to cooling performance in a conventional heat pump. Indeed, the heat-only heat pump, compared to conventional pumps, shows superior low- and high-temperature heating performance, as may be seen from Table 12.6.

The major difficulty with a heat-only heat pump approach is that the first cost, which for a conventional heat pump can be paid off through energy savings over the entire year, must be recovered over heating season operation only. Hence, the cost-effectiveness of this system in comparison to alternatives is open to question unless subsequent designs can effect substantial cost reductions.

12.2.5.   Air-Cycle Heat Pumps (ROVAC)

The electric heat pumps previously described utilize a fluorocarbon refrigerant-based Rankine cycle as the refrigeration system. However, other refrigeration cycles obviously also exist and are suitable for heat pumping, at least in theory. Some of them have been used in aircraft air-conditioning. The Brayton refrigeration cycle, for example, uses a permanent gas (e.g., air), which in the ideal cycle is compressed and expanded isentropically, the heat exchange processes being isobaric.

A rather novel air-cycle machine that is under development by Rovac Corporation for automobile air-conditioning applications is also under consideration as a residential heat pump [3]. Laboratory versions of the Rovac heat pump are driven by an electric motor through an external shaft, so the system is considered to be an electric heat pump. The original version of the Rovac system was designed as an automobile air conditioner to be driven by a power takeoff from the automobile engine. The compressor is a rotary vane, positive-displacement, constrained vane device as illustrated in Fig. 12.11. In the closed-system design version, a cooler-and-heater heat exchanger is employed as well as a regenerator. The stator-rotor system, called the *circulator,* forms a compressor-expander unit which, together with the heat exchangers, constitute the closed refrigeration loop. The rotor is shaft-

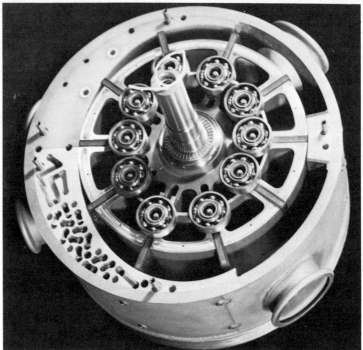

Fig. 12.11. The ROVAC air-cycle compressor/expander. Courtesy The ROVAC Corporation.

664

driven by an external power source—an automobile engine in the automobile air-conditioning version.

A novel feature of the Rovac system has been the injection of water in the compression and expansion sections of the circulator for added efficiency. Evidently the water vaporizes during the compression stage, keeping its temperature rise down; on entering the expander it condenses (and even freezes). Advantages claimed for this system (other than that fluorocarbons are not used) are a high COP for the electrically driven residential unit and the ease of capacity modulation by pressure ratio change. Possible disadvantages are (1) the nonhermetic refrigerant loop, (2) the necessity for more heat exchanger surface due to the sensible heat transfer processes of the air-cycle refrigeration machine, and (3) the problem of heat leakage, seals, and wear of the rotary compressor-expander. At present, however, the Rovac corporation is concentrating on the development of the compressor proper as a high-efficiency low-vapor-pressure nonfluorocarbon air-conditioning compressor [4].

12.3. OTHER HEAT PUMP CONCEPTS

In addition to the electric vapor-compression heat pump to which most of this book has been devoted, a number of nonconventional heat pump schemes were presented in Chapter 5. These include several thermal engine and absorption cycle types, solar-assisted or solar-supplemented units, add-on or hybrid systems, and chemical heat pumps. Besides these, other heat-pump devices have been devised or proposed. Since these are either in the very early experimental stage or, while technically feasible, appear to be uneconomical for space-conditioning applications, only brief descriptions of two selected systems will be presented in this chapter, the Peltier-effect (thermoelectric) heat pump and the magnetic heat pump.

12.3.1. Peltier-Effect Heat Pumps

The Peltier effect, the reverse of the thermocouple effect, is the development of a temperature gradient when electric current flows through two dissimilar materials that are joined thermally and electrically. Peltier heat pumps are inherently the simplest and most reliable of all electric heat pumps, consisting entirely of solid materials and requiring no moving parts or working fluids. The theory of the Peltier heat pump or refrigerator can be found in the ASHRAE *Handbook of Fundamentals—1977* [5]. Only a few considerations in developing potentially practical equipment will be summarized here.

First of all, it should be realized that the magnitude of the Peltier effect in most materials is relatively small, depending, quantitatively, on the relative magnitude of the Seebeck coefficient and the thermal conductivity of the materials involved. A high Seebeck coefficient implies a large temperature gradient for a given voltage drop; a low thermal conductivity is desirable, since thermal conduction from the hot to the cold end opposes the gradient set up by the potential difference. However, it is difficult to achieve the desired extreme values of both in real materials, since materials with a high Seebeck coefficient are also good thermal conductors. In this respect, semi-conductor materials are preferred over metals. Additional phenomena arising out of the effect are production of Joule (i^2R) heat and the second-order Thompson effect.

Although the Peltier effect is used in the cooling of electronic circuits and for portable cooler/heaters (Fig. 12.12), no known space-conditioning applications of it exist as far as is known. Several years ago in West Germany, Siemens conducted research in thermoelectric air conditioners for potential room use, but the program was discontinued due to their projected high first cost.

12.3.2. Magnetic Heat Pump

The magnetic heat pump uses a magnetic solid as the refrigerant. Its operation is based on the natural phenomenon that many magnetic materials become warmer when they are subjected to a magnetic field and cooler when that field is removed.

Scientists have been creating supercold temperatures (near absolute zero) magnetically in laboratories for the past 40 years, however they have never been able to produce useful cooling or heating at room temperatures. This difficulty was overcome by verifying that gadolinium (a rare-earth element discovered in 1880 and used in TV tubes and microwave applications) could produce a large change in temperatures when subjected to a magnetic field. In fact, gadolinium will change in temperature by as much as 14°C (25°F) when a strong magnetic field is applied to it and subsequently removed, and it will do this at room temperature. Furthermore, by using a technique called regeneration, a temperature change of over 60°C (140°F) has been achieved.

At the NASA Lewis Center a current model consists principally of a gadolinium refrigerant assembly, a fluid-filled tube called a regenerator, that is 100 cm (40 in.) long and 5 cm (2 in.) in diameter and an electromagnet, as shown in Fig 12.13.

The gadolinium refrigerator assembly is composed of 36 plates, 1-mm (0.04 in.) thick, assembled within a cylindrical stainless steel canister. The canister is suspended in a fixed position within the regenerator, which is filled with a mixture of 50% water and 50% alcohol. A circular-shaped magnet surrounds the regenerator and is located in the same fixed position as the refrigerant canister. The only moving part of the pump is the regenerator, which is driven up and down electrically, an action that serves to position the gadolinium canister alternately at either end of the regenerator. When the gadolinium is at the "hot" end of the regenerator, the magnet is turned on, heating the gadolinium, which in turn heats the surrounding fluid. The regenerator is then driven in the opposite direction so as to position the

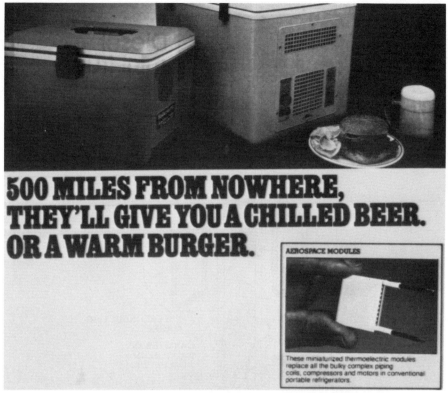

Fig. 12.12. Koolatron thermoelectric portable refrigerators and food warmers. (Courtesy Koolatron Industries Limited.

gadolinium at the "cold" end. At this point the magnet is turned off, and the gadolinium cools and absorbs heat from the surrounding fluid, thus cooling it.

Repetition of this process creates a cumulative effect at each end of the regenerator: The "hot" end gets hotter and the "cold" end gets cooler. In one recent test run, the magnetic heat pump achieved a temperature range from 55°C (131°F) at the hot end to −30°C (−22°F) at the cold end.

12.4. PRIMARY ENERGY BASIS COMPARISON OF CONVENTIONAL AND ADVANCED HEATING AND COOLING SYSTEMS

In order to compare the energy effectiveness of the electric heat pump with fossil-fuel heating equipment, it is necessary to trace the energy consumption of the heat pump back to the primary source energy used by the power plant. One way in which this may be done is by defining an effective heating COP, given by

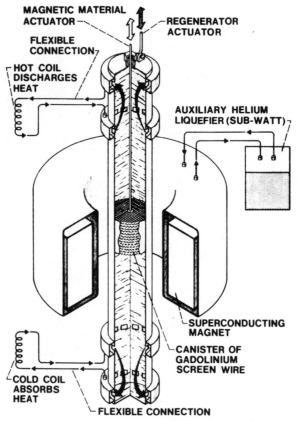

Fig. 12.13. A magnetic heat pump. Reprinted with permission from *ASHRAE Transactions 1981*.

Effective heating COP = (0.29)(heat pump's heating COP)

where 0.29 takes into account the fact that, on the average, only about 29% of the heat energy of the fuel burned at the power plant in making electricity will reach the heat pump.

Figure 12.14, prepared by Lennox, shows the relative efficiency of the use of primary energy by various systems. The graph is based on the following criteria: electric power generator and transmission energy requirements figured at 10,500 Btu/kWh; fossil-fuel furnace efficiency computed on a seasonal average of 62%.

In this section, a brief discussion of the impact of alternative space-conditioning systems on primary fuel resources based on the 1978 Gordian report [6] is presented. Although a reduction of total energy consumed on-site for space-conditioning is desirable, this is not an end in itself. Rather, the effect of increased heat pump market saturation on the types of fuels consumed to produce the energy for on-site use should be examined. This is particularly important in light of the increasing scarcity of such primary fuels

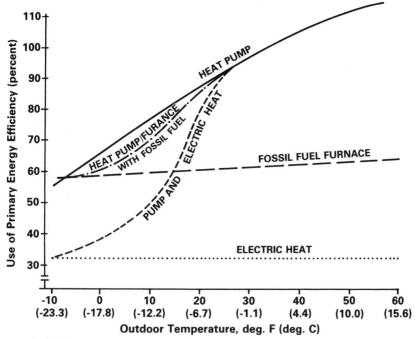

Fig. 12.14. Relative primary energy use efficiency. Courtesy Lennox Industries, Inc.

as oil and natural gas. Using on-site energy consumption data, primary energy consumption data were developed in the following manner:

Electric Energy. Average state plant heat rates were used to convert on-site consumption to resource energy consumption by the generating station; these heat rates are shown below for each city.

City	Plant Heat Rate (Btu/kWh)
Houston	10,304
Birmingham	10,194
Atlanta	10,696
Tulsa	10,283
Philadelphia	10,249
Seattle	9,717
Columbus	10,187
Cleveland	10,187
Concord	10,874

SOURCE: Edison Electric Institute.

In addition, transmission and distribution losses of 9% were assumed for electric energy.

Natural Gas. Heating value of natural gas was assumed to be constant throughout the nation at 1000 Btu/ft^3; 8% transmission and distribution losses were assumed.

Fuel Oil. Heating value taken as 139,000 Btu/gal. No distribution losses were assumed.

From the primary energy consumption data, primary energy seasonal performance factors were computed for residential and commercial HVAC equipment. These SPFs were defined as the ratio of total seasonal heating or cooling load satisfied by the HVAC system to the total primary energy consumption.

The results of the primary energy comparisons on an individual system basis are shown in Tables 12.7 through 12.10. The SPF data appear in Tables 12.7 and 12.8, for the conventional and advanced systems, respectively. As expected, electric heat pumps were found to be more efficient than any of the all-electric HVAC systems examined. High-efficiency residential electric heat pumps had performance factors in the 0.78–0.43 range for heating and the 0.61–0.82 range for cooling, constituting an average 9% improvement over the standard heat pump representing 1975 technology. In primary energy terms, electric heat pumps were comparable to or slightly less efficient than the combustion heating systems combined with electric cooling in regions with a significant heating load. In very cold regions, the combustion systems were more efficient. However, combustion system efficiencies were calculated based on insulated supply and return air ductwork. This is the observed practice with electric but not with combustion heating systems in general. Calculations showed that the performance of the combustion furnaces was considerably improved with duct insulation. With uninsulated ducts, electric heat pumps would probably have a slightly higher efficiency than the combustion systems.

Gas heat pumps now under development have a higher efficiency in a primary energy sense than electric heat pumps, combustion furnaces, or resistance heating systems. Of the residential gas heat pumps examined, the gas-fired free-piston and the V-type single-cylinder Stirling-Rankine heat pumps were found to have preformance factors in the 1.16–0.91 range for heating and in the 0.68–0.88 range for cooling. The corresponding results for the organic fluid absorption heat pump were lower: 0.95–0.87 for heating and 0.48–0.46 for cooling.

Total annual energy consumption, shown in Tables 12.9 and 12.10 for residential space-conditioning, is smallest for the combination of fossil-fuel furnaces and high-efficiency central air conditioners in colder parts of the country and is greatest for electric air conditioners. These results parallel heating-season energy consumption observations because of the relatively

TABLE 12.7. Conventional Electric Heat Pumps and Other Systems: Seasonal Performance Factors on a Primary Energy Basis*

Heating	Houston	Birmingham	Atlanta	Tulsa	Philadelphia	Seattle	Columbus	Cleveland	Concord
Electric Heat Pump - Standard	0.65 (0.72)*	0.59 (0.66)	0.58 (0.64)	0.51 (0.55)	0.57 (0.63)	0.59 (0.67)	0.52 (0.57)	0.52 (0.56)	0.41 (0.43)
Electric Heat Pump - High Efficiency I	0.78 (0.86)	0.69 (0.76)	0.68 (0.75)	0.59 (0.64)	0.66 (0.72)	0.65 (0.74)	0.58 (0.64)	0.57 (0.63)	0.43 (0.47)
Electric Heat Pump - High Efficiency II	0.73 (0.76)	0.67 (0.74)	0.61 (0.68)	0.60 (0.65)	0.64 (0.70)	0.68 (0.76)	0.58 (0.63)	0.57 (0.62)	0.48 (0.51)
Electric Heat Pump - High Efficiency III	0.69 (0.76)	0.62 (0.69)	0.61 (0.66)	0.52 (0.57)	0.60 (0.65)	0.67 (0.75)	0.54 (0.59)	0.53 (0.58)	0.46 (0.49)
Hybrid Electric Heat Pump-Gas Furnace	0.55 (0.61)	0.52 (0.57)	0.51 (0.55)	0.55 (0.60)	0.52 (0.56)	0.52 (0.55)	0.52 (0.55)	0.52 (0.55)	0.50 (0.53)
Hybrid Electric Heat Pump-Oil Furnace	0.55 (0.61)	0.53 (0.57)	0.51 (0.56)	0.56 (0.61)	0.52 (0.57)	0.53 (0.55)	0.53 (0.57)	0.54 (0.57)	0.52 (0.55)
Electric Heat Pump - Heat Only	0.78 (0.87)	0.72 (0.79)	0.69 (0.75)	0.65 (0.71)	0.75 (0.77)	0.75 (0.84)	0.63 (0.69)	0.62 (0.68)	0.51 (0.56)
Gas Warm Air Furnace	0.53	0.53	0.52	0.54	0.54	0.55	0.55	0.56	0.55
Oil Warm Air Furnace	0.59	0.59	0.59	0.61	0.60	0.61	0.61	0.62	0.62
Central Electric Furnace	0.30	0.30	0.30	0.30	0.30	0.30	0.30	0.30	0.28
Baseboard Convectors	0.30	0.31	0.29	0.30	0.31	0.32	0.31	0.31	0.29
Cooling									
Electric Heat Pump - Standard	0.60 (0.69)	0.60 (0.69)	0.57 (0.67)	0.58 (0.69)	0.60 (0.71)	0.67 (0.75)	0.61 (0.75)	0.61 (0.72)	0.57 (0.66)
Electric Heat Pump - High Efficiency I	0.70 (0.80)	0.70 (0.81)	0.67 (0.78)	0.70 (0.83)	0.71 (0.83)	0.82 (0.92)	0.71 (0.83)	0.73 (0.85)	0.69 (0.79)
Electric Heat Pump - High Efficiency II	0.65 (0.76)	0.65 (0.77)	0.61 (0.71)	0.66 (0.78)	0.66 (0.79)	0.73 (0.83)	0.66 (0.79)	0.67 (0.80)	0.61 (0.72)
Electric Heat Pump - High Efficiency III	0.69 (0.79)	0.68 (0.80)	0.66 (0.77)	0.58 (0.69)	0.69 (0.81)	0.79 (0.90)	0.69 (0.81)	0.70 (0.83)	0.71 (0.82)
Hybrid Electric Heat Pump	0.56 (0.65)	0.56 (0.66)	0.54 (0.64)	0.55 (0.67)	0.56 (0.67)	0.62 (0.72)	0.57 (0.67)	0.57 (0.68)	0.53 (0.64)
Central Air Conditioner - High Efficiency	0.66 (0.76)	0.66 (0.77)	0.63 (0.74)	0.60 (0.71)	0.67 (0.79)	0.75 (0.84)	0.67 (0.78)	0.68 (0.80)	0.64 (0.75)
Central Air Conditioner - Standard	0.55 (0.64)	0.55 (0.64)	0.53 (0.62)	0.57 (0.67)	0.56 (0.66)	0.63 (0.71)	0.56 (0.65)	0.57 (0.67)	0.57 (0.66)
Window Air Conditioners	0.67	0.71	0.67	0.69	0.70	0.68	0.71	0.70	0.67
Annual Heating and Cooling									
Electric Heat Pump - Standard	0.61 (0.69)	0.59 (0.68)	0.57 (0.65)	0.54 (0.60)	0.58 (0.65)	0.60 (0.68)	0.54 (0.59)	0.53 (0.58)	0.42 (0.45)
Electric Heat Pump - High Efficiency I	0.71 (0.81)	0.70 (0.79)	0.67 (0.77)	0.62 (0.70)	0.67 (0.75)	0.67 (0.75)	0.61 (0.67)	0.60 (0.66)	0.45 (0.49)
Electric Heat Pump - High Efficiency II	0.66 (0.78)	0.66 (0.75)	0.61 (0.70)	0.62 (0.69)	0.65 (0.72)	0.69 (0.77)	0.60 (0.66)	0.59 (0.65)	0.49 (0.53)
Electric Heat Pump - High Efficiency III	0.69 (0.78)	0.66 (0.75)	0.63 (0.72)	0.54 (0.61)	0.62 (0.69)	0.68 (0.77)	0.57 (0.63)	0.56 (0.61)	0.48 (0.51)
Hybrid Electric Heat Pump-Gas Furnace	0.56 (0.64)	0.54 (0.62)	0.52 (0.60)	0.55 (0.62)	0.53 (0.59)	0.53 (0.57)	0.53 (0.58)	0.53 (0.57)	0.50 (0.54)
Hybrid Electric Heat Pump-Oil Furnace	0.56 (0.65)	0.55 (0.62)	0.52 (0.60)	0.56 (0.63)	0.52 (0.59)	0.54 (0.57)	0.54 (0.59)	0.55 (0.58)	0.52 (0.56)
Gas Warm Air Furnace/Central Air Conditioner-High Efficiency	0.60	0.56	0.55	0.55	0.55	0.56	0.56	0.57	0.55
Oil Warm Air Furnace/Central Air Conditioner-High Efficiency	0.65	0.63	0.61	0.61	0.62	0.62	0.62	0.63	0.62
Central Electric Furnace/Central Air Conditioner-High Efficiency	0.54	0.44	0.41	0.37	0.36	0.35	0.35	0.33	0.30
Baseboard Convectors/Window Air Conditioners	0.49	0.41	0.38	0.35	0.34	0.34	0.34	0.33	0.30
Electric Heat Pump-Heat Only/Central Air Conditioner-High Efficiency	0.68 (0.78)	0.68 (0.78)	0.66 (0.74)	0.63 (0.71)	0.73 (0.78)	0.75 (0.84)	0.64 (0.71)	0.63 (0.70)	0.52 (0.57)

* Numbers in parenthese are on-site energy consumptions based on steady state or full load performance, i.e no dynamic efficiency losses are included.

SOURCE: Gordian and Associates [6].

671

TABLE 12.8. Advanced Heating and Cooling Systems: Seasonal Performance Factors on a Primary Energy Basis—Residential Space Conditioning*

	Houston	Birmingham	Atlanta	Tulsa	Philadelphia	Seattle	Columbus	Cleveland	Concord
HEATING									
Advanced Electric Heat Pump I	**	0.76	0.73	0.72	0.74	**	0.72	0.72	**
Advanced Electric Heat Pump II	1.04	0.86	0.86	0.69	0.81	0.96	0.74	0.73	0.58
Free Piston Stirling-Rankine Gas Heat Pump	1.10	1.07	1.06	1.04	1.05	1.09	1.03	1.03	0.98
V-Type Single Cylinder Stirling-Rankine Gas HP	1.16	1.11	1.10	1.01	1.10	1.11	1.07	1.07	0.91
Organic Fluid Absorption Gas Heat Pump	0.95	0.92	0.91	0.90	0.91	0.94	0.90	0.90	0.87
Pulse Combustion Furnace	0.78	0.79	0.79	0.81	0.81	0.82	0.82	0.82	0.82
COOLING									
Advanced Electric Heat Pump I	**	0.74	0.71	0.74	0.75	**	0.75	0.76	**
Advanced Electric Heat Pump II	1.00	1.01	0.96	0.98	1.02	1.15	1.02	1.04	0.97
Free Piston Stirling-Rankine Gas Heat Pump	0.72	0.69	0.69	0.71	0.70	0.71	0.69	0.71	0.68
V-Type Single Cylinder Stirling-Rankine Gas HP	0.81	0.80	0.81	0.73	0.84	0.88	0.82	0.85	0.82
Organic Fluid Absorption Gas Heat Pump	0.48	0.46	0.47	0.48	0.46	0.47	0.46	0.47	0.46
ANNUAL HEATING AND COOLING									
Advanced Electric Heat Pump I	**	0.75	0.72	0.73	0.74	**	0.73	0.73	**
Advanced Electric Heat Pump II	1.01	0.95	0.91	0.78	0.86	0.98	0.80	0.77	0.61
Free Piston Stirling-Rankine Gas Heat Pump	0.77	0.81	0.82	0.89	0.92	1.02	0.92	0.95	0.93
V-Type Single Cylinder Stirling-Rankine Gas HP	0.85	0.89	0.90	0.87	0.99	1.06	0.98	1.01	0.90
Organic Fluid Absorption Gas Heat Pump	0.53	0.59	0.60	0.68	0.71	0.83	0.73	0.78	0.80
Pulse Combustion Furnace/Central Air Conditioner-High Efficiency	0.67	0.71	0.70	0.72	0.76	0.81	0.78	0.79	0.80

* The energies presented may be underestimated for the advanced electric heat pump I, free piston Stirling-Rankine gas heat pump and the organic fluid absorption gas heat pump due to possible underestimation of cycling and defrost losses. If the dynamic efficiency losses of the advanced heat pumps were similar to those of the conventional electric (Rankine cycle) heat pump, then adjusted estimates of energies which include these effects can be obtained by multiplying the values in this table by a factor of 1.231 for the heating season, 1.181 for the cooling season and 1.252 for annual energy values.

** System is not sized for application in these cities.

SOURCE: Gordian and Associates [6].

TABLE 12.9. Conventional Electric Heat Pumps and Other Systems: Total Fuel And Electric Energy Consumption (Btu $\times 10^{-6}$) on a Primary Energy Basis*

Heating	Houston	Birmingham	Atlanta	Tulsa	Philadelphia	Seattle	Columbus	Cleveland	Concord
Electric Heat Pump - Standard	15.75 (14.16) *	42.08 (37.94)	44.17 (40.08)	88.30 (81.29)	85.05 (77.73)	72.67 (64.62)	114.79 (105.35)	137.25 (126.16)	199.09 (186.61)
Electric Heat Pump - High Efficiency I	13.17 (11.86)	36.11 (32.67)	37.72 (34.21)	76.84 (70.66)	73.90 (67.56)	66.07 (58.24)	102.29 (93.66)	123.00 (112.66)	185.06 (173.18)
Electric Heat Pump - High Efficiency II	14.09 (12.64)	37.55 (33.42)	41.53 (37.67)	75.56 (69.48)	75.97 (69.43)	63.36 (56.49)	103.08 (94.53)	123.39 (113.31)	168.61 (157.71)
Electric Heat Pump - High Efficiency III	14.95 (13.44)	40.04 (36.19)	42.12 (38.29)	86.16 (79.24)	81.37 (74.44)	64.62 (57.53)	110.69 (101.56)	132.46 (121.73)	176.50 (165.26)
Hybrid Electric Heat Pump-Gas Furnace	18.67 (16.82)	47.96 (43.91)	50.41 (46.12)	81.43 (75.35)	94.04 (86.91)	83.22 (78.75)	114.58 (107.73)	135.02 (127.83)	160.16 (152.15)
Hybrid Electric Heat Pump-Oil Furnace	18.63 (16.79)	47.58 (43.63)	50.15 (45.90)	80.22 (74.31)	92.95 (85.92)	81.80 (78.30)	111.77 (105.40)	131.34 (124.95)	155.41 (147.96)
Electric Heat Pump-Heat Only	13.16 (11.74)	34.87 (31.46)	37.14 (33.80)	69.51 (63.82)	69.00 (63.07)	57.67 (51.42)	94.48 (86.34)	113.50 (103.87)	154.42 (144.50)
Gas Warm Air Furnace	19.58	47.09	48.25	83.78	89.69	78.46	107.94	125.53	144.60
Oil Warm Air Furnace	17.29	41.97	43.03	74.05	80.62	70.51	97.40	111.55	130.95
Central Electric Furnace	34.69	82.68	88.64	149.54	161.47	136.69	196.28	231.35	284.62
Baseboard Convectors	33.40	81.06	86.75	147.30	159.11	133.76	193.41	221.94	250.46
Cooling									
Electric Heat Pump - Standard	81.04 (70.46)	55.87 (47.79)	54.80 (46.65)	43.41 (36.55)	33.58 (28.47)	10.39 (9.27)	31.10 (26.60)	23.91 (20.33)	16.20 (14.04)
Electric Heat Pump - High Efficiency I	66.61 (58.16)	47.07 (40.63)	46.39 (39.72)	35.40 (29.91)	28.23 (24.07)	8.49 (7.62)	26.11 (22.46)	19.89 (17.01)	13.37 (11.59)
Electric Heat Pump - High Efficiency II	73.41 (69.98)	50.69 (42.98)	51.08 (44.04)	38.66 (32.85)	30.45 (25.51)	9.59 (8.42)	28.24 (23.85)	21.60 (18.14)	14.96 (12.76)
Electric Heat Pump - High Efficiency III	70.61 (61.53)	48.55 (41.77)	47.80 (40.78)	43.60 (36.74)	29.29 (24.89)	8.86 (7.83)	27.11 (23.24)	20.85 (17.77)	13.01 (11.16)
Hybrid Electric Heat Pump	85.02 (73.08)	58.59 (49.73)	57.82 (48.71)	45.10 (37.32)	35.42 (29.62)	11.10 (9.68)	32.82 (27.76)	25.29 (21.27)	16.96 (14.11)
Central Air Conditioner - High Efficiency	73.12 (63.62)	50.27 (43.18)	49.46 (42.14)	42.46 (35.81)	30.29 (25.70)	9.42 (8.32)	28.06 (24.01)	21.58 (18.34)	14.04 (12.06)
Central Air Conditioner - Standard	87.20 (76.02)	60.06 (51.70)	59.13 (50.48)	44.66 (37.68)	36.23 (30.80)	11.00 (9.87)	33.53 (28.75)	25.78 (21.98)	16.27 (14.03)
Window Air Conditioners	38.21	26.51	26.68	21.40	17.63	8.14	16.73	1.84	9.53
Annual Heating and Cooling									
Electric Heat Pump - Standard	96.79 (84.62)	97.75 (85.73)	98.97 (86.73)	131.71 (117.84)	118.63 (106.21)	83.06 (73.89)	145.89 (131.95)	161.2 (146.50)	215.28 (200.65)
Electric Heat Pump - High Efficiency I	79.78 (70.02)	83.18 (73.30)	84.11 (73.93)	112.24 (100.51)	102.13 (91.63)	74.56 (65.86)	128.40 (116.12)	142.8 (129.67)	198.43 (184.78)
Electric Heat Pump - High Efficiency II	87.50 (75.62)	88.23 (76.91)	92.61 (81.72)	114.23 (102.33)	106.40 (94.94)	72.95 (64.91)	131.32 (118.38)	144.9 (131.45)	183.58 (170.47)
Electric Heat Pump - High Efficiency III	85.56 (74.96)	88.60 (77.96)	89.92 (79.07)	129.76 (115.97)	110.66 (99.33)	73.48 (65.36)	137.80 (124.80)	153.3 (139.50)	189.51 (176.43)
Hybrid Electric Heat Pump-Gas Furnace	103.70 (89.90)	106.55 (93.64)	108.23 (94.83)	126.53 (112.67)	129.46 (116.60)	94.31 (88.43)	147.40 (135.50)	160.6 (149.09)	177.13 (166.27)
Hybrid Electric Heat Pump-Oil Furnace	103.65 (89.86)	106.17 (93.36)	107.97 (94.61)	125.32 (111.63)	128.37 (115.61)	92.90 (87.98)	144.59 (133.16)	156.6 (146.22)	172.37 (162.07)
Gas Warm Air Furnace/Central Air Conditioner - High Efficiency	92.70	97.36	97.71	126.24	119.98	87.88	136.00	8.08	158.64
Oil Warm Air Furnace/Central Air Conditioner - High Efficiency	90.41	92.23	92.48	116.51	110.92	79.93	125.46	5.10	144.99
Central Electric Furnace/Central Air Conditioner - High Efficiency	107.81	133.05	138.10	192.00	191.76	146.11	224.34	2.90	298.66
Baseboard Convectors/Window Air Conditioners	71.61	107.57	113.43	168.70	176.74	141.90	210.14	3.78	259.99
Electric Heat Pump-Heat Only/Central Air Conditioner - High Efficiency	86.28 (75.41)	85.14 (74.64)	86.60 (75.94)	111.97 (99.63)	99.29 (88.77)	67.09 (59.74)	122.54 (110.35)	135.0 (122.21)	168.46 (156.56)

* Numbers in parentheses are on-site energy consumptions based on steady state or full load performance i.e., no dynamic efficiency losses are included.

SOURCE: Gordian and Associates [6].

TABLE 12.10. Advanced Heating and Cooling Systems: Total Fuel and Electric Energy Consumption (Btu $\times 10^{-6}$) on a Primary Energy Basis

	Houston	Birmingham	Atlanta	Tulsa	Philadelphia	Seattle	Columbus	Cleveland	Concord
HEATING									
Advanced Electric Heat Pump I	**	33.08	35.15	61.52	65.87	**	82.07	97.57	**
Advanced Electric Heat Pump II	8.11	22.16	23.33	49.97	45.91	36.45	61.24	73.26	108.05
Free Piston Stirling-Rankine Gas Heat Pump	9.32	23.47	24.17	43.30	46.43	39.43	59.05	68.97	83.05
V-Type Single Cylinder Stirling-Rankine Gas HP	8.83	22.51	23.12	44.46	44.51	38.77	55.67	66.02	87.19
Organic Fluid Absorption Gas Heat Pump	10.84	27.12	27.79	49.81	53.44	45.97	65.44	78.91	93.21
Pulse Combustion Furnace	13.09	31.34	32.21	55.97	60.37	53.39	73.04	86.34	98.31
COOLING									
Advanced Electric Heat Pump I	**	44.82	43.84	34.54	26.83	**	25.03	19.20	**
Advanced Electric Heat Pump II	40.68	27.89	27.22	21.33	16.57	5.41	15.52	11.87	8.19
Free Piston Stirling-Rankine Gas Heat Pump	66.82	48.00	45.31	35.59	28.87	9.99	26.95	20.64	13.41
V-Type Single Cylinder Stirling-Rankine Gas HP	75.64	55.71	51.72	45.36	32.62	11.20	30.89	23.11	15.48
Organic Fluid Absorption Gas Heat Pump	99.93	72.25	67.90	53.32	43.44	15.19	40.57	31.02	20.08
ANNUAL HEATING AND COOLING									
Advanced Electric Heat Pump I	**	77.90	78.99	96.06	92.70	**	107.10	116.77	**
Advanced Electric Heat Pump II	48.79	50.05	50.55	71.30	62.48	41.86	76.76	85.13	116.24
Free Piston Stirling-Rankine Gas Heat Pump	76.14	71.47	69.48	78.89	75.30	49.42	85.01	89.61	96.46
V-Type Single Cylinder Stirling-Rankine Gas HP	84.46	78.21	74.84	89.82	77.13	49.97	86.56	89.13	102.67
Organic Fluid Absorption Gas Heat Pump	110.77	99.37	95.69	103.13	96.88	61.16	107.01	109.93	113.29
Pulse Combustion Furnace/Central Air Conditioner-High Efficiency	76.71	74.53	74.34	91.77	86.08	61.72	97.06	104.68	110.36

* The performance factors presented may be underestimated for the advanced electric heat pump I, free piston Stirling-Rankine gas heat pump and organic fluid absorption gas heat pump due to possible underestimation of cycling and defrost losses. If the dynamic efficiency losses of the advanced heat pumps were similar to those of the conventional electric (Rankine cycle) heat pump, then adjusted estimates of performance factors which include these effects can be obtained by multiplying the values in this table by a factor of 1.087 for the heating season, 1.164 for the cooling season and 1.118 for annual performance factor values.

** System not sized for application in these cities.

SOURCE: Gordian and Associates [6].

674

small impact of air-conditioning requirements compared with heating-season energy needs in colder climates. The hybrid heat pump in colder regions consumes up to about 15% more total primary energy annually than fossil-fuel furnace/high-efficiency air conditioners, while the best high-efficiency electric heat pump uses up to 20% more.

On an annual basis, the free-piston Stirling-Rankine gas heat pump appears to consume the smallest amount of primary energy compared with the other heat-activated heat pump systems. In the warmer cities, the pulse combustion gas furnace/high-efficiency air conditioner combination is only slightly more energy-consuming than the free-piston Stirling-Rankine gas heat pump. The relative primary energy consumption of the combined system seems to increase in the colder cities, however, where the V-type single-cylinder Stirling-Rankine gas heat pump can be ranked second to the free-piston system. The advanced electric heat pump I also appears to consume more primary energy compared to other advanced systems as climatological conditions call for more heating in the colder cities. This behavior is experienced more dramatically with the advanced electric heat pump II, which is the most primary energy efficient system in the warm cities (Houston, Birmingham, etc.), but is the least efficient unit in the coldest city (Concord).

Commercial HVAC system primary energy SPF data are presented in Table 12.11. The results show that the Rankine-Rankine and Brayton-Rankine heat pumps have a higher primary energy efficiency (performance factors 1.43–0.76 heating; 0.91–0.72, cooling) than electric heat pumps or rooftop air conditioners with gas or electric resistance heat. Primary energy consumption data, presented in Table 12.12, clearly show that the new technology advanced unitary heat pump under development for light commercial building applications will consume less annual primary energy than any of the conventional heating and cooling systems examined. The Rankine-Rankine gas heat pump performs better than the Brayton-Rankine gas heat pump in all the cities examined. Among the conventional heating and cooling systems, the standard electric heat pump uses the least energy in all climates compared with the single-package rooftop gas furnace and electric resistance heating combined with electric air-conditioning. Generally, the Rankine-Rankine gas heat pump uses about 25–46% less energy for commercial building comfort-conditioning than the standard electric heat pump in all cities examined. The larger savings are associated with colder climates.

12.5. SUMMARY

Figure 12.15 presents the time schedule projected by Gordian [6] for the development of improved and advanced heat pump systems. The primary energy performances of these systems as determined in the Gordian Study [6] are summarized below.

TABLE 12.11. Electric Heat Pump, Conventional Systems, and Advanced Heat Pumps for Light Commercial Building Application: Average Heating, Cooling and Annual Performance Factor—Primary Energy Basis

Electric Heat Pump and Conventional Systems		Houston	Birmingham	Atlanta	Tulsa	Philadelphia	Seattle	Columbus	Cleveland	Concord
Electric Heat Pump - Standard										
	Heating	0.81	0.76	0.70	0.72	0.75	0.75	0.66	0.66	0.59
	Cooling	0.65	0.65	0.62	0.65	0.64	0.70	0.67	0.68	0.63
	Annual	0.66	0.68	0.64	0.67	0.69	0.73	0.66	0.66	0.60
Electric Resistance Heating/Electric Air Conditioning										
	Heating	0.29	0.29	0.28	0.29	0.29	0.31	0.30	0.30	0.28
	Cooling	0.65	0.65	0.62	0.65	0.54	0.71	0.68	0.68	0.61
	Annual	0.59	0.50	0.47	0.45	0.39	0.39	0.39	0.37	0.34
Rooftop Gas Furnace/Electric Air Conditioning										
	Heating	0.32	0.41	0.39	0.43	0.41	0.43	0.45	0.45	0.46
	Cooling	0.60	0.60	0.58	0.60	0.60	0.66	0.63	0.63	0.59
	Annual	0.56	0.54	0.51	0.53	0.50	0.49	0.51	0.50	0.50
Advanced Heat Pump Systems										
Brayton-Rankine Gas Heat Pump										
	Heating	0.94	0.92	0.86	0.87	0.91	0.93	0.84	0.84	0.76
	Cooling	0.78	0.76	0.76	0.74	0.77	0.73	0.77	0.77	0.72
	Annual	0.79	0.79	0.78	0.78	0.83	0.85	0.81	0.82	0.75
Rankine-Rankine Gas Heat Pump										
	Heating	1.43	1.35	1.32	1.30	1.34	1.35	1.28	1.27	1.27
	Cooling	0.85	0.86	0.86	0.88	0.85	0.91	0.88	0.90	0.91
	Annual	0.88	0.94	0.95	1.00	1.01	1.16	1.08	1.12	1.11

SOURCE: Gordian and Associates [6].

TABLE 12.12. Electric Heat Pump, Conventional Systems, and Advanced Heat Pumps for Light Commercial Building Application: Annual Primary Energy Consumption—Combined Heating and Cooling (MBtu)

Electric Heat Pump and Conventional Systems	Houston	Birmingham	Atlanta	Tulsa	Philadelphia	Seattle	Columbus	Cleveland	Concord
Electric Heat Pump - Standard	363.49	300.41	288.76	295.70	219.35	163.48	239.78	234.91	285.74
Electric Resistance Heating/Electric Air Conditioning	408.34	405.88	397.64	443.28	383.37	310.18	409.56	420.62	499.45
Rooftop Gas Furnace/Electric Air Conditioning	429.03	374.63	362.55	377.09	302.40	246.74	311.04	309.29	342.84
Advanced Heat Pump Systems									
Brayton-Rankine Gas Heat Pump	303.96	256.33	237.67	253.70	182.00	141.49	196.81	190.85	230.07
Rankine-Rankine Gas Heat Pump	274.23	215.46	195.96	199.18	148.46	103.73	147.31	139.30	154.27

SOURCE: Gordian and Associates [6].

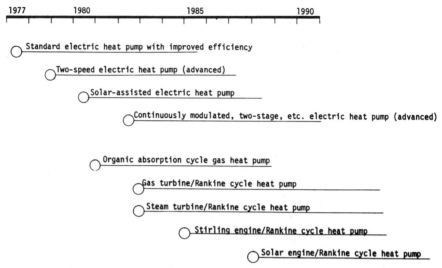

Fig. 12.15. Projected time scale of advanced electric and gas heat pump product introduction [6].

The greatest consumers of primary energy, on a total thermal equivalent basis, are electric resistance heating systems, represented by the central electric furnace and baseboard convectors. The smallest amount of total primary energy is consumed by fossil-fuel warm-air furnaces in mild winter areas of the country and by conventional electric heat pumps in colder winter areas. Conventional electric air conditioners are generally not as energy efficient with respect to primary energy usage as high-efficiency electric heat pumps for air cooling.

The Stirling-Rankine gas heat pump uses the least amount of total primary energy for both heating and cooling in all cities examined and is matched by a highly efficient advanced electric heat pump only in mild winter regions. The organic fluid absorption gas heat pump is not as efficient as the Stirling-Rankine gas heat pump.

The regional analysis was performed by selectively examining the impact on primary fuels of replacing conventional HVAC systems with advanced heating and cooling systems currently being developed. It was generally found that advanced systems replacing conventional systems using essentially the same type of on-site energy will generally conserve all types of primary fuel resources. Thus, an advanced electric heat pump replacing a conventional electric heat pump will conserve all forms of primary fuel resources. The quantities of primary fuels that could potentially be saved depend strongly on the relative efficiencies of the advanced and conventional systems being examined and the region of the country considered. Local conditions determine the fuel mix used to generate electric energy and the severity of winter and summer weather to which heat pumps are sensitive.

Advanced gas heat pumps examined were found to have potential for reducing all types of primary fuels, except natural gas, when replacing conventional electric heat pumps. However, this result is dependent upon the extent to which electrical energy is required for gas heat pump auxiliaries. Natural gas consumption is always found to increase for this replacement strategy. In replacing less-efficient gas equipment, advanced gas heat pumps were found to have a potential for conserving all primary fuels, including natural gas, but this result must be carefully qualified. Because heat pump performance is generally not as efficient for cooling as for heating, only very high efficiency gas heat pumps will be acceptable in warmer regions of the country where cooling season energy is a major part of annual system energy requirements. For the two different generic gas heat pump and gas heat pumps based on the Stirling-Rankine cycle, only the Stirling-Rankine system appears to have the potential for conserving natural gas in all regions of the country. The absorption gas heat pump would conserve natural gas (and other primary fuels) in colder regions of the country, but not in warmer regions where its cooling efficiency is too low. In any case, the conservation potential for natural gas is greater for the Stirling-Rankine gas heat pumps than for the absorption gas heat pump in all parts of the country.

The hybrid heat pump with fossil-fuel-fired furnace appears to have potential for conserving all primary fuel resources except natural gas or fuel oil compared with currently available conventional electric heat pumps.

The pulse combustion furnace appears to have the greatest potential for reducing natural gas use for residential heating of all the gas-fired advanced systems examined. Advanced electric heat pumps would probably reduce natural gas consumption even more dramatically and certainly to a greater extent than is estimated for advanced gas heat pumps.

The same general findings for residential HVAC systems may be applied to the light commercial HVAC systems. When compared to the all electric HVAC systems, gas heat pumps conserve all primary fuel resources except natural gas. When compared to the rooftop gas furnace/electric air conditioner system, gas heat pumps conserve all forms of primary fuel resources for heating, but consume more natural gas for cooling. Hence, in those regions where the primary HVAC service is for cooling, a net increase, on an annual basis, in natural gas consumption is observed.

REFERENCES

1. Christian, J. E., "Unitary Air-to-Air Heat Pumps," Argonne National Laboratory, ANL/ CES/TE 77-10, July 1977.
2. Hise, E. C., Moyers, J. C., and Fischer, H. C., "Design Report for the ACES Demonstration House," Oak Ridge National Laboratory, ORNL/CON-1, Oct. 1976.
3. Edwards, T. C., "The ROVAC Automotive Air Conditioning System," Society of Automotive Engineers, *SAE Tech. Paper* 750403, 1975.
4. Edwards, T., Clark, W., and Cavalleri, R., "Preliminary Development of a High Efficiency Low Vapor Pressure Non-Fluorocarbon Air Conditioning Compressor," Society of Automotive Engineers, *SAE Tech. Paper* 810505, 1981.
5. American Society of Heating, Refrigerating and Air Conditioning Engineers, *Handbook and Product Directory, 1977, Fundamentals,* ASHRAE, New York, 1977.
6. Gordian Associates, Inc., *Heat Pump Technology, A Survey of Technical Development Market Prospects,* report prepared for DOE, HCP/M2121-01, June 1978.

APPENDIX A

REFRIGERANT-12 TABLES

TABLE A.1. Refrigerant-12 (Dichlorodifluoromethane) Properties of Liquid and Saturated Vapor[a]

Temp F	Pressure		Volume cu ft/lb	Density lb/cu ft	Enthalpy Btu/lb		Entropy Btu/(lb)(°R)	
	psia	psig	Vapor v_g	Liquid $1/v_f$	Liquid h_f	Vapor h_g	Liquid s_f	Vapor s_g
-152	0.13799	29.64024*	197.58	104.52	-23.106	60.628	-0.063944	0.20818
-150	0.15359	29.60849*	178.65	104.36	-22.697	60.837	.062619	.20711
-145	0.19933	29.51537*	139.83	103.95	-21.674	61.365	.059344	.20452
-140	0.25623	29.39951*	110.46	103.54	-20.652	61.896	.056123	.20208
-135	0.32641	29.25663*	88.023	103.13	-19.631	62.430	.052952	.19978
-130	0.41224	29.08187*	70.730	102.71	-18.609	62.968	-0.049830	0.19760
-125	0.51641	28.86978*	57.283	102.29	-17.587	63.509	.046754	.19554
-120	0.64190	28.61429*	46.741	101.87	-16.565	64.052	.043723	.19359
-115	0.79200	28.30869*	38.410	101.45	-15.541	64.598	.040734	.19176
-110	0.97034	27.94558*	31.777	101.02	-14.518	65.145	.037786	.19002
-105	1.1809	27.5169*	26.458	100.59	-13.492	65.696	-0.034877	0.18838
-100	1.4280	27.0138*	22.164	100.15	-12.466	66.248	.032005	.18683
-95	1.7163	26.4268*	18.674	99.715	-11.438	66.801	.029169	.18536
-90	2.0509	25.7456*	15.821	99.274	-10.409	67.355	.026367	.18398
-85	2.4371	24.9593*	13.474	98.830	-9.3782	67.911	.023599	.18267
-80	2.8807	24.0560*	11.533	98.382	-8.3451	68.467	-0.020862	0.18143
-75	3.3879	23.0234*	9.9184	97.930	-7.3101	69.023	.018156	.18027
-70	3.9651	21.8482*	8.5687	97.475	-6.2730	69.580	.015481	.17916
-65	4.6139	20.5164*	7.4347	97.016	-5.2336	70.137	.012834	.17812
-60	5.3375	19.0133*	6.4774	96.553	-4.1919	70.693	.010214	.17714
-55	6.1874	17.3237*	5.6656	96.086	-3.1477	71.249	-0.007622	0.17621
-50	7.1168	15.4313*	4.9742	95.618	-2.1011	71.805	.005056	.17533
-45	8.1540	13.3196*	4.3828	95.141	-1.0519	72.358	.002516	.17451
-40	9.3076	10.9709*	3.8750	94.661	0.0000	72.913	.000000	.17373
-35	10.586	8.367*	3.4373	94.178	1.0546	73.464	.002492	.17299
-30	11.999	5.490*	3.0585	93.690	2.1120	74.015	0.004961	0.17229
-28	12.604	4.259*	2.9214	93.493	2.5358	74.234	.005942	.17203
-26	13.233	2.979*	2.7917	93.296	2.9601	74.454	.006919	.17177
-24	13.886	1.649*	2.6691	93.098	3.3848	74.673	.007894	.17151
-22	14.564	0.270*	2.5529	92.899	3.8100	74.891	.008864	.17126
-20	15.267	0.571	2.4429	92.699	4.2357	75.110	0.009831	0.17102
-18	15.996	1.300	2.3387	92.499	4.6618	75.328	.010795	.17078
-16	16.753	2.057	2.2399	92.298	5.0885	75.545	.011755	.17055
-14	17.536	2.840	2.1461	92.096	5.5157	75.762	.012712	.17032
-12	18.348	3.652	2.0572	91.893	5.9434	75.979	.013666	.17010
-10	19.189	4.493	1.9727	91.689	6.3716	76.196	0.014617	0.16989
-8	20.059	5.363	1.8924	91.485	6.8003	76.411	.015564	.16967
-6	20.960	6.264	1.8161	91.280	7.2296	76.627	.016508	.16947
-4	21.891	7.195	1.7436	91.074	7.6594	76.842	.017449	.16927
-2	22.854	8.158	1.6745	90.867	8.0898	77.055	.018388	.16907
0	23.849	9.153	1.6089	90.659	8.5207	77.271	0.019323	0.16888
2	24.878	10.182	1.5463	90.450	8.9522	77.485	.020255	.16868
4	25.939	11.243	1.4867	90.240	9.3843	77.698	.021184	.16851
† 5	26.483	11.787	1.4580	90.135	9.6005	77.805	.021647	.16842
6	27.036	12.340	1.4299	90.030	9.8169	77.911	.022110	.16833
8	28.167	13.471	1.3758	89.818	10.250	78.123	0.023033	0.16815
10	29.335	14.639	1.3241	89.606	10.684	78.335	.023954	.16798
12	30.539	15.843	1.2748	89.392	11.118	78.546	.024871	.16782
14	31.780	17.084	1.2278	89.178	11.554	78.757	.025786	.16765
16	33.060	18.364	1.1828	88.962	11.989	78.966	.026699	.16750
18	34.378	19.682	1.1399	88.746	12.426	79.176	0.027608	0.16734
20	35.736	21.040	1.0988	88.529	12.863	79.385	.028515	.16719
22	37.135	22.439	1.0596	88.310	13.300	79.593	.029420	.16704
24	38.574	23.878	1.0220	88.091	13.739	79.800	.030322	.16690
26	40.056	25.360	0.98612	87.870	14.178	80.007	.031221	.16676
28	41.580	26.884	0.95173	87.649	14.618	80.214	0.032118	0.16662
30	43.148	28.452	.91880	87.426	15.058	80.419	.033013	.16648
31	43.948	29.252	.90286	87.314	15.279	80.522	.033460	.16642
32	44.7C0	30.064	.88725	87.202	15.500	80.624	.033905	.16635
33	45.583	30.887	.87197	87.090	15.720	80.726	.034351	.16629
34	46.417	31.721	0.85702	86.977	15.942	80.828	0.034796	0.16622
35	47.263	32.567	.84237	86.865	16.163	80.930	.035240	.16616
36	48.120	33.424	.82803	86.751	16.384	81.031	.035683	.16610
37	48.989	34.293	.81399	86.638	16.606	81.133	.036126	.16604
38	49.870	35.174	.80023	86.524	16.828	81.234	.036569	.16598
39	50.763	36.067	0.78676	86.410	17.050	81.335	0.037011	0.16592
40	51.667	36.971	.77357	86.296	17.273	81.436	.037453	.16586
41	52.548	37.888	.76064	86.181	17.495	81.537	.037893	.16581
42	53.513	38.817	.74798	86.066	17.718	81.637	.038334	.16574
43	54.454	39.758	.73557	85.951	17.941	81.737	.038774	.16568
44	55.407	40.711	0.72341	85.836	18.164	81.837	0.039213	0.16562
45	56.373	41.677	.71149	85.720	18.387	81.937	.039652	.16557
46	57.352	42.656	.69982	85.604	18.611	82.037	.040091	.16551
47	58.343	43.647	.68837	85.487	18.835	82.136	.040529	.16546
48	59.347	44.651	.67715	85.371	19.059	82.236	.040966	.16541
49	60.364	45.668	0.66616	85.254	19.283	82.334	0.041403	0.16535
50	61.394	46.698	.65537	85.136	19.507	82.433	.041839	.16530
51	62.437	47.741	.64480	85.018	19.732	82.532	.042276	.16524
52	63.494	48.798	.63444	84.900	19.957	82.630	.042711	.16519
53	64.563	49.867	.62428	84.782	20.182	82.728	.043146	.16514
54	65.646	50.950	.61431	84.663	20.408	82.826	.043581	.16509
55	66.743	52.047	0.60453	84.544	20.634	82.924	0.044015	0.16504
56	67.853	53.157	.59495	84.425	20.859	83.021	.044449	.16499
57	68.977	54.281	.58554	84.305	21.086	83.119	.044883	.16494
58	70.115	55.419	.57632	84.185	21.312	83.215	.045316	.16489
59	71.267	56.571	.56727	84.065	21.539	83.312	.045748	.16484
60	72.433	57.737	0.55839	83.944	21.766	83.409	0.046180	0.16479
61	73.613	58.917	.54967	83.823	21.993	83.505	.046612	.16474
62	74.807	60.111	.54112	83.701	22.221	83.601	.047044	.16470
63	76.016	61.320	.53273	83.580	22.448	83.696	.047475	.16465
64	77.239	62.543	.52450	83.457	22.676	83.792	.047905	.16460
65	78.477	63.781	0.51642	83.335	22.905	83.887	0.048336	0.16456
66	79.729	65.033	.50848	83.212	23.133	83.982	.048765	.16451
67	80.996	66.300	.50070	83.089	23.362	84.077	.049195	.16446
68	82.279	67.583	.49305	82.965	23.591	84.171	.049624	.16442
69	83.576	68.880	.48555	82.841	23.821	84.266	.050053	.16438
70	84.888	70.192	0.47818	82.717	24.050	84.359	0.050482	0.16434
71	86.216	71.520	.47094	82.592	24.281	84.453	.050910	.16429
72	87.559	72.863	.46383	82.467	24.511	84.546	.051338	.16425
73	88.918	74.222	.45686	82.341	24.741	84.639	.051766	.16421
74	90.292	75.596	.45000	82.215	24.973	84.732	.052193	.16417
75	91.682	76.986	0.44327	82.089	25.204	84.825	0.052620	0.16412
76	93.087	78.391	.43666	81.962	25.435	84.916	.053047	.16408
77	94.509	79.813	.43016	81.835	25.667	85.008	.053473	.16404
78	95.946	81.250	.42378	81.707	25.899	85.100	.053900	.16400
79	97.400	82.704	.41751	81.579	26.132	85.191	.054326	.16396
80	98.870	84.174	0.41135	81.450	26.365	85.282	0.054751	0.16392
81	100.36	85.66	.40530	81.322	26.598	85.373	.055177	.16388
82	101.86	87.16	.39935	81.192	26.832	85.463	.055602	.16384
83	103.38	88.68	.39351	81.063	27.065	85.553	.056027	.16380
84	104.92	90.22	.38776	80.932	27.300	85.643	.056452	.16376
85	106.47	91.77	0.38212	80.802	27.534	85.732	0.056877	0.16372
† 86	108.04	93.34	.37657	80.671	27.769	85.821	.057301	.16368
87	109.63	94.93	.37111	80.539	28.005	85.910	.057725	.16364
88	111.23	96.53	.36575	80.407	28.241	85.998	.058149	.16360
89	112.85	98.15	.36047	80.275	28.477	86.086	.058573	.16357
90	114.49	99.79	0.35529	80.142	28.713	86.174	0.058997	0.16353
92	117.82	103.12	.34518	79.874	29.187	86.348	.059844	.16345
94	121.22	106.52	.33540	79.605	29.663	86.521	.060690	.16338
96	124.70	110.00	.32594	79.334	30.140	86.691	.061536	.16330
98	128.24	113.54	.31679	79.061	30.619	86.861	.062381	.16323
100	131.86	117.16	0.30794	78.785	31.100	87.029	0.063227	0.16315
102	135.56	120.86	.29937	78.508	31.583	87.196	.064072	.16308
104	139.33	124.63	.29106	78.228	32.067	87.360	.064916	.16301
106	143.18	128.48	.28303	77.946	32.553	87.523	.065761	.16293
108	147.11	132.41	.27524	77.662	33.041	87.684	.066606	.16286
110	151.11	136.41	0.26769	77.376	33.531	87.844	0.067451	0.16279
112	155.19	140.49	.26037	77.087	34.023	88.001	.068296	.16271
114	159.36	144.66	.25328	76.795	34.517	88.156	.069141	.16264
116	163.61	148.91	.24641	76.501	35.014	88.310	.069987	.16256
118	167.94	153.24	.23974	76.205	35.512	88.461	.070833	.16249
120	172.35	157.65	0.23326	75.906	36.013	88.610	0.071680	0.16241
122	176.85	162.15	.22698	75.604	36.516	88.757	.072528	.16234
124	181.43	166.73	.22089	75.299	37.021	88.902	.073376	.16226
126	186.10	171.40	.21497	74.991	37.529	89.044	.074225	.16218
128	190.86	176.16	.20922	74.680	38.040	89.184	.075075	.16210
130	195.71	181.01	0.20364	74.367	38.553	89.321	0.075927	0.16202
132	200.64	185.94	.19821	74.050	39.069	89.456	.076779	.16194
134	205.67	190.97	.19294	73.729	39.588	89.588	.077633	.16185
136	210.79	196.09	.18782	73.406	40.110	89.718	.078489	.16177
138	216.01	201.31	.18283	73.079	40.634	89.844	.079346	.16168
140	221.32	206.62	0.17799	72.748	41.162	89.967	0.080205	0.16159
145	235.00	220.30	.16644	71.904	42.495	90.261	.082361	.16135
150	249.31	234.61	.15564	71.035	43.850	90.534	.084531	.16110
155	264.24	249.54	.14552	70.137	45.229	90.783	.086719	.16083
160	279.82	265.12	.13604	69.209	46.633	91.006	.088927	.16053
165	296.07	281.37	0.12712	68.245	48.065	91.199	0.091159	0.16021
170	313.00	298.30	.11873	67.244	49.529	91.359	.093418	.15985
175	330.64	315.94	.11080	66.198	51.026	91.481	.095709	.15945
180	349.00	334.30	.10330	65.102	52.562	91.561	.098039	.15900
185	368.11	353.41	.096190	63.949	54.141	91.590	.10041	.15850
190	387.98	373.28	0.089418	62.728	55.769	91.561	0.10284	0.15793
195	408.63	393.93	.082946	61.426	57.453	91.462	.10532	.15727
200	430.09	415.39	.076728	60.026	59.203	91.278	.10789	.15651
205	452.38	437.68	.070714	58.502	61.032	90.987	.11055	.15561
210	475.52	460.82	.064843	56.816	62.959	90.558	.11332	.15453
215	499.53	484.83	0.059030	54.908	65.014	89.939	0.11626	0.15323
220	524.43	509.73	.053140	52.658	67.246	89.036	.11943	.15149
225	550.26	535.56	.046900	49.868	69.763	87.651	.12298	.14911
230	577.03	562.33	.039435	45.758	72.893	85.122	.12739	.14512
233.6 (Critical)	596.9	582.2	.02870	34.84	78.86	78.86	.1359	.1359

* From published data (1955 and 1956) of E. I. du Pont de Nemours & Co., Inc. Used by permission.
* Inches of mercury below one standard atmosphere.
** Based on 0 for the saturated liquid at -40 F.
† Standard cycle temperatures.

TABLE A.2. Superheated Freon-12[a]

Temp., F	v	h	s	v	h	s	v	h	s
		5 lbf/in.[2]			10 lbf/in.[2]			15 lbf/in.[2]	
0	8.0611	78.582	0.19663	3.9809	78.246	0.18471	2.6201	77.902	0.17751
20	8.4265	81.309	0.20244	4.1691	81.014	0.19061	2.7494	80.712	0.18349
40	8.7903	84.090	0.20812	4.3556	83.828	0.19635	2.8770	83.561	0.18931
60	9.1528	86.922	0.21367	4.5408	86.689	0.20197	3.0031	86.451	0.19498
80	9.5142	89.806	0.21912	4.7248	89.596	0.20746	3.1281	89.383	0.20051
100	9.8747	92.738	0.22445	4.9079	92.548	0.21283	3.2521	92.357	0.20593
120	10.234	95.717	0.22968	5.0903	95.546	0.21809	3.3754	95.373	0.21122
140	10.594	98.743	0.23481	5.2720	98.586	0.22325	3.4981	98.429	0.21640
160	10.952	101.812	0.23985	5.4533	101.669	0.22830	3.6202	101.525	0.22148
180	11.311	104.925	0.24479	5.6341	104.793	0.23326	3.7419	104.661	0.22646
200	11.668	108.079	0.24964	5.8145	107.957	0.23813	3.8632	107.835	0.23135
220	12.026	111.272	0.25441	5.9946	111.159	0.24291	3.9841	111.046	0.23614
		20 lbf/in.[2]			25 lbf/in.[2]			30 lbf/in.[2]	
20	2.0391	80.403	0.17829	1.6125	80.088	0.17414	1.3278	79.765	0.17065
40	2.1373	83.289	0.18419	1.6932	83.012	0.18012	1.3969	82.730	0.17671
60	2.2340	86.210	0.18992	1.7723	85.965	0.18591	1.4644	85.716	0.18257
80	2.3295	89.168	0.19550	1.8502	88.950	0.19155	1.5306	88.729	0.18826
100	2.4211	92.164	0.20095	1.9271	91.968	0.19704	1.5957	91.770	0.19379
120	2.5179	95.198	0.20628	2.0032	95.021	0.20240	1.6600	94.843	0.19918
140	2.6110	98.270	0.21149	2.0786	98.110	0.20763	1.7237	97.948	0.20445
160	2.7036	101.380	0.21659	2.1535	101.234	0.21276	1.7868	101.086	0.20960
180	2.7957	104.528	0.22159	2.2279	104.393	0.21778	1.8494	104.258	0.21463
200	2.8874	107.712	0.22649	2.3019	107.588	0.22269	1.9116	107.464	0.21957
220	2.9789	110.932	0.23130	2.3756	110.817	0.22752	1.9735	110.702	0.22440
240	3.0700	114.186	0.23602	2.4491	114.080	0.23225	2.0351	113.973	0.22915
		35 lbf/in.[2]			40 lbf/in.[2]			50 lbf/in.[2]	
40	1.1850	82.442	0.17375	1.0258	82.148	0.17112	0.80248	81.540	0.16655
60	1.2442	85.463	0.17968	1.0789	85.206	0.17712	0.84713	84.676	0.17271
80	1.3021	88.504	0.18542	1.1306	88.277	0.18292	0.89025	87.811	0.17862
100	1.3589	91.570	0.19100	1.1812	91.367	0.18854	0.93216	90.953	0.18434
120	1.4148	94.663	0.19643	1.2309	94.480	0.19401	0.97313	94.110	0.18988
140	1.4701	97.785	0.20172	1.2798	97.620	0.19933	1.0133	97.286	0.19527
160	1.5248	100.938	0.20689	1.3282	100.788	0.20453	1.0529	100.485	0.20051
180	1.5789	104.122	0.21195	1.3761	103.985	0.20961	1.0920	103.708	0.20563
200	1.6327	107.338	0.21690	1.4236	107.212	0.21457	1.1307	106.958	0.21064
220	1.6862	110.586	0.22175	1.4707	110.469	0.21944	1.1690	110.235	0.21553
240	1.7394	113.865	0.22651	1.5176	113.757	0.22420	1.2070	113.539	0.22032
260	1.7923	117.175	0.23117	1.5642	117.074	0.22888	1.2447	116.871	0.22502
		60 lbf/in.[2]			70 lbf/in.[2]			80 lbf/in.[2]	
60	0.69210	84.126	0.16892	0.58088	83.552	0.16556	...	...	...
80	0.72964	87.330	0.17497	0.61458	86.832	0.17175	0.52795	86.316	0.16885
100	0.76588	90.528	0.18079	0.64685	90.091	0.17768	0.55734	89.640	0.17489
120	0.80110	93.731	0.18641	0.67803	93.343	0.18339	0.58556	92.945	0.18070
140	0.83551	96.945	0.19186	0.70836	96.597	0.18891	0.61286	96.242	0.18629
160	0.86928	100.776	0.19716	0.73800	99.862	0.19427	0.63943	99.542	0.19170
180	0.90252	103.427	0.20233	0.76708	103.141	0.19948	0.66543	102.851	0.19696
200	0.93531	106.700	0.20736	0.79571	106.439	0.20455	0.69095	106.174	0.20207
220	0.96775	109.997	0.21229	0.82397	109.756	0.20951	0.71609	109.513	0.20706
240	0.99988	113.319	0.21710	0.85191	113.096	0.21435	0.74090	112.872	0.21193
260	1.0318	116.666	0.22182	0.87959	116.459	0.21909	0.76544	116.251	0.21669
280	1.0634	120.039	0.22644	0.90705	119.846	0.22373	0.78975	119.652	0.22135

Temp., F	v	h	s	v	h	s	v	h	s
		90 lbf/in.^{2}			100 lbf/in.^{2}			125 lbf/in.^{2}	
100	0.48749	89.175	0.17234	0.43138	88.694	0.16996	0.32943	87.407	0.16455
120	0.51346	92.536	0.17824	0.45562	92.116	0.17597	0.35086	91.008	0.17087
140	0.53845	95.879	0.18391	0.47881	95.507	0.18172	0.37098	94.537	0.17686
160	0.56268	99.216	0.18938	0.50118	98.884	0.18726	0.39015	98.023	0.18258
180	0.58629	102.557	0.19469	0.52291	102.257	0.19262	0.40857	101.484	0.18807
200	0.60941	105.905	0.19984	0.54413	105.633	0.19782	0.42642	104.934	0.19338
220	0.63213	109.267	0.20486	0.56492	109.018	0.20287	0.44380	108.380	0.19853
240	0.65451	112.644	0.20976	0.58538	112.415	0.20780	0.46081	111.829	0.20353
260	0.67662	116.040	0.21455	0.60554	115.828	0.21261	0.47750	115.287	0.20840
280	0.69849	119.456	0.21923	0.62546	119.258	0.21731	0.49394	118.756	0.21316
300	0.72016	122.892	0.22381	0.64518	122.707	0.22191	0.51016	122.238	0.21780
320	0.74166	126.349	0.22830	0.66472	126.176	0.22641	0.52619	125.737	0.22235
		150 lbf/in.^{2}			175 lbf/in.^{2}			200 lbf/in.^{2}	
120	0.28007	89.800	0.16629	...	...	...	...	...	...
140	0.29845	93.498	0.17256	0.24595	92.373	0.16859	0.20579	91.137	0.17480
160	0.31566	97.112	0.17849	0.26198	96.142	0.17478	0.22121	95.100	0.17130
180	0.33200	100.675	0.18415	0.27697	99.823	0.18062	0.23535	98.921	0.17737
200	0.34769	104.206	0.18958	0.29120	103.447	0.18620	0.24860	102.652	0.18311
220	0.36285	107.720	0.19483	0.30485	107.036	0.19156	0.26117	106.325	0.18860
240	0.37761	111.226	0.19992	0.31804	110.605	0.19674	0.27323	109.962	0.19387
260	0.39203	114.732	0.20485	0.33087	114.162	0.20175	0.28489	113.576	0.19896
280	0.40617	118.242	0.20967	0.34339	117.717	0.20662	0.29623	117.178	0.20390
300	0.42008	121.761	0.21436	0.35567	121.273	0.21137	0.30730	120.775	0.20870
320	0.43379	125.290	0.21894	0.36773	124.835	0.21599	0.31815	124.373	0.21337
340	0.44733	128.833	0.22343	0.37963	128.407	0.22052	0.32881	127.974	0.21793
		250 lbf/in.^{2}			300 lbf/in.^{2}			400 lbf/in.^{2}	
160	0.16249	92.717	0.16462	...	...	.*.	...	...	...
180	0.17605	96.925	0.17130	0.13482	94.556	0.16537	...	...	...
200	0.18824	100.930	0.17747	0.14697	98.975	0.17217	0.091005	93.718	0.16092
220	0.19952	104.809	0.18326	0.15774	103.136	0.17838	0.10316	99.046	0.16888
240	0.21014	108.607	0.18877	0.16761	107.140	0.18419	0.11300	103.735	0.17568
260	0.22027	112.351	0.19404	0.17685	111.043	0.18969	0.12163	108.105	0.18183
280	0.23001	116.060	0.19913	0.18562	114.879	0.19495	0.12949	112.286	0.18756
300	0.23944	119.747	0.20405	0.19402	118.670	0.20000	0.13680	116.343	0.19298
320	0.24862	123.420	0.20882	0.20214	122.430	0.20189	0.14372	120.318	0.19814
340	0.25759	127.088	0.21346	0.21002	126.171	0.20903	0.15032	124.235	0.20310
360	0.26639	130.754	0.21799	0.21770	129.900	0.21423	0.15668	128.112	0.20789
380	0.27504	134.423	0.22241	0.22522	133.624	0.21872	0.16285	131.961	0.21255
		500 lbf/in.^{2}			600 lbf/in.^{2}				
220	0.064207	92.397	0.15683	...	...	...			
240	0.077620	99.218	0.16672	0.047488	91.024	0.15335			
260	0.087054	104.526	0.17421	0.061922	99.741	0.16566			
280	0.094923	109.277	0.18072	0.070859	105.637	0.17374			
300	0.10190	113.729	0.18666	0.078059	110.729	0.18053			
320	0.10829	117.997	0.19221	0.084333	115.420	0.18663			
340	0.11426	122.143	0.19746	0.090017	119.871	0.19227			
360	0.11992	126.205	0.20247	0.095289	124.167	0.19757			
380	0.12533	130.207	0.20730	0.10025	128.355	0.20262			
400	0.13054	134.166	0.21196	0.10498	132.466	0.20746			
420	0.13559	138.096	0.21648	0.10952	136.523	0.21213			
440	0.14051	142.004	0.22087	0.11391	140.539	0.21664			

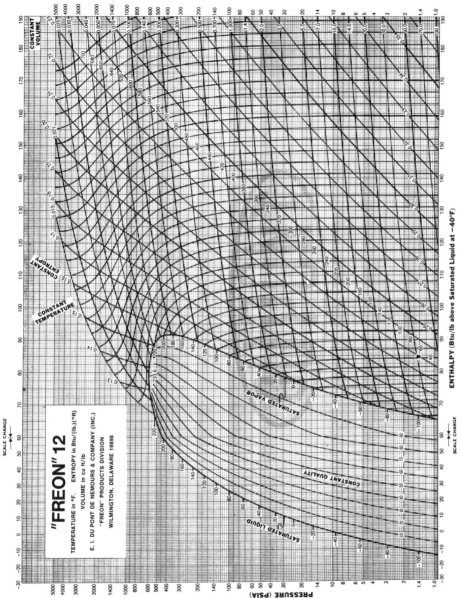

Figure A.1 Pressure-Enthalpy Diagram for R-12. Copyright © by Du Pont. Reproduced by permission.

685

APPENDIX B

REFRIGERANT-22 TABLES

TABLE B.1. Refrigerant-22 (Chlorodifluoromethane) Properties of Liquid and Saturated Vapor [a]

Temp F	Pressure psia	Pressure psig	Volume cu ft/lb Vapor v_g	Density lb/cu ft Liquid $1/v_f$	Enthalpy** Btu/lb Liquid h_f	Enthalpy** Btu/lb Vapor h_g	Entropy** Btu/(lb)(°R) Liquid s_f	Entropy** Btu/(lb)(°R) Vapor s_g
−150	0.27163	29.36816*	141.23	98.236	−25.974	87.521	−0.07147	0.29501
−145	0.34999	29.20861*	111.34	97.800	−24.851	88.100	−0.06787	0.29106
−140	0.44692	29.01126*	88.532	97.363	−23.725	88.681	−0.06432	0.28729
−135	0.56584	28.76914*	70.981	96.922	−22.596	89.263	−0.06082	0.28369
−130	0.71060	28.47441*	57.356	96.480	−21.463	89.848	−0.05736	0.28027
−125	0.88551	28.11829*	46.692	96.035	−20.326	90.433	−0.05394	0.27700
−120	1.0954	27.6910*	38.280	95.587	−19.185	91.020	−0.05055	0.27388
−115	1.3455	27.1818*	31.594	95.137	−18.038	91.608	−0.04720	0.27090
−110	1.6417	26.5788*	26.242	94.684	−16.886	92.196	−0.04389	0.26805
−105	1.9903	25.8689*	21.930	94.228	−15.728	92.783	−0.04060	0.26533
−100	2.3983	25.0383*	18.433	93.770	−14.564	93.371	−0.03734	0.26274
−98	2.5798	24.6688*	17.222	93.585	−14.097	93.606	−0.03605	0.26173
−96	2.7724	24.2765*	16.104	93.401	−13.628	93.840	−0.03476	0.26074
−94	2.9768	23.8604*	15.072	93.215	−13.158	94.075	−0.03347	0.25977
−92	3.1934	23.4193*	14.118	93.030	−12.688	94.309	−0.03219	0.25881
−90	3.4229	22.9522*	13.235	92.843	−12.216	94.544	−0.03091	0.25787
−88	3.6657	22.4579*	12.417	92.657	−11.743	94.777	−0.02963	0.25695
−86	3.9224	21.9382*	11.659	92.469	−11.268	95.011	−0.02836	0.25604
−84	4.1936	21.3829*	10.955	92.282	−10.793	95.244	−0.02709	0.25515
−82	4.4800	20.7998*	10.302	92.093	−10.316	95.478	−0.02583	0.25428
−80	4.7822	20.1846*	9.6949	91.905	−9.838	95.710	−0.02457	0.25342
−78	5.1007	19.5361*	9.1301	91.715	−9.359	95.943	−0.02331	0.25257
−76	5.4363	18.8528*	8.6043	91.525	−8.878	96.175	−0.02206	0.25174
−74	5.7896	18.1334*	8.1145	91.335	−8.397	96.406	−0.02081	0.25092
−72	6.1614	17.3766*	7.6579	91.144	−7.914	96.637	−0.01956	0.25012
−70	6.5522	16.5809*	7.2318	90.952	−7.429	96.868	−0.01832	0.24932
−68	6.9628	15.7449*	6.8339	90.760	−6.944	97.098	−0.01708	0.24855
−66	7.3939	14.8671*	6.4621	90.568	−6.457	97.328	−0.01584	0.24778
−64	7.8463	13.9460*	6.1144	90.374	−5.968	97.557	−0.01460	0.24703
−62	8.3208	12.9800*	5.7891	90.180	−5.479	97.786	−0.01337	0.24629
−60	8.8180	11.9677*	5.4844	89.986	−4.987	98.014	−0.01214	0.24556
−58	9.3388	10.9074*	5.1989	89.791	−4.495	98.241	−0.01092	0.24484
−56	9.8839	9.7975*	4.9312	89.595	−4.001	98.468	−0.00969	0.24414
−54	10.454	8.636*	4.6799	89.399	−3.506	98.694	−0.00847	0.24345
−52	11.051	7.422*	4.4440	89.202	−3.009	98.920	−0.00725	0.24276
−50	11.674	6.154*	4.2224	89.004	−2.511	99.144	−0.00604	0.24209
−48	12.324	4.820*	4.0140	88.806	−2.012	99.369	−0.00483	0.24143
−46	13.004	3.445*	3.8179	88.607	−1.511	99.592	−0.00361	0.24078
−44	13.712	2.002*	3.6334	88.407	−1.009	99.814	−0.00241	0.24014
−42	14.451	0.498*	3.4596	88.207	−0.505	100.036	−0.00120	0.23951
−40	15.222	0.526	3.2957	88.006	0.000	100.257	0.00000	0.23888
−38	16.024	1.328	3.1412	87.805	0.506	100.477	0.00120	0.23827
−36	16.859	2.163	2.9954	87.602	1.014	100.696	0.00240	0.23767
−34	17.728	3.032	2.8578	87.399	1.524	100.914	0.00359	0.23707
−32	18.633	3.937	2.7278	87.195	2.035	101.132	0.00479	0.23649
−30	19.573	4.877	2.6049	86.991	2.547	101.348	0.00598	0.23591
−28	20.549	5.853	2.4887	86.785	3.061	101.564	0.00716	0.23534
−26	21.564	6.868	2.3787	86.579	3.576	101.778	0.00835	0.23478
−24	22.617	7.921	2.2746	86.372	4.093	101.992	0.00953	0.23423
−22	23.711	9.015	2.1760	86.165	4.611	102.204	0.01072	0.23369
−20	24.845	10.149	2.0826	85.956	5.131	102.415	0.01189	0.23315
−18	26.020	11.324	1.9940	85.747	5.652	102.626	0.01307	0.23262
−16	27.239	12.543	1.9099	85.537	6.175	102.835	0.01425	0.23210
−14	28.501	13.805	1.8302	85.326	6.699	103.043	0.01542	0.23159
−12	29.809	15.113	1.7546	85.114	7.224	103.250	0.01659	0.23108
−10	31.162	16.466	1.6825	84.901	7.751	103.455	0.01776	0.23058
−8	32.563	17.867	1.6141	84.688	8.280	103.660	0.01892	0.23008
−6	34.011	19.315	1.5491	84.473	8.810	103.863	0.02009	0.22960
−4	35.509	20.813	1.4872	84.258	9.341	104.065	0.02125	0.22912
−2	37.057	22.361	1.4283	84.042	9.874	104.266	0.02241	0.22864
0	38.657	23.961	1.3723	83.825	10.409	104.465	0.02357	0.22817
2	40.309	25.613	1.3189	83.606	10.945	104.663	0.02472	0.22771
4	42.014	27.318	1.2680	83.387	11.483	104.860	0.02587	0.22725
5†	42.888	28.192	1.2434	83.277	11.752	104.958	0.02645	0.22703
6	43.775	29.079	1.2195	83.167	12.022	105.056	0.02703	0.22680
8	45.591	30.895	1.1732	82.946	12.562	105.250	0.02818	0.22636
10	47.464	32.768	1.1290	82.724	13.104	105.442	0.02932	0.22592
12	49.396	34.700	1.0869	82.501	13.648	105.633	0.03047	0.22548
14	51.387	36.691	1.0466	82.276	14.193	105.823	0.03161	0.22505
16	53.438	38.742	1.0082	82.051	14.739	106.011	0.03275	0.22463
18	55.551	40.855	0.97144	81.825	15.288	106.198	0.03389	0.22421
20	57.727	43.031	0.93631	81.597	15.837	106.383	0.03503	0.22379
22	59.967	45.271	0.90270	81.368	16.389	106.566	0.03617	0.22338
24	62.272	47.576	0.87055	81.138	16.942	106.748	0.03730	0.22297
26	64.644	49.948	0.83978	80.907	17.496	106.928	0.03844	0.22257
28	67.083	52.387	0.81031	80.675	18.052	107.107	0.03958	0.22218
30	69.591	54.895	0.78208	80.441	18.609	107.284	0.04070	0.22178
32	72.169	57.473	0.75503	80.207	19.169	107.459	0.04182	0.22139
34	74.818	60.122	0.72911	79.971	19.729	107.632	0.04295	0.22100
36	77.540	62.844	0.70425	79.733	20.292	107.804	0.04407	0.22062
38	80.336	65.640	0.68041	79.495	20.856	107.974	0.04520	0.22024
40	83.206	68.510	0.65753	79.255	21.422	108.142	0.04632	0.21986
42	86.153	71.457	0.63557	79.013	21.989	108.308	0.04744	0.21949
44	89.177	74.481	0.61448	78.770	22.558	108.472	0.04855	0.21912
46	92.280	77.584	0.59422	78.526	23.129	108.634	0.04967	0.21876
48	95.463	80.767	0.57476	78.280	23.701	108.795	0.05079	0.21839
50	98.727	84.031	0.55606	78.033	24.275	108.953	0.05190	0.21803
52	102.07	87.38	0.53808	77.784	24.851	109.109	0.05301	0.21768
54	105.50	90.81	0.52078	77.534	25.429	109.263	0.05412	0.21732
56	109.02	94.32	0.50414	77.282	26.008	109.415	0.05523	0.21697
58	112.62	97.93	0.48813	77.028	26.589	109.564	0.05634	0.21662
60	116.31	101.62	0.47272	76.773	27.172	109.712	0.05745	0.21627
62	120.09	105.39	0.45788	76.515	27.757	109.857	0.05855	0.21592
64	123.96	109.26	0.44358	76.257	28.344	110.000	0.05966	0.21558
66	127.92	113.22	0.42981	75.996	28.932	110.140	0.06076	0.21524
68	131.97	117.28	0.41653	75.733	29.523	110.278	0.06186	0.21490
70	136.12	121.43	0.40373	75.469	30.116	110.414	0.06296	0.21456
72	140.37	125.67	0.39139	75.202	30.710	110.547	0.06406	0.21422
74	144.71	130.01	0.37949	74.934	31.307	110.677	0.06516	0.21388
76	149.15	134.45	0.36800	74.664	31.906	110.805	0.06626	0.21355
78	153.69	138.99	0.35691	74.391	32.506	110.930	0.06736	0.21321
80	158.33	143.63	0.34621	74.116	33.109	111.052	0.06846	0.21288
82	163.07	148.37	0.33587	73.839	33.714	111.171	0.06956	0.21255
84	167.92	153.22	0.32588	73.560	34.322	111.288	0.07065	0.21222
86†	172.87	158.17	0.31623	73.278	34.931	111.401	0.07175	0.21188
88	177.93	163.23	0.30690	72.994	35.543	111.512	0.07285	0.21155
90	183.09	168.40	0.29789	72.708	36.158	111.619	0.07394	0.21122
92	188.37	173.67	0.28917	72.419	36.774	111.723	0.07504	0.21089
94	193.76	179.06	0.28073	72.127	37.394	111.824	0.07613	0.21056
96	199.26	184.56	0.27257	71.833	38.016	111.921	0.07723	0.21023
98	204.87	190.18	0.26467	71.536	38.640	112.015	0.07832	0.20989
100	210.60	195.91	0.25702	71.236	39.267	112.105	0.07942	0.20956
102	216.45	201.76	0.24962	70.933	39.897	112.192	0.08052	0.20923
104	222.42	207.72	0.24244	70.626	40.530	112.274	0.08161	0.20889
106	228.50	213.81	0.23549	70.317	41.166	112.353	0.08271	0.20855
108	234.71	220.02	0.22875	70.005	41.804	112.427	0.08381	0.20821
110	241.04	226.35	0.22222	69.689	42.446	112.498	0.08491	0.20787
112	247.50	232.80	0.21589	69.369	43.091	112.564	0.08601	0.20753
114	254.08	239.38	0.20974	69.046	43.739	112.626	0.08711	0.20718
116	260.79	246.10	0.20378	68.719	44.391	112.682	0.08821	0.20684
118	267.63	252.94	0.19800	68.388	45.046	112.735	0.08932	0.20649
120	274.60	259.91	0.19238	68.054	45.705	112.782	0.09042	0.20613
122	281.71	267.01	0.18692	67.714	46.368	112.824	0.09153	0.20578
124	288.95	274.25	0.18163	67.371	47.034	112.860	0.09264	0.20542
126	296.33	281.63	0.17648	67.023	47.705	112.891	0.09375	0.20505
128	303.84	289.14	0.17147	66.670	48.380	112.917	0.09487	0.20468
130	311.50	296.80	0.16661	66.312	49.059	112.936	0.09598	0.20431
132	319.29	304.60	0.16187	65.949	49.743	112.949	0.09711	0.20393
134	327.23	312.54	0.15727	65.581	50.432	112.955	0.09823	0.20354
136	335.32	320.63	0.15279	65.207	51.125	112.954	0.09936	0.20315
138	343.56	328.86	0.14843	64.826	51.824	112.947	0.10049	0.20275
140	351.94	337.25	0.14418	64.440	52.528	112.931	0.10163	0.20235
142	360.48	345.79	0.14004	64.047	53.238	112.908	0.10277	0.20194
144	369.17	354.48	0.13600	63.647	53.953	112.877	0.10391	0.20152
146	378.02	363.32	0.13207	63.240	54.677	112.836	0.10507	0.20109
148	387.03	372.33	0.12823	62.825	55.406	112.787	0.10622	0.20065
150	396.19	381.50	0.12448	62.402	56.143	112.728	0.10739	0.20020
152	405.52	390.83	0.12083	61.970	56.887	112.658	0.10856	0.19974
154	415.02	400.32	0.11726	61.529	57.638	112.577	0.10974	0.19926
156	424.68	409.99	0.11376	61.079	58.399	112.485	0.11093	0.19878
158	434.52	419.82	0.11035	60.617	59.168	112.381	0.11213	0.19828
160	444.53	429.83	0.10701	60.145	59.948	112.263	0.11334	0.19776
162	454.71	440.01	0.10374	59.660	60.737	112.131	0.11456	0.19723
164	465.07	450.37	0.10054	59.163	61.538	111.984	0.11580	0.19668
166	475.61	460.92	0.097393	58.651	62.351	111.820	0.11705	0.19611
168	486.34	471.65	0.094309	58.125	63.178	111.639	0.11831	0.19552
170	497.26	482.56	0.091279	57.581	64.019	111.439	0.11959	0.19490
172	508.37	493.67	0.088299	57.019	64.875	111.216	0.12089	0.19425
174	519.67	504.97	0.085365	56.438	65.750	110.970	0.12222	0.19358
176	531.17	516.47	0.082473	55.834	66.643	110.699	0.12356	0.19287
178	542.87	528.18	0.079616	55.205	67.558	110.400	0.12493	0.19212
180	554.78	540.09	0.076790	54.549	68.498	110.063	0.12635	0.19133
182	566.90	552.21	0.073987	53.861	69.465	109.700	0.12779	0.19050
184	579.24	564.54	0.071201	53.136	70.464	109.290	0.12928	0.18960
186	591.80	577.10	0.068421	52.370	71.500	108.832	0.13082	0.18864
188	604.58	589.88	0.065638	51.553	72.579	108.317	0.13242	0.18760
190	617.59	602.89	0.062837	50.677	73.711	107.734	0.13409	0.18646
192	630.84	616.14	0.059999	49.728	74.907	107.067	0.13585	0.18520
194	644.33	629.64	0.057096	48.685	76.184	106.294	0.13773	0.18380
196	658.08	643.38	0.054089	47.518	77.568	105.381	0.13972	0.18218
198	672.08	657.38	0.050912	46.178	79.102	104.270	0.14202	0.18029
200	686.36	671.66	0.047438	44.571	80.862	102.833	0.14460	0.17794
202	700.91	686.21	0.043375	42.476	83.030	100.870	0.14779	0.17475
204	715.75	701.05	0.037545	38.991	86.309	97.260	0.15264	0.16914
204.81	721.91	707.21	0.030525	32.760	91.329	91.329	0.16016	0.16016

* From published data (1964) of E. I. du Pont de Nemours & Co., Inc. Used by permission.
* Inches of mercury below one standard atmosphere.
** Based on 0 for the saturated liquid at −40 F.
† Standard cycle temperatures.

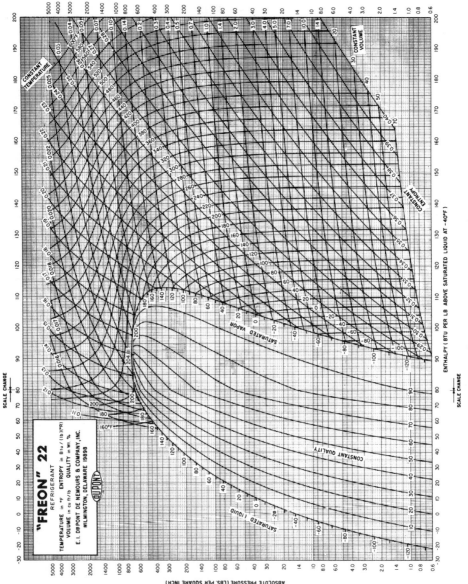

Figure B.1 Pressure-Enthalpy Diagram for R-22. Copyright © by Du Pont. Reproduced by permission.

688

APPENDIX C

COMPOUND INTEREST TABLES

TABLE C.1. 6% Compound Interest Factors[a,b]

	Single Payment		Uniform Annual Series				
	Compound Amount Factor caf'	Present Worth Factor pwf'	Sinking Fund Factor sff	Capital Recovery Factor crf	Compound Amount Factor caf	Present Worth Factor pwf	
	F/P	P/F	A/F	A/P	F/A	P/A	
	$F = P \times \mathrm{caf}'$	$P = F \times \mathrm{pwf}'$	$A = F \times \mathrm{sff}$	$A = P \times \mathrm{crf}$	$F = A \times \mathrm{caf}$	$P = A \times \mathrm{pwf}$	
n	$(1+i)^n$	$\dfrac{1}{(1+i)^n}$	$\dfrac{i}{(1+i)^n - 1}$	$\dfrac{i(1+i)^n}{(1+i)^n - 1}$	$\dfrac{(1+i)^n - 1}{i}$	$\dfrac{(1+i)^n - 1}{i(1+i)^n}$	n
1	1.0600	0.9434	1.00000	1.06000	1.000	0.943	1
2	1.1236	.8900	0.48544	0.54544	2.060	1.833	2
3	1.1910	.8396	.31411	.37411	3.184	2.673	3
4	1.2625	.7921	.22859	.22859	4.375	3.465	4
5	1.3382	.7473	.17740	.27340	5.637	4.212	5
6	1.4185	.7050	.14336	.20336	6.975	4.917	6
7	1.5036	.6651	.11914	.17914	8.394	5.582	7
8	1.5938	.6274	.10104	.16104	9.897	6.210	8
9	1.6895	.5919	.08702	.14702	11.491	6.802	9
10	1.7908	.5584	.07587	.13587	13.181	7.360	10
11	1.8983	.5268	.06679	.12679	14.972	7.887	11
12	2.0122	.4970	.05928	.11928	16.870	8.384	12
13	2.1329	.4688	.05296	.11296	18.882	8.853	13
14	2.2609	.4423	.04758	.10758	21.015	9.295	14
15	2.3966	.4173	.04296	.10296	23.276	9.712	15
16	2.5404	.3936	.03895	.09895	25.673	10.106	16
17	2.6928	.3714	.03544	.09544	28.213	10.477	17
18	2.8543	.3503	.03236	.09236	30.906	10.828	18
19	3.0256	.3305	.02962	.08962	33.760	11.158	19

n							n
20	3.2071	.3118	.02718	.08718	36.786	11.470	20
21	3.3996	.2942	.02500	.08500	39.993	11.764	21
22	3.6035	.2775	.02305	.08305	43.392	12.042	22
23	3.8197	.2618	.02128	.08128	46.996	12.303	23
24	4.0489	.2470	.01968	.07968	50.816	12.550	24
25	4.2919	.2330	.01823	.07823	54.865	12.783	25
26	4.5494	.2198	.01690	.07690	59.156	13.003	26
27	4.8223	.2074	.01570	.07570	63.706	13.211	27
28	5.1117	.1956	.01459	.07459	68.528	13.406	28
29	5.4184	.1846	.01358	.07358	73.640	13.591	29
30	5.7435	.1741	.01265	.07265	79.058	13.765	30
31	6.0881	.1643	.01179	.07179	84.802	13.929	31
32	6.4534	.1550	.01100	.07100	90.890	14.084	32
33	6.8406	.1462	.01027	.07027	97.343	14.230	33
34	7.2510	.1379	.00960	.06960	104.184	14.368	34
35	7.6861	.1301	.00897	.06897	111.435	14.498	35
40	10.2857	.0972	.00646	.06646	154.762	15.046	40
45	13.7646	.0727	.00470	.06470	212.744	15.456	45
50	18.4202	.0543	.00344	.06344	290.336	15.762	50
55	24.6503	.0406	.00254	.06254	394.172	15.991	55
60	32.9877	.0303	.00188	.06188	533.128	16.161	60
65	44.1450	.0227	.00139	.06139	719.083	16.289	65
70	59.0759	.0169	.00103	.06103	967.932	16.385	70
75	79.0569	.0126	.00077	.06077	1300.949	16.456	75
80	105.7960	.0095	.00057	.06057	1746.600	16.509	80
85	141.5789	.0071	.00043	.06043	2342.982	16.549	85
90	189.4645	.0053	.00032	.06032	3141.075	16.579	90
95	253.5463	.0039	.00024	.06024	4209.104	16.601	95
100	339.3021	.0029	.00018	.06018	5638.368	16.618	100

[a] The letter notations, e.g., F/P, P/F, P/F, indicate the operation to be performed with the factors in that column. For example, F/P indicates that the single compound amount factors are used to find the future value of a given present amount.

[b] P = present sum; F = sum at end of n interest periods; A = end-of-period payment in a uniform series running for n periods.

TABLE C.2. 8% Compound Interest Factors[a,b]

| | Single Payment | | Uniform Annual Series | | | |
	Compound Amount Factor caf'	Present Worth Factor pwf'	Sinking Fund Factor sff	Capital Recovery Factor crf	Compound Amount Factor caf	Present Worth Factor pwf
	$F = P \times \text{caf}'$	$P = F \times \text{pwf}'$	$A = F \times \text{sff}$	$A = P \times \text{crf}$	$F = A \times \text{caf}$	$P = A \times \text{pwf}$
	F/P	P/F	A/F	A/P	F/A	P/A
n	$(1+i)^n$	$\dfrac{1}{(1+i)^n}$	$\dfrac{i}{(1+i)^n - 1}$	$\dfrac{i(1+i)^n}{(1+i)^n - 1}$	$\dfrac{(1+i)^n - 1}{i}$	$\dfrac{(1+i)^n - 1}{i(1+i)^n}$	n
1	1.0800	0.9259	1.00000	1.08000	1.000	0.926	1
2	1.664	.8573	0.48077	0.56077	2.080	1.783	2
3	1.2597	.7938	.30803	.38803	3.246	2.577	3
4	1.3605	.7350	.22192	.30192	4.506	3.312	4
5	1.4693	.6806	.17046	.25046	5.867	3.993	5
6	1.5869	.6302	.13632	.21632	7.336	4.623	6
7	1.7138	.5835	.11207	.19207	8.923	5.206	7
8	1.8509	.5403	.09401	.17401	10.637	5.747	8
9	1.9990	.5002	.08008	.16008	12.488	6.247	9
10	2.1589	.4632	.06903	.14903	14.487	6.710	10
11	2.3316	.4289	.06008	.14008	16.645	7.139	11
12	2.5182	.3971	.05270	.13270	18.977	7.536	12
13	2.7196	.3677	.04652	.12652	21.495	7.904	13
14	2.9372	.3405	.04130	.12130	24.215	8.244	14
15	3.1722	.3152	.03683	.11683	27.152	8.559	15
16	3.4259	.2919	.03298	.11298	30.324	8.851	16
17	3.7000	.2703	.02963	.10963	33.750	9.122	17
18	3.9960	.2502	.02670	.10670	37.450	9.372	18
19	4.3157	.2317	.02413	.10413	41.446	9.604	19

692

n							n
20	4.6610	.2145	.02185	.10185	45.762	9.818	20
21	5.0338	.1987	.01983	.09983	50.423	10.017	21
22	5.4365	.1839	.01803	.09803	55.457	10.201	22
23	5.8715	.1703	.01642	.09642	60.893	10.371	23
24	6.3412	.1577	.01498	.09498	66.765	10.529	24
25	6.8485	.1460	.01368	.09368	73.106	10.675	25
26	7.3964	.1352	.01251	.09251	79.954	10.810	26
27	7.9881	.1252	.01145	.09145	87.351	10.935	27
28	8.6271	.1159	.01049	.09049	95.339	11.051	28
29	9.3173	.1073	.00962	.08962	103.966	11.158	29
30	10.0627	.0994	.00883	.08883	113.283	11.258	30
31	10.8677	.0920	.00811	.08811	123.346	11.350	31
32	11.7371	.0852	.00745	.08745	134.214	11.435	32
33	12.6760	.0789	.00685	.08685	145.951	11.514	33
34	13.6901	.0730	.00630	.08630	158.627	11.587	34
35	14.7853	.0676	.00580	.08580	172.317	11.655	35
40	21.7245	.0460	.00386	.08386	259.057	11.925	40
45	31.9204	.0313	.00259	.08259	386.506	12.108	45
50	46.9016	.0213	.00174	.08174	573.770	12.233	50
55	68.9139	.0145	.00118	.08118	848.923	12.319	55
60	101.2571	.0099	.00080	.08080	1253.213	12.377	60
65	148.7798	.0067	.00054	.08054	1847.248	12.416	65
70	218.6064	.0046	.00037	.08037	2720.080	12.443	70
75	321.2045	.0031	.00025	.08025	4002.557	12.461	75
80	471.9548	.0021	.00017	.08017	5886.935	12.474	80
85	693.4565	.0014	.00012	.08012	8655.706	12.482	85
90	1018.9151	.0010	.00008	.08008	12723.939	12.488	90
95	1497.1205	.0007	.00005	.08005	18701.507	12.492	95
100	2199.7613	.0005	.00004	.08004	27484.516	12.494	100

[a]The letter notations, e.g., F/P, P/F, indicate the operation to be performed with the factors in that column. For example, F/P indicates that the single compound amount factors are used to find the future value of a given present amount.

[b]P = present sum; F = sum at end of n interest periods; A = end-of-period payment in a uniform series running for n periods.

TABLE C.3. 10% Compound Interest Factors[a,b]

	Single Payment			Uniform Annual Series			
	Compound Amount Factor caf'	Present Worth Factor pwf'	Sinking Fund Factor sff	Capital Recovery Factor crf	Compound Amount Factor caf	Present Worth Factor pwf	
	F/P	P/F	A/F	A/P	F/A	P/A	
	$F = P \times \text{caf}'$	$P = F \times \text{pwf}'$	$A = F \times \text{sff}$	$A = P \times \text{crf}$	$F = A \times \text{caf}$	$P = A \times \text{pwf}$	
n	$(1+i)^n$	$\dfrac{1}{(1+i)^n}$	$\dfrac{i}{(1+i)^n - 1}$	$\dfrac{i(1+i)^n}{(1+i)^n - 1}$	$\dfrac{(1+i)^n - 1}{i}$	$\dfrac{(1+i)^n - 1}{i(1+i)^n}$	n
1	1.1000	0.9091	1.00000	1.10000	1.000	0.909	1
2	1.2100	.8264	0.47619	0.57619	2.100	1.736	2
3	1.3310	.7513	.30211	.40211	3.310	2.487	3
4	1.4641	.6830	.21547	.31547	4.641	3.170	4
5	1.6105	.6209	.16380	.26380	6.105	3.791	5
6	1.7716	.5645	.12961	.22961	7.716	4.355	6
7	1.9487	.5132	.10541	.20541	9.487	4.868	7
8	2.1436	.4665	.08744	.18744	11.436	5.335	8
9	2.3579	.4241	.07364	.17364	13.579	5.759	9
10	2.5937	.3855	.06275	.16275	15.937	6.144	10
11	2.8531	.3505	.05396	.15396	18.531	6.495	11
12	3.1384	.3186	.04676	.14676	21.384	6.814	12
13	3.4523	.2897	.04078	.14078	24.523	7.103	13
14	3.7975	.2633	.03575	.13575	27.975	7.367	14
15	4.1772	.2394	.03147	.13147	31.772	7.606	15
16	4.5950	.2176	.02782	.12782	35.950	7.824	16
17	5.0545	.1978	.02466	.12466	40.545	8.022	17
18	5.5599	.1799	.02193	.12193	45.599	8.201	18
19	6.1159	.1635	.01955	.11955	51.159	8.365	19

n							n
20	6.7275	.1486	.01746	.11746	57.275	8.514	20
21	7.4002	.1351	.01562	.11562	64.002	8.649	21
22	8.1403	.1228	.01401	.11401	71.403	8.772	22
23	8.9543	.1117	.01257	.11257	79.543	8.883	23
24	9.8497	.1015	.01130	.11130	88.497	8.985	24
25	10.8347	.0923	.01017	.11017	98.347	9.077	25
26	11.9182	.0839	.00916	.10916	109.182	9.161	26
27	13.1100	.0763	.00826	.10826	121.100	9.237	27
28	14.4210	.0693	.00745	.10745	134.210	9.307	28
29	15.8631	.0630	.00673	.10673	148.631	9.370	29
30	17.4494	.0573	.00608	.10608	164.494	9.427	30
31	19.1943	.0521	.00550	.10550	181.943	9.479	31
32	21.1138	.0474	.00497	.10497	201.138	9.526	32
33	23.2252	.0431	.00450	.10450	222.252	9.569	33
34	25.5477	.0391	.00407	.10407	245.477	9.609	34
35	28.1024	.0356	.03369	.10369	271.024	9.644	35
40	45.2593	.0221	.00226	.10226	442.593	9.779	40
45	72.8905	.0137	.00139	.10139	718.905	9.863	45
50	117.3909	.0085	.00086	.10086	1163.909	9.915	50
55	189.0591	.0053	.00053	.10053	1880.591	9.947	55
60	304.4816	.0033	.00033	.10033	3034.816	9.967	60
65	490.3707	.0020	.00020	.10020	4893.707	9.980	65
70	789.7470	.0013	.00013	.10013	7887.470	9.987	70
75	1271.8952	.0008	.00008	.10008	12708.954	9.992	75
80	2048.4002	.0005	.00005	.10005	20474.002	9.995	80
85	3298.9690	.0003	.00003	.10003	32979.690	9.997	85
90	5313.0226	.0002	.00002	.10002	53120.226	9.998	90
95	8556.6760	.0001	.00001	.10001	85556.760	9.999	95
100	13780.6123	.0001	.00001	.10001	137796.123	9.999	100

[a] The letter notations, e.g., F/P, P/F, indicate the operation to be performed with the factors in that column. For example, F/P indicates that the single compound amount factors are used to find the future value of a given present amount.

[b] P = present sum; F = sum at end of n interest periods; A = end-of-period payment in a uniform series running for n periods.

TABLE C.4. 12% Compound Interest Factors[a,b]

	Single Payment		Uniform Annual Series				
	Compound Amount Factor caf'	Present Worth Factor pwf'	Sinking Fund Factor sff	Capital Recovery Factor crf	Compound Amount Factor caf	Present Worth Factor pwf	
	F/P	P/F	A/F	A/P	F/A	P/A	
	$F = P \times caf'$	$P = F \times pwf'$	$A = F \times sff$	$A = P \times crf$	$F = A \times caf$	$P = A \times pwf$	
n	$(1+i)^n$	$\dfrac{1}{(1+i)^n}$	$\dfrac{i}{(1+i)^n - 1}$	$\dfrac{i(1+i)^n}{(1+i)^n - 1}$	$\dfrac{(1+i)^n - 1}{i}$	$\dfrac{(1+i)^n - 1}{i(1+i)^n}$	n
1	1.1200	0.8929	1.00000	1.12000	1.000	0.893	1
2	1.2544	.7972	0.47170	0.59170	2.120	1.690	2
3	1.4049	.7118	.29635	.41635	3.374	2.402	3
4	1.5735	.6355	.20923	.32923	4.779	3.037	4
5	1.7623	.5674	.15741	.27741	6.353	3.605	5
6	1.9738	.5066	.12323	.24323	8.115	4.111	6
7	2.2107	.4523	.09912	.21912	10.089	4.564	7
8	2.4760	.4039	.08130	.20130	12.300	4.968	8
9	2.7731	.3606	.06768	.18768	14.776	5.328	9
10	3.1058	.3220	.05698	.17698	17.549	5.650	10
11	3.4785	.2875	.04842	.16842	20.655	5.938	11
12	3.8960	.2567	.04144	.16144	24.133	6.194	12
13	4.3635	.2292	.03568	.15568	28.029	6.424	13
14	4.8871	.2046	.03087	.15087	32.393	6.628	14
15	5.4736	.1827	.02682	.14682	37.280	6.811	15

n							n
16	6.1304	.1631	.02339	.14339	42.753	6.974	16
17	6.8660	.1456	.02046	.14046	48.884	7.120	17
18	7.6900	.1300	.01794	.13794	55.750	7.250	18
19	8.6128	.1161	.01576	.13576	63.440	7.366	19
20	9.6463	.1037	.01388	.13388	72.052	7.469	20
21	10.8038	.0926	.01224	.13224	81.699	7.562	21
22	12.1003	.0826	.01081	.13081	92.503	7.645	22
23	13.5523	.0738	.00956	.12956	104.603	7.718	23
24	15.1786	.0659	.00846	.12846	118.155	7.784	24
25	17.0001	.0588	.00750	.12750	133.334	7.843	25
26	19.0401	.0525	.00665	.12665	150.334	7.896	26
27	21.3249	.0469	.00590	.12590	169.374	7.943	27
28	23.8839	.0419	.00524	.12524	190.699	7.984	28
29	26.7499	.0374	.00466	.12466	214.583	8.022	29
30	29.9599	.0334	.00414	.12414	241.333	8.055	30
31	33.5551	.0298	.00369	.12369	271.292	8.085	31
32	37.5817	.0266	.00328	.12328	304.847	8.112	32
33	42.0915	.0238	.00292	.12292	342.429	8.135	33
34	47.1425	.0212	.00260	.12260	384.520	8.157	34
35	52.7996	.0189	.00232	.12232	431.663	8.176	35
40	93.0510	.0107	.00130	.12130	767.091	8.244	40
45	163.9876	.0061	.00074	.12074	1358.230	8.283	45
50	289.0022	.0035	.00042	.12042	2400.018	8.305	50
∞				.12000		8.333	∞

[a] The letter notations, e.g., F/P, P/F, indicate the operation to be performed with the factors in that column. For example, F/P indicates that the single compound amount factors are used to find the future value of a given present amount.

[b] P = present sum; F = sum at end of n interest periods; A = end-of-period payment in a uniform series running for n periods.

TABLE C.5. 15% Compound Interest Factors[a,b]

	Single Payment		Uniform Annual Series				
	Compound Amount Factor caf'	Present Worth Factor pwf'	Sinking Fund Factor sff	Capital Recovery Factor crf	Compound Amount Factor caf	Present Worth Factor pwf	
	F/P	P/F	A/F	A/P	F/A	P/A	
	$F = P \times \text{caf}'$	$P = F \times \text{pwf}'$	$A = F \times \text{sff}$	$A = P \times \text{crf}$	$F = A \times \text{caf}$	$P = A \times \text{pwf}$	
n	$(1+i)^n$	$\dfrac{1}{(1+i)^n}$	$\dfrac{i}{(1+i)^n - 1}$	$\dfrac{i(1+i)^n}{(1+i)^n - 1}$	$\dfrac{(1+i)^n - 1}{i}$	$\dfrac{(1+i)^n - 1}{i(1+i)^n}$	n
1	1.1500	0.8696	1.00000	1.15000	1.000	0.870	1
2	1.3225	.7561	.46512	.61512	2.150	1.626	2
3	1.5209	.6575	.28798	.43798	3.472	2.283	3
4	1.7490	.5718	.20026	.35027	4.993	2.855	4
5	2.0114	.4972	.14832	.29832	6.742	3.352	5
6	2.3131	.4323	.11424	.26424	8.754	3.784	6
7	2.6600	.3759	.09036	.24036	11.067	4.160	7
8	3.0590	.3269	.07285	.22285	13.727	4.487	8
9	3.5179	.2843	.05957	.20957	16.786	4.772	9
10	4.0456	.2472	.04925	.19925	20.304	5.019	10
11	4.6524	.2149	.04107	.19107	24.349	5.234	11
12	5.3503	.1869	.03448	.18448	29.002	5.421	12
13	6.1528	.1625	.02911	.17911	34.352	5.583	13
14	7.0757	.1413	.02469	.17469	40.505	5.724	14
15	8.1371	.1229	.02102	.17102	47.580	5.847	15

n	F/P	P/F	A/F	A/P	F/A	P/A	n
16	9.3576	.1069	.01795	.16795	55.717	5.954	16
17	10.7613	.0929	.01537	.16537	65.075	6.047	17
18	12.3755	.0808	.01319	.16319	75.836	6.128	18
19	14.2318	.0703	.01134	.16134	88.212	6.198	19
20	16.3665	.0611	.00976	.15976	102.444	6.259	20
21	18.8215	.0531	.00842	.15842	118.810	6.312	21
22	21.6447	.0462	.00727	.15727	137.632	6.359	22
23	24.8915	.0402	.00628	.15628	159.276	6.399	23
24	28.6252	.0349	.00543	.15543	184.168	6.434	24
25	32.9190	.0304	.00470	.15470	212.793	6.464	25
26	37.8568	.0264	.00407	.15407	245.712	6.491	26
27	45.5353	.0230	.00353	.15353	283.569	6.514	27
28	50.0656	.0200	.00306	.15306	327.104	6.534	28
29	57.5755	.0174	.00265	.15265	377.170	6.551	29
30	66.2118	.0151	.00230	.15230	434.745	6.566	30
31	76.1435	.0131	.00200	.15200	500.957	6.579	31
32	87.5651	.0114	.00173	.15173	577.100	6.591	32
33	100.6998	.0099	.00150	.15150	664.666	6.600	33
34	115.8048	.0086	.00131	.15131	765.365	6.609	34
35	133.1755	.0075	.00113	.15113	881.170	6.617	35
40	267.8635	.0037	.0056	.15056	1779.090	6.642	40
45	538.7693	.0019	.00028	.15028	3585.128	6.654	45
50	1083.6574	.0009	.00014	.15014	7217.716	6.661	50
∞				.15000		6.667	∞

[a]The letter notations, e.g., F/P. P/F, indicate the operation to be performed with the factors in that column. For example, F/P indicates that the single compound amount factors are used to find the future value of a given present amount.

[b]P = present sum; F = sum at end of n interest periods; A = end-of-period payment in a uniform series running for n periods.

TABLE C.6. 20% Compound Interest Factors[a,b]

	Single Payment		Uniform Annual Series				
	Compound Amount Factor caf'	Present Worth Factor pwf'	Sinking Fund Factor sff	Capital Recovery Factor crf	Compound Amount Factor caf	Present Worth Factor pwf	
	F/P	P/F	A/F	A/P	F/A	P/A	
	$F = P \times caf'$	$P = F \times pwf'$	$A = F \times sff$	$A = P \times crf$	$F = A \times caf$	$P = A \times pwf$	
n	$(1+i)^n$	$\dfrac{1}{(1+i)^n}$	$\dfrac{i}{(1+i)^n - 1}$	$\dfrac{i(1+i)^n}{(1+i)^n - 1}$	$\dfrac{(1+i)^n - 1}{i}$	$\dfrac{(1+i)^n - 1}{i(1+i)^n}$	n
1	1.2000	0.8333	1.00000	1.20000	1.000	0.833	1
2	1.4400	.6944	0.45455	0.65455	2.200	1.528	2
3	1.7280	.5787	.27473	.47473	3.640	2.106	3
4	2.0736	.4823	.18629	.38629	5.368	2.589	4
5	2.4883	.4019	.13438	.33438	7.442	2.991	5
6	2.9860	.3349	.10071	.30071	9.930	3.326	6
7	3.5832	.2791	.07742	.27742	12.916	3.605	7
8	4.2998	.2326	.06061	.26061	16.499	3.837	8
9	5.1598	.1938	.04808	.24808	20.799	4.031	9
10	6.1917	.1615	.03852	.23852	25.959	4.192	10
11	7.4301	.1346	.03110	.23110	32.150	4.327	11
12	8.9161	.1122	.02526	.22526	39.581	4.439	12
13	10.6993	.0935	.02062	.22062	48.497	4.533	13
14	12.8392	.0779	.01689	.21689	59.196	4.611	14
15	15.4070	.0649	.01388	.21388	72.035	4.675	15

n							n
16	18.4884	.0541	.01144	.21144	87.442	4.730	16
17	22.1861	.0451	.00944	.20944	105.931	4.775	17
18	26.6233	.0376	.00781	.20781	128.117	4.812	18
19	31.9480	.0313	.00646	.20646	154.740	4.844	19
20	38.3376	.0261	.00536	.20536	186.688	4.870	20
21	46.0051	.0217	.00444	.20444	225.026	4.891	21
22	55.2061	.0181	.00369	.20369	271.031	4.909	22
23	66.2474	.0151	.00307	.20307	326.237	4.925	23
24	79.4968	.0126	.00255	.20255	392.484	4.937	24
25	95.3962	.0105	.00212	.20212	471.981	4.948	25
26	114.4755	.0087	.00176	.20176	567.377	4.956	26
27	137.3706	.0073	.00147	.20147	681.853	4.964	27
28	164.8447	.0061	.00122	.20122	819.223	4.970	28
29	197.8136	.0051	.00102	.20102	984.068	4.975	29
30	237.3763	.0042	.00085	.20085	1181.882	4.979	30
31	284.8516	.0035	.00070	.20070	1419.258	4.982	31
32	341.8219	.0029	.00059	.20059	1704.109	4.985	32
33	410.1863	.0024	.00049	.20049	2045.931	4.988	33
34	492.2235	.0020	.00041	.20041	2456.118	4.990	34
35	590.6682	.0017	.00034	.20034	2948.341	4.992	35
40	1469.7716	.0007	.00014	.20014	7343.858	4.997	40
45	3657.2620	.0003	.00005	.20005	18281.310	4.999	45
50	9100.4382	.0001	.00002	.20002	45497.191	4.999	50
∞				.20000		5.000	∞

[a] The letter notations, e.g., F/P, P/F, indicate the operation to be performed with the factors in that column. For example, F/P indicates that the single compound amount factors are used to find the future value of a given present amount.

[b] P = present sum; F = sum at end of n interest periods; A = end-of-period payment in a uniform series running for n periods.

701

TABLE C.7. 25% Compound Interest Factors[a,b]

	Single Payment		Uniform Annual Series				
	Compound Amount Factor caf′	Present Worth Factor pwf′	Sinking Fund Factor sff	Capital Recovery Factor crf	Compound Amount Factor caf	Present Worth Factor pwf	
	F/P	P/F	A/F	A/P	F/A	P/A	
	$F = P \times \text{caf}′$	$P = F \times \text{pwf}′$	$A = F \times \text{sff}$	$A = P \times \text{crf}$	$F = A \times \text{caf}$	$P = A \times \text{pwf}$	
n	$(1+i)^n$	$\dfrac{1}{(1+i)^n}$	$\dfrac{i}{(1+i)^n - 1}$	$\dfrac{i(1+i)^n}{(1+i)^n - 1}$	$\dfrac{(1+i)^n - 1}{i}$	$\dfrac{(1+i)^n - 1}{i(1+i)^n}$	n
1	1.2500	0.8000	1.00000	1.25000	1.000	0.800	1
2	1.5625	.6400	.44444	.69444	2.250	1.440	2
3	1.9531	.5120	.26230	.51230	3.813	1.952	3
4	2.4414	.4096	.17344	.42344	5.766	2.362	4
5	3.0518	.3277	.12185	.37185	8.207	2.689	5
6	3.8147	.2621	.08882	.33882	11.259	2.951	6
7	4.7684	.2097	.06634	.31634	15.073	3.161	7
8	5.9605	.1678	.05040	.30040	19.842	3.329	8
9	7.4506	.1342	.03876	.28876	25.802	2.463	9
10	9.3132	.1074	.03007	.28007	33.253	3.571	10
11	11.6415	.0859	.02349	.27349	42.566	3.656	11
12	14.5519	.0687	.01845	.26845	54.208	3.725	12
13	18.1899	.0550	.01454	.26454	68.760	3.780	13
14	22.7374	.0440	.01150	.26150	86.949	3.824	14
15	28.4217	.0352	.00912	.25912	109.687	3.859	15

n							n
16	3.887	138.109	.25724	.00724	.0281	35.5271	16
17	3.910	173.636	.25576	.00576	.0225	44.4089	17
18	3.928	218.045	.25459	.00459	.0180	55.5112	18
19	3.942	273.556	.25366	.00366	.0144	69.3889	19
20	3.954	342.945	.25292	.00292	.0115	86.7362	20
21	3.963	429.681	.25233	.00233	.0092	108.4202	21
22	3.970	538.101	.25186	.00186	.0074	135.5253	22
23	3.976	673.626	.25148	.00148	.0059	169.4066	23
24	3.981	843.033	.25119	.00119	.0047	211.7582	24
25	3.985	1054.791	.25095	.00095	.0038	264.6978	25
26	3.988	1319.489	.25076	.00076	.0030	330.8722	26
27	3.990	1650.361	.25061	.00061	.0024	413.5903	27
28	3.992	2063.952	.25048	.00048	.0019	516.9879	28
29	3.994	2580.939	.25039	.00039	.0015	646.2349	29
30	3.995	3227.174	.25031	.00031	.0012	807.7936	30
31	3.996	4034.968	.25025	.00025	.0010	1009.7420	31
32	3.997	5044.710	.25020	.00020	.0008	1262.1774	32
33	3.997	6306.887	.25016	.00016	.0006	1577.7218	33
34	3.998	7884.609	.25013	.00013	.0005	1972.1523	34
35	3.998	9856.761	.25010	.00010	.0004	2465.1903	35
40	3.999	30088.655	.25003	.00003	.0001	7523.1638	40
45	4.000	91831.496	.25001	.00001	.0001	22958.8740	45
50	4.000	280255.693	.25000	.00000	.0000	70064.9232	50
∞	4.000		.25000				∞

[a] The letter notations, e.g., F/P, P/F, indicate the operation to be performed with the factors in that column. For example, F/P indicates that the single compound amount factors are used to find the future value of a given present amount.

[b] P = present sum; F = sum at end of n interest periods; A = end-of-period payment in a uniform series running for n periods.

TABLE C.8. Capital Recovery Factors for Interest Rates from 25 to 50%

n	25%	30%	35%	40%	45%	50%	n
1	1.25000	1.30000	1.35000	1.40000	1.45000	1.50000	1
2	0.69444	0.73478	0.77553	0.81667	0.85816	0.90000	2
3	0.51230	0.55063	0.58966	0.62936	0.66966	0.71053	3
4	0.42344	0.46163	0.50076	0.54077	0.58156	0.62308	4
5	0.37185	0.41058	0.45046	0.49136	0.53318	0.57583	5
6	0.33882	0.37839	0.41926	0.46126	0.50426	0.54812	6
7	0.31634	0.35687	0.39880	0.44192	0.48607	0.53108	7
8	0.30040	0.34192	0.38489	0.42907	0.47427	0.52030	8
9	0.28876	0.33124	0.37519	0.42034	0.46646	0.51335	9
10	0.28007	0.32346	0.36832	0.41432	0.46123	0.50882	10
11	0.27349	0.31773	0.36339	0.41013	0.45768	0.50585	11
12	0.26845	0.31345	0.35982	0.40718	0.45527	0.50388	12
13	0.26454	0.31024	0.35722	0.40510	0.45362	0.50258	13
14	0.26150	0.30782	0.35532	0.40363	0.45249	0.50172	14
15	0.25912	0.30598	0.35393	0.40259	0.45172	0.50114	15
16	0.25724	0.30458	0.35290	0.40185	0.45118	0.50076	16
17	0.25576	0.30351	0.35214	0.40132	0.45081	0.50051	17
18	0.25459	0.30269	0.35158	0.40094	0.45056	0.50034	18
19	0.25366	0.30207	0.35117	0.40067	0.45039	0.50023	19
20	0.25292	0.30159	0.35087	0.40048	0.45027	0.50015	20
21	0.25233	0.30122	0.35064	0.40034	0.45018	0.50010	21
22	0.25186	0.30094	0.35048	0.40024	0.45013	0.50007	22
23	0.25148	0.30072	0.35035	0.40017	0.45009	0.50004	23
24	0.25119	0.30055	0.35026	0.40012	0.45006	0.50003	24
25	0.25095	0.30043	0.35019	0.40009	0.45004	0.50002	25
26	0.25076	0.30033	0.35014	0.40006	0.45003	0.50001	26
27	0.25061	0.30025	0.35011	0.40005	0.45002	0.50001	27
28	0.25048	0.30019	0.35008	0.40003	0.45001	0.50001	28
29	0.25039	0.30015	0.35006	0.40002	0.45001	0.50000	29
30	0.25031	0.30011	0.35004	0.40002	0.45001	0.50000	30
31	0.25025	0.30009	0.35003	0.40001	0.45000	0.50000	31
32	0.25020	0.30007	0.35002	0.40001	0.45000	0.50000	32
33	0.25016	0.30005	0.35002	0.40001	0.45000	0.50000	33
34	0.25013	0.30004	0.35001	0.40000	0.45000	0.50000	34
35	0.25010	0.30003	0.35001	0.40000	0.45000	0.50000	35
∞	0.25000	0.30000	0.35000	0.40000	0.45000	0.50000	∞

TABLE C.9. **Single-Payment Present Worth Factors for Interest Rates from 25 to 50%**

n	25%	30%	35%	40%	45%	50%	n
1	0.8000	0.7692	0.7407	0.7143	0.6897	0.6667	1
2	0.6400	0.5917	0.5487	0.5102	0.4756	0.4444	2
3	0.5120	0.4552	0.4064	0.3644	0.3280	0.2963	3
4	0.4096	0.3501	0.3011	0.2603	0.2262	0.1975	4
5	0.3277	0.2693	0.2230	0.1859	0.1560	0.1317	5
6	0.2621	0.2072	0.1652	0.1328	0.1076	0.0878	6
7	0.2097	0.1594	0.1224	0.0949	0.0742	0.0585	7
8	0.1678	0.1226	0.0906	0.0678	0.0512	0.0390	8
9	0.1342	0.0943	0.0671	0.0484	0.0353	0.0260	9
10	0.1074	0.0725	0.0497	0.0346	0.0243	0.0173	10
11	0.0859	0.0558	0.0368	0.0247	0.0168	0.0116	11
12	0.0687	0.0429	0.0273	0.0176	0.0116	0.0077	12
13	0.0550	0.0330	0.0202	0.0126	0.0080	0.0051	13
14	0.0440	0.0254	0.0150	0.0090	0.0055	0.0034	14
15	0.0352	0.0195	0.0111	0.0064	0.0038	0.0023	15
16	0.0281	0.0150	0.0082	0.0046	0.0026	0.0015	16
17	0.0225	0.0116	0.0061	0.0033	0.0018	0.0010	17
18	0.0180	0.0089	0.0045	0.0023	0.0012	0.0007	18
19	0.0144	0.0068	0.0033	0.0017	0.0009	0.0005	19
20	0.0115	0.0053	0.0025	0.0012	0.0006	0.0003	20
21	0.0092	0.0040	0.0018	0.0009	0.0004	0.0002	21
22	0.0074	0.0031	0.0014	0.0006	0.0003	0.0001	22
23	0.0059	0.0024	0.0010	0.0004	0.0002	0.0001	23
24	0.0047	0.0018	0.0007	0.0003	0.0001	0.0001	24
25	0.0038	0.0014	0.0006	0.0002	0.0001	......	25
26	0.0030	0.0011	0.0004	0.0002	0.0001	......	26
27	0.0024	0.0008	0.0003	0.0001	......	......	27
28	0.0019	0.0006	0.0002	0.0001	......	......	28
29	0.0015	0.0005	0.0002	0.0001	......	......	29
30	0.0012	0.0004	0.0001	......	......	......	30
31	0.0010	0.0003	0.0001	......	......	......	31
32	0.0008	0.0002	0.0001	......	......	......	32
33	0.0006	0.0002	0.0001	......	......	......	33
34	0.0005	0.0001	......	......	......	......	34
35	0.0004	0.0001	......	......	......	......	35

TABLE C.10. Series Present Worth Factors for Interest Rates from 25 to 50%

n	25%	30%	35%	40%	45%	50%	n
1	0.800	0.769	0.741	0.714	0.690	0.667	1
2	1.400	1.361	1.289	1.224	1.165	1.111	2
3	1.952	1.816	1.696	1.589	1.493	1.407	3
4	2.362	2.166	1.997	1.849	1.720	1.605	4
5	2.689	2.436	2.220	2.035	1.876	1.737	5
6	2.951	2.643	2.385	2.168	1.983	1.824	6
7	3.161	2.802	2.507	2.263	2.057	1.883	7
8	3.329	2.925	2.598	2.331	2.109	1.922	8
9	3.463	3.019	2.665	2.379	2.144	1.948	9
10	3.571	3.092	2.715	2.414	2.168	1.965	10
11	3.656	3.147	2.752	2.438	2.185	1.977	11
12	3.725	3.190	2.779	2.456	2.196	1.985	12
13	3.780	3.223	2.799	2.469	2.204	1.990	13
14	3.824	3.249	2.814	2.478	2.210	1.993	14
15	3.859	3.268	2.825	2.484	2.214	1.995	15
16	3.887	3.283	2.834	2.489	2.216	1.997	16
17	3.910	3.295	2.840	2.492	2.218	1.998	17
18	3.928	3.304	2.844	2.494	2.219	1.999	18
19	3.942	3.311	2.848	2.496	2.220	1.999	19
20	3.954	3.316	2.850	2.497	2.221	1.999	20
21	3.963	3.320	2.852	2.498	2.221	2.000	21
22	3.970	3.323	2.853	2.498	2.222	2.000	22
23	3.976	3.325	2.854	2.499	2.222	2.000	23
24	3.981	3.327	2.855	2.499	2.222	2.000	24
25	3.985	3.329	2.856	2.499	2.222	2.000	25
26	3.988	3.330	2.856	2.500	2.222	2.000	26
27	3.990	3.331	2.856	2.500	2.222	2.000	27
28	3.992	3.331	2.857	2.500	2.222	2.000	28
29	3.994	3.332	2.857	2.500	2.222	2.000	29
30	3.995	3.332	2.857	2.500	2.222	2.000	30
31	3.996	3.332	2.857	2.500	2.222	2.000	31
32	3.997	3.333	2.857	2.500	2.222	2.000	32
33	3.997	3.333	2.857	2.500	2.222	2.000	33
34	3.998	3.333	2.857	2.500	2.222	2.000	34
35	3.998	3.333	2.857	2.500	2.222	2.000	35
∞	4.000	3.333	2.857	2.500	2.222	2.000	∞

TABLE C.11. Factors to Convert a Gradient Series to an Equivalent Uniform Annual Series[a]

n	1%	2%	3%	4%	5%	6%	7%	8%	10%	12%	15%	20%	25%	30%	35%	40%	45%	50%	n
2	0.50	0.50	0.49	0.49	0.49	0.49	0.48	0.48	0.48	0.47	0.47	0.45	0.44	0.43	0.43	0.42	0.41	0.40	2
3	0.99	0.99	0.98	0.97	0.97	0.96	0.95	0.95	0.94	0.92	0.91	0.88	0.85	0.83	0.80	0.78	0.76	0.74	3
4	1.49	1.48	1.46	1.45	1.44	1.43	1.42	1.40	1.38	1.36	1.33	1.27	1.22	1.18	1.13	1.09	1.05	1.02	4
5	1.98	1.96	1.94	1.92	1.90	1.88	1.86	1.85	1.81	1.77	1.72	1.64	1.56	1.49	1.42	1.36	1.30	1.24	5
6	2.47	2.44	2.41	2.39	2.36	2.33	2.30	2.28	2.22	2.17	2.10	1.98	1.87	1.77	1.67	1.58	1.50	1.42	6
7	2.96	2.92	2.88	2.84	2.81	2.77	2.73	2.69	2.62	2.55	2.45	2.29	2.14	2.01	1.88	1.77	1.66	1.56	7
8	3.45	3.40	3.34	3.29	3.24	3.20	3.15	3.10	3.00	2.91	2.78	2.58	2.39	2.22	2.06	1.92	1.79	1.68	8
9	3.93	3.87	3.80	3.74	3.68	3.61	3.55	3.49	3.37	3.26	3.09	2.84	2.60	2.40	2.21	2.04	1.89	1.76	9
10	4.42	4.34	4.26	4.18	4.10	4.02	3.95	3.87	3.73	3.58	3.38	3.07	2.80	2.55	2.33	2.14	1.97	1.82	10
11	4.90	4.80	4.70	4.61	4.51	4.42	4.33	4.24	4.06	3.90	3.65	3.29	2.97	2.68	2.44	2.22	2.03	1.87	11
12	5.38	5.26	5.15	5.03	4.92	4.81	4.70	4.60	4.39	4.19	3.91	3.48	3.11	2.80	2.52	2.28	2.08	1.91	12
13	5.86	5.72	5.59	5.45	5.32	5.19	5.06	4.94	4.70	4.47	4.14	3.66	3.24	2.89	2.59	2.33	2.12	1.93	13
14	6.34	6.18	6.02	5.87	5.71	5.56	5.42	5.27	5.00	4.73	4.36	3.82	3.36	2.97	2.64	2.37	2.14	1.95	14
15	6.81	6.63	6.45	6.27	6.10	5.93	5.76	5.59	5.28	4.98	4.56	3.96	3.45	3.03	2.69	2.40	2.17	1.97	15
16	7.29	7.08	6.87	6.67	6.47	6.28	6.09	5.90	5.55	5.21	4.75	4.09	3.54	3.09	2.72	2.43	2.18	1.98	16
17	7.76	7.52	7.29	7.07	6.84	6.62	6.41	6.20	5.81	5.44	4.93	4.20	3.61	3.13	2.75	2.44	2.19	1.98	17
18	8.23	7.97	7.71	7.45	7.20	6.96	6.72	6.49	6.05	5.64	5.08	4.30	3.67	3.17	2.78	2.46	2.20	1.99	18
19	8.70	8.41	8.12	7.83	7.56	7.29	7.02	6.77	6.29	5.84	5.23	4.39	3.72	3.20	2.79	2.47	2.21	1.99	19
20	9.17	8.84	8.52	8.21	7.90	7.61	7.32	7.04	6.51	6.02	5.37	4.46	3.77	3.23	2.81	2.48	2.21	1.99	20
21	9.63	9.28	8.92	8.58	8.24	7.92	7.60	7.29	6.72	6.19	5.49	4.53	3.80	3.25	2.82	2.48	2.21	2.00	21
22	10.10	9.70	9.32	8.94	8.57	8.22	7.87	7.54	6.92	6.35	5.60	4.59	3.84	3.26	2.83	2.49	2.22	2.00	22
23	10.56	10.13	9.71	9.30	8.90	8.51	8.14	7.78	7.11	6.50	5.70	4.65	3.86	3.28	2.83	2.49	2.22	2.00	23
24	11.02	10.55	10.10	9.65	9.21	8.80	8.39	8.01	7.29	6.64	5.80	4.69	3.89	3.29	2.84	2.49	2.22	2.00	24
25	11.48	10.97	10.48	9.99	9.52	9.07	8.64	8.23	7.46	6.77	5.88	4.74	3.91	3.30	2.84	2.49	2.22	2.00	25
26	11.94	11.39	10.85	10.33	9.83	9.34	8.88	8.44	7.62	6.89	5.96	4.77	3.92	3.30	2.85	2.50	2.22	2.00	26
27	12.39	11.80	11.23	10.66	10.12	9.60	9.11	8.64	7.77	7.00	6.03	4.80	3.94	3.31	2.85	2.50	2.22	2.00	27

TABLE C.11. (Continued)

n	1%	2%	3%	4%	5%	6%	7%	8%	10%	12%	15%	20%	25%	30%	35%	40%	45%	50%	n
28	12.85	12.21	11.59	10.99	10.41	9.86	9.33	8.83	7.91	7.11	6.10	4.83	3.95	3.32	2.85	2.50	2.22	2.00	28
29	13.30	12.62	11.96	11.31	10.69	10.10	9.54	9.01	8.05	7.21	6.15	4.85	3.96	3.32	2.85	2.50	2.22	2.00	29
30	13.75	13.02	12.31	11.63	10.97	10.34	9.75	9.19	8.18	7.30	6.21	4.87	3.96	3.32	2.85	2.50	2.22	2.00	30
31	14.20	13.42	12.67	11.94	11.24	10.57	9.95	9.36	8.30	7.38	6.25	4.89	3.97	3.32	2.85	2.50	2.22	2.00	31
32	14.65	13.82	13.02	12.24	11.50	10.80	10.14	9.52	8.41	7.46	6.30	4.91	3.97	3.33	2.85	2.50	2.22	2.00	32
33	15.10	14.22	13.36	12.54	11.76	11.02	10.32	9.67	8.52	7.53	6.34	4.92	3.98	3.33	2.86	2.50	2.22	2.00	33
34	15.54	14.61	13.70	12.83	12.01	11.23	10.50	9.82	8.61	7.60	6.37	4.93	3.98	3.33	2.86	2.50	2.22	2.00	34
35	15.98	15.00	14.04	13.12	12.25	11.43	10.67	9.96	8.71	7.66	6.40	4.94	3.99	3.33	2.86	2.50	2.22	2.00	35
40	18.18	16.89	15.65	14.48	13.38	12.36	11.42	10.57	9.10	7.90	6.52	4.97	4.00	3.33	2.86	2.50	2.22	2.00	40
50	22.44	20.44	18.56	16.81	15.22	13.80	12.53	11.41	9.57	8.16	6.62	4.99	4.00	3.33	2.86	2.50	2.22	2.00	50
60	26.53	23.70	21.07	18.70	16.61	14.79	13.23	11.90	9.80	8.27	6.65	5.00	4.00	3.33	2.86	2.50	2.22	2.00	60
70	30.47	26.66	23.21	20.20	17.62	15.46	13.67	12.18	9.91	8.31	6.66	5.00	4.00	3.33	2.86	2.50	2.22	2.00	70
80	34.25	29.36	25.04	21.37	18.35	15.90	13.93	12.33	9.96	8.32	6.67	5.00	4.00	3.33	2.86	2.50	2.22	2.00	80
90	37.87	31.79	26.57	22.28	18.87	16.19	14.08	12.41	9.98	8.33	6.67	5.00	4.00	3.33	2.86	2.50	2.22	2.00	90
100	41.34	33.99	27.84	22.98	19.23	16.37	14.17	12.45	9.99	8.33	6.67	5.00	4.00	3.33	2.86	2.50	2.22	2.00	100

[a]This table contains multipliers for a gradient g to convert the n-year end-of-year series $0, g, 2g, \ldots, (n-1)g$ to an equivalent uniform annual series for n years.

TABLE C.12.   Factors to Compute the Present Worth of a Gradient Series, Interest Rates from 3 to 20% [a]

n	3%	4%	5%	6%	7%	8%	10%	12%	15%	20%	n
2	0.94	0.92	0.91	0.89	0.87	0.86	0.83	0.80	0.76	0.69	2
3	2.77	2.70	2.63	2.57	2.51	2.45	2.33	2.22	2.07	1.85	3
4	5.44	5.27	5.10	4.95	4.79	4.65	4.38	4.13	3.79	3.30	4
5	8.89	8.55	8.24	7.93	7.65	7.37	6.86	6.40	5.78	4.91	5
6	13.08	12.51	11.97	11.46	10.98	10.52	9.68	8.93	7.94	6.58	6
7	17.95	17.06	16.23	15.45	14.71	14.02	12.76	11.64	10.19	8.26	7
8	23.48	22.18	20.97	19.84	18.79	17.81	16.03	14.47	12.48	9.88	8
9	29.61	27.80	26.13	24.58	23.14	21.81	19.42	17.36	14.75	11.43	9
10	36.31	33.88	31.65	29.60	27.72	25.98	22.89	20.25	16.98	12.89	10
11	43.53	40.38	37.50	34.87	32.47	30.27	26.40	23.13	19.13	14.23	11
12	51.25	47.25	43.62	40.34	37.35	34.63	29.90	25.95	21.18	15.47	12
13	59.42	54.45	49.99	45.96	42.33	39.05	33.38	28.70	23.14	16.59	13
14	68.01	61.96	56.55	51.71	47.37	43.47	36.80	31.36	24.97	17.60	14
15	77.00	69.73	63.29	57.55	52.45	47.89	40.15	33.92	26.69	18.51	15
16	86.34	77.74	70.16	63.46	57.53	52.26	43.42	36.37	28.30	19.32	16
17	96.02	85.96	77.14	69.40	62.59	56.59	46.58	38.70	29.78	20.04	17
18	106.01	94.35	84.20	75.36	67.62	60.84	49.64	40.91	31.16	20.68	18
19	116.27	102.89	91.33	81.31	72.60	65.01	52.58	43.00	32.42	21.24	19
20	126.79	111.56	98.49	87.23	77.51	69.09	55.41	44.97	33.58	21.74	20
21	137.54	120.34	105.67	93.11	82.34	73.06	58.11	46.82	34.64	22.17	21
22	148.51	129.20	112.85	98.94	87.08	76.93	60.69	48.55	35.62	22.55	22
23	159.65	138.13	120.01	104.70	91.72	80.67	63.15	50.18	36.50	22.89	23
24	170.97	147.10	127.14	110.38	96.25	84.30	65.48	51.69	37.30	23.18	24
25	182.43	156.10	134.23	115.97	100.68	87.80	67.70	53.11	38.03	23.43	25
26	194.02	165.12	141.26	121.47	104.98	91.18	69.79	54.42	38.69	23.65	26
27	205.73	174.14	148.22	126.86	109.17	94.44	71.78	55.64	39.29	23.84	27
28	217.53	183.14	155.11	132.14	113.23	97.57	73.65	56.77	39.83	24.00	28
29	229.41	192.12	161.91	137.31	117.16	100.57	75.41	57.81	40.31	24.14	29
30	241.36	201.06	168.62	142.36	120.97	103.46	77.08	58.78	40.75	24.26	30
31	253.35	209.95	175.23	147.29	124.66	106.22	78.64	59.68	41.15	24.37	31
32	265.40	218.79	181.74	152.09	128.21	108.86	80.11	60.50	41.50	24.46	32
33	277.46	227.56	188.13	156.77	131.64	111.38	81.49	61.26	41.82	24.54	33
34	289.54	236.26	194.42	161.32	134.95	113.79	82.78	61.96	42.10	24.60	34
35	301.62	244.88	200.58	165.74	138.13	116.09	83.99	62.61	42.36	24.66	35

[a] This table contains multipliers for a gradient g to find the present worth of the n-year end-of-year series $0, g, 2g, \ldots (n - 1)g$.

CONVERSION TABLES

TABLE D.1. Conversion Factors

Multiply	By	To Obtain	Multiply	By	To Obtain
acre	.405	ha	EDR steam (240 Btu/hr)	70.3	W
bar	100	kPa	ft²·hr·°F/Btu (R, thermal resistance)	0.176	m²·°C/W
barrel (42 gal)	159	l	ft	0.3048	m
Btu, IT	1.055	kJ	ft	304.8	mm
Btu/ft³	37.3	kJ/m³, J/l	ft/min, fpm	0.00508	m/s
Btu/gal (US)	0.279	kJ/l	ft/s, fps	0.3048	m/s
Btu·in/(ft²·hr·°F) (k, thermal conductivity)	144	W·mm/(m²·°C)	ft of water	2.99	kPa
Btu/h	0.293	W	ft²	0.0929	m²
Btu/ft²	11.4	kJ/m²	ft²/s, (μ/ρ, kinematic viscosity)	92900	mm²/s
Btu/(h·ft²)	3.15	W/m²	ft³	28.3	l
Btu/(ft²·hr·°F) (U, overall heat trans coeff) (C, thermal conductance)	5.68	W/(m²·°C)	ft³	0.0283	m³
Btu/lb	2.33	kJ/kg	ft³/h, cfh	7.87	ml/s
Btu/(lb·°F) (c, specific heat)	4.19	kJ/(kg·°C)	ft³/min, cfm	0.472	l/s
bushel	0.0352	m³	ft³/s, cfs	28.3	l/s
calorie, gram	4.19	J	ft-lb (work)	1.36	J
calorie, kilogram; kilocalorie	4.19	kJ	ft-lb/min (power)	0.0226	W
centipoise (μ, dynamic viscosity)	1.00	mPa·s	gallon (US)	3.79	l
centistoke (μ/ρ, kinematic viscosity)	1.00	mm²/s	gallon (US)	0.00379	m³
cents per gallon	0.264	¢/l	gph (US)	1.05	ml/s
cents per gallon (no. 2 fuel oil)	0.0677	$/GJ	gpm (US)	0.0631	l/s
cents per gallon (no. 6 fuel oil)	0.0632	$/GJ	grain (1/7000 lb)	0.648	g
cents per gallon (propane)	0.112	$/GJ	gr/gal	17.1	mg/l
cents per kWh	2.78	$/GJ	gr/lb	0.143	g/kg
cents per therm	0.0948	$/GJ	horsepower (boiler)	9.81	kW
cost, $ per square (100 sq ft)	0.108	$/m²	horsepower	0.746	kW
cost, $ per square foot	10.8	$/m²	inch	25.4	mm
cost, $ per pound	2.205	$/kg	in. of mercury	3.38	kPa
cost, $ per ton (refrigeration)	0.284	$/kW	in. of water	249	Pa
EDR hot water (150 Btu/hr)	44.0	W	in./100 ft, thermal expansion	0.833	mm/m
			in.²	645	mm²

TABLE D.1. (Continued)

Multiply	By	To Obtain	Multiply	By	To Obtain
in.³ (volume)	16.4	ml	lb (mass)	454	g
in.³/min (SCIM)	0.273	ml/s	lb/ft (uniform load)	1.49	kg/m
in.³ (section modulus)	16400	mm³	lb/(ft·hr) (μ, dynamic viscosity)	0.413	mPa·s
in.⁴ (section moment)	416000	mm⁴	lb/(ft·s) (μ, dynamic viscosity)	1488	mPa·s
km/hr	0.278	m/s	lb/hr	0.126	g/s
kWh	3.60	MJ	lb/min	0.00756	kg/s
kilopound (kg force)	9.81	N	lb of steam per hour @ 212°F (100°C)	0.284	kW
kip	4.45	kN	lbf/ft²	47.9	Pa
kip/in.² (ksi)	6.90	MPa	lbf·s/ft² (μ, dynamic viscosity)	47900	mPa·s
liter	0.001	m³	lbm/ft²	4.88	kg/m²
micron of mercury	133	mPa	lb/ft³ (ρ, density)	16.0	kg/m³
mile	1.61	km	lb/gal	120	kg/m³
mile, nautical	1.85	km	lb-ft (torque or moment)	1.36	N·m
mph	1.61	km/h	lb-in. (torque or moment)	113	mN·m
mph	0.447	m/s	ppm	1.00	mg/kg
millibar	0.100	kPa	psi	6.89	kPa
mm of mercury (torr)	0.133	kPa	quart (liquid)	0.946	l
mm of water (20°C)	9.79	Pa	square (100 ft²)	9.29	m²
meter of water	9.79	kPa	tablespoon	15	ml
ounce (mass, avoir.)	28.3	g	teaspoon	5	ml
ounce (force or thrust)	0.278	N	therm	106	MJ
ounce (liquid)	29.6	ml	ton, long (2240 lb)	1.02	Mg (tonne)
ounce-inch (torque, moment)	7.06	mN·m	ton, short (2000 lb)	0.907	Mg (tonne)
ounce (avoir.) per gallon	7.49	g/l	ton, refrigeration	3.52	kW
perm (permeance)	57.4	μg/(kPa·s·m²)	W/ft²	10.8	W/m²
pint (liquid)	473	ml	yd	0.914	m
pound			yd²	0.836	m²
lb (mass)	0.454	kg	yd³	0.765	m³
lb (force or thrust)	4.45	N			

TABLE D.2. Conversion Factors

Physical Quantity	Symbol	Conversion Factor	Physical Quantity	Symbol	Conversion Factor
Area	A	1 ft² = 0.0929 m² 1 in.² = 6.452 × 10⁻⁴ m²	Pressure	P	1 lbf/in.² = 6894.8 N/m² 1 lbf/ft² = 47.88 N/m² 1 atm = 101,325 N/m²
Density	ρ	1 lbm/ft³ = 16.018 kg/m³ 1 slug/ft³ = 515.379 kg/m³	Specific heat	c	1 Btu/(lbm·°F) = 4187 J/(kg·K)
Energy	Q or W	1 Btu = 1055.1 J 1 cal = 4.186 J 1 (ft)(lbf) = 1.3558 J 1 (hp)(hr) = 2.685 × 10⁶ J	Specific energy	Q/m	1 Btu/lbm = 2326.1 J/kg
			Temperature	T	$T(°R) = (9/5)T(K)$ $T(°F) = [T(°C)](9/5) + 32$ $T(°F) = [T(K) - 273.15](9/5) + 32$
Force	F	1 lbf = 4.448 N	Thermal conductivity	k	1 Btu/(hr·ft·°F) = 1.731 W/(m·K)
Heat flow rate	q	1 Btu/hr = 0.2931 W 1 Btu/s = 1055.1 W	Thermal diffusivity	α	1 ft²/s = 0.0929 m²/s 1 ft²/hr = 2.581 × 10⁻⁵ m²/s
Heat flux	q''	1 Btu/(hr·ft²) = 3.1525 W/m²	Thermal resistance	R_t	1 (hr·°F)/Btu = 1.8958 K/W
Heat generation per unit volume	q'''_G	1 Btu/(hr·ft³) = 10.343 W/m³	Velocity	V	1 ft/s = 0.3048 m/s 1 mph = 0.44703 m/s
Heat transfer coefficient	h_c	1 Btu/(hr·ft²·°F) = 5.678 W/(m²·K)	Viscosity, dynamic	μ	1 lbm/(ft·s) = 1.488 N·s/m² 1 centipoise = 0.00100 N·s/m²
Length	L	1 ft = 0.3048 m 1 in. = 2.54 cm = 0.0254 m 1 mile = 1.6093 km = 1609.3 m	Viscosity, kinematic	ν	1 ft²/s = 0.0929 m²/s 1 ft²/hr = 2.581 × 10⁻⁵ m²/s
Mass	m	1 lbm = 0.4536 kg 1 slug = 14.594 kg	Volume	V	1 ft³ = 0.02832 m³ 1 in.³ = 1.6387 × 10⁻⁵ m³ 1 gal (U.S. liq.) = 0.003785 m³
Mass flow rate	$\dot{m}$	1 lbm/hr = 0.000126 kg/s 1 lbm/s = 0.4536 kg/s			
Power	$\dot{W}$	1 hp = 745.7 W 1 (ft·lbf)/s = 1.3558 W 1 Btu/s = 1055.1 W 1 Btu/hr = 0.293 W			

713

TABLE D.3. Pressure Conversion Chart

in/H₂O	P.S.I.	in/Hg	mm/H₂O	mm/Hg	kg/cm²	bar	mbar	Pa	kPa
.1	.0036	.0073	2.534	.1863	.0002	.0002	.2482	24.82	.0248
.2	.0072	.0146	5.067	.3726	.0005	.0005	.4964	49.64	.0496
.4	.0144	.0293	10.13	.7452	.0010	.0010	.9928	99.28	.0993
.6	.0216	.0440	15.20	1.118	.0015	.0015	1.489	148.9	.1489
.8	.0289	.0588	20.34	1.496	.0020	.0020	1.992	199.2	.1992
1.0	.0361	.0735	25.41	1.868	.0025	.0025	2.489	248.9	.2489
2	.0722	.1470	50.81	3.736	.0051	.0050	4.978	497.8	.4978
3	.1083	.2205	76.22	5.604	.0076	.0075	7.467	746.7	.7467
4	.1444	.2940	101.62	7.472	.0102	.0099	9.956	995.6	.9956
5	.1804	.3673	127.0	9.335	.0127	.0124	12.44	1244	1.244
6	.2165	.4408	152.4	11.203	.0152	.0149	14.93	1493	1.493
7	.2526	.5143	177.8	13.072	.0178	.0174	17.42	1742	1.742
8	.2887	.5878	203.2	14.940	.0203	.0199	19.90	1990	1.990
9	.3248	.6613	228.6	16.808	.0228	.0224	22.39	2239	2.239
10	.3609	.7348	254.0	18.676	.0254	.0249	24.88	2488	2.488
11	.3970	.8083	279.4	20.544	.0279	.0274	27.37	2737	2.737
12	.4331	.8818	304.8	22.412	.0304	.0298	29.86	2986	2.986
13	.4692	.9553	330.2	24.280	.0330	.0323	32.35	3235	3.235
14	.5053	1.029	355.6	26.148	.0355	.0348	34.84	3484	3.484
15	.5414	1.102	381.0	28.016	.0381	.0373	37.33	3733	3.733
16	.5774	1.176	406.4	29.879	.0406	.0398	39.81	3981	3.981
17	.6136	1.249	431.8	31.752	.0431	.0423	42.31	4231	4.231
18	.6496	1.322	457.2	33.616	.0457	.0448	44.79	4479	4.479
19	.6857	1.396	482.6	35.484	.0482	.0472	47.28	4728	4.728
20	.7218	1.470	508.0	37.352	.0507	.0497	49.77	4977	4.977
21	.7579	1.543	533.4	39.22	.0533	.0522	52.26	5226	5.226
22	.7940	1.616	558.8	41.09	.0558	.0547	54.74	5474	5.474
23	.8301	1.690	584.2	42.96	.0584	.0572	57.23	5723	5.723
24	.8662	1.764	609.6	44.82	.0609	.0597	59.72	5972	5.972
25	.9023	1.837	635.0	46.69	.0634	.0622	62.21	6221	6.221
26	.9384	1.910	660.4	48.56	.0660	.0646	64.70	6470	6.470
27	.9745	1.984	685.8	50.43	.0685	.0671	67.19	6719	6.719
28	1.010	2.056	710.8	52.26	.0710	.0696	69.64	6964	6.964
29	1.047	2.132	736.8	54.18	.0736	.0721	72.19	7219	7.219
30	1.083	2.205	762.2	56.04	.0761	.0746	74.67	7467	7.467
31	1.119	2.278	787.5	57.91	.0787	.0771	77.15	7715	7.715
32	1.155	2.352	812.8	59.77	.0812	.0796	79.63	7963	7.963
33	1.191	2.425	838.2	61.63	.0837	.0820	82.12	8212	8.212
34	1.227	2.498	863.5	63.49	.0862	.0845	84.60	8460	8.460
35	1.263	2.571	888.9	65.36	.0888	.0870	87.08	8708	8.708
36	1.299	2.645	914.2	67.22	.0913	.0895	89.56	8956	8.956
37	1.335	2.718	939.5	69.08	.0938	.0920	92.04	9204	9.204
38	1.371	2.791	964.9	70.95	.0964	.0945	94.53	9453	9.453
39	1.408	2.867	990.9	72.86	.0990	.0970	97.08	9708	9.708
40	1.444	2.940	1016	74.72	.1015	.0995	99.56	9956	9.956
41	1.480	3.013	1042	76.59	.1040	.1020	102.0	10204	10.20
42	1.516	3.086	1067	78.45	.1066	.1044	104.5	10452	10.45

P.S.I.	in/H₂O	in/Hg	mm/H₂O	mm/Hg	kg/cm²	bar	mbar	Pa	kPa
1.0	27.71	2.036	703.1	51.75	.0703	.0689	68.95	6895	6.895
1.1	30.45	2.240	773.4	56.89	.0773	.0758	75.84	7584	7.584
1.2	33.22	2.443	843.7	62.06	.0844	.0827	82.74	8274	8.274
1.3	35.98	2.647	914.0	67.23	.0914	.0896	89.63	8963	8.963
1.4	38.75	2.850	984.3	72.40	.0984	.0965	96.52	9652	9.652
1.5	41.52	3.054	1055	77.57	.1055	.1034	103.4	10340	10.34
1.6	44.29	3.258	1125	82.74	.1125	.1103	110.3	11030	11.03
1.7	47.06	3.461	1195	87.92	.1195	.1172	117.2	11720	11.72
1.8	49.82	3.665	1266	93.09	.1266	.1241	124.1	12410	12.41
1.9	52.59	3.868	1336	98.26	.1336	.1310	131.0	13100	13.10
2.0	55.36	4.072	1406	103.4	.1406	.1379	137.9	13790	13.79
2.1	58.13	4.276	1476	108.6	.1476	.1448	144.8	14480	14.48
2.2	60.90	4.479	1547	113.8	.1547	.1517	151.7	15170	15.17
2.3	63.67	4.683	1617	118.9	.1617	.1586	158.6	15860	15.86
2.4	66.43	4.886	1687	124.1	.1687	.1655	165.5	16550	16.55
2.5	69.20	5.090	1758	129.3	.1758	.1724	172.4	17240	17.24
2.6	71.97	5.294	1828	134.5	.1828	.1793	179.3	17930	17.93
2.7	74.74	5.497	1898	139.6	.1898	.1862	186.2	18620	18.62
2.8	77.51	5.701	1969	144.8	.1968	.1930	193.0	19300	19.30
2.9	80.27	5.904	2039	150.0	.2039	.1999	199.9	19990	19.99
3.0	83.04	6.108	2109	155.1	.2109	.2068	206.8	20680	20.68
3.1	85.81	6.312	2180	160.3	.2180	.2137	213.7	21370	21.37
3.2	88.58	6.515	2250	165.5	.2250	.2206	220.6	22060	22.06
3.3	91.35	6.719	2320	170.7	.2320	.2275	227.5	22750	22.75
3.4	94.11	6.922	2390	175.8	.2390	.2344	234.4	23440	23.44
3.5	96.88	7.126	2461	181.0	.2461	.2413	241.3	24130	24.13
3.6	99.65	7.330	2531	186.2	.2531	.2482	248.2	24820	24.82
3.7	102.4	7.533	2601	191.3	.2601	.2551	255.1	25510	25.51
3.8	105.2	7.737	2672	196.5	.2672	.2620	262.0	26200	26.20
3.9	108.0	7.940	2742	201.7	.2742	.2689	268.9	26890	26.89
4.0	110.7	8.144	2812	206.9	.2812	.2758	275.8	27580	27.58
4.1	113.5	8.348	2883	212.0	.2883	.2827	282.7	28270	28.27
4.2	116.3	8.551	2953	217.2	.2953	.2896	289.6	28960	28.96
4.3	119.0	8.775	3023	222.4	.3023	.2965	296.5	29650	29.65
4.4	121.8	8.958	3094	227.5	.3094	.3034	303.4	30340	30.34
4.5	124.6	9.162	3164	232.7	.3164	.3103	310.3	31030	31.03
4.6	127.3	9.366	3234	237.9	.3234	.3172	317.2	31720	31.72
4.7	130.1	9.569	3304	243.1	.3304	.3240	324.0	32400	32.40
4.8	132.9	9.773	3375	248.2	.3375	.3310	331.0	33000	33.10
4.9	135.6	9.976	3445	253.4	.3445	.3378	337.8	33780	33.78
5.0	138.4	10.18	3515	258.6	.3515	.3447	344.7	34470	34.47
5.1	141.2	10.38	3586	263.7	.3586	.3516	351.6	35160	35.16
5.2	143.9	10.59	3656	268.9	.3656	.3585	358.5	35850	35.85
5.3	146.7	10.79	3726	274.1	.3726	.3654	365.4	36540	36.54
5.4	149.5	10.99	3797	279.3	.3797	.3723	372.3	37230	37.23
5.5	152.2	11.20	3867	284.4	.3867	.3792	379.2	37920	37.92
5.6	155.0	11.40	3937	289.6	.3937	.3861	386.1	38610	38.61

Pressure Conversion Tables

Conversion Factors

Conversion	Conversion
P.S.I. x .0689 = bar	P.S.I. x 27.71 = in. H₂O
P.S.I. x 68.95 = mbar	P.S.I. x 2.036 = in. Hg
P.S.I. x 6895 = Pa	P.S.I. x 703.1 = mm/H₂O
P.S.I. x 6.895 = kPa	P.S.I. x 51.75 = mm/Hg
	P.S.I. x .0703 = kg/cm²

NOTE: CONVERSION FACTORS ROUNDED

Table 1 (P.S.I. index)

P.S.I.	kPa	Pa	mbar	bar	kg/cm²	mm/Hg	mm/H₂O	in. Hg	in. H₂O
5.7	39.30	39300	393.0	.3930	.4007	294.8	4008	11.60	157.8
5.8	39.99	39990	399.9	.3999	.4078	299.9	4078	11.81	160.5
5.9	40.68	40680	406.8	.4068	.4148	305.1	4148	12.01	163.3
6.0	41.37	41370	413.7	.4137	.4218	310.3	4218	12.22	166.1
6.1	42.06	42060	420.6	.4206	.4289	315.5	4289	12.42	168.8
6.2	42.75	42750	427.5	.4275	.4359	320.6	4359	12.62	171.6
6.3	43.44	43440	434.4	.4344	.4429	325.8	4429	12.83	174.4
6.4	44.13	44130	441.3	.4413	.4500	331.0	4500	13.03	177.2
6.5	44.82	44820	448.2	.4482	.4570	336.1	4570	13.23	179.9
6.6	45.50	45500	455.0	.4550	.4640	341.3	4640	13.44	182.7
6.7	46.19	46190	461.9	.4619	.4711	346.5	4711	13.64	185.5
6.8	46.88	46880	468.8	.4688	.4781	351.7	4781	13.84	188.2
6.9	47.57	47570	475.7	.4757	.4851	356.8	4851	14.05	191.0
7.0	48.26	48260	482.6	.4826	.4921	362.0	4922	14.25	193.8
7.1	48.95	48950	489.5	.4895	.4992	367.2	4992	14.46	196.5
7.2	49.64	49640	496.4	.4964	.5062	372.3	5062	14.66	199.3
7.3	50.33	50330	503.3	.5033	.5132	377.5	5132	14.86	202.1
7.4	51.02	51020	510.2	.5102	.5203	382.7	5203	15.07	204.8
7.5	51.71	51710	517.1	.5171	.5273	387.9	5273	15.27	207.6
7.6	52.40	52400	524.0	.5240	.5343	393.0	5343	15.47	210.4
7.7	53.09	53090	530.9	.5309	.5414	398.2	5414	15.68	213.1
7.8	53.78	53780	537.8	.5378	.5484	403.4	5484	15.88	215.9
8.0	55.16	55160	551.6	.5516	.5625	413.7	5625	16.29	221.4
8.2	56.54	56540	565.4	.5654	.5765	424.1	5765	16.70	227.0
8.4	57.92	57920	579.2	.5792	.5906	434.4	5906	17.10	232.5
8.6	59.29	59290	592.9	.5929	.6046	444.7	6047	17.51	238.0
8.8	60.67	60670	606.7	.6067	.6187	455.1	6187	17.92	243.6
9.0	62.05	62050	620.5	.6205	.6328	465.4	6328	18.32	249.1
9.2	63.43	63430	634.3	.6343	.6468	475.8	6468	18.73	254.7
9.4	64.81	64810	648.1	.6481	.6609	486.1	6609	19.14	260.2
9.6	66.19	66190	661.9	.6619	.6749	496.5	6750	19.54	265.7
9.8	67.57	67570	675.7	.6757	.6890	506.8	6890	19.95	271.3
10.0	68.95	68950	689.5	.6895	.7031	517.1	7031	20.36	276.8
11.0	75.84	75840	758.4	.7584	.7734	568.9	7734	22.40	304.5
12.0	82.74	82740	827.4	.8274	.8437	620.6	8437	24.43	332.2
13.0	89.63	89630	896.3	.8963	.9140	672.3	9140	26.47	359.8
14.0	96.52	96520	965.2	.9652	.9843	724.0	9843	28.50	387.5
14.7	101.4	101400	1014	1.014	1.033	760.2	10340	29.93	406.9
15.0	103.4	103400	1034	1.034	1.055	775.7	10550	30.54	415.2
16.0	110.3	110300	1103	1.103	1.125	827.4	11250	32.58	442.9
17.0	117.2	117200	1172	1.172	1.195	879.1	11950	34.61	470.6
18.0	124.1	124100	1241	1.241	1.265	930.9	12660	36.65	498.2
19.0	131.0	131000	1310	1.310	1.336	982.6	13360	38.68	525.9
20.0	137.9	137900	1379	1.379	1.406	1034	14060	40.72	553.6
21.0	144.8	144800	1448	1.448	1.476	1086	14770	42.76	581.3
22.0	151.7	151700	1517	1.517	1.547	1138	15470	44.79	609.0
23.0	158.6	158600	1586	1.586	1.617	1189	16170	46.83	636.7
24.0	165.5	165500	1655	1.655	1.687	1241	16870	48.86	664.3
25.0	172.4	172400	1724	1.724	1.758	1293	17580	50.90	692.0

Table 2 (in. H₂O index)

in. H₂O	psi	in. Hg	kg/cm²	bar	mbar	Pa	kPa	mm/Hg	mm/H₂O
43	1.552	3.160	.1091	.1069	107.0	10701	10.70	80.31	1092
44	1.588	3.233	.1116	.1094	109.5	10949	10.95	82.18	1118
45	1.624	3.306	.1142	.1119	112.0	11197	11.20	84.04	1143
46	1.660	3.378	.1167	.1144	114.5	11445	11.44	85.90	1168
47	1.696	3.453	.1192	.1168	116.9	11694	11.69	87.76	1194
48	1.732	3.526	.1218	.1193	119.4	11942	11.94	89.63	1219
49	1.768	3.600	.1243	.1218	121.9	12190	12.19	91.49	1244
50	1.804	3.673	.1268	.1243	124.4	12438	12.44	93.35	1270
51	1.841	3.748	.1294	.1268	126.9	12690	12.69	95.27	1296
52	1.877	3.822	.1319	.1293	129.4	12938	12.94	97.13	1321
53	1.913	3.895	.1345	.1318	131.9	13190	13.19	98.99	1346
54	1.949	3.968	.1370	.1343	134.4	13438	13.44	100.8	1372
55	1.985	4.041	.1395	.1368	136.9	13686	13.69	102.7	1397
56	2.021	4.115	.1421	.1392	139.3	13934	13.93	104.6	1422
57	2.057	4.188	.1446	.1417	141.8	14182	14.18	106.4	1448
58	2.093	4.261	.1471	.1442	144.3	14431	14.43	108.3	1473
59	2.129	4.335	.1497	.1467	146.8	14679	14.68	110.2	1498
60	2.165	4.408	.1522	.1492	149.3	14927	14.93	112.0	1524
61	2.202	4.483	.1548	.1517	151.8	15182	15.18	113.9	1550
62	2.238	4.556	.1573	.1542	154.3	15430	15.43	115.8	1575
63	2.274	4.630	.1599	.1567	156.8	15679	15.68	117.7	1600
64	2.310	4.703	.1624	.1592	159.3	15927	15.93	119.5	1626
65	2.346	4.776	.1649	.1616	161.7	16175	16.18	121.4	1651
66	2.382	4.850	.1674	.1641	164.2	16423	16.42	123.3	1676
67	2.418	4.923	.1700	.1666	166.7	16672	16.67	125.1	1702
68	2.454	4.996	.1725	.1691	169.2	16920	16.92	127.0	1727
69	2.490	5.070	.1750	.1716	171.7	17168	17.17	128.8	1752
70	2.526	5.143	.1776	.1740	174.2	17416	17.42	130.7	1778
71	2.562	5.216	.1801	.1765	176.6	17664	17.66	132.6	1803
72	2.598	5.290	.1826	.1790	179.1	17912	17.91	134.4	1828
73	2.635	5.365	.1852	.1816	181.7	18161	18.17	136.4	1854
74	2.671	5.438	.1878	.1840	184.2	18416	18.42	138.2	1880
75	2.707	5.511	.1903	.1865	186.6	18664	18.66	140.1	1905
76	2.743	5.585	.1928	.1890	189.1	18912	18.91	141.9	1930
77	2.779	5.658	.1954	.1915	191.6	19160	19.16	143.8	1956
78	2.815	5.731	.1979	.1940	194.1	19409	19.41	145.7	1981
79	2.851	5.805	.2004	.1964	196.6	19657	19.66	147.5	2006
80	2.887	5.878	.2030	.1989	199.0	19905	19.90	149.4	2032
81	2.923	5.951	.2055	.2014	201.5	20153	20.15	151.2	2057
82	2.959	6.024	.2080	.2039	204.0	20402	20.40	153.1	2082
83	2.996	6.097	.2106	.2064	206.6	20657	20.66	155.0	2108
84	3.032	6.173	.2131	.2089	209.0	20905	20.90	156.9	2134
85	3.068	6.246	.2157	.2114	211.5	21153	21.15	158.8	2159
86	3.104	6.320	.2182	.2139	214.0	21401	21.40	160.6	2184
87	3.140	6.393	.2207	.2163	216.5	21650	21.65	162.5	2210
88	3.176	6.466	.2233	.2188	219.0	21898	21.90	164.4	2235
89	3.212	6.540	.2258	.2213	221.4	22146	22.15	166.2	2260
90	3.248	6.613	.2283	.2238	223.9	22394	22.39	168.1	2286
91	3.284	6.686	.2309	.2263	226.4	22642	22.64	169.9	2311
92	3.320	6.760	.2334	.2287	228.9	22890	22.89	171.8	2336
93	3.356	6.833	.2359	.2312	231.4	23139	23.14	173.7	2362
94	3.392	6.906	.2384	.2337	233.9	23387	23.39	175.5	2387
95	3.429	6.981	.2410	.2362	236.4	23642	23.64	177.4	2413
96	3.465	7.055	.2436	.2387	238.9	23890	23.89	179.3	2438
97	3.501	7.128	.2461	.2412	241.4	24138	24.14	181.2	2464
98	3.537	7.201	.2486	.2437	243.9	24387	24.39	183.0	2489
99	3.573	7.275	.2512	.2462	246.3	24635	24.64	184.9	2514
100	3.609	7.348	.2537	.2487	248.8	24883	24.88	186.8	2540

SOURCE: Dwyer Instrument, Inc.

TABLE D.4.   Temperature Conversion

Temperature Conversion

The numbers in bold-face type in the center column refer to the temperature, either in Celsius or Fahrenheit, which is to be converted to the other scale. If converting Fahrenheit to Celsius, the equivalent temperature will be found in the left column. If converting Celsius to Fahrenheit, the equivalent temperature will be found in the column on the right.

Celsius	C or F	Fahr	Celsius	C or F	Fahr	Celsius	C or F	Fahr	Celsius	C or F	Fahr
−40.0	**−40**	−40.0	−6.7	**+20**	+68.0	+26.7	**+80**	+176.0	+60.0	**+140**	+284.0
−39.4	**−39**	−38.2	−6.1	**+21**	+69.8	+27.2	**+81**	+177.8	+60.6	**+141**	+285.8
−38.9	**−38**	−36.4	−5.5	**+22**	+71.6	+27.8	**+82**	+179.6	+61.1	**+142**	+287.6
−38.3	**−37**	−34.6	−5.0	**+23**	+73.4	+28.3	**+83**	+181.4	+61.7	**+143**	+289.4
−37.8	**−36**	−32.8	−4.4	**+24**	+75.2	+28.9	**+84**	+183.2	+62.2	**+144**	+291.2
−37.2	**−35**	−31.0	−3.9	**+25**	+77.0	+29.4	**+85**	+185.0	+62.8	**+145**	+293.0
−36.7	**−34**	−29.2	−3.3	**+26**	+78.8	+30.0	**+86**	+186.8	+63.3	**+146**	+294.8
−36.1	**−33**	−27.4	−2.8	**+27**	+80.6	+30.6	**+87**	+188.6	+63.9	**+147**	+296.6
−35.6	**−32**	−25.6	−2.2	**+28**	+82.4	+31.1	**+88**	+190.4	+64.4	**+148**	+298.4
−35.0	**−31**	−23.8	−1.7	**+29**	+84.2	+31.7	**+89**	+192.2	+65.0	**+149**	+300.2
−34.4	**−30**	−22.0	−1.1	**+30**	+86.0	+32.2	**+90**	+194.0	+65.6	**+150**	+302.0
−33.9	**−29**	−20.2	−0.6	**+31**	+87.8	+32.8	**+91**	+195.8	+66.1	**+151**	+303.8
−33.3	**−28**	−18.4	.0	**+32**	+89.6	+33.3	**+92**	+197.6	+66.7	**+152**	+305.6
−32.8	**−27**	−16.6	+0.6	**+33**	+91.4	+33.9	**+93**	+199.4	+67.2	**+153**	+307.4
−32.2	**−26**	−14.8	+1.1	**+34**	+93.2	+34.4	**+94**	+201.2	+67.8	**+154**	+309.2
−31.7	**−25**	−13.0	+1.7	**+35**	+95.0	+35.0	**+95**	+203.0	+68.3	**+155**	+311.0
−31.1	**−24**	−11.2	+2.2	**+36**	+96.8	+35.6	**+96**	+204.8	+68.9	**+156**	+312.8
−30.6	**−23**	−9.4	+2.8	**+37**	+98.6	+36.1	**+97**	+206.6	+69.4	**+157**	+314.6
−30.0	**−22**	−7.6	+3.3	**+38**	+100.4	+36.7	**+98**	+208.4	+70.0	**+158**	+316.4
−29.4	**−21**	−5.8	+3.9	**+39**	+102.2	+37.2	**+99**	+210.2	+70.6	**+159**	+318.2
−28.9	**−20**	−4.0	+4.4	**+40**	+104.0	+37.8	**+100**	+212.0	+71.1	**+160**	+320.0
−28.3	**−19**	−2.2	+5.0	**+41**	+105.8	+38.3	**+101**	+213.8	+71.7	**+161**	+321.8
−27.8	**−18**	−0.4	+5.5	**+42**	+107.6	+38.9	**+102**	+215.6	+72.2	**+162**	+323.6
−27.2	**−17**	+1.4	+6.1	**+43**	+109.4	+39.4	**+103**	+217.4	+72.8	**+163**	+325.4
−26.7	**−16**	+3.2	+6.7	**+44**	+111.2	+40.0	**+104**	+219.2	+73.3	**+164**	+327.2
−26.1	**−15**	+5.0	+7.2	**+45**	+113.0	+40.6	**+105**	+221.0	+73.9	**+165**	+329.0
−25.6	**−14**	+6.8	+7.8	**+46**	+114.8	+41.1	**+106**	+222.8	+74.4	**+166**	+330.8
−25.0	**−13**	+8.6	+8.3	**+47**	+116.6	+41.7	**+107**	+224.6	+75.0	**+167**	+332.6
−24.4	**−12**	+10.4	+8.9	**+48**	+118.4	+42.2	**+108**	+226.4	+75.6	**+168**	+334.4
−23.9	**−11**	+12.2	+9.4	**+49**	+120.2	+42.8	**+109**	+228.2	+76.1	**+169**	+336.2
−23.3	**−10**	+14.0	+10.0	**+50**	+122.0	+43.3	**+110**	+230.0	+76.7	**+170**	+338.0
−22.8	**−9**	+15.8	+10.6	**+51**	+123.8	+43.9	**+111**	+231.8	+77.2	**+171**	+339.8
−22.2	**−8**	+17.6	+11.1	**+52**	+125.6	+44.4	**+112**	+233.6	+77.8	**+172**	+341.6
−21.7	**−7**	+19.4	+11.7	**+53**	+127.4	+45.0	**+113**	+235.4	+78.3	**+173**	+343.4
−21.1	**−6**	+21.2	+12.2	**+54**	+129.2	+45.6	**+114**	+237.2	+78.9	**+174**	+345.2
−20.6	**−5**	+23.0	+12.8	**+55**	+131.0	+46.1	**+115**	+239.0	+79.4	**+175**	+347.0
−20.0	**−4**	+24.8	+13.3	**+56**	+132.8	+46.7	**+116**	+240.8	+80.0	**+176**	+348.8
−19.4	**−3**	+26.6	+13.9	**+57**	+134.6	+47.2	**+117**	+242.6	+80.6	**+177**	+350.6
−18.9	**−2**	+28.4	+14.4	**+58**	+136.4	+47.8	**+118**	+244.4	+81.1	**+178**	+352.4
−18.3	**−1**	+30.2	+15.0	**+59**	+138.2	+48.3	**+119**	+246.2	+81.7	**+179**	+354.2
−17.8	**0**	+32.0	+15.6	**+60**	+140.0	+48.9	**+120**	+248.0	+82.2	**+180**	+356.0
−17.2	**+1**	+33.8	+16.1	**+61**	+141.8	+49.4	**+121**	+249.8	+82.8	**+181**	+357.8
−16.7	**+2**	+35.6	+16.7	**+62**	+143.6	+50.0	**+122**	+251.6	+83.3	**+182**	+359.6
−16.1	**+3**	+37.4	+17.2	**+63**	+145.4	+50.6	**+123**	+253.4	+83.9	**+183**	+361.4
−15.6	**+4**	+39.2	+17.8	**+64**	+147.2	+51.1	**+124**	+255.2	+84.4	**+184**	+363.2
−15.0	**+5**	+41.0	+18.3	**+65**	+149.0	+51.7	**+125**	+257.0	+85.0	**+185**	+365.0
−14.4	**+6**	+42.8	+18.9	**+66**	+150.8	+52.2	**+126**	+258.8	+85.6	**+186**	+366.8
−13.9	**+7**	+44.6	+19.4	**+67**	+152.6	+52.8	**+127**	+260.6	+86.1	**+187**	+368.6
−13.3	**+8**	+46.4	+20.0	**+68**	+154.4	+53.3	**+128**	+262.4	+86.7	**+188**	+370.4
−12.8	**+9**	+48.2	+20.6	**+69**	+156.2	+53.9	**+129**	+264.2	+87.2	**+189**	+372.2
−12.2	**+10**	+50.0	+21.1	**+70**	+158.0	+54.4	**+130**	+266.0	+87.8	**+190**	+374.0
−11.7	**+11**	+51.8	+21.7	**+71**	+159.8	+55.0	**+131**	+267.8	+88.3	**+191**	+375.8
−11.1	**+12**	+53.6	+22.2	**+72**	+161.6	+55.6	**+132**	+269.6	+88.9	**+192**	+377.6
−10.6	**+13**	+55.4	+22.8	**+73**	+163.4	+56.1	**+133**	+271.4	+89.4	**+193**	+379.4
−10.0	**+14**	+57.2	+23.3	**+74**	+165.2	+56.7	**+134**	+273.2	+90.0	**+194**	+381.2
−9.4	**+15**	+59.0	+23.9	**+75**	+167.0	+57.2	**+135**	+275.0	+90.6	**+195**	+383.0
−8.9	**+16**	+60.8	+24.4	**+76**	+168.8	+57.8	**+136**	+276.8	+91.1	**+196**	+384.8
−8.3	**+17**	+62.6	+25.0	**+77**	+170.6	+58.3	**+137**	+278.6	+91.7	**+197**	+386.6
−7.8	**+18**	+64.4	+25.6	**+78**	+172.4	+58.9	**+138**	+280.4	+92.2	**+198**	+388.4
−7.2	**+19**	+66.2	+26.1	**+79**	+174.2	+59.4	**+139**	+282.2	+92.8	**+199**	+390.2

SOURCE: *ASHRAE Handbook of Fundamentals 1977.*

716

INDEX